Analytical Chemistry

Analytical Chemistry

PRINCIPLES AND TECHNIQUES

Larry G. Hargis

UNIVERSITY OF NEW ORLEANS

PRENTICE HALL
Englewood Cliffs, New Jersey 07632

Library of Congress Cataloging-in-Publication Data

Hargis, Larry G.
  Analytical chemistry.

  Bibliography: p.
  Includes index.
  1. Chemistry, Analytic. I. Title.
QD75.2.H365 1988   543   87-29270
ISBN 0-13-033507-X

Editorial/production supervision: Zita de Schauensee
Cover, title page, and chapter-opening design: Lorraine Mullaney
Manufacturing buyer: Paula Benevento
Cover photo by Denis Hart
Courtesy Vestec Corporation
        9299 Kirby Drive,
        Houston, Texas

 © 1988 by Prentice-Hall, Inc.
A Division of Simon & Schuster
Englewood Cliffs, New Jersey 07632

Printed in the United States of America
10 9 8 7 6 5 4 3 2 1

ISBN   0-13-033507-X

Prentice-Hall International (UK) Limited, *London*
Prentice-Hall of Australia Pty. Limited, *Sydney*
Prentice-Hall Canada Inc., *Toronto*
Prentice-Hall Hispanoamericana, S.A., *Mexico*
Prentice-Hall of India Private Limited, *New Delhi*
Prentice-Hall of Japan, Inc., *Tokyo*
Simon & Schuster Southeast Asia Pte. Ltd., *Singapore*
Editora Prentice-Hall do Brasil, Ltda., *Rio de Janeiro*

CONTENTS

Contents

Contents

23

LABORATORY OPERATIONS AND PRACTICES

24

REPRESENTATIVE EXPERIMENTS

APPENDIXES

PREFACE

Analytical chemistry in its broadest sense encompasses the theory and practice of all means of acquiring information about the composition of matter. Quantitative analysis constitutes the largest part of analytical chemistry and is devoted to the techniques, methods, and instrumentation involved in determining the amounts or concentrations of constituents in samples. Many disciplines, including agriculture, biology, engineering, environmental science, geology, and medicine, rely heavily on quantitative analysis in solving routine problems and in research. As our ability to synthesize and formulate new substances grows and we become more knowledgeable about their properties, the demands on analytical chemistry grow. The instruments and techniques used in modern analytical chemistry laboratories change constantly but many principles and much underlying theory remain fundamentally important to the practice of analytical chemistry. This book is devoted largely to the study of those principles and theories.

I began this book with four major goals in mind: (1) to develop and explain the theories upon which the principles of various analytical procedures are based, (2) to instill in students an appreciation of the diversity of chemical analyses and to demonstrate the importance of being able to discern and evaluate the relative merits and limitations of analytical methods, (3) to familiarize students with important details of specific methods of analysis, and (4) to instruct students in the techniques of solving equilibrium and stoichiometry problems. A book in which these goals are achieved can serve the needs of a diverse group of students whose programs of study require an introductory course in analytical chemistry. The book contains more material than is usually covered in a one-semester course. As a result, instructors will have some latitude in selecting topics that best serve the particular needs of their students. The presence of a liberal number of instrumental topics makes the book appropriate for those programs that offer only a single course in analytical chemistry, as instructors may include sufficient instrumental material to insure that students become familiar with the range of modern analytical methods.

The book begins in a traditional manner. Chapter 1 provides a brief overview of the unit operations that comprise most analysis procedures and sets the tone for the remainder of the book. Chapter 2 is a review of numerous concepts and terms used routinely in quantitative analysis. Most or all of these topics should have been learned in the general chemistry course, and if they have not already done so, students should seek to master this material very early in the course. The handling of experimental data is covered in Chapter 3. The material is presented at a fairly advanced level, which reflects my opinion that a sound understanding of error and its effect on the reliability of experimental results is crucial to the practice of analytical chemistry.

The traditional topics of gravimetry and titrimetry, along with the associated equilibrium concepts, are presented in Chapters 4–11. Chapter 4 on gravimetry is self-contained and can be taught later, or omitted, if the instructor wishes. The general features of titration methods are described in Chapter 5, which is followed by chapters devoted to specific types of titrations. Acid-base equilibria and titrations are covered first because they are so important to the other types. Chapters 8 and 12, on nonaqueous acid-base and precipitation titrations respectively, may be omitted without losing any continuity.

The "instrumental" topics are covered in Chapters 13–22 and include two chapters on electrochemistry (13, 14), four chapters of spectroscopy (15–18), and four chapters on separations (19–22). The emphasis in these chapters is on the underlying principles and capabilities of instruments for solving analytical problems rather than on instrument design and construction. Chapter 15 covers the basic principles of spectrometry and should precede the other three chapters in this group. Likewise, Chapter 20 is a general introduction to chromatography and should precede Chapters 21 and 22, which are devoted to specific chromatographic methods. The last two chapters contain material relevant to the quantitative analysis laboratory. The basic tools and laboratory operations are discussed in Chapter 23, which also contains sections on safety and recordkeeping. Chapter 24 is a collection of specific procedures for determining various substances using the techniques described in the book.

Two features of this book merit special mention. A large number of figures, including nearly all the titration graphs, were drawn using data generated from a computer solution of the appropriate theoretical equation. As a result, there should be few, if any, of the inconsistencies or errors that occur when data are obtained from simplified, approximate equations. The problems at the end of the chapters cover a wide range of difficulty and include many examples of current applications to which students can relate. I strongly urge students not to avoid attempting the more difficult problems on the premise that they are too long or complicated to be on an exam. While this may be true, there is often significant learning value in attempting to solve difficult problems. Even though you may be unsuccessful in solving such a problem, you may begin to see which particular concepts or operations are causing you difficulty, and you can then direct your study toward overcoming those difficulties.

Two supplements accompany this book: a laboratory manual and a problem solutions manual. The laboratory manual contains 50 percent more experiments than are given in Chapter 24 and they cover a broader range of techniques and sample types. The solutions manual is much more than a simple answer book; it contains complete step-by-step solutions to the problems at the end of the chapters, including brief explanations of how to approach the solution, why certain assumptions can be made, and so forth. This manual can be made available to students if the instructor wishes.

A great many colleagues aided in the development of this book by reviewing the manuscript at different stages. They were a most conscientious group, catching numerous errors and taking the time to enlighten me on a number of important matters. I am truly grateful for their efforts. Especially, I would like to thank the following:

Jon Carnahan, Northern Illinois University
Gordon Ewing, Pennsylvania State University
Edward Gray, Jr., University of Hartford
George Harrington, Temple University
James Hicks, University of Chicago
Raymond Hunt, University of Illinois
Linda McGown, University of Washington
John Marquart, Eastern Illinois University
Paul Rosenberg, University of New Hampshire
Joseph Sherma, Lafayette College
James Winefordner, University of Florida

I owe special thanks to my colleague Professor Ronald F. Evilia, who listened willingly when I was beset with uncertainties and helped me to understand and find the direction I needed. He read much of the early drafts and the entire final manuscript, providing valuable criticism at every stage. I wish also to express my appreciation to all of the Prentice-Hall staff who assisted in the development and production of this book. Without exception they carried out their duties with the highest degree of professionalism. Finally, to my wife, Caroly, I give my deepest thanks for her understanding and forbearance while I was engaged in this extensive project.

Larry G. Hargis
New Orleans, Louisiana

Analytical Chemistry

INTRODUCTION

Before beginning a serious study of quantitative analysis it is important that students have a general understanding of analytical chemistry, of which quantitative analysis is the major part. As the book delves into specific theories and practices, it becomes important for one to be aware of how such things fit into the general goals of analytical chemistry. When deeply immersed in the chemical and mathematical procedures used to obtain data and calculate results, it is easy to lose sight of why the answer was desired and to what use it will be put. These general considerations often influence what methods and procedures are used and how mathematical calculations are performed. In this chapter we describe what analytical chemistry is, the role played by quantitative analysis, and the general methodology used to carry out quantitative determinations.

ANALYTICAL CHEMISTRY

Analytical chemistry is a branch of chemistry involved with the analysis of chemical substances. Analysis is used here in a broad sense to include identifying substances, called *qualitative analysis*; determining the concentration or amount of substances, called *quantitative analysis*; and determining the structure of substances. The theory on which the processes of analysis are based and the instrumentation or tools with which measurements are made are integral parts of the analytical chemist's domain.

The variety of problems in which analysis plays a role is indeed great. Below are listed six broad areas in which analytical chemistry is commonly involved. With a little thought you should be able to add to this list.

1. *Establishing economic value.* Determining the amount of silver in a coin, the amount of oil in shale, and the amount of protein in animal feedstock are examples.

2. *Determining health hazards.* Establishing the concentration of sulfur dioxide in emissions from a coal-burning power plant and the amount of pesticide residue on fresh fruit and vegetables are two diverse examples.

3. *Diagnosing disease.* Clinical determinations such as those for glucose and urea are absolutely essential to the physician in making a proper diagnosis and making it quickly. Many clinical laboratories are set up to determine more than 50 substances routinely.

4. *Controlling quality.* Virtually all manufacturers try to achieve a constant, predetermined quality of product. To do this they often need to know the quality of their raw materials. Drug and processed-food manufacturers must analyze their starting materials, additives, and finished products regularly to ensure that their product meets acceptable standards.

5. *Relating properties to composition or structure.* Both the physical and chemical properties of alloys, adhesives, lubricants, plastics, and so on, depend on their chemical composition. Similarly, the activity or effectiveness of pharmaceuticals, pesticides, and herbicides depends largely on their chemical structure.

6. *Conducting research.* Analytical chemistry plays a major role in many research projects. Following the accumulation of pesticides in the food chain or the metabolic fate of drugs in the human body, determining the nature of catalytic surfaces necessary to convert coal to natural gas, looking for new ways to separate, identify, and determine the concentration of important enzymes and proteins, and developing new and better instruments for making quantitative measurements are but a few examples.

It should be obvious that analytical chemists often work in close cooperation with other types of chemists and scientists. Consider, for example, a chemical manufacturing plant that suffers a large explosion. Management wants to know the cause of the explosion. Engineers will be called upon to assess the damage, estimate the force of the explosion, and try to pinpoint its source. They may also establish the effect of different types of equipment failure or malfunction. Organic or inorganic chemists will be called upon to determine what improper conditions or impurities could have resulted in undesirable reactions leading to explosive products. Physical chemists will be needed to calculate the amount of explosive energy that could result from the various possible reactions. Analytical chemists may work with all of these scientists, analyzing residue samples from around the explosion site to identify the combustion products in hopes that they will provide a clue as to the nature of the explosion reaction and/or the location of the explosion. Also, they might decide to analyze the raw materials being used to determine if impurities are present that could have been responsible for an undesirable reaction or condition that might lead to an explosion. The effectiveness of the scientists and engineers in solving this problem will be diminished if their knowledge is limited *solely* to their own disciplines. The extent to which we know things beyond our own discipline is important in determining our effectiveness in many endeavors both scientific and nonscientific. And it is, of course, the reason many of you are enrolled in an analytical chemistry course and reading this book.

QUANTITATIVE ANALYSIS

Quantitative analysis, the part of analytical chemistry that deals with the determination of amounts or concentrations of constituents, is the topic of this book. An extremely large number of methods have been developed for use in quantitative analysis, each with certain capabilities and limitations. Because there are so many methods, it is often helpful to organize them into groups based on similar capabilities or characteristics. Unfortunately, there is no single classification that is suitable for all purposes. A *complete* (or ultimate) analysis means that the amounts of all of the constituents in the sample were determined, while a *partial* analysis means that only some were determined. *Elemental* and *molecular* analysis refer to the determination of the amounts of elements and molecules, respectively. *Major-constituent* (macro) analysis implies that the constituent determined was present in high concentration, whereas *trace* analysis means just the opposite. Another classification scheme used in this and many other textbooks is based on the type of final measurement. In this scheme, *gravimetric* methods are those in which the weight of a substance is measured, *titrimetric* methods refer to measurement of a volume, and *physicochemical* (instrumental) methods are based on the measurement of some physical or chemical property. There are, of course, subdivisions within each of these categories, some of which are discussed in the various chapters of this book.

Care should be exercised in using the terms "analyze" and "determine." Samples are analyzed, but specific constituents are determined. This book will follow the modern practice of referring to the constituent of interest as the *analyte*.

STEPS IN AN ANALYSIS

In selecting a method and performing an analysis, several important decisions must be made. It is worthwhile to consider briefly some of the common questions and problems that arise in carrying out the individual operations of an analysis.

Overall Plan or Goal

The first step in any analysis is to learn as much as possible about the overall problem that is being faced. Specifically, one needs to know what questions are being asked in order to determine what information is needed to answer those questions, what part of that information can be supplied by analysis, and what methods should be used for the analyses. The choice of a particular method is determined largely by the diverse substances present in the sample, the expected concentration level of the analyte, the accuracy needed, the number of samples to be analyzed, the cost, and the time, equipment, and technical expertise available.

Collecting the Sample

The normal laboratory sample weighs less than 1 gram, yet the analysis results of such a sample are generally used to characterize the entire material from which the sample was taken. It is essential therefore that the laboratory sample accurately represents the composition of the bulk material. Sampling homogeneous material is easy since, by definition, the composition is uniform throughout. One needs only to select the desired size of sample for the laboratory. Sampling heterogeneous material is more difficult, generally requiring carefully prescribed procedures that are designed to ensure proper representation in the final sample.

Methods of sampling bulk materials gain widespread acceptance only after careful, intensive studies have been conducted to evaluate and test their accuracy. Acceptable methods for sampling bulk materials have been specified by various scientific organizations concerned with the use of these materials. The American Society for Testing and Materials (ASTM), the National Bureau of Standards (NBS), the Association of Official Analytical Chemists (AOAC), and the American Oil Chemists Society (AOCS) are a few such organizations. The exact procedures used vary depending on the nature of the sample, but often they are comprised of three steps: obtaining a gross sample, reducing the gross sample to a transportable size, and obtaining a laboratory sample.

The gross sample is obtained by random selection. For example, in unloading ore from a ship, a chute may be designed that siphons off a certain fraction of the material—perhaps as much as a ton—which, taken as a whole, must be representative of the entire load of ore. Since a ton of material is too large and heavy a load to bring into the laboratory, a second step is performed to reduce this sample to a size suitable for transportation. In the case of our shipload of ore, this may require some crushing to get a more uniform size, followed by some sampling procedure that removes a fixed portion of the gross sample. One such procedure is the *cone and quarter* method, where the gross sample is piled into a cone which is flattened and quartered. Opposite quarters are retained, mixed, and the procedure repeated until an appropriate sample (perhaps 5 to 25 lb) is obtained.

In the laboratory, the same or a similar procedure is used to further reduce the sample to a size that can be analyzed. Usually, three or more samples will be prepared and analyzed simultaneously because we can place more confidence in a number representing the results of three measurements than we can in one representing the result of a single measurement. Also, in case a sample is ruined during the analysis, the chemist can continue with the remaining two rather than having to start over.

The chemist must be constantly on guard against contamination during the sample collection process. This is especially important when the constituents to be determined are present in very low concentrations (trace analysis). It is good practice to make certain that all sample-handling equipment is cleaned before use. Special care is required when handling corrosive substances since they can often react with their containers and the equipment used to handle them.

It may not always be possible to analyze samples immediately after they are brought to the laboratory. Fortunately, most samples are stable and can be stored easily and safely. There are instances, however, when this is not the case, especially with samples taken from or containing living organisms, where ongoing chemical reactions can change the composition. To store such samples without deterioration, the rate of these chemical reactions must be stopped or at least decreased. One useful technique is to freeze the sample, thereby depriving the molecules of the mobility and energy they require to react.

Sometimes it is not only important how the sample is taken but *when* it is taken. Many clinical procedures give quite different results when a blood sample is taken immediately after eating as opposed to after 12 hours of fasting. Similarly, impurities in air and natural water samples may vary considerably over a period of a day, a week, or a month.

Measuring the Sample

The result of a quantitative analysis is usually reported in relative terms, that is, the amount of substance per unit amount of sample. The amount of sample is expressed most often as a weight, but volume is used also. The weighing of solids and nonvolatile liquids is fairly routine and can be done with great accuracy using modern analytical balances. Gases and volatile liquids are difficult to weigh accurately, and usually are measured in terms of their volume. If the density of a liquid is known, its weight can be determined from the volume.

It is usually not acceptable to weigh solid samples for analysis without some treatment to remove "occasional" water—that is, water that can be removed from the sample without causing structural changes. In solids, occasional water may be hydrated or simply adsorbed on the surface. Unfortunately, the water content of most samples is not constant, varying with changes in atmospheric humidity, temperature, particle size, and sample-handling procedures. These things are not easy to control, and to obtain reproducible results, samples commonly are dried *before* being weighed. The usual procedure is to heat the sample in an oven to volatilize the water. The best temperature for accomplishing this depends on how firmly the water is attached to the sample. After a sufficient time, usually 1 to 2 hours, the sample is removed and allowed to cool in a dry atmosphere. There are instances when samples cannot be dried, often because they decompose (sometimes explosively!) at the temperatures necessary to drive off the water. In such cases, the sample can be analyzed on a "wet" or "as received" basis and the water content determined by analyzing another portion of the same sample.

The size of the sample taken for analysis depends on the concentration of the analyte and the specific steps and equipment that will be used. The sample weight must be large enough to make the relative weighing error insignificant and small enough so that the operations can be performed easily.

Dissolving the Sample

Most methods, particularly noninstrumental ones, are designed for use with liquid solutions of the sample, partly because solutions are homogeneous and easy to handle. As a rule, it is best to dissolve the entire sample using the mildest conditions possible.

Water
Samples composed entirely of soluble salts may be dissolved directly in water. Some salts, although soluble, dissolve slowly at room temperature and mild heating may be advisable.

Acids

In many instances, at least a portion of the sample will not dissolve in water. In such cases, the addition of strong acids will usually render the sample soluble. Hydrochloric, dilute sulfuric, and dilute perchloric acid, all considered nonoxidizing acids, are tried first. Oxidizing acids such as hot, concentrated sulfuric acid, nitric acid, and aqua regia (a mixture of hydrochloric and nitric acids) are used with more stubborn samples. Hydrofluoric acid, although considered a weak acid, is uniquely suited for dissolving silicate ores because it forms a volatile fluorosilicate, H_2SiF_6.

Organic solvents

Samples containing organic compounds generally do not dissolve in water and require the use of organic solvents. There are many organic solvents and the choice of a particular one depends on its compatibility with later sample treatment and measurement steps.

Fusion

Samples that fail to dissolve in aqueous or organic solvents must be *fused*—dissolved in a molten solvent (called a *flux*) that has acidic, basic, or oxidizing properties. Potassium pyrosulfate ($K_2S_2O_7$), sodium carbonate, and sodium peroxide are common examples of an acidic, basic, and oxidizing flux, respectively. Fusion is carried out by mixing finely ground sample with solid flux in an inert metal crucible and heating until the flux melts, at which time it becomes very reactive and dissolves the sample. After a period of time, the melt is cooled and dissolved in water or dilute aqueous acid.

The reasons why molten solvents exhibit much greater dissolving power than aqueous solvents are threefold. First, the temperature of a molten flux is several hundred degrees higher than the temperature of a boiling aqueous solution. Second, since the solvent itself is the acid, base, or oxidizing agent in a fusion, its concentration is 2 to 10 times larger than it would be in an aqueous solution. Finally, as you will learn in Chapter 8, the absence of water can greatly increase the acid or base strength of a substance.

Dealing with Potential Interferences

The ideal quantitative method would allow the analyte to be determined in the presence of any possible combination of other substances. If such methods existed, the practice of quantitative analysis would be simple indeed! Chemists use the word "selectivity" in a relative sense to indicate how close a particular method comes to being ideal. Thus a method that has few interferences is said to be very selective. In applying a method to a given sample it is necessary to know what interfering substances are possibly present in the sample and how they will interfere. Armed with this knowledge, the analytical chemist can develop a plan for dealing with the problem. There are three general ways to eliminate interferences, and they are discussed below in order of increasing difficulty.

Selecting the method

A great many interferences can be avoided by judicious selection of the method to be used. Several methods exist for determining most substances, and while each is subject to interferences, they are seldom the same interferences. If the chemist is not familiar with the general composition of the sample to be analyzed, a qualitative study may have to be made to determine what constituents are present, or some time spent in the library learning about the sample material. Equally important to knowing the general composition of the sample is knowing which methods possibly can be used. As they practice their profession, analytical chemists become familiar with the capabilities and limitations of a great variety of methods. For those not blessed with 30 years of experience, most libraries contain numerous collections of "accepted" or "standard" methods of analysis. These are methods that have been carefully tested and evaluated under controlled conditions and are recommended by various organizations of chemists for use with specific substances and samples. In addition, there are review articles published periodically that evaluate and summarize new methods and applications.

Masking

The elimination of an interference from a substance by converting it to a noninterfering form is called *masking*. For example, in a method for determining copper based on measuring the amount of iodine produced when Cu^{2+} reacts with iodide ion, Fe^{3+} interferes by also oxidizing iodide to iodine.

$$2Fe^{3+} + 2I^- \longrightarrow 2Fe^{2+} + I_2$$

Ferric ion can be masked by adding fluoride ion, which forms a stable complex ion with iron(III), FeF_6^{3-}, and prevents its reduction.

$$FeF_6^{3-} + I^- \longrightarrow \text{no reaction}$$

Obviously, to plan and execute successful masking of diverse substances, an extensive knowledge of chemical properties is required.

Separation

Separating the analyte and interfering constituents from each other is an effective solution to the problem of interferences, although generally it is considered a last resort because it is time consuming and can add significantly to the analysis error. Sometimes it is easier to separate the analyte from the sample containing the interference; other times it is best to remove only the interfering substance. Separations can be effected by precipitation, electrodeposition, extraction, ion exchange, volatilization, and chromatography. Some of these techniques are discussed in later chapters. The combination of chromatographic separation and continuous detection has proven to be a very powerful analysis tool and is discussed extensively in Chapters 20 through 22.

Treating the Sample and Measuring the Analyte

Most methods require the analyte to be in a particular form that is suitable for the measurement. For example, in a redox titration the analyte must be in a particular oxidation state before the titration begins. Similarly, in a colorimetric method a colorless analyte must be converted to a colored product for the measurement. The techniques, reactions, and conditions used to accomplish such transformations are diverse and learned mostly through experience.

The exact property or substance measured depends on the type of method selected. There are a great diversity of types, far too many to explore in an introductory course.

In this book we discuss the theory, methodology, and selected applications of the common, simpler types of analytical methods.

Calculating the Amount and Evaluating the Results

The procedure used to calculate the amount or concentration of an analyte depends on the analysis method. Some methods rely on absolute measurements. That is, the amount or concentration of analyte is determined directly from the measured value. Others rely on relative measurements, where the amount or concentration is determined by comparison of the measured value for the unknown with similar values obtained for known amounts of analyte. Both approaches are explored in this book. Absolute measurements are inherently simpler than relative measurements, but uncertainty in the values used to relate the measured quantity to amount or concentration often forces us to employ a relative measurement approach.

Once a result has been obtained, someone is going to want to know how "good" it is or how certain you are that it is the "correct" answer. Therefore, you will need to be familiar with the things that could go wrong and the extent of their influence on the final result. There are two ways to improve your confidence that a final result is "correct"; improve the quality of the apparatus, solutions, and techniques used to obtain the data from which the value is calculated, and increase the number of replicate determinations. As you will learn in Chapter 3, statistics can be very helpful in evaluating the quality of experimental data and showing how it might be improved.

TOOLS AND TECHNIQUES

The range of applications of analytical chemistry is truly enormous and has led to the development of a vast array of tools and techniques with which analytical chemists must be familiar. Some consider the prospect of learning about so many different things forbidding and frustrating; others see it as an exciting challenge.

It is not possible in any textbook to discuss and illustrate all of the important tools and techniques used in analytical chemistry. The topics discussed in this book include both classical and modern methods of analysis and should serve to familiarize the reader with many of the common tools and techniques.

2

REVIEW OF FUNDAMENTALS

As mentioned in Chapter 1, most noninstrumental methods of analysis are designed for use with solutions of the sample. We will, therefore, be making use of various solution concepts, many of which are commonly introduced in general chemistry courses. Any new subject is likely to be challenging enough in its own right without the added burden of constantly being unsure of certain terms and concepts that are assumed to be already a part of one's knowledge. In this chapter we review the concepts and terms that must be mastered prior to studying the principles and applications of quantitative analysis. Most of the topics will be familiar ones, but a few may be new. Your ability to answer the questions and work the problems at the end of this chapter is a good measure of your preparation for the new material in this book.

ELECTROLYTES AND IONIZATION

The word "electrolyte" is used to describe a substance that dissolves in a solvent to produce an electrically conducting solution. Since ions are responsible for conducting current in a liquid, substances that ionize completely are called *strong electrolytes* and those that ionize partly are called *weak electrolytes*. Fortunately, in water, the large majority of substances ionize more than 90% or less than 10%, making the strong–weak classification a useful one. A discussion of the factors influencing the ionization process is beyond the scope of this book but students who are unsure of their ability to predict whether a substance is a strong or weak electrolyte should review their general chemistry notes on bonding, dissociation, and solute–solvent interactions. It is convenient to remember that virtually all sodium, potassium, and ammonium salts, as well as all nitrate and most halide salts, are strong electrolytes in water.

TABLE 2-1 SOLUBILITIES OF COMMON INORGANIC SALTS IN WATER[a]

Ion	Compound	Solubility
NO_3^-	Nitrates	All soluble
NO_2^-	Nitrites	All soluble except Ag^+
$CH_3CO_2^-$	Acetates	All soluble except Ag^+, Hg_2^{2+}, Bi^{3+}
Cl^-	Chlorides	All soluble except Ag^+, Hg_2^{2+}, Bi^{3+}
Br^-	Bromides	All soluble except Ag^+, Hg_2^{2+}, Pb^{2+}
I^-	Iodides	All soluble except Ag^+, Hg_2^{2+}, Pb^{2+}, Bi^{3+}
SO_4^{2-}	Sulfates	All soluble except Pb^{2+}, Ba^{2+}, Sr^{2+}, Ca^{2+}
SO_3^{2-}	Sulfites	All insoluble except Na^+, K^+, NH_4^+
S^{2-}	Sulfides	All insoluble except Na^+, K^+, NH_4^+, Ba^{2+}, Sr^{2+}, Ca^{2+}
PO_4^{3-}	Phosphates	All insoluble except Na^+, K^+, NH_4^+
CO_3^{2-}	Carbonates	All insoluble except Na^+, K^+, NH_4^+
$C_2O_4^{2-}$	Oxalates	All insoluble except Na^+, K^+, NH_4^+
O^{2-}	Oxides	All insoluble except Na^+, K^+, Ba^{2+}, Sr^{2+}, Ca^{2+}
OH^-	Hydroxides	All insoluble except Na^+, K^+, NH_4^+, Ba^{2+}, Sr^{2+}, Ca^{2+}

[a]The compounds listed here include only those of the common metals of groups IA, IB, IIA, and IIB, and Cr, Mn, Fe, Co, Ni, Al, Sn, Pb, Sb, and Bi. The polyatomic ion NH_4^+ is included because of its importance and similarity to group IA metals.

SOLUBILITY

The concentration of dissolved solute at equilibrium with its undissolved form is the substance's *solubility*. Most substances have a finite solubility in water, although some values are so small that currently they cannot be measured. A few compounds, such as ethanol, are infinitely soluble. In much the same manner as electrolytes, a large number of simple inorganic compounds fall into one of two groups: those having a solubility greater than 1 molar (called *soluble*) and those having a solubility of less than 0.01 molar (called *insoluble* or *slightly soluble*). Compounds with solubilities between these two extremes are referred to as being *moderately* soluble.

A solution at equilibrium containing the maximum amount of dissolved solute is said to be *saturated*, and the addition of any more solute or the evaporation of any solvent can lead to separation of the excess solute. It is possible to prepare solutions that contain more solute than that required for saturation and these are called *supersaturated* solutions. They are not stable (although some will last a long time if carefully stored) and ultimately enough solute will separate to produce a saturated solution.

There are many occasions when we prefer to express chemical processes in terms of *ionic* reactions. Consequently, it is important to know whether or not a given substance dissolves and, if it does, how it dissociates. The solubilities of common inorganic salts fall into groups that are fairly easy to remember and are summarized in Table 2-1.

CONCENTRATION

Concentration is a general term expressing the amount of solute contained in a given amount of material. Chemists express concentration in different ways to suit particular needs.

Molarity

The number of moles of solute divided by the number of liters of solution containing the solute is called the *molarity, M,* of the solute. One can also express the amount in millimoles and the volume in milliliters. Thus

$$\text{molarity} = \frac{\text{amount solute (moles)}}{\text{volume solution (liters)}} = \frac{\text{amount solute (millimoles)}}{\text{volume solution (milliliters)}} \quad (2\text{-}1)$$

Recalling that the number of moles of a substance is related to its weight in grams through the molecular weight (MW), we have

$$\text{amount (moles)} = \frac{\text{weight (grams)}}{\text{MW}} \quad (2\text{-}2)$$

or

$$\text{amount (millimoles)} = \frac{\text{weight (milligrams)}}{\text{MW}} \quad (2\text{-}3)$$

It is important to remember that the amount of a substance in moles describes the *number* of particles or molecules present. Thus Equations 2-2 and 2-3 are used to calculate how many particles or molecules are contained in a particular weight of a pure substance.

Example 2-1

What is the molarity of a sodium hydroxide solution prepared by dissolving 2.40 g of NaOH and diluting to 500 mL with water?

Solution   According to Equation 2-1,

$$\text{molarity} = \frac{\text{amount NaOH (mol)}}{\text{volume of solution (L)}}$$

where

$$\text{volume of solution} = \frac{500 \text{ mL}}{1000 \text{ mL/L}} = 0.500 \text{ L}$$

and

$$\text{amount NaOH} = \frac{2.40 \text{g NaOH}}{40.0 \ \dfrac{\text{g NaOH}}{\text{mol NaOH}}} = 0.0600 \text{ mol NaOH}$$

Substitution yields

$$\text{molarity} = \frac{0.0600 \text{ mol NaOH}}{0.500 \text{ L solution}} = 0.120 \ M$$

Example 2-2

How many grams of potassium permanganate are needed to prepare 250 mL of a 0.100 *M* solution?

Solution   The weight of $KMnO_4$ can be calculated from its amount in moles using Equation 2-2.

$$\text{amount KMnO}_4 \text{ (mol)} = \frac{\text{weight KMnO}_4 \text{ (g)}}{\text{MW KMnO}_4 \text{ (g/mol)}}$$

or

$$\text{weight KMnO}_4 \text{ (g)} = \text{amount KMnO}_4 \text{ (mol)} \times \text{MW KMnO}_4$$

According to Equation 2-1,

$$\text{amount KMnO}_4 = 0.100 \text{ mol/L} \times 0.250 \text{ L} = 0.0250 \text{ mol}$$

Substituting this value and the MW for $KMnO_4$ in the equation above produces the desired answer:

$$\text{wt KMnO}_4 = 0.0250 \text{ mol KMnO}_4 \times 158 \frac{\text{g KMnO}_4}{\text{mol KMnO}_4} = 3.95 \text{ g}$$

Normality

The use of normality as an expression of concentration is a subject of some controversy among chemists. The trend seems to be in favor of avoiding its use. Accordingly, normality is not used in solving problems in this book. However, because this once popular unit of concentration is still encountered in the workplace and in the literature, some familiarity with its basis and use is necessary.

The driving force behind the development of the normality system was a desire to simplify the calculations involved with certain types of problems. As will be learned shortly, one common approach to determining the amount of an analyte is to measure the amount of a substance required to react with it completely. For example, we may wish to determine the amount of calcium hydroxide in a solution by measuring the amount of hydrochloric acid needed to react with it according to the following reaction:

$$Ca(OH)_2 + 2HCl \longrightarrow CaCl_2 + 2H_2O \qquad (2\text{-}4)$$

We measure the amount of HCl but want to know the amount of $Ca(OH)_2$. Obviously, we need a ''bridge'' to get from one substance to the other. This bridge is the reacting ratio of 1 molecule of $Ca(OH)_2$ per 2 molecules of HCl or 1 mole of $Ca(OH)_2$ per 2 moles of HCl. To determine the reacting ratio the chemical reaction must be written and balanced. Some chemists find it desirable to work problems as if this ''bridge'' were always 1 to 1. That is, the $Ca(OH)_2$ present consumes an *equivalent* amount of HCl and produces an *equivalent* amount of $CaCl_2$. The reacting ratio still must be taken into account, but it is done by combining it with the molarity to produce a new unit of concentration called *normality*.

Normality is defined as the number of equivalents (eq) of solute divided by the number of liters of solution containing the solute. As with molarity, milliequivalents and milliliters can be used also. Thus

$$\text{normality} = \frac{\text{amount solute (eq)}}{\text{volume solution (L)}} = \frac{\text{amount solute (meq)}}{\text{volume solution (mL)}} \qquad (2\text{-}5)$$

An equivalent is a number-based unit similar to a mole and is related to the weight of a substance through its equivalent weight (EW).

$$\text{amount (equivalents)} = \frac{\text{weight (grams)}}{\text{EW}} \tag{2-6}$$

or

$$\text{amount (milliequivalents)} = \frac{\text{weight (milligrams)}}{\text{EW}} \tag{2-7}$$

The equivalent weight is chosen so that the number of equivalents of one substance used or produced in a reaction *equals* the number of equivalents of every other substance used or produced in that reaction. For Reaction 2-4 one could say that 1 eq of $Ca(OH)_2$ reacts with 1 eq of HCl and produces 1 eq of $CaCl_2$.

Equivalent weight is related to molecular weight according to a simple formula:

$$\text{EW} = \frac{\text{MW}}{h} \tag{2-8}$$

where h has units of eq/mol. The numerical value of h depends on the chemical reaction in which a substance is involved. The same substance undergoing different reactions can have different values of h and, therefore, different equivalent weights. It is *imperative* that normality, equivalents, or equivalent weight always be referred to in terms of a specified or implied reaction.

Normality is related to molarity in just the same way that equivalent weight is related to molecular weight. From Equations 2-1 and 2-2, we have

$$\text{molarity} = \frac{\text{weight}}{\text{MW} \times \text{volume}} \quad \text{or} \quad \text{MW} = \frac{\text{weight}}{\text{molarity} \times \text{volume}} \tag{2-9}$$

Similarly, from Equations 2-5 and 2-7,

$$\text{normality} = \frac{\text{weight}}{\text{EW} \times \text{volume}} \quad \text{or} \quad \text{EW} = \frac{\text{weight}}{\text{normality} \times \text{volume}} \tag{2-10}$$

Substituting Equations 2-9 and 2-10 for MW and EW in Equation 2-8 gives

$$\frac{\cancel{\text{weight}}}{\text{normality} \times \cancel{\text{volume}}} = \frac{\left(\dfrac{\cancel{\text{weight}}}{\text{molarity} \times \cancel{\text{volume}}}\right)}{h}$$

or

$$\text{normality} = \text{molarity} \times h \tag{2-11}$$

Since h is *almost* always greater than or equal to 1, normality is almost always greater than or equal to molarity.

Chemists seem to agree that the easiest way to define reactivity, and thus h, is in terms of unit electrical charge. How this charge is established depends on the type of reaction.

Concentration

Oxidation-reduction (redox) reactions

The value of h for a reactant or product in a redox reaction equals the number of electrons lost or gained in the reaction by *one* ion or molecule of the substance. The easiest way to determine this value is to write and balance the redox half-reaction for the substance of interest. Consider the oxidation of iodide ion by ferric ion:

$$2I^- + 2Fe^{3+} \longrightarrow I_2 + 2Fe^{2+}$$

The balanced half-reaction for the oxidation of iodide ion is

$$2I^- \rightleftharpoons I_2 + 2e^-$$

This equation says that *one* I_2 is worth, or is equivalent to, two e^- and thus h for I_2 is 2. Similarly, *one* I^- is equivalent to one e^- and h for I^- is 1. The balanced half-reaction for the reduction of iron is

$$Fe^{3+} + e^- \rightleftharpoons Fe^{2+}$$

and h is 1 for both Fe^{3+} and Fe^{2+}. Since success in determining h for redox reactions depends almost entirely on your ability to balance half-reactions, and since half-reactions are used extensively in electrochemistry, it is a skill worth developing. A detailed procedure for balancing redox reactions using the method of half-reactions is given in Appendix B.

Example 2-3

Calculate the normality of a solution prepared by dissolving 220.0 mg of $K_2Cr_2O_7$ in 100.0 mL of water that will be used to oxidize ferrous chloride according to the following reaction (unbalanced):

$$K_2Cr_2O_7 + FeCl_2 + HCl \longrightarrow CrCl_3 + FeCl_3 + KCl + H_2O$$

Solution   The normality can be calculated from Equation 2-10 once the equivalent weight of $K_2Cr_2O_7$ is determined.

$$\text{normality} = \frac{\text{weight } K_2Cr_2O_7}{\text{EW } K_2Cr_2O_7 \times \text{volume}}$$

To find h and the EW we need the balanced half-reaction for $K_2Cr_2O_7$. It is simpler to use ionic reactions; thus

$$Cr_2O_7{}^{2-} + 14H^+ + 6e^- \rightleftharpoons 2Cr^{3+} + 7H_2O$$

and h is 6 for $Cr_2O_7{}^{2-}$ or $K_2Cr_2O_7$, giving us an equivalent weight of

$$EW = \frac{MW}{h} = \frac{294.2 \text{ mg/mmol}}{6 \text{ meq/mmol}} = 49.03 \text{ mg/meq}$$

Substituting in the equation above gives us the answer:

$$\text{normality} = \frac{220.0 \text{ mg } K_2Cr_2O_7}{49.03 \text{ mg/meq} \times 100.0 \text{ mL}} = 0.04487 \text{ meq/mL}$$

Example 2-4

What weight of ferrous chloride is needed to react completely with 50.00 mL of the $K_2Cr_2O_7$ solution from Example 2-3?

Solution We know that the number of milliequivalents of $K_2Cr_2O_7$ and $FeCl_2$ are equal by definition. Thus, from Equation 2-5,

$$\text{amount } FeCl_2 = \text{amount } K_2Cr_2O_7 = N_{K_2Cr_2O_7} \times V_{soln}$$

$$= 0.04487 \text{ meq/mL} \times 50.00 \text{ mL} = 2.244 \text{ meq}$$

Now that the amount of $FeCl_2$ is known, we can calculate its weight from Equation 2-7:

$$2.244 \text{ meq } FeCl_2 = \frac{\text{wt } FeCl_2}{\text{EW } FeCl_2}$$

We need the balanced half-reaction for iron to find h and the EW:

$$Fe^{2+} \rightleftharpoons Fe^{3+} + e^-$$

Thus h is 1 for Fe^{2+} or $FeCl_2$ and the EW is

$$EW = \frac{MW}{h} = \frac{126.8 \text{ mg/mmol}}{1 \text{ meq/mmol}} = 126.8 \text{ mg/meq}$$

Substituting in the equation above, we obtain

$$2.244 \text{ meq } FeCl_2 = \frac{\text{wt } FeCl_2}{126.8 \text{ mg/meq } FeCl_2}$$

or

$$\text{wt } FeCl_2 = 2.244 \text{ meq} \times 126.8 \text{ mg/meq} = 284.5 \text{ mg}$$

Nonredox reactions

In nonredox reactions the unit electrical charge is taken to be a unit of *ionic charge*, either positive or negative. The ionic charge, and therefore h, is best determined by counting the number of univalent cations *or their equivalent* consumed or produced by *one* ion or molecule of substance. For the purpose of this calculation, an anion is considered equivalent to a cation, a divalent ion is equivalent to two univalent cations, and so forth. The application of this definition to acid-base reactions is quite simple because the primary reactants are usually hydrogen or hydroxide ions which are univalent and therefore represent one unit of charge each. Suppose that you wanted to know the value of h for sulfuric acid used in the following reaction:

$$H_2SO_4 + 2NaOH \longrightarrow Na_2SO_4 + 2H_2O$$

It is easy to see that one sulfuric acid molecule consumes *two* univalent hydroxide ions or two units of charge, making h equal to 2. Beware of the tendency simply to look at the acid (or base) and count the number of hydrogens (or hydroxides). This simple approach does not always lead to the correct answer. The definition clearly states that one must determine not what the substance of interest *has*, but what it *reacts with or produces*. Phosphoric acid has 3 acidic hydrogens, but as you can see from the following reactions, it may not use all of them.

$$H_3PO_4 + NaOH \longrightarrow NaH_2PO_4 + H_2O$$

$$H_3PO_4 + 2NaOH \longrightarrow Na_2HPO_4 + 2H_2O$$

$$H_3PO_4 + 3NaOH \longrightarrow Na_3PO_4 + 3H_2O$$

The value of h for phosphoric acid used according to the first reaction is 1 because it reacts with one hydroxide ion, representing one unit of charge. The values of h for phosphoric acid used in the second and third reactions are 2 and 3, respectively.

Example 2-5

What is the normality of a 0.05140 M $Ca(OH)_2$ solution that reacts with HCl according to Equation 2-4?

Solution Since one $Ca(OH)_2$ molecule reacts with two hydrogen ions from the HCl, h is 2 and

$$\text{normality} = \text{molarity} \times h = 0.05140 \times 2 = 0.1028$$

The definition is also valid for reactions in which precipitates and complex ions are formed, but it is often awkward to apply. For this reason, and because many of the important precipitation and complexation reactions take place with one-to-one stoichiometry, most chemists prefer not to use the normality system with these types of reactions.

Percent

The most common manner of reporting the final result of a quantitative determination is by percent (parts per hundred) of analyte. The percentage can be expressed in terms of weight or volume.

$$\text{weight percent (w/w)} = \frac{\text{weight analyte}}{\text{weight sample}} \times 100$$

$$\text{volume percent (v/v)} = \frac{\text{volume analyte}}{\text{volume sample}} \times 100$$

Both weight and volume percent are relative values and, as such, do not depend on the units of weight or volume used, provided that both numerator and denominator have the same units. Ordinarily, if the type of percent is not specified, we assume it to be a weight percent. The bridge between weight and volume is density, which can be used to convert one type of percent to another, as illustrated in the following example.

Example 2-6

The weight percent of commercial concentrated perchloric acid is 70.0%. If its density is 1.664 g/mL, calculate (a) the weight of perchloric acid in 1.00 mL of the solution, and (b) the molarity of the perchloric acid.

Solution This type of problem is best solved by paying close attention to the units of the information available and sought.

(a) What we want is weight of $HClO_4$/volume of solution. What we have is %(w/w), which has units of g $HClO_4$/g solution, and density, which has units of g solution/mL solution. Thus

$$\text{wt } HClO_4/\text{vol solution} = 0.700 \, \frac{\text{g } HClO_4}{\text{g solution}} \times 1.664 \, \frac{\text{g solution}}{\text{mL solution}}$$

$$= 1.16 \, \frac{\text{g } HClO_4}{\text{mL solution}}$$

(b) Molarity has units of mol/L, so we must convert g of $HClO_4$ to mol using the molecular weight:

$$\frac{1.16 \ \dfrac{\text{g } HClO_4}{\text{mL solution}}}{100.5 \ \dfrac{\text{g } HClO_4}{\text{mol } HClO_4}} = 0.0115 \ \frac{\text{mol } HClO_4}{\text{mL solution}}$$

and

$$\text{molarity} = 0.0115 \ \frac{\text{mol } HClO_4}{\text{mL solution}} \times 1000 \ \frac{\text{mL}}{\text{L}} = 11.5 \ \frac{\text{mol } HClO_4}{\text{L}}$$

Percentage is seldom used to express very small concentrations, presumably because of the inconvenience of using zeros or powers of 10 to keep track of the decimal point. To avoid this inconvenience, chemists often change the multiplier of the weight or volume ratio.

Parts per Million

Recognizing that weight percent is also called parts per hundred, the obvious definition of parts per million (ppm) is

$$\text{ppm} = \frac{\text{weight analyte}}{\text{weight sample}} \times 10^6 \tag{2-12}$$

Occasionally, parts per thousand (ppt) and parts per billion (ppb) are used as well. The only difference is the number by which the weight ratio is multiplied.

When the analyte and other solute concentrations are of the order of a few parts per million or less, the solution is nearly pure solvent and will have a density essentially equal to that of the solvent. If the solvent is *water*, its density is 1.00 g solution/mL solution. This means that 1.0 L of solution will weigh 1.0 kg and Equation 2-12 could be written

$$\text{ppm} = \frac{\text{kg analyte}}{\text{kg sample}} \times 10^6 = \frac{\text{kg analyte}}{\text{L solution}} \times 10^6 = \frac{\text{mg analyte}}{\text{L solution}}$$

When this equation is valid, the conversion of parts per million to molarity is straightforward, as illustrated in the following example.

Example 2-7

Calculate the molarity of a 5.00-ppm $Ca(NO_3)_2$ solution.

Solution Since 5.00 ppm means 5.00 mg $Ca(NO_3)_2$/L solution, if milligrams of $Ca(NO_3)_2$ can be converted to moles, the problem is solved. Recalling that molecular weight is the bridge between weight and moles, we have

$$\text{molarity} = \frac{5.00 \ \dfrac{\text{mg } Ca(NO_3)_2}{\text{L solution}} \times \dfrac{1.000 \ \text{g } Ca(NO_3)_2}{1000 \ \text{mg } Ca(NO_3)_2}}{164 \ \dfrac{\text{g } Ca(NO_3)_2}{\text{mol } Ca(NO_3)_2}}$$

$$= 3.05 \times 10^{-5} \ \frac{\text{mol}}{\text{L}}$$

Since parts per million is a weight-based unit of concentration, comparisons between solutions of different substances must be made carefully. Ten-ppm solutions of Ca^{2+} and Mg^{2+} do not contain the same number of calcium and magnesium ions per unit volume.

Types of Concentration

When substances dissolve they frequently undergo chemical changes such as dissociation into ions. As a result, the amount of a substance added to the solution may not be the same as the amount of *that same substance* in the solution. We will find it absolutely essential to distinguish between the two amounts. The *total* number of moles of a particular substance X, irrespective of its state of dissociation or association, contained in a liter of solution is referred to as the molar *analytical concentration* and is denoted as C_x. The term *formal concentration* also has been used to express this quantity. The *actual* concentration of a particular ionic or molecular substance X in solution is called its *equilibrium concentration* and is denoted as [X]. The distinction between these two types of concentration is illustrated by examining a solution prepared by dissolving 60 g (1 mol) of acetic acid, CH_3CO_2H, in water and diluting to 1.0 L. The analytical concentration of CH_3CO_2H is 1.0 *M* but the equilibrium concentration is less, about 0.98 *M*, because some of the acetic acid molecules dissociate into H^+ and $CH_3CO_2^-$. To determine the exact equilibrium concentration, one must know the extent of the dissociation. How this is done will be one of the main topics of discussion in several later chapters.

ACTIVITY

The molar equilibrium concentration describes the number of a particular species present per liter of solution. Circumstances can arise, however, where certain physical properties are representative of a solution of different concentration than actually exists. This different or effective concentration is called *activity*, a_x, and results mostly from transient, electrostatic interactions between ions in solution. To see how this can come about, consider a 1.0-L solution containing 1.0 mol of sodium chloride. The electrical conductivity of this solution depends on the concentration of ions, which would seem to be 1.0 *M* Na^+ and 1.0 *M* Cl^-. But sodium and chloride ions are electrostatically attracted to one another, and at any given instant, some of the ions may be so close together that they behave as a neutral molecule. As a result, the "effective" concentration of ions is different from the actual concentration and it is this effective value that determines the magnitude of the conductivity.

The mathematical relationship between concentration and activity appears very simple:

$$a_x = \gamma [X] \tag{2-13}$$

where γ is called the *activity coefficient*. Estimating activity from concentration, or vice versa, depends on our ability to calculate the activity coefficient. The value of γ depends on the concentration of *all* ions in the solution. Since the probability of interactions decreases with a decrease in the concentration of ions, the activity coefficient approaches unity as the ionic concentration approaches zero. When dealing with a nonionic substance or when the total ionic concentration is low, the difference between activity and concentration may be small enough to be ignored, allowing concentration to be used in place of activity.

Since the activity coefficient depends on the ionic concentration in the solution, it is essential that we have some way of calculating its value for a particular set of solution conditions. In 1923, Debye and Hückel reported the derivation of the following expression, relating the activity coefficient to the charge on an ion and the ionic strength of the solution:

$$-\log \gamma_i = \frac{0.5 Z_i^2 \sqrt{\mu}}{1 + \sqrt{\mu}} \tag{2-14}$$

where γ_i is the activity coefficient for ion i, Z_i the charge on ion i, and μ the ionic strength of the solution. The ionic strength is a measure of the number and charge of all the ions in the solution and is calculated from the equation

$$\mu = 0.5 \sum (C_i Z_i^2) \tag{2-15}$$

where C_i is the molar concentration of ion i. The Greek letter sigma, Σ, means "the sum of" and Equation 2-15 says that the concentration times the charge squared for *each* ion in the solution is summed before multiplying by 0.5.

Example 2-8

What is the calcium-ion activity of a solution containing 0.010 M $CaCl_2$ and 0.20 M $MgCl_2$?

Solution The activity can be calculated from Equation 2-13:

$$a_{Ca^{2+}} = \gamma_{Ca^{2+}} [Ca^{2+}]$$

The calcium-ion concentration is known, but $\gamma_{Ca^{2+}}$ must be calculated using Equation 2-14:

$$-\log \gamma_{Ca^{2+}} = \frac{0.5 Z_{Ca^{2+}}^2 \sqrt{\mu}}{1 + \sqrt{\mu}}$$

The ionic strength is calculated using Equation 2-15:

$$\mu = 0.5 \sum (C_i Z_i^2)$$
$$= 0.5 (C_{Ca^{2+}} Z_{Ca^{2+}}^2 + C_{Mg^{2+}} Z_{Mg^{2+}}^2 + C_{Cl^-} Z_{Cl^-}^2)$$

Since both $CaCl_2$ and $MgCl_2$ are strong electrolytes, the concentration of each ion is

$$[Cl^-] = 2(0.010) + 2(0.20) = 0.42 \ M$$
$$[Mg^{2+}] = 0.20 \ M$$
$$[Ca^{2+}] = 0.010 \ M$$

and

$$\mu = 0.5[(0.01)(+2)^2 + (0.20)(+2)^2 + (0.42)(-1)^2] = 0.63$$

Then

$$-\log \gamma_{Ca^{2+}} = \frac{0.5(+2)^2 \sqrt{0.63}}{1 + \sqrt{0.63}} = 0.89$$

Activity

and

$$\gamma_{Ca^{2+}} = 0.13$$

Finally,

$$a_{Ca^{2+}} = (0.13)(0.010) = 0.0013$$

The Debye–Hückel equation is not an exact equation for determining activity coefficients, but it does give good estimates when the ionic strength is not substantially greater than about 1.0.

The activities of pure liquids, solids, and gases are defined differently from the activity of dissolved solutes. Pure liquids or solids at the standard temperature and pressure are assigned an activity of exactly 1.0. The activity of a gaseous substance is equal to its partial pressure in atmospheres.

EXPONENTIAL AND LOGARITHMIC FUNCTIONS

The use of exponents and logarithms is very important in science and historically, students have had to be proficient with both of these mathematical functions. Modern electronic calculators have largely eliminated the drudgery of looking up log and antilog values in tables, but they have not lessened the need to understand and appreciate the properties and characteristics of these functions. A brief review of the expression and manipulation of numbers using exponents and logarithms is given in Appendix A.

p-VALUES

Scientists often find it convenient to express a molar concentration or equilibrium constant in terms of its p-value, which is defined as the negative logarithm (base 10) of the value. Thus, for a value X,

$$pX = -\log X \quad \text{or} \quad \log \frac{1}{X}$$

By using this function, a very wide range of numerical values can be expressed in terms of small, usually positive numbers. In most cases, p-values are defined to represent the *activities* of dilute solutes, but as pointed out in the preceding section, the difference between activity and concentration often is negligible in problems of quantitative analysis. p-values are used with concentrations throughout this text with the assumption that such a substitution introduces no appreciable error.

Example 2-9

Represent a hydrogen-ion concentration of $7.4 \times 10^{-5}\ M$, a sodium-ion concentration of $2.5\ M$, and an equilibrium constant of 2.6×10^{-14} in terms of their p-values.

Solution

$$pH = -\log [H^+] = -\log (7.4 \times 10^{-5})$$
$$= -\log 7.4 - \log 10^{-5}$$
$$= -0.87 - (-5) = 4.13$$
$$pNa = -\log [Na^+] = -\log 2.5 = -0.40$$
$$pK_{eq} = -\log K_{eq} = -\log (2.6 \times 10^{-14})$$
$$= -\log 2.6 - \log 10^{-14}$$
$$= -0.41 - (-14) = 13.59$$

Example 2-10

Calculate the molar concentration of H^+ in a solution whose pH is (a) 3.71, (b) 7.00, and (c) 12.84.

Solution

(a)
$$pH = -\log [H^+] = 3.71$$
$$\log [H^+] = -3.71$$
$$[H^+] = 10^{-3.71} = 1.9 \times 10^{-4}\ M$$

(b)
$$pH = -\log [H^+] = 7.00$$
$$\log [H^+] = -7.00$$
$$[H^+] = 10^{-7.00} = 1.0 \times 10^{-7}\ M$$

(c)
$$pH = -\log [H^+] = 12.84$$
$$\log [H^+] = -12.84$$
$$[H^+] = 10^{-12.84} = 1.45 \times 10^{-13}\ M$$

Note, as indicated by the sodium-ion concentration in Example 2-9, that the p-values of numbers greater than unity are negative.

CHEMICAL EQUILIBRIUM

Chemical equilibrium is a consequence of the fact that many reactions can proceed simultaneously in opposite directions. This interesting observation plays an extremely important role in both the methodology and calculations used in quantitative analysis, and it is therefore necessary to have a good understanding of how and why it comes about. To review this phenomenon called equilibrium, consider the general reaction

$$A + B \rightleftharpoons C + D$$

If C and D are the first and only products formed by the reaction of A and B, the rate of the reaction depends on the concentrations of A and B. As the reaction proceeds the concentrations decrease, causing a simultaneous decrease in the rate of the reaction, as shown by the top line in Figure 2-1. From this part of the figure two important observations can be made: first, the rate does not decrease to zero; and second, after a time the rate levels off to a constant value.

Accepting the premise that the opposite (right to left) reaction can also occur, consider what should happen to its rate. Initially, no C or D is present and there can be no opposite or reverse reaction. When C and D are formed and their concentrations increase as a result of the forward reaction, the rate of the reverse reaction increases as shown by the bottom line in Figure 2-1. If the forward rate is decreasing and the reverse rate increasing, eventually they will become equal. The condition that exists when this occurs is called *equilibrium* and it is characterized by the fact that while both forward and reverse reactions are taking place, no *net* change occurs in the concentrations of any of the species. The time required to establish equilibrium, t_{eq} in Figure 2-1, depends on the rates of the forward and reverse reactions, which, in turn, depend on the nature of the reactants and the conditions of the reaction. Simple neutralization reactions can reach equilibrium in less than a nanosecond; other reactions may take days or even years to reach such a state.

Since the rate of a reaction depends on the concentration of reacting substances, we can also examine the equilibrium process by plotting the concentration of reactants and products versus time as shown in Figure 2-2. As the forward reaction proceeds, the concentration of the reactants A and B decrease while the concentrations of the products C and D increase. A careful comparison of curves such as those in Figures 2-1 and 2-2 for a given chemical system would show that the concentrations and rates of reactions reach their constant values simultaneously. The fact that a reaction is reversible and reaches a condition of equilibrium is depicted by the use of a double arrow ($\rightleftharpoons$) in the chemical equation for the reaction. In dealing with reactions where equilibrium exists but is not an important consideration, chemists frequently omit the double arrow. Consequently, a single arrow does not unequivocally indicate an irreversible reaction.

The position of the equilibrium, which is indicated by the concentrations of prod-

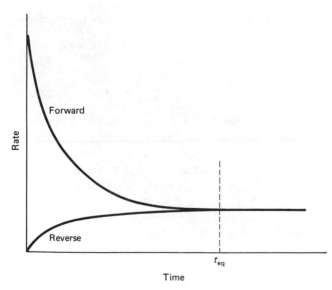

Figure 2-1 Rate versus time profiles for a reversible reaction.

Chapter 2   Review of Fundamentals

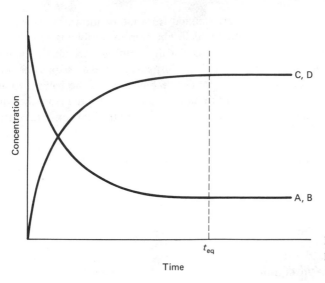

Figure 2-2 Concentration versus time profiles for a reversible reaction: A + B $\rightleftharpoons$ C + D.

ucts and reactants, can vary from almost no reaction to almost complete reaction. It is important to recognize that the *time* required to establish equilibrium and the *position* of the equilibrium are not directly related, each being determined by different properties of the system.

The idea that reactions do not always go to completion—which is to say that back or reverse reactions can exist—complicates the way in which the concentrations of substances in solution are calculated. Not only must the amounts of substances taken be considered, but also the extent to which the reaction occurs. It is essential therefore that we be able to describe the equilibrium condition mathematically.

The Equilibrium Constant

A consequence of the equilibrium condition is that the activities of the reactants are always related in the same way to the activities of the products. For the general reaction

$$m\text{A} + n\text{B} \quad \rightleftharpoons \quad p\text{C} + q\text{D} \tag{2-16}$$

the mathematical description of the equilibrium process is given by the expression

$$K_{eq} = \frac{(a_C)^p (a_D)^q}{(a_A)^m (a_B)^n}$$

where K_{eq} is the equilibrium constant. This equation says that regardless of how the equilibrium is established, the ratio of the activities of *products to reactants*, when each term is raised to the power of its coefficient in the equilibrium reaction, is always the same value, K_{eq}. Although the equilibrium constant is independent of the starting activities used, it does depend on other solution parameters such as temperature and solvent. Unless otherwise specified, the equilibrium constants used in this book are for water solutions and temperatures of 25°C.

Equilibrium expressions are simplified by the fact that the activity of pure solids and liquids is fixed and defined as unity. Consequently, these substances do not affect the value of the equilibrium constant and are omitted from the equilibrium expression. Furthermore, many of the solutions encountered in equilibrium problems have solute concentrations less than about 1 M, which means that the solution is comprised mainly of solvent; and pure solvent has an activity of 1.0. Normally, we can assume the presence of small concentrations of solutes does not alter the activity of the solvent to any appreciable extent, which means that the solvent term also can be deleted from the equilibrium expression. Keeping these assumptions in mind, the equilibrium constant expression for the reaction

$$2Sn^{2+} + H_2SO_3 + 4H^+ \rightleftharpoons 2Sn^{4+} + S(s) + 3H_2O$$

would be written

$$K_{eq} = \frac{\left(a_{Sn^{4+}}\right)^2}{\left(a_{Sn^{2+}}\right)^2 \left(a_{H_2SO_3}\right) \left(a_{H^+}\right)^4}$$

The use of activities in equilibrium expressions complicates most calculations. However, insofar as *solutes* are concerned, the difference between activity and equilibrium concentration may be sufficiently small to ignore if the total ionic concentration of the solution is low (see Equations 2-14 and 2-15). Such an assumption is made routinely throughout this book. Furthermore, the activity of gaseous substances may be given by their partial pressures in atmospheres. Although equilibrium expressions written in terms of concentration are not exactly correct, they generally provide answers that are sufficiently accurate for most applications. The equilibrium expression for Reaction 2-16, written in terms of concentration, is

$$K_{eq} = \frac{[C]^p [D]^q}{[A]^m [B]^n} \tag{2-17}$$

The magnitude of the equilibrium constant is a measure of the extent to which a reaction occurs. For a reaction that goes nearly to completion, the concentration of products will be high relative to the concentration of reactants remaining, and therefore the equilibrium constant is large. Exactly the opposite is expected for a reaction that proceeds only slightly to completion.

A chemical equilibrium can be attained through either of two opposite reactions. Equation 2-17 represents an equilibrium condition attained by reacting A with B according to Reaction 2-16. The same equilibrium, established by the opposite reaction,

$$pC + qD \rightleftharpoons mA + nB$$

is represented by the expression

$$K_{eq} = \frac{[A]^m [B]^n}{[C]^p [D]^q} \tag{2-18}$$

Equilibrium constants for reactions that are opposites of each other are reciprocals. Thus

$$K_{eq(2-17)} = \frac{1}{K_{eq(2-18)}}$$

Use in problem solving

Before specific problems using the equilibrium constant expression are examined, it is worthwhile to consider how we go about or should go about solving problems. All too often, the approach is to look for a single formula that all of the data can be "plugged" into and solved for the final answer. Although this approach may work with elementary, one-step problems, it is a disaster with more complicated, multistep problems. Learning to solve chemical problems—and more specifically, equilibrium problems—is somewhat like a carpenter learning to build houses. A good carpenter can build a house that he has never built or seen before because he is familiar and experienced with many simpler operations, such as how to frame a door, construct a wall, and pitch a roof, and is able to select and perform these operations in a sequence that will lead to the desired house. Good scientists use a similar approach in solving scientific problems. Experience in problem solving (meaning homework) is just as important to the quantitative analysis student as experience in building is to the apprentice carpenter.

One of the most common uses of equilibrium constants lies in the determination of equilibrium concentrations of substances from their analytical concentrations. To illustrate, consider a substance AB that dissociates into A and B when dissolved in water:

$$AB \rightleftharpoons A + B \tag{2-19}$$

The equilibrium concentration of A, B, or AB can be calculated from the equilibrium constant equation for the dissociation reaction:

$$K_{eq} = \frac{[A][B]}{[AB]} \tag{2-20}$$

To find the concentration of A, the quantities [B] and [AB] must be represented in terms of the unknown. If the dissociation of AB is the only source of A and B, their equilibrium concentrations must be equal, or

$$[B] = [A] \tag{2-21}$$

The equilibrium concentration of AB is the analytical concentration minus the concentration of AB lost through dissociation. But the concentration of dissociated AB is the same as the concentration of A (or B); thus

$$[AB] = C_{AB} - [A] \tag{2-22}$$

Substituting these values in the expression for the equilibrium constant gives

$$K_{eq} = \frac{[A][A]}{C_{AB} - [A]} = \frac{[A]^2}{C_{AB} - [A]} \tag{2-23}$$

or

$$[A]^2 + K_{eq}[A] - K_{eq}C_{AB} = 0 \tag{2-24}$$

If the equilibrium constant for the reaction and the analytical concentration of AB are known, Equation 2-24 can be solved using the quadratic formula (see Appendix C).

It is often possible to make assumptions that will simplify Equation 2-23 without changing the final answer significantly. We are concerned specifically with the denominator in which one value is being subtracted from another. It is often the case that $[A]$ is much smaller than C_{AB}, which means that

$$C_{AB} - [A] \simeq C_{AB} \qquad (2\text{-}25)$$

and Equation 2-23 simplifies to a nonquadratic form:

$$K_{eq} = \frac{[A]^2}{C_{AB}} \qquad (2\text{-}26)$$

or

$$[A] = \sqrt{K_{eq} C_{AB}}$$

Whether or not Equation 2-25 can be accepted as valid depends on the use to which the final answer will be put. Generally, such assumptions are less likely to be made when they will become the major source of error in the final answer. In this book such assumptions are considered acceptable if they do not change the calculated value by more than 2%, relative. Remember that it is not a question of whether the concentration of A is small, but whether it is small *compared to* the number from which it is being subtracted (or added). It is crucial that you learn to recognize and evaluate assumptions of this type because they are encountered again and again in the problems of quantitative analysis.

Students frequently ask how they can decide *in advance* whether a simplifying assumption should be made. The value of the equilibrium constant is the key: when K_{eq} is "large," it means that the equilibrium represented by Equation 2-19 lies far to the right and the equilibrium concentration of AB will be significantly less than its analytical concentration. When K_{eq} is "small," the equilibrium lies far to the left and the equilibrium and analytical concentration of AB will be essentially the same. What constitutes a "small" or "large" K_{eq} is somewhat difficult to specify. Generally, if the value of K_{eq} is about three orders of magnitude (1000 times) less than the analytical concentration, it can be considered "small," making Equation 2-25 valid.

Example 2-11

A solution is prepared by dissolving 0.200 mol of $NaHSO_4$ in water and diluting to 1.00 L. If the equilibrium constant for the dissociation of HSO_4^- is 1.02×10^{-2}, calculate the concentration of hydrogen ion (a) without and (b) with the assumption shown by Equation 2-25.

Solution

(a) When dissolved in water, $NaHSO_4$ dissociates completely into Na^+ and HSO_4^- (typical of sodium salts). The further dissociation of HSO_4^- can be written as

$$HSO_4^- \rightleftharpoons H^+ + SO_4^{2-}$$

for which

$$K_{eq} = \frac{[H^+][SO_4^{2-}]}{[HSO_4^-]}$$

where

$$[H^+] = [SO_4^{2-}]$$

$$[HSO_4^-] = C_{HSO_4^-} - [H^+]$$

Substituting these quantities in the equilibrium expression gives the equation

$$K_{eq} = \frac{[H^+]^2}{C_{HSO_4^-} - [H^+]}$$

where

$$C_{HSO_4^-} = \frac{0.200 \text{ mol}}{1.00 \text{ L}} = 0.200 \text{ } M$$

With the known values, the equilibrium equation can be solved using the quadratic formula:

$$1.02 \times 10^{-2} = \frac{[H^+]^2}{0.200 - [H^+]}$$

$$[H^+]^2 + (1.02 \times 10^{-2})[H^+] - (2.04 \times 10^{-3}) = 0$$

$$[H^+] = 4.04 \times 10^{-2} \text{ } M$$

(b) If we make the assumption that

$$C_{HSO_4^-} - [H^+] \simeq C_{HSO_4^-}$$

the equilibrium equation reduces to

$$K_{eq} = \frac{[H^+]^2}{C_{HSO_4^-}}$$

or

$$1.02 \times 10^{-2} = \frac{[H^+]^2}{0.200}$$

and

$$[H^+] = 4.51 \times 10^{-2} \text{ } M$$

It is always good practice to check the validity of any assumptions made in solving equilibrium problems once the final answer has been computed. In Example 2-11(b), it is clear that the calculated hydrogen-ion concentration of 4.51×10^{-2} M is not small enough compared to the analytical concentration of HSO_4^- to accept the assumption represented by Equation 2-25.

Example 2-12

A solution of acetic acid whose analytical concentration was 0.100 M was found to have a hydrogen-ion concentration of 1.32×10^{-3} M. Calculate the equilibrium constant for the dissociation of this acid.

Solution   The dissociation reaction is

$$CH_3CO_2H \rightleftharpoons H^+ + CH_3CO_2^-$$

for which

$$K_{eq} = \frac{[H^+][CH_3CO_2^-]}{[CH_3CO_2H]}$$

Since the dissociation is the sole source of both H^+ and $CH_3CO_2^-$,

$$[H^+] = [CH_3CO_2^-] = 1.32 \times 10^{-3} \text{ } M$$

Chemical Equilibrium

and

$$[CH_3CO_2H] = C_{CH_3CO_2H} - [H^+]$$

$$= 0.100 - (1.32 \times 10^{-3}) = 0.099 \ M$$

Substituting these values in the equilibrium expression gives

$$K_{eq} = \frac{(1.32 \times 10^{-3})(1.32 \times 10^{-3})}{0.099} = 1.76 \times 10^{-5}$$

The Common-Ion Effect

Although the value of K_{eq} is constant at a given temperature and pressure, the *position* of an equilibrium, as indicated by the concentrations of reactants and products, can vary. For example, suppose that solutions of A and B are mixed and Reaction 2-16 proceeds until an equilibrium is established. The addition of more A will lead to an increase in the rate of the forward reaction. There is no *immediate* change in the rate of the reverse reaction because the concentrations of C and D were not changed. Since the forward and reverse reaction rates are no longer equal, a net reaction takes place until a new condition of equilibrium is established. The concentrations of A, B, C, and D will be different from their original values, but their ratio according to Equation 2-17 will remain the same—that is, K_{eq} does not change.

This so-called *common-ion effect* is encountered often in solution chemistry and it influences the way we solve equilibrium problems. Suppose that a weak electrolyte AB is dissolved in water along with some of its common ion B and we wish to find the equilibrium concentration of A. The dissociation reaction and equilibrium constant equation are given by Equations 2-19 and 2-20:

$$AB \; \rightleftharpoons \; A + B$$

$$K_{eq} = \frac{[A][B]}{[AB]}$$

To calculate the concentration of A, values for K_{eq}, [B], and [AB] are needed. Previously, we were able to say that the concentrations of A and B were equal, but this is no longer the case because there are now two sources of B: that added to the solution, whose concentration is C_B, and that formed upon dissociation of AB, whose concentration is unknown but equal to that of A. Thus the equilibrium concentration of B is

$$[B] = C_B + [A]$$

The equilibrium concentration of AB is its analytical concentration minus the concentration of ionized AB or of A formed, as given by Equation 2-22:

$$[AB] = C_{AB} - [A]$$

Substituting these values into the equilibrium constant equation gives

$$K_{eq} = \frac{[A](C_B + [A])}{C_{AB} - [A]} \tag{2-27}$$

or

$$[A]^2 + (C_B + K_{eq})[A] - K_{eq}C_{AB} = 0$$

Although Equation 2-27 is solvable, often it can be simplified. If $[A] \ll C_B$ and C_{AB},

$$C_B + [A] \simeq C_B \tag{2-28}$$

$$C_{AB} - [A] \simeq C_{AB} \tag{2-29}$$

and Equation 2-27 reduces to

$$K_{eq} = \frac{[A]C_B}{C_{AB}} \tag{2-30}$$

Example 2-13

The 0.100 M acetic acid solution in Example 2-12 also contains 0.200 M sodium acetate. Calculate the equilibrium concentration of hydrogen ion in the solution.

Solution Hydrogen ion is formed by the dissociation of CH_3CO_2H,

$$CH_3CO_2H \rightleftharpoons H^+ + CH_3CO_2^-$$

for which

$$K_{eq} = \frac{[H^+][CH_3CO_2^-]}{[CH_3CO_2H]} = \frac{[H^+](C_{CH_3CO_2^-} + [H^+])}{C_{CH_3CO_2H} - [H^+]}$$

Since K_{eq} is small, very little dissociation occurs and we can assume, at least temporarily, that Equations 2-28 and 2-29 are valid (remember, sodium acetate is a strong electrolyte and dissociates completely into Na^+ and $CH_3CO_2^-$):

$$C_{CH_3CO_2^-} + [H^+] \simeq C_{CH_3CO_2^-} = 0.200 \ M$$

$$C_{CH_3CO_2H} - [H^+] \simeq C_{CH_3CO_2H} = 0.100 \ M$$

and Equation 2-30 is valid:

$$K_{eq} = \frac{[H^+]C_{CH_3CO_2^-}}{C_{CH_3CO_2H}}$$

$$1.76 \times 10^{-5} = \frac{[H^+](0.200)}{0.100}$$

$$[H^+] = 8.80 \times 10^{-6} \ M$$

The concentration of hydrogen ion is much smaller than the analytical concentrations of acetic acid or sodium acetate, so the assumptions made earlier are indeed justified.

Example 2-14

What weight of sodium acetate must be added to 250 mL of 0.100 M acetic acid to obtain a hydrogen-ion concentration of $2.25 \times 10^{-4} \ M$?

Solution The relationship between the concentrations of hydrogen ion, acetate ion, and acetic acid is given by the equilibrium expression for the dissociation of the acid,

$$K_{eq} = \frac{[H^+][CH_3CO_2^-]}{[CH_3CO_2H]}$$

If we make the usual assumptions, we can use analytical concentrations in place of the equilibrium concentrations for CH_3CO_2H and $CH_3CO_2^-$:

$$K_{eq} = \frac{[H^+]C_{CH_3CO_2^-}}{C_{CH_3CO_2H}}$$

Solving this equation for the concentration of acetate ion, we obtain

$$1.76 \times 10^{-5} = \frac{(2.25 \times 10^{-4})C_{CH_3CO_2^-}}{0.100}$$

$$C_{CH_3CO_2^-} = 7.82 \times 10^{-3} \, M$$

Since sodium acetate is a strong electrolyte and is considered the sole source of acetate ion,

$$C_{CH_3CO_2^-} = C_{CH_3CO_2Na}$$

The amount of this salt in millimoles is

$$\text{amount } CH_3CO_2Na = (7.82 \times 10^{-3} \, M)(250 \text{ mL}) = 1.96 \text{ mmol}$$

Converting this quantity to weight gives

$$\text{weight } CH_3CO_2Na = (1.96 \text{ mmol})(82.0 \text{ mg/mmol}) = 161 \text{ mg}$$

Simultaneous Equilibria

Once you develop a good understanding of the equilibrium concept, you will be able to appreciate the fact that a substance can participate in several equilibrium systems simultaneously. The following pairs of reactions all describe cases where X is a part of two equilibrium systems.

$$MX_2 \rightleftharpoons MX^+ + X^-$$
$$MX^+ \rightleftharpoons M^{2+} + X^-$$

$$A^+ + X^- \rightleftharpoons AX$$
$$B^+ + X^- \rightleftharpoons BX$$

$$HX \rightleftharpoons H^+ + X^-$$
$$M^+ + X^- \rightleftharpoons MX$$

The correct concentration of X cannot be calculated by solving independently the equilibrium equation for each reaction and simply adding the two results, because the equilibrium systems are not independent of each other. The correct mathematical approach involves the *simultaneous* solution of two equations, representing the two equilibria. The solution of simultaneous equations is tedious when done by hand, and often requires more information than may readily be available. Fortunately, a simple alternative solution exists when it can be established that the concentration of analyte determined by one equilibrium (call this the *principal* equilibrium) is significantly larger than its concentration determined by any of the other equilibria. When such a condition exists, the other equilibria do not significantly affect the principal equilibrium and can be ignored, thereby reducing the problem to a single equilibrium type.

The difficulty is shifted from the mathematical solution of simultaneous equations to determining whether a principal equilibrium exists and, if so, which one. It may be possible to solve independently each pertinent equilibrium equation for the desired concentration and then compare the results, but it is much simpler and quicker to recognize insignificant equilibria by the values of their equilibrium constants. For example, sulfurous acid can ionize in two steps:

$$H_2SO_3 \rightleftharpoons H^+ + HSO_3^- \qquad K_{eq(1)} = 1.2 \times 10^{-2} \qquad (2\text{-}31)$$

$$HSO_3^- \rightleftharpoons H^+ + SO_3^{2-} \qquad K_{eq(2)} = 6.6 \times 10^{-8} \qquad (2\text{-}32)$$

The large difference between the values of the equilibrium constants tells us that when H_2SO_3 is dissolved in water, the first reaction supplies most of the H^+ to the solution and its concentration could be calculated from the K_{eq} equation for this reaction alone,

$$K_{eq(1)} = \frac{[H^+][HSO_3^-]}{[H_2SO_3]}$$

The hydrogen-ion concentration of a 0.10 M H_2SO_3 solution, calculated assuming Reaction 2-31 to be the only source of hydrogen ion, is 3.5×10^{-2} M. When both Reactions 2-31 and 2-32 are included as sources of hydrogen ion, the calculated concentration is the same through the first six digits. As a general rule, if $K_{eq(1)}$ is at least 100 times larger than $K_{eq(2)}$, the second equilibrium can be safely neglected for the purpose of such a calculation.

PROBLEMS

0.0015

2-1. Indicate which of the following substances are considered soluble in water. For those indicated, write the ions that will exist in the solution.

(a) NaBr   (e) $BaSO_4$ (h) Na_3PO_4
(b) NH_4Cl (f) $(NH_4)_2CO_3$ (i) $NiSO_4$
(c) HNO_3 (g) $Fe(OH)_3$ (j) $Ba(OH)_2$
(d) C_2H_6

2-2. How many millimoles of reagent are contained in each of the following?

(a) 41.5 mg of Na_2CO_3 (d) 12.0 mL of 0.180 M $CuSO_4$
(b) 6.00 g of NaOH   (e) 25.0 mL of 0.644 ppm Cu^{2+}
(c) 1.50 mL of pure H_2O (f) 10.0 g of 3.00%(w/w) H_2O_2

2-3. How many millimoles of reagent are contained in each of the following?

(a) 1.76 g of $AgNO_3$ (d) 2.00 L of 4.50×10^{-3} M CH_3CO_2H
(b) 800 mg of Na_3PO_4 (e) 5.00 mL of 31.4 ppm Cr^{3+}
(c) 1.00 mL of 5.00×10^{-2} M HCl   (f) 2500 mg of 6.97%(w/w) Na_2SO_3

2-4. What is the weight in milligrams of the following reagents?

(a) 5.50 mmol of I_2 (d) 500 mL of 7.50×10^{-3} M $K_2Cr_2O_7$
(b) 1.42×10^{-1} mol of $NaNO_2$ (e) 100 mL of 216 ppm Ca^{2+}
(c) 1.00 mL of 18.0 M H_2SO_4 (f) 1.00 g of 4.00%(w/w) NaCl

2-5. What is the weight in milligrams of the following reagents?

(a) 3.94 mmol of C_2H_5OH (d) 0.350 L of 0.114 M $KMnO_4$
(b) 3.00 mol of NH_3 (e) 500 mL of 91.4 ppb Pb
(c) 2.50 mL of 6.00 M KOH   (f) 100 mg of 34.1%(w/w) NH_3

2-6. Calculate the molar analytical concentration of each of the following solutes dissolved in 500 mL of solution.

(a) 4.18 g of NH_4HF_2

(b) 5.00 mL of 12.0 M HCl

(c) 14.7 g of $H_2C_2O_4 \cdot 2H_2O$

(d) 1200 mg of NaOH

(e) 3.85 mmol of $NaHCO_3$

(f) 50.0 mL of 200 ppm NaCN

2-7. Calculate the molar analytical concentration of each of the following solutes dissolved in 250 mL of solution.

(a) 9.00 g of KI

(b) 750 mg of $BaCl_2$

(c) 0.0168 mol of $Zn(NO_3)_2$

(d) 25.0 mL of 15.0 M NH_3

(e) 844 mg of $Na_2B_4O_7 \cdot 10H_2O$

(f) 50.0 mL of 9.88 ppm Hg

2-8. Calculate the equivalent weight of each underlined substance in the following reactions. The reactions are *not* all balanced.

(a) $\underline{Ca(OH)_2}$ + HCl $\longrightarrow$ $CaCl_2$ + H_2O

(b) $\underline{H_3PO_4}$ + NaOH $\longrightarrow$ NaH_2PO_4 + H_2O

(c) CaC_2O_4 + $\underline{H_2SO_4}$ $\longrightarrow$ $CaSO_4$ + $H_2C_2O_4$

(d) $MgCO_3$ + HCl $\longrightarrow$ $\underline{MgCl_2}$ + H_2O + $CO_2(g)$

(e) Fe^{3+} + $\underline{I^-}$ $\longrightarrow$ Fe^{2+} + $\underline{I_2}$

(f) $\underline{MnO_4^-}$ + Fe^{2+} + H^+ $\longrightarrow$ Mn^{2+} + Fe^{3+} + H_2O

(g) $\underline{H_3AsO_3}$ + I_2 + H_2O $\longrightarrow$ H_3AsO_4 + HI

(h) $\underline{Al(OH)_3}$ + KOH $\longrightarrow$ $KAl(OH)_4$

2-9. Balance the following oxidation-reduction reactions.

(a) $H_2C_2O_4$ + $KMnO_4$ + H_2SO_4 $\longrightarrow$ $MnSO_4$ + K_2SO_4 + H_2O + $CO_2(g)$

(b) $FeCl_3$ + $SnCl_2$ $\longrightarrow$ $FeCl_2$ + $SnCl_4$

(c) $K_2Cr_2O_7$ + KI + HCl $\longrightarrow$ $CrCl_3$ + KCl + I_2 + H_2O

(d) $Na_2S_2O_3$ + I_2 $\longrightarrow$ $Na_2S_4O_6$ + NaI

(e) H_2O_2 + $K_4Fe(CN)_6$ $\longrightarrow$ KOH + $K_3Fe(CN)_6$

(f) $U(SO_4)_2$ + $Ce(SO_4)_2$ + H_2O $\longrightarrow$ UO_2SO_4 + $Ce_2(SO_4)_3$ + H_2SO_4

2-10. If 50.4 mmol of a pure organic compound weighs 7.13 g, what is its molecular weight?

2-11. Calculate the normality of each of the following.

(a) a 0.100 M H_3PO_4 solution reacting as shown in Problem 2-8(b)

(b) a 0.100 M $KMnO_4$ solution reacting as shown in Problem 2-8(f)

(c) 450 mg of pure $K_2Cr_2O_7$ dissolved in 250 mL of solution and used to produce iodine as shown in Problem 2-9(c)

2-12. Barium sulfate dissolves in water to the extent of 2.46 mg $BaSO_4$ per liter of solution. What is the molar analytical concentration of dissolved $BaSO_4$?

2-13. The concentration of dissolved calcium oxalate in a saturated solution is 5.24×10^{-5} M. What is the weight in milligrams of dissolved CaC_2O_4 in 1.00 L of this solution?

2-14. Calculate the molar analytical concentration of each of the following.

(a) 0.947 g of Ag dissolved in HNO_3 and diluted to 500 mL

(b) 7.74 g of KH_2PO_4 per 100 mL of solution

(c) 4.06 mg of urea, $CO(NH_2)_2$, per mL of solution

2-15. Calculate the molar analytical concentration of each of the following.

(a) 5.00 g of 2.50%(w/w) Na_3PO_4 diluted to 100 mL

(b) 12.4 g of iodine per liter of ethanol.

(c) 1.63 g of iron wire dissolved in HNO_3 and diluted to 250 mL

2-16. A sample of copper ore weighing 1.35 g contains 62.0 mg of copper. What is the weight percent Cu in the ore?

2-17. What weight of limestone containing 23.0%(w/w) Ca must be taken in order to have 750 mg of Ca present?

2-18. A 37.0%(w/w) solution of hydrochloric acid has a density of 1.18 g/mL.
  (a) Calculate the molar analytical concentration of HCl in this solution.
  (b) What volume of this solution should be taken to make 500 mL of a 1.00 M HCl solution?

2-19. A portion of pure magnesium chloride weighing 4.00 g is dissolved in water and diluted to 500 mL. Determine the following concentrations.
  (a) molar analytical concentration of $MgCl_2$
  (b) molar equilibrium concentration of Mg^{2+}
  (c) molar equilibrium concentration of Cl^-

2-20. A sample of Mississippi River water is found to have a calcium concentration of 183 ppm. Calculate the following.
  (a) milligrams of Ca contained in 100 mL of the water
  (b) molar concentration of Ca in the water
  (c) % Ca (w/w) in the water

2-21. Calculate the ionic strength of each of the following solutions.
  (a) 0.10 M NaCl
  (b) 0.10 M Na_2SO_4
  (c) 0.10 M $Al_2(SO_4)_3$
  (d) 0.20 M $CaCl_2$ + 0.10 M Na_2SO_4
  (e) 0.15 M HCl + 0.15 M $NiCl_2$

2-22. Calculate the activity coefficient and the activity of $SO_4{}^{2-}$ in each of the following solutions.
  (a) 0.0750 M $CuSO_4$
  (b) 0.0750 M Na_2SO_4
  (c) 0.0750 M $Fe_2(SO_4)_3$

2-23. What weight of pure potassium chloride is needed to make 1.00 L of a solution whose analytical concentration is 0.120 M?

2-24. To what volume must 8.32 g of NaOH be diluted to make its analytical concentration 0.200 M?

2-25. What is the molar equilibrium concentration of Cl^- in 250 mL of solution containing 1.39 g of $MgCl_2$?

2-26. A 500-mL sample of effluent from a chemical plant contains 26.1 μg of mercury. If the density of the effluent liquid is 1.073 g/mL, calculate the concentration of mercury in the given units.
  (a) weight percent
  (b) parts per million
  (c) parts per billion
  (d) molarity

2-27. An aqueous solution of 6.00 M ammonia has a density of 0.935 g/mL. What is the %(w/w) NH_3 in the solution?

2-28. Calculate the molar equilibrium concentration of Li^+ in a 12.0%(w/w) Li_2SO_4 solution whose density is 1.067 g/mL.

2-29. What weight of urea, $CO(NH_2)_2$, would you need to prepare 500 mL of a 1.50 M solution?

2-30. What weight of sodium carbonate must be taken to make 3.00 L of a solution whose equilibrium concentration of Na^+ is 0.120 M?

2-31. Describe how you would prepare each of the following solutions.
  (a) 1.00 L of 0.193 M $BaCl_2$ from solid $BaCl_2 \cdot 2H_2O$
  (b) 500 mL of 3.0%(w/w) H_2O_2 from 30%(w/w) H_2O_2
  (c) 2.00 L of 750 ppm $Zn(NO_3)_2$ from solid $Zn(NO_3)_2$
  (d) 250 mL of a solution containing 1.00 mg $AgNO_3$/mL solution from a 0.350 M solution of $AgNO_3$
  (e) 300 mL of 0.0247 M $K_2Cr_2O_7$ from a solution of 1.31 M $K_2Cr_2O_7$
  (f) 5.00 L of 6.00 M H_3PO_4 from a concentrated solution that is 85.1%(w/w) H_3PO_4 and has a density of 1.69 g/mL

2-32. Write the equilibrium constant expression for each of the following.

(a) $2Fe^{3+} + Sn^{2+} \rightleftharpoons 2Fe^{2+} + Sn^{4+}$

(b) $AgCl(s) \rightleftharpoons Ag^+ + Cl^-$

(c) $Ni^{2+} + 4CN^- \rightleftharpoons Ni(CN)_4^{2-}$

(d) $H_3PO_4 + H_2O \rightleftharpoons H_3O^+ + H_2PO_4^-$

(e) $H_3PO_4 + 2H_2O \rightleftharpoons 2H_3O^+ + HPO_4^{2-}$

(f) $Cr_2O_7^{2-} + 6I^- + 14H^+ \rightleftharpoons 2Cr^{3+} + 3I_2(s) + 7H_2O$

2-33. Write the equilibrium constant expression for each of the following. Are the values of the equilibrium constants different for the two expressions?

$$2Ce^{4+} + Sn^{2+} \rightleftharpoons 2Ce^{3+} + Sn^{4+}$$

$$Ce^{4+} + \tfrac{1}{2}Sn^{2+} \rightleftharpoons Ce^{3+} + \tfrac{1}{2}Sn^{4+}$$

2-34. If the equilibrium constant for the reaction in Problem 2-32(a) is 9.20×10^{20}, what is the equilibrium constant for the following reaction?

$$2Fe^{2+} + Sn^{4+} \rightleftharpoons 2Fe^{3+} + Sn^{2+}$$

2-35. The equilibrium constant for the dissociation of dichloroacetic acid, Cl_2CHCO_2H, into H^+ and $Cl_2CHCO_2^-$ is 5.00×10^{-3}. If a solution of this acid has an analytical concentration of $0.160\ M$, calculate the equilibrium concentration of H^+, both with and without using the assumption illustrated by Equation 2-25.

2-36. The equilibrium constant for the reaction

$$Fe^{3+} + 3C_2O_4^{2-} \rightleftharpoons Fe(C_2O_4)_3^{3-}$$

is 1.58×10^{20}. Calculate the equilibrium concentration of all three ions in a solution prepared by mixing 1.00 mmol of Fe^{3+} with 3.00 mmol of $C_2O_4^{2-}$ and diluting to 500 mL.

2-37. If the equilibrium constant for the dissociation of solid silver chloride,

$$AgCl(s) \rightleftharpoons Ag^+ + Cl^-$$

is 1.8×10^{-10}, what is the weight in milligrams of dissolved Ag^+ in equilibrium with the solid in 4.00 L of solution?

2-38. Water can dissociate into H^+ and OH^-. What is the equilibrium constant for this reaction if pure water has a hydrogen-ion concentration of $1.00 \times 10^{-7}\ M$?

2-39. A solution of nitrous acid (HNO_2) prepared by dissolving 1.93 g of HNO_2 in 500 mL of solution has a hydrogen-ion concentration of $7.63 \times 10^{-3}\ M$. Calculate the equilibrium constant for the dissociation of HNO_2 into H^+ and NO_2^-.

2-40. Suppose the solution described in Problem 2-37 contains enough NaCl to make the chloride-ion concentration $0.340\ M$ and there are no competing simultaneous equilibria. Calculate the equilibrium concentration of silver ion.

2-41. What weight of NaCl must be dissolved in the solution described in Problem 2-37 to produce a silver-ion concentration of $8.44 \times 10^{-9}\ M$?

2-42. Copper forms a complex ion with ammonia that can be represented with the following reaction:

$$Cu^{2+} + 4NH_3 \rightleftharpoons Cu(NH_3)_4^{2+}$$

If the equilibrium constant for the reaction is 3.98×10^{12}, calculate the equilibrium concentration of Cu^{2+} in a solution containing $2.50 \times 10^{-2}\ M\ Cu(NH_3)_4^{2+}$ and $0.140\ M$ excess NH_3.

2-43. The equilibrium constant for the dissociation of pyruvic acid, CH_3COCO_2H, is $3.24 \times$

10^{-3}. Calculate the equilibrium concentration of this acid in a solution whose analytical concentration is 0.100 M.

2-44. What volume of 6.00 M acetic acid must be added to 200 mmol of sodium acetate so that, upon dilution to 500 mL with water, the solution will have a hydrogen-ion concentration of 3.50×10^{-5} M?

2-45. The following equilibrium exists between NO_2 and its dimer N_2O_4:

$$2NO_2(g) \rightleftharpoons N_2O_4(g)$$

What happens (quantitatively) to the equilibrium concentration of N_2O_4 if the equilibrium concentration of NO_2 is doubled?

2-46. What molar ratio of HF and NaF must be mixed to prepare 500 mL of a solution whose hydrogen-ion concentration is 6.18×10^{-4} M?

2-47. A solution is prepared by mixing 4.00 mmol of $SnCl_2$ with 8.00 mmol of $FeCl_3$ and diluting to 500 mL. The reagents react according to the following equation:

$$2Fe^{3+} + Sn^{2+} \rightleftharpoons 2Fe^{2+} + Sn^{4+} \qquad K_{eq} = 9.20 \times 10^{20}$$

Calculate the concentration of Fe^{3+} remaining in the solution at equilibrium.

2-48. The equilibrium constant for the dissociation of $PbCl_2$ into its ions

$$PbCl_2(s) \rightleftharpoons Pb^{2+} + 2Cl^-$$

is 1.7×10^{-5}. Calculate the following.
  (a) concentration of Pb^{2+} in a solution where such an equilibrium has been established
  (b) concentration of Pb^{2+} in a solution containing 0.100 M Cl^- in equilibrium with solid $PbCl_2$
  (c) percentage decrease in the concentration of Pb^{2+} caused by the addition of the Cl^-

REPRESENTING AND MANIPULATING DATA

Despite our very best efforts, no physical measurement is ever completely free of experimental error or uncertainty. The effective use of any scientific data depends on knowing the degree of uncertainty associated with the data. Unfortunately, determining this uncertainty often is neither routine nor easy, and it is not uncommon to spend more time evaluating the data than it took to acquire them. It is a common mistake for beginners, and even experienced professionals, to seek a greater degree of certainty than is necessary to answer the question at hand. This pitfall can be avoided by always remembering the larger question that the results of the experiment will be used to answer. For example, a scrap metal dealer who uses a vat of hydrochloric acid solution for cleaning iron may want to know whether the concentration has become too low for a particular cleaning job. It would be expensive and time consuming to carry out a sophisticated analysis capable of measuring the concentration with an uncertainty of a few thousandths of a percent when a quick, simple test could tell the metal dealer what he or she wants to know.

SIGNIFICANT FIGURES

In literature it is well understood that a precise choice of words is necessary to make one's meaning clear. The same can be said for the digits used in a number representing a physical measurement. Consider a volume of solution measured with two different graduated cylinders as shown in Figure 3-1. Cylinder (a) tells us that the volume is more than 3 but less than 4 mL. How should it be reported? The rule used by most scientists is to include all digits that are known with certainty plus the *first* uncertain digit. These digits are called *significant figures*. We *know* the volume is at least 3 mL. Measuring

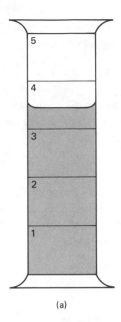
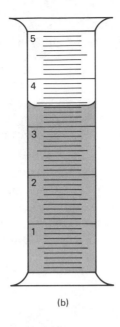

Figure 3-1 A volume of liquid measured with two different graduated cylinders: (a) volume = 3.4 mL; (b) volume = 3.43 mL.

from the bottom of the meniscus, we *think* it is 0.4 of the distance between the 3- and 4-mL marks on the graduated cylinder. Thus the correct representation is 3.4 mL, a number with two significant figures. Obviously, if the 4 is an uncertain digit, any digits following it must also be uncertain and, according to our rule, should not be included.

If we desire to decrease the uncertainty of the final value, a better graduated cylinder must be used. Consider the same volume being measured with cylinder (b) shown in Figure 3-1. Now it is clear that the volume lies between 3.4 and 3.5 mL. We *know* that it is at least 3.4 mL and *think* it is 0.2 of the distance between the 3.4- and 3.5-mL marks on the cylinder. The volume reported correctly is 3.42 mL, a number with three significant figures.

When values are reported with the proper number of significant figures, we know something of the quality of the measurements. The general rule adopted for this chapter is that, unless otherwise specified, the last digit in a number is assumed to be uncertain by ±1. Thus, the volume of solution in cylinder (a) of Figure 3-1 has an uncertainty of ±0.1 mL and lies between 3.3 and 3.5 mL. According to the measurement in cylinder (b) the uncertainty is ±0.01 mL and the volume is between 3.41 and 3.43 mL.

Counting Rules

In counting significant figures you must be aware of the fact that there are two different types of zeros: those that are regular digits, which must be treated the same as any other digit, and those that are used *only* to show the position of the decimal point. Leading zeros are present only to show the position of the decimal point. Thus 0.06307 contains *four* significant figures. Trailing zeros are significant. It is often necessary to set the decimal point using the power-of-10 notation to avoid introducing the appearance of unwanted significant figures. For example, the number twelve hundred, written with two significant figures, is 12×10^2 or 1.2×10^3. Written as "1200" it contains not two but *four* significant figures. Not all scientists use the same rule concerning trailing zeros, so check with your instructor to see if a different rule will be used in your course.

Arithmetic Rules

Often, you will have to carry out arithmetic operations with measured quantities. In doing so, it is easy to generate meaningless digits, especially with electronic calculators, and it is important that you learn to recognize and eliminate these digits.

Rounding off

There are several different rules commonly used for discarding unwanted digits in a number. The following rule, which is used in this book, is the simplest and most common. If the digit to be discarded is 5 or greater, increase the retained preceding digit by 1. If it is less than 5, do not change the preceding digit. For example,

5.175 rounded off to three digits becomes 5.18.

7.009 rounded off to three digits becomes 7.01.

1.083 rounded off to three digits becomes 1.08.

When more than one digit is to be dropped, the rule is applied *one time* to the first digit following the last retained digit. Under no circumstances should the rounding off be done sequentially. Thus 9.1647 rounded off to three digits becomes 9.16 because 4 is less than 5. You should *not* round off the 7, making the number 9.165, and then round off the 5, making it 9.17.

Addition and subtraction

The rule here is that the *absolute* uncertainty of the answer must not exceed that of the most uncertain value. In other words, retain as many digits to the right of the decimal as is found in the number with the fewest digits to the right of the decimal. Adding 12.2, 0.365, and 1.04 on your calculator gives 13.605, which should be rounded off to 13.6. It is preferable to add the numbers as they are and round off the answer rather than round off each value and then add.

When the numbers to be added or subtracted have exponents, they must be made to have the same exponent before the addition or subtraction is carried out. Thus

$$4.16 \times 10^{-3} \longrightarrow 0.0416 \times 10^{-1}$$
$$1.724 \times 10^{-1} \longrightarrow 1.724 \ \times 10^{-1}$$
$$\underline{3.86 \times 10^{-4} \longrightarrow 0.00386 \times 10^{-1}}$$
$$1.76946 \times 10^{-1} \longrightarrow 1.769 \times 10^{-1}$$

Multiplication and division

There are two different rules commonly used in performing multiplication and division with significant figures. One is easier to apply than the other but is less valid scientifically. The simplest rule is to keep as many significant figures in the final answer as is found in the value with the least number of significant figures. The shortcoming of this simple rule is readily apparent when the first digit of the value with the least number of significant digits is a 9. For example, according to the simple rule,

$$1.074 \times 0.993 = 1.07$$

because there are three significant figures in the value, 0.993. Yet if this value were just 1% larger or 1.002, the answer would be 1.076, a value with four significant figures. The simple rule is used in the examples and problems of this chapter, but you may refer to Appendix D for the description of a more rigorous rule.

Example 3-1

What is the answer, with the correct number of significant figures, to the following arithmetic expression?

$$\frac{40.1 \times 0.1633}{204.228}$$

Solution The value 40.1 contains the *least* number of significant figures, with three. The result computed with an electronic calculator is 0.0320638208. Rounded off to three significant figures, the correct answer is 0.0321 or 3.21×10^{-2}.

The rules for problems combining addition/subtraction and multiplication/division are applied individually to each operation as it is performed.

Example 3-2

Compute the answer to the following expression using the correct number of significant figures.

$$\frac{21.6 \times 0.317}{4.1} + 16.037$$

Solution The result of the multiplication/division should contain two significant figures (same as the value 4.1). This rounded-off result is then added to 16.037, with the answer rounded off according to the rules of addition.

$$\frac{21.6 \times 0.317}{4.1} + 16.037$$

$$\downarrow$$

$$1.67004\ldots \quad + 16.037$$

$$\downarrow$$

$$1.7 \quad + 16.037 = 17.737$$

$$\downarrow$$

$$17.7$$

It is very important to distinguish between experimental values and *integers* or *pure* numbers. Integers are known with absolute certainty and contain as many significant figures as necessary, even though they are usually not written out. For example, the average of two experimental values is computed from the expression

$$\frac{62.31 + 62.47}{2}$$

The 2 is an integer and the final answer should contain four significant figures.

Logarithms

A log quantity should contain as many digits to the *right of the decimal* as there are significant figures in the number used to calculate the log quantity. Thus

$$\log (3.67 \times 10^{-3}) = -2.435$$

It appears that an additional significant figure has been added during this mathematical transformation, but the digits to the left of the decimal, the 2 in this case, only represent the position of the decimal point in the original number.

FUNDAMENTAL TERMS

It is common practice among analytical chemists to carry out two or more determinations of the analyte in a given sample. The results of such replicate determinations are seldom exactly the same and the analyst is faced with the problem of how to determine the "best" answer. The results of replicate measurements tend to cluster about a central value, and this is the value most often reported.

We benefit in two ways from making replicate determinations or measurements. First, the central value is likely to be closer to the true value than any single value, and second, the variation in the replicate values may tell us something about the reliability of the measurements.

Central Value

The most common central value used by chemists is the *arithmetic mean*, $\bar{x}$, which is obtained by dividing the sum of the individual values by the number of values. Mathematically,

$$\bar{x} = \frac{x_1 + x_2 + x_3 + \cdots + x_n}{n} = \frac{\sum x_i}{n}$$

where $x_1, x_2, x_3, \ldots, x_n$ are the individual values, n the number of values, and $\sum x_i$ the sum of values of x. You will recognize immediately that the arithmetic mean is nothing more than the *average* value. Cynics are certain that scientists periodically invent new words for old terms simply to confuse "outsiders." Actually, there are two types of mean: arithmetic and geometric. Only the arithmetic mean is important to introductory quantitative analysis. Another central value, less commonly used by chemists, is the *median*. It is the middle *numerical* value in a set of values. Thus the median of the five values 20.4, 20.6, 20.1, 20.7, and 20.0 is 20.4. Do not make the mistake of selecting 20.1 because it is *physically* located in the middle of the data set. When the data set contains an even number of values, the median is the average of the two middle numerical values. For a data set consisting of 20.4, 20.6, 20.1, and 20.7, the median is calculated as follows:

$$\text{middle numerical values} = 20.4 \text{ and } 20.6$$

$$\text{median} = \frac{20.4 + 20.6}{2} = 20.5$$

Since both the mean and the median of a single set of data represent the central or "best" value, one might expect them to be the same. They are often not for small data sets, but as the size of the data set increases, both values numerically approach the same value, which in the absence of any systematic error is the true value.

Generally, the mean is a better central value than the median for small data sets. The main advantage of the median is its relative insensitivity to a single divergent value in the data set, as is seen in the following example.

Example 3-3

Calculate the mean and the median for each of the following sets of data. Set A: 6.37, 6.33, 6.41, 6.80; set B: 6.37, 6.33, 6.41, 6.93.

Solution According to the definitions stated above, for data set A,

$$\bar{x} = \frac{6.37 + 6.33 + 6.41 + 6.80}{4} = \frac{25.91}{4} = 6.48$$

$$\text{median} = \frac{6.37 + 6.41}{2} = 6.39$$

For data set B,

$$\bar{x} = \frac{6.37 + 6.33 + 6.41 + 6.93}{4} = \frac{26.04}{4} = 6.51$$

$$\text{median} = \frac{6.37 + 6.41}{2} = 6.39$$

Accuracy

The term *accuracy* describes the nearness of an experimental value, x_i, or a mean, $\bar{x}$, to the true value, μ. It is expressed as error, where

$$\text{error} = x_i - \mu \quad \text{or} \quad \bar{x} - \mu$$

An error calculated this way is called an *absolute* error and carries the units of x_i and μ. When comparing measurement errors of different quantities, it is more useful to use *relative* error, which is calculated by dividing the absolute error by the true value:

$$\text{relative error} = \frac{\text{error}}{\mu}$$

Parts per hundred (pph) or percent error is the relative error multiplied by 100. Parts per thousand (ppt) error is the relative error multiplied by 1000, and so on.

Example 3-4

Calculate the absolute error, percent error, and parts per thousand error for the mean of the following data set.

x_i (mg)

8.33
8.29 $\mu = 8.27$ mg
8.28
8.34
8.36

Solution To determine the error, the mean must be calculated:

$$\bar{x} = \frac{\sum x_i}{n} = \frac{41.6 \text{ mg}}{5} = 8.32 \text{ mg}$$

$$\text{error} = \bar{x} - \mu = 8.32 \text{ mg} - 8.27 \text{ mg} = 0.05 \text{ mg}$$

$$\% \text{ error} = \frac{\text{error}}{\mu} \times 100 = \frac{0.05 \text{ mg}}{8.27 \text{ mg}} \times 100 = 0.6$$

$$\text{ppt error} = \frac{\text{error}}{\mu} \times 1000 = \frac{0.05 \text{ mg}}{8.27 \text{ mg}} \times 1000 = 6$$

The *average error* is calculated like the average value or arithmetic mean except that the individual errors rather than the individual values are used.

It should be noted that the concept of true value in science is not the same as that used in philosophy. Many philosophers believe that there is no absolute truth, or true value. In science we use the term "true value" synonymously with *accepted value*. For example, the true value of the amount of analyte in an unknown may be the value determined from the weighed amount of pure material used to make the unknown or even the result of an analysis performed by your instructor.

Precision

Precision, a term often mistakenly used in place of accuracy, refers to the agreement between values in a set of data. The fact that the values of replicate measurements all agree well does not necessarily mean that they are close to the true value. Why this may be so is discussed in a later section. There are several popular ways to express the precision of data.

Average deviation

Individual deviations are simply the difference, taken *without regard to sign*, between the experimental values and the central value. The average deviation, $\bar{d}$, is found by summing the individual deviations and dividing by the number of measurements. Thus the average deviation from the mean is given by

$$\bar{d} = \frac{\sum |x_i - \bar{x}|}{n} \tag{3-1}$$

Historically, average deviation has been used extensively as a measure of precision, but it is not preferred, primarily because, unlike other estimates of precision, it is not statistically interpretable and it gives equal weight to large and small deviations, which are not equally probable.

Standard deviation

The standard deviation, s, or root-mean-square deviation as it is sometimes called, is the preferred measure of precision and is calculated from the equation

$$s = \sqrt{\frac{\sum (x_i - \bar{x})^2}{n - 1}} \tag{3-2}$$

The most common mistake made by students using this equation is to "square the sum of the deviations" rather than "sum the squares of the deviations." Look carefully at Example 3-5 to make sure that you learn to use the formula correctly.

Example 3-5

Quantitative analysis student Anna Litical obtained the following results for the determination of isooctane in gasoline.

Determination number	Percent isooctane
1	3.83
2	3.97
3	3.94
4	3.88
5	3.94
6	3.90

Calculate the standard deviation from the mean

Solution According to Equation 3-2,

$$s = \sqrt{\frac{\sum (x_i - \bar{x})^2}{n - 1}}$$

$$\bar{x} = \frac{3.83 + 3.97 + 3.94 + 3.88 + 3.94 + 3.90}{6} = 3.91\%$$

x_i	$|x_i - \bar{x}|$	$(x_i - \bar{x})^2$
3.83	0.08	0.0064
3.97	0.06	0.0036
3.94	0.03	0.0009
3.88	0.03	0.0009
3.94	0.03	0.0009
3.90	0.01	0.0001

$$0.0128 = \sum (x_i - \bar{x})^2$$

$$s = \sqrt{\frac{0.0128}{6 - 1}} = 0.051\%$$

Both average and standard deviation can also be expressed in relative terms to facilitate comparison between data sets:

$$\text{relative average deviation} = \frac{\bar{d}}{\bar{x}}$$

$$\text{relative standard deviation} = \frac{s}{\bar{x}}$$

Chemists sometimes complicate matters by referring to the relative standard deviation as the *coefficient of variation*. As was pointed out in the section on error, relative values

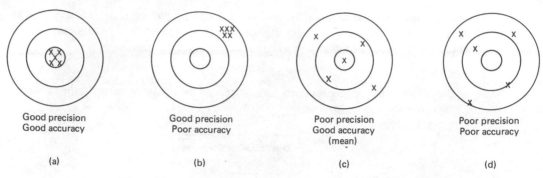

Good precision Good accuracy	Good precision Poor accuracy	Poor precision Good accuracy (mean)	Poor precision Poor accuracy
(a)	(b)	(c)	(d)

Figure 3-2 Targets illustrating the difference between accuracy and precision.

can be expressed fractionally (as above), as parts per hundred or percent (multiplied by 100), as parts per thousand (multiplied by 1000), and so forth.

Range

The *range*, R, is the absolute difference between the largest and smallest values in the data set. It is generally not a good estimate of precision in large data sets, becoming increasingly better in smaller sets until for two observations it, together with the mean, describes the data completely.

Proper Use of Accuracy and Precision

The terms "accuracy" and "precision" often are used incorrectly, by interchanging their meanings. Error and deviation, the respective measures of accuracy and precision, are not always numerically related to one another. That is, the precision of the data in a set can be excellent while the overall accuracy is terrible. Examine the rifle targets in Figure 3-2 and assume that the holes (marked by x's) are equivalent to individual values and the bull's-eye is the true value. Target (a) of Figure 3-2 represents good precision (holes close to each other) and good accuracy (holes close to true value); target (b) represents good precision but poor accuracy. The precision in target (c) is poor because the individual holes are not close to each other but the accuracy of the *mean* is good. Both the precision and the accuracy are poor in target (d). The reasons for the different distributions and placements of shots in these targets are important because they indicate whether a numerical relationship exists between the accuracy and the precision and how the accuracy might be improved.

TYPES OF ERRORS

It is easily apparent that different types of error were responsible for the distribution of shots in targets (b) and (c) of Figure 3-2. A rifle with an improperly aligned sight is a logical explanation for the distribution in target (b) but could not explain target (c). The apparent random nature of the distribution in target (c) could be the result of very slight rifle movements that occur when the trigger is being pulled.

Errors in chemical analyses fall into either of two categories based on the characteristics exemplified in targets (b) and (c) of Figure 3-2. The manner in which errors are decreased or evaluated is determined partly by the category to which they belong. Unfortunately, determining the source and category of an error is frequently a difficult task, often requiring additional experiments, comparisons with other data, and statistical analysis of the data.

Determinate Errors

Errors that can be determined and eliminated are called *determinate* or *systematic errors* and are usually constant in both magnitude and direction. The sources of determinate errors can be organized into four groups.

1. *Methods.*   Errors in this category arise from undesirable behavior of the sample or reagents used in the analysis. For example, in a gravimetric method, analyte is precipitated, separated from the solution, and weighed. If the precipitate is not sufficiently insoluble, a weight less than the correct one will be obtained, introducing a negative error in the final result. It is also possible that other substances will precipitate along with the analyte, causing a weight greater than the correct one to be obtained, leading to a positive error.

2. *Equipment and materials.*   All measuring devices are sources of determinate error. For example, a voltmeter draws a small current while operating, thereby changing the voltage it is measuring; a piece of volumetric glassware, such as a pipet or buret, whose volume markings were made at 20°C is used at 25°C; or a weighed precipitate is thought to be pure but contains a small amount of an impurity. Note that each time the voltmeter is used to measure a particular voltage, the result will be low by the same amount. Thus the error is constant in both magnitude and direction.

3. *Personal judgments.*   Few if any methods can be performed without making judgments. For example, weighing a sample may require estimating which of two divisions line up on a vernier scale. Similarly, transferring a portion of a dissolved sample may entail deciding when the liquid level matches the fill line on a pipet. Such judgments contain both determinate and indeterminate errors. Indeterminate errors are discussed a little later in the chapter. One of the most common types of judgment errors is an unjustified preference or dislike of certain digits. You may be madly in love with a girl 5 feet 5 inches tall, born the 5th child on May 5, and living at 555 Fifth Street, which has caused you to look very fondly upon the digit 5. Thus anytime a pointer appears to lie 0.4 to 0.6 of the way between two divisions, subconsciously you favor choosing 0.5. Number biases are very common, with zero and five being favored digits. It is a difficult bias to overcome, but it can be decreased through self-discipline.

In another situation you may be called on to turn off a buret stopcock when a certain color change occurs in a solution, but if you are color blind or have a decreased sensitivity to one of the colors, you may consistently wait too long, thereby introducing a determinate error. There is, of course, little one can do about errors that are the result of physical limitations or handicaps.

One of the worst types of bias you will encounter is "final answer bias." This occurs when there is a particular answer you want or hope to get. It is deceptively easy to make decisions in the direction that leads to a result closer to the "desired" one. In the quantitative analysis laboratory many determinations are performed in triplicate to provide a more reliable estimate of the true value. It is common for beginners to want to use three samples of identical weight so that when the measurements with the first sample are completed, they will "know" what the measurement values "should be" for the next two samples.

4. *Mistakes.* These are errors resulting from carelessness or ignorance and can be decreased by being more careful and informed. Mathematical errors, improper reading of instrument scales and recordings, and misrecorded data are the three major sources of such mistakes.

Effect on results

There are two types of determinate error, proportional and constant, each with a different effect on the results. A *proportional* error is one whose magnitude varies in proportion to the amount of analyte and, therefore, the amount of sample. Note that we are talking here about absolute error, not relative error. In the gravimetric determination of sulfur, a precipitate of barium sulfate is isolated and weighed. The weight of sulfur is computed by multiplying the weight of the precipitate by a gravimetric factor consisting of the atomic weight of sulfur divided by the molecular weight of barium sulfate. If an incorrect gravimetric factor is used, it will lead to a proportional error. Suppose, for example, that a sample yielded a precipitate weighing 800.0 mg. The correct weight of sulfur is given by

$$\text{wt S} = \text{wt BaSO}_4 \times \frac{\text{AW of S}}{\text{MW of BaSO}_4}$$

$$= 800.0 \text{ mg} \times \frac{32.06}{233.4} = 109.9 \text{ mg}$$

If an incorrect gravimetric factor of $32.06/223.4$ is used, the calculated weight of sulfur is

$$\text{wt S} = 800.0 \text{ mg} \times \frac{32.06}{223.4} = 114.8 \text{ mg}$$

and the error is

$$\text{error} = 114.8 \text{ mg} - 109.9 \text{ mg} = 4.9 \text{ mg}$$

Now examine what happens if a sample twice as large is analyzed similarly. The weight of the barium sulfate precipitate also will be twice as large. Repeating the calculations gives

$$\text{wt S} = 1600 \text{ mg} \times \frac{32.06}{233.4} = 219.8 \text{ mg}$$

$$= 1600 \text{ mg} \times \frac{32.06}{223.4} = 229.6 \text{ mg}$$

resulting in an error of

$$\text{error} = 229.6 \text{ mg} - 219.8 \text{ mg} = 9.8 \text{ mg}$$

which, like the sample weight, has doubled.

A *constant error*, as the name implies, does not depend on the magnitude of the measured quantity or the amount of analyte. In the gravimetric determination just described, a small amount of barium sulfate will not be collected because it remains dissolved in equilibrium with the solid. This dissolved barium sulfate will not depend on the quantity of solid barium sulfate, and therefore is considered a source of constant error. Constant errors become more serious as the measured quantity gets smaller. For example, if 0.3 mg of $BaSO_4$ remains dissolved due to its solubility, look at what happens to the absolute and relative errors as the amount of precipitate doubles.

Weight $BaSO_4$ precipitate (mg)	Weight $BaSO_4$ dissolved (mg)	Absolute error (mg)	Relative error (error/wt ppt $\times$ 100)
800.0	0.3	0.3	0.04
1600.0	0.3	0.3	0.02

Detection, reduction, and compensation

Identifying that a determinate error exists and tracing it to a particular source or sources is sometimes a devilishly difficult task requiring great ingenuity and perserverance, and at other times certain sources of error are simply assumed. The tracing process is complicated by the fact that an observed error in a determination is the accumulation of many individual determinate and indeterminate errors. A discussion of the procedures used to identify and locate sources of errors is beyond the scope of this book, but some of the common approaches to reducing and compensating for errors will be described.

Method errors are obviously reduced by the judicious *selection of the method* itself. Some method errors are inherent—that is, they are always present when the method is used—while others are dependent on the particular sample being analyzed. A method might be quite suitable for determining phosphate in fresh water but not in seawater, due to its high salt content.

Method errors can sometimes be reduced or eliminated by carefully *controlling the variables* of the method. For example, iodine can be determined by measuring the amount of arsenite required to react with it according to the reaction

$$I_2 + AsO_3^{3-} + H_2O \longrightarrow 2I^- + AsO_4^{3-} + 2H^+$$

The reaction is slow and incomplete in acidic solution and an unfavorable side reaction occurs in highly basic solution. Consequently, the acidity of the solution must be maintained within prescribed limits to avoid large errors.

As mentioned in Chapter 1, *separation* of potentially interfering substances is sometimes necessary to avoid serious errors in chemical measurements. Whether or not a separation should be attempted is determined largely by the difficulty of the separation procedure and the availability of other suitable methods.

A different approach to the problem of determinate error is to evaluate its magnitude, then *compensate* for the error rather than try to eliminate it. Compensation techniques have an advantage that the effort expended in determining the correction factor is independent of the number of samples to be analyzed.

Standard samples can be used to compensate for both proportional and constant errors. A standard sample is one for which the concentration of analyte is known accurately. Such a sample may be prepared from known amounts of pure reagents or it may be a natural substance that has been very carefully analyzed using procedures that are known to be free of significant errors. The National Bureau of Standards (NBS) has available for purchase several hundred common substances that contain certified concentrations of one or more constituents. Proportional errors are compensated for using the proportion

$$\frac{\text{amount analyte in unknown}}{\text{amount analyte in standard}} = \frac{\text{result for unknown}}{\text{result for standard}}$$

The success of this approach depends on the error in the determination of the unknown being reproduced in the determination of the standard. To be certain of this it is necessary that the standard be very similar in composition to the unknown. Unfortunately, the complexity of many natural substances can make this a formidable and often impossible task. When a determinate error is constant and reproducible, its magnitude may be found by analyzing the standard and thereafter subtracting the error from each unknown result. Sometimes, when the error is positive, it can be found by carrying out a determination with a sample containing all reagents except the analyte. Such a determination is called a *blank determination* and the result or error is called the *blank.*

Certain equipment and apparatus errors can be determined by direct experimental *calibration*. The measured error of the device is then added to its normal experimental value to give the correct value.

On occasion, when a determinate error can be attributed largely to a single cause, its magnitude can be calculated theoretically and a correction applied accordingly. For example, the change in volume of a pipet with a change in temperature can be calculated from the coefficient of expansion of the glass. This approach to error compensation is not commonly applicable.

Indeterminate Errors

Errors that cannot be determined and controlled are called *indeterminate* or *random errors*. Both the magnitude and direction of indeterminate errors vary nonreproducibly from one measurement to the next and are never the same, except by chance. These errors are the cumulative effect of many small, uncontrollable variables and personal judgments that lead to uncertainty in a measured value.

Effect on results

The presence of indeterminate errors is manifested in the scatter of replicate experimental values. At first thought, one might conclude that random errors in a set of data should cancel one another, and in the absence of determinate error, the mean should equal the true value. Such a conclusion is not incorrect, but it is incomplete. The net random error is zero when the number of values in the data set is infinitely large, but it may be greater than zero for any finite number of values. We are very interested in how the errors (or the values) are distributed, because this distribution gives us valuable information about the reliability of the mean as an estimate of the true value. This same distribution can provide information on the likelihood of any particular error being caused by a random process. What we are saying, then, is that indeterminate errors cannot be eliminated or reduced for a given procedure but can be evaluated to supply information about the reliability of the data.

Distribution of measurements: The error curve

The distribution of indeterminate error (or of data containing only indeterminate error) is determined by the ways in which the small uncertainties making up the error can combine. Three such uncertainties might all be positive for one measurement, two positive and one negative for the next measurement, and so on. Suppose that a large number of measurements are made on a sample whose true value is known. The error in each measurement could be calculated and tabulated according to how frequently each error occurred. The error distribution most commonly (but not always) encountered in chemical analysis approximates a Gaussian distribution, shown graphically in Figure 3-3. Four observations are immediately obvious:

1. An equal number and type of positive and negative errors occur.
2. Small errors occur much more frequently than large errors.
3. Very large errors occur very infrequently.
4. The value that occurs most often is the one with no error, namely, the true value.

Detailed mathematical analysis of the error curve has provided us with certain generalizations that are helpful in obtaining, treating, and interpreting data. We have no difficulty accepting the fact that the mean, $\bar{x}$, is the best estimate of the true value, μ, that can be obtained from a finite set of measurements. The symmetry of the error curve suggests that positive and negative errors tend to cancel one another in the averaging process and this cancellation becomes more perfect as the number of measurements increases. Mathematically, it can be shown that the reliability, Re, of a mean as an estimate of the true value increases as the *square root* of the number of measurements, or

$$Re = k\sqrt{n}$$

where k is a proportionality constant. If Re_1 is the reliability of the mean of n_1 values,

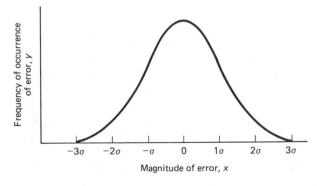

Figure 3-3 Theoretical distribution of indeterminate error for a large number of measurements.

the number of values n_2 needed to obtain a reliability of Re_2 can be calculated using a simple ratio and proportion equation:

$$\frac{Re_1}{Re_2} = \frac{\cancel{k}\sqrt{n_1}}{\cancel{k}\sqrt{n_2}}$$

To double the reliability of a mean calculated from 5 measurements one must acquire 20 measurements. To increase the reliability by a factor of 4 would require a total of 80 measurements. Obviously, a condition of limiting return is reached quickly in trying to increase the reliability of the mean simply by taking more measurements.

Whereas the general shape of the error curve is the same for any Gaussian-like distribution of data, the width or spread of the curve varies according to the precision of the data. In the mathematical description of a Gaussian distribution the scatter (or precision) is represented by the *standard deviation of an infinitely large data set, σ*. It is convenient to think of σ as the "true" standard deviation (the standard deviation you would get if you took an infinite number of measurements) and s as an approximation of σ, based on a finite number of measurements. Although the standard deviation can and does vary according to the data, it can be shown that 68.3% of the area beneath the curve in Figure 3-3 lies within $\pm 1\sigma$ of the mean, while 95.4% lies within $\pm 2\sigma$ and 99.7% lies within $\pm 3\sigma$. Obviously, a large value of σ means of broad error curve, while a small value means a narrow curve. This property of the Gaussian error curve can be stated another way: The chances are 68.3 out of 100 that the indeterminate error of a given measurement in a set is less than $\pm 1\sigma$ or the chances are 95.4 out of 100 that the error is less than $\pm 2\sigma$, and so on. We must, however, know the value of the "true" standard deviation. Like the true value, σ is never known with absolute certainty and must be estimated from the values in the data set using Equation 3-2. The use of $n - 1$ in the denominator compensates for a small negative bias that occurs when n alone is used. The bias is apparent only for small values of n, as the difference between n and $n - 1$ for large values of n is negligible. Such bias-correcting terms are called *degrees of freedom*. The rationale for their use is not important at this point, but they do appear in equations and tables of statistical constants in later portions of this chapter. The reliability of s as an estimate of σ, although not very good for $n \leq 3$, increases with n.

Statistics applied to small data sets

It should now be apparent that σ and μ are important quantities associated with any measurement and scientists would like to know their values. Because we deal with finite data sets, we can only acquire s and $\bar{x}$, which are *estimates* of the standard deviation and true value, respectively. However, the reliability of these estimates can be calculated from certain statistical equations.

Confidence intervals. How "good" the mean is as an estimate of the true value may be answered statistically by defining an interval about the mean, within which one expects to find the true value. The size of the interval is determined in part by the statistical probability (called confidence level) of being correct. If you wish a 100% probability of being correct, the interval would have to be infinitely large to protect against a very large random error, even though such an error is not very likely. Such an interval has no value to the scientist. On the other hand, if we are willing to settle for a *finite* probability of being correct, we can have a finite confidence interval.

In the absence of determinate error, the confidence interval about the mean, within which the true value lies, is calculated from the equation

$$\mu = \bar{x} \pm \frac{ts}{\sqrt{n}} \qquad (3\text{-}3)$$

where t is a statistical constant that depends both on the confidence level and the number of measurements involved. The values of t for three different confidence levels are given in Table 3-1.

Example 3-6

Cal Culator obtained the following results for replicate determinations of calcium in limestone: 14.35%, 14.41%, 14.40%, 14.32%, and 14.37%. Calculate the confidence interval at the 95% confidence level.

Solution The confidence interval is calculated from Equation 3-3.

$$\bar{x} = \frac{\sum x_i}{n} = \frac{14.35 + 14.41 + 14.40 + 14.32 + 14.37}{5} = 14.37\%$$

$$s = \sqrt{\frac{\sum (x_i - \bar{x})^2}{n - 1}} = \sqrt{\frac{(0.02)^2 + (0.04)^2 + (0.03)^2 + (0.05)^2 + (0.00)^2}{5 - 1}}$$

$$= 0.037\%$$

Using $t = 2.776$ for $n = 5$ from Table 3-1 gives us

$$\bar{x} \pm \frac{ts}{\sqrt{n}} = 14.37\% \pm \frac{(2.776)(0.037\%)}{\sqrt{5}} = 14.37\% \pm 0.05$$

TABLE 3-1 VALUES OF t FOR CALCULATING CONFIDENCE INTERVALS

Number of observations, n	Degrees of freedom, $n - 1$	Confidence level		
		90%	95%	99%
2	1	6.314	12.706	63.657
3	2	2.920	4.303	9.925
4	3	2.353	3.182	5.841
5	4	2.132	2.776	4.604
6	5	2.015	2.571	4.032
7	6	1.943	2.447	3.707
8	7	1.895	2.365	3.499
9	8	1.860	2.306	3.355
10	9	1.833	2.262	3.250
11	10	1.812	2.228	3.169
12	11	1.796	2.201	3.106
13	12	1.782	2.179	3.055
14	13	1.771	2.160	3.012
15	14	1.761	2.145	2.977
20	19	1.729	2.093	2.861
∞	∞	1.645	1.960	2.576

Comparing values: Tests of significance.   Chemists and other scientists often find themselves posing questions about the significance or validity of data that can be answered with a simple yes or no. The posing and answering of such questions is called *hypothesis testing*. A common procedure called the null hypothesis is to state that there is no significant difference between two numbers or sets of numbers, or that a variable exerts no significant effect on the result, and then statistically prove or disprove the statement. It is important to recognize that the proof will be qualified by a confidence level indicating the degree of certainty of the proof. Three common questions that can be answered using the null hypothesis approach are:

1. Is the mean of a data set significantly different from the true value?
2. Are the means of two different data sets significantly different?
3. Are the precisions of two different data sets significantly different?

Each of these questions is examined below.

Comparing a mean with a true value.   A common way to test a method, procedure, instrument, or laboratory technician as a source of determinate error is to perform the determination on a sample whose "true value" is known. How the true value gets known is not important here, but it could be the NBS certified value of a standard sample or the mean of a very large number of careful measurements on the same sample. The test is based on a comparison of the actual difference between the mean and the true value with the largest difference that could be expected as a result of indeterminate error. The latter value is given by $\pm ts/\sqrt{n}$. Thus if $(\bar{x} - \mu) > \pm\, ts/\sqrt{n}$, the hypothesis that no significant difference exists is rejected.

Example 3-7

A new procedure for determining trace amounts of zinc in vegetables was evaluated by using it to determine the zinc content of an NBS standard sample, with the following results: 0.083, 0.088, 0.087, and 0.086 ppm Zn. The certified value of the standard sample is 0.082 ppm Zn. Is there a significant difference between the mean of the results and the certified or "true" value?

Solution   The first step is to calculate the standard deviation needed to compute the confidence interval, $\pm ts/\sqrt{n}$:

$$\bar{x} = \frac{\sum x_i}{n} = \frac{0.083 + 0.088 + 0.087 + 0.086}{4} = 0.086 \text{ ppm}$$

$$s = \sqrt{\frac{\sum (x_i - \bar{x})^2}{n - 1}} = \sqrt{\frac{(0.003)^2 + (0.002)^2 + (0.001)^2 + (0.000)^2}{4 - 1}}$$

$$= 0.0022 \text{ ppm}$$

From Table 3-1 we find $t = 3.182$ for $n = 4$ at the 95% confidence level; thus

$$\frac{ts}{\sqrt{n}} = \frac{(3.182)(0.0022 \text{ ppm})}{\sqrt{4}} = 0.003 \text{ ppm}$$

Also,

$$(\bar{x} - \mu) = 0.086 \text{ ppm} - 0.082 \text{ ppm} = 0.004 \text{ ppm}$$

Since $(\bar{x} - \mu) > \pm ts/\sqrt{n}$, the hypothesis is rejected and the conclusion is that a significant difference *does* exist.

Comparing two means. There are many occasions when chemists wish to determine if two independently obtained results are essentially the same. For example, an oil refiner that purchases a large amount of high-grade crude oil as it is being loaded aboard a tanker may wish to have it analyzed at the time of purchase and again upon its arrival at the refinery to ensure that the oil being delivered is indeed the same oil that was purchased. If n_1 replicate determinations were made during loading and n_2 determinations on arrival, we may write

$$\mu_1 = \bar{x}_1 \pm \frac{ts_1}{\sqrt{n_1}} \tag{3-4}$$

and

$$\mu_2 = \bar{x}_2 \pm \frac{ts_2}{\sqrt{n_2}} \tag{3-5}$$

If we assume that $\mu_1 = \mu_2$ and $\sigma_1 = \sigma_2$, Equations 3-4 and 3-5 can be combined to give

$$\bar{x}_1 - \bar{x}_2 = \pm ts_p \sqrt{\frac{n_1 + n_2}{n_1 n_2}} \tag{3-6}$$

where s_p is the *pooled* standard deviation of the two data sets and is calculated from the equation

$$s_p = \sqrt{\frac{(n_1 - 1)s_1^2 + (n_2 - 1)s_2^2}{n_1 + n_2 - 2}}$$

The value for t is based on $n_1 + n_2 - 2$ degrees of freedom. If

$$(\bar{x}_1 - \bar{x}_2) > \pm ts_p \sqrt{\frac{n_1 + n_2}{n_1 n_2}}$$

the null hypothesis that no significant difference exists is rejected.

It is not necessary to *assume* that σ_1 equals σ_2, as this can be tested using the standard deviations, s_1 and s_2, as shown in the next section.

Example 3-8

A ship of copper ore from Chile was purchased by a local metal refiner. The analysis certificate, made out while the ship was being loaded, showed that % Cu = 14.66 with a standard deviation of 0.07% for 5 measurements. When the ore arrived at the refinery, it was analyzed with the following results: % Cu = 14.58, 14.61, 14.69, and 14.64. Should the refiner accept the ore?

Types of Errors

Solution To compare the two means, use Equation 3-6. To calculate the pooled standard deviation, we must first calculate s_2.

$$\bar{x}_2 = \frac{14.58 + 14.61 + 14.69 + 14.64}{4} = 14.63\%$$

$$s_2 = \sqrt{\frac{(0.05)^2 + (0.02)^2 + (0.06)^2 + (0.01)^2}{4 - 1}} = 0.05\%$$

Then

$$s_p = \sqrt{\frac{(5 - 1)(0.07)^2 + (4 - 1)(0.05)^2}{5 + 4 - 2}} = 0.04\%$$

Using the value of t from Table 3-1 for 7 degrees of freedom $(n_1 + n_2 - 2)$ at the 95% confidence level gives us

$$\pm ts_p \sqrt{\frac{n_1 + n_2}{n_1 n_2}} = (2.37)(0.04)\sqrt{\frac{5 + 4}{(5)(4)}} = 0.06\%$$

$$\bar{x}_1 - \bar{x}_2 = 14.66\% - 14.63\% = 0.03\%$$

Since $(\bar{x}_1 - \bar{x}_2)$ is not greater than $\pm ts_p \sqrt{(n_1 + n_2)/n_1 n_2}$, the hypothesis that no significant difference between means exists is accepted and the refiner should accept the shipment of ore.

Comparing two precisions: The F test. It is sometimes desired to determine if the standard deviation, s_1, from one data set is significantly different from the standard deviation, s_2, from another data set. The test is made by comparing the *square* of each standard deviation, called the *variance*, V. The larger variance is always the numerator.

$$F_c = \frac{s_1^2}{s_2^2} = \frac{V_1}{V_2} \tag{3-7}$$

If the calculated value, F_c, exceeds a tabulated, statistical value, F_t, the hypothesis of no difference is rejected. An abbreviated compilation of F_t values is given in Table 3-2.

TABLE 3-2 VALUES OF F_t FOR COMPARING VARIANCES AT THE 95% CONFIDENCE LEVEL

Number of observations in denominator	Number of observations in numerator						
	3	4	5	6	7	10	∞
3	19.00	19.16	19.25	19.30	19.33	19.38	19.50
4	9.55	9.28	9.12	9.01	8.94	8.81	8.53
5	6.94	6.59	6.39	6.26	6.16	6.00	5.63
6	5.79	5.41	5.19	5.05	4.95	4.78	4.36
7	5.14	4.76	4.53	4.39	4.28	4.10	3.67
10	4.26	3.86	3.63	3.48	3.37	3.18	2.71
∞	2.99	2.60	2.37	2.21	2.09	1.88	1.00

Example 3-9

The director of a hospital clinical laboratory was trying to decide whether or not to keep a young, recently hired technician. The director decided to see if the new technician's work was of the same quality as that of the other staff. She asked both a senior technician and the new technician to analyze the same sample, using the same procedure, reagents, and instruments. They obtained the following results:

Senior technician	New technician
1.38%	1.28%
1.33%	1.36%
1.34%	1.35%
1.35%	1.40%
1.30%	1.31%

Use the F test to determine if there is a significant difference in the precision of the data.

Solution The standard deviation is a measure of precision. Computing the standard deviation for use in Equation 3-7 gives

$$ s_{\text{senior}} = \sqrt{\frac{(0.04)^2 + (0.01)^2 + (0.00)^2 + (0.01)^2 + (0.04)^2}{5-1}} $$

$$ = 0.029\% $$

and

$$ s_{\text{new}} = \sqrt{\frac{(0.06)^2 + (0.02)^2 + (0.01)^2 + (0.06)^2 + (0.03)^2}{5-1}} $$

$$ = 0.046\% $$

Then

$$ F_c = \frac{s_{\text{new}}^2}{s_{\text{senior}}^2} = \frac{(0.046)^2}{(0.029)^2} = 2.5 $$

The tabular value, F_t, from Table 3-2 is 6.39. Since F_c does not exceed this value, the hypothesis of no difference is accepted.

Rejecting data.   Occasionally in a set of results we find that one value is much larger or smaller than the others, and we must decide whether to retain or reject the value. Let's follow the case of quantitative analysis student Ty Trate to see an example of such a situation and how to deal with it. Ty obtained the following results for % Cl in analyzing a laboratory unknown:

Trial	% Cl
1	12.69
2	12.58
3	13.02
4	12.63

If he uses all four values, the mean is

$$\bar{x} = \frac{50.92}{4} = 12.73\%$$

whereas if trial number 3 is rejected, the mean is

$$\bar{x} = \frac{37.90}{3} = 12.63\%$$

Ty cannot recall any unique event that occurred during trial 3 that might be responsible for the high value. Although his intuition is to reject the value, Ty knows that intuition is seldom a good basis for judgment. In preparing for the experiment he remembers reading about the Q test for discordant values. In this test, Q_c is calculated from the equation

$$Q_c = \frac{\left|\text{questionable value} - \text{nearest numerical value}\right|}{\text{range}}$$

If Q_c is found to be less than Q_t from a table of statistical constants, the hypothesis of no significant difference is accepted and the value is retained. In Ty's case,

$$Q_c = \frac{\left|13.02 - 12.69\right|}{13.02 - 12.58} = \frac{0.33}{0.44} = 0.75$$

The value for Q_t at the 90% confidence level, taken from Table 3-3, is 0.76. Since Q_c is less than Q_t, Ty is probably better off retaining the value.

Ty's story could go on. Suppose that his grade on this laboratory experiment determines whether he passes or fails the course. With so much at stake, Ty wisely decides to repeat the analysis three more times and obtains the following results:

Trial	% Cl
5	12.81
6	13.04
7	12.37

TABLE 3-3 VALUES OF Q_t FOR REJECTING DATA

Number of observations, n	Degrees of freedom, $n - 1$	Confidence level		
		90%	96%	99%
3	2	0.94	0.98	0.99
4	3	0.76	0.85	0.93
5	4	0.64	0.73	0.82
6	5	0.56	0.64	0.74
7	6	0.51	0.59	0.68
8	7	0.47	0.54	0.63
9	8	0.44	0.51	0.60
10	9	0.41	0.48	0.57

indicating that it was merely fortuitous that three of the first four results were so closely grouped in value. Using all seven results, he gets a mean of

$$\bar{x} = \frac{89.14}{7} = 12.73\%$$

which is the same mean obtained from the first four results.

If time permits, the acquisition of more results is the surest way to a more informed decision. In Ty's case, it might have turned out that his additional three measurements were 12.61, 12.66, and 12.60, which would have clearly established 13.02 as a value that should have been rejected.

PROPAGATION OF INDETERMINATE ERRORS

The final result of a determination is always computed from two or more measurements, each of which has an error associated with it. The way in which the individual errors accumulate depends on the type of arithmetic operation performed. It also depends on whether the error is determinate or indeterminate in nature. Since determinate errors are of known direction and magnitude, each individual measurement can be corrected *before* any arithmetic is performed. The accumulation of indeterminate errors is based on the fact that the individual *variances* or uncertainties are additive.

Addition and Subtraction

Suppose that a result, R, is to be obtained from the algebraic equation

$$R = A + B - C$$

where A, B, and C are experimentally measured quantities. If each of these quantities has associated with it an indeterminate error, the equation we really care about is

$$(R \pm r) = (A \pm a) + (B \pm b) - (C \pm c)$$

where the lowercase letters refer to the error or uncertainty of the measured or calculated value. The uncertainty is usually expressed as the standard deviation, but the confidence interval, $\pm ts/\sqrt{n}$, may also be used.

Statistical theory tells us that the squares of the uncertainties are additive; thus

$$r^2 = a^2 + b^2 + c^2$$

or

$$r = \sqrt{a^2 + b^2 + c^2} \tag{3-8}$$

Recalling that indeterminate errors are randomly positive or negative, we must, in any estimate of the overall error, account for the possibility that all of the individual errors were of the same sign (in the same direction). As a result, the uncertainty of C is additive even though C is subtracted from $A + B$.

Example 3-10

Calculate the error in the molecular weight of FeS from the following atomic weights: Fe = 55.847 $\pm$ 0.004, S = 32.064 $\pm$ 0.003.

Propagation of Indeterminate Errors   **57**

Solution According to Equation 3-8

$$r = \sqrt{(\pm 0.004)^2 + (\pm 0.003)^2} = 0.005$$

$$MW = 87.911 \pm 0.005$$

Multiplication and Division

The error in the result of a multiplication and/or division is calculated in a similar way except that it is the squares of the *relative* uncertainties that are additive. Hence

$$\left(\frac{r}{R}\right)^2 = \left(\frac{a}{A}\right)^2 + \left(\frac{b}{B}\right)^2 + \left(\frac{c}{C}\right)^2$$

or

$$\frac{r}{R} = \sqrt{\left(\frac{a}{A}\right)^2 + \left(\frac{b}{B}\right)^2 + \left(\frac{c}{C}\right)^2} \qquad (3\text{-}9)$$

The absolute error in R can be found by multiplying both sides of Equation 3-9 by R:

$$r = R \sqrt{\left(\frac{a}{A}\right)^2 + \left(\frac{b}{B}\right)^2 + \left(\frac{c}{C}\right)^2}$$

Example 3-11

In a titration, the % analyte is computed from the following equation:

$$\% \text{ analyte} = \frac{\text{vol}_{titrant} \times M_{titrant} \times MW_{analyte}}{\text{wt}_{sample}} \times 100$$

Calculate the relative error and the absolute error in the % analyte from the following data.

$$\text{vol}_{titrant} = 38.04 \pm 0.02 \text{ mL}$$

$$M_{titrant} = 0.1137 \pm 0.0003 \text{ mmol/mL}$$

$$MW_{analyte} = 74.116 \pm 0.005 \text{ mg/mmol}$$

$$\text{wt}_{sample} = 800.0 \pm 0.2 \text{ mg}$$

Solution The individual relative errors are

$$\frac{\pm 0.02}{38.04} = \pm 5.3 \times 10^{-4}$$

$$\frac{\pm 0.0003}{0.1137} = \pm 2.6 \times 10^{-3}$$

$$\frac{\pm 0.005}{74.116} = \pm 6.7 \times 10^{-5}$$

$$\frac{\pm 0.2}{800.0} = \pm 2.5 \times 10^{-4}$$

The total relative error is calculated using Equation 3-9:

$$\frac{r}{R} = \sqrt{(\pm 5.3 \times 10^{-4})^2 + (\pm 2.6 \times 10^{-3})^2 + (\pm 6.7 \times 10^{-5})^2 + (\pm 2.5 \times 10^{-4})^2}$$

$$= 2.7 \times 10^{-3}$$

The absolute error is given by

$$r = \frac{r}{R} \times R$$

where

$$R = \% \text{ analyte} = \frac{(38.04)(0.1137)(74.116)}{800.0} \times 100 = 40.07$$

Thus

$$r = (2.7 \times 10^{-3})(40.07) = 0.11$$

CONSTRUCTING AND INTERPRETING GRAPHS

Line graphs are an extremely efficient way to present large amounts of data and, at the same time, permit important conclusions to be drawn easily and with confidence. Consequently, they are used extensively in all sciences. Although many different relationships exist between experimental variables, scientists are especially attracted to linear relationships, partly because straight lines are easier to draw than curved lines, but more so because the calculations associated with straight lines are easier to perform than those for curved lines.

Ideally, when linearly related data are plotted, a single straight line can be drawn through all the points. In reality, such a situation rarely occurs because indeterminate errors associated with the experimental variables being plotted cause some deviation from strictly linear behavior. In such cases the chemist must try to find the "best" straight line described by the data points on the graph. Intuitively, you know that when the points are scattered, the line should be drawn so that there are about as many points above the line as below. It is more satisfying, however, if we do not have to rely entirely on our intuition and it is here that statistics can help. The techniques used for obtaining the "best" line from a set of data points and for specifying the uncertainties of that line are called *regression analyses*. In this section we examine only the simplest regression analysis for a straight line, called the *method of least squares*.

In applying the method of least squares, the following two fundamental assumptions are made:

1. A linear relationship exists between the plotted variables.
2. One of the plotted variables is independent of the other one and is known with a comparatively high degree of accuracy.

A nonlinear relationship often can be transformed into a linear one by plotting the logarithm, reciprocal, or some root of one or both of the variables. Such transformations

Constructing and Interpreting Graphs

must be used with caution, however, because they can also convert a distribution from Gaussian to non-Gaussian, and certain statistical calculations are based on data having a Gaussian distribution.

In the second assumption above, the term *independent variable* refers to one whose value is not influenced by the other variable. Conversely, a *dependent variable* is one whose value is influenced by the other variable. In reaction-rate studies it is common to measure the concentration of a reactant or product as a function of time. Time is the independent variable, as it is not influenced by the progress of the reaction. Concentration is the dependent variable since its value depends on the time. Also note that the independent variable, time, can be measured much more accurately than the dependent variable, concentration, and is therefore assumed to contain none of the measurement uncertainty.

Fitting the Least-Squares Line

The equation for a straight-line relationship between a dependent variable y and an independent variable x is

$$y = a + bx$$

where a is the intercept (value of y when x is zero) and b is the slope of the line. Recalling that there is little or no uncertainty in the x direction, the deviation of a given data point (x_i, y_i) from the line in the y direction is given by

$$y_i - (a + bx_i)$$

The method of least squares is based on finding values of a and b that make the sum of the squares of the deviations, Q, a minimum, where

$$Q = \sum [y_i - (a + bx_i)]^2$$

Values of a and b for Q_{min} are determined using differential calculus. It is not the derivation that interests us, but the results, which are

$$a = \bar{y} - b\bar{x} \tag{3-10}$$

$$b = \frac{\sum [(x_i - \bar{x})(y_i - \bar{y})]}{\sum (x_i - \bar{x})^2} \tag{3-11}$$

where $\bar{x}$ and $\bar{y}$ are the means of all x's and y's. Once the "best" slope and intercept have been determined, a line with those values can be put on the graph along with the original data points to complete the plot.

It is important to remember that like the standard deviation and the mean, the quantities a and b are statistical *estimates* of the true intercept, α, and slope, β. As such, we are naturally interested in how close these estimates are to the true values. You have already learned how the confidence interval can be used to describe the range of values within which the true value is likely to be found. In the same way, it can be shown that the confidence interval for the slope and the intercept are given by

$$\beta = b \pm \frac{ts}{\sqrt{\sum (x_i - \bar{x})^2}} \tag{3-12}$$

$$\alpha = a \pm ts \sqrt{\frac{1}{n} + \frac{\bar{x}^2}{\sum (x_i - \bar{x})^2}} \qquad (3\text{-}13)$$

where

$$s = \sqrt{\frac{\sum (y_i - \bar{y})^2 - b^2 \sum (x_i - \bar{x})^2}{n - 2}} \qquad (3\text{-}14)$$

Values of t from Table 3-1 are for $n - 2$ degrees of freedom. These equations, when seen for the first time, may seem overwhelming. However, they are not difficult, only tedious, and you can take heart in the fact that you will probably not have to remember the exact formulas, but only know how to apply them.

Example 3-12

As pointed out in chapter 1, many instrumental methods of analysis utilize calibration plots of measured signal versus concentration. Below are shown the data collected for such a plot. Assume that a linear relationship exists and that the concentration is the independent variable, known with a high degree of certainty.

Concentration (ppm)	Signal
1.00	0.116
2.50	0.281
5.00	0.567
7.50	0.880
10.00	1.074

Calculate, using the least-squares method, the slope and intercept of the "best-fit" line along with the confidence interval of each at the 95% level.

Solution Let the independent variable of concentration equal x and the dependent variable of signal equal y. Then

$$\bar{x} = \frac{1.00 + 2.50 + 5.00 + 7.50 + 10.00}{5} = 5.20$$

$$\bar{y} = \frac{0.116 + 0.281 + 0.567 + 0.880 + 1.074}{5} = 0.584$$

Data necessary for the equations are:

$|x_i - \bar{x}|$	$(x_i - \bar{x})^2$	$|y_i - \bar{y}|$	$(y_i - \bar{y})^2$	$|x_i - \bar{x}||y_i - \bar{y}|$
4.20	17.64	0.468	0.219	1.966
2.70	7.29	0.303	0.091	0.818
0.02	0.00	0.017	0.000	0.000
2.30	5.29	0.296	0.088	0.681
4.80	23.04	0.490	0.240	2.352
$\sum = \overline{14.02}$	$\sum = \overline{53.26}$	$\sum = \overline{1.574}$	$\sum = \overline{0.638}$	$\sum = \overline{5.817}$

Constructing and Interpreting Graphs

From Equation 3-11,

$$\text{slope} = b = \frac{5.817}{53.26} = 0.1092$$

and from Equation 3-10,

$$\text{intercept} = a = 0.584 - (0.1092)(5.20) = 0.016$$

In calculating the confidence interval for the slope, s is obtained from Equation 3-14 and t is for $5 - 2$ or 3 degrees of freedom.

$$s = \sqrt{\frac{0.638 - (0.1092)^2 (53.26)}{5 - 2}} = 0.0316$$

$$\beta = 0.1092 \pm \frac{(3.18)(0.0316)}{53.26} = 0.1092 \pm 0.00189$$

Finally, the confidence interval of the intercept is calculated from Equation 3-13:

$$\alpha = 0.016 \pm (3.18)(0.0316)\sqrt{\frac{1}{5} + \frac{(5.20)^2}{53.26}}$$

$$= 0.016 \pm 0.085$$

PROBLEMS

3-1. How many significant figures are contained in each of the following values?
(a) 6.317
(b) 0.08091
(c) 4.30×10^2
(d) 1200
(e) 1.0×10^{-5}

3-2. How many significant figures are contained in each of the following values?
(a) 200
(b) 31.70
(c) 0.03066
(d) 4.17×10^4
(e) 6.00483

3-3. Write the following values with three significant figures.
(a) two thousand
(b) the decimal equivalent of $\frac{1}{64}$
(c) pi (the geometric constant)
(d) one millionth of one

3-4. Write the following values with two significant figures.
(a) one pound, in grams
(b) the height in centimeters of a person who is 5 feet 6 inches tall
(c) the number of centimeters in a yard
(d) the length of a football field in feet

3-5. What is the implied absolute and relative uncertainty in each of the following experimental values?
(a) 16
(b) 2.037
(c) 1.00×10^{-3}
(d) 0.060714

3-6. To deliver a certain volume from a buret requires two readings: initial and final. If each reading is made with an uncertainty of ± 0.02 mL, what is the relative uncertainty in dispensing 30.00 mL from a buret?

3-7. Round off the following values to three significant figures.
(a) 1.6194
(b) 0.0316477
(c) 1437
(d) 8.6153×10^{-3}

3-8. Report the results of the following arithmetic equations with the appropriate number of significant figures. Assume that all values are experimental.

(a) $1.1037 + 0.914 + 8.62 =$

(b) $3.1994 \times 10^{-3} + 6.145 \times 10^{-1} + 0.00317 =$

(c) $(106.8)(21.4) =$

(d) $(1.447 \times 10^{-3})(8.19 \times 10^{-1}) =$

(e) $(4.18 \times 10^{-4})^3 =$

(f) $\dfrac{8.11}{1.6} =$

(g) $\dfrac{14.16 + 3.179 - 9.08}{3.159 + 4.4} =$

(h) $\log(3.162 \times 10^4) =$

3-9. Report the results of the following arithmetic equations with the appropriate number of significant figures. Assume that all values are experimental.

(a) $9.388 + 1.4066 - 5.2 =$

(b) $4.16 \times 10^3 - 3.79 \times 10^2 =$

(c) $(6.4)(1.741) =$

(d) $(7.72 \times 10^{-2})(4.116 \times 10^{-3}) =$

(e) $\sqrt{2.164 \times 10^5} =$

(f) $\dfrac{1.6}{4.1793} =$

(g) $\dfrac{(6.02 \times 10^1) + 4.4711}{12.833} + 2.649 =$

(h) $\log(8.063 \times 10^{-4} + 1.17 \times 10^{-2}) =$

3-10. An analysis of city drinking water for total hardness produced the following results (in ppm $CaCO_3$): 228.3, 226.4, 226.9, 227.1, and 228.6. Calculate the following.

(a) mean   (c) range   (e) standard deviation

(b) median   (d) average deviation

3-11. The following pH data were collected in the analysis of water from an upstate New York lake as part of an acid rain study: 4.17, 4.20, 4.19, 4.23, 4.22, 4.14, 4.20, and 4.15. Calculate the following.

(a) mean   (c) percent average deviation

(b) median   (d) percent standard deviation

3-12. If the true concentration of $CaCO_3$ in the drinking water described in Problem 3-10 is 225.9 ppm, calculate the absolute and parts per thousand error.

3-13. An alloy from the National Bureau of Standards was analyzed for its chromium content and the following results were obtained: 2.61, 2.66, 2.61, 2.70, and 2.68% Cr. If the NBS certified value is 2.69% Cr, what is the absolute error? What is the percent error?

3-14. Indicate which of the following are determinate and which are indeterminate errors.

(a) A hygroscopic sample is not dried in the oven prior to weighing.

(b) A buret reading of 28.43 mL is written down as 28.34 mL.

(c) The bottom of the meniscus in a buret is estimated to be at the zero-milliliter mark but in fact is very slightly below this mark.

(d) The liquid in a 10-mL TD (to deliver) pipet is blown out by mouth.

(e) The rest point of a meter needle, oscillating due to some desktop vibration, is estimated incorrectly.

3-15. Quantitative analysis student Kenny Duit analyzed an ore sample for its copper content and obtained a mean of 4.67% Cu for three measurements. How many additional measurements must Kenny make to obtain a mean that is five times more reliable?

3-16. A chemist attempting to determine the benzene content of a commercial gasoline obtained a mean of 1.73% for four determinations. How much would the reliability of the mean be increased if an additional six determinations were made?

3-17. Which of the following three data sets has the best precision? Justify your answer.

A	B	C
2.31	11.74	56.33
2.33	11.82	56.21
2.30	11.79	56.27
2.30	11.80	56.16

3-18. An analytical chemist working for the Environmental Protection Agency obtained the following results for the determination of dioxin in contaminated soil: 28.4, 26.1, 26.7, 29.0, 28.3, 27.5, and 27.7 ppb.
  (a) Calculate the confidence interval of the mean at the 95% confidence level.
  (b) How many measurements must be obtained for the confidence interval to be ± 0.77? Assume that there is no change in the standard deviation.

3-19. What must the standard deviation of eight measurements be to produce a confidence interval of ± 0.075 at the following confidence levels?
  (a) 90%
  (b) 95%

3-20. A new procedure for determining cholesterol in blood serum was used with an NBS standard sample, producing the following results: 2.24, 2.21, 2.16, 2.21, and 2.19 mg/mL. If the NBS certified value is 2.12 mg/mL, determine if there is a significant difference between the mean of the results and the certified or "true" value at the 95% confidence level.

3-21. Suppose that a young chemist working in your group proposes a new method for determining polychlorinated biphenyls (PCBs) in wastewater. You give her a sample whose concentration is known to be 0.238 ppm and she makes five replicate determinations with a mean of 0.230 ppm. What is the maximum standard deviation of the data that will still allow you to conclude that no significant difference exists between the reported mean and the true value at the 95% confidence level?

3-22. Determine if there is any significant difference at the 90% confidence level between the means of the following two data sets.

Set 1	Set 2
92.61	93.08
92.84	92.87
92.77	92.91
92.61	93.03
92.65	93.06
92.69	

3-23. Determine if there is any significant difference between the means of two data sets that are described below. Use a value of 1.708 for t.

	Set A	Set B
n	12	15
s	0.84	0.75
$\bar{x}$	34.61	34.99

3-24. A research chemist has proposed a new method for determining phenylalanine in the whole blood of newborns. A clinical laboratory evaluating the method obtained replicate values of 0.0682, 0.0677, 0.0685, 0.0685, and 0.0679 mg/mL. The same sample analyzed by an accepted standard method produced a mean of 0.0687 and standard deviation of 0.00037 mg/mL based on seven measurements. Can the laboratory director be 90% confident that the new method will give results that are not significantly different from the presently accepted standard method? Can he be 99% confident?

3-25. The air inside a nickel refinery was analyzed for gaseous SO_2 on two successive days, with the following results:

Day	Mean concentration $SO_2(g)$ (ppm)	Number of measurements
1	8.63	6
2	8.84	5

The method used is known to yield a standard deviation of 0.093 ppm SO_2. Are the two means significantly different at the 95% confidence level? Are they different at the 99% confidence level?

3-26. In trying to decide whether an old instrument should be retired from service, the director of a pharmaceutical laboratory asked a technician to determine the cortisone content of a drug sample using the same procedure but making the final measurements on both the old instrument and on a new instrument. The following data were obtained. Is there any significant difference at the 95% confidence level between the precisions of the two data sets?

	Percent cortisone	
Trial	Old instrument	New instrument
1	12.7	12.3
2	12.3	12.3
3	12.4	12.5
4	12.0	12.0
5	11.9	12.6

3-27. Use the Q test at the 96% confidence level to determine if the last value of the following data should be omitted before computing the mean: 2.93, 3.08, 3.11, 3.04, and 2.70% Fe.

3-28. A chemist working in a hospital clinical laboratory obtained the following values for uric acid in blood serum: 4.63, 4.58, 4.82, 4.31, 4.67, and 4.60 mg-% (mg × 100/mL). What value should be reported as the mean? Use the 90% confidence level for testing any dubious values.

3-29. Quantitative analysis student I. M. Smart obtained the following results for his determination of copper in an unknown: 4.86, 4.86, 4.77, and 4.86% Cu. Is this a case where the Q test can be used to determine if a value should be discarded? Explain your answer.

3-30. In the gravimetric determination of sulfur as barium sulfate, % S is calculated from the equation

$$\% \text{ S} = \frac{\text{wt BaSO}_4 \times \dfrac{\text{AW S}}{\text{MW BaSO}_4} \times 100}{\text{wt sample}}$$

Calculate the absolute and relative error in % S from the following data:

$$\text{wt BaSO}_4 = 655.4 \pm 0.2 \text{ mg}$$

$$\text{wt sample} = 1.4336 \pm 0.0002 \text{ g}$$

$$\text{AW S} = 32.064 \pm 0.008$$

$$\text{AW Ba} = 137.34 \pm 0.01$$

$$\text{AW O} = 15.9994 \pm 0.0003$$

3-31. There is a constant loss of 1.5 mg of nickel due to its solubility in a gravimetric method. What is the percent error due to this loss for the following weights of nickel in a sample?
(a) 50 mg
(b) 250 mg

3-32. A gravimetric method for Al was found to give results that are constantly low by 0.70 mg Al. Calculate the parts per thousand error when a 750-mg sample containing approximately 12% Al is analyzed by this method.

3-33. A linear relationship exists between the concentration of a colored substance and the amount of light it absorbs (as measured by the absorbance). The following data were collected for different concentrations of permanganate ion:

$C_{\text{MnO}_4^-}$ (ppm)	Absorbance	$C_{\text{MnO}_4^-}$ (ppm)	Absorbance
1.00	0.030	10.00	0.301
2.50	0.072	15.00	0.442
5.00	0.147	20.00	0.577
7.50	0.217	25.00	0.738

(a) Plot the data on linear graph paper and fit a straight line to the points by eye.
(b) Use a least-squares regression analysis of the data to calculate a slope and intercept. Plot the line with these values on the same graph used for part (a).
(c) Calculate the confidence intervals at the 95% confidence level for the slope and intercept.

GRAVIMETRY

In gravimetry, or gravimetric analysis as it is sometimes called, the analyte is *converted* to an insoluble substance (precipitate) which is *isolated* and *weighed*. Knowing the weight of the precipitate and its composition, the weight of the analyte can be calculated. The italicized words in the first sentence represent the three fundamental steps of every gravimetric determination. The weighing process is usually simple and accurate using modern balances, leaving the success of the determination dependent on the ability to convert the analyte to an insoluble substance of known, uniform composition and to effect a quantitative recovery of that insoluble substance in a form suitable for weighing. All of the following discussion in this chapter deals with the conversion and isolation steps.

MECHANISM OF PRECIPITATION

The conversion of a constituent to an insoluble substance and the isolation of that substance cannot, insofar as gravimetry is concerned, be regarded as independent topics. The *manner* in which a precipitate is formed often determines how it can or cannot be isolated and also its purity. To learn why this is so, we need to examine in some detail how precipitates form. Broadly speaking, there are three important stages of precipitate formation: nucleation, crystal growth, and aging.

Nucleation

Nucleation is the formation, in a supersaturated solution, of the smallest aggregate of molecules capable of growing into a large precipitate particle. The physical properties and even to some extent the chemical composition of the final precipitate particles are largely dependent on the nucleation process. It is no wonder then that chemists are interested in knowing how to control nucleation. Unfortunately, the nucleation process is extraordinarily difficult to study experimentally and considerable uncertainty and disagreement exists over the meaning of the available data. The number of molecules comprising the nucleation aggregate depends both on the substance itself and the conditions of the precipitation. Although accepted specific values are not available, most scientists involved in such studies believe that it is usually less than 100.

The time between mixing and the visual appearance of a precipitate is called the *induction period* and it varies with the nature of the substance being precipitated, the concentration of the reagents being mixed, and even the order of addition of the reagents. When concentrated solutions are mixed, the induction periods usually are so short that precipitation appears to occur instantly upon mixing. On the other hand, induction periods of several minutes have been observed when mixing dilute solutions.

Crystal Growth

Once a nucleation aggregate has formed, it begins to grow as ions or molecules from the solution deposit on the surface in a regular, geometric pattern. The types of crystal growth, that is, the geometric manners in which crystals expand, are not an important consideration at this point. Some precipitates consist almost entirely of tiny single crystals, each of which grew from a nucleation aggregate, while others consist of aggregates of single crystals that have merged during the growth process. The aggregation process is very important because it determines the final size of the precipitate particles and therefore the ease with which they can be separated from the solution.

Aggregate (Particle) Growth

Natural cohesive forces exist between small particles having the same composition and, as a result, one would expect most precipitates to consist of a relatively few large aggregates of crystals. However, small particles can have other properties that tend to counteract the natural forces of aggregation and lead to the formation of colloidal suspensions. The individual *particles* in such a suspension are so small that they are not retained on ordinary filter paper and will not settle out on standing. Obviously, this condition cannot be tolerated in a gravimetric determination, where the precipitate must be separated. To learn how it can be prevented, we must understand why it occurs.

Silver halide precipitates are well known for their tendency to form colloidal suspensions, so let us examine the process that occurs when sodium chloride is treated with an excess of silver nitrate in dilute nitric acid solution. As small crystals and aggregates of silver chloride form they attract to their surface, and hold by normal bonding forces, new silver and chloride ions in proportion to their availability in the bulk solution. Since they are not yet a part of the regular crystal lattice, these ions are said to be adsorbed on the surface and comprise what is called the *primary adsorption layer*. Because there are more silver ions than chloride ions in the solution, the primary adsorption layer will contain an excess of silver ions and be positively charged. Had the solution contained an excess of chloride ion, the primary adsorption layer would have been negatively charged. These two conditions are shown in Figure 4-1. The *net* charge of the primary

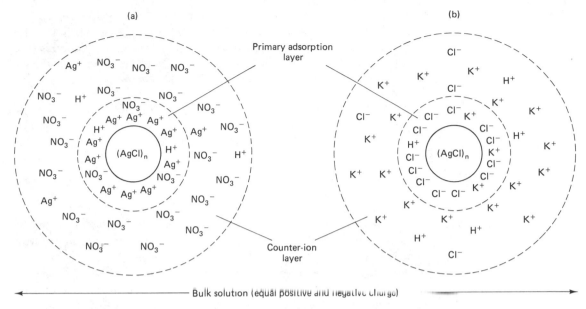

(a)   (b)

Primary adsorption
layer

Counter-ion
layer

Bulk solution (equal positive and negative charge)

Figure 4-1 Colloidal silver chloride particles: (a) in a solution of $AgNO_3$; (b) in a solution of KCl.

adsorption layer depends on *both* the absolute and relative silver-ion and chloride-ion concentrations in the bulk solution. A positively charged primary adsorption layer attracts an excess of negative ions into an adjacent secondary or *counter-ion layer*. These ions are held purely by electrostatic forces and are much more mobile than the ions in the primary adsorption layer.

The two charged layers of the particle constitute an *electrical double layer* that exerts a repulsive force toward other similarly charged particles. This force may be greater than the normal cohesive forces that exist between the particles, preventing them from getting close enough to coalesce or coagulate into larger aggregates. How this problem is dealt with experimentally is discussed later in the section "Methodology."

Purity

One of the basic requirements of gravimetric analysis is the formation of a precipitate of *known composition*, which means that the isolated precipitate must be pure. Impurities can be incorporated into a precipitate during its formation, called *coprecipitation*, or after its formation while still in contact with the precipitating solution, called *postprecipitation*.

Coprecipitation

The term "coprecipitation" is generally meant to include only the contamination of a precipitate by normally soluble substances. A diverse substance can be coprecipitated by the process of *adsorption, inclusion,* or *occlusion*. The adsorption process has already been discussed in the preceding section. Since it is a surface phenomenon, the surface area of the precipitate, and thus the particle size, plays a vital role in determining

Mechanism of Precipitation   **69** the extent of contamination by adsorption. It is important to realize that once precipitation is complete and the particles have stopped growing, precipitating ions in the primary adsorption layer are in *excess* of the analyte ion and are therefore impurities that will add to the final weight. Precipitating ions that become adsorbed during crystal growth are eventually converted into regular lattice ions and become a uniform part of the crystal.

Diverse ions in the primary adsorption layer that are not desorbed during growth are also converted into lattice ions—a process called *inclusion*. If the diverse ion replaces an analyte ion, which will simply find another lattice site, the effect is to increase the weight of the precipitate in proportion to the number of diverse ions (plus precipitating ions of opposite charge) that are included. When the diverse ion replaces the precipitating ion, it is a simple substitution making the precipitate heavier if the atomic mass of the diverse ion is greater than that of the analyte ion and lighter if the opposite is true. Since the coprecipitated substance is technically soluble, its inclusion in a precipitate is due largely to its inability to escape the growing crystal. If crystal growth can be made to occur more slowly, this type of coprecipitation can be reduced.

Growing precipitate particles also can entrap diverse substances that do not become a part of the crystal lattice but are simply surrounded by the crystalline material. This process is called *occlusion* and occurs most often when fairly soluble, hydrated salts are precipitated. It is seldom a problem in precipitating the highly insoluble substances used for gravimetric determinations.

Postprecipitation

The formation of a second insoluble substance on an existing precipitate is called postprecipitation. It is always the result of a difference in the rates of precipitation of the analyte and contaminant. For example, calcium in the presence of magnesium can be precipitated with oxalate as relatively pure calcium oxalate, even though the solubility of magnesium oxalate may be exceeded, because magnesium oxalate precipitates very slowly using the calcium oxalate particles as nucleation sites. Obviously, the sooner the calcium oxalate precipitate can be separated from the solution, the less it will be contaminated by magnesium oxalate.

METHODOLOGY

Most gravimetric methods consist of only a few fundamental steps, but the manner in which these steps are performed varies depending on the chemical and physical properties of the substance. In this section of the chapter we describe some of the factors that must be considered in deciding how to accomplish these steps.

Forming the Precipitate

The principal goals of every gravimetric procedure are to produce a precipitate that is pure and that can be filtered easily. The procedures used to accomplish these goals are the result of our understanding of the physical properties of precipitates and how they form.

Favoring growth over nucleation

The ease with which a precipitate is isolated, and often its purity, depend on the size of the particles making up the precipitate. Large particles are retained by porous filters that permit rapid flow of liquid. Small particles require small-pore filters through which liquid flows slowly and that clog easily. The question, then, is how we can produce a relatively few large crystals rather than many small ones in a precipitation.

Experimentally, it has been observed that the particle size of a precipitate is influenced partly by such experimental variables as precipitate solubility, reactant concentrations in the precipitating solution, the rate of addition and mixing of reactants, and the temperature. To account for the effect of these variables on particle size, chemists have theorized that particle size is related to a single solution property called the *supersaturation ratio*, or relative supersaturation, given by

$$\text{supersaturation ratio} = \frac{Q - S}{S}$$

where Q is the concentration of the solute at any given instant and S is its equilibrium solubility. The numerator, $Q - S$, is a measure of the degree of supersaturation. Careful studies have led most chemists to believe that the rate of nucleation increases exponentially with the supersaturation ratio, while the rate of growth increases linearly, as shown in Figure 4-2.

According to this theory, the most favorable relationship between growth and nucleation occurs when the supersaturation ratio is at its smallest finite value. When a precipitating reagent and sample are mixed, the solution becomes momentarily supersaturated, a condition that usually is quickly relieved by precipitation. Dilute solutions, slow addition of the precipitating reagent, and rapid mixing all have the effect of keeping Q, and thus the supersaturation ratio, small. Increasing the solution temperature may have the same final effect because generally it increases the equilibrium solubility S. Some of these variables have other effects that must be considered before reaching a judgment on the "best" or "right" conditions for a precipitation. For example, the use of dilute solutions (and therefore greater volumes) and increased temperature not only

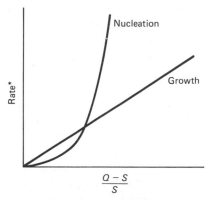

*The numerical values of the rate for the two processes may differ substantially.

Figure 4-2 Effect of supersaturation ratio on rate of nucleation and crystal growth.

may help produce larger particles but also may increase the amount of dissolved precipitate.

Coagulating colloids

Sometimes, despite our best efforts to the contrary, the supersaturation ratio cannot be maintained small enough to prevent the formation of a colloidal suspension. Fortunately, colloidal particles usually can be encouraged to coagulate into larger, filterable particles by reducing the size, and therefore the influence, of their electrical double layer. The size or volume of the double layer is essentially the size of the counter-ion layer. The primary adsorption layer is small, consisting of a volume about one ion thick directly adjacent to the particle surface. The counter-ion layer has a *net* charge equal to that of the primary layer, but a volume that is much larger because the difference between the concentration of cations and anions in this layer is not as great.

One way to keep the volume of the counter-ion layer small is to keep the charge of the primary adsorption layer small by avoiding an excess of the precipitating ion. This is, of course, difficult to do in quantitative analysis where the concentration of analyte is not known beforehand. A more practical approach is to increase the ionic strength by adding a soluble electrolyte. This has the effect of increasing the concentration of excess charge in the counter-ion layer, which is another way of saying that a smaller volume of the counter-ion layer will contain the same net charge. The effectiveness of this approach increases with the charge on the ion; thus A^{3-} is more effective than A^{2-}, which is more effective than A^-.

Heating and stirring, while generally less effective than addition of an electrolyte, are also good means of reducing the volume of the electrical double layer. The higher kinetic energies of the ions in the counter-ion layer allow them to approach the oppositely charged ions of the primary layer more closely.

Minimizing impurities

Crystalline precipitates have a natural tendency to rid themselves of included and occluded impurities. Often, the impurities are there because the precipitation process took place too fast to allow impurities to get out of the way. Heating these precipitates in the precipitating solution, a process called *digestion*, usually results in larger and purer particles by giving the crystals a chance to dissolve and reprecipitate under equilibrium conditions. The crystals or parts of crystals most likely to dissolve during digestion are those with the greatest surface area per volume (smallest particles) and those with lattice impurities or imperfections. As the average crystal size increases, filterability improves. It might be noted also that larger particles have less surface area per unit volume and, therefore, fewer adsorbed impurities. Although digestion is an important aid, it cannot be used in situations where postprecipitation can occur. In such cases, the damaging effects of a postprecipitated contaminant almost always outweigh the benefits of digestion, and the best thing to do is remove the precipitate from the solution as soon as possible.

In certain cases, where neither digestion nor immediate filtering will yield a sufficiently pure precipitate, it may be possible to recrystallize the filtered precipitate. To do this, it is necessary that some experimental variable such as pH can be changed in one direction to make the precipitate dissolve and returned to the original condition to cause the reprecipitation. Only a small fraction of the diverse substance present in the solution actually gets carried down with the precipitate. When the precipitate is dissolved, the resulting solution contains much less diverse substance than the original precipitating solution, and a proportionately smaller amount of this substance is carried down in the second precipitation. Multiple precipitation sounds like an excellent way to achieve the desired purity, but the errors associated with the additional filtering and washing steps are usually quite large, making the technique a choice of "last resort."

Separating and Rinsing Precipitates

A precipitate may be separated by filtering it through paper, sintered glass, or sintered porcelain. The choice depends on the nature of the precipitate and on the temperature to which it will be heated after filtering.

Paper filters are inexpensive and readily available in a broad range of porosities, but they have many disadvantages: They react with strong acids, strong bases, and some oxidizing agents; break under mild stress, preventing the use of suction to speed filtration; and cannot be dried to constant weight, requiring that they be burned away. This last restriction prevents their use with precipitates that are not stable at the ignition temperature of the paper or that react with the hot gases liberated during burning.

Sintered glass filters are made by heating a thin round disk of tiny glass beads until they just begin to fuse together. After cooling, the disk is made the bottom of a Pyrex glass crucible. Such filters are available in coarse, medium, and fine porosity as determined by the size of the glass beads used. Sintered glass filters have the advantages of sufficient mechanical strength to allow the use of suction to speed the filtering process and being able to withstand temperatures up to about 500°C before the sintered glass begins to soften. Sintered porcelain crucibles are constructed in a fashion similar to their glass counterparts and, like glass, are available in various degrees of porosity. Their single advantage over sintered glass is the ability to withstand temperatures up to 1200°C.

Once the precipitate has been collected on the filter, traces of the precipitating solution with its dissolved substances must be rinsed away. With crystalline precipitates, several rinsings with small amounts of pure water generally will remove the impurities. When coagulated colloids are washed with pure water the ions in the primary adsorption layer are largely unaffected because of their strong attraction to ions in the crystal, but much of the electrolyte in the counter-ion layer is removed. As this happens, the counter-ion layer increases in volume and the strong repulsive forces responsible for the original colloidal state are reestablished. This process is called *peptization* and results in the loss of colloidal precipitate through the filter.

The dilemma of whether or not to wash coagulated colloids commonly is resolved by washing, not with pure water, but with a solution of a volatile electrolyte. Instead of the electrolyte in the counter-ion layer being removed, it is *replaced* by the volatile electrolyte, which is removed later when the precipitate is dried or ignited.

Drying and Igniting Precipitates

After a precipitate has been filtered it is heated over a burner or in an oven until it reaches a constant weight. During the heating process, moisture and volatile electrolytes are removed and, in some cases, the chemical form of the precipitate is changed. The temperatures to which precipitates are heated vary considerably and are determined largely from data obtained with an automatic thermobalance, a device that continuously records the weight of a substance in a furnace while its temperature is increased at a constant rate.

The curves for $BaSO_4$ and $Al_2O_3 \cdot xH_2O$ in Figure 4-3a are typical of precipitates losing water or volatile electrolytes in that they show a slow weight decrease over a range of temperatures. Changes in composition of the precipitated molecules also may occur over a range of temperatures or at a single temperature as exemplified by the thermogram of calcium oxalate in Figure 4-3b. At temperatures below 100°C, $CaC_2O_4 \cdot H_2O$ is stable and only unbound moisture is lost. Above 100°C the hydrated water is lost until at 226°C anhydrous CaC_2O_4 is formed. Although this form is stable up to almost 400°C, it is not a suitable weighing form because it is very hygroscopic. Slightly above 400°C, CO is evolved leaving $CaCO_3$, which is stable up to 635°C, where CO_2 evolution begins leading ultimately to CaO. Although both $CaCO_3$ or CaO are suitable weighing forms, $CaCO_3$ is preferred because it is heavier, thereby producing a smaller relative weighing error.

Cooling and Weighing Precipitates

Once the precipitate has been heated to the desired temperature and kept there for the prescribed time, it must be cooled to room temperature before weighing. Even slightly warm objects placed in the enclosed pan compartment of an analytical balance will set up air currents that cause very erratic and incorrect weights to be recorded. Since many gravimetric precipitates have a tendency to pick up moisture while cooling, they are cooled and stored in a moisture-free container called a desiccator. To ensure that *all* of the precipitate is converted to the desired weighing form, the process of heating, cooling, and weighing is generally repeated until a constant weight is achieved. Usually, this means three successive weights that are the same, within experimental error.

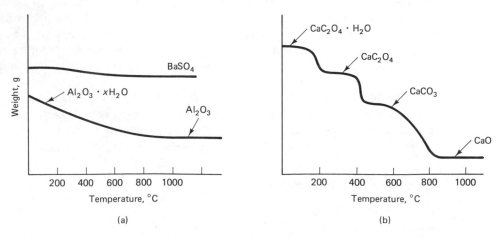

Figure 4-3 Thermogravimetric curves.

Calculating the Results

The result of a gravimetric determination usually is reported as a percentage of analyte. From the equation

$$\% \text{ analyte} = \frac{\text{weight of analyte}}{\text{weight of sample}} \times 100 \qquad (4\text{-}1)$$

it is obvious that only two values are needed: the weight of the analyte and the weight of the sample. The data available from a gravimetric determination are the weight of the precipitate and the weight of the sample. The precipitate is seldom the analyte itself, but a definite relationship between the two always exists. This relationship is called the *gravimetric factor* and is used to convert the known weight of the precipitate to the corresponding weight of analyte. Suppose that the analyte is calcium chloride and it is precipitated as silver chloride by the addition of excess silver nitrate. The reaction leading to the formation of the precipitate is

$$CaCl_2 + 2AgNO_3 \longrightarrow 2AgCl(s) + Ca(NO_3)_2$$

The amount of $CaCl_2$ in moles or millimoles can be obtained from the amount of $AgCl(s)$ if we take into account the reaction stoichiometry. Thus

$$\text{amount } CaCl_2 = \text{amount } AgCl \times \frac{1 \text{ mol } CaCl_2}{2 \text{ mol } AgCl} \qquad (4\text{-}2)$$

where the amounts of $CaCl_2$ and $AgCl$ are given by

$$\text{amount } CaCl_2 = \frac{\text{wt } CaCl_2}{\text{MW } CaCl_2}$$

$$\text{amount } AgCl = \frac{\text{wt } AgCl}{\text{MW } AgCl}$$

When these expressions are substituted in Equation 4-2, we obtain

$$\frac{\text{wt } CaCl_2}{\text{MW } CaCl_2} = \frac{\text{wt } AgCl}{\text{MW } AgCl} \times \frac{1 \text{ mol } CaCl_2}{2 \text{ mol } AgCl}$$

or

$$\text{wt } CaCl_2 = \text{wt } AgCl \times \left(\frac{\text{MW } CaCl_2}{\text{MW } AgCl} \times \frac{1 \text{ mol } CaCl_2}{2 \text{ mol } AgCl} \right) \qquad (4\text{-}3)$$
$$\underbrace{}_{\text{gravimetric factor}}$$

The portion of Equation 4-3 in parentheses is called the *gravimetric factor*. In many cases it is not necessary to write reactions to determine the stoichiometry involved, but only to look at what is desired and what is weighed. For example, suppose that the amount of photographic ''hypo'' or sodium thiosulfate ($Na_2S_2O_3$) in a sample is being determined by converting the sulfur to sulfate and precipitating it as barium sulfate, $BaSO_4$. It is easy to see that two molecules or moles of $BaSO_4$ will be produced for each molecule or mole of $Na_2S_2O_3$ present in the sample. We do not need to know *how*

$Na_2S_2O_3$ gets converted to $BaSO_4$. Because gravimetric factors are intuitively easy to establish, students often prefer not to go through the individual steps shown above. However, it may be a good idea to use the stepwise approach the first few times in order to reinforce the fundamental relationships that are involved.

Example 4-1

A 0.8310-g sample of waste material from a zinc smelter was treated with concentrated HNO_3 and $KClO_3$ to convert the sulfur to sulfate. After removing the nitrate and chlorate by repeated fuming with concentrated HCl, the sulfate was precipitated as $BaSO_4$. After ignition and cooling the $BaSO_4$ precipitate weighed 0.2997 g. What is the % S in the waste material?

Solution From Equation 4-1,

$$\% \text{ S} = \frac{\text{wt S}}{\text{wt sample}} \times 100$$

The weight of S is calculated from the known weight of $BaSO_4$. Since each $BaSO_4$ molecule contains one S atom, the equivalent of Equation 4-3 is

$$\text{wt S} = \text{wt BaSO}_4 \times \frac{\text{AW S}}{\text{MW BaSO}_4} \times \frac{1 \text{ mol S}}{1 \text{ mol BaSO}_4}$$

$$= 0.2997 \text{ g} \times \frac{32.06 \text{ g S/mol S}}{233.4 \text{ g BaSO}_4/\text{mol BaSO}_4} \times \frac{1 \text{ mol S}}{1 \text{ mol BaSO}_4}$$

$$= 0.04117 \text{ g}$$

Substituting in the equation for percent S gives us

$$\% \text{ S} = \frac{0.04117 \text{ g S}}{0.8310 \text{ g sample}} \times 100 = 4.954$$

Example 4-2

A 644.0-mg sample containing magnesium sulfate and inert material was dissolved and treated with $(NH_4)_2HPO_4$ to precipitate the magnesium as $MgNH_4PO_4 \cdot 6H_2O$. After being filtered and washed, the precipitate was ignited at 1050°C to $Mg_2P_2O_7$ that, upon cooling, weighed 293.0 mg. Calculate the % Mg in the sample.

Solution From Equation 4-1,

$$\% \text{ Mg} = \frac{\text{wt Mg}}{\text{wt sample}} \times 100$$

Each molecule of $Mg_2P_2O_7$ contains two atoms of magnesium; thus

$$\text{wt Mg} = \text{wt Mg}_2\text{P}_2\text{O}_7 \times \frac{\text{AW Mg}}{\text{MW Mg}_2\text{P}_2\text{O}_7} \times \frac{2 \text{ mol Mg}}{1 \text{ mol Mg}_2\text{P}_2\text{O}_7}$$

$$= 0.2930 \text{ g Mg}_2\text{P}_2\text{O}_7 \times \frac{24.31 \text{ g Mg/mol Mg}}{222.6 \text{ g Mg}_2\text{P}_2\text{O}_7/\text{mol Mg}_2\text{P}_2\text{O}_7} \times \frac{2 \text{ mol Mg}}{1 \text{ mol Mg}_2\text{P}_2\text{O}_7}$$

$$= 0.06400 \text{ g}$$

Substituting in the equation above yields

$$\% \ Mg = \frac{0.06400 \ g \ Mg}{0.6440 \ g \ sample} \times 100 = 9.938$$

PRECIPITATING REAGENTS

Precipitating reagents (precipitants) are chosen for their ability to be selective and to form highly insoluble precipitates that are easily filtered and of reproducible stoichiometry. Good selectivity ensures that other substances in the sample will not precipitate along with the analyte and a low solubility means losses during precipitation and rinsing will be small. In addition to loss of analyte due to the finite solubility of the precipitate, physical or mechanical losses are incurred during the filtering process. Some precipitates exhibit an annoying tendency to adhere to the wall of the beaker, making a quantitative transfer to the filter medium very difficult. A precipitate of unknown or nonreproducible stoichiometry prevents the analyst from relating the weight of the precipitate to a weight of analyte. On occasion, this problem can be overcome simply by heating the precipitate to change its composition to one that is known and reproducible. For example, several divalent metal ions can be precipitated by the reaction shown below with magnesium:

$$Mg^{2+} + NH_4^+ + PO_4^{3-} + 6H_2O \longrightarrow MgNH_4PO_4 \cdot 6H_2O(s)$$

The composition of the product is not sufficiently reproducible to use as a final weighing form but upon heating to about 1000°C, it is converted to magnesium pyrophosphate $Mg_2P_2O_7$, which is a suitable weighing form.

Inorganic Reagents

Table 4-1 lists some of the common inorganic precipitants used in gravimetric determinations. Most of the early work in gravimetry was done with inorganic precipitants because they were available in a pure state and their chemical and physical properties had been studied and were well understood. Although it is difficult to generalize, inorganic precipitates tend to be more reproducible in composition and more easily filtered than metal-organic precipitates. Their biggest drawback, and it is a severe one, is lack of specificity. Much of the older literature on gravimetric analysis is concerned with ways in which potentially interfering substances can be chemically masked or separated from the analyte.

Organic Reagents

As organic synthesis as a science grew, along with our understanding of bonding between metal ions and organic compounds, new gravimetric procedures were developed using organic precipitants. The ability to design and synthesize a precipitant for a spe-

TABLE 4-1 SELECTED INORGANIC PRECIPITANTS USED IN GRAVIMETRY

Precipitating reagent	Element precipitated[a]	Precipitated form	Weighed form
NH_3	Al	$Al(OH)_3$	Al_2O_3
	Be	$Be(OH)_2$	BeO
	Fe	$Fe(OH)_3$	Fe_2O_3
$(NH_4)_2Cr_2O_7$	Ba	$BaCrO_4$	$BaCrO_4$
	Pb	$PbCrO_4$	$PbCrO_4$
$(NH_4)_2C_2O_4$	Ca	$CaC_2O_4 \cdot H_2O$	$CaCO_3$
	Mg	MgC_2O_4	$MgCO_3$
	Zn	ZnC_2O_4	$ZnCO_3$
$(NH_4)_2HPO_4$	Al	$AlPO_4$	$AlPO_4$
	Ca	$Ca_3(PO_4)_2$	$CaSO_4$
	Mg	$MgNH_4PO_4 \cdot 6H_2O$	$Mg_2P_2O_7$
	Mn	$MnNH_4PO_4 \cdot H_2O$	$Mn_2P_2O_7$
	Zn	$ZnNH_4PO_4 \cdot H_2O$	$Zn_2P_2O_7$
	Zr	$Zr(HPO_4)_2$	ZrP_2O_7
$BaCl_2$	S	$BaSO_4$	$BaSO_4$
HCl	Ag	$AgCl$	$AgCl$
	Bi	$BiOCl$	$BiOCl$
H_2S	Bi	Bi_2S_3	Bi_2O_3
	Cd	CdS	$CdSO_4$
	Cu	CuS	CuO
	Hg	HgS	HgS
	Mo	MoS_3	MoO_3
	Sn	SnS	SnO_2
	Zn	ZnS	ZnO
$HClO_4$	Nb	Nb_2O_5	Nb_2O_5
	Si	SiO_2	SiO_2
	Ta	Ta_2O_5	Ta_2O_5
$AgNO_3$	Cl	$AgCl$	$AgCl$
	Br	$AgBr$	$AgBr$
	I	AgI	AgI
H_2SO_4	Ba	$BaSO_4$	$BaSO_4$
	Pb	$PbSO_4$	$PbSO_4$
	Sr	$SrSO_4$	$SrSO_4$
	W	WO_3	WO_3

[a] Elements precipitated must be in specific oxidation states.

cific metal ion has improved both the sensitivity and selectivity of gravimetric methods. In many cases the selectivity comes about because the organic precipitants are weak acids whose ability to react with metal ions is governed partly by the pH of the solution. By controlling the pH, the analyst can sometimes "tune" a reagent so that it will react with one metal ion but not another.

It is regrettable that many metal-organic precipitates do not have uniform, reproducible compositions and must be ignited to the much lighter metal oxide for weighing. A few common organic precipitants are listed in Table 4-2.

TABLE 4-2 SELECTED ORGANIC PRECIPITANTS USED IN GRAVIMETRY

Name	Structure	Elements precipitated[a,b]
α-Benzoin oxime (cuproin)		Bi, **Cu**, Mo, Zn
Dimethylglyoxime	$CH_3 - C = N - OH$ $\mid$ $CH_3 - C = N - OH$	**Ni**, Pd
8-Hydroxyquinoline (8-quinolinol or oxine)		**Al**, Bi, Cd, Cu, Fe, **Mg**, Pb, Ti, U, **Zn**
Nitron		ClO_4^-, **NO_3^-**
1-Nitroso-2-naphthol		Bi, Cr, **Co**, Hg, Fe
Sodium tetraphenylborate	$Na^+B(C_6H_5)_4^-$	Ag^+, Cs^+, **K^+**, NH^+, Rb^+, large univalent cations
Tetraphenylarsonium chloride	$(C_6H_5)_4As^+Cl^-$	**ClO_4^-**, MnO_4^-, MoO_4^{2-}, **ReO_4^{2-}**, WO_4^{2-}

[a] Reagents are especially useful for determining the elements shown in boldface.
[b] Elements precipitated must be in specific oxidation states.

SPECIAL TECHNIQUES

Although gravimetry is among the oldest of quantitative analysis methods, the general techniques used have not diversified to a very great extent. The very large majority of gravimetric methods involve direct precipitation: that is, addition of the precipitating agent directly to the solution containing the analyte. In this section we briefly describe a few other approaches to obtaining a weighable precipitate.

Homogeneous Precipitation

It was pointed out earlier that an important goal of every gravimetric determination is to produce a precipitate with large, easily filtered particles. The surest way to accomplish this is to maintain a very low supersaturation ratio during the precipitation process, thereby favoring crystal growth over nucleation. When the precipitating reagent is added to a sample, its concentration at the interface of the two mixing solutions is large, causing the supersaturation ratio to be large. Such a condition favors nucleation over growth. Using dilute solutions, slow addition, and rapid mixing only partly overcomes this condition. A much more efficient way to avoid these local regions of high supersaturation is to slowly generate the precipitating reagent directly in the solution by means of a chemical reaction—a process called *homogeneous precipitation*. The slow, homogeneous appearance of the precipitating reagent results in less nucleation and more growth, leading ultimately to a purer and more easily filtered precipitate.

The slow hydrolysis of potassium pyrosulfate has been used to generate sulfate ion homogeneously for the precipitation of barium.

$$S_2O_7{}^{2-} + 3H_2O \longrightarrow 2SO_4{}^{2-} + 2H_3O^+$$

A slight variation of this approach is to change the pH of the solution homogeneously, thereby causing an insoluble substance to form. For example, when oxalate is added to a slightly acidic solution of calcium ion, no precipitate occurs because the oxalate is present as $HC_2O_4{}^-$ or $H_2C_2O_4$ and Ca^{2+} cannot displace the hydrogens to form insoluble CaC_2O_4. Addition of urea to the solution and heating to 90°C will cause a gradual increase in pH due to hydrolysis:

$$\underset{\text{urea}}{CO(NH_2)_2} + H_2O \xrightarrow{90°C} CO_2(g) + 2NH_3$$

The ammonia will neutralize the $H_2C_2O_4$ and $HC_2O_4{}^-$, thereby freeing the oxalate anion to react with Ca^{2+}. This entire process takes place homogeneously throughout the solution. Table 4-3 lists several reagents that have proven useful in homogeneous precipitations.

Direct Volatilization

Volatile components in a sample often can be separated by boiling and collecting the vapor in a form suitable for weighing. Water and carbon dioxide are the two substances most commonly determined in this manner. Water is quantitatively volatilized from many inorganic samples simply by heating to an appropriate temperature. The water vapor in a closed system is passed through a weighed portion of desiccant, often magnesium perchlorate, which is then reweighed. The gain in weight is the weight of water from the sample. Inorganic carbonates, when treated with a strong nonvolatile acid such as sulfuric acid and then heated, will evolve carbon dioxide. The gas is adsorbed on Ascarite II, which consists of sodium hydroxide dispersed on a nonfibrous silicate. It retains the carbon dioxide by the reaction

$$2NaOH(s) + CO_2(g) \longrightarrow Na_2CO_3(s) + H_2O(g)$$

TABLE 4-3 SELECTED REAGENTS USED IN HOMOGENEOUS PRECIPITATIONS

Reagent	Precipitating ion generated	Elements precipitated[a]
8-Acetoxyquinoline	O^- (8-hydroxyquinolate)	Al, Bi, Cd, Cu, Fe, Mg, Pb, Ti, U, Zr
Diethyloxalate	$C_2O_4^{2-}$	Mg, Ca, Zn
Dimethylsulfate, sulfamic acid	SO_4^{2-}	Ba, Pb, Sr, W
Thioacetamide	S^{2-}	Bi, Cu, Cd, Mo, Sn, Zn
Trimethylphosphate	PO_4^{3-}	Al, Hf, Zr
Urea	OH^-	Al, Be, Fe

[a] Elements precipitated must be in specific oxidation states.

To avoid the loss of gaseous water produced in the reaction, which would decrease the weight gain, the Ascarite II is combined with a desiccant.

Perhaps the most troublesome aspect of the direct volatilization technique is the necessity of carrying out the steps in a closed apparatus to avoid contamination from the atmosphere. With some types of samples this inconvenience can be avoided by using an indirect volatilization technique.

Indirect Volatilization

The indirect procedure is based on the weight loss of a sample resulting from heating. Its significant advantage lies in the simplicity of the procedure: One needs merely to weigh, heat, cool, and reweigh the sample. For the procedure to work, a single, volatile component must be lost completely during the heating step. This severe restriction limits the application of the technique almost exclusively to the determination of water in certain types of samples.

APPLICATIONS

Over the years, the importance of gravimetry in quantitative analysis has declined, not so much because of changes in the things we desire analyzed but because of the development of other methods that are superior in some particular way. One must not conclude, however, that gravimetry has no place in modern analysis. In this section we describe the present scope of gravimetric analysis along with several common applications.

Scope

The scope of applications of any method depends on how well the method can meet certain goals. Let us look at the capabilities of gravimetric analysis in six areas: sensitivity, accuracy, selectivity, ease of operation, time, and cost.

Sensitivity

The sensitivity, or ability to distinguish between two nearly identical amounts or concentrations, of most methods is limited by the measurement device, but not for gravimetry where the measurement device is an analytical balance. The balance is capable of accurately weighing less material than can be precipitated, filtered, and rinsed without significant loss. Consequently, the sensitivity is limited by some aspect of the latter three operations, such as solubility losses occurring during precipitation and rinsing, the inability of the filter to retain the smallest precipitate particles, or our inability to transfer every single precipitate particle from the precipitation beaker onto the filter device. Increasing the sample size will produce a larger amount of precipitate, which, in theory, should decrease the relative error due to precipitate loss, but usually it also means a larger volume, which, in turn, further complicates the precipitate handling operations. Gravimetry is best suited to samples having a concentration of analyte greater than about 1%, but may in certain instances be suitable for samples with concentrations as low as 0.1%.

Accuracy

The *inherent* accuracy of gravimetric methods is among the best of all quantitative methods, with relative errors as low as 0.1%. Unfortunately, this degree of accuracy is attainable only with fairly simple samples that have been handled very carefully by experienced analysts.

Selectivity

Gravimetric methods are limited mainly to inorganic substances and generally are not very selective. As a result, many samples require extensive pretreatment to remove potential interferences or to change them to a noninterfering form. When separation is necessary, the accuracy of the method always suffers because the extra sample manipulation invariably results in the loss of some analyte.

Ease of operations

Gravimetric methods require considerable manipulative skills that are obtained only through practice. Furthermore, the operations tend to be time consuming, tedious, and impossible to automate. Without question, gravimetric analyses are among the most difficult of all quantitative methods to perform.

Time

A typical gravimetric determination may require from 3 to 6 hours to complete, although the analyst often has time for other work during digestion, ignition, and cooling of the precipitate. The lengthy analysis time and inability to automate the procedures makes gravimetric methods undesirable when results must be obtained quickly or when large numbers of samples must be analyzed.

Cost

The reagent and equipment expense for most gravimetric determinations is small, but the personnel cost is high because of the long analysis times and required technical skills.

Selected Examples

Up to this point various topics important in gravimetric analysis have been discussed somewhat independently. It is instructive to see how the type of sample and the specific properties of a precipitate influence the details of a gravimetric procedure. In this section we describe gravimetric methods for several common substances.

Determination of sulfur

Sulfur is determined in the form of sulfate by precipitating it with barium chloride:

$$Ba^{2+} + SO_4^{2-} \longrightarrow BaSO_4(s)$$

Samples containing other forms of sulfur are first treated with a strong oxidizing agent to convert the sulfur to sulfate. Barium sulfate is precipitated from acidic solution, even though its solubility can be as much as 20 times larger than in neutral solution, to avoid the simultaneous precipitation of other barium salts. Anions forming insoluble barium salts in neutral solution include AsO_3^{3-}, AsO_4^{3-}, CO_3^{2-}, $C_2O_4^{2-}$, F^-, and PO_4^{3-}. Unlike sulfate, these are anions of weak acids and form strong covalent bonds with hydrogen ions in acidic solution, which reduces their ability to precipitate barium ion. Unless they are present in only trace amounts, PO_4^{3-} and F^- will precipitate with barium even in dilute acid solution and must therefore be removed from the sample.

Coprecipitation is a serious problem with barium sulfate precipitates, even in acidic solution. The coprecipitation of various anions decreases in the order: $BrO_3^- >$ $Fe(CN)_6^{4-} > NO_3^- > MnO_4^- > Fe(CN)_6^{3-} > Cl^- > Br^- > I^- > SCN^- >$ $CH_3CO_2^-$ (acetate). Although modest amounts of chloride can be tolerated, common ions such as nitrate and chlorate must be removed, usually by fuming (almost boiling) the sample with concentrated hydrochloric acid.

$$NO_3^- + 4H^+ + 3Cl^- \xrightarrow{\Delta} Cl_2(g) + NOCl(g) + 2H_2O$$

$$ClO_3^- + 6H^+ + 5Cl^- \xrightarrow{\Delta} 3Cl_2(g) + 3H_2O$$

Coprecipitation of cations decreases in the order $Cd^{2+} > Mn^{2+} > Cu^{2+} > Zn^{2+} >$ $Al^{3+} > Fe^{2+} > K^+ > Ni^{2+} > Na^+ > Li^+ > Mg^{2+}$. Both Fe^{3+} and H^+ are also coprecipitated, Fe^{3+} quite strongly, but their position in the series is not certain. The interference of Fe^{3+} is especially troublesome in many samples and usually is minimized by reducing it with hydroxylamine to Fe^{2+}, which can be tolerated in larger amounts, or by adding tartrate or citrate ions which react with Fe^{3+} to give a product that does not coprecipitate.

Barium sulfate is a finely divided, crystalline precipitate that is usually filtered through fine, ashless filter paper. Sintered porcelain filter crucibles give slightly more accurate results but are only seldom used, perhaps because they are fairly expensive.

The precipitate shows no tendency to peptize, so it is washed with pure water. The volume of the rinse water is kept small because $BaSO_4$ is only moderately insoluble.

The filter paper containing the precipitate is placed in a regular porcelain crucible and burned off at a low temperature in the presence of fresh air—a process called *ashing*. If ashing is done in an oxygen-deficient atmosphere, carbon from the paper can reduce the barium sulfate to barium sulfide:

$$BaSO_4(s) + 4C(s) \xrightarrow{\Delta} BaS(s) + 4CO(g)$$

The formation of a small amount of BaS can be ignored because it is slowly oxidized during continued heating after the carbon is burned off:

$$BaS(s) + 2O_2(g) \xrightarrow{\Delta} BaSO_4(s)$$

When large amounts of BaS are formed, this oxidation may be incomplete and lead to low results. In such cases, the sulfide ion can be displaced by adding a drop of concentrated sulfuric acid before continued heating:

$$BaS(s) + H_2SO_4 \xrightarrow{\Delta} BaSO_4(s) + H_2S(g)$$

The excess H_2SO_4 is volatilized as $SO_3(g)$ and $H_2O(g)$. When ashing is complete, the precipitate is ignited to about $800°C$ to remove traces of water that will not volatilize at lower temperatures.

Anions that are coprecipitated as barium salts are almost always oxidized to BaO during ignition. This makes the precipitate heavier than it should be, producing a positive error. Cations that are coprecipitated as sulfate salts usually cause negative errors because the coprecipitated cation is usually lighter than the barium ion it replaces in the crystal lattice.

Example 4-3

An 800.0-mg sample containing sulfate was treated with a slight excess of barium chloride, yielding a precipitate that contained 4.3 mg of coprecipitated $BaCO_3$. After ignition and cooling, the precipitate weighed 377.0 mg. Calculate (a) the apparent % SO_4, (b) the true % SO_4, and (c) the absolute error in % SO_4.

Solution

(a) The apparent % SO_4 is calculated from the equation

$$\% \ SO_4 = \frac{\text{wt } SO_4 \ (\text{apparent})}{\text{wt sample}} \times 100$$

The weight of the sample is known, but the weight of SO_4 must be calculated from the weight of the $BaSO_4$ precipitated. In calculating the *apparent* % SO_4, the precipitate is assumed to be *pure* $BaSO_4$ and

$$\text{wt } SO_4 = \text{wt } BaSO_4 \times \frac{\text{MW } SO_4}{\text{MW } BaSO_4} \times \frac{1 \text{ mmol } SO_4}{1 \text{ mmol } BaSO_4}$$

$$= 377.0 \text{ mg} \times \frac{96.06 \text{ mg } SO_4/\text{mmol } SO_4}{233.4 \text{ mg } BaSO_4/\text{mmol } BaSO_4} \times \frac{1 \text{ mmol } SO_4}{1 \text{ mmol } BaSO_4}$$

$$= 155.2 \text{ mg}$$

Then

$$\% \ SO_4 = \frac{155.2 \text{ mg } SO_4}{800.0 \text{ mg sample}} \times 100 = 19.40$$

(b) The coprecipitated $BaCO_3$ is converted to BaO when the precipitate is ignited. Thus the weight of the precipitate is the sum of two weights.

weight precipitate = weight $BaSO_4$ + weight BaO
(known)   (want to find)   (need to know)

The weight of BaO can be found from the weight of the $BaCO_3$ using the gravimetric factor:

$$\text{wt } BaO = \text{wt } BaCO_3 \times \frac{MW \ BaO}{MW \ BaCO_3} \times \frac{1 \text{ mmol } BaO}{1 \text{ mmol } BaCO_3}$$

$$= 4.3 \text{ mg} \times \frac{153.3 \text{ mg } BaO/\text{mmol } BaO}{197.4 \text{ mg } BaCO_3/\text{mmol } BaCO_3} \times \frac{1 \text{ mmol } BaO}{1 \text{ mmol } BaCO_3}$$

$$= 3.3 \text{ mg}$$

Substituting the known values in the equation above gives us

$$377.0 \text{ mg precipitate} = \text{wt } BaSO_4 + 3.3 \text{ mg } BaO$$

or

$$\text{wt } BaSO_4 = 377.0 \text{ mg} - 3.3 \text{ mg} = 373.7 \text{ mg}$$

As in part (1) the weight of SO_4 is found from the weight of $BaSO_4$:

$$\text{wt } SO_4 = 373.7 \text{ mg } BaSO_4 \times \frac{96.06 \text{ mg } SO_4/\text{mmol } SO_4}{233.4 \text{ mg } BaSO_4/\text{mmol } BaSO_4} \times \frac{1 \text{ mmol } SO_4}{1 \text{ mmol } BaSO_4}$$

$$= 153.8 \text{ mg}$$

Finally,

$$\% \ SO_4(\text{true}) = \frac{153.8 \text{ mg } SO_4}{800.0 \text{ mg sample}} \times 100 = 19.23$$

(c) The absolute error is the difference between the actual value and the true value:

$$\text{error} = 19.40\% - 19.23\% = 0.17\%$$

Example 4-4

Barium sulfate precipitated from a 1.300-g sample was contaminated with 9.4 mg of $Fe_2(SO_4)_3$ and weighed a total of 314.0 mg. Calculate (a) the apparent $\%$ $BaSO_4$ and (b) the true $\%$ $BaSO_4$.

Solution

(a) The apparent $\%$ $BaSO_4$ is calculated from the weights given:

$$\% \ BaSO_4 = \frac{\text{wt } BaSO_4 \text{ apparent}}{\text{wt sample}} \times 100$$

$$= \frac{314.0 \text{ mg } BaSO_4}{1300 \text{ mg sample}} \times 100 = 24.15\%$$

(b) If no contamination had occurred, the sulfate in $Fe_2(SO_4)_3$ would have been present as $BaSO_4$. Since the amount of coprecipitated $Fe_2(SO_4)_3$ is known, it can be converted to an equivalent amount of $BaSO_4$* using a gravimetric factor:

$$\text{wt } BaSO_4^* = \text{wt } Fe_2(SO_4)_3 \times \frac{\text{MW } BaSO_4}{\text{MW } Fe_2(SO_4)_3} \times \frac{3 \text{ mmol } BaSO_4}{1 \text{ mmol } Fe_2(SO_4)_3}$$

$$= 9.4 \text{ mg } Fe_2(SO_4)_3 \times \frac{233.4 \text{ mg } BaSO_4/\text{mmol } BaSO_4}{399.9 \text{ mg } Fe_2(SO_4)_3/\text{mmol } Fe_2(SO_4)_3}$$

$$\times \frac{3 \text{ mmols } BaSO_4}{1 \text{ mmol } Fe_2(SO_4)_3}$$

$$= 16 \text{ mg}$$

If the precipitate had been pure, the 9.4 mg of $Fe_2(SO_4)_3$ would have been present as 16 mg of $BaSO_4$. Thus

$$\text{weight } BaSO_4(\text{true}) = 314.0 \text{ mg} - 9.4 \text{ mg} + 16 \text{ mg} = 321 \text{ mg}$$

and

$$\% \ BaSO_4 = \frac{321 \text{ mg } BaSO_4}{1300 \text{ mg sample}} \times 100 = 24.7\%$$

Determination of chloride

Chloride is precipitated from solution as silver chloride by the addition of silver nitrate.

$$Ag^+ + Cl^- \longrightarrow AgCl(s)$$

The precipitation is performed in moderately acidic solution ($0.1 - 0.3 \ M \ HNO_3$) to avoid the simultaneous precipitation of insoluble silver salts of OH^-, CN^-, CO_3^{2-}, PO_4^{3-} and AsO_4^{3-}. Strongly acidic solutions should be avoided because they increase the solubility of silver chloride. Unfortunately, numerous anions form insoluble compounds with silver ion even in acidic solution, including Br^-, I^-, SCN^-, S^{2-}, $S_2O_3^{2-}$, $Fe(CN)_6^{4-}$. These ions cannot be tolerated in the sample and must be removed or destroyed prior to precipitation. Cations that interfere the most seriously are those forming soluble complex ions with chloride, thereby competing with Ag^+ for the chloride in solution. Such cations include Cr^{3+}, Cd^{2+}, Sn^{2+}, Pt^{4+}, and Hg^{2+}.

Silver chloride is a colloidal-type precipitate whose major source of contamination is via surface adsorption. A large excess of the silver nitrate precipitating reagent must be avoided to prevent the formation of an electrical double layer that will make the colloid difficult to coagulate. Most procedures call for addition of a 10% excess of silver nitrate. If the amount of chloride in the unknown sample can be roughly estimated, the amount of silver nitrate needed can be calculated. Alternatively, a quick preliminary determination may be performed to establish the amount needed. Controlling the amount of excess silver nitrate and boiling the solution after precipitation is sufficient to coagulate the colloid. Since postprecipitation is not a problem, freshly precipitated silver chloride solutions may be digested overnight to improve the particle size and decrease the amount of adsorbed impurities.

Example 4-5

The chloride in a sample containing approximately 3.5% NaCl is to be determined gravimetrically as AgCl. What weight of sample must be taken to obtain a precipitate that will weigh about 0.10 g?

Solution Using the gravimetric factor, we can calculate the weight of NaCl necessary to produce the desired weight of AgCl.

$$\text{wt NaCl} = \text{wt AgCl} \times \frac{\text{MW NaCl}}{\text{MW AgCl}} \times \frac{1 \text{ mol NaCl}}{1 \text{ mol AgCl}}$$

$$= 0.10 \text{ g AgCl} \times \frac{58.44 \text{ g NaCl/mol NaCl}}{143.3 \text{ g AgCl/mol AgCl}} \times \frac{1 \text{ mol NaCl}}{1 \text{ mol AgCl}}$$

$$= 0.041 \text{ g}$$

Substituting this weight in the equation for % NaCl gives us the desired answer:

$$\% \text{ NaCl} = \frac{\text{wt NaCl}}{\text{wt sample}} \times 100$$

$$3.5 = \frac{0.041 \text{ g NaCl}}{\text{wt sample}} \times 100$$

or

$$\text{wt sample} = 1.2 \text{ g}$$

Because they can be dried at low temperatures and are easily dissolved for cleanup after the determination is completed, silver chloride precipitates are usually filtered through sintered glass crucibles. The precipitate is rinsed sparingly with dilute nitric acid (0.01 M) to minimize peptization and replace adsorbed, nonvolatile electrolytes, such as $AgNO_3$, KNO_3, and $NaNO_3$, with HNO_3 that is volatilized during the drying step.

Silver chloride precipitates are dried to constant weight at 110 to 120°C. A few hundredths of a percent of water is retained at this temperature and is volatilized only by heating to the melting point (460°C). Such drastic heating is not recommended, however, because silver chloride is somewhat volatile at this temperature and also is easily decomposed.

Once formed, silver chloride precipitates should be kept away from intense light to prevent photodecomposition:

$$2\text{AgCl(s)} \xrightarrow{\text{light}} 2\text{Ag(s)} + \text{Cl}_2\text{(g)} \tag{4-4}$$

The progress of photodecomposition can be readily seen by the darkening (purple) of the normally white precipitate that is due to the presence of finely divided silver metal. The direction of the error caused by photodecomposition depends on whether or not the silver chloride is in the presence of excess silver ion. When no excess silver ion is present, only Reaction 4-4 occurs and the net result is that silver chloride molecules are converted to an equal number of silver atoms. The precipitate gets lighter and the error is negative. When excess silver ion is present, the chlorine generated in the photodecomposition reaction reacts with water, forming Cl^-, which combines with the excess Ag^+ to produce more AgCl:

$$Cl_2 + H_2O \rightleftharpoons H^+ + Cl^- + HOCl \qquad (4\text{-}5)$$

$$Ag^+ + Cl^- \rightleftharpoons AgCl(s) \qquad (4\text{-}6)$$

Adding Reactions 4-4 through 4-6, we obtain

$$AgCl(s) + H_2O + Ag^+ \rightleftharpoons 2Ag(s) + H^+ + HOCl \qquad (4\text{-}7)$$

Since two silver atoms are formed for each silver chloride lost, the precipitate gets heavier and the error is positive.

Example 4-6

A sample of freshly precipitated and filtered silver chloride weighs 459.0 mg. If 1.00% of the silver chloride becomes photodecomposed, what will the precipitate weigh?

Solution The weight of AgCl lost due to photodecomposition is

$$459.0 \text{ mg AgCl} \times 0.0100 = 4.59 \text{ mg AgCl}$$

The weight of Ag(s) formed is calculated from the weight of AgCl lost by using the gravimetric factor:

$$\text{wt Ag formed} = \text{wt AgCl lost} \times \frac{\text{AW Ag}}{\text{MW AgCl}} \times \frac{1 \text{ mmol Ag}}{1 \text{ mmol AgCl}}$$

$$= 4.59 \text{ mg} \times \frac{107.9 \text{ mg Ag/mmol Ag}}{143.3 \text{ mg AgCl/mmol AgCl}} \times \frac{1 \text{ mmol Ag}}{1 \text{ mmol AgCl}}$$

$$= 3.46 \text{ mg}$$

Finally, the new weight of the precipitate is given by

$$\text{new wt ppt} = \text{old wt ppt} - \text{wt AgCl lost} + \text{wt Ag formed}$$

$$= 459.0 \text{ mg} - 4.59 \text{ mg} + 3.46 \text{ mg} = 457.9 \text{ mg}$$

Example 4-7

If the silver chloride precipitate in Example 4-6 had not been filtered prior to the photodecomposition, what would the precipitate weigh?

Solution In this case the net result of the photodecomposition is given by Equation 4-7, which says that 2 mmol of silver is formed for each mmol of AgCl lost. Thus

$$\text{wt Ag formed} = \text{wt AgCl lost} \times \frac{\text{AW Ag}}{\text{MW AgCl}} \times \frac{2 \text{ mmol Ag}}{1 \text{ mmol AgCl}}$$

$$= 4.59 \text{ mg} \times \frac{107.9 \text{ mg Ag/mmol Ag}}{143.3 \text{ mg AgCl/mmol AgCl}} \times \frac{2 \text{ mmol Ag}}{1 \text{ mmol AgCl}}$$

$$= 6.91 \text{ mg}$$

and the new weight of precipitate is

$$\text{new wt ppt} = 459.0 \text{ mg} - 4.59 \text{ mg} + 6.91 \text{ mg} = 461.3 \text{ mg}$$

Determination of nickel

Nickel ion is precipitated with a very selective organic precipitating reagent called dimethylglyoxime (DMGH).

$$Ni^{2+} + 2\;\begin{array}{l} CH_3-C=N-OH \\ | \\ CH_3-C=N-OH \end{array} \longrightarrow 2H^+ +$$

or

$$Ni^{2+} + 2DMGH \longrightarrow 2H^+ + Ni(DMG)_2(s)$$

The dotted lines represent hydrogen bonds. Only Ni^{2+}, Pd^{2+}, and Bi^{3+} form insoluble compounds with DMGH. Palladium precipitates from dilute HCl solution only, nickel from solution above pH 5, and bismuth from solution above pH 11. Consequently, the method is selective for nickel if the pH is between 5 and 11. Nickel is usually precipitated by adding the DMGH reagent to an acidified solution of Ni^{2+}, which is then made basic with ammonia. The addition of ammonia serves two purposes: It makes the solution basic, thereby lowering the solubility of $Ni(DMG)_2$, and it forms soluble complex ions with several transition metal ions, thereby preventing their simultaneous precipitation as metal hydroxides.

Neither coprecipitation nor postprecipitation are serious problems and the precipitate is usually digested at 60°C for an hour or two before it is filtered through a sintered glass crucible. The precipitate does not peptize and can be washed with pure water. Some care is required during washing because the precipitate has an unfortunate tendency to creep up the side of the crucible. The precipitate loses its water above 80°C but should not be heated above 175°C, where it begins to decompose.

Determination of carbon and hydrogen

Determining the carbon and hydrogen content of pure organic compounds is very common because the results can be used to establish the empirical ratio of atoms in the compound. The determinations are based on a direct volatilization procedure and require only 5 to 10 mg of sample. The sample is heated in a stream of oxygen in the presence of catalysts, causing it to decompose into CO_2 and H_2O. Typically,

$$C_{10}H_8 + 12O_2 \longrightarrow 10CO_2 + 4H_2O$$

The combustion usually is performed in a long tube like that shown in Figure 4-4. The copper oxide–lead chromate mixture is a catalyst for the combustion reaction, the metallic silver removes any gaseous sulfur and halogen compounds formed during combustion, and the lead dioxide removes any nitrogen oxides.

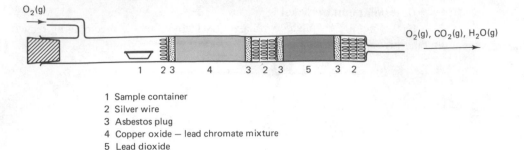

1 Sample container
2 Silver wire
3 Asbestos plug
4 Copper oxide – lead chromate mixture
5 Lead dioxide

Figure 4-4 Pregl-type combustion tube for a carbon–hydrogen analyzer.

Water and carbon dioxide are carried out of the tube by the oxygen and into a carefully weighed tube of desiccant (usually magnesium perchlorate). The desiccant absorbs the water but passes the carbon dioxide onto a second weighed tube containing Ascarite, where it is absorbed. The tubes are sealed, removed from the apparatus, and weighed. The weight gains of the tubes, which are typically only a few milligrams, are the weights of H_2O and CO_2 evolved from the sample. Because such small weight differences are involved, special microbalances must be used in weighing to attain the desired accuracy. On paper, the procedure appears quite simple, but in practice it requires great manipulative skills and some chemists like to refer to it as an art rather than a science. For example, some procedures call for the removal of dust and fingerprints from the glass tubes containing the desiccant and Ascarite by wiping with a particular type of cloth a fixed number of times always in the same direction. The necessity of observing such details has prevented automation of the procedure.

Gas chromatography (Chapter 21) can be used to determine both CO_2 and H_2O with the significant advantages of being much faster and easily automated. Unfortunately, gas chromatography cannot match the precision or the accuracy of the gravimetric method.

PROBLEMS

4-1. Write the gravimetric factors (using chemical formulas) for each of the following.

Sought	Weighed	Sought	Weighed
(a) Al	Al_2O_3	**(e)** $Na_2B_4O_7 \cdot 10H_2O$	B_2O_3
(b) Cl	AgCl	**(f)** P_2O_5	$Mg_2P_2O_7$
(c) Zn	$Zn_2P_2O_7$	**(g)** $Pb(C_2H_5)_4$	$PbCrO_4$
(d) MoS_3	$BaSO_4$	**(h)** SO_3	$BaSO_4$

4-2. The formation of insoluble $MgNH_4PO_4 \cdot 6H_2O$, which is ignited to $Mg_2P_2O_7$ for weighing, can be used for the determination of Mg, N, or P. Write the appropriate gravimetric factor that would apply to each determination.

4-3. What weight of silver chloride precipitate can be produced from an 800-mg sample containing 14.5% $CaCl_2$?

4-4. Calculate the weight of $CaCO_3$ produced from heating 400 mg of $CaC_2O_4 \cdot H_2O$ at 600°C.

4-5. What will 628 mg of $Zr(HPO_4)_2$ weigh after ignition to ZrP_2O_7?

4-6. Calculate the percentage of chlorine in a swimming pool chlorinating agent that is 99.4% trichloro-s-triazine, $C_3Cl_3N_3$.

4-7. A fertilizer advertised as 8-8-8 generally means that it contains 8% total nitrogen, 8% P_2O_5, and 8% K_2O. Calculate the percentage of P and of K in the fertilizer.

4-8. Calculate the percentage of iron in each of the following.
    (a) FeO   **(b)** $FeSO_4$ **(c)** Fe_3O_4

4-9. A 1.0000-g sample of ultrapure silicon was treated with an oxidizing agent and the resulting SiO_2, after proper collection and drying, weighed 2.1387 g. Calculate the % Si in the sample.

4-10. A 350-mg sample containing $KClO_3$ was carefully reduced and treated with excess $AgNO_3$. The resulting AgCl weighed 185 mg. Calculate the percentage of $KClO_3$ in the sample.

4-11. A sample of brass weighing 712 mg was dissolved and treated in such a way as to successively precipitate SnS_2, CuS, $ZnNH_4PO_4 \cdot H_2O$, and $PbSO_4$. Each precipitate was collected and ignited, producing the following weights: 83.2 mg of SnO_2, 700.5 mg of CuO, 50.6 mg of $Zn_2P_2O_7$, and 36.3 mg of $PbSO_4$. Calculate the percentage of each metal in the brass. Does this account for all of the sample?

4-12. A commercial algaecide containing an organocopper compound was treated with concentrated nitric acid and evaporated to dryness. After dissolution of the residue, the copper was precipitated with α-benzoin oxime. If the sample weighed 15.443 g and the precipitate of $Cu(C_{14}H_{12}NO_2)_2$ weighed 0.6314 g, calculate the % Cu in the algaecide.

4-13. The calcium from a sample of limestone weighing 607.4 mg was precipitated as calcium oxalate and ignited to calcium carbonate weighing 246.7 mg.
    (a) Calculate the % Ca in the sample.
    (b) If the precipitate had been ignited at a higher temperature giving CaO, what would be its weight?
    (c) If a constant error of 0.1 mg existed each time the balance was read, calculate the relative weighing error (in parts per thousand) in determining the weight of $CaCO_3$. Remember, two balance readings are required to weigh an object.
    (d) Repeat part (c) if the precipitate was ignited to CaO instead of $CaCO_3$.

4-14. A sample of hydrated barium chloride weighed 428.3 mg before and 356.2 mg after heating to constant weight. What is the average number of waters of hydration per molecule of $BaCl_2$?

4-15. A sample of inorganic desiccant weighing 6.187 g was placed in an oven at 160°C and left overnight. The sample was removed the next morning and after cooling to room temperature in a moisture-free environment was found to weigh 6.133 g. What is the % H_2O in the sample?

4-16. A 2.446-g sample thought to contain only one volatile substance, $MgCO_3$, was heated in a nitrogen atmosphere for 30 min. After cooling, the residue weighed 2.216 g. Calculate the percentage of magnesium carbonate in the sample.

4-17. Manganese dioxide in acidic solution can oxidize chloride ion to chlorine:

$$MnO_2(s) + 2Cl^- + 4H^+ \longrightarrow Mn^{2+} + Cl_2(g) + 2H_2O$$

Calculate the weight of MnO_2 consumed in the formation of 100 mL of chlorine gas at STP.

4-18. A sample weighs 850.4 mg and contains 3.52% S. Calculate the absolute (% S) and percent relative error for a gravimetric sulfate determination if 4.70 mg of the precipitated $BaSO_4$ is converted to BaS during ignition.

4-19. The sulfate in a 674.3-mg sample was determined by a gravimetric method. If the precipitate consisted of 214.4 mg of $BaSO_4$ and 9.7 mg of $Ba_3(PO_4)_2$ before ignition, calculate the following.

(a) apparent % S (c) relative error in parts per thousand

(b) true % S

4-20. The precipitate in a gravimetric method for sulfate weighed 622.7 mg of which 13.6 mg was a contaminant, $Fe_2(SO_4)_3$.

(a) What would the precipitate weigh if *all* of the sulfate was present as $BaSO_4$?

(b) What will the precipitate weigh after ignition at 800°C?

(c) What is the relative error (in percent) in the ignited precipitate due to the fact that some of the sulfate precipitated as $Fe_2(SO_4)_3$ rather than $BaSO_4$?

4-21. In a gravimetric determination of sulfur as $BaSO_4$, 0.8863 g of the ignited precipitate is found to contain 7.71 mg of coprecipitated Na_2SO_4. Calculate the percentage error of the determination resulting from the presence of the Na_2SO_4.

4-22. A sample of freshly precipitated silver chloride weighs 704.7 mg. Before the solution can be filtered and washed, 10.0 mg of AgCl is photodecomposed. After the precipitate is filtered and washed, another 10.0 mg of AgCl is photodecomposed. Calculate the weight of the precipitate.

4-23. What weight of sample containing 8.00% Fe_3O_4 must be taken to obtain a precipitate of $Fe(OH)_3$ that, when ignited to Fe_2O_3, weighs 150 mg?

4-24. What volume of 0.250 *M* NaOH is needed to precipitate the iron in Problem 4-23?

4-25. What weight of $AgNO_3$ is required to precipitate the chloride in 750 mg of 14.0% $BaCl_2$?

4-26. A sample weighing 640 mg is comprised of 12.0% KCl, 8.71% $CaCl_2$, and inert material. What weight of AgCl can be precipitated from this sample?

4-27. What volume of 0.214 *M* $(NH_4)_2HPO_4$ is necessary to precipitate the calcium as $Ca_3(PO_4)_2$ from 838 mg of a sample that contains 9.74% Ca?

4-28. A pure organic compound weighing 947.4 mg was heated in a stream of oxygen. The effluent gas was swept successively through a tube of magnesium perchlorate desiccant weighing 18.6137 g and a tube of Ascarite weighing 16.8425 g. After several minutes the tubes were sealed and reweighed: The tube containing the magnesium perchlorate weighed 18.6448 g and the tube containing the Ascarite weighed 16.9098 g.

(a) Calculate the % H and % C in the sample.

(b) If the compound consisted only of C and H, calculate its empirical formula.

4-29. An organic pesticide (MW 183.7) is 8.43% Cl. A 0.627-g sample containing the pesticide plus inert material containing no chloride was decomposed with metallic sodium in alcohol. The liberated chloride ion was precipitated as AgCl weighing 0.0831 g. Calculate the percentage of pesticide in the sample.

4-30. In the gravimetric determination of barium as $BaCrO_4$, what weight of sample must be taken so that the weight of the precipitate in milligrams multiplied by 100 will equal the % Ba in the sample?

4-31. A sample containing $BaCl_2 \cdot 2H_2O$, KCl, and inert material weighed 0.8417 g. After heating the sample at 160°C for 45 min, it weighed 0.8076 g. The sample was then dissolved in water and treated with a slight excess of $AgNO_3$. The resulting precipitate was collected and found to weigh 0.5847 g. Calculate the % $BaCl_2 \cdot 2H_2O$ and % KCl in the sample.

4-32. The nitrogen content of a 3.6342-g sample of undried spinach leaves was determined to be 4.63%. If the moisture content of the leaves was 8.68%, what is the percentage of nitrogen in an oven-dried sample of the leaves?

4-33. An antacid tablet weighing 3.084 g was dissolved and diluted to 100.0 mL in a volumetric flask. A 25.00-mL aliquot was removed and treated with sufficient 8-hydroxyquinoline to precipitate all of the aluminum and magnesium. After proper collection and drying, this precipitate weighed 1.7748 g. Another 25.00-mL aliquot of the sample was treated with enough ammonia to precipitate just $Al(OH)_3$. After ignition at $800°C$, this precipitate weighed 0.1167 g. Calculate the % Al and % Mg in the antacid.

4-34. In the gravimetric method for cobalt using 1-nitroso-2-naphthol, about 0.8 mg of $Co(C_{10}H_6O_2N)_3$ is lost due to solubility and handling of the precipitate. If this method is to be used to determine cobalt at the 7% level in samples, what is the minimum sample weight that must be taken to ensure that the relative error due to these losses will not exceed 0.30%?

5

TITRIMETRY

A quantitative determination performed by measuring the amount of some substance required to react with the analyte is called a *titrimetric determination*. The substance reacting with the analyte is called the *titrant* or *standard solution* and its amount is usually determined by measuring the volume of a solution of known concentration needed to react completely with the analyte. Such a procedure is referred to as a *volumetric titration*. If the stoichiometry of the reaction is known, the amount of analyte can be calculated from the amount of titrant used.

The titrimetric method has been applied to many different chemical systems using all four of the major types of reactions: precipitation, acid-base, complexation, and oxidation-reduction. Titrimetry is used much more often than gravimetry for routine determinations because it offers superior speed and convenience with little sacrifice in accuracy and precision.

BASIC REQUIREMENTS

All successful titrations are based on reactions that are stoichiometric, quantitative, fast, and for which there is a suitable means for estimating the equivalence point. The *equivalence point* of a titration refers to that point when just enough titrant has been added to react completely with the analyte. That is, an amount of titrant has been added *equivalent* to the amount of analyte present. The *end point* refers to the experimental estimation of the equivalence point. If the equivalence point is considered to be the true value, the end point is the experimental value, and any difference between them is called the *titration error*. There are numerous ways in which end points are determined, but their discussion must be postponed until certain important characteristics of titrations can be discussed.

A *stoichiometric reaction* is one in which a definite, reproducible relationship exists between the reacting substances. This relationship is the "bridge" used by the chemist to get from what is measured (the amount of titrant) to what is desired (the amount of analyte). Most chemical reactions are stoichiometric and therefore potentially useful for titrations. A quantitative reaction is one that goes to completion when an amount of titrant equivalent to the amount of analyte present is added. The terms "completion" and "quantitative" are arbitrary but usually taken to mean that the reaction is at least 99.9% complete. Less complete reactions are unsatisfactory because of difficulty in estimating the equivalence point with an acceptable degree of accuracy and precision. Slow reactions are not suitable for the same reason. Slow, like quantitative, is a relative term that has different meanings when used in different contexts. For the purposes of a titration, the reaction is considered slow if its rate is less than the normal rate of mixing of the sample and titrant.

TITRANTS

The titrant is chosen partly on the basis of its reaction with the analyte. As already pointed out, the titration reaction must be stoichiometric, quantitative, and fast. In addition, the titrant must be stable and of known concentration. The exact concentration of a titrant can be determined by dissolving a weighed portion of the pure reagent in a known volume of solvent or by titration with another solution of known concentration.

Concentration Determined Directly from Weight

Substances whose exact solution concentration can be determined by transferring a known weight of the reagent to a volumetric flask and diluting to the mark with solvent are called *primary standards*. The key word in the preceding sentence is *exact*. The inherent accuracy of the titrimetric method demands that the concentration of the titrant is known within a few parts per thousand of the true value. This places stringent requirements on substances that are to be called primary standards:

1. The material must be of very high purity, preferably 99.9% or better.
2. It should be stable in solid form and be able to withstand mild heating to remove adsorbed moisture or water of hydration. In solution, it should not react with normal atmospheric substances or the solvent.
3. It should have a high equivalent weight in order to minimize the relative error of the weighing process.

Relatively few substances can meet all of these criteria. As a result, chemists often must use an alternative method for determining the concentration of titrants.

Concentration Determined by Standardization

While the list of primary standards is quite short, there are many substances that meet all of the requirements except that of purity. Consequently, in many cases a titrant is prepared in an approximate concentration and then titrated against a primary standard to determine its exact concentration. This process, called *standardization*, makes possible the use of a variety of titrants, which, in turn, expands the range of titrimetric determinations.

DETERMINING THE END POINT

There are several ways to determine the end point of a titration, some of which will be discussed in later chapters. The objective here is not to learn the different ways but rather to examine the specific characteristics of a titration on which the end-point determinations are based. All such determinations are based directly or indirectly on changes that occur in a concentration of the reactants or products during the titration. The specific changes vary depending on the particular reaction and starting concentrations but the general behavior is the same for all titrations and is illustrated by the data in the first two columns of Table 5-1. These data were calculated for a hypothetical reaction

$$A \quad + \quad T \rightleftharpoons P$$

analyte   titrant   product

having an equilibrium constant of 1.00×10^{10}.

Using Data near the Equivalence Point

It is clear from the data in Table 5-1 that the relative change in the concentration of analyte is small in the beginning and final stages of the titration but very large in the

TABLE 5-1 CONCENTRATION CHANGES DURING A TITRATION[a]

Volume 0.1000 M T added (mL)	Concentration of A (mol/L)	pA
0	1.00×10^{-1}	1.00
5.00	6.67×10^{-2}	1.18
10.00	4.29×10^{-2}	1.37
24.00	2.04×10^{-3}	2.69
24.90	2.00×10^{-4}	3.70
24.99	2.00×10^{-5}	4.70
25.00	2.24×10^{-6}	5.65
25.01	2.50×10^{-7}	6.60
25.10	2.50×10^{-8}	7.60
26.00	2.50×10^{-10}	9.60
40.00	1.67×10^{-10}	9.78
45.00	1.25×10^{-10}	9.90
50.00	1.00×10^{-10}	10.00

[a]The initial sample consists of 25.00 mL of 0.1000 M analyte.

vicinity of the equivalence point. In fact, the maximum relative rate of change of concentration corresponds exactly to the equivalence point. Unfortunately, the concentrations extend over several orders of magnitude and when they are displayed in a linear plot as shown in Figure 5-1, the behavior in the vicinity of the equivalence point is greatly obscured.

To emphasize the *rate of change of concentration*, and, at the same time, compress the values to a manageable range, a log function of the concentration is plotted, $-\log A$ or pA. The third column in Table 5-1 shows the pA values that are plotted in Figure 5-2. This plot is much more descriptive of the changes that occur in the vicinity of the equivalence point. Had the concentrations of titrant been calculated and pT rather than pA values plotted, the curve would have the same shape but the change in values would be in the opposite direction, as shown by the dashed line in Figure 5-2. These types of log plots are called *titration curves*.

Graphic end points

It should be obvious that if a graph similar to that shown in Figure 5-2 can be constructed from experimental (measured) data, the equivalence point can be estimated by locating the point of maximum slope. One simple method, illustrated in Figure 5-3, involves three steps:

1. Fit a straight line to the nearly linear portions of the curve before and after the equivalence point.
2. Fit a straight line to the rising portion of the curve.
3. Locate the midpoint of line 2 between the lower and upper lines. This point is taken as the end point.

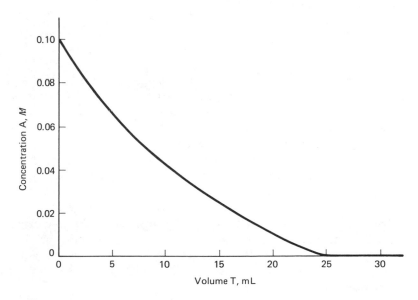

Figure 5-1 Changes in analyte concentration during the titration of 25.00 mL of 0.1000 *M* A with 0.1000 *M* T.

Determining the End Point

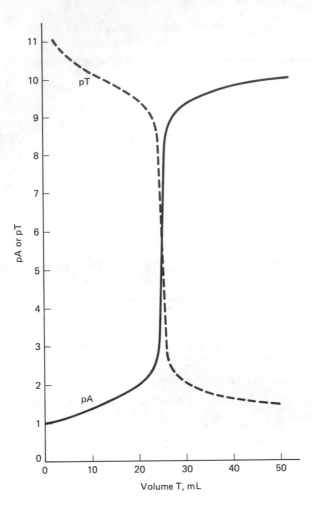

Figure 5-2 Titration curve for 25.00 mL of A with 0.1000 M T.

Any error in locating the midpoint along the pA or pT axis will produce an error in its location along the volume axis. The extent to which this occurs is determined by the steepness of the rising portion of the curve. The error on the pA axis is the same for the two titration curves shown in Figure 5-4, but the error on the volume axis is much smaller for the titration with the steeper end-point break.

Although the negative logarithm of the concentration has been plotted in the examples, the negative logarithm of any experimental quantity that is proportional to concentration can be used just as well. In later chapters you will learn that a measure of color intensity called *absorbance* is directly proportional to the concentration of the colored substance. As a result, the negative logarithm of the absorbance can be used in the construction of a titration curve. Other physical properties, such as electrode potential (discussed in Chapters 10 and 11), are related to the concentration in a logarithmic manner. In such cases, the measured property itself is plotted versus the volume of titrant to produce the titration curve.

There are two reasons why the graphic method is not the one used most often to

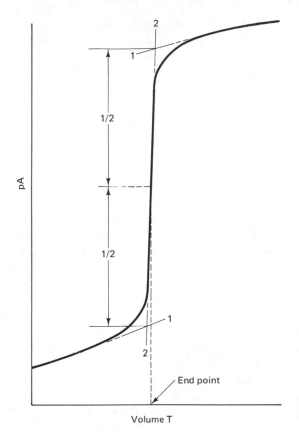

Figure 5-3 Determining the point of maximum slope.

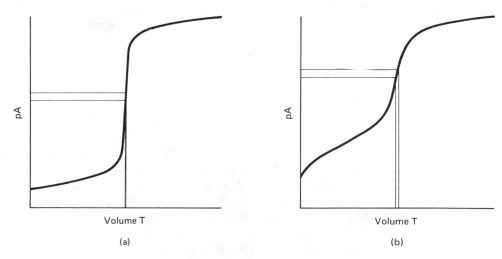

Figure 5-4 Effect of steepness on end-point error.

determine titration end points. First, most titrations are performed manually, and collecting the data and graphing the results are tedious and time-consuming operations. Second, we may not have a simple means of measuring the concentration or some property proportional to the concentration of analyte or titrant during a titration.

Visual end points

The end points of titrations are most often determined by noting some visual change in the solution, usually color. These changes are always based on the fact that pA or pT changes dramatically and rapidly in the vicinity of the equivalence point. There are three ways that color changes can occur near the equivalence point of a titration. Regardless of the mechanism by which it functions, the substance undergoing the color change is called an *indicator*.

Colored titrants or analytes. Since the color intensity of a solution is directly proportional to the concentration of colored substance, it should change during a titration in the same way that the concentration changes. Potassium permanganate is an intensely colored (purple) titrant commonly used in oxidation-reduction titrations. Prior to the equivalence point virtually all of the titrant added is consumed by the analyte and the sample solution remains nearly colorless. As the equivalence point is reached, or rather slightly passed, the concentration of permanganate in the solution increases dramatically and the solution becomes purple, signaling that the titration is finished. When the analyte rather than the titrant is the colored substance, a similar change in color is expected but in the opposite direction—from colored to colorless. This method of determining end points is limited to those few occasions when we have a suitable colored titrant or analyte and no other strongly colored substances present that might mask the color change.

Specific compound formers. When neither the titrant nor the analyte is colored, it may be possible to add a substance that will react with one of them to form a colored product. For example, the chromate ion serves as an indicator for the titration of chloride ion with silver nitrate titrant. Prior to the equivalence point, added titrant reacts with chloride, forming a white precipitate of silver chloride. When the chloride is "gone" (the equivalence point) the silver nitrate titrant reacts with the chromate ion forming orange-red silver chromate. The sudden appearance of the orange-red color signals the end point.

Certain restrictions apply to the selection of these indicators that severely limit the available choices. The specific properties of such indicators will be discussed later in the chapters on complexation, oxidation-reduction, and precipitation titrations.

Nonspecific, equilibrium-dependent compounds. Indicators of this type are characterized by two fundamental properties: (1) they exist in two equilibrium forms (i.e., protonated–unprotonated or oxidized–reduced), and (2) each form has a different color. The position of the equilibrium between the two forms of the indicator, and therefore its color, depends in some way on the concentration of analyte or titrant in the solution. The large relative and rapid change in analyte or titrant concentrations in the vicinity of the equivalence point causes a large shift in the position of the indicator equilibrium. When the concentrations of the two forms of the indicator change in order to coincide with the new equilibrium, the observed color also changes. In general, about a 100-fold change in concentration (two-fold change in pA or pT) is necessary to convert an indicator from predominately one form to the other.

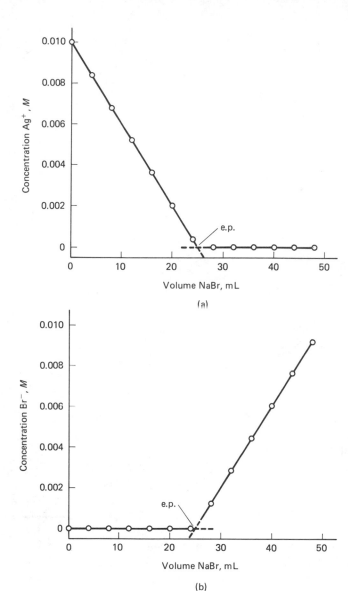

Figure 5-5 Linear titration curves for 25 mL of 0.010 M $AgNO_3$ titrated with 0.010 M NaBr. Concentration of Ag^+ and Br^- corrected for dilution.

Using Data Away from the Equivalence Point

During a titration, the concentration of analyte decreases for two reasons: analyte is consumed by reaction with the titrant, and the solution is diluted by the titrant. If dilution is eliminated or corrected for, the decrease in analyte concentration will be linear with the titrant added, *as long as the reaction goes to completion.* We can expect the reaction to go to completion if the equilibrium constant is very large, or in the case of a not-so-large value, there is a substantial excess of one reactant. Typically, a plot of analyte concentration (corrected for dilution) versus the volume of titrant added will look like the example shown in Figure 5-5a. The end point is found by extrapolating the straight-

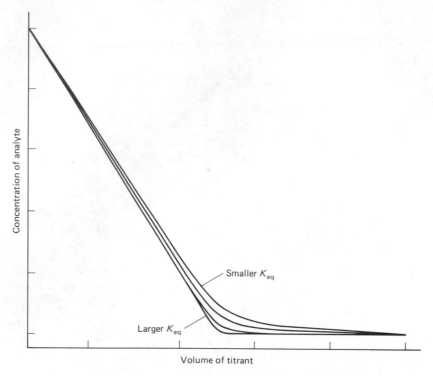

Figure 5-6 Effect of K_{eq} on shape of linear titration plot.

line portions of the plot until they intersect. Figure 5-5b is for the same titration when the titrant concentration is plotted.

In practice, actual concentrations are seldom plotted because they are either not known or must be calculated from other experimental data. Instead, some property *directly* proportional to concentration is measured and plotted. Although this will affect the numerical values on the ordinate of the graph, it will not alter the shape or the position of the end point. Color intensity and electrical conductivity are two physical properties sometimes used for constructing linear titration graphs.

Curvature near the equivalence point is the result of an incomplete reaction. In this region, there is no large excess of either analyte or titrant to force the reaction to completion. The smaller the equilibrium constant for the reaction, the greater the amount of curvature. Linear titration plots for a family of reactions differing only in their equilibrium constants (degree of completeness) are shown in Figure 5-6. Obviously, too small an equilibrium constant will result in a curvature so great that an accurate extrapolation is not possible.

The dilution resulting from the addition of titrant is a serious problem because it causes the analyte concentration (or any instrument signal proportional to concentration) to change in a nonlinear manner as shown in Figure 5-1. The measured signal can be corrected for dilution using the following equation:

$$S_{\text{corrected}} = S_{\text{measured}} \times \frac{V_{\text{original}} + V_{\text{added}}}{V_{\text{original}}}$$

Chapter 5 Titrimetry

where S stands for signal and V for volume. It is often possible to make the dilution effect negligibly small by using a titrant that is much more concentrated than the analyte, as illustrated in Example 5-1.

Example 5-1

Exactly 50 mL of a solution containing 0.0400 M Cl^- is being titrated with a 0.0500 M $AgNO_3$ solution. (a) Calculate the concentration of Cl^- when the titration is 70% complete, first neglecting and then including dilution. Assume that the reaction goes to completion. (b) Repeat the calculations for a 0.500 M $AgNO_3$ titrant.

Solution

(a) The reaction between Cl^- and Ag^+ is

$$Cl^- + Ag^+ \longrightarrow AgCl(s)$$

The amount of $AgNO_3$ added is 70% of the amount of Cl^- present. Thus

$$\text{amount } AgNO_3 \text{ added} = \text{amount } Cl^- \text{ present} \times 0.70$$

$$= (50.0 \text{ mL} \times 0.0400 \text{ } M) \times 0.70 = 1.40 \text{ mmol}$$

The amount of Cl^- left is found by difference:

$$\text{amount } Cl^- \text{ initial} = 50.0 \text{ mL} \times 0.0400 \text{ } M = 2.00 \text{ mmol}$$

$$- \left\{ \begin{array}{c} \text{amount } Ag^+ \text{added} \\ \| \\ \text{amount } Cl^- \text{reacted} \end{array} \right\} \qquad = 1.40 \text{ mmol}$$

$$\overline{\text{amount } Cl^- \text{ remaining}} \qquad \qquad = \overline{0.60 \text{ mmol}}$$

Neglecting dilution, we have

$$[Cl^-] = \frac{0.60 \text{ mmol}}{50.0 \text{ mL}} = 0.012 \text{ } M$$

To account for dilution, the *volume* of $AgNO_3$ added must be determined.

$$\text{amount } AgNO_3 \text{ added} = 1.40 \text{ mmol}$$

$$\text{vol} \times 0.0500 \text{ } M = 1.40 \text{ mmol}$$

$$\text{vol} = 28.0 \text{ mL}$$

Accounting for dilution gives us

$$[Cl^-] = \frac{0.60 \text{ mmol}}{50.0 \text{ mL} + 28.0 \text{ mL}} = 0.0077 \text{ } M$$

(b) Neglecting dilution yields

$$[Cl^-] = \frac{0.60 \text{ mmol}}{50.0 \text{ mL}} = 0.012 \text{ } M$$

The volume of $AgNO_3$ added is found in the same manner as before:

$$\text{amount } AgNO_3 \text{ added} = 1.40 \text{ mmol}$$

$$\text{vol} \times 0.500 \text{ } M = 1.40 \text{ mmol}$$

$$\text{vol} = 2.8 \text{ mL}$$

Determining the End Point

Accounting for dilution, we obtain

$$[Cl^-] = \frac{0.60 \text{ mmol}}{50.0 \text{ mL} + 2.8 \text{ mL}} = 0.011 \; M$$

Obviously, the more concentrated the titrant, the less volume is used and the smaller the dilution becomes. Our ability to measure accurately the volume of the titrant being added represents a practical limit on using more concentrated titrants. For example, 50 mL of sample solution being titrated with a titrant 100 times more concentrated than the analyte may require only 0.5 mL of titrant to reach the equivalence point, a volume too small to dispense and measure with the required accuracy and precision.

Relative Merits of Methods for Determining End Points

End-point methods relying on data near the equivalence point are potentially suitable only for reactions that "go to completion" readily (have a large equilibrium constant). There are some otherwise suitable reactions for which this is not true. Clearly, the great advantages of visual indicators lie in their simplicity and speed. The titrant can be added more or less continuously without the need for periodic volume measurements and the human eye serves as an instantaneous detector. There are chemical reactions, however, for which no suitable indicators are readily available. Furthermore, visual indicators cannot be used with highly colored samples or titrants, or when diverse substances are present that can destroy their indicating properties.

Graphic methods using data away from the equivalence point have the advantage of being applicable to chemical reactions with smaller equilibrium constants than are required for indicator methods, but they take considerably more time and require some sort of instrumental measurement device. All such methods require the collection of data at periodic intervals and subsequent preparation of a graph to determine the end point.

CALCULATING AMOUNT AND CONCENTRATIONS

There are three different ways in which titrations are used to determine the amount of an analyte; a direct titration method, a replacement or indirect titration method, and a back-titration method. In this section we describe the manner in which each method is carried out together with a description of the type of calculations that must be performed.

Direct Titration Method

In a direct titration method, the titrant reacts directly with the analyte and a simple relationship exists between the amount of titrant used and the amount of analyte present:

amount analyte = amount titrant × reacting ratio

where the amount of titrant used is expressed in units of moles or millimoles. In the titration of phosphoric acid with sodium hydroxide according to the reaction

$$\underset{\text{analyte}}{H_3PO_4} + \underset{\text{titrant}}{2NaOH} \longrightarrow Na_2HPO_4 + 2H_2O \tag{5-1}$$

2 molecules or moles of sodium hydroxide are needed to react with each molecule or mole of phosphoric acid, and

$$\text{amount } H_3PO_4 \text{ (mmol)} = \text{amount NaOH (mmol)} \times \frac{1H_3PO_4}{2NaOH}$$

Once the amount of phosphoric acid has been calculated it can be converted to a weight and a percentage as follows:

$$\text{wt } H_3PO_4 = \text{amount } H_3PO_4 \times \text{MW } H_3PO_4$$

$$\% \ H_3PO_4 = \frac{\text{wt } H_3PO_4}{\text{wt sample}} \times 100$$

Example 5-2

A 300.0-mg sample containing phosphoric acid and inert material was diluted with water and titrated with 0.05000 M NaOH according to Reaction 5-1. The end point was reached after 29.00 mL of titrant was added. Calculate the % H_3PO_4 in the sample.

Solution The percent of H_3PO_4 is given by

$$\% \ H_3PO_4 = \frac{\text{wt } H_3PO_4}{\text{wt sample}} \times 100$$

The amount of H_3PO_4 is found from the amount of titrant:

$$\text{amount } H_3PO_4 = \text{amount NaOH} \times \frac{1H_3PO_4}{2NaOH}$$

$$= (29.00 \text{ mL} \times 0.05000 \ M) \times \frac{1H_3PO_4}{2NaOH} = 0.7250 \text{ mmol}$$

Converting to weight, we have

$$\text{wt } H_3PO_4 = \text{amount } H_3PO_4 \times \text{MW } H_3PO_4$$

$$= 0.7250 \text{ mmol} \times 98.00 \ \frac{\text{mg}}{\text{mmol}} = 71.05 \text{ mg}$$

Finally,

$$\% \ H_3PO_4 = \frac{71.05 \text{ mg } H_3PO_4}{300.0 \text{ mg sample}} \times 100 = 23.68$$

It may be more difficult and tedious to determine the stoichiometry between the titrant and analyte in more complicated systems, when several reactions are coupled. Direct titration methods are not always possible due to the lack of a suitable indicator or titrant. In such instances, one of the other titration methods may be applicable.

As was pointed out earlier, many titrants are not available as primary standards, and their concentrations must be determined by titration with some other primary standard. The problem is somewhat the opposite of the one just described in Example 5-2; that is, the amount of analyte (standard) is known and the concentration of titrant is the unknown.

Example 5-3

Exactly 600.0 mg of pure sodium oxalate, $Na_2C_2O_4$, was dissolved in acid and titrated with a potassium permanganate solution according to the reaction

$$2MnO_4^- + 5C_2O_4^{2-} + 16H^+ \longrightarrow 2Mn^{2+} + 10CO_2(g) + 8H_2O$$

The end point was reached after adding 34.00 mL of titrant. Calculate the molarity of the $KMnO_4$.

Solution The molar concentration is given by

$$C_{KMnO_4} = \frac{\text{amount } KMnO_4 \, (\text{mmol})}{\text{volume solution } (\text{mL})}$$

The amount of $KMnO_4$ can be found from the amount of $Na_2C_2O_4$ with which it reacted:

$$\text{amount } KMnO_4 = \text{amount } Na_2C_2O_4 \times \frac{2KMnO_4}{5Na_2C_2O_4}$$

$$= \frac{600.0 \text{ mg } Na_2C_2O_4}{134.0 \, \dfrac{\text{mg } Na_2C_2O_4}{\text{mmol } Na_2C_2O_4}} \times \frac{2}{5} = 1.791 \text{ mmol}$$

Then

$$C_{KMnO_4} = \frac{1.791 \text{ mmol}}{34.00 \text{ mL}} = 0.05268 \; M$$

Replacement or Indirect Titration Method

This titration method employs a preliminary reaction in which the analyte is replaced by an equivalent amount of another substance which is then determined by titration. For example, iron can be determined this way by treating a solution of ferric ions with excess potassium iodide and titrating the liberated iodine with a sodium thiosulfate titrant. The reactions are

$$\underset{\text{analyte}}{2Fe^{3+}} + 2I^- \longrightarrow 2Fe^{2+} + I_2 \tag{5-2}$$

$$I_2 + \underset{\text{titrant}}{2S_2O_3^{2-}} \longrightarrow 2I^- + S_4O_6^{2-} \tag{5-3}$$

Note that the titrant and analyte do not react with each other, but they are related through the iodine. To develop the quantitative relationship between titrant and analyte, the stoichiometries of both Reactions 5-2 and 5-3 must be taken into account.

Example 5-4

A 750.0-mg sample of iron ore was dissolved in acid and treated to oxidize all of the iron to ferric ion. After destroying any remaining oxidizing agent, excess KI was added. The liberated I_2 required 28.50 mL of 0.07500 M $Na_2S_2O_3$ for titration. What is the % Fe in the sample?

Solution Using the stoichiometries of Reactions 5-2 and 5-3 we can work backward to find the amount of Fe^{3+}.

$$\text{amount } I_2 = \text{amount } S_2O_3^{2-} \times \frac{1I_2}{2S_2O_3^{2-}}$$

$$\text{amount } Fe^{3+} = \text{amount } I_2 \times \frac{2Fe^{3+}}{1I_2}$$

where amount is in moles or millimoles. Combining these gives

$$\text{amount } Fe^{3+} = \text{amount } S_2O_3^{2-} \times \frac{1I_2}{2S_2O_3^{2-}} \times \frac{2Fe^{3+}}{1I_2}$$

$$= (18.50 \text{ mL})(0.07500 \text{ } M)(1/2)(2/1) = 1.388 \text{ mmol}$$

Converting this to weight and then % Fe yields

$$\text{wt Fe} = \text{amount Fe} \times \text{AW Fe}$$

$$= 1.388 \text{ mmol} \times 55.85 \frac{\text{mg}}{\text{mmol}} \quad 77.52 \text{ mg}$$

$$\% \text{ Fe} = \frac{\text{wt Fe}}{\text{wt sample}} \times 100$$

$$= \frac{77.52 \text{ mg}}{750.0 \text{ mg}} \times 100 = 10.34$$

Determining the molarity of a titrant being standardized by a replacement titration is done similarly except that we work backward from the known amount of standard to the unknown concentration of titrant. This type of problem is illustrated in the following example.

Example 5-5

A sodium thiosulfate solution is standardized by weighing and dissolving 250.0 mg of pure copper metal. The copper solution is treated with excess KI and the liberated iodine requires 44.90 mL of the thiosulfate titrant to reach the end point. What is the molarity of the sodium thiosulfate?

Solution The pertinent reactions are

$$2Cu^{2+} + 4I^- \longrightarrow 2CuI(s) + I_2$$
$$\text{excess}$$

$$I_2 + 2S_2O_3^{2-} \longrightarrow 2I^- + S_4O_6^{2-}$$

From the stoichiometries of these reactions,

$$\text{amount } S_2O_3^{2-} = \text{amount } I_2 \times \frac{2S_2O_3^{2-}}{1I_2}$$

$$\text{amount } I_2 = \text{amount Cu} \times \frac{1I_2}{2Cu^{2+}}$$

Combining these relationships gives

$$\text{amount } S_2O_3^{2-} = \text{amount Cu} \times \frac{1I_2}{2Cu^{2+}} \times \frac{2S_2O_3^{2-}}{1I_2}$$

$$= \frac{250.0 \text{ mg}}{63.54 \dfrac{\text{mg}}{\text{mmol}}} \times \frac{1}{2} \times \frac{2}{1} = 3.935 \text{ mmol}$$

and

$$c_{S_2O_3^{2-}} = \frac{\text{amount of } S_2O_3^{2-}}{\text{volume of solution}} = \frac{3.935 \text{ mmol}}{44.90 \text{ mL}} = 0.08764 \ M$$

Back-Titration Method

In a back-titration method the analyte is found by difference. A known excess of some reagent is added to the analyte solution. A portion of this reagent reacts with the analyte; the rest is determined by titration. A long-used back-titration method for chloride involves the addition of a known excess of silver nitrate which reacts to form insoluble silver chloride, followed by titration of the remaining excess silver nitrate with potassium thiocyanate. The ionic reactions for this process are

$$\underset{\text{analyte}}{Cl^-} + \underset{\text{known amount}}{Ag^+} \longrightarrow AgCl(s)$$

$$\underset{\text{excess}}{Ag^+} + \underset{\text{titrant}}{SCN^-} \longrightarrow AgSCN(s)$$

The direct titration of Cl^- in acidic solution using $AgNO_3$ is not entirely satisfactory, due to the lack of a good indicator (there are some available but they are only moderately effective). On the other hand, we have an excellent indicator, Fe^{3+}, for the back titration with thiocyanate (Chapter 12).

Most procedures call for adding a definite volume of a known concentration of the reagent used to react with the analyte. The amount remaining is determined from the amount of titrant used and the stoichiometry of the titration reaction. The difference between the amount of reagent added and the amount left over, that is, the amount used, can be related to the amount of analyte by the stoichiometry of the analyte reaction. Thus, like a replacement titration method, two potentially different reaction stoichiometries must be taken into account. The pertinent relationships are summarized in the following equations:

$$\text{amount analyte} = \text{amount reagent used} \times \text{reacting ratio (analyte/reagent)}$$

$$\text{amount reagent used} = \text{amount reagent added} - \text{amount reagent remaining}$$

$$\text{amount reagent remaining} = \text{amount titrant} \times \text{reacting ratio (reagent/titrant)}$$

Example 5-6

An 800.0-mg sample of chromium ore was dissolved and the chromium oxidized to chromate ion. The solution was treated with 10.00 mL of 0.2000 M $AgNO_3$. The resulting precipitate of Ag_2CrO_4 was removed and discarded. The excess $AgNO_3$ required 14.50 mL of 0.1200 M KSCN for titration. Calculate the % Cr_2O_3 in the ore.

Solution   The pertinent ionic reactions are

$$\underset{\substack{\text{analyte} \quad \text{known amount}}}{CrO_4{}^{2-} + \quad 2Ag^+} \longrightarrow Ag_2CrO_4(s)$$

$$\underset{\substack{\text{excess} \quad \text{titrant}}}{Ag^+ + \quad SCN^-} \longrightarrow AgSCN(s)$$

To obtain the amount of Ag^+ used to react with the chromate, we determine the amount added and subtract that which is left over:

$$\text{amount } Ag^+ \text{ added} \qquad = 10.00 \text{ mL} \times 0.2000\ M \qquad = 2.000 \text{ mmol}$$

$$-\left\{ \begin{array}{c} \text{amount } Ag^+ \text{ left} \\ \| \\ \text{amount } SCN^- \times \dfrac{1\ Ag^+}{1\ SCN^-} \end{array} \right\} = 14.50 \text{ mL} \times 0.1200\ M \times \frac{1}{1} = 1.740 \text{ mmol}$$

$$\overline{\text{amount } Ag^+ \text{ used}} \qquad\qquad\qquad\qquad\qquad\qquad\quad = \overline{0.260 \text{ mmol}}$$

From the stoichiometry of the reaction with chromate,

$$\text{amount } CrO_4{}^{2-} = \text{amount } Ag^+ \text{ used} \times \frac{1CrO_4{}^{2-}}{2Ag^+}$$

$$= 0.260 \text{ mmol} \times \frac{1}{2} = 0.130 \text{ mmol}$$

Finally, we get the equivalent amount of Cr_2O_3, convert to weight and then to percentage.

$$\text{amount } Cr_2O_3 = \text{amount } CrO_4{}^{2-} \times \frac{1Cr_2O_3}{2CrO_4{}^{2-}}$$

$$= 0.130 \text{ mmol} \times \frac{1}{2} = 0.0650 \text{ mmol}$$

$$\text{wt } Cr_2O_3 = \text{amount } Cr_2O_3 \times \text{MW } Cr_2O_3$$

$$= 0.0650 \text{ mmol} \times 152.0 \frac{\text{mg}}{\text{mmol}} = 9.88 \text{ mg}$$

$$\% \ Cr_2O_3 = \frac{\text{wt } Cr_2O_3}{\text{wt sample}} \times 100 = \frac{9.88 \text{ mg}}{800.0 \text{ mg}} \times 100 = 1.24$$

APPLICATIONS

Titrimetry is used more widely than gravimetry in quantitative analysis. In this section we discuss the scope of titrimetry as a general technique. There are many specific features that are unique to one class of titrations (i.e., acid-base) and these will be discussed, together with selected examples, in the individual chapters on acid-base, complexation, precipitation, and redox titrations.

Scope

It was pointed out in Chapter 4 that the range of applications of a method depends on how well the method can meet certain goals. Let us examine the capabilities of titrimetry in the same six areas considered for gravimetry. It is a most worthwhile exercise to make comparisons between the two techniques and to learn the strengths and weaknesses of each. In doing so you will find yourself better prepared to make sound judgments about how to solve analytical problems.

Limit of detection

The limit of detection for a titration is almost always determined by the increasing difficulty in establishing when the equivalence point has been reached as the concentration of analyte gets smaller. It is somewhat difficult to specify a particular concentration value as the typical limit of detection because the magnitude of the equivalence point break in a titration, and therefore the ease or difficulty of detecting it, depends on both the concentration of the reactants and the equilibrium constant of the reaction. For those reactions whose equilibrium constants are very favorable, it is often possible to determine concentrations as low as 1 millimolar. Graphic methods for finding end points, especially those using data away from the vicinity of the equivalence point, often are more sensitive than visual indicator methods and do not require as favorable an equilibrium constant for the titration reaction.

Accuracy

The accuracy of titrimetric methods is probably not quite as good as it is for gravimetric methods when the samples are fairly simple and of comparable size. Nonetheless, they are still among the most accurate methods available.

Selectivity

Titrimetric methods are applicable to both inorganic and organic substances and in some instances may be more selective than gravimetric methods. For example, in the gravimetric determination of calcium as CaC_2O_4, silica contaminates the precipitate, causing serious error. This same silica contamination has no effect on a replacement titration method for Ca using potassium permanganate to titrate the oxalate from CaC_2O_4.

Ease of operation

Titrations are relatively easy to perform and this, perhaps as much as anything else, is the reason they are used so often in analytical determinations. The manipulative operations are simple and quite easily automated. Perhaps the only advantage that an experienced chemist may have over a beginner with a few days' practice is a familiarity with the subtleties of various indicators.

Time

Titrimetric determinations usually are performed more quickly than gravimetric determinations. Unless extensive sample pretreatment is necessary, procedures seldom take more than an hour to complete.

Cost

The tools of titrimetry are simple and the cost per determination is small. When large numbers of samples have to be determined, the personnel costs can become an important consideration; consequently, laboratories that consistently handle large numbers of similar samples almost always use automated systems.

PROBLEMS

5-1. Calculate the molar analytical concentration of solute when 2.500 g of each of the following primary standard reagents is dissolved and diluted to 500.0 mL.
  (a) $K_2Cr_2O_7$ (d) HSO_3NH_2 (g) $(NH_4)_2Ce(NO_3)_6$
  (b) $AgNO_3$ (e) $KHC_8H_4O_4$ (h) $Na_2C_2O_4$
  (c) $KH(IO_3)_2$ (f) Na_2CO_3

5-2. Calculate the weight of the primary standard shown in parentheses needed to prepare each of the following solutions.
  (a) 500 mL of 0.100 M Ca^{2+} ($CaCO_3$)
  (b) 1.00 L of 0.200 M $H_2C_2O_4$ ($Na_2C_2O_4$)
  (c) 0.250 L of 0.125 M IO_3^- ($KH(IO_3)_2$)
  (d) 500 mL of 0.200 M Cl^- ($BaCl_2 \cdot 2H_2O$)
  (e) 250 mL of 0.250 M Ag^+ ($AgNO_3$)
  (f) 500 mL of 0.100 M H_3BO_3 ($Na_2B_4O_7 \cdot 10H_2O$)

5-3. Arsenic(III) oxide dissolves in base to form Na_3AsO_3, which, upon acidification, is converted to arsenious acid, H_3AsO_3. What weight of primary standard As_2O_3 is required to prepare 250.0 mL of 0.2000 M H_3AsO_3?

5-4. A standard solution of iodine can be prepared by treating a known amount of KIO_3 with excess KI in acidic solution:

$$IO_3^- + 5I^- + 6H^+ \longrightarrow 3I_2 + 3H_2O$$

Calculate the weight of primary standard KIO_3 needed to make 500.0 mL of 0.1000 M I_2.

5-5. The following data were collected for the titration of iron with permanganate according to the reaction

$$5Fe^{2+} + MnO_4^- + 8H^+ \longrightarrow 5Fe^{3+} + Mn^{2+} + 4H_2O$$

The sample weighed 0.8037 g, the concentration of the $KMnO_4$ titrant was 0.1034 M, and the original volume was 20.00 mL. The measured signal is known to be directly proportional to the concentration of Fe^{2+}.

Volume $KMnO_4$ added (mL)	Signal	Volume $KMnO_4$ added (mL)	Signal
0	1.00×10^{-1}	9.40	8.84×10^{-4}
2.00	7.18×10^{-2}	9.60	4.22×10^{-5}
4.00	4.83×10^{-2}	9.80	1.16×10^{-5}
6.00	2.85×10^{-2}	10.00	6.67×10^{-6}
8.00	1.14×10^{-2}	11.00	2.08×10^{-6}
9.00	3.79×10^{-3}	12.00	1.20×10^{-6}
9.20	2.33×10^{-3}	14.00	6.26×10^{-7}

(a) Sketch the linear titration curve, with and without accounting for dilution.

(b) Sketch the logarithmic titration curve, accounting for dilution.

(c) Determine the end point from each titration curve and use it to calculate the % Fe in the sample.

5-6. Potassium hydrogen phthalate, $KHC_8H_4O_4$, is a primary standard acid that reacts with sodium hydroxide on a 1:1 molar basis. A sample of the acid weighing 0.5893 g was titrated with NaOH, requiring 22.49 mL to reach the end point. Calculate the molar concentration of the NaOH solution.

5-7. Sodium carbonate is a primary standard base that reacts with hydrochloric acid as follows:

$$Na_2CO_3 + 2HCl \longrightarrow 2NaCl + H_2O + CO_2(g)$$

If 40.37 mL of an HCl solution were required to titrate a solution containing 221.4 mg of primary standard Na_2CO_3, calculate the molar concentration of the HCl solution.

5-8. Potassium dichromate is an excellent primary standard that is often used to standardize reducing agents such as Fe^{2+}:

$$Cr_2O_7^{2-} + 6Fe^{2+} + 14H^+ \longrightarrow 2Cr^{3+} + 6Fe^{3+} + 7H_2O$$

If 33.06 mL of an Fe^{2+} solution were required to titrate a sample containing 322.8 mg of pure $K_2Cr_2O_7$, what is the molar concentration of the Fe^{2+} solution?

5-9. A solution of iodine was standardized by titration with H_3AsO_3 prepared from weighing and dissolving 141.6 mg of pure As_2O_3. The titration reaction is

$$H_3AsO_3 + I_2 + H_2O \longrightarrow H_3AsO_4 + 2I^- + 2H^+$$

What is the molar concentration of I_2 if 19.62 mL of the solution were required to reach the end point?

5-10. A 0.5537-g sample containing oxalic acid required 21.62 mL of 0.09377 M NaOH for titration. If the reaction is

$$H_2C_2O_4 + 2NaOH \longrightarrow Na_2C_2O_4 + 2H_2O$$

calculate the % $H_2C_2O_4$ in the sample.

5-11. A sodium thiosulfate solution can be standardized by using it to titrate the iodine liberated by the action of excess KI on a known weight of primary standard $K_2Cr_2O_7$:

$$Cr_2O_7^{2-} + 6I^- + 14H^+ \longrightarrow 2Cr^{3+} + 3I_2 + 7H_2O$$

$$I_2 + 2S_2O_3^{2-} \longrightarrow 2I^- + S_4O_6^{2-}$$

Calculate the molar concentration of the sodium thiosulfate solution if 31.47 mL of this solution were required to titrate a sample prepared using 0.2177 g of pure $K_2Cr_2O_7$:

5-12. A sample of phosphate detergent weighing 0.6637 g was dissolved in water and titrated with 0.1216 M HCl according to the reaction

$$PO_4^{3-} + 2HCl \longrightarrow H_2PO_4^- + 2Cl^-$$

The end point was observed after the addition of 28.33 mL of the HCl titrant. Calculate the amount of phosphorus present as % PO_4^{3-} and % P_2O_5.

5-13. A sample of pure $CaCl_2 \cdot xH_2O$ weighing 0.6370 g was dissolved in water containing a few drops of concentrated HNO_3. It took 24.04 mL of 0.3606 M $AgNO_3$ to titrate the chloride in the sample:

$$Ag^+ + Cl^- \longrightarrow AgCl(s)$$

What is the degree of hydration of the $CaCl_2$?

5-14. A 250.0-mg sample of a pure chlorinated aromatic compound with a molecular weight of 261.4 was decomposed, converting the bound chlorine to Cl^-. The resulting solution required 21.70 mL of 0.1322 M $AgNO_3$ for titration of the chloride ion. How many chlorine atoms does one molecule of the aromatic compound contain?

5-15. A sample of iron ore weighing 916 mg was dissolved in acid and pretreated to oxidize all of the iron to Fe^{3+}. After removing any remaining oxidizing agent, an excess of KI was added to the solution. The liberated iodine required 22.9 mL of 0.0934 M $Na_2S_2O_3$ for titration. Calculate the % Fe_3O_4 in the sample (see Reactions 5-2 and 5-3).

5-16. A 5.00-mL aliquot of bleach (density = 1.16 g/mL) was diluted and treated with excess KI, leading to the reaction

$$OCl^- + 2I^- + 2H^+ \longrightarrow Cl^- + I_2 + H_2O$$

The liberated I_2 required 22.8 mL of 0.214 M $Na_2S_2O_3$ for titration (see Reaction 5-3). Calculate the % OCl^- (hypochlorite ion) in the bleach.

5-17. A sample of an antihistamine, brompheniramine maleate, weighing 4.6330 g was dissolved in alcohol and decomposed with metallic sodium. The resulting solution was treated with 10.00 mL of 0.2500 M $AgNO_3$, an amount sufficient to precipitate all of the liberated bromide ion as AgBr. The excess $AgNO_3$ was titrated with 0.1214 M KSCN, requiring 14.42 mL to reach the end point:

$$Ag^+ + SCN^- \longrightarrow AgSCN(s)$$

Calculate the % Br in the sample. What additional piece of information is needed to determine the percentage of the antihistamine in the sample?

5-18. During the titration of an $HClO_4$ solution with 0.1032 M NaOH, quantitative analysis student Al Cohol became distracted and overshot the end point. A fellow student, I. M. Smart, suggested that he simply record the present volume of NaOH added and titrate the excess with a standard acid solution. If the original sample volume was 25.00 mL, the volume of NaOH added was 28.06 mL, and it took 3.47 mL of 0.1094 M HCl to back titrate the NaOH, calculate the molar concentration of the original $HClO_4$ solution.

5-19. A 0.1428-g sample of carbonate rock was pulverized and heated in a closed system causing the evolution of CO_2:

$$M_xCO_3 \longrightarrow M_xO + CO_2(g)$$

A stream of nitrogen was used to sweep the CO_2 into 100.0 mL of 0.05172 M NaOH solution:

$$CO_2(g) + 2NaOH \longrightarrow Na_2CO_3 + H_2O$$

The excess NaOH required 28.14 mL of 0.1788 M HCl for titration. Calculate the amount of carbonate in the sample as % $CaCO_3$.

5-20. What weight of sample containing sodium carbonate must be used so that 10 times the volume of 0.2269 M HCl (in milliliters) will equal the % Na_2CO_3 in the sample? The titration reaction is

$$Na_2CO_3 + 2HCl \longrightarrow 2NaCl + H_2O + CO_2(g)$$

5-21. An 850-mg sample containing 4.79% La was dissolved and titrated with 0.100 M NaF:

$$La^{3+} + 3F^- \longrightarrow LaF_3(s)$$

What fraction of La^{3+} remains untitrated after the addition of 7.50 mL of the sodium fluoride titrant?

5-22. A sample containing both HCl and HNO_3 had a density of 1.083 g/mL. A 10.00-mL aliquot of this sample required 28.44 mL of 0.1163 M NaOH for titration. Another 10.00-mL aliquot of the sample required 16.31 mL of 0.1216 M $AgNO_3$ for titration. Calculate the % HCl and % HNO_3 in the sample.

ACID-BASE EQUILIBRIA

It is difficult to overstate the importance of acid-base equilibria in chemistry and related sciences such as agriculture, biology, geology, and medicine. Such topics as the production of chlorophyll in a plant, the formation of a limestone cave, the deterioration of a marble statue in polluted air, or the activity of a life-sustaining enzyme cannot be discussed intelligently without an understanding of acid-base behavior. The quantitative aspects of acid-base chemistry described in this chapter should broaden and strengthen both your understanding of the principles of equilibrium and your ability to apply those principles to the solution of a variety of common problems.

DEFINING ACIDS AND BASES

Acids and bases are recognized as specific classes of compounds because of certain distinctive properties they exhibit. It should not be surprising to discover that chemists have not always entirely agreed on the particular properties that should define such classes. Early chemists defined acids as substances that taste sour or turn litmus red and bases as substances that taste bitter or turn litmus blue. These definitions are far from ideal because not all compounds exhibiting these properties are chemically similar.

With acceptance in the 1880s of the Arrhenius theory of ionization, acids became defined as substances that dissociate into hydrogen ions and anions when dissolved in water. Since bases were known to react with acids to form a salt and water, they became defined as substances that dissociate into hydroxide ions and cations when dissolved in water. The water produced in an acid-base reaction was considered to be the result of hydrogen ions combining with hydroxide ions. Although the Arrhenius definitions were a significant advance, they could not explain the role played by the solvent in determining the strength of an acid or base, nor could they explain why certain salts exhibit acid-base properties.

Brønsted–Lowry Definition

In 1923, J. N. Brønsted in Denmark and T. M. Lowry in England, independently, recognized that both acid and base behavior could be described in terms of a single substance—the hydrogen ion (or proton). They defined an acid as any substance capable of *donating* a proton in a reaction. A base is defined in terms of an opposite to an acid; that is, as any substance capable of *accepting* a proton in a reaction. The Brønsted–Lowry concept of acid-base behavior has proven especially convenient when using water and water-like (protophilic) solvents.

Role of the solvent

One of the most significant features of the Brønsted–Lowry concept of acid-base behavior is the incorporation of the solvent as a vital part of the acid-base system. Protons, like electrons, are not thought to exist in a free state in solution. As a result, the proton-donating ability of an acid or the proton-accepting ability of a base can be realized only in the presence of a substance of the opposite ability. The solvent can be that substance, either accepting or donating protons, and therefore must be considered a participant in the dissociation or ionization of acids and bases. For example, when formic acid dissolves in water, ionization occurs because the solvent acts as a base or proton acceptor:

$$\overset{\frown}{HCO_2H} + H_2O \; \rightleftharpoons \; H_3O^+ + HCO_2^- \qquad (6\text{-}1)$$

$$\text{acid}_1 \qquad \text{base}_2 \qquad \quad \text{acid}_2 \qquad \text{base}_1$$

The H_3O^+ is merely a solvated proton called the *hydronium ion*. Present data indicate that several molecules of water are attached to each proton but H_3O^+ is written for the sake of simplicity. When ammonia is dissolved in water, the solvent acts as an acid or proton donor:

$$\overset{\frown}{NH_3} + H_2O \; \rightleftharpoons \; NH_4^+ + OH^- \qquad (6\text{-}2)$$

$$\text{base}_1 \quad \text{acid}_2 \qquad \quad \text{acid}_1 \qquad \text{base}_2$$

Solvents like water that can exhibit both acidic and basic character, depending on the solute, are called *amphiprotic* solvents. Low-molecular-weight alcohols and acetic acid are other common amphiprotic solvents. When formic acid or ammonia is dissolved in ethanol, the following reactions occur:

$$\overset{\frown}{HCO_2H} + C_2H_5OH \; \rightleftharpoons \; C_2H_5OH_2^+ + HCO_2^-$$

$$\overset{\frown}{NH_3} + C_2H_5OH \; \rightleftharpoons \; NH_4^+ + C_2H_5O^-$$

Conjugate acids and bases

The reversible behavior of acid-base reactions leads to an interesting and quite useful observation. When an acid donates a proton in a reaction, it becomes a substance that is capable of accepting a proton to re-form the original acid. Such a substance is a

Brønsted–Lowry base. Similarly, when a base accepts a proton in a reaction, it becomes a substance capable of donating a proton. Brønsted and Lowry referred to such substances as *conjugate pairs*. Accordingly, every Brønsted–Lowry acid has a conjugate base and vice versa. Reactions 6-1 and 6-2 consist simply of two conjugate pairs, labeled 1 and 2. The strengths of conjugate acids and bases are related to each other in an inverse manner; that is, the stronger the acid, the weaker the conjugate base.

IONIZATION OF WATER

Since water can act as a proton donor in the presence of a base and as a proton acceptor in the presence of an acid, it should not be surprising to learn that it can undergo an acid-base reaction with itself:

$$H_2O + H_2O \rightleftharpoons H_3O^+ + OH^- \tag{6-3}$$

$$\text{acid}_1 \quad \text{base}_2 \qquad \text{acid}_2 \quad \text{base}_1$$

conjugates

conjugates

All amphiprotic solvents undergo such self-ionization or *autoprotolysis* reactions. The extent to which this reaction occurs is represented by the expression

$$K_{eq} = \frac{(a_{H_3O^+})(a_{OH^-})}{(a_{H_2O})} = \frac{(a_{H_3O^+})(a_{OH^-})}{1}$$

Written in terms of concentrations, we have

$$K_{eq} = K_w = [H_3O^+][OH^-] \tag{6-4}$$

or, taking the negative logarithm of both sides,

$$pK_w = pH + pOH \tag{6-5}$$

The autoprotolysis constant for water, K_w, has a value of 1.00×10^{-14} at 24°C (about room temperature). According to Reaction 6-3, the concentrations of hydronium and hydroxide ions in pure water must be equal and Equation 6-4 can be used to calculate the hydronium-ion concentration of pure water:

$$K_w = [H_3O^+]^2$$

$$[H_3O^+] = \sqrt{1.00 \times 10^{-14}} = 1.00 \times 10^{-7} M$$

$$pH = -\log(1.00 \times 10^{-7}) = 7.000$$

Likewise, the pOH of pure water is 7.000. This is the neutral pH for water (it will be different for other solvents because their autoprotolysis constants are different). At 24°C, any solution whose hydronium-ion concentration exceeds $1.00 \times 10^{-7} M$ is acidic and any solution whose hydronium-ion concentration is less than $1.00 \times 10^{-7} M$ is basic.

According to Equation 6-4, an increase in the hydronium-ion concentration resulting from the addition of an acid to water must be accompanied by an equal decrease in the hydroxide-ion concentration. Conversely, an increase in the hydroxide-ion concentration must be accompanied by a decrease in the hydronium-ion concentration. If the concentration of either H_3O^+ or OH^- is known, the other can be calculated using Equation 6-4.

Example 6-1

What is the concentration of H_3O^+ in a 1.00×10^{-2} M NaOH solution?

Solution The H_3O^+ comes from the ionization of water:

$$2H_2O \rightleftharpoons H_3O^+ + OH^-$$

The mathematical equation describing this equilibrium is

$$K_w = [H_3O^+][OH^-]$$

The OH^- comes from the NaOH, which ionizes completely, and from water, which ionizes only slightly. Accordingly, the concentration of OH^- from water is small compared to that from the NaOH. Thus

$$[OH^-] \simeq C_{NaOH} = 1.00 \times 10^{-2}\ M$$

Substituting in the K_w expression gives us

$$1.00 \times 10^{-14} = [H_3O^+](1.00 \times 10^{-2})$$

and

$$[H_3O^+] = 1.00 \times 10^{-12}\ M$$

Example 6-2

Calculate the pOH of a 0.020 M HCl solution.

Solution To calculate pOH we must find first the OH^- concentration. The source of hydroxide ion is the ionization of water:

$$2H_2O \rightleftharpoons H_3O^+ + OH^-$$

The equation describing this equilibrium is

$$K_w = [H_3O^+][OH^-]$$

The H_3O^+ comes both from the HCl (which ionizes completely) and water (which ionizes only slightly). Accordingly, the amount from water is small compared to that from the HCl. Thus

$$[H_3O^+] \simeq C_{HCl} = 2.0 \times 10^{-2}\ M$$

Substituting in the K_w expression yields

$$1.00 \times 10^{-14} = (2.0 \times 10^{-2}\ M)[OH^-]$$

and

$$[OH^-] = 5.0 \times 10^{-13}\ M$$

Finally,

$$pOH = -\log(5.0 \times 10^{-13}) = 12.30$$

IONIZATION OF ACIDS AND BASES

Chemists find it convenient to classify acids and bases as "strong" or "weak" to indicate their approximate degree of ionization (reaction with the solvent). Strong acids and bases are those which ionize essentially 100% in dilute solution. Acids and bases that ionize less than 10% in dilute solutions are said to be weak. Relatively few common acids and bases fall between these two limits, making the distinction between strong and weak a useful one. Strong acids and bases that are encountered frequently in the laboratory are listed in Table 6-1. Lists of weak acids and bases and their ionization constants are given in Appendices G and H.

Strong Acids and Bases

Equilibrium expressions are not a part of the calculations involved with the ionization of strong acids and strong bases because the equilibrium constants are either very large or undefined. A strong acid such as hydrochloric acid is considered to be completely ionized in water:

$$HCl + H_2O \xrightarrow{\sim 100\%} H_3O^+ + Cl^- \qquad (6\text{-}6)$$

The equilibrium constant for this reaction is given by

$$K_{eq} = \frac{(a_{H_3O^+})(a_{Cl^-})}{(a_{HCl})(a_{H_2O})} = \frac{(a_{H_3O^+})(a_{Cl^-})}{(a_{HCl})(1)}$$

If Reaction 6-6 is complete, no molecular HCl exists in the solution and the equilibrium constant expression becomes

$$K_{eq} = \frac{(a_{H_3O^+})(a_{Cl^-})}{0}$$

making K_{eq} undefined. Even if the acid is not quite 100% ionized, the activity (concentration) of molecular HCl is very small, making K_{eq} a correspondingly large number. In either case, there is never any significant difference between the analytical concentration of the acid or base and the equilibrium concentration of the ionization products.

TABLE 6-1   COMMON STRONG ACIDS AND BASES

Acids		Bases	
HCl	Hydrochloric acid	LiOH	Lithium hydroxide
HBr	Hydrobromic acid	NaOH	Sodium hydroxide
HI	Hydriodic acid	KOH	Potassium hydroxide
HClO$_4$	Perchloric acid	Ba(OH)$_2$	Barium hydroxide
HNO$_3$	Nitric acid		
H$_2$SO$_4$[a]	Sulfuric acid		

[a]Only the first hydrogen is completely ionized.

Weak Acids and Bases

Equilibrium expressions are fundamental to solving concentration problems involving weak acids and bases. Using HA to represent a weak acid, the ionization in water is

$$HA + H_2O \rightleftharpoons H_3O^+ + A^-$$

The equilibrium constant for this reaction is called the *acid dissociation* or *acidity* constant, K_a.

$$K_a = \frac{(a_{H_3O^+})(a_{A^-})}{(a_{HA})(a_{H_2O})} = \frac{(a_{H_3O^+})(a_{A^-})}{(a_{HA})(1)} \simeq \frac{[H_3O^+][A^-]}{[HA]}$$

The value of K_a is small when the concentration terms in the numerator are small compared to that in the denominator. Such a situation exists when the position of equilibrium for the ionization lies far to the left.

A weak base, B, can be treated in a similar manner:

$$B + H_2O \rightleftharpoons BH^+ + OH^-$$

$$K_b = \frac{(a_{BH^+})(a_{OH^-})}{(a_B)(a_{H_2O})} = \frac{(a_{BH^+})(a_{OH^-})}{(a_B)(1)} \simeq \frac{[BH^+][OH^-]}{[B]}$$

where K_b is referred to as the *basicity* or *base dissociation* constant.

Polyprotic acids and polyequivalent bases

Acids that can donate more than one proton are called polyprotic acids, and bases that can accept more than one proton are called polyequivalent bases. The ionization of such compounds occurs in steps, as illustrated below by the triprotic acid H_3PO_4 and the diequivalent base CO_3^{2-}.

$$H_3PO_4 + H_2O \rightleftharpoons H_3O^+ + H_2PO_4^- \qquad K_{a_1} = \frac{[H_3O^+][H_2PO_4^-]}{[H_3PO_4]}$$

$$H_2PO_4^- + H_2O \rightleftharpoons H_3O^+ + HPO_4^{2-} \qquad K_{a_2} = \frac{[H_3O^+][HPO_4^{2-}]}{[H_2PO_4^-]}$$

$$HPO_4^{2-} + H_2O \rightleftharpoons H_3O^+ + PO_4^{3-} \qquad K_{a_3} = \frac{[H_3O^+][PO_4^{3-}]}{[HPO_4^{2-}]}$$

$$CO_3^{2-} + H_2O \rightleftharpoons HCO_3^- + OH^- \qquad K_{b_1} = \frac{[HCO_3^-][OH^-]}{[CO_3^{2-}]}$$

$$HCO_3^- + H_2O \rightleftharpoons H_2CO_3 + OH^- \qquad K_{b_2} = \frac{[H_2CO_3][OH^-]}{[HCO_3^-]}$$

The subscripts on the acid and base dissociation constants are used to identify the individual ionization steps. The numerical values of the dissociation constants always decrease with each dissociation step; thus $K_1 > K_2 > K_3 \cdots$. A simple explanation of this trend can be made in terms of electrostatic forces. The ionization of the first proton of H_3PO_4 involves removing H^+ from a monovalent anion, $H_2PO_4^-$. Ionization of the second proton is more difficult because it involves removal of H^+ from a divalent anion, HPO_4^{2-}. The magnitude of the difference between successive dissociation constants for polyprotic acids (and for polyequivalent bases) is determined by the structure of the molecule or ion and the nature of the bond holding the proton. A brief look at Appendix G will show that the differences vary from less than one to more than six orders of magnitude.

Relation between K_a and K_b for conjugate pairs

In the Brønsted–Lowry system of acidity and basicity, the stronger an acid, the weaker its conjugate base. Since K_a and K_b are used as measures of acid and base strength, it is useful to examine their quantitative relationship. Consider the base ammonia and its conjugate acid ammonium ion. The ionization reactions and equilibrium expressions are as follows:

$$NH_3 + H_2O \rightleftharpoons NH_4^+ + OH^- \qquad K_b = \frac{[NH_4^+][OH^-]}{[NH_3]}$$

$$NH_4^+ + H_2O \rightleftharpoons H_3O^+ + NH_3 \qquad K_a = \frac{[H_3O^+][NH_3]}{[NH_4^+]}$$

Multiplying the two equilibrium constant expressions gives

$$K_a K_b = \frac{[H_3O^+][NH_3]}{[NH_4^+]} \times \frac{[NH_4^+][OH^-]}{[NH_3]} = [H_3O^+][OH^-]$$

Since

$$[H_3O^+][OH^-] = K_w$$

then

$$K_a K_b = K_w \qquad\qquad (6\text{-}7)$$

Equation 6-7 is applicable to any *acid-base conjugate pair* in aqueous solution. This simple equation is worth remembering, as you will encounter many problems in which the K_a or K_b needed is not immediately available and must be calculated from the value for the conjugate base or acid.

Example 6-3

What is the base dissociation constant for the nitrite ion?

Solution Appendix H contains no K_b for NO_2^-, but Appendix G lists a K_a of 7.1×10^{-4} for the conjugate acid, HNO_2. Accordingly,

$$K_a K_b = K_w$$

$$(7.1 \times 10^{-4}) K_b = 1.0 \times 10^{-14}$$

$$K_b = \frac{1.0 \times 10^{-14}}{7.1 \times 10^{-4}} = 1.4 \times 10^{-11}$$

In the case of polyprotic acids and polyequivalent bases it somehow seems natural that the desired K_b should be calculated from the K_a with the same subscript: that is,

Ionization of Acids and Bases

K_{b_1} from K_{a_1}, K_{b_2} from K_{a_2}, and so on. Resist this temptation; it is not correct! The K_a and K_b used in Equation 6-7 must be for a *conjugate* pair. Suppose that we desire to calculate the K_b for PO_4^{3-}. The ionization reaction for this base is

$$PO_4^{3-} + H_2O \rightleftharpoons HPO_4^{2-} + OH^-$$

Since this is the *first* proton acquired, the equilibrium constant is represented as K_{b_1}. The conjugate acid of PO_4^{3-} is HPO_4^{2-}, whose acid dissociation constant is K_{a_3} (ionization of the *third* proton). Thus

$$K_{a_3} K_{b_1} = K_w$$

Example 6-4

What is the base dissociation constant for the hydrogen carbonate (bicarbonate) ion?

Solution There is no listing for HCO_3^- in Appendix H. As a base, bicarbonate would ionize according to the reaction

$$HCO_3^- + H_2O \rightleftharpoons H_2CO_3 + OH^-$$

The reaction describes the addition of the *second* proton and would therefore be represented by K_{b_2}. The conjugate acid of HCO_3^- is H_2CO_3, which is represented by K_{a_1}. Consequently, we write

$$K_{a_1} K_{b_2} = K_w$$

and the value of K_{a_1} is available from Appendix G.

$$(4.45 \times 10^{-7}) K_{b_2} = 1.00 \times 10^{-14}$$

$$K_{b_2} = 2.25 \times 10^{-8}$$

CALCULATION OF pH

The solutions for which chemists want to calculate pH are often quite complex, with the hydrogen-ion concentration being controlled by the combined effect of several equilibria. In those cases where an exact answer is even possible, the calculations can be lengthy and tedious. Although there may be several different substances contributing to the overall hydronium-ion or hydroxide-ion concentration in a solution, it is frequently the case that one substance is the major supplier. If we are willing to settle for a good approximation of the exact concentration of H_3O^+ or OH^-, the minor suppliers can be neglected, thereby greatly simplifying the required calculations. A good understanding of the principles of chemical equilibrium is essential when trying to identify the major source or sources of a particular substance in solution.

In each of the following sections on strong and weak acids and bases, the same multiple equilibrium situation arises. When an acid HA dissolves in water, both the acid and water are suppliers of H_3O^+:

$$HA + H_2O \rightleftharpoons H_3O^+ + A^- \tag{6-8}$$

$$H_2O + H_2O \rightleftharpoons H_3O^+ + OH^- \tag{6-9}$$

Chapter 6   Acid-Base Equilibria

When a base B dissolves in water, both the base and water are suppliers of OH^-.

$$B + H_2O \rightleftharpoons BH^+ + OH^- \qquad (6\text{-}10)$$

$$H_2O + H_2O \rightleftharpoons H_3O^+ + OH^- \qquad (6\text{-}11)$$

How the pH or pOH is calculated depends on which of *three* situations applies. It is worthwhile to examine these situations in a general sense before proceeding to the sections describing the specific treatment of each type of acid and base.

Case 1. The acid is the major supplier of H_3O^+. This means that Reaction 6-9 contributes a negligible amount of H_3O^+ to the solution and can be ignored. The hydronium-ion concentration can be determined by solving the equilibrium expression for Reaction 6-8. This is the most commonly encountered case and prevails as long as the acid is neither very dilute nor very weak.

Case 2. Water is the major supplier of H_3O^+. This means that Reaction 6-8 contributes a negligible amount of H_3O^+ to the solution and can be ignored. The hydronium-ion concentration can be determined by solving the equilibrium expression for Reaction 6-9. This is equivalent to having a container of pure water. Such a situation exists when the acid is either very weak, very dilute, or both

Case 3. Both the acid and water are major suppliers of H_3O^+. Since both Reactions 6-8 and 6-9 contribute significant amounts of H_3O^+ to the solution, both must be considered simultaneously in determining the hydronium-ion concentration. This is the intermediate situation between the extremes defined by cases 1 and 2.

The same three cases apply to bases except that OH^- is the ion of interest and Reactions 6-10 and 6-11 are the sources of the ion.

Strong Acids and Bases

Strong acids and bases, by definition, ionize completely in water, which is another way of saying that Reactions 6-8 and 6-10 go to completion. To calculate the pH of a solution of strong acid or strong base, we must first decide which of the three cases applies.

We know that the hydronium-ion and hydroxide-ion concentrations in pure water are both $1.0 \times 10^{-7}\ M$. Since strong acids ionize completely, any such acid whose analytical concentration is significantly greater than $1.0 \times 10^{-7}\ M$ can be considered the major supplier of H_3O^+ and case 1 applies. If the concentration of strong acid is significantly less than $1.0 \times 10^{-7}\ M$, water will be the major supplier of H_3O^+ and case 2 applies. Obviously, case 3 applies when the analytical concentration of strong acid is not significantly different from $1.0 \times 10^{-7}\ M$. Exactly what constitutes a "significant" difference depends on how the answer will be used. *For the purpose of pH calculations in this chapter, two orders of magnitude will be considered "significant."* Thus $1.0 \times 10^{-7}\ M\ H_3O^+$ is negligible compared to $1.0 \times 10^{-5}\ M\ H_3O^+$. The three situations can be summarized as follows:

Case	Major supplier of H_3O^+	Necessary condition
1	Strong acid	$C_{HX} \gg 10^{-7}\ M$
2	Water	$C_{HX} \ll 10^{-7}\ M$
3	Both	$C_{HX} \simeq 10^{-7}\ M$

where HX represents a strong acid that dissociates completely. Analogous arguments apply to strong bases when referring to the hydroxide-ion concentration. Now let us consider the approximate pH calculations for each case.

Case 1. The strong acid is the major supplier of H_3O^+. Consequently,

$$[H_3O^+] = C_{HX}$$

Only the analytical concentration of HX in the solution must be known to obtain the concentration of H_3O^+.

Case 2. Water is the major supplier of H_3O^+. Since water is also the only major source of OH^-,

$$[H_3O^+] \simeq [OH^-]$$

and the autoprotolysis equilibrium expression becomes

$$K_w = [H_3O^+]^2$$

or

$$[H_3O^+] = \sqrt{K_w} = \sqrt{1.0 \times 10^{-14}} = 1.0 \times 10^{-7}\ M$$

Case 3. Both the strong acid and water are major suppliers of H_3O^+. The equilibrium concentration of H_3O^+ is the sum of the concentrations produced by both sources:

$$[H_3O^+] = [H_3O^+]\ \text{from HX} + [H_3O^+]\ \text{from } H_2O$$
$$\| \qquad\qquad\qquad \|$$
$$C_{HX} \qquad\qquad\qquad [OH^-]$$

or

$$[H_3O^+] = C_{HX} + [OH^-] \tag{6-12}$$

From the autoprotolysis equilibrium for water,

$$[OH^-] = \frac{K_w}{[H_3O^+]} \tag{6-13}$$

Substituting Equation 6-13 for $[OH^-]$ in Equation 6-12 gives

$$[H_3O^+] = C_{HX} + \frac{K_w}{[H_3O^+]}$$

which is a quadratic equation that can be rearranged and written as

$$[H_3O^+]^2 - C_{HX}[H_3O^+] - K_w = 0 \tag{6-14}$$

This equation can be solved for the hydronium-ion concentration using the quadratic formula.

It should be noted that Equation 6-14 is an exact solution to the problem of determining the hydronium-ion concentration of a solution of strong acid, *regardless* of the concentration of the acid. Cases 1 and 2 are limiting cases where Equation 6-14 can be simplified for an easier and quicker mathematical solution.

Strong bases are examined in an analogous manner except that OH^- is the ion of interest. Once the hydroxide-ion concentration is known, the hydronium-ion concentration can be calculated from the water autoprotolysis equation.

Example 6-5

Calculate the pH of an aqueous HCl solution whose analytical concentration is (a) $1.0 \times 10^{-1}\ M$, (b) $1.0 \times 10^{-7}\ M$, and (c) $1.0 \times 10^{-10}\ M$.

Solution

(a) At a concentration of $1.0 \times 10^{-1}\ M$, the HCl is clearly the major supplier of H_3O^+. Accordingly, we neglect the contribution of the minor supplier, H_2O, and write

$$[H_3O^+] = C_{HCl} = 1.0 \times 10^{-1}\ M$$

and

$$pH = -\log(1.0 \times 10^{-1}) = 1.00$$

(b) At a concentration of $1.0 \times 10^{-7}\ M$, the amount of H_3O^+ coming from the HCl is comparable to that coming from the water and the two sources must be considered together. Using Equation 6-14 gives us

$$[H_3O^+]^2 - C_{HCl}[H_3O^+] \quad K_w = 0$$

$$[H_3O^+]^2 - 1.0 \times 10^{-7}\ [H_3O^+] - 1.0 \times 10^{-14} = 0$$

Solving this quadratic equation gives

$$[H_3O^+] = 1.62 \times 10^{-7}\ M$$

and

$$pH = 6.79$$

(c) At a concentration of $1.0 \times 10^{-10}\ M$, the HCl must be considered the *minor* supplier of H_3O^+. Thus, its contribution is neglected and only the ionization of water is considered:

$$2H_2O \rightleftharpoons H_3O^+ + OH^-$$

$$K_w = [H_3O^+][OH^-]$$

Since H_3O^+ and OH^- are produced together in the ionization of water and there is no other *important* source of either ion, their concentrations are the same and

$$K_w = [H_3O^+]^2 = 1.0 \times 10^{-14}\ M$$

or

$$[H_3O^+] = \sqrt{1.0 \times 10^{-14}} = 1.0 \times 10^{-7}\ M$$

and

$$pH = 7.00$$

Weak Monoprotic Acids

For the purpose of calculating pH it is convenient to subdivide weak acids into one of three cases, as we did with strong acids and bases. Unfortunately, the incomplete ionization of weak acids makes it more difficult to decide which case applies in addition to complicating the actual pH calculation. In an aqueous solution of a weak acid the amount of H_3O^+ coming from the acid is directly related to the acid dissociation constant and the concentration of the acid, while the amount of H_3O^+ coming from water is directly related to the water autoprotolysis constant, K_w. Therefore, the major supplier of H_3O^+ can be determined by comparing the product of $K_a \times C_{HA}$ with K_w. The three possible situations are summarized as follows:

Case	Major supplier of H_3O^+	Necessary condition
1	Weak acid	$K_a \times C_{HA} \gg K_w$
2	Water	$K_a \times C_{HA} \ll K_w$
3	Both	$K_a \times C_{HA} \simeq K_w$

Only the calculations for case 1 will be discussed further. Case 2 involves calculating the pH of pure water, which has already been described. The calculations for case 3 are complicated and beyond the scope of this book. Fortunately, we are not often interested in calculating the pH of solutions fitting case 3.

The dissociation of a weak acid in water may be described by the following equilibrium:

$$HA + H_2O \rightleftharpoons H_3O^+ + A^- \qquad K_a = \frac{[H_3O^+][A^-]}{[HA]}$$

For each molecule of acid that dissociates, one H_3O^+ and one A^- ion are produced. As long as there is no other important source of either ion (which is not always the case), their concentrations must be equal; that is,

$$[H_3O^+] = [A^-] \qquad (6\text{-}15)$$

Furthermore, the sum of the molar concentration of the weak acid and its conjugate base must equal the analytical concentration of the acid:

$$C_{HA} = [HA] + [A^-] \qquad (6\text{-}16)$$

Substituting Equations 6-15 in 6-16 and rearranging gives

$$[HA] = C_{HA} - [H_3O^+] \qquad (6\text{-}17)$$

When Equations 6-15 and 6-17 are substituted for $[A^-]$ and $[HA]$ in the equilibrium expression for the weak acid, we obtain

$$K_a = \frac{[H_3O^+]^2}{C_{HA} - [H_3O^+]} \qquad (6\text{-}18)$$

This is a quadratic equation and can be rearranged to the form

$$[H_3O^+]^2 + K_a[H_3O^+] - K_a C_{HA} = 0 \qquad (6\text{-}19)$$

If both the ionization constant and the analytical concentration of the weak acid are known, Equation 6-19 can be solved for $[H_3O^+]$ using the quadratic formula. It is possible to avoid the task of solving a quadratic equation when certain conditions are met. It is often the case that $[H_3O^+]$ is much smaller than C_{HA}, which is another way of saying that

$$C_{HA} - [H_3O^+] \simeq C_{HA}$$

Substituting this expression in Equation 6-18 gives

$$K_a = \frac{[H_3O^+]^2}{C_{HA}} \tag{6-20}$$

which can be rearranged to

$$[H_3O^+] = \sqrt{K_a C_{HA}} \tag{6-21}$$

The magnitude of the error introduced by using Equation 6-21 instead of 6-19 to calculate $[H_3O^+]$ is directly proportional to the ionization constant and inversely proportional to the concentration of acid, as can be seen from the data in Table 6-2. When C_{HA}/K_a is 10^3, the relative error is 1.6%. Unless otherwise stated, this will be the maximum acceptable error for such problems in this book. That is, if $C_{HA}/K_a \geqslant 10^3$, Equation 6-21 will be accepted as valid. In general, if there is some uncertainty as to the validity of Equation 6-21, a trial value for $[H_3O^+]$ can be calculated using Equation 6-21 and compared with C_{HA} in Equation 6-17. If the value for $[HA]$ is altered by an amount less than the allowable error, the trial value may be accepted as the final value. If the allowable error is exceeded, the quadratic Equation 6-19 must be solved.

Example 6-6

Calculate the hydronium-ion concentration of an aqueous, 0.150 M acetic acid solution. The K_a for acetic acid is 1.76×10^{-5}.

Solution   The equilibrium principally responsible for the concentration of H_3O^+ is

$$CH_3CO_2H + H_2O \rightleftharpoons H_3O^+ + CH_3CO_2^-$$

TABLE 6-2   ERRORS RESULTING FROM USING THE APPROXIMATE EQUATION 6-21 INSTEAD OF THE EXACT EQUATION 6-19

K_a	C_{HA}	$[H_3O^+]$ from Equation 6-21	$[H_3O^+]$ from Equation 6-19	Relative error (%)
1.00×10^{-1}	1.00×10^{-1}	1.00×10^{-1}	6.18×10^{-2}	62
	1.00×10^{-3}	1.00×10^{-2}	9.90×10^{-4}	910
	1.00×10^{-5}	1.00×10^{-3}	1.00×10^{-5}	9900
1.00×10^{-3}	1.00×10^{-1}	1.00×10^{-2}	9.51×10^{-3}	5.1
	1.00×10^{-3}	1.00×10^{-3}	6.18×10^{-4}	62
	1.00×10^{-5}	1.00×10^{-4}	9.90×10^{-6}	910
1.00×10^{-5}	1.00×10^{-1}	1.00×10^{-3}	9.95×10^{-4}	0.50
	1.00×10^{-3}	1.00×10^{-4}	9.51×10^{-5}	5.1
	1.00×10^{-5}	1.00×10^{-5}	6.18×10^{-6}	62
1.00×10^{-7}	1.00×10^{-1}	1.00×10^{-4}	1.00×10^{-4}	0.0050
	1.00×10^{-3}	1.00×10^{-5}	9.95×10^{-6}	0.50
	1.00×10^{-5}	1.00×10^{-6}	9.56×10^{-7}	5.1

for which

$$K_a = \frac{[H_3O^+][CH_3CO_2{}^-]}{[CH_3CO_2H]}$$

Since the ionization of acetic acid is the only important source of either hydronium or acetate ions,

$$[CH_3CO_2{}^-] = [H_3O^+]$$

$$[CH_3CO_2H] = C_{CH_3CO_2H} - [H_3O^+]$$

and

$$1.76 \times 10^{-5} = \frac{[H_3O^+]^2}{0.150 - [H_3O^+]}$$

To determine if the denominator can be simplified, the ratio C_{HA}/K_a must be calculated and compared with its minimum value, 10^3.

$$\frac{C_{HA}}{K_a} = \frac{0.150}{1.76 \times 10^{-5}} = 8.5 \times 10^3$$

Since $C_{HA}/K_a > 10^3$, we can assume that $0.150 - [H_3O^+] \simeq 0.150$. Thus

$$1.76 \times 10^{-5} = \frac{[H_3O^+]^2}{0.150}$$

$$[H_3O^+] = \sqrt{(1.76 \times 10^{-5})(0.150)} = 1.62 \times 10^{-3} \ M$$

Example 6-7

Calculate the hydrogen-ion concentration of an aqueous, 0.150 M chloroacetic acid solution. The K_a for chloroacetic acid is 1.36×10^{-3}.

Solution The equilibrium principally responsible for the concentration of H_3O^+ is

$$CH_2ClCO_2H + H_2O \ \rightleftharpoons \ H_3O^+ + CH_2ClCO_2{}^-$$

for which

$$K_a = \frac{[H_3O^+][CH_2ClCO_2{}^-]}{[CH_2ClCO_2H]}$$

Since the ionization of chloroacetic acid is the only important source of either hydronium or chloroacetate ions,

$$[CH_2ClCO_2{}^-] = [H_3O^+]$$

$$[CH_2ClCO_2H] = C_{CH_2ClCO_2H} - [H_3O^+]$$

and

$$1.36 \times 10^{-3} = \frac{[H_3O^+]^2}{0.150 - [H_3O^+]}$$

To determine if the denominator can be simplified, we write

$$\frac{C_{HA}}{K_a} = \frac{0.150}{1.36 \times 10^{-3}} = 110$$

Since $110 < 10^3$, the denominator cannot be simplified without introducing an unacceptable error. Therefore, the equilibrium equation is rearranged and solved for $[H_3O^+]$ using the quadratic formula.

$$[H_3O^+]^2 + 1.36 \times 10^{-3}[H_3O^+] - (1.36 \times 10^{-3})(0.150) = 0$$

$$[H_3O^+] = \frac{-1.36 \times 10^{-3} \pm \sqrt{(1.36 \times 10^{-3})^2 - (4)(1)(-2.04 \times 10^{-4})}}{2}$$

$$= 0.0136 \ M$$

Weak Monoequivalent Bases

The techniques described in the preceding section on weak acids are the same ones that are used to calculate the hydroxide-ion concentration in solutions of weak bases. Following is a comparison of the important equations:

Acids	*Bases*
$HA + H_2O \rightleftharpoons H_3O^+ + A^-$	$B + H_2O \rightleftharpoons BH^+ + OH^-$
$K_a = \dfrac{[H_3O^+][A^-]}{[HA]}$	$K_b = \dfrac{[BH^+][OH^-]}{[B]}$
$[H_3O^+] = [A^-]$	$[BH^+] = [OH^-]$
$[HA] = C_{HA} - [H_3O^+]$	$[B] = C_B - [OH^-]$
$K_a = \dfrac{[H_3O^+]^2}{C_{HA} - [H_3O^+]}$	$K_b = \dfrac{[OH^-]^2}{C_B - [OH^-]}$
If $C_{HA}/K_a \geqslant 10^3$,	If $C_B/K_b \geqslant 10^3$,
$K_a = \dfrac{[H_3O^+]^2}{C_{HA}}$	$K_b = \dfrac{[OH^-]^2}{C_B}$

Once the hydroxide-ion concentration is known, the hydronium-ion concentration can be calculated from the water autoprotolysis equilibrium.

Example 6-8

Calculate the pH of a 0.0750 M solution of ammonia in water. The K_b for ammonia is 1.75 $\times 10^{-5}$.

Solution   Although it is the pH that is desired, the hydroxide-ion concentration must be found first. The equilibrium controlling the concentration of OH^- is

$$NH_3 + H_2O \rightleftharpoons NH_4^+ + OH^-$$

for which

$$K_b = \frac{[NH_4^+][OH^-]}{[NH_3]}$$

Since the ionization of ammonia is the only important source of either ammonium ion or hydroxide ion,

$$[OH^-] = [NH_4^+]$$

$$[NH_3] = C_{NH_3} - [OH^-]$$

and

$$1.75 \times 10^{-5} = \frac{[OH^-]^2}{0.0750 - [OH^-]}$$

To determine if the denominator can be simplified:

$$\frac{C_B}{K_b} = \frac{0.0750}{1.75 \times 10^{-5}} = 4.29 \times 10^3$$

Since the ratio is greater than 1.0×10^3,

$$0.0750 - [OH^-] \simeq 0.0750$$

and

$$1.75 \times 10^{-5} = \frac{[OH^-]^2}{0.0750}$$

$$[OH^-] = \sqrt{(1.75 \times 10^{-5})(0.0750)} = 1.15 \times 10^{-3} \, M$$

Finally, from the water autoprotolysis equilibrium,

$$K_w = [H_3O^+][OH^-]$$

$$1.00 \times 10^{-14} = [H_3O^+](1.15 \times 10^{-3})$$

$$[H_3O^+] = 8.73 \times 10^{-12} \, M$$

and

$$pH = 11.06$$

Conjugate Acid-Base Pairs

A solution containing a conjugate acid-base pair may be acidic, neutral, or basic, depending on the strengths and concentrations of the acid and base. The general approach used to calculate the pH of a solution containing an acid-base conjugate pair is similar to that used to calculate the pH of a weak acid or a weak base. For a solution of the weak acid HA and the sodium salt of its conjugate base NaA, there are two pertinent equilibria:

$$HA + H_2O \rightleftharpoons H_3O^+ + A^- \tag{6-22}$$

$$A^- + H_2O \rightleftharpoons HA + OH^- \tag{6-23}$$

If we can neglect the ionization of water, Equation 6-22 represents the only source of H_3O^+. To solve the equilibrium equation

$$K_a = \frac{[H_3O^+][A^-]}{[HA]} \tag{6-24}$$

for $[H_3O^+]$, concentrations of HA and A^- must be determined. To be exact, these concentrations must account for the HA and A^- that are formed and consumed in the ionization reactions. Reaction 6-22 causes the concentration of HA to decrease and the concentration of A^- to increase, both by an amount equal to the concentration of H_3O^+. Similarly, Reaction 6-23 increases the concentration of HA and decreases the concentration of A^-, both by an amount equal to the concentration of OH^-. Thus the equilibrium concentrations of HA and A^- are given by

$$[HA] = C_{HA} - [H_3O^+] + [OH^-] \qquad (6\text{-}25)$$

$$[A^-] = C_{NaA} + [H_3O^+] - [OH^-] \qquad (6\text{-}26)$$

Ordinarily, C_{HA} and C_{NaA} are much larger than the difference between the concentration of H_3O^+ and OH^-, and Equations 6-25 and 6-26 simplify to

$$[HA] \simeq C_{HA} \qquad (6\text{-}27)$$

$$[A^-] \simeq C_{NaA} \qquad (6\text{-}28)$$

Substituting these quantities in Equation 6-24 gives an equation that can be solved for $[H_3O^+]$:

$$K_a = \frac{[H_3O^+] C_{NaA}}{C_{HA}} \qquad (6\text{-}29)$$

It is difficult to determine exactly when the assumptions leading to Equation 6-29 break down and produce an unacceptable error. The concentrations of H_3O^+ and OH^- decrease as the acid and base ionization constants K_a and K_b decrease. Consequently, Equations 6-27 and 6-28 are most valid when the analytical concentrations are large and the ionization constants are small. For problems in this book, Equation 6-29 will be accepted as valid if the following conditions are met.

$$C_{HA} \text{ and } C_{NaA} \geqslant 10^{-3}$$

$$K_a \text{ and } K_b \leqslant 10^{-3}$$

If the simplifying assumptions are not allowable and a more accurate solution is required, Equations 6-25 and 6-26 may be substituted directly into Equation 6-24, giving

$$K_a = \frac{[H_3O^+] \left(C_{NaA} + [H_3O^+] - [OH^-] \right)}{C_{HA} - [H_3O^+] + [OH^-]}$$

Replacing $[OH^-]$ with $K_w/[H_3O^+]$ and rearranging gives a complicated but solvable equation:

$$[H_3O^+]^3 + (C_{NaA} + K_a)[H_3O^+]^2 - (K_a C_{HA} + K_w)[H_3O^+] - K_a K_w = 0$$

Example 6-9

What is the pH of a solution prepared by mixing 3.00 g of sodium acetate with 5.00 mL of 12.0 M acetic acid and diluting to 2.00 L? The K_a for acetic acid is 1.76×10^{-5}.

Solution The first question to be asked when substances are mixed is: Do they react? In the laboratory, this question should always be asked *before* mixing substances! The answer here is "no" because acetic acid and acetate ion (from the sodium acetate) are conjugates. The solution after mixing will contain acetic acid, sodium acetate, and water. The analytical concentration of acetic acid and sodium acetate are

$$C_{CH_3CO_2H} = \frac{5.00 \text{ mL} \times 12.0 \text{ mmol/mL}}{2000 \text{ mL}} = 0.0300 \text{ } M$$

$$C_{CH_3CO_2Na} = \frac{3.00 \text{ g}}{82.0 \text{ g/mol} \times 2.00 \text{ L}} = 0.0183 \text{ } M$$

Acetic acid is the major source of H_3O^+,

$$CH_3CO_2H + H_2O \rightleftharpoons H_3O^+ + CH_3CO_2^-$$

for which

$$K_a = \frac{[H_3O^+][CH_3CO_2^-]}{[CH_3CO_2H]}$$

Since

$$C_{CH_3CO_2H} \text{ and } C_{CH_3CO_2Na} > 10^{-3}$$

and

$$K_a \text{ and } K_b < 10^{-3}$$

the following approximations are valid:

$$[CH_3CO_2H] \simeq C_{CH_3CO_2H} = 0.0300 \text{ } M$$

$$[CH_3CO_2^-] \simeq C_{CH_3CO_2Na} = 0.0183 \text{ } M$$

Substituting these values in the equilibrium constant expression gives

$$K_a = 1.76 \times 10^{-5} = \frac{[H_3O^+](0.0183)}{0.0300}$$

$$[H_3O^+] = 2.89 \times 10^{-5} \text{ } M$$

and

$$pH = -\log(2.89 \times 10^{-5}) = 4.540$$

In dealing with a conjugate pair it does not matter which substance is considered the conjugate of the other; that is, in a mixture of NH_4^+ and NH_3, NH_4^+ can be the weak acid and NH_3 its conjugate base or NH_3 can be the weak base and NH_4^+ its conjugate acid. If we had chosen to solve for $[OH^-]$ directly, rather than $[H_3O^+]$, Equation 6-24 would be replaced with the equilibrium equation for Reaction 6-23:

$$K_b = \frac{C_{HA}[OH^-]}{C_{NaA}} \tag{6-30}$$

Example 6-10

Calculate the pH of a solution that is 0.120 M in NH_3 and 0.0750 M in NH_4Cl. The K_b for ammonia is 1.75×10^{-5}.

Solution

Calculating $[H_3O^+]$ *directly.* Since NH_3 and NH_4^+ are conjugates, there is no net reaction. Ammonium ion is an acid and the principal source of H_3O^+ is the reaction

$$NH_4^+ + H_2O \rightleftharpoons H_3O^+ + NH_3$$

for which

$$K_a = \frac{[H_3O^+][NH_3]}{[NH_4^+]}$$

where

$$K_a = \frac{K_w}{K_b} = \frac{1.00 \times 10^{-14}}{1.75 \times 10^{-5}} = 5.71 \times 10^{-10}$$

Since

$$C_{NH_4Cl} \text{ and } C_{NH_3} > 10^{-3}$$

and

$$K_a \text{ and } K_b < 10^{-3}$$

the following approximations are valid:

$$[NH_4^+] \simeq C_{NH_4Cl} = 0.0750 \ M$$

$$[NH_3] \simeq C_{NH_3} = 0.120 \ M$$

Using Equation 6-29, we have

$$5.71 \times 10^{-10} = \frac{[H_3O^+](0.120)}{0.0750}$$

$$[H_3O^+] = 3.57 \times 10^{-10} \ M$$

and

$$pH = -\log (3.57 \times 10^{-10}) = 9.448$$

Calculating $[OH^-]$ *initially.* Ammonia is the base and principal source of OH^-:

$$NH_3 + H_2O \rightleftharpoons NH_4^+ + OH^-$$

for which

$$K_b = \frac{[NH_4^+][OH^-]}{[NH_3]}$$

Since

$$C_{NH_4Cl} \text{ and } C_{NH_3} > 10^{-3}$$

and

$$K_b \text{ and } K_a < 10^{-3}$$

the following approximations are valid:

$$[NH_4^+] \simeq C_{NH_4Cl} = 0.0750 \ M$$

$$[NH_3] \simeq C_{NH_3} = 0.120 \ M$$

Substituting in Equation 6-30 gives us

$$K_b = 1.75 \times 10^{-5} = \frac{(0.0750)[OH^-]}{0.120}$$

$$[OH^-] = 2.80 \times 10^{-5} \ M$$

Converting to $[H_3O^+]$ yields

$$[H_3O^+] = \frac{K_w}{[OH^-]} = \frac{1.00 \times 10^{-14}}{2.80 \times 10^{-5}} = 3.57 \times 10^{-10} \ M$$

and

$$pH = -\log (3.57 \times 10^{-10}) = 9.448$$

pH buffers

A mixture of a weak acid or base and its conjugate is called a *buffer*, and it has the interesting ability to resist changes in pH upon dilution or addition of small amounts of acid or base. This property is important in all areas of science, but especially so in those disciplines concerned with the chemistry of living things. Virtually all living matter depends on enzymes to control and direct the chemical reactions that are fundamental to life itself. The activity of these life-sustaining enzymes is critically dependent on pH, as illustrated in Figure 6-1. Living organisms are constantly coming into contact with acids and bases, and to survive they must be able to control the pH of the cellular solutions where the enzymes are located.

Effect of dilution. The pH of a buffer solution remains essentially independent of dilution until the concentrations of the weak acid and base are decreased to the point where Equations 6-27 and 6-28 are no longer valid. The reason for this constancy is more apparent when Equation 6-24 is viewed in a different form. Solving the equation for $[H_3O^+]$ and taking the negative logarithm of both sides gives

$$-\log [H_3O^+] = -\log K_a + \left(-\log \frac{[HA]}{[A^-]} \right)$$

or

$$pH = pK_a - \log \frac{[HA]}{[A^-]} \qquad (6\text{-}31)$$

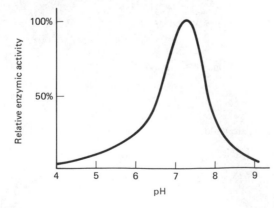

Figure 6-1 Activity of the enzyme ribonuclease A as a function of pH. [From W. Ferdinand, *The Enzyme Molecule* (New York: John Wiley & Sons, 1976), p. 133. Copyright © by John Wiley & Sons Ltd. Reprinted by permission of John Wiley & Sons Ltd.]

Chapter 6   Acid-Base Equilibria

From this equation it is clear that the pH depends on the *ratio* of the concentrations of acid and conjugate base rather than their absolute values. The addition of water to a buffer solution will dilute both HA and A$^-$ equally, leaving their concentration ratio unchanged. This behavior is illustrated in the following example.

Example 6-11

(a) Calculate the pH of a buffer prepared by mixing 0.250 mol of acetic acid with 0.100 mol of sodium acetate and diluting to 1.00 L. (b) Calculate the pH when 10.0 mL of this buffer is diluted to 250 mL with water.

Solution

(a) The dissociation of acetic acid is the major source of H$_3$O$^+$:

$$CH_3CO_2H + H_2O \;\rightleftharpoons\; H_3O^+ + CH_3CO_2^-$$

Using Equation 6-31, we have

$$pH = pK_a - \log \frac{[CH_3CO_2H]}{[CH_3CO_2^-]} \approx pK_a - \log \frac{C_{CH_3CO_2H}}{C_{CH_3CO_2Na}}$$

The pK_a for acetic acid is

$$pK_a = -\log(1.76 \times 10^{-5}) = 4.754$$

Thus

$$pH = 4.754 - \log \frac{0.250 \text{ mol}/1.00 \text{ L}}{0.100 \text{ mol}/1.00 \text{ L}} = 4.356$$

(b) After dilution, the analytical concentrations of CH$_3$CO$_2$H and CH$_3$CO$_2$Na are

$$C_{CH_3CO_2H} = \frac{10.0 \text{ mL} \times 0.250 \; M}{250 \text{ mL}} = 0.0100 \; M$$

$$C_{CH_3CO_2Na} = \frac{10.0 \text{ mL} \times 0.100 \; M}{250 \text{ mL}} = 0.00400 \; M$$

Using Equation 6-31 yields

$$pH = 4.754 - \log \frac{0.0100}{0.00400} = 4.356$$

Effect of addition of acids or bases.   When a strong acid such as HCl is added to water, the quantity of H$_3$O$^+$ is increased by an amount equal to the quantity of the strong acid. When that same strong acid is added to a buffer solution, it is consumed by reaction with the basic component of the buffer. Similarly, a strong base such as NaOH is consumed by reaction with the acidic component of the buffer. If the buffer consists of HA and NaA, the reactions are

$$HCl + NaA \;\longrightarrow\; HA + NaCl$$

$$NaOH + HA \;\longrightarrow\; NaA + H_2O$$

In the process of reacting with the added strong acid or base, the relative concentrations of the buffer components are changed, but the *effect* of these changes is not very large because the pH depends on the logarithm of the *ratio* of the concentrations. For example, consider a buffer consisting of 0.20 M HA and 0.10 M NaA. The pH of this buffer can be calculated using Equation 6-31:

$$pH = pK_a - \log \frac{0.20}{0.10} = pK_a - 0.30 \qquad (6\text{-}32)$$

If sufficient strong base is added to react with half of the HA present, the new concentration of HA will be 0.10 M. The reaction of the strong base with HA will produce some NaA, making its new concentration 0.20 M:

$$HA + NaOH \longrightarrow NaA + H_2O$$

The pH of the new solution is given by

$$pH = pK_a - \log \frac{0.10}{0.20} = pK_a + 0.30 \qquad (6\text{-}33)$$

Subtracting Equation 6-32 from 6-33 gives

$$\Delta pH = pK_a + 0.30 - (pK_a - 0.30) = 0.60$$

which is a relatively small change in pH considering the amount of strong base added. Had the same amount of strong base been added to the same volume of pure water, the pH would have increased from 7.00 to 13.00, a change of 6 units.

Buffer capacity. While the pH of a buffer depends on the ratio of the concentrations of the acid-base conjugate pair, the capacity of a buffer to resist a change in pH depends on both the individual concentrations and the concentration ratio. The *buffer capacity*, β, is defined as the quantity (in equivalents) of strong acid or strong base needed to cause 1.0 L of the buffer to undergo a pH change of 1.0 unit. Since a buffer will resist changes in pH only as long as there is weak acid or weak base left to react, the greater the concentration of buffer components, the greater the buffer capacity. The buffer capacity also increases as the concentration ratio of the acid-base pair approaches unity. It is usually not possible to have a concentration ratio greater than about $10/1$ or less than about $1/10$ and still have a sufficient amount of either buffer component present to react with the added base or acid. If these ratios are accepted as practical limits, Equation 6-31 can be used to determine the useful pH range of a buffer:

$[HA]/[A^-]$	pH *from Equation 6-31*
$1/10$	$pH_{max} = pK_a - \log 1/10 = pK_a + 1$
$10/1$	$pH_{min} = pK_a - \log 10/1 = pK_a - 1$

The useful pH range is $pK_a \pm 1$.

Example 6-12

Calculate the change in pH that occurs when 0.100 mol of solid NaOH is added to 1.00 L of each of the following buffers: (a) $[HCO_2H] = 0.200 \ M$; $[HCO_2^-] = 0.100 \ M$ and (b) $[HCO_2H] = 0.0200 \ M$; $[HCO_2^-] = 0.0100 \ M$.

Solution

(a) The initial pH is calculated using Equation 6-31:

$$pH_{initial} = pK_a - \log \frac{[HCO_2H]}{[HCO_2^-]} = 3.75 - \log \frac{0.200}{0.100} = 3.45$$

When NaOH is added to the buffer, the following reaction takes place,

$$HCO_2H + NaOH \longrightarrow HCO_2Na + H_2O$$

thereby decreasing the concentration of HCO_2H and simultaneously increasing the concentration of HCO_2^-. The new composition of the solution is

amount HCO_2H initial $\quad = 1.00 \text{ L} \times 0.200 \ M \quad = 0.200$ mol

$$-\left\{\begin{array}{c} \text{amount NaOH added} \\ \parallel \\ \text{amount } HCO_2H \text{ reacted} \end{array}\right\} = \text{amount } HCO_2^- \text{ formed} = 0.100 \text{ mol}$$

amount HCO_2H remaining $\qquad\qquad\qquad = \overline{0.100 \text{ mol}}$

The present amount of HCO_2^- is given by

$$\text{amount } HCO_2^- = \text{amount } HCO_2^- \text{ (initial)} + \text{amount } HCO_2^- \text{ (formed)}$$

$$= (1.00 \text{ L} \times 0.100 \ M) + 0.100 = 0.200 \text{ mol}$$

and

$$C_{HCO_2^-} = \frac{0.200 \text{ mol}}{1.00 \text{ L}} = 0.200 \ M$$

$$C_{HCO_2H} = \frac{0.100 \text{ mol}}{1.00 \text{ L}} = 0.100 \ M$$

Using Equation 6-31 to calculate the new pH, we have

$$pH_{final} = 3.75 - \log \frac{0.100}{0.200} = 4.05$$

Then

$$\Delta pH = pH_{final} - pH_{initial} = 4.05 - 3.45 = 0.60$$

(b) The initial pH is calculated using Equation 6-31:

$$pH_{initial} = pK_a - \log \frac{[HCO_2H]}{[HCO_2^-]} = 3.75 - \log \frac{0.0200}{0.0100} = 3.45$$

The composition of the solution after addition of the NaOH is

$$-\left\{\begin{array}{c} \text{amount } HCO_2H \text{ initial} \\ \parallel \\ \text{amount } HCO_2^- \text{ formed} \end{array}\right\} = 1.00 \text{ L} \times 0.0100 \ M = 0.0100 \text{ mol}$$

amount NaOH added $\qquad\qquad\qquad\qquad = 0.100$ mol

amount NaOH remaining $\qquad\qquad\qquad = \overline{0.090 \text{ mol}}$

The solution contains a strong base NaOH and a weak base HCO_2^-. Neglecting the OH^- produced by the weak base yields

$$[OH^-] \simeq C_{NaOH} = \frac{0.090 \text{ mol}}{1.00 \text{ L}} = 0.090 \ M$$

$$pOH = -\log(0.090) = 1.05$$

TABLE 6-3 COMMON pH BUFFERS

Name	pK_a
Phosphoric acid/potassium dihydrogen phosphate	2.15
Citric acid/sodium dihydrogen citrate	3.13
Sodium dihydrogen citrate/disodium hydrogen citrate	4.76
Acetic acid/sodium acetate	4.76
Disodium hydrogen citrate/trisodium citrate	6.40
Potassium dihydrogen phosphate/disodium hydrogen phosphate	7.20
Tris(hydroxymethyl)aminomethane hydrochloride/ tris(hydroxymethyl)aminomethane	8.08
Boric acid/sodium borate	9.23
Ammonium chloride/ammonia	9.25
Sodium bicarbonate/sodium carbonate	10.33
Disodium hydrogen phosphate/trisodium phosphate	12.4

and

$$pH = pK_w - pOH = 14.00 - 1.05 = 12.95$$

Thus,

$$\Delta pH = pH_{final} - pH_{initial} = 12.95 - 3.45 = 9.50$$

Note that the added sodium hydroxide reacted with all of the formic acid present, thereby destroying the buffer and causing a large change in the pH.

Selecting a buffer. Table 6-3 is a list of some common buffers along with the pK_a for the acid component. There are actually as many buffers as there are weak acid–weak base conjugate pairs. The selection of a particular buffer for a given application is based on two considerations: the desired pH and the chemical compatibility of the buffer components with the sample. The first consideration is satisfied by selecting a buffer whose acid component has a pK_a as close as possible to the desired pH. This will permit the concentration ratio of the acid-base pair to be as close to unity as possible, thereby maximizing the buffer capacity. The consideration of chemical compatibility is much more complicated and requires some knowledge of the possible interactions between the buffer components and the constituents in the chemical system being studied.

Preparing a buffer. Equation 6-31 suggests that a buffer of any desired pH can be prepared by combining the calculated quantities of an acid-base conjugate pair. Occasionally, one of the conjugate pair is not available or cannot be weighed easily. In such cases the buffer can be made by combining an excess of the available weak acid or base with an appropriate amount of strong base or acid. The other buffer component is produced in the ensuing acid-base reaction.

Example 6-13

Calculate the pH of the buffer prepared by mixing 500 mL of 0.200 M acetic acid with 1.00 g of sodium hydroxide. The K_a for acetic acid is 1.76×10^{-5}.

Solution When acetic acid and sodium hydroxide are mixed, they react with one another:

$$CH_3CO_2H + NaOH \longrightarrow CH_3CO_2Na + H_2O$$

The first order of business is to determine the approximate composition of the solution after the reaction is complete.

$$\begin{array}{lr} \text{amount CH}_3\text{CO}_2\text{H added} & = 500 \text{ mL} \times 0.200\ M = 100 \text{ mmol} \\ -\begin{cases} \text{amount NaOH added} \\ \qquad \| \\ \text{amount CH}_3\text{CO}_2\text{H used} \\ \qquad \| \\ \text{amount CH}_3\text{CO}_2\text{Na formed} \end{cases} = \dfrac{1000 \text{ mg}}{40.0 \text{ mg/mmol}} & = 25 \text{ mmol} \\ \hline \text{amount CH}_3\text{CO}_2\text{H remaining} & = \overline{75 \text{ mmol}} \end{array}$$

Thus

$$[\text{CH}_3\text{CO}_2\text{H}] \simeq C_{\text{CH}_3\text{CO}_2\text{H}} = \frac{75 \text{ mmol}}{500 \text{ mL}} = 0.15\ M$$

$$[\text{CH}_3\text{CO}_2{}^-] \simeq C_{\text{CH}_3\text{CO}_2\text{Na}} = \frac{25 \text{ mmol}}{500 \text{ mL}} = 0.050\ M$$

and

$$\text{pH} = \text{p}K_a - \log \frac{[\text{CH}_3\text{CO}_2\text{H}]}{[\text{CH}_3\text{CO}_2{}^-]} = 4.75 - \log \frac{0.15}{0.050} = 4.28$$

The measured pH of buffers prepared by combining calculated quantities as described above are often found to differ from the predicted value. These differences, which can be as much as 0.5 pH unit, are due mainly to uncertainties in the values of the equilibrium constants, the fact that activities rather than concentrations should be used in the equilibrium equations, and simplifications used in the calculations. To avoid these discrepancies, buffers often are prepared with the aid of a pH meter (Chapter 13). For example, the buffer described in Example 6-13 could be prepared by placing the monitoring electrode of a pH meter in the 500 mL of acetic acid and then slowly adding the sodium hydroxide until the meter shows the desired pH. Many reference works and handbooks of chemistry contain experimentally derived recipes for the preparation of buffer solutions of known pH.

Polyprotic Acids and Polyequivalent Bases

Calculating the *exact* pH of solutions containing polyprotic acids or polyequivalent bases is very complicated. Fortunately, assumptions often can be made that greatly simplify the necessary calculations. To examine the general methods used, let us consider the five different compositions that comprise a diprotic acid or diequivalent base system: (1) H_2A, (2) $H_2A + HA^-$, (3) HA^-, (4) $HA^- + A^{2-}$, and (5) A^{2-}. In all cases we shall assume that the H_3O^+ and OH^- contributed by water are negligible.

Case 1. A solution containing H_2A. The stepwise ionization of H_2A provides two sources of H_3O^+:

$$H_2A + H_2O \rightleftharpoons H_3O^+ + HA^- \qquad K_{a_1} = \frac{[H_3O^+][HA^-]}{[H_2A]}$$

$$HA^- + H_2O \rightleftharpoons H_3O^+ + A^{2-} \qquad K_{a_2} = \frac{[H_3O^+][A^{2-}]}{[HA^-]}$$

If K_{a_1} is "much larger" than K_{a_2}, the first ionization is the major source of H_3O^+ and the effect of the second ionization on the concentration of H_3O^+ is negligible. The problem becomes one of calculating the pH of a weak acid. The meaning of "much larger" is arbitrary, but for the purpose of solving problems in this book it will mean 100 times larger. Thus, if $K_{a_1}/K_{a_2} \geqslant 100$, the second ionization can be neglected *for the purpose of calculating the pH*. A different value for this ratio is used to determine the feasibility of carrying out a stepwise titration and it is important not to confuse the two values.

Example 6-14

Calculate the pH of a 0.0100 M solution of carbonic acid. The ionization constants are $K_{a_1} = 4.45 \times 10^{-7}$ and $K_{a_2} = 4.69 \times 10^{-11}$.

Solution Since the ratio of K_{a_1}/K_{a_2} is greater than 10^2,

$$\frac{K_{a_1}}{K_{a_2}} = \frac{4.45 \times 10^{-7}}{4.69 \times 10^{-11}} = 9.49 \times 10^3$$

$$9.49 \times 10^3 > 10^2$$

only the first ionization needs to be considered. Thus

$$H_2CO_3 + H_2O \ \rightleftharpoons \ H_3O^+ + HCO_3^-$$

for which

$$K_{a_1} = \frac{[H_3O^+][HCO_3^-]}{[H_2CO_3]} \simeq \frac{[H_3O^+]^2}{C_{H_2CO_3}}$$

$$4.45 \times 10^{-7} = \frac{[H_3O^+]^2}{0.0100}$$

$$[H_3O^+] = 6.67 \times 10^{-5} \ M$$

and

$$pH = 4.176$$

Case 2. A solution containing $H_2A + HA^-$. This is very similar to case 1: if $K_{a_1}/K_{a_2} \geqslant 100$, the second ionization can be neglected and the problem becomes one of calculating the pH of a weak acid in the presence of its conjugate base—a buffer problem.

Example 6-15

Calculate the pH of a solution containing 0.100 M o-phthalic acid and 0.250 M potassium hydrogen o-phthalate. For $C_6H_4(CO_2H)_2$: $K_{a_1} = 1.12 \times 10^{-3}$ and $K_{a_2} = 3.91 \times 10^{-6}$.

Solution Since the ratio of K_{a_1}/K_{a_2} exceeds 10^2,

$$\frac{K_{a_1}}{K_{a_2}} = \frac{1.12 \times 10^{-3}}{3.91 \times 10^{-6}} = 286$$

$$286 > 100$$

the second ionization can be neglected. Using H_2P to represent o-phthalic acid,

$$H_2P + H_2O \ \rightleftharpoons \ H_3O^+ + HP^-$$

for which

$$K_{a_1} = \frac{[H_3O^+][HP^-]}{[H_2P]} \simeq \frac{[H_3O^+]\,C_{KHP}}{C_{H_2P}}$$

$$1.12 \times 10^{-3} = \frac{[H_3O^+](0.250)}{(0.100)}$$

$$[H_3O^+] = 4.48 \times 10^{-4}\ M$$

and

$$pH = 3.349$$

Case 3. A solution containing HA^-. Substances such as HA^- that exhibit both acidic and basic character when dissolved in water are called *amphiprotic* substances. When a salt such as NaHA is dissolved in water, it dissociates completely into Na^+ and HA^-. The HA^- can undergo acid ionization,

$$HA^- + H_2O \;\rightleftharpoons\; H_3O^+ + A^{2-} \qquad K_{a_2} = \frac{[H_3O^+][A^{2-}]}{[HA^-]} \tag{6-34}$$

and base ionization,

$$HA^- + H_2O \;\rightleftharpoons\; H_2A + OH^- \qquad K_{b_2} = \frac{K_w}{K_{a_1}} = \frac{[H_2A][OH^-]}{[HA^-]} \tag{6-35}$$

Whether the solution is acidic or basic is determined by the relative magnitude of the equilibrium constants for Reactions 6-34 and 6-35. If K_{a_2} is greater than K_{b_2}, the solution will be acidic; if it is less than K_{b_2}, the solution will be basic.

It is important to realize that the equilibrium represented by Reactions 6-34 and 6-35 occur *simultaneously* and both must be considered in deriving an equation for calculating the concentration of H_3O^+. To solve the K_{a_2} expression for $[H_3O^+]$, a suitable substitution for the concentration of A^{2-} must be found. Considering that some of the H_3O^+ formed in Reaction 6-34 will be lost by reaction with the OH^- formed in Reaction 6-35, we can write

$$[H_3O^+] = [H_3O^+]_{formed} - [H_3O^+]_{lost} \tag{6-36}$$

But

$$[H_3O^+]_{formed} = [A^{2-}]$$

$$[H_3O^+]_{lost} = [OH^-]_{formed} = [H_2A]$$

Substituting these expressions in Equation 6-36 gives

$$[H_3O^+] = [A^{2-}] - [H_2A]$$

or

$$[A^{2-}] = [H_3O^+] + [H_2A] \tag{6-37}$$

Calculation of pH

Solving the K_{a_1} expression for $[H_2A]$ and substituting the result in Equation 6-37 gives

$$[A^{2-}] = [H_3O^+] + \frac{[H_3O^+][HA^-]}{K_{a_1}} \qquad (6\text{-}38)$$

When Equation 6-38 is substituted for $[A^{2-}]$ in the K_{a_2} equilibrium expression (Equation 6-34), we obtain

$$K_{a_2} = \frac{[H_3O^+]\left([H_3O^+] + \dfrac{[H_3O^+][HA^-]}{K_{a_1}}\right)}{[HA^-]}$$

which can be rearranged to give the complicated-looking equation

$$[H_3O^+]^2 = \frac{K_{a_1}K_{a_2}[HA^-]}{K_{a_1} + [HA^-]} \qquad (6\text{-}39)$$

It is frequently the case that

$$K_{a_1} \ll [HA^-]$$

which is another way of saying that

$$K_{a_1} + [HA^-] \simeq [HA^-] \qquad (6\text{-}40)$$

Under such conditions Equation 6-39 reduces to

$$[H_3O^+]^2 = \frac{K_{a_1}K_{a_2}[HA^-]}{[HA^-]} = K_{a_1}K_{a_2}$$

and

$$[H_3O^+] = \sqrt{K_{a_1}K_{a_2}} \quad \text{or} \quad pH = \frac{K_{a_1} + pK_{a_2}}{2} \qquad (6\text{-}41)$$

It is interesting to note that as long as the assumption shown by Equation 6-40 remains valid and the ionization of water has a negligible effect, the pH is independent of the concentration of HA^-. An equimolar mixture of a weak acid and a weak base, that are not conjugates, is equivalent to an amphiprotic substance such as HA^-, and the pH of such a solution can be calculated using the following equation, which is virtually the same as Equation 6-41:

$$[H_3O^+] = \sqrt{K_{a(1)}K_{a(2)}}$$

where $K_{a(1)}$ is the dissociation constant for the weak acid and $K_{a(2)}$ is the dissociation constant for the conjugate acid of the weak base.

A triprotic acid such as H_3PO_4 can give rise to two amphiprotic substances: $H_2PO_4^-$ and HPO_4^{2-}. The only problem with calculating the pH of a solution of either of these is to select the correct K_a values for use with Equation 6-41. The following rule applies: Use the K_a of the substance itself plus the K_a of its *conjugate* acid. Thus

For $H_2PO_4^-$: $[H_3O^+] = \sqrt{K_{a_1}K_{a_2}}$

For HPO_4^{2-}: $[H_3O^+] = \sqrt{K_{a_2}K_{a_3}}$

Chapter 6 Acid-Base Equilibria

Example 6-16

Calculate the pH of a 0.0250 M solution of sodium bicarbonate. For H_2CO_3, $K_{a_1} = 4.45 \times 10^{-7}$ and $K_{a_2} = 4.69 \times 10^{-11}$.

Solution Since K_{a_1} is much smaller than the concentration of HA^-, Equation 6-41 is valid, and

$$[H_3O^+] = \sqrt{(4.45 \times 10^{-7})(4.69 \times 10^{-11})} = 4.57 \times 10^{-9} \ M$$

$$pH = 8.340$$

Case 4. A solution containing $HA^- + A^{2-}$. This solution is the base equivalent of case 2. If $K_{b_1}/K_{b_2} \geq 100$, the second ionization may be neglected and the problem reduces to finding the pH of a weak base in the presence of its conjugate acid.

Case 5. A solution containing A^{2-}. This solution is the base equivalent of case 1 and $[OH^-]$ is calculated from the K_{b_1} equilibrium expression in the same way that $[H_3O^+]$ was calculated from the K_{a_1} equilibrium expression.

Example 6-17

Calculate the pH of a 0.150 M sodium oxalate solution. For $H_2C_2O_4$, $K_{a_1} = 5.60 \times 10^{-2}$ and $K_{a_2} = 5.42 \times 10^{-5}$.

Solution First determine the base dissociation constants:

$$K_{b_1} = \frac{K_w}{K_{a_2}} = \frac{1.00 \times 10^{-14}}{5.42 \times 10^{-5}} = 1.85 \times 10^{-10}$$

$$K_{b_2} = \frac{K_w}{K_{a_1}} = \frac{1.00 \times 10^{-14}}{5.60 \times 10^{-2}} = 1.79 \times 10^{-13}$$

Since the ratio of ionization constants exceeds 10^2,

$$\frac{K_{b_1}}{K_{b_2}} = \frac{1.85 \times 10^{-10}}{1.79 \times 10^{-13}} = 1.03 \times 10^3$$

$$1.03 \times 10^3 > 10^2$$

only the first ionization must be considered. Thus

$$C_2O_4^{2-} + H_2O \rightleftharpoons HC_2O_4^- + OH^-$$

for which

$$K_{b_1} = \frac{[HC_2O_4^-][OH^-]}{[C_2O_4^{2-}]} \approx \frac{[OH^-]^2}{C_{Na_2C_2O_4}}$$

$$1.85 \times 10^{-10} = \frac{[OH^-]^2}{0.150}$$

$$[OH^-] = 5.27 \times 10^{-6} \ M$$

and

$$pOH = 5.28$$

Finally,

$$pH = pK_w - pOH = 14.00 - 5.28 = 8.72$$

Calculation of pH

Composition as a function of pH

It is often necessary to know what effect the pH will have on the composition of a polyprotic acid or polyequivalent base solution. As you will see in Chapter 9, many complexing agents are also polyequivalent bases and their ability to react with metal ions depends on the pH. One easy way to see the effect of pH on the composition of a polyprotic acid is in terms of the *relative* concentrations of the various possible species. A weak diprotic acid, H_2A, can exist in three different forms (or degrees of protonation): H_2A, HA^-, and A^{2-}. The relative concentration of each, defined here as α, is its concentration divided by the total concentration:

$$\alpha_{H_2A} = \frac{[H_2A]}{[H_2A] + [HA^-] + [A^{2-}]}$$

$$\alpha_{HA^-} = \frac{[HA^-]}{[H_2A] + [HA^-] + [A^{2-}]}$$

$$\alpha_{A^{2-}} = \frac{[A^{2-}]}{[H_2A] + [HA^-] + [A^{2-}]}$$

Since the sum of the fractions must equal unity,

$$\alpha_{H_2A} + \alpha_{HA^-} + \alpha_{A^{2-}} = 1$$

Each α can be expressed in terms of the equilibrium constants, K_{a_1} and K_{a_2}, and the concentration of H_3O^+. For example, the concentrations of HA^- and A^{2-} from the K_{a_1} and K_{a_2} equilibrium expressions are

$$[HA^-] = \frac{K_{a_1}[H_2A]}{[H_3O^+]} \tag{6-42}$$

$$[A^{2-}] = \frac{K_{a_2}[HA^-]}{[H_3O^+]} \tag{6-43}$$

Substituting Equation 6-42 for $[HA^-]$ in Equation 6-43 gives

$$[A^{2-}] = \frac{K_{a_2}K_{a_1}[H_2A]}{[H_3O^+]^2} \tag{6-44}$$

When Equations 6-42 and 6-44 are substituted in the equation defining α_{H_2A}, the following equation is obtained:

$$\alpha_{H_2A} = \frac{[H_2A]}{[H_2A] + \dfrac{K_{a_1}[H_2A]}{[H_3O^+]} + \dfrac{K_{a_1}K_{a_2}[H_2A]}{[H_3O^+]^2}}$$

which can be rearranged and simplified to give

$$\alpha_{H_2A} = \frac{[H_3O^+]^2}{[H_3O^+]^2 + K_{a_1}[H_3O^+] + K_{a_1}K_{a_2}} \tag{6-45}$$

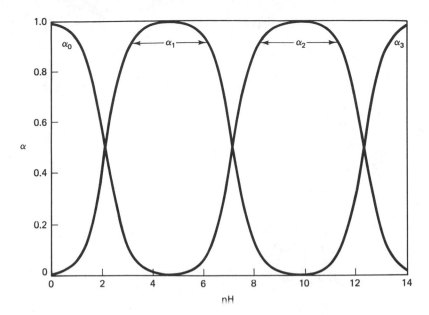

Figure 6-2   Composition of a phosphate solution as a function of pH.

In an analogous fashion, it can be shown that

$$\alpha_{HA^-} = \frac{K_{a_1}[H_3O^+]}{[H_3O^+]^2 + K_{a_1}[H_3O^+] + K_{a_1}K_{a_2}} \qquad (6\text{-}46)$$

$$\alpha_{A^{2-}} = \frac{K_{a_1}K_{a_2}}{[H_3O^+]^2 + K_{a_1}[H_3O^+] + K_{a_1}K_{a_2}} \qquad (6\text{-}47)$$

Note that Equations 6-45, 6-46, and 6-47 all have the same denominator, making them somewhat easier to remember. More important, the equations indicate that for a *given* polyprotic acid, the fractional amount of each species present depends only on the concentration of H_3O^+. Figure 6-2 illustrates the common practice of plotting all α-values on the same graph, in order to show at a glance how the solution composition changes with pH. The general equations for the various forms of any polyprotic acid are

$$\alpha_{H_nA} = \frac{[H_3O^+]^n}{[H_3O^+]^n + K_{a_1}[H_3O^+]^{n-1} + K_{a_1}K_{a_2}[H_3O^+]^{n-2} + \cdots + K_{a_1}K_{a_2}\cdots K_{a_n}} \qquad (6\text{-}48)$$

$$\alpha_{H_{n-1}A^-} = \frac{K_{a_2}[H_3O^+]^{n-1}}{D} \qquad (6\text{-}49)$$

Calculation of pH

$$\alpha_{H_{n-2}A^{2-}} = \frac{K_{a_1}K_{a_2}[H_3O^+]^{n-2}}{D} \qquad (6\text{-}50)$$

$$\vdots \qquad \vdots$$

$$\alpha_{A^{n-}} = \frac{K_{a_1}K_{a_2}\cdots K_{a_n}}{D} \qquad (6\text{-}51)$$

where D is the denominator in Equation 6-48.

PROBLEMS

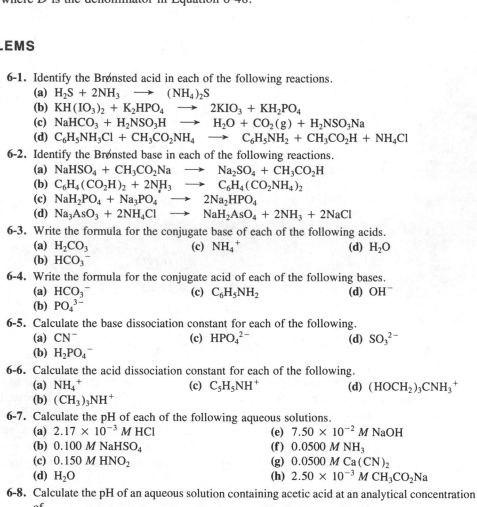

6-1. Identify the Brønsted acid in each of the following reactions.
  (a) $H_2S + 2NH_3 \longrightarrow (NH_4)_2S$
  (b) $KH(IO_3)_2 + K_2HPO_4 \longrightarrow 2KIO_3 + KH_2PO_4$
  (c) $NaHCO_3 + H_2NSO_3H \longrightarrow H_2O + CO_2(g) + H_2NSO_3Na$
  (d) $C_6H_5NH_3Cl + CH_3CO_2NH_4 \longrightarrow C_6H_5NH_2 + CH_3CO_2H + NH_4Cl$

6-2. Identify the Brønsted base in each of the following reactions.
  (a) $NaHSO_4 + CH_3CO_2Na \longrightarrow Na_2SO_4 + CH_3CO_2H$
  (b) $C_6H_4(CO_2H)_2 + 2NH_3 \longrightarrow C_6H_4(CO_2NH_4)_2$
  (c) $NaH_2PO_4 + Na_3PO_4 \longrightarrow 2Na_2HPO_4$
  (d) $Na_3AsO_3 + 2NH_4Cl \longrightarrow NaH_2AsO_4 + 2NH_3 + 2NaCl$

6-3. Write the formula for the conjugate base of each of the following acids.
  (a) H_2CO_3 (c) NH_4^+ (d) H_2O
  (b) HCO_3^-

6-4. Write the formula for the conjugate acid of each of the following bases.
  (a) HCO_3^- (c) $C_6H_5NH_2$ (d) OH^-
  (b) PO_4^{3-}

6-5. Calculate the base dissociation constant for each of the following.
  (a) CN^- (c) HPO_4^{2-} (d) SO_3^{2-}
  (b) $H_2PO_4^-$

6-6. Calculate the acid dissociation constant for each of the following.
  (a) NH_4^+ (c) $C_5H_5NH^+$ (d) $(HOCH_2)_3CNH_3^+$
  (b) $(CH_3)_3NH^+$

6-7. Calculate the pH of each of the following aqueous solutions.
  (a) $2.17 \times 10^{-3}\ M$ HCl   (e) $7.50 \times 10^{-2}\ M$ NaOH
  (b) $0.100\ M$ NaHSO$_4$ (f) $0.0500\ M$ NH$_3$
  (c) $0.150\ M$ HNO$_2$ (g) $0.0500\ M$ Ca(CN)$_2$
  (d) H_2O (h) $2.50 \times 10^{-3}\ M$ CH$_3$CO$_2$Na

6-8. Calculate the pH of an aqueous solution containing acetic acid at an analytical concentration of
  (a) $3.00 \times 10^{-1}\ M$ (b) $3.00 \times 10^{-10}\ M$

6-9. What is the minimum analytical concentration of salicylic acid for which a quadratic equation does not have to be used in solving the equilibrium expression for the hydrogen-ion concentration?

6.10. Calculate the pH of the solution produced when 25.0 mL of 0.100 M NaOH are mixed with 50.0 mL of each of the following.
(a) water   (c) 0.150 M HCl   (e) 0.0100 M H_2SO_4
(b) 0.100 M NaOH   (d) 0.0500 M HCl   (f) 0.200 M CH_3CO_2H

6-11. Calculate the pH of the solution produced when 20.0 mL of 0.125 M HCl are mixed with 25.0 mL of each of the following.
(a) 0.0500 M $Ba(OH)_2$ (c) 0.0800 M Na_2CO_3 (e) 0.0800 M NaCl
(b) 0.120 M NH_3 (d) 0.200 M $HClO_4$ (f) water

6-12. Calculate the pH of the following buffers.
(a) 0.100 M CH_3CO_2H + 0.100 M $(CH_3CO_2)_2Ca$
(b) 0.0500 M KH_2PO_4 + 0.250 M Na_2HPO_4
(c) 0.250 M NH_3 + 0.150 M NH_4Cl
(d) 0.0400 M Tris + 0.0600 M Tris · HCl [Tris = tris(hydroxymethyl)aminomethane]

6-13. What weight of the first component listed must be mixed with 10.0 mL of a 5.00 M solution of the second component to make 500 mL of each of the following buffers?
(a) $NaHCO_3$ + Na_2CO_3 at pH 10.000
(b) HCO_2H + HCO_2Na at pH 4.300
(c) $CHCl_2CO_2H$ + $CHCl_2CO_2Na$ at pH 2.200

6-14. Calculate the volume of 2.50 M base (first component listed) that must be mixed with 5.00 g of conjugate acid (second component listed) to prepare 1.00 L of each of the following buffers.
(a) NH_3 + NH_4NO_3 at pH 8.700   (c) NaH_2BO_3 + H_3BO_3 at pH 10.000
(b) $HONH_2$ + $HONH_3Cl$ at pH 6.000

6-15. For each of the following, calculate the molar ratio of acid to conjugate base necessary to produce a pH of 5.00.
(a) acetic acid/sodium acetate
(b) benzoic acid/sodium benzoate
(c) phenylacetic acid/sodium phenylacetate
(d) sodium hydrogen oxalate/disodium oxalate

6-16. For each of the following, calculate the molar ratio of base to conjugate acid that would produce a pH of 9.00.
(a) ammonia/ammonium nitrate
(b) methyl amine/methyl ammonium chloride
(c) sodium dihydrogen borate/boric acid
(d) sodium sulfite/sodium hydrogen sulfite

6-17. A 0.0722 M solution of a weak acid has a pH of 3.11. Calculate the K_a of this acid.

6-18. A 0.250 M solution of a weak base has a pH of 9.91. Calculate the K_b of this base.

6-19. A solution prepared by dissolving 2.25 g of HA (MW = 51.0) and 4.00 g of NaA in 500 mL of solution has a pH of 6.02. Calculate the K_a of the acid.

6-20. Calculate the change in pH that occurs when 1.00 g of NaOH is added to 250 mL of each of the following solutions.
(a) pure H_2O (c) 0.100 M NaOH   (e) 0.100 M $HClO_4$
(b) 0.0200 M NH_3 (d) 0.250 M HCl

6-21. Calculate the change in pH that occurs when 1.00 g of NaOH is added to 250 mL of each of the following buffers.
(a) 0.200 M CH_3CO_2H + 0.200 M CH_3CO_2Na
(b) 0.250 M Na_2HPO_4 + 0.100 M Na_3PO_4

(c) $0.150\ M\ H_2C_2O_4 + 0.0500\ M\ NaHC_2O_4$

(d) $0.100\ M\ NH_4Cl + 0.100\ M\ NH_3$

(e) $0.0500\ M\ NaHCO_3 + 0.0250\ M\ Na_2CO_3$

6-22. Calculate the buffer capacity, β, in terms of addition of HCl, of each of the following buffers.

(a) $0.150\ M$ lactic acid $+ 0.250\ M$ sodium lactate

(b) $0.0150\ M$ lactic acid $+ 0.0250\ M$ sodium lactate

(c) $0.150\ M$ lactic acid $+ 0.150\ M$ sodium lactate

6-23. Suggest a conjugate acid–base pair that could be used to prepare a buffer at each of the following pH values.

(a) 0.75   (c) 5.00   (e) 9.50

(b) 2.00   (d) 7.00   (f) 12.50

6-24. Calculate the pH of the buffer prepared by mixing each of the following with 5.00 mL of 6.00 M HCl and diluting to 500 mL with water.

(a) 5.00 g of sodium acetate

(b) 8.00 g of sodium tartrate

(c) 7.50 g of disodium hydrogen phosphate

(d) 65.0 mmol of NH_3

6-25. Calculate the pH of the buffer prepared by mixing 2.00 g of NaOH with each of the following and diluting to 500 mL with water.

(a) 12.0 g of NH_4Cl

(b) 7.50 g of disodium hydrogen phosphate

(c) 25.0 mL of 3.00 M chloroacetic acid

(d) 80.0 mmol of salicylic acid

6-26. What volume of 6.00 M HCl must be added to 15.0 g of Na_3PO_4 to yield a solution that has a pH of 2.500 when diluted to 1.00 L with water?

6-27. Calculate the pH of a solution prepared by mixing 15.0 g of $H_2C_2O_4$ with 5.00 g of NaOH and diluting to 250 mL with water?

6-28. Calculate the pH of each of the following.

(a) $0.100\ M\ H_3PO_4$ (c) $0.100\ M\ Na_2HPO_4$ (d) $0.100\ M\ Na_3PO_4$

(b) $0.100\ M\ NaH_2PO_4$

6-29. The acid dissociation constants for sulfurous acid are: $K_{a_1} = 1.2 \times 10^{-2}$ and $K_{a_2} = 6.6 \times 10^{-8}$.

(a) Calculate the pH of a solution of $0.100\ M\ H_2SO_3$.

(b) Calculate the pH of a solution of $0.100\ M\ Na_2SO_3$.

(c) Write the reaction that occurs when H_2SO_3 and Na_2SO_3 are mixed.

(d) Calculate the pH of the solution resulting when equal volumes of the solutions described in parts (a) and (b) are mixed.

6-30. A solution of sodium o-phthalate is adjusted to pH 5.00 with hydrochloric acid. Calculate the fraction of the phthalate that exists as H_2P, HP^-, and P^{2-} (where P = phthalate).

7

ACID-BASE TITRATIONS

Acid-base titrations are used routinely in virtually all fields of chemistry and in related areas, such as biology, pharmacy, medicine, and geology. In addition to common inorganic substances, thousands of organic compounds exhibit sufficient acidity or basicity that they can be determined by titration. Our excellent understanding of acid-base properties, together with the relative ease, speed, and low cost of performing titrations, are major factors contributing to the popularity of acid-base titration methods. For most applications, water is a convenient and suitable solvent. In those instances where the analyte is too weak to be titrated in water, a nonaqueous solvent may be used to enhance the acidity or basicity to a degree that permits its titration. Such nonaqueous titrations are discussed in Chapter 8.

A successful titration depends on the availability of a suitable titrant, a fast and quantitative reaction, and a means of estimating the equivalence point. The shape of a titration curve is critical in determining the success or failure of a titration. In this chapter we discuss various factors that influence the shapes of titration curves and correlate the equilibrium theory presented in Chapter 6 with specific acid-base methodology.

TITRATION CURVES

The course of an acid-base titration is best followed by examining the pH as the titration progresses; that is, plotting pH versus volume of titrant added. The data for such a plot can be measured experimentally with a pH electrode and meter (Chapter 13) or it can be calculated using equations derived from the principles of equilibrium. The ability to *calculate* titration curves is significant because it permits the chemist to study the effects of many parameters, to determine the conditions necessary for an acceptable titration, and to choose proper titrants and indicators without having to resort to laboratory trial-and-error procedures. All of the pH calculations necessary to construct a theoretical titration curve have been discussed in Chapter 6. One of the main objectives of this chapter is to illustrate how it is decided which pH calculations should be used in a specific situation. Also, some of the common reagents and methods used in acid-base titrimetry will be discussed.

Strong Acid Titrated with Strong Base

Consider HCl as a typical strong acid being titrated with the strong base, NaOH. The titration reaction is

$$HCl + NaOH \longrightarrow NaCl + H_2O$$

or, written in terms of the participating ions,

$$H^+ + OH^- \longrightarrow H_2O \tag{7-1}$$

Reaction 7-1 is simply the reverse of the water dissociation reaction, and therefore has an equilibrium constant of $1/K_w$ or 1.0×10^{14}. With such a large equilibrium constant, it is possible to say that Reaction 7-1 goes to completion and any amount of NaOH added prior to the equivalence point will consume a stoichiometric amount of HCl.

The specific equations used to calculate the hydrogen-ion concentration depend on the composition of the solution, which in turn depends on the stage or region of the titration. A strong acid-strong base titration can be divided into three distinct regions:

Region	Major constituents	Comments
1. Before the equivalence point	HCl + NaCl	Treat as strong acid
2. At the equivalence point	NaCl	Treat as pure solvent
3. After the equivalence point	NaCl + NaOH	Treat as strong base

Let us consider each region separately, concentrating on how to determine the approximate composition of the solution.

Region 1: Before the equivalence point. Since the reaction stoichiometry is $1:1$ and the equilibrium constant is large, the addition of any NaOH consumes an equal amount of HCl.

$$
\begin{array}{l}
\text{amount HCl initial} \\
-\left\{
\begin{array}{l}
\text{amount NaOH added} \\
\quad\quad \| \\
\text{amount HCl reacted}
\end{array}
\right. \\
\overline{\text{amount HCl remaining}}
\end{array}
$$

where the amount is in moles or millimoles. The analytical concentration of HCl can be determined by dividing the amount by the new volume of the solution. Since HCl is a strong acid,

$$[H_3O^+] = C_{HCl}$$

and the calculation is finished.

Region 2: At the equivalence point. The product of the titration, NaCl, is a neutral salt and the H_3O^+ results only from the ionization of water. Thus

$$[H_3O^+] = \sqrt{K_w} = 1.00 \times 10^{-7}\ M$$

$$pH = 7.000$$

Resist the temptation to conclude that the pH is 7.000 at the equivalence point of *every* acid-base titration—it is not.

Region 3: After the equivalence point. The amount of NaOH in the solution is determined by comparing the amount of HCl present initially with the amount of NaOH added:

$$\begin{array}{c} \text{amount NaOH added} \\ -\left\{ \begin{array}{c} \text{amount HCl initial} \\ \| \\ \text{amount NaOH reacted} \end{array} \right\} \\ \hline \text{amount NaOH remaining} \end{array}$$

Again, the concentration of NaOH is determined by dividing the amount by the new volume. The problem is now reduced to simply calculating the pH of a strong base solution.

Occasions will arise when you are uncertain as to whether the equivalence point has yet been reached or exceeded. A simple mathematical comparison of the amount of analyte present initially with the amount of titrant added should quickly clear up any uncertainty. One word of caution: Remember that not every titration will have a 1 : 1 reaction stoichiometry! We will learn how to deal with such a situation in a later section on polyprotic acids and polyequivalent bases.

Example 7-1

Calculate the pH of the solution when 25.0 mL of 0.0920 *M* HCl has been titrated with 0, 15.0, 23.0, and 30.0 mL of 0.100 *M* NaOH.

Solution

Before the addition of any titrant. Since HCl is a strong acid:

$$[H_3O^+] \simeq C_{HCl} = 0.0920\ M$$

$$pH = 1.036$$

After 15.0 mL of NaOH *added.* The titration reaction is

$$HCl + NaOH \longrightarrow NaCl + H_2O$$

Comparing the amount of HCl present initially with the amount of NaOH added,

$$\begin{array}{ll} \text{amount HCl initial} & = 25.0\ \text{mL} \times 0.0920\ M = 2.30\ \text{mmol} \\ -\left\{ \begin{array}{c} \text{amount NaOH added} \\ \| \\ \text{amount HCl used} \end{array} \right\} & = 15.0\ \text{mL} \times 0.100\ M\ \ = 1.50\ \text{mmol} \\ \hline \text{amount HCl remaining} & = \overline{0.80\ \text{mmol}} \end{array}$$

Since HCl is the major source of H_3O^+,

$$[H_3O^+] \simeq C_{HCl} = \frac{0.80\ \text{mmol}}{40.0\ \text{mL}} = 0.020\ M$$

and

$$pH = 1.70$$

After 23.0 mL of NaOH *added.* Again, comparing the amount of analyte present initially with the amount of titrant added:

amount HCl initial = 25.0 mL × 0.0920 M = 2.30 mmol
amount NaOH added = 23.0 mL × 0.100 M = 2.30 mmol

Since neither analyte nor titrant is left over, we are at the equivalence point and the flask contains only the products of the reaction, NaCl and H_2O. Since water is the major supplier of both H_3O^+ and OH^-,

$$2H_2O \rightleftharpoons H_3O^+ + OH^-$$

$$K_w = [H_3O^+][OH^-] = [H_3O^+]^2$$

$$[H_3O^+] = \sqrt{K_w} = \sqrt{1.00 \times 10^{-14}} = 1.00 \times 10^{-7} M$$

$$pH = 7.000$$

After 30.0 mL of NaOH *added.*

$$-\begin{Bmatrix} \text{amount HCl initial} \\ \parallel \\ \text{amount NaOH used} \end{Bmatrix} = 25.0 \text{ mL} \times 0.0920 \ M = 2.30 \text{ mmol}$$

$$\frac{\text{amount NaOH added}}{\text{amount NaOH remaining}} = \frac{30.0 \text{ mL} \times 0.100 \ M}{} = \frac{3.00 \text{ mmol}}{0.70 \text{ mmol}}$$

The hydroxide-ion concentration is determined by the excess NaOH:

$$[OH^-] \simeq C_{NaOH} = \frac{0.70 \text{ mmol}}{55.0 \text{ mL}} = 0.013 \ M$$

$$pOH = 1.89$$

From Equation 6-5,

$$pH = 14.00 - 1.89 = 12.11$$

Titration curve and effect of concentration

The concentration of both analyte and titrant affect the shape of the titration curve, predominately in terms of the magnitude of the pH change near the equivalence point, as shown in Figure 7-1. The pH *before* the equivalence point is determined by the concentration of untitrated strong acid: For more dilute acids, the pH is larger (higher on graph). The pH *after* the equivalence point is determined by the concentration of excess strong base titrant: For more dilute bases, the pH is smaller (lower on the graph). The combination of these two effects diminishes the size of the equivalence-point break. Obviously, there is a point where the break will not be large enough for a chemical indicator to function properly. In general, the steeply rising portion of the pH curve must exceed 2 pH units for an indicator to function well. This limits the smallest concentration of strong acid that can be titrated to about $5 \times 10^{-4} M$.

Clearly, it is an advantage to use as concentrated a titrant as possible. However, one must remember that the *volume* of titrant must be accurately measured in a titration. The titration of 25 mL of $5 \times 10^{-4} M$ HCl will require only 0.125 mL of a 0.1 M NaOH

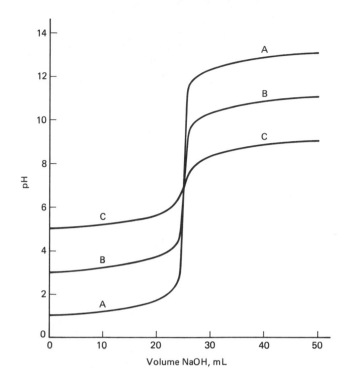

Figure 7-1 Effect of concentration on the titration curve for a strong acid.

A: 25.0 mL of 1.0×10^{-1} *M* HCl titrated with 1.0×10^{-1} *M* NaOH.

B: 25.0 mL of 1.0×10^{-3} *M* HCl titrated with 1.0×10^{-3} *M* NaOH.

C: 25.0 mL of 1.0×10^{-5} *M* HCl titrated with 1.0×10^{-5} *M* NaOH.

titrant to reach the equivalence point! The relative error involved in measuring such a small volume with an ordinary buret will be quite large.

Strong Base Titrated with Strong Acid

The curve for the titration of a strong base is derived in an analogous manner to that for a strong acid. The solution is basic before the equivalence point and acidic afterward. If the titration curve is plotted using pOH, it will have the same general shape as those in Figure 7-1. It is common practice, however, to plot all acid-base titration curves in terms of pH. Under these circumstances, the curve has an inverse shape, as shown in Figure 7-2. The effect of concentration on the shape of the titration curve is the same as it was for a strong acid being titrated with a strong base.

Weak Acid Titrated with Strong Base

Acetic acid titrated with hydroxide is a typical example of a weak acid-strong base titration. The titration reaction is

$$CH_3CO_2H + OH^- \longrightarrow CH_3CO_2^- + H_2O \qquad (7\text{-}2)$$

Reaction 7-2 is simply the reverse of the ionization reaction for the base, $CH_3CO_2^-$. Consequently, the equilibrium constant is $1/K_b$ or 1.76×10^9. As is the case with every successful titration, the equilibrium constant is so large that any amount of titrant added prior to the equivalence point will consume a stoichiometric amount of analyte.

Titration Curves

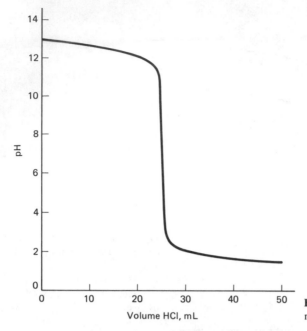

Figure 7-2 Curve for the titration of 25.0 mL of 0.100 M NaOH with 0.100 M HCl.

A weak acid–strong base titration can be divided into four regions, each corresponding to a different solution composition:

Region	Major constituents	Comment
1. Before the addition of titrant	CH_3CO_2H	Treat as weak acid
2. Before the equivalence point	$CH_3CO_2H + CH_3CO_2Na$	Treat as buffer
3. At the equivalence point	CH_3CO_2Na	Treat as weak base
4. After the equivalence point	$CH_3CO_2Na + NaOH$	Treat as strong base

As was done in the preceding section on strong acid–strong base titrations, we will examine each region separately, concentrating on how to determine the composition of the solution.

Region 1: Before the addition of titrant. The solution contains a weak acid plus solvent and the $[H_3O^+]$ is calculated using Equation 6-21:

$$[H_3O^+] = \sqrt{K_a[CH_3CO_2H]} \simeq \sqrt{K_a C_{CH_3CO_2H}}$$

Region 2: Before the equivalence point. The solution contains untitrated acetic acid together with the titration product sodium acetate. Since these two substances are acid-base conjugates, we have a buffer solution and Equation 6-29 is used to calculate the concentration of H_3O^+:

$$K_a = \frac{[H_3O^+][CH_3CO_2^-]}{[CH_3CO_2H]} \simeq \frac{[H_3O^+] C_{CH_3CO_2Na}}{C_{CH_3CO_2H}}$$

The analytical concentrations of CH_3CO_2Na and CH_3CO_2H are found by comparing the amount of titrant added with the amount of acid present initially,

$$\underbrace{\text{amount } CH_3CO_2H \text{ initial} - \left\{ \begin{array}{c} \text{amount NaOH added} \\ \| \\ \text{amount } CH_3CO_2H \text{ used} \end{array} \right\}}_{\text{amount } CH_3CO_2H \text{ remaining}} = \text{amount } CH_3CO_2Na \text{ formed}$$

and

$$C_{CH_3CO_2Na} = \frac{\text{amount } CH_3CO_2Na \ (\text{mmol})}{\text{vol} \ (\text{mL})}$$

$$C_{CH_3CO_2H} = \frac{\text{amount } CH_3CO_2H \ (\text{mmol})}{\text{vol} \ (\text{mL})}$$

Region 3: At the equivalence point. The solution consists of sodium acetate, which ionizes completely into Na^+ and the weak base, $CH_3CO_2^-$. The concentration of OH^- is obtained from the equilibrium expression for a weak base,

$$[OH^-] = \sqrt{K_b[CH_3CO_2^-]} \simeq \sqrt{K_b C_{CH_3CO_2Na}}$$

and converted to $[H_3O^+]$ using the water autoprotolysis expression. The concentration of sodium acetate can be found quite simply. At the equivalence point,

$$\text{amount } CH_3CO_2Na \text{ formed} = \text{amount } CH_3CO_2H \text{ present initially}$$

and

$$C_{CH_3CO_2Na} = \frac{\text{amount } CH_3CO_2Na \ (\text{mmol})}{\text{volume at e.p.} \ (\text{mL})}$$

Region 4: After the equivalence point. The solution contains two bases, NaOH and CH_3CO_2Na. Sodium hydroxide is a much stronger base than sodium acetate and, as such, is responsible for the existing OH^- in the solution. In other words, the problem is one of calculating the pH of a strong base solution. The procedure is identical to that used for region 3 of a strong acid–strong base titration.

Example 7-2

Calculate the pH of the solution resulting when 20.0 mL of 0.100 M CH_3CO_2H has been titrated with 0, 10.0, 20.0, and 30.0 mL of 0.100 M NaOH. The K_a for CH_3CO_2H is 1.76 $\times 10^{-5}$.

Solution

Before the addition of any titrant. Acetic acid is the major supplier of both H_3O^+ and $CH_3CO_2^-$,

$$CH_3CO_2H + H_2O \; \rightleftharpoons \; H_3O^+ + CH_3CO_2^-$$

and

$$K_a = \frac{[H_3O^+][CH_3CO_2^-]}{[CH_3CO_2H]} = \frac{[H_3O^+]^2}{[CH_3CO_2H]} \simeq \frac{[H_3O^+]^2}{C_{CH_3CO_2H}}$$

Solving for $[H_3O^+]$, we have

$$1.76 \times 10^{-5} = \frac{[H_3O^+]^2}{0.100}$$

$$[H_3O^+] = 1.33 \times 10^{-3} \, M$$

$$pH = 2.876$$

After 10.0 mL of NaOH *added.* The titration reaction is

$$CH_3CO_2H + NaOH \longrightarrow CH_3CO_2Na + H_2O$$

amount CH_3CO_2H initial $= 20.0$ mL $\times 0.100 \, M = 2.00$ mmol

$-\left\{ \begin{array}{c} \text{amount NaOH added} \\ \| \\ \text{amount } CH_3CO_2H \text{ used} \end{array} \right\} = 10.0$ mL $\times 0.100 \, M = 1.00$ mmol

amount CH_3CO_2H remaining $= 1.00$ mmol

amount CH_3CO_2Na formed $=$ amount CH_3CO_2H used $= 1.00$ mmol

The solution contains both CH_3CO_2H and CH_3CO_2Na corresponding to region 2. Acetic acid is the major supplier of H_3O^+:

$$CH_3CO_2H + H_2O \rightleftharpoons H_3O^+ + CH_3CO_2^-$$

Making the usual assumptions gives us

$$K_a \simeq \frac{[H_3O^+]\, C_{CH_3CO_2Na}}{C_{CH_3CO_2H}}$$

where

$$C_{CH_3CO_2Na} = \frac{1.00 \text{ mmol}}{30.0 \text{ mL}}$$

$$C_{CH_3CO_2H} = \frac{1.00 \text{ mmol}}{30.0 \text{ mL}}$$

Substituting these concentrations in the equilibrium expression gives

$$1.76 \times 10^{-5} = \frac{[H_3O^+](1.00/30.0)}{(1.00/30.0)}$$

$$[H_3O^+] = 1.76 \times 10^{-5} \, M$$

$$pH = 4.754$$

After 20.0 mL of NaOH *added.* To determine the composition, we use

amount CH_3CO_2H initial $= 20.0$ mL $\times 0.100 \, M$ $= 2.00$ mmol

$-\left\{ \begin{array}{c} \text{amount NaOH added} \\ \| \\ \text{amount } CH_3CO_2H \text{ used} \end{array} \right\} = 20.0$ mL $\times 0.100 \, M$ $= 2.00$ mmol

nothing left over

amount CH_3CO_2Na formed $=$ amount CH_3CO_2H used $= 2.00$ mmol

The solution contains only the weak base CH_3CO_2Na, corresponding to region 3. The weak base is the major supplier of both OH^- and CH_3CO_2H;

$$CH_3CO_2^- + H_2O \rightleftharpoons CH_3CO_2H + OH^-$$

and

$$K_b = \frac{K_w}{K_a} = \frac{[OH^-]^2}{[CH_3CO_2^-]} \approx \frac{[OH^-]^2}{C_{CH_3CO_2Na}}$$

where

$$C_{CH_3CO_2Na} = \frac{2.00 \text{ mmol}}{40.0 \text{ mL}} = 0.0500 \ M$$

Substituting in the equilibrium expression yields

$$\frac{1.00 \times 10^{-14}}{1.76 \times 10^{-5}} = \frac{[OH^-]^2}{0.0500}$$

$$[OH^-] = 5.33 \times 10^{-6} \ M$$

$$pOH = 5.273$$

and

$$pH = 14.000 - 5.273 = 8.727$$

After 30.0 mL of NaOH *added.* To determine the composition, we use

$$-\left\{ \begin{array}{l} \text{amount CH}_3\text{CO}_2\text{H initial} \\ \parallel \\ \text{amount NaOH used} \end{array} \right\} = 20.0 \text{ mL} \times 0.100 \ M = 2.00 \text{ mmol}$$

$$\frac{\text{amount NaOH added}}{\text{amount NaOH remaining}} \quad \begin{array}{l} = 30.0 \text{ mL} \times 0.100 \ M = 3.00 \text{ mmol} \\ = 1.00 \text{ mmol} \end{array}$$

The solution contains the strong base NaOH and the weak base CH_3CO_2Na, corresponding to region 4. The pH is determined by the strong base.

$$[OH^-] \approx C_{NaOH} = \frac{1.00 \text{ mmol}}{50.0 \text{ mL}} = 0.0200 \ M$$

$$pOH = -\log(0.0200) = 1.699$$

and

$$pH = 14.000 - 1.699 = 12.301$$

Titration curve

Figure 7-3 shows the curve for the titration of acetic acid, a typical weak acid, with sodium hydroxide. For the purpose of comparison, the titration curve for hydrochloric acid at the same concentration is also shown. There are important differences between the curves at and before but not after the equivalence point. We said earlier that, in the titration of a strong acid, the pH before the equivalence point is determined by the concentration of untitrated acid present. In the titration of a weak acid this is still true but the acid is only partly dissociated, which leads to a lower concentration of H_3O^+ (higher pH). Furthermore, the product of a weak acid–strong base titration is a weak base, making the pH basic at the equivalence point. The product of a strong acid–strong base titration is a neutral salt, making the pH 7.0 at the equivalence point. After the equivalence point, the pH is determined by the concentration of excess titrant, which is the same for both titrations. This explains why the two titration curves coincide after their equivalence points.

Titration Curves

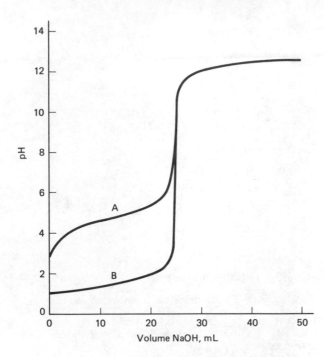

Figure 7-3 Titration curves for a weak and strong acid.

A: 25.0 mL of 0.100 M CH$_3$CO$_2$H with 0.100 M NaOH.

B: 25.0 mL of 0.100 M HCl with 0.100 M NaOH.

The easiest pH value on the titration curve to calculate is the one when the volume of titrant added is one-half the equivalence point volume, that is, when the titration is 50% complete. At this point the concentration of untitrated acid equals the concentration of product (conjugate base) formed,

$$[HA] = [A^-]$$

and Equation 6-31 reduces to

$$pH = pK_a - \log \frac{[HA]}{[A^-]} = pK_a - \log 1$$

$$= pK_a$$

This is a convenient way to determine the pK_a of an acid.

Effect of concentration

Titration curves for different concentrations of acetic acid are shown in Figure 7-4. In contrast to the same set of titration curves for HCl, the concentration has very little effect on the shape of the curve *prior* to the equivalence point. The reason for this lies in the fact that in this region the hydronium-ion concentration is calculated from the following equilibrium expression:

$$[H_3O^+] = \frac{K_a[HA]}{[A^-]}$$

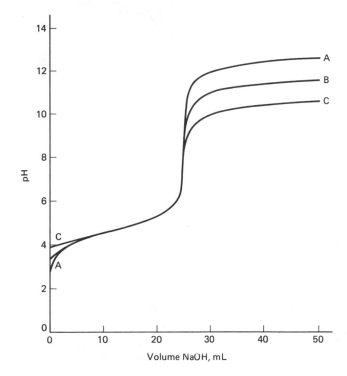

or

$$pH = pK_a - \log \frac{[HA]}{[A^-]} \qquad (7\text{-}3)$$

According to Equation 7-3, the pH depends only on the *ratio* of $[HA]$ to $[A^-]$, not on the individual concentrations. The concentration effect after the equivalence point is the same for a weak acid titration as it is for a strong acid titration because the pH in this region is controlled by the titrant.

Effect of ionization constant

The ionization constant of the weak acid has a significant effect on the shape of the titration curve, as shown by the family of curves in Figure 7-5. Note that the pH values before the equivalence point become larger as the acid becomes weaker, resulting in a smaller pH change near the equivalence point. The equivalence point pH also changes because the weaker the acid being titrated, the stronger the conjugate base formed, and it is this base that is responsible for the pH at the equivalence point.

There is obviously some minimum set of values for K_a and the concentration of the weak acid beyond which the pH change near the equivalence point is too small to permit an accurate estimate of the equivalence point. The exact minimum is arbitrary, depending somewhat on the required accuracy. One commonly used rule states that $K_a \times C_{HA}$ must exceed 10^{-9} for a successful titration.

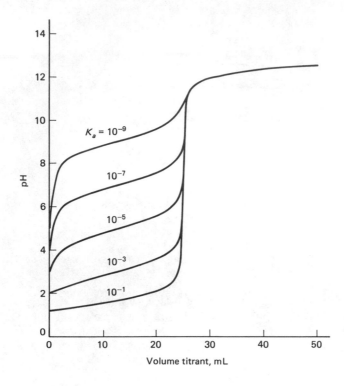

Figure 7-5 Effect of ionization constant on the curve for the titration of $0.10\ M$ weak acid with $0.10\ M$ strong base.

Weak Base Titrated with Strong Acid

The derivation of the curve for the titration of a weak base is analogous to that for a weak acid. The solution is basic initially, becoming less so as the titration proceeds. The pH at the end point will be below 7 because the product of the titration reaction is the conjugate *acid* of the initial weak base. Using ammonia as a typical weak base, the four regions of the titration with HCl are:

Region	Major constituents	Comment
1. Before the addition of titrant	NH_3	Treat as weak base
2. Before the equivalence point	$NH_3 + NH_4Cl$	Treat as buffer
3. At the equivalence point	NH_4Cl	Treat as weak acid
4. After the equivalence point	$NH_4Cl + HCl$	Treat as strong acid

The concentration and ionization constant affect the shape of the titration curve for a weak base the same way they do for a weak acid. Figure 7-6 shows a family of curves for bases of differing ionization constants.

Polyprotic Acids

Deriving the curve for the titration of a weak polyprotic acid with a strong base requires a little more consideration than was given to deriving the curve for a weak monoprotic acid. To illustrate, consider a diprotic acid H_2A that can react with NaOH in two steps:

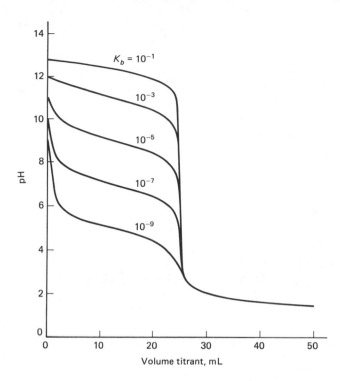

Figure 7-6 Effect of ionization constant on the curve for the titration of 0.10 M weak base with 0.10 M strong acid.

$$H_2A + NaOH \longrightarrow NaHA + H_2O \qquad (7\text{-}4)$$

$$NaHA + NaOH \longrightarrow Na_2A + H_2O \qquad (7\text{-}5)$$

If H_2A is a much stronger acid than NaHA (that is, $K_{a_1} \gg K_{a_2}$) we can expect Reaction 7-4 to be nearly complete before Reaction 7-5 begins. As a result, there are two equivalence points and the titration curve can be divided into six distinct regions.

Region	Major Constituents	Comment
1. Before the addition of any titrant	H_2A	Treat as weak monoprotic acid
2. Before the first equivalence point	H_2A + NaHA	Treat as buffer
3. At the first equivalence point	NaHA	Treat as amphiprotic substance
4. Between the first and second equivalence points	NaHA + Na_2A	Treat as buffer
5. At the second equivalence point	Na_2A	Treat as weak monoequivalent base
6. After the second equivalence point	Na_2A + NaOH	Treat as strong base

The specific calculations used to determine the pH in regions 1 and 2 are the same as for a weak monoprotic acid. At the first equivalence point, only the amphiprotic substance NaHA is present and Equation 6-41 or 6-39 is used to calculate the pH.

In region 4, between the first and second equivalence points, the solution consists of a buffer and the pH calculation is routine. However, determining the concentrations of NaHA and Na_2A requires some care because the *overall* stoichiometry between titrant and analyte is no longer $1:1$. The matter is further complicated if there is uncertainty as to whether this stage of the titration has been reached. Probably the simplest approach to solving these problems is to maintain a $1:1$ stoichiometry for the calculations by carrying them out in a stepwise fashion like the reaction itself. For example, suppose that 25.0 mL of 0.100 M H_2A has been partially titrated with 10.0 mL of 0.150 M NaOH. The first step of the reaction is

$$H_2A + NaOH \longrightarrow NaHA + H_2O$$

and we can write

$$-\left\{\begin{array}{c}\text{amount } H_2A \text{ initial} \\ \parallel \\ \text{amount NaOH used}\end{array}\right\} = 25.0 \text{ mL} \times 0.100 \ M = 2.50 \text{ mmol}$$

$$\frac{\text{amount NaOH added}}{\text{amount NaOH remaining}} \quad \frac{= 20.0 \text{ mL} \times 0.150 \ M = 3.00 \text{ mmol}}{0.50 \text{ mmol}}$$

$$\text{amount NaHA formed} = \text{amount } H_2A \text{ initial} = 2.50 \text{ mmol}$$

Clearly, more than enough NaOH has been added to reach the first equivalence point. Recognizing that the NaHA formed in this step can react with the remaining NaOH,

$$NaHA + NaOH \longrightarrow Na_2A + H_2O$$

the final composition can be obtained by comparing the amounts of each:

$$\text{amount NaHA formed at first equivalence point} \qquad = 2.50 \text{ mmol}$$

$$-\left\{\begin{array}{c}\text{amount NaOH remaining at first equivalence point} \\ \parallel \\ \text{amount NaHA consumed}\end{array}\right\} = 0.50 \text{ mmol}$$

$$\overline{\text{amount NaHA remaining in final solution}} \qquad = \overline{2.00 \text{ mmol}}$$

$$\text{amount } Na_2A \text{ formed} = \text{amount NaHA used} = 0.50 \text{ mmol}$$

Note that the individual reaction steps involve $1:1$ stoichiometry. A similar stepwise calculation can be used for region 6 to determine the amount of sodium hydroxide remaining after the second equivalence point.

Example 7-3

Calculate one pH value in each of the six regions for the titration of 20.0 mL of 0.100 M diprotic acid H_2A with 0.100 M NaOH. For H_2A: $K_{a_1} = 1.00 \times 10^{-4}$ and $K_{a_2} = 1.00 \times 10^{-8}$.

Solution

Region 1: Before the addition of any titrant. Since $K_{a_1}/K_{a_2} > 100$, only the first ionization need be considered.

$$H_2A + H_2O \rightleftharpoons H_3O^+ + HA^-$$

$$K_{a_1} = \frac{[H_3O^+][HA^-]}{[H_2A]} \simeq \frac{[H_3O^+]^2}{C_{H_2A}}$$

Substituting the known values gives us

$$1.00 \times 10^{-4} = \frac{[H_3O^+]^2}{0.100}$$

$$[H_3O^+] = 3.16 \times 10^{-3}\ M$$

$$pH = 2.500$$

Region 2: Before the first equivalence point. Suppose that 5.0 mL of titrant has been added. The titration reaction is

$$H_2A + NaOH \longrightarrow NaHA + H_2O$$

The composition of the solution is determined by comparing the amounts of analyte and titrant:

amount H_2A initial = 20.0 mL × 0.100 M = 2.00 mmol

$-\left\{\begin{array}{l} \text{amount NaOH added} \\ \qquad\qquad \| \\ \text{amount } H_2A \text{ used} \end{array}\right\}$ = 5.0 mL × 0.100 M = 0.50 mmol

amount H_2A remaining   1.50 mmol

amount NaHA formed = amount H_2A used = 0.50 mmol

and

$$C_{H_2A} = \frac{1.50\ \text{mmol}}{25.0\ \text{mL}} = 0.0600\ M$$

$$C_{NaHA} = \frac{0.50\ \text{mmol}}{25.0\ \text{mL}} = 0.020\ M$$

Using Equation 6-29 for a buffer solution yields

$$K_{a_1} \simeq \frac{[H_3O^+]C_{NaHA}}{C_{H_2A}}$$

$$1.00 \times 10^{-4} = \frac{[H_3O^+](0.020)}{0.0600}$$

$$[H_3O^+] = 3.0 \times 10^{-4}\ M$$

$$pH = -\log(3.0 \times 10^{-4}) = 3.52$$

Region 3: At the 1st equivalence point. The solution contains only solvent and NaHA. The NaHA dissociates completely into Na^+ and HA^-. Since HA^- is an amphiprotic substance, Equation 6-41 can be used to calculate the concentration of H_3O^+.

$$[H_3O^+] = \sqrt{K_{a_1}K_{a_2}} = \sqrt{(1.00 \times 10^{-4})(1.00 \times 10^{-8})} = 1.00 \times 10^{-6}\ M$$

$$pH = -\log(1.00 \times 10^{-6}) = 6.000$$

Region 4: Between the first and second equivalence points. Suppose that 35.0 mL of titrant has been added. Assume that the neutralization of H_2A takes place stepwise. The reaction for the first step is

$$H_2A + NaOH \longrightarrow NaHA + H_2O$$

and

$$-\left\{ \begin{array}{l} \text{amount } H_2A \text{ initial} \\ \quad\quad \| \\ \text{amount NaOH used} \end{array} \right\} = 20.0 \text{ mL} \times 0.100 \ M = 2.00 \text{ mmol}$$

$$\frac{\text{amount NaOH added}}{\text{amount NaOH remaining}} = 35.0 \text{ mL} \times 0.100 \ M = \frac{3.50 \text{ mmol}}{1.50 \text{ mmol}}$$

$$\text{amount NaHA formed} = \text{amount } H_2A \text{ initial} = 2.00 \text{ mmol}$$

Since there is NaOH remaining, it will react with the NaHA formed:

$$NaHA + NaOH \longrightarrow Na_2A + H_2O$$

and

$$\begin{array}{l} \text{amount NaHA initial} \quad\quad = 2.00 \text{ mmol} \\ -\left\{ \begin{array}{l} \text{amount NaOH added} \\ \quad\quad \| \\ \text{amount NaHA used} \end{array} \right\} = 1.50 \text{ mmol} \\ \hline \text{amount NaHA remaining} = 0.50 \text{ mmol} \end{array}$$

$$\text{amount } Na_2A \text{ formed} = \text{amount NaHA used} = 1.50 \text{ mmol}$$

Converting to concentrations, we obtain

$$C_{Na_2A} = \frac{1.50 \text{ mmol}}{55.0 \text{ mL}} = 2.73 \times 10^{-2} \ M$$

$$C_{NaHA} = \frac{0.50 \text{ mmol}}{55.0 \text{ mL}} = 9.1 \times 10^{-3} \ M$$

Using the K_{a_2} equilibrium expression yields

$$K_{a_2} = \frac{[H_3O^+][A^{2-}]}{[HA^-]} \approx \frac{[H_3O^+]C_{Na_2A}}{C_{NaHA}}$$

and substituting the known values gives

$$1.00 \times 10^{-8} = \frac{[H_3O^+](2.73 \times 10^{-2})}{9.1 \times 10^{-3}}$$

$$[H_3O^+] = 3.3 \times 10^{-9} \ M$$

$$\text{pH} = -\log(3.3 \times 10^{-9}) = 8.48$$

Region 5: At the second equivalence point. The amount of diequivalent base formed is equal to the amount of acid present initially.

$$\text{amount } Na_2A \text{ formed} = \text{amount } H_2A \text{ initially} = 2.00 \text{ mmol}$$

The volume of NaOH added is twice that needed to reach the first equivalence point, or 40.0 mL, making the total volume 60.0 mL. Thus

$$C_{Na_2A} = \frac{2.00 \text{ mmol}}{60.0 \text{ mL}} = 0.0333 \ M$$

Since K_{b_1}/K_{b_2} is greater than 10^2, the first ionization of A^{2-} is considered to be the sole source of OH^-:

$$A^{2-} + H_2O \rightleftharpoons HA^- + OH^-$$

for which

$$K_{b_1} = \frac{K_w}{K_{a_2}} = \frac{[HA^-][OH^-]}{[A^{2-}]} \simeq \frac{[OH^-]^2}{C_{Na_2A}}$$

Substituting the known values gives

$$\frac{1.00 \times 10^{-14}}{1.00 \times 10^{-8}} = \frac{[OH^-]^2}{0.0333}$$

$$[OH^-] = 1.82 \times 10^{-4}\ M$$

$$pOH = -\log(1.82 \times 10^{-4}) = 3.740$$

and

$$pH = 14.000 - 3.740 = 10.260$$

Region 6: After the second equivalence point. Suppose that 50.0 mL of NaOH has been added. The composition is determined in the usual way:

$$
\begin{array}{ll}
\text{amount NaOH added} & = 50.0\ \text{mL} \times 0.100\ M = 5.00\ \text{mmol} \\
-\left\{\begin{array}{c}\text{amount H}_2\text{A initial} \\ \| \\ \text{amount NaOH used}\end{array}\right\} & = 20.0\ \text{mL} \times 0.100\ M = 2.00\ \text{mmol} \\
\hline
\text{amount NaOH remaining} & \hspace{2cm} 3.00\ \text{mmol}
\end{array}
$$

$$\text{amount NaHA formed} = \text{amount H}_2\text{A initial} = 2.00\ \text{mmol}$$

$$
\begin{array}{ll}
\text{amount NaOH remaining at first e.p.} & = 3.00\ \text{mmol} \\
-\left\{\begin{array}{c}\text{amount NaHA formed at first e.p.} \\ \| \\ \text{amount NaOH used}\end{array}\right\} & = 2.00\ \text{mmol} \\
\hline
\text{amount NaOH remaining} & = 1.00\ \text{mmol}
\end{array}
$$

and

$$[OH^-] \simeq C_{NaOH} = \frac{1.00\ \text{mmol}}{70.0\ \text{mL}} = 0.0143\ M$$

$$pOH = -\log(0.0143) = 1.845$$

Finally,

$$pH = 14.000 - 1.845 = 12.155$$

Figure 7-7 shows the curve for the titration described in Example 7-3. Two end points occur, equally spaced on the volume axis. The titration could be stopped at either end point by using the appropriate indicator.

In trying to establish the general shape of the titration curve for a polyprotic acid, it is necessary to estimate the size of the equivalence point breaks. The magnitude of each break depends not only on the value of $K_a \times C$, but also on the relative acidities of the acid being titrated and the product of the reaction. For example, in the first step of the titration of H_2A with NaOH,

Titration Curves

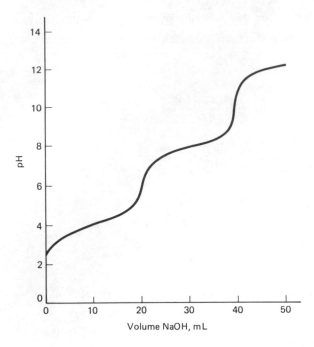

Figure 7-7 Curve for the titration of 20.0 mL of 0.100 M H_2A with 0.100 M NaOH. $K_{a_1} = 1.00 \times 10^{-4}$ and $K_{a_2} = 1.00 \times 10^{-8}$.

$$H_2A + NaOH \longrightarrow NaHA + H_2O$$

if the acidity of HA^- is not very different from that of H_2A, the pH cannot rise much, as the last of the H_2A is converted to HA^- at the equivalence point. On the other hand, if HA^- is a much weaker acid than H_2A, a larger change in pH is possible. Unless the ratio of successive ionization constants equals or exceeds about 10^4, the first end-point break *cannot* be large enough to be of practical value. If the ratio *does* exceed 10^4, the size of the break will *still* depend on the value of $K_a \times C$ (must exceed 10^{-9} to be of value). Titration curves for three hypothetical diprotic acids, all with the same value of K_{a_1}, are shown in Figure 7-8. In curve A, the ratio of K_{a_1} to K_{a_2} for the acid is 10^5, and assuming the concentration to be about 0.1 M, both $K_{a_1} \times C$ and $K_{a_2} \times C$ exceed 10^{-9}. Thus two well-defined end-point breaks are observed, either of which could be used in a titration. In curve B, the ratio is 10^3 and only the second break is large enough to be of use in estimating the equivalence point. The ratio is only 10^1 in curve C and virtually no trace of the first equivalence point can be seen.

Example 7-4

Sketch the titration curve for 0.25 M phosphoric acid with 0.25 M NaOH. For H_3PO_4, $K_{a_1} = 7.1 \times 10^{-3}$, $K_{a_2} = 6.3 \times 10^{-8}$, and $K_{a_3} = 4.2 \times 10^{-13}$.

Solution   First, determine if the three hydrogens react stepwise:

$$\frac{K_{a_1}}{K_{a_2}} = \frac{7.1 \times 10^{-3}}{6.3 \times 10^{-8}} = 1.1 \times 10^5$$

$$\frac{K_{a_2}}{K_{a_3}} = \frac{6.3 \times 10^{-8}}{4.2 \times 10^{-13}} = 1.5 \times 10^5$$

Both ratios exceed 10^4; therefore, the titration reaction occurs in three discrete steps. The

Chapter 7   Acid-Base Titrations

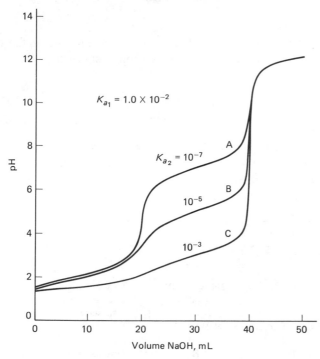

$K_{a_1} = 1.0 \times 10^{-2}$

$K_{a_2} = 10^{-7}$

A

B

10^{-5}

C

10^{-3}

Volume NaOH, mL

Figure 7-8 Effect of the difference between K_{a_1} and K_{a_2} on the curve for the titration 0.10 M diprotic acid with 0.10 M NaOH.

pH break in the vicinity of the third equivalence point is not large enough to be of practical value because

$$K_{a_3} \times C_{H_3PO_4} = 4.2 \times 10^{-13} \times 0.25 = 1.1 \times 10^{-13}$$

which is less than 10^{-9}. The curve will have two breaks, evenly spaced (at 25 and 50 mL) on the volume axis. Placing the first two breaks on the pH axis is relatively easy because both equivalence points are characterized by the formation of amphiprotic substances. Thus

At the first e.p.: $\quad pH = \dfrac{pK_{a_1} + pK_{a_2}}{2} = \dfrac{-\log\,(7.1 \times 10^{-3}) - \log\,(6.3 \times 10^{-8})}{2} = 4.67$

At the second e.p.: $\quad pH = \dfrac{pK_{a_2} + pK_{a_3}}{2} = \dfrac{-\log\,(6.3 \times 10^{-8}) - \log\,(4.2 \times 10^{-13})}{2} = 9.79$

The following sketch can now be made:

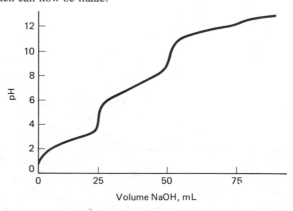

Volume NaOH, mL

Polyequivalent Bases

Titration curves for weak polyequivalent bases are derived in the same manner as those for polyprotic acids. A diequivalent base can react with a strong acid such as HCl in two steps:

$$Na_2A + HCl \longrightarrow NaHA + NaCl \qquad (7\text{-}6)$$
$$NaHA + HCl \longrightarrow H_2A + NaCl \qquad (7\text{-}7)$$

The extent to which Reaction 7-6 is completed before Reaction 7-7 begins, and therefore the degree to which two distinct end-point breaks may exist, depends on the difference between K_{b_1} and K_{b_2} for the diequivalent base, A^{2-}. The titration curve can be divided into the same six regions as those used for a diprotic acid:

Region	Major constituent	Comment
1. Before the addition of any titrant	Na_2A	Treat as weak monoequivalent base
2. Before the first equivalence point	$Na_2A + NaHA$	Treat as buffer
3. At the first equivalence point	$NaHA$	Treat as amphiprotic substance
4. Between the first and second equivalence point	$NaHA + H_2A$	Treat as buffer
5. At the second equivalence point	H_2A	Treat as weak monoprotic acid
6. After the second equivalence point	$H_2A + HCl$	Treat as strong acid

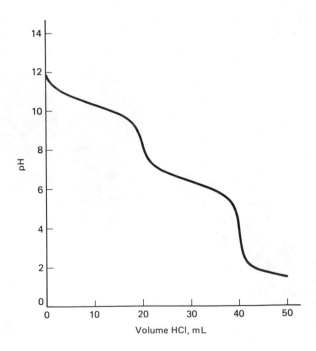

Figure 7-9 Curve for the titration of 20.0 mL of 0.200 M Na_2CO_3 with 0.200 M HCl.

Sodium carbonate is a diequivalent base that is often determined by titration with HCl. Two distinct end-point breaks are observed in the titration curve because

$$\frac{K_{b_1}}{K_{b_2}} = \frac{K_w/K_{a_2}}{K_w/K_{a_1}} = \frac{2.13 \times 10^{-4}}{2.25 \times 10^{-8}} = 0.95 \times 10^4$$

and assuming a concentration of about $0.1\ M$,

$$K_{b_2} \times C_B = 2.25 \times 10^{-8} \times 0.1 = 2.25 \times 10^{-9}$$

The ratio of K_{b_1} to K_{b_2} is just close enough to 10^4 to make the first equivalence-point break large enough to use. The second equivalence-point break meets the $K_{b_2} \times C$ limit of 10^{-9} and will be large enough to use in the titration. The titration curve is shown in Figure 7-9.

DETERMINING THE EQUIVALENCE POINT

Generally, titrations are performed to determine the amount of analyte present in a sample. To do this the volume of titrant needed to reach the equivalence point must be measured. The various techniques for estimating equivalence points were described in Chapter 5. Equivalence points in acid-base titrations are determined most often using chemical indicators, primarily because of the inherent simplicity of this technique, but also because of the availability of many excellent indicators. Although they are not used as often, the graphic methods described in Chapter 5 are applicable to acid-base titrations because of the feasibility of making pH measurements directly with the glass electrode.

Theory of Acid-Base Indicators

Acid-base indicators are weak organic acids or bases whose conjugate forms have different colors. The indicator acts as a second acid or base in the solution being titrated and must be *weaker* than the analyte acid or base so that it reacts last with the titrant. The amount of indicator added must be kept small compared to the amount of analyte present so that it does not consume an appreciable amount of titrant in the process of "indicating" (changing from one conjugate form or color to the other). For this reason, indicators must be very intensely colored so that only a few drops of a dilute solution are needed to produce a color that is observed easily by the eye.

To determine the conditions necessary to cause an indicator to change color near the equivalence point of a titration, it is necessary to examine the equilibrium behavior of the indicator. The ionization equilibrium for a typical indicator can be written as

$$\underset{\text{(color A)}}{\text{HIn}} + H_2O \rightleftharpoons H_3O^+ + \underset{\text{(color B)}}{\text{In}^-} \qquad (7\text{-}8)$$

for which

$$K_a = \frac{[H_3O^+][In^-]}{[HIn]}$$

Rearranging this equation to the logarithmic form gives

$$pH = pK_a - \log \frac{[HIn]}{[In^-]} \quad (7-9)$$

According to strict equilibrium principles, we would never expect to have *only* HIn or In^- in solution, and the color of the solution will always be intermediate between the colors of HIn and In^-. The human eye, however, is not good at discerning slight variations in color and a large excess of one indicator form appears as only that form. Once this occurs, an even larger excess of that one form will have no observable effect on the color of the solution.

We are very interested in knowing the change in pH that must occur in order to bring about a distinct, easily recognizable indicator color change. Suppose that for a typical indicator, when $[HIn]/[In^-] \geqslant 10/1$ we "see" only color A and when $[HIn]/[In^-] \leqslant 1/10$ we "see" only color B. Equation 7-9 can be used to calculate the pH range over which the indicator will change color:

$$pH_{acid\ color} = pK_a - \log \frac{10}{1} = pK_a - 1$$

$$pH_{base\ color} = pK_a - \log \frac{1}{10} = pK_a + 1$$

The indicator transition range is the difference between these values:

$$\Delta pH = pH_{base\ color} - pH_{acid\ color} = (pK_a + 1) - (pK_a - 1) = 2$$

Not all indicators have transition ranges of exactly 2 pH units. If the color intensity of the two indicator forms is different or if the eye is more sensitive to one color, the ratio of indicator forms needed for one color to appear predominant may not be 10/1 or 1/10. Most indicator transition ranges based on visual observations are between 1 and 2 pH units.

Types of Acid-Base Indicators

A great number of compounds have properties that make them suitable for use as acid-base indicators. Consequently, indicators are available for almost any desired pH range. An abbreviated list of some common indicators is given in Table 7-1.

TABLE 7-1 SELECTED pH INDICATORS

Common name	pK_a	Indicator range
Methyl yellow	3.3	2.9–4.0
Methyl orange	4.2	3.1–4.4
Methyl red	5.0	4.2–6.2
Chlorophenol red	6.0	4.8–6.4
Bromothymol blue	7.1	6.0–7.6
Cresol purple	8.3	7.4–9.0
Phenolphthalein	9.7	8.0–9.8
Thymolphthalein	9.9	9.3–10.5

Chapter 7   Acid-Base Titrations

Despite the large number of acid-base indicators, they can be divided into relatively few groups based on their structural characteristics. Two important groups are described briefly in the following paragraphs.

Phthalein indicators

The best known indicator in this group is *phenolphthalein*, whose structure is

acid form	base form
(colorless)	(red)

When phenolphthalein is used in the titration of an acid the solution begins colorless and the equivalence point is signaled by the first appearance of a red color, usually around pH 8.0 to 8.5. At this point only a small portion of the indicator may be in the colored or basic form, but it is easily observed nonetheless because the acidic form is colorless. Phenolphthalein and the very similar thymolphthalein are not very soluble in water, so their solutions are usually prepared with ethanol.

The solubility of phthalein indicators can be increased significantly by replacing the weak carboxylic acid functional group with the much stronger sulfonic acid group. Such a change often affects the colors of the indicator as well as its solubility. Thus phenol red,

yellow	red

is much more soluble in water than phenolphthalein and has a yellow acid form. Other indicators that differ in color and pH transition range are obtained by substituting halogens and various alkyl groups on the benzene rings of the parent structure.

Determining the Equivalence Point

Azo indicators

Methyl orange and methyl red are the two best known azo-type indicators and they are related to one another in the same way that phenolphthalein and phenol red are related; methyl orange contains a sulfonic acid group where methyl red has a carboxylic acid group. The ionization of methyl red is described by the equation

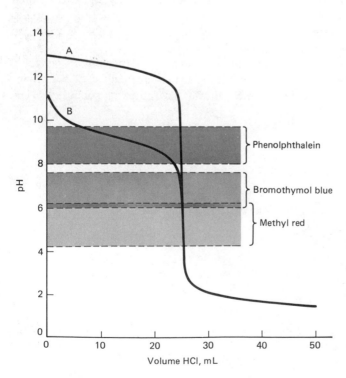

Placing different substituents on the amino nitrogen and the benzene rings yields indicators of different pH ranges and colors.

Selecting the Proper Indicator

The role of a chemical indicator is to change color at or very near the equivalence point of a titration. Therefore, the transition range of the indicator should overlap the steepest part of the titration curve; which is another way of saying that the equivalence point pH should fall within the indicator transition range. The titration curve for ammonia, shown by curve B in Figure 7-10, has a break that nicely encompasses the transition range for

Figure 7-10 Titration curves for weak and strong bases along with various indicator transition ranges.

A: 25.0 mL of 0.100 M NH$_3$ titrated with 0.100 M HCl.

B: 25.0 mL of 0.100 M NaOH titrated with 0.100 M HCl.

Chapter 7   Acid-Base Titrations

methyl red. If bromothymol blue was selected as the indicator, the color transition would begin when 20 mL of HCl had been added (pH 7.6) and end when 25 mL had been added (pH 6.0). Obviously, such a gradual color change is useless for estimating the arrival of the equivalence point. The titration curve for hydrochloric acid, shown as curve A in Figure 7-10, has a very large break that spans the transition range of several indicators, any one of which can be used to signify the end point of the titration. It should be clear now why very weak acids and bases cannot be successfully titrated. Their equivalence-point breaks are too small and gradual, resulting in a slow, rather than rapid, change in the indicator color.

Indicator Errors

There are three sources of error associated with the use of chemical indicators. The first is a determinate error resulting from the fact that the added indicator is an acid or base and must be titrated together with the analyte to produce the desired color change. By requiring indicators to be very intensely colored, they need be present only in tiny amounts, thereby making this error very small.

A second error occurs when the pH at which the indicator completes its color change is not the same as the pH at the equivalence point. This also is a determinate error and is best minimized through careful selection of the indicator. If the expected error is still too large, an *indicator blank* correction may be made. The indicator blank is the volume of titrant needed only to cause the indicator to change color and is determined by titrating a solution containing the indicator but no analyte. The volume required for the blank is subtracted from the volume required for the analyte. Such a correction has the added advantage of eliminating the first determinate error.

Our inability to decide reproducibly when the indicator color change occurs is responsible for the third error. This is an indeterminate error and can become quite serious when the pH break near the equivalence point is not very steep. Under such conditions the color transition tends to be gradual, thereby making it harder to decide when the desired change has occurred. One way of minimizing this error is to match the color of the solution being titrated to the color of a reference solution containing the same amount of indicator at the proper pH.

TITRANTS

As noted earlier, the largest end-point breaks occur when the equilibrium constants for the titration reactions are large. For this reason, titrants for neutralization reactions are always strong acids or bases.

Preparing Acid Titrants

Hydrochloric acid is, by far, the acid titrant used most commonly. It has most of the important properties we seek in a titrant; it is a strong acid, its dilute solutions are very stable, it has very little oxidizing or reducing power, and the chloride ion does not form complicating precipitates with most cations. Although pure hydrogen chloride is a gas, its high solubility and complete dissociation in water make dilute solutions of the acid quite stable. Solutions that are 0.1 M in HCl can be boiled for at least an hour without loss of acid, provided that the evaporated water is replaced periodically.

Concentrated hydrochloric acid is not a primary standard. Consequently, the acid titrant is normally prepared by diluting the concentrated reagent and standardizing it with a primary standard base. It is possible to prepare a standard hydrochloric acid solution directly by weighing constant-boiling HCl. The procedure consists of distilling a dilute acid solution, discarding the first 75%, and collecting the next 10 to 20% of the distillate. The composition of the collected distillate, which is known as *constant-boiling HCl*, is about 20% HCl by weight. The exact concentration depends slightly on the atmospheric pressure, as shown in Table 7-2.

Dilute solutions of perchloric and sulfuric acids are also stable and can be used as acid titrants. Sulfuric acid suffers the disadvantages of having a second hydrogen that is only partially dissociated ($pK_{a_2} = 1.92$) and forming insoluble sulfates with a number of common cations. Dilute nitric acid is not very stable and can undergo reactions in which it behaves as an oxidizing agent.

Standardizing Acids

Although the direct preparation of a standard HCl solution is possible, the problems of generating and handling constant-boiling HCl leads most chemists to use a standardization procedure. Fortunately, there are several good primary-standard bases available for standardizing strong acids.

Sodium carbonate, Na₂CO₃

This is the reagent used most often for the standardization of acid solutions. Although it is available commercially in a very high state of purity, sodium carbonate is somewhat hygroscopic; that is, it picks up moisture from the air. For this reason, the reagent should be dried prior to weighing by heating it in an oven at 200°C for about 30 minutes.

Sodium carbonate can be titrated to sodium bicarbonate using phenolphthalein as the indicator or it may be titrated to carbonic acid using methyl orange or methyl red as the indicator. The second end point is preferred for two reasons: The pH break is larger, leading to a smaller indicator error, and the volume of titrant required is larger, leading to a smaller volume measurement error. The sharpness of the pH break can be enhanced

TABLE 7-2 COMPOSITION
OF CONSTANT-BOILING
HYDROCHLORIC ACID

Pressure (mm Hg)	% HCl (vacuum-weight basis)
730	20.293
740	20.269
750	20.245
760	20.221

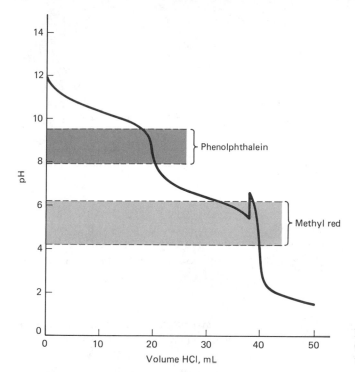

Figure 7-11 Effect of boiling to remove CO_2 in the titration of sodium carbonate with hydrochloric acid.

by boiling the solution briefly just prior to the methyl red end point to remove the titration product, carbonic acid:

$$H_2CO_3 \xrightarrow{\Delta} H_2O + CO_2(g)$$

This produces two beneficial effects. Just prior to the equivalence point the solution contains carbonic acid plus a little untitrated bicarbonate ion, a buffer that changes pH only gradually on addition of a strong acid. Removing the carbonic acid not only destroys the buffer, it also raises the pH by as much as 2 units, thereby increasing the size of the pH break when the last of the titrant is added. Figure 7-11 shows the titration curve that can be expected when the boiling step is incorporated in the procedure.

Example 7-5

About 10 mL of concentrated HCl was transferred to a 1-L bottle and diluted with water. When this solution was used to titrate a sample of pure sodium carbonate weighing 0.3054 g, it took 35.09 mL to reach the methyl red end point. Calculate the molarity of the HCl solution.

Solution In the presence of methyl red indicator the titration reaction is

$$2HCl + Na_2CO_3 \longrightarrow H_2CO_3 + 2NaCl$$

and

$$\text{amount HCl} = \text{amount Na}_2\text{CO}_3 \times \frac{2HCl}{1Na_2CO_3}$$

Titrants

where the amount of Na_2CO_3 is given by

$$\text{amount } Na_2CO_3 = \frac{\text{wt } Na_2CO_3}{\text{MW } Na_2CO_3}$$

Thus

$$\text{amount HCl} = \frac{305.4 \text{ mg}}{106.0 \text{ mg/mmol}} \times \frac{2}{1} = 5.762 \text{ mmol}$$

Finally,

$$C_{HCl} = \frac{\text{amount HCl}}{\text{volume solution}} = \frac{5.762 \text{ mmol}}{35.09 \text{ mL}} = 0.1642 \ M$$

Sodium tetraborate, $Na_2B_4O_7 \cdot 10H_2O$

This compound, sometimes called borax, is a suitable primary-standard base for standardizing acid solutions. It behaves as a diequivalent base whose stoichiometry may be represented as

$$B_4O_7^{2-} + 2H^+ + 5H_2O \longrightarrow 4H_3BO_3$$

The pH break at the equivalence point nicely overlaps the transition range for methyl red. Sodium tetraborate decahydrate has the advantage of a high equivalent weight (190.69) that leads to a small relative weighing error. Its principal disadvantage is the specific precautions that must be taken to ensure an exact, reproducible composition in regard to the hydrated water. Sodium tetraborate cannot be heated to remove this water. Instead, it is stored in a desiccator over an aqueous solution saturated with sodium chloride and sucrose. This maintains a relative humidity of about 70% and ensures that no dehydration of the decahydrate occurs.

Tris(hydroxymethyl)aminomethane, $(HOCH_2)_3CNH_2$

The long name of this reagent is often shortened to TRIS or THAM. It is readily available in a very high degree of purity, can be dried at 110°C, is stable in air and in solution, and has a relatively high equivalent weight (121.06).

Preparing Base Titrants

Sodium hydroxide is used almost exclusively for titrating acids in aqueous solution. On the rare occasions when sodium ions cannot be tolerated, potassium hydroxide may be substituted. Neither base is available in primary-standard-grade purity and their solutions must be standardized.

Dilute solutions of sodium hydroxide are reasonably stable when stored in plastic containers and protected from exposure to carbon dioxide. There is a slow reaction with glass to form soluble silicates, and solutions that are to be kept for more than a week should not be stored in glass containers. This is especially true of containers with ground-glass stoppers as the stoppers often "freeze" so tightly that their removal is impossible without destroying the container.

Clearly, the most troublesome aspect of NaOH as a titrant is its propensity for reacting with atmospheric carbon dioxide:

$$CO_2(g) + 2OH^- \longrightarrow CO_3^{2-} + H_2O \qquad (7\text{-}10)$$

Both the solid reagent and its solutions can react this way; consequently, an already standardized sodium hydroxide solution must be protected from atmospheric CO_2 if its concentration is to remain unchanged. Although the concentration of NaOH is reduced when it comes in contact with CO_2, its reactive capacity toward acids *may* not change. From Equation 7-10 it is seen that for every two hydroxide ions that react, one carbonate ion is formed. If the sodium hydroxide solution is used in a titration with *an acidic range indicator* such as methyl orange or methyl red, each contaminating carbonate ion will react with two hydronium ions from the titrant,

$$CO_3^{2-} + 2H_3O^+ \longrightarrow H_2CO_3 + 2H_2O$$

and is, therefore, equivalent to the amount of hydroxide used in its formation. Consequently, a small amount of carbonate contamination will cause no significant error and may go unnoticed.

Unfortunately, the most common applications of sodium hydroxide titrants are for the titration of weak acids, which require an indicator with a basic transition range such as phenolphthalein. In these cases, each carbonate ion will be titrated to bicarbonate, reacting with only one hydronium ion from the analyte:

$$CO_3^{2-} + H_3O^+ \longrightarrow HCO_3^- + H_2O$$

As a result, the effective concentration of the base solution is decreased and a determinate error will be incurred.

Since solid sodium hydroxide also reacts with atmospheric carbon dioxide, even freshly prepared solutions of the reagent will be contaminated with significant amounts of sodium carbonate. If the solution is protected from *further* contamination, its presence sometimes can be tolerated *if* the same indicator is used in both the standardization and unknown titrations.

Carbonate-free hydroxide solutions are usually prepared in one of two ways. The easiest way, and one that works with both NaOH and KOH, is to prepare an aqueous solution of the base and add enough $BaCl_2$ or $Ba(NO_3)_2$ to precipitate the carbonate as $BaCO_3$, which can be removed by filtering. An excess of $BaCl_2$ or $Ba(NO_3)_2$ is not harmful in an acid-base way because they are neutral salts. Barium ion does, however, form insoluble salts with many anions and may be undesirable for this reason.

Another method for removing carbonate makes use of the fact that sodium carbonate is relatively insoluble in a 50%(w/w) solution of sodium hydroxide. After settling, filtering, or centrifuging, a portion of the clear supernatant liquid can be withdrawn and diluted with freshly boiled distilled water. The boiling step is important because distilled water is often saturated in carbon dioxide, which should be removed prior to dissolving the sodium hydroxide. It is interesting to note that CO_2-free potassium hydroxide cannot be prepared this way because potassium carbonate is quite soluble in concentrated potassium hydroxide.

Standardizing Bases

Several good primary standards are available for the standardization of bases. Two reagents serve perhaps 90% of the applications.

Potassium hydrogen phthalate, $KHC_8H_4O_4$

This reagent, often abbreviated KHP, is an excellent primary standard. It is readily available in a very pure state, is soluble in water, has a high equivalent weight, is not hygroscopic, and can be heated to 135°C without decomposition. Its only shortcoming is that, being a weak acid ($K_a = 3.91 \times 10^{-6}$), the pH at the equivalence point of its titration with a strong base is in the alkaline region. Consequently, the base must be free of carbonate if a sharp end-point break is to be achieved. As potassium hydrogen phthalate is also used as a reference standard for pH measurements, it can be obtained from the National Bureau of Standards in an exceptionally high, certified degree of purity for the most accurate work.

Sulfamic acid, H_2NSO_3H

Sulfamic acid is quite a strong acid ($K_a = 1.03 \times 10^{-1}$) and can be titrated using virtually any indicator whose transition range is between pH 4 and 9. In solution, it undergoes a slow hydrolysis, forming HSO_4^- and NH_4^+:

$$H_2NSO_3H + H_2O \xrightarrow{\text{slow}} HSO_4^- + NH_4^+ \tag{7-11}$$

If the hydrolysis is allowed to occur to any appreciable extent, indicators with transition ranges in the basic region cannot be used because *both* HSO_4^- and NH_4^+ will react with the base titrant before the end point is observed, thereby changing the overall stoichiometry. Indicators whose transition ranges are in the acidic region can still be used because HSO_4^- is quite a strong acid ($K_a = 1.0 \times 10^{-2}$) and will react in place of the sulfamic acid that is destroyed during its formation.

Example 7-6

Two 25.00-mL aliquots of an old sulfamic acid solution were titrated with 0.1000 M NaOH. When phenolphthalein was used as the indicator, 22.00 mL is required to reach the end point. When methyl red was used, 20.50 mL was required. Calculate the percent of sulfamic acid that had hydrolyzed.

Solution According to Reaction 7-11, at the phenolphthalein end point,

amount NaOH = amount H_2NSO_3H left + amount HSO_4^- + amount NH_4^+

while at the methyl red end point,

amount NaOH = amount H_2NSO_3H left + amount HSO_4^-

Subtracting gives the amount of NH_4^+ formed, which is equal to the amount of H_2NSO_3H hydrolyzed:

amount NaOH to phenolphthalein = 22.00 mL × 0.1000 M = 2.200 mmol

$$-\left\{ \begin{matrix} \text{amount NaOH to methyl red} \\ \| \\ \text{amount } H_2NSO_3H \text{ initial} \end{matrix} \right\} = 20.50 \text{ mL} \times 0.1000 \ M = 2.050 \text{ mmol}$$

amount H_2NSO_3H hydrolyzed = 0.150 mmol

and

$$\% \text{ hydrolyzed} = \frac{0.150 \text{ mmol}}{2.050 \text{ mmol}} \times 100 = 7.32$$

Other primary standard acids

Benzoic acid is available in sufficient purity to be considered a primary standard, but it is not very soluble in water. It is used most often in mixed water–alcohol and nonaqueous solvents.

Potassium hydrogen iodate, $KH(IO_3)_2$, is an excellent standard, being a strong acid with a high equivalent weight. Its good oxidizing ability prevents its use in the presence of oxidizable substances.

APPLICATIONS

Acid-base titrimetry will always play an important role in quantitative analysis because of the extremely large number of inorganic, organic, and biochemical compounds that possess inherent acidic or basic properties. The titrimetric technique is very old, but the chemical problems to which it is applied are constantly changing. We have discussed many of the principles and techniques important in titrimetry and it is instructive to see how they are applied to specific methods. The following examples were chosen not because they are the best or most important acid-base methods, but because they illustrate the range of applications and provide some insight into the specific details that often must be considered.

Determination of Carbonate and Carbonate Mixtures

Carbonate, bicarbonate, and hydroxide ions are all bases that can be titrated with a strong acid such as HCl. The determination of these constituents, alone or in permissible combinations, in a sample provides several interesting examples of acid-base titrations. First, it is important to note that appreciable amounts of bicarbonate and hydroxide ions cannot coexist in the same solution because they react:

$$HCO_3^- + OH^- \longrightarrow CO_3^{2-} + H_2O$$

Any attempt to prepare a mixture of these two substances will yield a solution containing carbonate plus whichever reactant is in excess. This makes for five possible sample compositions: (1) Na_2CO_3 alone, (2) Na_2CO_3 + NaOH, (3) Na_2CO_3 + $NaHCO_3$, (4) NaOH alone, and (5) $NaHCO_3$ alone. Titration curves representative of each composition are shown in Figure 7-12. When sodium carbonate is present, the curve has two breaks; the first at about pH 8.4, in the phenolphthalein transition range, and the second at about pH 4.7, in the methyl red transition range. The data necessary to identify the constituents present and determine their amounts can be obtained by attempting, and when possible, performing a stepwise titration first using phenolphthalein and then methyl red indicator. Consider each of the sample compositions listed above:

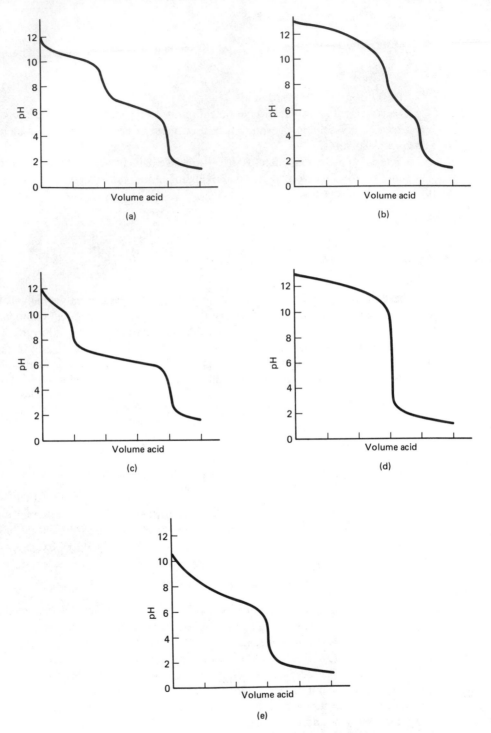

Figure 7-12 Titration curves for carbonate–bicarbonate–hydroxide samples: (a) Na_2CO_3;
(b) Na_2CO_3 + NaOH; (c) Na_2CO_3 + $NaHCO_3$; (d) NaOH; (e) $NaHCO_3$.

Chapter 7   Acid-Base Titrations

Case 1: Na_2CO_3 *alone.* Two end points can be observed and the volume of titrant required to reach the phenolphthalein end point will *equal* (within experimental error) the volume required to titrate from the phenolphthalein (Ph) to the methyl red (MR) end point (Figure 7-12a). If the titration is stopped at the phenolphthalein end point,

$$\text{amount HCl titrant} = \text{amount } Na_2CO_3$$

If the titration is stopped at the methyl red end point,

$$\text{amount HCl titrant} = 2 \times \text{amount } Na_2CO_3$$

Case 2: Na_2CO_3 + NaOH. Sodium hydroxide will titrate along with the sodium carbonate *prior to* the phenolphthalein color change, thereby lengthening the first leg of the titration. That is, the volume of titrant needed to reach the phenolphthalein end point will be *greater than* the volume needed to titrate from the phenolphthalein to the methyl red end point (Figure 7-12b). At the phenolphthalein end point,

$$\text{amount HCl to Ph} = \text{amount NaOH} + \text{amount } Na_2CO_3$$

In titrating from the phenolphthalein to the methyl red end point the sodium carbonate is in the second step of its reaction with the titrant and

$$\text{amount HCl from Ph to MR} = \text{amount } Na_2CO_3$$

The amount of sodium hydroxide can be found by difference.

Case 3: Na_2CO_3 + $NaHCO_3$. Sodium bicarbonate does not begin to titrate until *after* the phenolphthalein color change and the volume of titrant needed to reach the phenolphthalein end point will be *less than* the volume needed to titrate from the phenolphthalein to the methyl red end point (Figure 7-12c). At the phenolphthalein end point,

$$\text{amount HCl to Ph} = \text{amount } Na_2CO_3$$

In titrating from the phenolphthalein to the methyl red end point, the second step of the sodium carbonate reaction is being completed *and* the original sodium bicarbonate is reacting. Thus

$$\text{amount HCl from Ph to MR} = \text{amount } Na_2CO_3 + \text{amount } NaHCO_3$$

The amount of sodium bicarbonate present in the original sample is found by difference.

Case 4: NaOH *alone.* Only one break will be observed and it will be typical of a strong base–strong acid titration, encompassing the transition range of both phenolphthalein and methyl red as illustrated in Figure 7-12d. At either end point,

$$\text{amount HCl to Ph or MR} = \text{amount NaOH}$$

Case 5: $NaHCO_3$ *alone.* The titration curve is that for a weak base with one endpoint break encompassing the methyl red transition range (Figure 7-12e). At the end point,

$$\text{amount HCl to MR} = \text{amount } NaHCO_3$$

Table 7-3 summarizes the relative volumes of titrant needed to reach the end points for each type of sample.

Applications

Sample composition	Volumes of titrant
Na_2CO_3	$V_{0 \to Ph} = V_{Ph \to MR}$
$Na_2CO_3 + NaOH$	$V_{0 \to Ph} > V_{Ph \to MR}$
$Na_2CO_3 + NaHCO_3$	$V_{0 \to Ph} < V_{Ph \to MR}$
NaOH	$V_{0 \to Ph} = V_{0 \to MR} > 0$
$NaHCO_3$	$V_{0 \to Ph} = 0, V_{Ph \to MR} > 0$

Example 7-7

A 300.0-mg sample containing Na_2CO_3, $NaHCO_3$, NaOH, and inert material either alone or in some combination was dissolved and titrated with 0.1000 M HCl. The titration required 24.41 mL to reach the phenolphthalein end point and an additional 8.67 mL to reach the methyl red end point. Determine the composition of the sample and calculate the percent of each titrated component.

Solution Since two end points are observed and $V_{0 \to Ph} > V_{Ph \to MR}$, the sample contains Na_2CO_3 plus NaOH.

$$\left.\begin{array}{c} \text{amount HCl to reach Ph e.p.} \\ \parallel \\ \text{amount } Na_2CO_3 + \text{amount NaOH} \end{array}\right\} = 24.41 \text{ mL} \times 0.1000 \ M = 2.441 \text{ mmol}$$

$$-\left\{\begin{array}{c} \text{amount HCl from Ph to MR e.p.} \\ \parallel \\ \text{amount } Na_2CO_3 \text{ (as } NaHCO_3) \end{array}\right\} = 8.67 \text{ mL} \times 0.1000 \ M = 0.867 \text{ mmol}$$

$$\overline{\text{amount NaOH present}} \qquad\qquad\qquad\qquad\qquad\qquad = \overline{1.574 \text{ mmol}}$$

Then

$$\% \ Na_2CO_3 = \frac{0.867 \text{ mmol} \times 106.0 \text{ mg/mmol}}{300.0 \text{ mg}} \times 100 = 30.6$$

$$\% \ NaOH = \frac{1.574 \text{ mmol} \times 40.00 \text{ mg/mmol}}{300.0 \text{ mg}} \times 100 = 20.99$$

Example 7-8

A solution prepared by mixing NaOH with $NaHCO_3$ was diluted to 500 mL with water. A 25.00-mL aliquot was titrated with 0.1202 M HCl. The buret read 14.37 mL at the phenolphthalein end point and 38.52 mL at the methyl red end point. What were the concentrations of the substances in the 500-mL solution?

Solution When NaOH and $NaHCO_3$ are mixed they react,

$$NaHCO_3 + NaOH \longrightarrow Na_2CO_3 + H_2O$$

Since $V_{0 \to Ph} < V_{Ph \to MR}$ for the titration, the solution after mixing must contain Na_2CO_3 plus $NaHCO_3$. Accordingly,

$$-\left\{\begin{array}{c}\text{amount HCl to reach Ph e.p.} \\ \parallel \\ \text{amount } Na_2CO_3 \text{ present}\end{array}\right\} = 14.37 \text{ mL} \times 0.1202 \text{ } M \qquad = 1.727 \text{ mmol}$$

$$\underline{\left\{\begin{array}{c}\text{amount HCl from Ph to MR e.p.} \\ \parallel \\ \text{amount } Na_2CO_3 + NaHCO_3\end{array}\right\}} = (38.52 \text{ mL} - 14.37 \text{ mL}) \times 0.1202 \text{ } M = 2.903 \text{ mmol}$$

$$\text{amount } NaHCO_3 \text{ present} \qquad\qquad\qquad\qquad\qquad\qquad = \overline{1.176 \text{ mmol}}$$

Finally,

$$C_{Na_2CO_3} = \frac{1.727 \text{ mmol}}{25.00 \text{ mL}} = 0.06908 \text{ } M$$

$$C_{NaHCO_3} = \frac{1.176 \text{ mmol}}{25.00 \text{ mL}} = 0.04704 \text{ } M$$

Although the procedures used in the two preceding examples are simple and straightforward, they are not entirely satisfactory for the determination of such mixtures because of a somewhat gradual pH change at the first equivalence point, corresponding to the formation of bicarbonate. Methods have been developed that do not use the bicarbonate end point. In the titration of a sodium hydroxide–sodium carbonate mixture, a sample is titrated to the methyl red end point to determine the total amount of the two analytes:

$$\text{amount HCl to MR} = \text{amount NaOH} + (2 \times \text{amount } Na_2CO_3)$$

A second sample of the same size is dissolved and treated with excess barium chloride to precipitate the carbonate ion:

$$BaCl_2 + Na_2CO_3 \longrightarrow BaCO_3(s) + 2NaCl$$

The remaining solution, now containing only the NaOH plus some NaCl and $BaCl_2$, is titrated with HCl to the phenolphthalein end point, where

$$\text{amount HCl to Ph} = \text{amount NaOH}$$

The solid barium carbonate need not be removed, as it does not interfere in the titration. The amount of Na_2CO_3 is found by subtracting the amount of NaOH from the total amount.

This method cannot be used with carbonate–bicarbonate mixtures because barium bicarbonate is not very soluble and will coprecipitate with the barium carbonate. The procedure can be modified by treating a sample with a known excess of sodium hydroxide to convert the bicarbonate to carbonate:

$$NaHCO_3 + NaOH \longrightarrow Na_2CO_3 + H_2O$$

This carbonate is precipitated by the addition of excess barium chloride and the remaining excess sodium hydroxide is back titrated with HCl to a phenolphthalein end point. The amount of bicarbonate is given by

$$\text{amount NaHCO}_3 = \text{amount NaOH added} - \left\{\begin{array}{c} \text{amount NaOH remaining} \\ \| \\ \text{amount HCl back titration} \end{array}\right\}$$

A second sample of the same size is titrated to the methyl red end point to obtain the total amount of the two analytes, and the amount of Na_2CO_3 is found by difference.

Example 7-9

A 0.5000-g sample containing Na_2CO_3, $NaHCO_3$, and inert material was dissolved and required 38.00 mL of 0.1100 M HCl to reach the methyl red end point. A second 0.5000-g sample was dissolved and treated with 10.00 mL of 0.5000 M NaOH. A large excess of $BaCl_2$ was added, precipitating $BaCO_3$. The remaining solution required 17.00 mL of the HCl to reach the phenolphthalein end point. Calculate the % Na_2CO_3 and % $NaHCO_3$ in the sample.

Solution For the titration to the methyl red end point:

$$\left\{\begin{array}{c} \text{amount HCl to MR e.p.} \\ \| \\ 2 \times \text{amount Na}_2\text{CO}_3 + \text{amount NaHCO}_3 \end{array}\right\} = 38.00 \text{ mL} \times 0.1100 \ M = 4.180 \text{ mmol}$$

From the back titration:

$$\begin{aligned} \text{amount NaHCO}_3 &= \text{amount NaOH added} - \text{amount NaOH left} \\ &= (10.00 \text{ mL} \times 0.5000 \ M) - (17.00 \text{ mL} \times 0.1100 \ M) \\ &= 3.130 \text{ mmol} \end{aligned}$$

Subtracting this from the total gives

$$2 \times \text{amount Na}_2\text{CO}_3 = 4.180 \text{ mmol} - 3.130 \text{ mmol} = 1.050 \text{ mmol}$$
$$\text{amount Na}_2\text{CO}_3 = 1.050 \text{ mmol}/2 = 0.5250 \text{ mmol}$$

Finally,

$$\% \text{ NaHCO}_3 = \frac{3.130 \text{ mmol} \times 84.01 \text{ mg/mmol}}{500.0 \text{ mg}} \times 100 = 52.59$$

$$\% \text{ Na}_2\text{CO}_3 = \frac{0.5250 \text{ mmol} \times 106.0 \text{ mg/mmol}}{500.0 \text{ mg}} \times 100 = 11.13$$

Determination of Nitrogen

Nitrogen is found in many important substances, including proteins, fertilizers, explosives, drugs, pesticides, and natural waters. As such, it is one of the most often determined elements. The protein content of grain, meat, and biological material in general is usually estimated by a nitrogen determination. This is possible because most proteins contain nearly the same percentage of nitrogen.

The most popular method for determining nitrogen is the Kjeldahl method, developed in 1883. It is based on the conversion of the bound nitrogen to ammonia, which is then separated by distillation and determined by titration. The important steps are as follows:

1. *Digestion.* In this step the sample is oxidized in hot, concentrated sulfuric acid. The carbon and hydrogen in the sample are converted to carbon dioxide and water. If the nitrogen exists as an amine or an amide as it does in proteins, it will be converted to the desired ammonium ion. If it exists in a higher oxidation state such as an azo or nitro group as it does in some explosives and fertilizers, it will be converted to elemental nitrogen or nitrogen oxides, leading to low results. When such groups are present they must be reduced prior to digestion.

The digestion process can be quite slow, requiring more than an hour with some samples. To speed the process, potassium sulfate may be added to increase the boiling point of the sulfuric acid solution. Many published procedures also recommend the addition of a catalyst such as mercury or mercuric oxide.

2. *Distillation.* When oxidation is complete, the solution is cooled, diluted somewhat with water, cooled again, and finally made basic with a sodium hydroxide solution to yield ammonia:

$$NH_4^+ + OH^- \longrightarrow NH_3(g) + H_2O$$

The liberated ammonia is distilled into another container for titration. At first this seems like an easy task. In fact, it requires considerable skill and great care together with some special apparatus. The addition of a concentrated sodium hydroxide solution to a concentrated sulfuric acid solution generates a great deal of heat, enough to boil the resulting solution if the mixing is not done properly. Ordinarily, a special long-necked container called a Kjeldahl flask is used for both the digestion and distillation steps. The more dense sodium hydroxide solution is poured carefully down the side of the flask with a minimum of mixing to form a separate lower layer. A spray trap is fitted to the top of the flask and to a condenser. Only then is the Kjeldahl flask gently swirled to mix the two layers. A Kjeldahl distillation apparatus is shown in Figure 7-13.

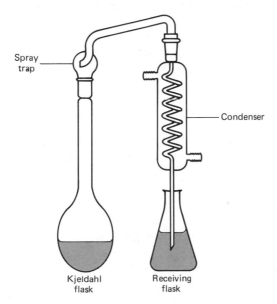

Spray trap

Condenser

Kjeldahl flask

Receiving flask

Figure 7-13 Kjeldahl distillation apparatus.

Applications

3. *Titration.* The distilled ammonia cannot be collected in water without significant loss due to its high volatility. One procedure calls for an excess of boric acid in the collection flask. As the ammonia is collected, it reacts to form the borate ion $H_2BO_3^-$,

$$H_3BO_3 + NH_3 \longrightarrow NH_4^+ + H_2BO_3^-$$

which is a sufficiently strong base ($K_b = 1.72 \times 10^{-5}$) to be titrated with HCl using methyl red indicator.

An alternative method employs a back titration. The collection flask contains a known excess of standard HCl, and as the ammonia enters it reacts with some of the HCl,

$$HCl + NH_3 \longrightarrow NH_4Cl$$

The remaining HCl is titrated with a standard NaOH solution. An indicator with an acidic transition range, such as methyl red, is required to avoid having any of the ammonium ion react with the NaOH prior to the color change.

Example 7-10

A 1.500-g sample of dried soil was analyzed for nitrogen using a Kjeldahl procedure. The receiving flask contained 75.00 mL of 0.04000 *M* HCl. After the ammonia was collected, the solution was titrated with 0.1000 *M* NaOH, requiring 23.70 mL to reach the methyl red end point. Calculate the percent nitrogen in the sample.

Solution   For the back titration,

$$
\begin{aligned}
\text{amount HCl initial} &= 75.00 \text{ mL} \times 0.04000 \ M = 3.000 \text{ mmol} \\
-\left\{ \begin{array}{c} \text{amount NaOH used} \\ \| \\ \text{amount HCl remaining} \end{array} \right\} &= 23.70 \text{ mL} \times 0.1000 \ M \ = 2.370 \text{ mmol} \\
\hline
\text{amount HCl used} = \text{amount NH}_3 = \text{amount N} &\quad= 0.630 \text{ mmol}
\end{aligned}
$$

and

$$\% \text{ N} = \frac{0.630 \text{ mmol} \times 14.0 \text{ mg/mmol}}{1500 \text{ mg}} \times 100 = 0.588$$

Determination of Esters

A large variety of esters can be determined by an acid-base back-titration technique. Almost all esters react with hydroxide to give a carboxylate anion and an alcohol—a process called *saponification*:

$$RCO_2R' + OH^- \longrightarrow RCO_2^- + R'OH$$

Relatively few esters react fast enough with OH^- to permit their direct titration, so most procedures call for refluxing the ester with a known excess of KOH for $\frac{1}{2}$ to 2 hours, after which the remaining KOH is determined by titration with a standard acid.

PROBLEMS

7-1. A 25.00-mL sample of 0.103 M Ba(OH)$_2$ was titrated with 0.188 M HClO$_4$. Calculate the pH of the solution before addition of any titrant and after the addition of 15.6, 27.4, and 38.1 mL of titrant.

7-2. Repeat the calculations in Problem 7-1 using concentrations of 1.03×10^{-3} M for Ba(OH)$_2$ and 1.88×10^{-3} M for HClO$_4$.

7-3. A sample of 0.100 M HCl is being titrated with 0.100 M NaOH. Calculate the pH when the volume of titrant added is 95%, 100%, and 105% of the equivalence-point volume.

7-4. Repeat the calculations in Problem 7-3 using concentrations of 1.00×10^{-3} M for both HCl and NaOH. What is the effect of the lower concentrations on the pH at the equivalence point? What is the effect on the *size* of the pH break surrounding the equivalence point?

7-5. For the titration of 20.00 mL of 0.1026 M NH$_3$ with 0.0947 M HCl, calculate the pH before the addition of any titrant, at 10.00 mL prior to the equivalence point, at the equivalence point, and at 10.00 mL after the equivalence point.

7-6. Repeat the calculations in Problem 7-5 using concentrations of 0.01026 M NH$_3$ and 0.00947 M HCl. In which regions of the titration is the pH affected by the change in concentration of analyte and titrant?

7-7. Calculate the pH of the solution when the volume of 0.100 M sodium hydroxide added in the titration of 0.100 M lactic acid is 50%, 95%, 100%, and 105% of the equivalence-point volume.

7-8. Perform the calculations in Problem 7-7 for the titration of hypochlorous acid in place of lactic acid. In which regions of the titration is the pH affected by the change in ionization constant of the acid?

7-9. The unprotonated form of histadine can behave as a diequivalent base when titrated with a strong acid. The values of K_{b_1} and K_{b_2} are 1.2×10^{-5} and 1.1×10^{-8}, respectively. If 25.00 mL of 0.1000 M unprotonated histadine is titrated with 0.2000 M HCl, calculate the pH after the addition of the following volumes of titrant.

(a) 0 mL   (d) 12.50 mL   (g) 25.00 mL

(b) 6.250 mL   (e) 15.00 mL   (h) 30.00 mL

(c) 10.00 mL   (f) 18.75 mL

7-10. Sketch the curve for the titration described in Problem 7-9.

7-11. Sketch the expected titration curves when each of the following acids is titrated with NaOH. Assume that the concentration of NaOH is approximately the same as the concentration of the acid being titrated.

(a) 1.0×10^{-2} M formic acid   (e) 1.0×10^{-1} M sulfurous acid

(b) 1.0×10^{-1} M boric acid   (f) 1.0×10^{-1} M fumaric acid

(c) 1.0×10^{-1} M sulfamic acid   (g) 1.0×10^{-1} M arsenious acid

(d) 1.0×10^{-2} M hydrocyanic acid   (h) 1.0×10^{-1} M arsenic acid

7-12. Sketch the expected titration curves when each of the following bases is titrated with HCl. Assume that the concentration of HCl is approximately equal to that of the base.

(a) 1.0×10^{-2} M trimethylamine   (d) 1.0×10^{-1} M sodium carbonate

(b) 1.0×10^{-2} M pyridine   (e) 1.0×10^{-1} M sodium sulfide

(c) 1.0×10^{-1} M sodium borate   (f) 1.0×10^{-1} M sodium malonate

7-13. Sketch the expected titration curves for each of the following.

(a) 0.10 M HCl + 0.10 M HClO$_4$ with 0.10 M NaOH

(b) 0.10 M HCl + 0.10 M HCO$_2$H with 0.10 M NaOH

(c) 0.10 M HCl + 0.20 M HCO$_2$H with 0.10 M NaOH

(d) 0.10 M H$_3$PO$_4$ + 0.10 M KH$_2$PO$_4$ with 0.10 M NaOH

(e) 0.10 M C$_6$H$_5$ONa (sodium phenoxide) + 0.30 M HONH$_2$ with 0.10 M HCl

(f) 0.10 M NaOH + 0.20 M Na$_2$CO$_3$ with 0.10 M HCl

7-14. Select an indicator that would be suitable for each of the following titrations.

(a) 0.010 M Ba(OH)$_2$ with 0.010 M HCl

(b) 0.10 M HCO$_2$H with 0.10 M NaOH

(c) 0.10 M HONH$_2$ with 0.10 M HCl

(d) 0.10 M NaHCO$_3$ with 0.10 M HCl

(e) 0.10 M H$_3$AsO$_4$ with 0.050 M NaOH (second equivalence point)

(f) 0.020 M Cl$_3$CCO$_2$H with 0.020 M NaOH

7-15. What is the pH transition range of an indicator that appears to the eye to be pure yellow when the ratio of H$_2$In to HIn$^-$ is at least 6.2 : 1 and pure green when the ratio of HIn$^-$ to H$_2$In is at least 8.6 : 1?

7-16. Suppose the indicator described in Problem 7-15 was used for the titration of 20.00 mL of 0.1000 M HA with 0.1000 M NaOH. If the indicator concentration was 4.0 $\times$ 10^{-5} M, calculate the titration error resulting from the need for extra titrant to convert the indicator to its base-form color. Assume that the equilibrium constant for the titration reaction is very large.

7-17. Approximately 6 mL of concentrated perchloric acid (72%) was transferred to a bottle and diluted with about 1 L of water. A sample containing 251.5 mg of primary standard Na$_2$B$_4$O$_7 \cdot$10H$_2$O required 27.41 mL of the HClO$_4$ solution to reach the methyl red end point. Calculate the molar concentration of the HClO$_4$ solution.

7-18. What weight of pure sodium carbonate should be taken so that about 25 mL of approximately 0.1 M HCl will be used in a standardization titration using a methyl red end point?

7-19. A 2.0284-g sample of primary standard potassium hydrogen phthalate was dissolved and diluted to 250.0 mL. A 20.00-mL aliquot of this solution required 32.19 mL of a sodium hydroxide solution for titration to the phenolphthalein end point. Calculate the molar concentration of the sodium hydroxide solution.

7-20. A sulfamic acid solution, prepared by dissolving 417.4 mg of the solid reagent in 250.0 mL of solution, has hydrolyzed to the extent of 2.10%. If this solution is used to determine the concentration of a sodium hydroxide solution by titration to the phenolphthalein end point, calculate the absolute concentration error.

7-21. A 5.00-mL aliquot of wine vinegar was diluted and titrated with 0.1104 M NaOH, requiring 32.88 mL to reach the phenolphthalein end point. If the vinegar has a density of 1.055 g/mL, calculate its acidity as % acetic acid.

7-22. A 2.000-g sample of juice from a fresh lime was diluted with water and filtered to remove the suspended matter. The clear liquid required 40.39 mL of 4.022 $\times$ 10^{-2} M NaOH for titration to the phenolphthalein end point. Calculate the acidity as % citric acid.

7-23. A sample of powdered cleanser weighing 3.500 g was diluted with water and filtered to remove the suspended matter. The solution, containing sodium tetraborate, required 34.81 mL of 0.1009 M HCl for titration to the methyl red end point. Calculate the percentage of Na$_2$B$_4$O$_7$ in the sample.

7-24. The owner of a swimming pool supply company is offered what appears to be an exceptionally good deal on some concentrated muriatic acid (hydrochloric acid). The solution is supposed to be 31.25% HCl by weight. The store owner has some suspicions about the seller and decides to have a sample analyzed before agreeing to make a purchase. The chemist he hired determined the density of the solution to be 1.367 g/mL. Then a

5.00-mL aliquot of the acid was diluted to 250 mL. A 50.0-mL aliquot of the diluted sample required 22.5 mL of 0.300 M NaOH for titration to the phenolphthalein end point. Calculate the percentage of HCl in the original acid and tell the store owner what he should do.

7-25. The "total alkalinity" of boiler feedwater often is determined by titrating the various metal bicarbonates with a standard acid. Calculate the total alkalinity as milligrams of $CaCO_3/$ 100 mL in a 50.00-mL water sample that required 40.73 mL of 0.08074 M HCl for titration to the methyl red end point.

7-26. A sodium hydroxide solution contains a small amount of sodium carbonate as a result of contamination by atmospheric carbon dioxide. A 25.00-mL aliquot requires 26.47 mL of 0.1021 M HCl to reach the phenolphthalein end point and an additional 0.87 mL to reach the methyl red end point.
   (a) Calculate the molar concentration of NaOH and Na_2CO_3.
   (b) What weight of CO_2 has been absorbed per milliliter of solution?

7-27. A sample may contain NaOH, Na_2CO_3, and $NaHCO_3$, alone or in some combination, along with inert material. A portion of the sample weighing 857.6 mg was dissolved and titrated with 0.1163 M HCl. It took 21.64 mL to reach the phenolphthalein end point and an additional 14.90 mL to reach the methyl red end point.
   (a) What is the composition of the sample?
   (b) Calculate the percentage of each titrated component.

7-28. Suppose that the sample in Problem 7-27 required 15.37 mL to reach the phenolphthalein end point and an additional 20.88 mL to reach the methyl red end point.
   (a) What is the composition of the sample?
   (b) Calculate the percentage of each titrated component.

7-29. A 3.750-g sample of baking soda containing Na_2CO_3, $NaHCO_3$, and inert material was dissolved in water and diluted to 500.0 mL. A 50.00-mL aliquot of this solution required 46.93 mL of 0.1208 M HCl to reach the methyl red end point. Another 50.00-mL aliquot was treated with 10.00 mL of 0.2506 M NaOH to convert the bicarbonate to carbonate ion, which then was precipitated by the addition of excess $BaCl_2$. After removal of the precipitated $BaCO_3$, the excess NaOH in the solution required 14.09 mL of 0.09783 M HCl for titration. Calculate the % Na_2CO_3 and % $NaHCO_3$ in the sample.

7-30. Five samples, each weighing 5.041 g, contain $HClO_4$, H_3PO_4, and KH_2PO_4, alone or in some combination, along with inert material. When each sample was titrated with 0.1533 M NaOH, the following end points were observed:

Sample	Volume NaOH to methyl orange e.p. (mL)	Volume NaOH to phenolphthalein e.p. (mL)
1	20.67	20.67
2	12.06	33.90
3	18.74	37.48
4	0.00	31.49
5	15.41	26.39

Determine the composition of each sample and calculate the percentage of each titrated component.

7-31. A 0.643-g sample of a water-soluble fertilizer was dissolved and analyzed for nitrogen using the Kjeldahl method. After digestion, the distilled ammonia was collected in 100 mL of

0.1334 M boric acid. This solution required 21.4 mL of 0.202 M HCl for titration to the methyl red end point. Calculate the % N in the fertilizer.

7-32. A 1.00-mL aliquot of fish oil was analyzed for nitrogen using the Kjeldahl method. After digestion, the distilled ammonia was collected in 100 mL of 0.0503 M HCl. The excess HCl required 28.3 mL of 0.124 M NaOH for titration. Calculate the amount of nitrogen in the sample as mg N/mL.

7-33. The protein content of wheat flour can be determined reasonably accurately by multiplying the percentage of nitrogen present by 5.7. A 2.06-g sample of flour was taken through a Kjeldahl procedure and the ammonia produced was distilled into a boric acid solution. If this solution required 34.7 mL of 0.174 M HCl for titration to the methyl red end point, what is the percentage of protein in the flour?

7-34. The carbon dioxide evolved on heating a 2.407-g sample of steel in a closed system was swept into a flask containing 50.00 mL of 0.05081 M barium hydroxide. After filtering the solution to remove the precipitated barium carbonate, the remaining base consumed 14.87 mL of 0.1125 M hydrochloric acid for titration to the phenolphthalein end point. Calculate the percentage of carbon in the steel.

7-35. The purity of a sample of ethyl acetate, $CH_3CO_2C_2H_5$, was determined by treating 2.011 g of the material with 75.00 mL of 0.3861 M KOH, heating the mixture below the boiling point for about 2 hours to saponify the ester, and titrating the excess KOH with 32.53 mL of 0.2066 M HCl. Calculate the percent purity of the ester.

ACIDS AND BASES IN NONAQUEOUS SOLVENTS

Although water is an extraordinarily versatile solvent in which to carry out acid-base titrations, there are occasions when a nonaqueous solvent may be necessary or preferred, such as when the analyte is not water soluble and the neutralization reaction is not sufficiently complete in water. The completeness of a neutralization reaction depends, in part, on the acid or base strength of the analyte. There is, of course, little we can do about the *intrinsic* acidity or basicity of a substance, but Brønsted and Lowry have made it clear that the *observed* acidity or basicity depends on the solvent because it participates in the ionization. An acid too weak to titrate in water sometimes can be titrated in a nonaqueous solvent, where its observed acidity is greater.

Acid-base equilibrium in nonaqueous solvents is considerably more complicated than in water and many of the important equilibrium constants are not known. As a result, it is difficult to calculate the data needed to construct theoretical titration curves as was done in Chapter 7. It is possible, however, to measure pH in many nonaqueous solvents and thereby construct experimental titration curves. Finally, it should be remembered that the terms "pH" and "hydrogen-ion concentration" actually refer to the concentration of solvated protons. In water, the "hydrogen ion" is H_3O^+; in acetic acid it is $CH_3CO_2H_2^+$; and so forth.

SOLVENTS

Insofar as acid-base equilibria are concerned, solvents may be classified as either amphiprotic, basic, or aprotic. *Amphiprotic solvents* exhibit both acidic and basic character and undergo self-ionization or autoprotolysis, as illustrated by the following reactions:

$$2H_2O \rightleftharpoons H_3O^+ + OH^-$$

$$2C_2H_5OH \rightleftharpoons C_2H_5OH_2^+ + C_2H_5O^-$$

$$2CH_3CO_2H \rightleftharpoons CH_3CO_2H_2^+ + CH_3CO_2^-$$

$$2NH_3 \rightleftharpoons NH_4^+ + NH_2^-$$

Autoprotolysis may be represented in general by the reaction

$$2HS \rightleftharpoons H_2S^+ + S^- \tag{8-1}$$

where H_2S^+ and S^- are sometimes referred to as the lyonium and lyate ions, respectively.

Basic solvents can accept but not donate protons. Pyridine is considered a basic solvent because it has no proton to donate but can accept a proton according to the reaction

Obviously, such solvents cannot undergo autoprotolysis.

Aprotic solvents exhibit no appreciable acidic or basic character and, like basic solvents, do not undergo autoprotolysis. Such solvents are of limited importance in acid-base chemistry because of their inability to dissolve many common acids and bases. Table 8-1 lists a few solvents belonging to each of the three classifications.

TABLE 8-1 PROPERTIES OF COMMON SOLVENTS

Solvent	Autoprotolysis constant, pK_{HS}	Dielectric constant
Amphiprotic		
Acetic acid	14.45	6.1
Acetonitrile	32.2	36.0
Ammonia[a]	33	22
Ethanol	19.5	24.3
Ethylenediamine	15.3	12.9
Methanol	16.7	32.6
Water	14.00	78.5
Aprotic or basic		
Benzene	—	2.3
Dimethylformamide	—	36.7
Dioxane	—	2.2
n-Hexane	—	1.9
Methyl isobutylketone	—	13.1
Pyridine	—	12.3

[a] At $-50°C$.

Important Characteristics of Amphiprotic Solvents

The completeness of a neutralization reaction in an amphiprotic solvent depends not only on the acid or base strength of the analyte, but also on three properties of the solvent. This section is devoted to a discussion of those properties and their influence on the behavior of acid-base equilibria.

Acid-base properties

According to the Brønsted–Lowry concept, the ionization of a weak acid HA in a solvent HS is the sum of two half-reactions:

$$HA \;\rightleftharpoons\; H^+ + A^- \tag{8-2}$$

$$\underline{HS + H^+ \;\rightleftharpoons\; H_2S^+} \tag{8-3}$$

$$HA + HS \;\rightleftharpoons\; H_2S^+ + A^- \tag{8-4}$$

The equilibrium constant for Reaction 8-2 can be considered a measure of the *intrinsic acidity* of HA. Similarly, the equilibrium constant for Reaction 8-3 can be considered a measure of the *intrinsic basicity* of HS. The product of these two values is the equilibrium constant for Reaction 8-4:

$$K_{\text{acidity, HA}} = \frac{[H^+][A^-]}{[HA]}$$

$$K_{\text{basicity, HS}} = \frac{[H_2S^+]}{[HS][H^+]}$$

$$(K_{\text{acidity, HA}})(K_{\text{basicity, HS}}) = \frac{[\cancel{H^+}][A^-]}{[HA]} \times \frac{[H_2S^+]}{[HS][\cancel{H^+}]} = K_{\text{eq}} \tag{8-5}$$

Equation 8-5 is of little practical benefit because we have no means of determining equilibrium constants for half-reactions. However, the equation does make clear the importance of the solvent basicity in determining the overall ionization constant of a weak acid: The acidic behavior of a solute, represented by K_{eq} in Equation 8-5, is directly proportional to the intrinsic basicity, $K_{\text{basicity, HS}}$, of the solvent. Solvents with a large intrinsic basicity will enhance the observed acidity of weak acids. Similarly, acidic solvents will enhance the observed basicity of weak bases. Phenol is too weak an acid to titrate in water, but it is easily titrated in a more basic solvent such as ethylenediamine, as illustrated in Figure 8-1.

The acidic and basic character of the solvent are responsible for another important effect—the maximum acidity or basicity a solute can exhibit in the solvent. If a solute acid is a much stronger proton donor than the solvated proton (lyonium ion), the ionization reaction

$$HA + HS \;\rightleftharpoons\; H_2S^+ + A^-$$

goes essentially to completion. The net effect is that all of the solute acid HA is replaced with the weaker acid H_2S^+. Chemists refer to this phenomenon as the *leveling effect*. Strong bases are similarly leveled to the strength of the solvent anion S^-. An example of this effect can be seen in the behavior of perchloric, hydrochloric, and nitric acids.

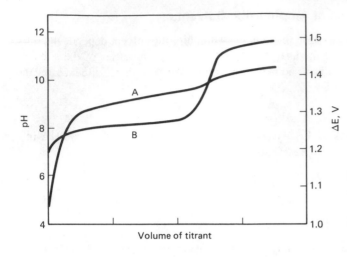

Figure 8-1 Titration of phenol:

A: In water with sodium hydroxide titrant (pH axis).

B: In ethylenediamine with sodium aminoethoxide titrant (ΔE axis).

All three acids ionize to the same extent in water—100%. In a solvent less willing to accept protons (a weaker base), such as glacial (concentrated) acetic acid, none of the acids ionize completely, although perchloric acid ionizes to the greatest extent and therefore is the stronger acid. It can be said that acetic acid exhibits less leveling power toward acids than does water. In general, as the intrinsic basicity (or acidity) of the solvent decreases, its leveling power for acids (or bases) also decreases.

Autoprotolysis

The degree to which a pure solvent ionizes is represented by its *autoprotolysis constant*, K_{HS}. For Reaction 8-1 the constant is defined by

$$K_{HS} = [H_2S^+][S^-]$$

Autoprotolysis is an acid-base reaction between identical solvent molecules in which some act as an acid and others as a base. Consequently, the extent of an autoprotolysis reaction depends both on the intrinsic acidity and the intrinsic basicity of the solvent.

The importance of the autoprotolysis constant in titrations lies in its effect on the completeness of a titration reaction. To illustrate, consider a weak acid HA being titrated in water with the strong base NaOH. The titration reaction is

$$HA + NaOH \longrightarrow NaA + H_2O$$

or, as an ionic reaction,

$$HA + OH^- \longrightarrow A^- + H_2O \tag{8-6}$$

For this same acid being titrated in ethanol with the strong base sodium ethoxide, C_2H_5ONa, the reactions are

$$HA + C_2H_5ONa \longrightarrow NaA + C_2H_5OH$$

$$HA + C_2H_5O^- \longrightarrow A^- + C_2H_5OH \tag{8-7}$$

Reaction 8-6 is simply the reverse of the ionization reaction for the base A^-. Thus

$$K_{eq} = \frac{1}{K_b} = \frac{K_a}{K_w} \qquad (8\text{-}8)$$

Similarly, for Reaction 8-7:

$$K_{eq} = \frac{1}{K_b'} = \frac{K_a'}{K_{HS}} \qquad (8\text{-}9)$$

where K_b' and K_a' are the base and acid ionization constants in ethanol. According to Equation 8-9, as K_{HS} increases in value, K_{eq} becomes smaller, which means that the titration reaction is less complete. A similar set of equations can be derived for a weak base being titrated with a strong acid in each solvent. In Chapter 7 it was shown that the most complete reactions yield the best titration curves (largest and steepest end-point breaks). For the same reason, solvents with small autoprotolysis constants are preferred in nonaqueous titrations. Values of the autoprotolysis constants for several common amphiprotic solvents are given in Table 8-1.

Example 8-1

Acetic acid has an autoprotolysis constant of 3.55×10^{-15}. Calculate the pH of pure acetic acid.

Solution The autoprotolysis reaction is the source of "hydrogen ions":

$$2CH_3CO_2H \rightleftharpoons CH_3CO_2H_2^+ + CH_3CO_2^-$$

for which

$$K_{HS} = [CH_3CO_2H_2^+][CH_3CO_2^-]$$

Since there is no other source of either ion,

$$[CH_3CO_2^-] = [CH_3CO_2H_2^+]$$

and

$$K_{HS} = [CH_3CO_2H_2^+]^2$$

$$[CH_3CO_2H_2^+] = \sqrt{K_{HS}} = \sqrt{3.55 \times 10^{-15}} = 5.96 \times 10^{-8}\ M$$

Finally,

$$pH = -\log(5.96 \times 10^{-8}) = 7.225$$

Dielectric constant

The behavior of acids and bases in nonaqueous solvents suggests that ionization should be considered a two-step process: proton transfer (ion-pair formation) followed by dissociation (ion-pair separation). For a weak acid HA in a solvent HS, we would write

$$\text{Proton transfer: } HA + HS \rightleftharpoons H_2S^+A^- \qquad (8\text{-}10)$$

$$\text{Dissociation: } \quad H_2S^+A^- \rightleftharpoons H_2S^+ + A^- \qquad (8\text{-}11)$$

The overall ionization reaction is the sum of these two reactions:

$$HA + HS \rightleftharpoons H_2S^+ + A^- \tag{8-12}$$

The extent of the proton transfer step is determined by the acid-base properties of the solute and solvent and by K_{HS}. The extent of the dissociation step is determined by the ionic charges and the dielectric constant of the solvent.

The *dielectric constant* is a measure of a solvent's capacity for separating oppositely charged particles. A large dielectric constant means good separating power. In water, which has the largest dielectric constant of the common solvents (see Table 8-1), Reaction 8-11 is virtually complete and the extent of the overall ionization is determined solely by the extent of the proton transfer step. Acetic acid, whose dielectric constant is only 6.2, is not very efficient at separating ion pairs. When perchloric acid is dissolved in acetic acid solvent, we may write

$$HClO_4 + CH_3CO_2H \rightleftharpoons CH_3CO_2H_2^+ClO_4^-$$
$$\underline{CH_3CO_2H_2^+ClO_4^- \rightleftharpoons CH_3CO_2H_2^+ + ClO_4^-}$$
$$HClO_4 + CH_3CO_2H \rightleftharpoons CH_3CO_2H_2^+ + ClO_4^-$$

The equilibrium constant for the overall process is about 10^{-5} and one is tempted to conclude that perchloric acid is a weak acid in this solvent. Research has shown, however, that the proton transfer reaction is essentially complete; that is, perchloric acid is a strong acid. The small equilibrium constant is due to the inability of acetic acid to separate the ion pair formed in the proton transfer reaction.

Some acids and bases undergo proton transfer reactions that do not produce ion pairs. For example,

$$HA^+ + HS \rightleftharpoons H_2S^+ + A$$
$$B^- + HS \rightleftharpoons HB + S^-$$

For such reactions, the strength of the acid or base (the extent of the overall ionization) is not affected significantly by the dielectric constant of the solvent.

Meaning of pH

A solution of pH 3 is 10 times more acidic than another solution of pH 4, as long as *the same solvent* is used in both solutions. But what if different solvents are involved? There is no universal scale of pH that exists for all solvents. Thus we cannot assume that an aqueous solution of pH 6 is more acidic than an ethanolic solution of pH 7 simply because 6 is smaller than 7. Perhaps the easiest way to recognize this fact is to compare the neutral pH for several solvents (Table 8-2). This value is easily calculated because, for a pure solvent, the "hydrogen-ion" concentration is the square root of the autoprotolysis constant.

Important Characteristics of Aprotic Solvents

Aprotic solvents exhibit no acid-base character of their own and therefore do not undergo autoprotolysis or compete with the reactants for protons in a titration. As a result, leveling does not occur and neutralization reactions often are more complete. Acids that are

TABLE 8-2 NEUTRAL pH OF VARIOUS
SOLVENTS AT 25°C

Solvent	pK_{HS}	Neutral pH
Acetic acid	14.45	7.23
Ethanol	19.5	9.8
Ethylenediamine	15.3	7.7
Methanol	16.7	8.4
Water	14.00	7.00

leveled in water can be titrated stepwise in the aprotic solvent methyl isobutyl ketone, as illustrated in Figure 8-2.

It is truly regretable that many aprotic solvents do not dissolve common acids and bases very well and/or have very low dielectric constants. Some effort has been made to use mixtures of aprotic and amphiprotic solvents, such as benzene/methanol and hexane/ethylene glycol, but they seldom offer significant advantages over pure amphiprotic solvents.

Choosing a Solvent

It is instructive at this point to consider what properties are desired of a solvent for an acid-base titration. They can be divided into two groups: those that improve the shape

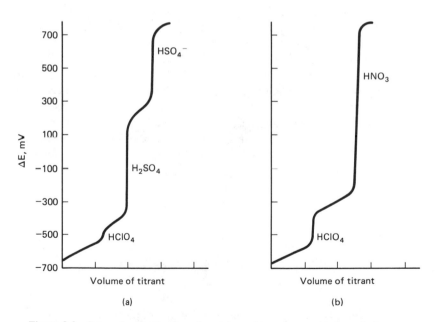

Figure 8-2 Curves for the titration of a mixture of two strong acids in methyl isobutylketone with tetrabutylammonium hydroxide: (a) $HClO_4$ + H_2SO_4; (b) $HClO_4$ + HNO_3. (Reprinted with permission from D.B. Bruss and G.E.A. Wyld, *Anal. Chem.* **1957,** *29*, 232–235. Copyright 1957 American Chemical Society.)

of the titration curve, thereby making it easier to locate the equivalence point; and those that improve our ability to actually perform the titration. It should be clear from the preceding discussion that insofar as the shape of the titration curve is concerned, a solvent should:

1. Be a good proton acceptor if the analyte is an acid or a good proton donor if the analyte is a base. This will enhance the observed acidity or basicity of the analyte.
2. Have a small autoprotolysis constant to minimize leveling.
3. Have a large dielectric constant to facilitate ion-pair dissociation.

These properties define the theoretical limitations imposed on a titration by the solvent. From a practical standpoint, it is desirable that the solvent:

1. Dissolves the reactants, products, and indicator.
2. Is easy to work with: that is, has a low toxicity and vapor pressure, has no color that will obscure the indicator color change, and so on.

Something else that may be of considerable importance when a method has to be modified to suit a particular problem is our understanding of the equilibrium processes governing the titration reaction and the availability of pertinent equilibrium constants.

No single solvent is best in every category but water comes pretty close. As a result, the large majority of applications employ water as the solvent. Nonaqueous solvents are used primarily when the analyte is either too weak or too insoluble in water to permit its titration. Occasionally, a solvent more discriminating than water is needed to perform a stepwise titration of two acids (or bases) of similar strength.

Selected solvents

Glacial acetic acid is a good solvent for titrating very weak bases because of its strong proton-donating ability. It has an autoprotolysis constant (pK_{HS} for $CH_3CO_2H = 14.45$) similar to water's, but its much lower dielectric constant ($D = 6.1$) is an unfortunate disadvantage. It is used frequently in the titration of anionic bases where a low dielectric constant is less significant in affecting the degree of ionization. Many different acids and bases dissolve in acetic acid and it is not a difficult solvent to work with.

Ethylenediamine, $H_2NCH_2CH_2NH_2$, commonly abbreviated en, is the basic counterpart to acetic acid. Its dielectric constant is twice that for acetic acid but still much smaller than that for water. The strongly basic character of this solvent makes it desirable for titrating very weak acids. Like acetic acid, it is not a difficult substance to handle in the laboratory.

Dimethylformamide, $HCON(CH_3)_2$, is a much weaker base than ethylenediamine, but it is a popular solvent because of its large dielectric constant ($D = 36.7$) and ability to dissolve many organic acids, including salts and polymers.

Pure aprotic solvents are seldom useful due to their inability to dissolve common acids and bases. There are a number of applications using a mixed aprotic–amphiprotic solvent such as benzene–methanol or hexane–ethylene glycol. The idea is to combine the nonleveling property of the aprotic solvent with the dissolving ability of amphiprotic solvent. The results are only moderately successful.

TITRANTS AND STANDARDS

Nonaqueous titrations tend to be more highly specialized than aqueous titrations. As a result, many different combinations of titrant, solvent, and indicator have been used. It is not uncommon, for example, to discover that a solvent used to dissolve an excellent titrant will not dissolve the analyte, forcing the analyst to look for another titrant or solvent. Sometimes it is necessary to use different solvents for the titrant and analyte.

Acidic Titrants

Perchloric acid is one of the strongest acids known and is by far the most widely used of the acidic titrants listed in Table 8-3. Acetic acid and dioxane are the usual solvents for nonaqueous perchloric acid titrants. Although anhydrous dioxane yields sharper end-point breaks than acetic acid, it is more difficult to remove water from this solvent. Concentrated perchloric acid is 72% $HClO_4$ and 28% H_2O; thus water is automatically introduced into the diluted titrant solution. When acetic acid is the solvent, the water is easily removed by addition of an equivalent amount of acetic anhydride, $(CH_3CO_2)_2O$, with which it reacts to form acetic acid.

$$(CH_3CO_2)_2O + H_2O \longrightarrow 2CH_3CO_2H$$

The perchlorate salts formed in the titration often are insoluble in acetic acid and can obscure a visual end point. Also, they may interfere in the functioning of the electrode when the end point is determined potentiometrically (Chapter 13).

Standards for acidic titrants

Potassium hydrogen phthalate and tris(hydroxymethyl)aminomethane are two primary standard bases used to standardize acetic acid solutions of perchloric acid. You may recall from Chapter 7 that potassium hydrogen phthalate was used to standardize *aqueous base* solutions. The hydrogen phthalate ion is an amphiprotic substance and in glacial acetic acid solvent it is sufficiently basic to be used for the standardization of *acids*, the reaction with perchloric acid being

$$HClO_4 + KHC_8H_4O_4 \xrightarrow[\text{Solvent}]{CH_3CO_2H} H_2C_8H_4O_4 + KClO_4(s)$$

TABLE 8-3 TITRANTS USED IN NONAQUEOUS TITRIMETRY

Acidic titrants	Basic titrants
Perchloric acid	Tetrabutylammonium hydroxide
p-Toluenesulfonic acid	Sodium acetate
2,4-Dinitrobenzenesulfonic acid	Potassium methoxide
	Sodium aminoethoxide

Basic Titrants

Some common basic titrants used in nonaqueous titrations are listed in Table 8-3. Tetrabutylammonium hydroxide is probably the most widely used, even though it is restricted to alcohol and alcohol/benzene solvents, where its solubility is great enough to permit solutions of sufficient concentration to be prepared. There are two properties of this titrant that are primarily responsible for its widespread use: It is a strong base and it generally forms soluble titration products. Two methods of preparation are used. In the first, tetrabutylammonium iodide is shaken with silver oxide in alcohol (ROH), producing the reaction

$$2(C_4H_9)_4N^+I^- + Ag_2O(s) + ROH \longrightarrow$$
$$(C_4H_9)_4N^+OH^- + (C_4H_9)_4N^+OR^- + 2AgI(s)$$

The AgI and excess Ag_2O are removed by filtration and the remaining solution diluted with benzene. The diluted titrant solution is usually about 10% alcohol and 90% benzene. The second method involves passing an alcoholic solution of tetrabutylammonium iodide through an ion-exchange resin (Chapter 19) in the hydroxide form.

Sodium and potassium methoxide in a mixed methanol/benzene solvent have been used to titrate weak acids in several basic solvents, although neither titrant appears to be as strongly basic as tetrabutylammonium hydroxide. Furthermore, the salts produced in the titration reactions are usually insoluble and often complicate the end-point determination.

TITRATION CURVES

The equilibrium behavior of acids and bases in nonaqueous solvents can be very complex. We have already seen that ion-pair formation is a complication encountered in solvents with low dielectric constants. Other types of association reactions also occur, such as self-association

$$2HA \rightleftharpoons (HA)_2$$

In solvents that are not very polar, ionic solutes may be poorly solvated and seek stabilization via homoconjugation:

$$A^- + HA \rightleftharpoons HA_2^-$$
$$BH^+ + B \rightleftharpoons B_2H^+$$

Even when "quantitative" equilibrium calculations are possible, they tend to include more assumptions and be less exact than the corresponding calculations for an aqueous system. Because of the difficulty of calculating theoretical titration curves, much of what is known about nonaqueous titrations has been learned by trial and error. It can reasonably be assumed that the parameters that affect the shapes of aqueous titration curves also affect, in the same general way, the shapes of nonaqueous titration curves.

Effect of Water

Water behaves as a weak base in acidic solvents and a weak acid in basic solvents. In doing so it competes with the solvent for protons, thereby exerting a leveling effect. The

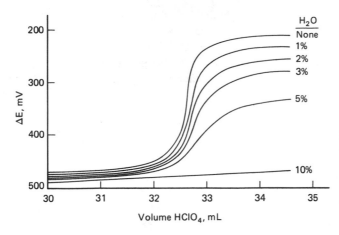

Figure 8-3 Effect of water on the titration of an amine acetate salt (Asterol) in glacial acetic acid with $HClO_4$. (Reprinted with permission from C.W. Pifer and E.G. Wollish, *Anal. Chem.* **1952**, *24*, 300–306. Copyright 1952 American Chemical Society.)

net result is a less satisfactory titration, owing to a smaller pH break near the equivalence point. The amount of water that can be tolerated depends in part on the strength of the acid or base being titrated. Many amines are moderately strong bases in acetic acid and can be titrated when the solvent contains as much as 2 to 3% water. On the other hand, as little as 0.2% water can prevent the successful titration of a very weak base such as caffeine. The effect of water on the titration of Asterol in acetic acid is shown in Figure 8-3.

Water is such a common chemical in our environment that it is often present as a contaminant in other reagents. Unfortunately, its removal is not always simple. The analyst is well advised to consider the extent to which water can be tolerated in a particular application before starting complicated procedures for its removal.

DETERMINING THE EQUIVALENCE POINT

Many of the indicators used in aqueous acid-base titrimetry can be used in nonaqueous titrations. Although a great deal of equilibrium data for aqueous solutions of indicators is available, it cannot be used, even in a qualitative way, to predict indicator behavior in nonaqueous solvents. Unfortunately, comparable equilibrium data for nonaqueous solutions generally are not available and indicators for specific applications usually are selected on the basis of careful empirical (trial and error) studies. A few of the more common indicators are listed in Table 8-4. Remember, however, that these indicators are not applicable in all solvents or with every titrant.

TABLE 8-4 INDICATORS USED IN NONAQUEOUS ACID-BASE TITRATIONS

For titrating acids	For titrating bases
Azo violet	Methyl orange
o-Nitroaniline	Methyl red
Thymol blue	Methyl violet

Because of the difficulties encountered with visual indicators in nonaqueous solvents, potentiometric detection with the glass electrode (Chapter 13) is commonly employed to construct an experimental titration curve—the end point then being obtained graphically. Occasionally, end points are determined from spectrophotometric, amperometric, or conductometric measurements using data away from the vicinity of the equivalence point (see Chapter 5).

APPLICATIONS

After reading this chapter, it should be apparent that far less is known about the quantitative behavior of acids and bases in nonaqueous solvents than in water. The lack of fundamental equilibrium data has made it difficult to predict the behavior of a solute in one solvent from its observed behavior in another solvent. As a result, each new application has to be studied and evaluated independently, even though it may use the same titrant or solvent employed in other applications. Having done so much work on a new application it is understandable that the chemist should want to have it published! Thus an enormous literature exists dealing with applications of nonaqueous acid-base titrimetry.

More care is necessary in performing nonaqueous than aqueous titrations. One problem is getting water out of the solutions and keeping it out. We do not realize how prevalent water is in our environment until we don't want it! Surface moisture on glassware must be removed (but remember, volumetric glassware should *never* be dried in an oven) and anhydrous solutions must be protected and handled in such a way that they do not acquire significant amounts of water from the air. Most organic solvents have coefficients of expansion that are significantly larger than that for water—0.11% per degree Celsius for anhydrous acetic acid compared to 0.025% for water. Consequently, greater care must be exercised in eliminating or compensating for temperature changes that occur while making volume measurements.

Carboxylic Acids

Most carboxylic acids are sufficiently acidic to make their titration comparatively easy. Simple low-molecular-weight acids can be titrated in alcohol, benzene–methanol, or dimethylformamide with potassium methoxide. Phenolphthalein can be used as the indicator in alcoholic solvents. Polyprotic acids generally form troublesome insoluble salts with potassium methoxide and are better titrated in dimethylformamide or acetonitrile with a tetraalkylammonium hydroxide titrant.

Phenols

Most phenols are too weakly acidic to titrate in alcoholic solvents, requiring instead a basic solvent such as ethylenediamine. Sodium aminoethoxide has proved to be an excellent titrant in this solvent. It is interesting to note that phenol esters of carboxylic acids can be titrated in this solvent, apparently because the ester reacts with ethylenediamine to yield the phenol, which can be titrated.

$$\underset{\text{O}}{\overset{\text{O}}{\text{RC}}}-\text{OAr} + \text{H}_2\text{NCH}_2\text{CH}_2\text{NH}_2 \longrightarrow \underset{\text{O \ H}}{\overset{\text{O \ H}}{\text{RC}}}-\text{NCH}_2\text{CH}_2\text{NH}_2 + \text{ArOH}$$

Example 8-2

A 3.000-g sample containing phenol and other inert material was dissolved in ethylenediamine and titrated with 0.04000 M sodium aminoethoxide, requiring 16.40 mL to reach the end point. Calculate the percent phenol in the sample.

Solution The titration reaction is

$$\text{C}_6\text{H}_5\text{OH} + \text{NaOCH}_2\text{CH}_2\text{NH}_2 \longrightarrow \text{C}_6\text{H}_5\text{ONa} + \text{HOCH}_2\text{CH}_2\text{NH}_2$$
(phenol)

From the balanced reaction

$$\text{amount phenol} = \text{amount titrant} \times \frac{1 \text{ phenol}}{1 \text{ titrant}}$$

$$= (16.40 \text{ mL})(0.04000 \, M) \times \frac{1}{1} = 0.6560 \text{ mmol}$$

Converting to weight gives us

$$\text{wt phenol} = 0.6560 \text{ mmol} \times 94.11 \text{ mg/mmol} = 61.74 \text{ mg}$$

Finally,

$$\% \text{ phenol} = \frac{\text{wt phenol}}{\text{wt sample}} \times 100 = \frac{61.74 \text{ mg}}{3000 \text{ mg}} \times 100 = 2.058$$

Amines

Most aliphatic amines are about as basic as ammonia and their titrations are not difficult. Acetic acid is the most common solvent, but dioxane, nitromethane, acetonitrile, and benzene have also been used. Aromatic and certain substituted alkyl amines are much less acidic, having a value of pK_b in water in the range 9 to 12. Even so, they can be titrated in acetic acid if steps are taken to remove even traces of water.

Some amino acids are not soluble enough in acetic acid to permit their direct titration. In such cases, a determination may be carried out by dissolving the sample in acetic acid containing a known amount of perchloric acid and then back titrating the excess acid with a standard base such as potassium acetate.

Example 8-3

A 500.0-mg sample of a sugar substitute containing amino acids as the active ingredients was dissolved in 25.00 mL of 0.02000 M $HClO_4$ in glacial acetic acid. The excess $HClO_4$ required 14.62 mL of 0.01000 M CH_3CO_2K for titration to the methyl violet end point. If the average molecular weight of the amino acid mixture is 90.00, calculate the percentage present.

Solution Using the formula $\underset{\overset{|}{\text{R}}}{\text{HO}_2\text{CCHNH}_2}$ to represent the amino acids, the reaction with $HClO_4$ is

$$\mathrm{HO_2C\overset{\displaystyle R}{\overset{|}{C}}HNH_2 + HClO_4 \xrightarrow{CH_3CO_2H} HO_2C\overset{\displaystyle R}{\overset{|}{C}}HNH_3{}^+ClO_4{}^-}$$

The back-titration reaction also has a 1:1 stoichiometry:

$$\mathrm{HClO_4 + CH_3CO_2K \xrightarrow{CH_3CO_2H} CH_3CO_2H + K^+ClO_4{}^-}$$

The amount of amino acid present is taken as the difference between the amount of $HClO_4$ in the initial solution and that remaining after reaction:

$$\begin{array}{lll} & \text{amount } \mathrm{HClO_4} \text{ initial} & = 25.00\text{ mL} \times 0.02000\ M = 0.5000\text{ mmol} \\ - & \left\{\begin{array}{c}\text{amount } \mathrm{HClO_4} \text{ left} \\ \| \\ \text{amount } \mathrm{CH_3CO_2K} \text{ required}\end{array}\right\} & = 14.62\text{ mL} \times 0.01000\ M = 0.1462\text{ mmol} \\ \hline & \text{amount } \mathrm{HClO_4} \text{ used} = \text{amount amino acid} & = 0.3538\text{ mmol} \end{array}$$

Converting to weight and then percent gives us

$$\text{wt amino acid} = 0.3538\text{ mmol} \times 90.00\text{ mg/mmol} = 31.84\text{ mg}$$

$$\%\text{ amino acid} = \frac{31.84\text{ mg}}{500.0\text{ mg}} \times 100 = 6.368$$

PROBLEMS

8-1. Write the autoprotolysis reactions and equilibrium equations for each of the following solvents.

(a) H_2O

(b) CH_3OH

(c) CH_3CO_2H

(d) NH_3

(e) $H_2NCH_2CH_2NH_2$

8-2. The concentration of S^- in a solvent HS was found to be 6.30×10^{-8} *M*. Calculate the autoprotolysis constant for the solvent.

8-3. Calculate the concentration of protonated solvent for each of the solvents in Problem 8-1.

8-4. Write two-step reactions (proton transfer/ion-pair separation) for the ionization of each of the following substances.

(a) perchloric acid in methanol

(b) hydrofluoric acid in ammonia

(c) ammonia in acetic acid

(d) urea in phosphoric acid

8-5. The equilibrium concentration of $CH_3CO_2H_2{}^+$ in 0.100 *M* HCl in acetic acid is about 1.7×10^{-5} *M*. Even so, HCl is considered a *strong* acid in this solvent.

(a) Calculate the equilibrium constant for the equilibrium between HCl and its ions in this solvent.

(b) Explain how HCl can be considered a strong acid in this solvent.

8-6. The pK_a for HCl in acetic acid is 8.55. Calculate the pH of the following solutions (in acetic acid).

(a) 0.026 *M* HCl

(b) 0.026 *M* HCl + 0.040 *M* NaCl

(c) 0.040 *M* NaCl

8-7. The K_a for HBr in ethylenediamine is 4.2×10^{-3}. Calculate the pH of the following solutions (in ethylenediamine).

(a) 9.50×10^{-3} *M* HBr

(b) 9.50×10^{-3} *M* HBr + 6.00×10^{-3} *M* $CaBr_2$

(c) 6.00×10^{-3} *M* $CaBr_2$

8-8. As an estimate of the size of the end-point break in an acid-base titration, calculate the amount by which the pH changes in the transition from 10^{-3} M strong acid (typical of a point just before the end point) to 10^{-3} M strong base (typical of a point just after the end point) in each of the following solvents.

(a) water   (c) ethanol   (e) ammonia

(b) methanol   (d) acetonitrile   (f) ethylenediamine

8-9. Calculate the pH of the solution after addition of the following volumes of 0.1047 M perchloric acid in methanol to 25.00 mL of 0.1103 M sodium methoxide in methanol.

(a) 0 mL   (c) 26.34 mL   (d) 35.00 mL

(b) 10.00 mL

8-10. A solution containing 183.6 mg of primary standard potassium hydrogen phthalate in glacial acetic acid required 21.63 mL of $HClO_4$ in glacial acetic acid to reach the methyl violet end point. Calculate the molar concentration of the perchloric acid solution.

8-11. A 0.2104-g sample containing *p*-bromoaniline was dissolved in glacial acetic acid and titrated with 0.01250 M perchloric acid in the same solvent, requiring 16.47 mL to reach the end point. Calculate the percentage of C_6H_6NBr in the sample.

8-12. A sample weighing 450.3 mg, containing the amino acid phenylalanine, along with inert material, was dissolved in glacial acetic acid containing 10.00 mL of 0.1034 M $HClO_4$. After the reaction was complete, the excess perchloric acid required 12.33 mL of 0.05525 M CH_3CO_2K for titration to the methyl violet end point. Calculate the percentage of phenylalanine, $H_2NCHCH_2C_6H_5$, in the sample.

$$\overset{|}{CO_2H}$$

COMPLEXATION EQUILIBRIA AND TITRATIONS

Most metal ions can accept unshared pairs of electrons from an anion or molecule to form coordinate covalent bonds. The word *coordinate* specifies that both bonding electrons were contributed by one of the two atoms involved. The molecule or ion containing the donor atom is called a *ligand* or *coordinating agent* and the product resulting from a reaction between a metal ion and a ligand is referred to as a *coordination compound* or *complex ion*.

Metal ions have the capacity to accept at least four and often six pairs of electrons. The actual number accepted, which depends mainly on the metal ion but partly on the ligand, is called the *coordination number*. Ligands containing a single donor atom are called monodentate; those containing two donor atoms are bidentate; and so forth. When a polydentate ligand coordinates with a metal ion, a ring structure is formed as shown in the following reaction between cupric ion and ethylenediamine:

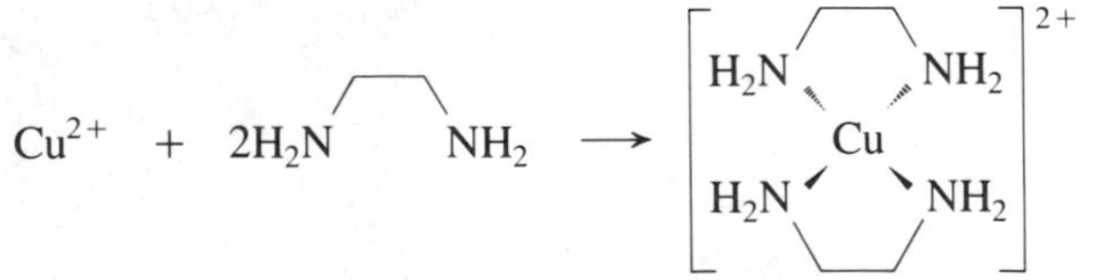

$$Cu^{2+} \quad + \quad 2H_2N \qquad NH_2 \quad \longrightarrow \quad \begin{bmatrix} H_2N & & NH_2 \\ & Cu & \\ H_2N & & NH_2 \end{bmatrix}^{2+} \qquad (9\text{-}1)$$

Such compounds are called *chelates* and they have proven to be very useful in analytical chemistry.

The explanation of why coordination compounds form and the structural characteristics of such compounds are outside the scope of this book. In this chapter we are concerned mainly with the equilibrium behavior of coordination compounds in aqueous solution and with their applications in titrimetry.

FORMATION OF COMPLEXES

It is not possible to have an uncoordinated metal ion in aqueous solution. Water is an excellent ligand and will react to form an aquo ion. As a result, complexation reactions in aqueous solution are actually ligand replacement reactions, as illustrated by the following examples:

$$Cu(H_2O)_4^{2+} + 4NH_3 \rightleftharpoons Cu(NH_3)_4^{2+} + 4H_2O$$

$$Al(H_2O)_6^{3+} + 6F^- \rightleftharpoons AlF_6^{3-} + 6H_2O$$

It is common practice to omit the water and write the reactions in the following, simplified form:

$$Cu^{2+} + 4NH_3 \rightleftharpoons Cu(NH_3)_4^{2+} \tag{9-2}$$

$$Al^{3+} + 6F^- \rightleftharpoons AlF_6^{3-}$$

Such a practice will be followed in this book, but students should remember that uncoordinated metal ions do not exist in polar solvents.

Monodentate Ligands

Most monodentate ligands are simple inorganic anions or molecules containing a single donor atom. As you can see from the examples in Table 9-1, nitrogen, oxygen, sulfur, and halogens are the most common donor atoms. Some ligands, such as the carboxylate ion, may appear to be bidentate because they contain two *potential* donor atoms (the oxygens). However, the geometry of the bonding orbitals in the coordination compound requires the donor atoms to be spaced at certain distances and angles from each other. The two oxygens in the carboxylate ion cannot satisfy this geometrical requirement, so only one oxygen acts as a donor.

Polydentate Ligands

Polydentate ligands possess a variety of chemical and physical properties that make them very useful in analytical chemistry. They are, for example, often more selective than monodentate ligands in their reactions with metal ions. Reaction selectivity is largely a

TABLE 9-1 COMMON MONODENTATE LIGANDS

Neutral	Anionic
H_2O	F^-, Cl^-, Br^-, I^-
NH_3	SCN^-
RNH_2 (aliphatic amines)	CN^-
	OH^-
	RCO_2^- (carboxylate)
	S^{2-}

TABLE 9-2 COMMON POLYDENTATE LIGANDS

Type	Structure	Name
Bidentate		Ethylenediamine (en)
Tetradentate		Triethylenetetraamine (Trien)
		Nitrilotriacetic acid (NTA)
Hexadentate		Ethylenediaminetetra-acetic acid (EDTA)
		Cyclohexanediaminetetraacetic acid (CDTA)
Octadentate		Diethylenetriaminepenta-acetic acid (DTPA)

function of the geometry of the ligand and the type of donor atoms it contains. Some metal ions prefer to coordinate with oxygen, while others favor nitrogen or sulfur. Aminopolycarboxylic acids, containing both nitrogen and oxygen donors, form especially stable complexes with a large number of metal ions. Table 9-2 lists a few examples of the many polydentate ligands having applications in complexation titrimetry.

Metal complexes with polydentate ligands generally are more stable than those with similar monodentate ligands. For example, the equilibrium constant for Reaction 9-1 ($K_{eq} = 1.1 \times 10^{20}$) is 7 orders of magnitude larger than that for Reaction 9-2 ($K_{eq} = 1.1 \times 10^{13}$). Thermodynamics tells us that there are two driving forces for a chemical reaction: decreasing the enthalpy (liberation of heat, ΔH) and increasing the entropy (more disorder, ΔS). The enthalpy change is about the same for both reactions because the same number and types of bonds are formed in both cases, but the entropy change is much greater in Reaction 9-1 because it involves the combining of three particles to form the product, whereas Reaction 9-2 involves the combining of five particles (bringing more order or less disorder to the system).

USES OF COORDINATION COMPOUNDS

It is difficult to overstate the importance of coordination compounds in analytical chemistry. Organic chemists often are able to synthesize ligands that are designed to have the specific physical and chemical properties needed for a particular application. A few general uses are described briefly in this section. Specific applications are discussed in some of the later chapters.

Formation of Colored Substances

The formation of a colored compound can be used to indicate the presence of a specific metal ion and/or determine its concentration in a sample. For example, ferrous ion forms a highly colored complex with the bidentate ligand, 1,10-phenanthroline:

$$Fe^{2+} + 3 \quad \longrightarrow \quad Fe$$

colorless   red

The complex ion, whose common name is ferroin, is used both as an indicator for oxidation-reduction titrations and for the determination of trace concentrations of iron based on its intense red color. As little as 0.1 ppm Cu^{2+} has been determined based on the formation of reddish-brown bis-diethyldithiocarbamatecopper(II) in chloroform:

$$Cu^{2+} + 2(C_2H_5)_2N-C\overset{\displaystyle S}{\underset{\displaystyle S^-NH_4^+}{\diagdown}} \longrightarrow \left[(C_2H_5)_2N-C\overset{\displaystyle S}{\underset{\displaystyle S}{\diagdown}}\right]_2 Cu + 2NH_4^+$$

The development of color via a complexation reaction and subsequent measurement of the "intensity" or "degree" of that color (Chapters 15 and 16) represents an extremely important use of coordination compounds in analytical chemistry.

Chemical Masking

A *masking agent* decreases the concentration of a free metal ion to a level where a particular interfering reaction will not occur. For example, in the popular titrimetric method for determining copper, based on the titration of I_2 formed in the reaction between Cu^{2+} and I^-, ferric ion interferes by undergoing a similar redox reaction with iodide. The interference is eliminated by adding a soluble fluoride salt to the solution, prior to the addition of iodide ion, to complex the iron and prevent its reaction with the iodide.

$$Fe^{3+} + 6F^- \rightleftharpoons FeF_6^{3-}$$

$$FeF_6^{3-} + I^- \longrightarrow \text{N.R.}$$

Similarly, cyanide ion can be used to mask certain metals that interfere in the titration of Ca^{2+} or Mg^{2+} with a polydentate ligand called EDTA. In the absence of cyanide, transition metals such as Zn^{2+}, Cd^{2+}, and Hg^{2+} interfere by reacting simultaneously with the EDTA. When cyanide ion is added, these cations form stable complexes and are prevented from reacting with the EDTA.

Titrating Metal Ions

The titration of metals with monodentate ligands usually is not very successful. The large majority of metal ions have a coordination number of 4 or 6, which means they will combine with that many monodentate ligands. As we might expect, the ligands may be added sequentially, with each step being represented by a separate equilibrium constant. For example, $BiCl_4^-$ is formed in the following steps:

$$Bi^{3+} + Cl^- \rightleftharpoons BiCl^{2+} \qquad K_1 = \frac{[BiCl^{2+}]}{[Bi^{3+}][Cl^-]} = 1.6 \times 10^2$$

$$BiCl^{2+} + Cl^- \rightleftharpoons BiCl_2^+ \qquad K_2 = \frac{[BiCl_2^+]}{[BiCl^{2+}][Cl^-]} = 2.0 \times 10^1$$

$$BiCl_2^+ + Cl^- \rightleftharpoons BiCl_3 \qquad K_3 = \frac{[BiCl_3]}{[BiCl_2^+][Cl^-]} = 2.0 \times 10^2$$

$$BiCl_3 + Cl^- \rightleftharpoons BiCl_4^- \qquad K_4 = \frac{[BiCl_4^-]}{[BiCl_3][Cl^-]} = 1.0 \times 10^1$$

The overall reaction is

$$Bi^{3+} + 4Cl^- \rightleftharpoons BiCl_4^-$$

whose equilibrium constant is given by

$$K_{overall} = K_1 K_2 K_3 K_4 = 6.4 \times 10^6$$

If we consider the idea of using chloride ion as a titrant for bismuth, it must be discarded for two reasons. First, the equilibrium constants for the individual steps are too similar to allow any step to be "complete" before the next one begins. That is, a stepwise titration cannot be done. Second, while the overall equilibrium constant may be large enough for a successful titration, there will be no appreciable amount of free Bi^{3+} remaining after 25% titration due to the formation of the first complex, $BiCl^{2+}$. As a result, pBi^{3+} becomes large well before the equivalence point. The undesirable equilibrium effects due to stepwise reactions generally preclude the use of monodentate ligands as titrants for metal ions. The few successful applications usually involve metal ions that have a coordination number of 2.

In contrast to their monodentate cousins, polydentate ligands are exceptionally useful titrants for metal ions. Such ligands often react with metals in a single step, thereby avoiding the complications of stepwise reactions. Furthermore, they are often more selective than monodentate ligands in their reactions with metal ions. The selectivity is determined largely by the geometry of the ligand and the type of donor atoms it contains. One ligand, ethylenediaminetetraacetic acid, has proven to be exceptionally useful as a titrant, and the remainder of this chapter is devoted mostly to the behavior of this reagent.

COMPLEXES OF EDTA AND METAL IONS

Ethylenediaminetetraacetic acid, often abbreviated EDTA, is a hexadentate ligand, containing four oxygen and two nitrogen donor atoms (see Table 9-2). In strongly basic solution (pH > 12), all four carboxylic acid groups are unprotonated and EDTA forms stable, 1:1 complexes with virtually all multivalent metal ions. The structure of a typical metal–EDTA complex is shown in Figure 9-1. In addition to being a hexadentate ligand,

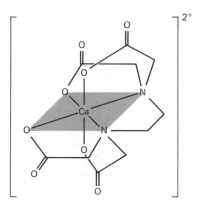

Figure 9-1 Structure of calcium–EDTA complex. Bond lengths are not to scale.

EDTA is also a tetraprotic acid and can exist in a variety of protonated forms, represented by H_4Y, H_3Y^-, H_2Y^{2-}, HY^{3-}, and Y^{4-}. Any or all of these forms may react with a given metal to yield a metal–EDTA complex, and each reaction would be represented with a different equilibrium constant. Chemists have arbitrarily chosen to use the reaction with the tetraanion (Y^{4-}) and its accompanying equilibrium constant to describe the formation of a metal–EDTA complex:

$$M^{n+} + Y^{4-} \rightleftharpoons MY^{n-4} \qquad K_{MY} = \frac{[MY^{n-4}]}{[M^{n+}][Y^{4-}]} \qquad (9\text{-}3)$$

Understand, however, that the hydrogen-ion concentration will affect the equilibrium position of Reaction 9-3 by influencing the concentration of Y^{4-} in the solution.

For EDTA to react with a metal ion, the hydrogens attached to the carboxylate groups must be removed. In strongly basic solution, these hydrogens are removed by reaction with hydroxide ion. In more acidic solutions, metal ions must be able to displace the hydrogens if a complex is to be formed. Since metal ions differ significantly in their ability to displace these hydrogens, the solution acidity can be used to "regulate" the reactivity of EDTA toward metal ions. For example, most metal ions react quantitatively with a stoichiometric amount of EDTA at pH 10, but only a few, such as Fe^{3+} and Hg^{2+}, also react quantitatively at pH 2.

Effect of pH on the Composition of EDTA

Equations that describe the effect of pH on the composition of a diprotic acid were derived in Chapter 6. You may recall that we found it convenient to describe the composition in terms of the relative concentrations (α) of the various species. A tetraprotic acid, such as EDTA, can exist in five different states of protonation: H_4Y, H_3Y^-, H_2Y^{2-}, HY^{3-}, and Y^{4-}. For the tetraanion,

$$\alpha_{Y^{4-}} = \frac{[Y^{4-}]}{[Y']} \qquad (9\text{-}4)$$

where $[Y']$ is the total concentration of uncomplexed EDTA or

$$[Y'] = [H_4Y] + [H_3Y^-] + [H_2Y^{2-}] + [HY^{3-}] + [Y^{4-}]$$

The value of $\alpha_{Y^{4-}}$ depends only on the concentration of H_3O^+ and the acid dissociation constants for EDTA. The expressions for calculating α values are easily derived using the method described in Chapter 6. For Y^{4-},

$$\alpha_{Y^{4-}} = \frac{K_{a_1}K_{a_2}K_{a_3}K_{a_4}}{[H_3O^+]^4 + K_{a_1}[H_3O^+]^3 + K_{a_1}K_{a_2}[H_3O^+]^2 + K_{a_1}K_{a_2}K_{a_3}[H_3O^+] + K_{a_1}K_{a_2}K_{a_3}K_{a_4}}$$

$$(9\text{-}5)$$

Equations for the other values are

$$\alpha_{HY^{3-}} = \frac{[HY^{3-}]}{[Y']} = \frac{K_{a_1} K_{a_2} K_{a_3}[H_3O^+]}{D}$$

$$\alpha_{H_2Y^{2-}} = \frac{[H_2Y^{2-}]}{[Y']} = \frac{K_{a_1} K_{a_2}[H_3O^+]^2}{D}$$

$$\alpha_{H_3Y^-} = \frac{[H_3Y^-]}{[Y']} = \frac{K_{a_1}[H_3O^+]^3}{D}$$

$$\alpha_{H_4Y} = \frac{[H_4Y]}{[Y']} = \frac{[H_3O^+]^4}{D}$$

where D is the denominator of Equation 9-5. The effect of pH on the fraction of EDTA existing in its various forms is shown graphically in Figure 9-2. Note that Y^{4-} is present at very low concentrations in acidic solution and does not become the predominant species until the pH exceeds a value of 10.

As will be seen shortly, values of $\alpha_{Y^{4-}}$ often are needed to calculate the metal-ion concentration of an EDTA solution. The values may be estimated from a graph such as Figure 9-2 or calculated using Equation 9-5. Selected values calculated from this equation are given in Table 9-3.

Example 9-1

Calculate the fraction of EDTA present as Y^{4-} in solution at pH 8.0 and at pH 11.0.

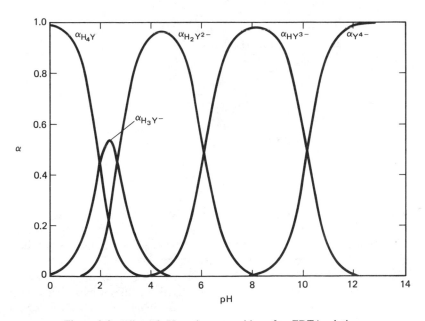

Figure 9-2 Effect of pH on the composition of an EDTA solution.

TABLE 9-3 VALUES OF $\alpha_{Y^{4-}}$ FOR EDTA AT DIFFERENT VALUES OF pH

pH	$\alpha_{Y^{4-}}$	pH	$\alpha_{Y^{4-}}$
2.0	5.0×10^{-14}	8.0	6.9×10^{-3}
3.0	3.4×10^{-11}	9.0	6.5×10^{-2}
4.0	5.0×10^{-9}	10.0	4.1×10^{-1}
5.0	4.8×10^{-7}	11.0	8.5×10^{-1}
6.0	3.0×10^{-5}	12.0	9.8×10^{-1}
7.0	6.1×10^{-4}		

Solution The fraction present as Y^{4-} is $\alpha_{Y^{4-}}$, which can be calculated from Equation 9-5. It is convenient first to calculate the products of the acid dissociation constants:

$$K_{a_1} = 1.0 \times 10^{-2}$$

$$K_{a_1}K_{a_2} = (1.0 \times 10^{-2})(2.1 \times 10^{-3}) = 2.1 \times 10^{-5}$$

$$K_{a_1}K_{a_2}K_{a_3} = (2.1 \times 10^{-5})(7.8 \times 10^{-7}) = 1.6 \times 10^{-11}$$

$$K_{a_1}K_{a_2}K_{a_3}K_{a_4} = (1.6 \times 10^{-11})(6.8 \times 10^{-11}) = 1.1 \times 10^{-21}$$

At pH *8.0.* $[H_3O^+] = 1.0 \times 10^{-8}$ *M* and substituting in Equation 9-5 gives

$$\alpha_{Y^{4-}} = \frac{1.1 \times 10^{-21}}{1.6 \times 10^{-19}} = 6.9 \times 10^{-3}$$

At pH *11.0.* $[H_3O^+] = 1.0 \times 10^{-11}$ *M* and substituting in Equation 9-5 gives

$$\alpha_{Y^{4-}} = \frac{1.1 \times 10^{-21}}{1.3 \times 10^{-21}} = 0.85$$

Effect of Auxiliary Complexing Agents on Metal-Ion Concentrations

Metals react most completely with EDTA in basic solution where many of them also form insoluble hydroxides or basic oxides. Once formed, these insoluble products will react only slowly with EDTA, making a titration impossible. To avoid this problem, an auxiliary complexing agent may be added to react with the metal ion and prevent its precipitation when the solution is made basic. The complex must be intermediate in stability between the metal hydroxide and the metal–EDTA complex. In this way it will form in preference to the hydroxide, but will give up the metal ion to the EDTA titrant as it is added. Ammonia is especially useful for this purpose because it forms soluble complexes with many transition metals, and when mixed with its conjugate acid, ammonium ion, it forms a basic pH buffer. Zinc forms four different ammonia complexes:

$$Zn^{2+} + NH_3 \rightleftharpoons Zn(NH_3)^{2+} \qquad K_{f_1} = \frac{[Zn(NH_3)^{2+}]}{[Zn^{2+}][NH_3]}$$

$$Zn(NH_3)^{2+} + NH_3 \rightleftharpoons Zn(NH_3)_2^{2+} \qquad K_{f_2} = \frac{[Zn(NH_3)_2^{2+}]}{[Zn(NH_3)^{2+}][NH_3]}$$

$$Zn(NH_3)_2^{2+} + NH_3 \rightleftharpoons Zn(NH_3)_3^{2+} \qquad K_{f_3} = \frac{[Zn(NH_3)_3^{2+}]}{[Zn(NH_3)_2^{2+}][NH_3]}$$

$$Zn(NH_3)_3^{2+} + NH_3 \rightleftharpoons Zn(NH_3)_4^{2+} \qquad K_{f_4} = \frac{[Zn(NH_3)_4^{2+}]}{[Zn(NH_3)_3^{2+}][NH_3]}$$

The relative concentrations of the different forms of zinc can be expressed in the same way they were for EDTA. Thus

$$\alpha_{Zn^{2+}} = \frac{[Zn^{2+}]}{[Zn']} \tag{9-6}$$

where $[Zn']$ is the total concentration of zinc not complexed to EDTA, or

$$[Zn'] = [Zn^{2+}] + [Zn(NH_3)^{2+}] + [Zn(NH_3)_2^{2+}]$$
$$+ [Zn(NH_3)_3^{2+}] + [Zn(NH_3)_4^{2+}]$$

The value of $\alpha_{Zn^{2+}}$ depends only on the concentration of NH_3 and the formation constants for the zinc–ammonia complexes. The expression for calculating $\alpha_{Zn^{2+}}$ is derived in the same manner as the expression for $\alpha_{Y^{4-}}$. The final equation is slightly different because it is derived using *formation* rather than dissociation constants for the zinc–ammonia complexes.

$$\alpha_{Zn^{2+}} = \frac{1}{1 + K_{f_1}[NH_3] + K_{f_1}K_{f_2}[NH_3]^2 + K_{f_1}K_{f_2}K_{f_3}[NH_3]^3 + K_{f_1}K_{f_2}K_{f_3}K_{f_4}[NH_3]^4} \tag{9-7}$$

Example 9-2

Calculate the concentration of Zn^{2+} in a 0.100 M zinc nitrate solution buffered at pH 9.15 with an NH_3/NH_4Cl solution. The free ammonia concentration in the buffer is 0.0800 M. The formation constants for the zinc–ammonia complexes are: $K_{f_1} = 1.62 \times 10^2$, $K_{f_2} = 1.95 \times 10^2$, $K_{f_3} = 2.29 \times 10^2$, and $K_{f_4} = 1.07 \times 10^2$.

Solution Since the total concentration of uncomplexed zinc, $[Zn']$, is known, the concentration of Zn^{2+} can be calculated if the value for $\alpha_{Zn^{2+}}$ is found:

$$\alpha_{Zn^{2+}} = \frac{[Zn^{2+}]}{[Zn']} \quad \text{or} \quad [Zn^{2+}] = \alpha_{Zn^{2+}}[Zn']$$

Complexes of EDTA and Metal Ions

Calculating the terms for the denominator of Equation 9-7 gives us

$$K_{f_1}[NH_3] = (1.62 \times 10^2)(0.0800) = 13.0$$

$$K_{f_1}K_{f_2}[NH_3]^2 = (1.62 \times 10^2)(1.95 \times 10^2)(0.0800)^2 = 202$$

$$K_{f_1}K_{f_2}K_{f_3}[NH_3]^3 = (1.62 \times 10^2)(1.95 \times 10^2)(2.29 \times 10^2)(0.0800)^3 = 3.70 \times 10^3$$

$$K_{f_1}K_{f_2}K_{f_3}K_{f_4}[NH_3]^4 = (1.62 \times 10^2)(1.95 \times 10^2)(2.29 \times 10^2)(1.07 \times 10^2)(0.0800)^4$$
$$= 3.17 \times 10^4$$

Substituting in Equation 9-7 yields

$$\alpha_{Zn^{2+}} = \frac{1}{1 + 13.0 + 202 + (3.70 \times 10^3) + (3.17 \times 10^4)} = 2.66 \times 10^{-5}$$

Finally,

$$[Zn^{2+}] = (2.66 \times 10^{-5})(0.0100) = 2.66 \times 10^{-7}\ M$$

Conditional Formation Constants

The general approach to solving metal–ligand equilibrium problems is not much different from that used to solve acid-base equilibrium problems. In fact, chemists generally try to make the approaches as similar as possible. You already know how to calculate the hydronium-ion concentration in an acetic acid solution. Now let us see what happens when we try to apply that same approach to finding the zinc-ion concentration in a ZnY^{2-} solution. The steps might be summarized as follows:

The equilibrium:
$$HOAc + H_2O \rightleftharpoons H_3O^+ + OAc^- \qquad Zn^{2+} + Y^{4-} \rightleftharpoons ZnY^{2-}$$

Initial equation: $\qquad K_a = \dfrac{[H_3O^+][OAc^-]}{[HOAc]} \qquad K_{ZnY} = \dfrac{[ZnY^{2-}]}{[Zn^{2+}][Y^{4-}]}$

Assumption 1: $\qquad [H_3O^+] \simeq [OAc^-] \qquad [Zn^{2+}] \simeq [Y^{4-}]$

Assumption 2: $\qquad [HOAc] \simeq C_{HOAc} \qquad [ZnY^{2-}] \simeq C_{ZnY^{2-}}$

Final equation: $\qquad K_a = \dfrac{[H_3O^+]^2}{C_{HOAc}} \qquad K_{ZnY^{2-}} = \dfrac{C_{ZnY^{2-}}}{[Zn^{2+}]^2}$

Solution: $\qquad [H_3O^+] = \sqrt{K_a C_{HOAc}} \qquad [Zn^{2+}] = \sqrt{C_{ZnY^{2-}}/K_{ZnY^{2-}}}$

Aside from the fact that chemists prefer to use *formation* rather than dissociation constants when dealing with metal complexes, the steps are the same. Unfortunately, assumption 1 is seldom valid for EDTA complexes because both the metal ion and the EDTA are usually involved in other competing equilibria. The assumption that *can* be made is that

$$[Zn'] = [Y']$$

because these concentrations are the sums of all the different equilibrium forms of zinc and EDTA that are not complexed to each other. If an expression can be developed containing these concentrations rather than the equilibrium concentrations of Zn^{2+} and Y^{4-}, the general problem-solving approach can stay the same. Such an expression can be derived quite simply. From Equations 9-6 and 9-4 it is known that

$$[Zn^{2+}] = \alpha_{Zn^{2+}}[Zn'] \quad \text{and} \quad [Y^{4-}] = \alpha_{Y^{4-}}[Y']$$

Substituting for these values in the equilibrium expression gives

$$K_{ZnY} = \frac{[ZnY^{2-}]}{(\alpha_{Zn^{2+}}[Zn'])(\alpha_{Y^{4-}}[Y'])}$$

or

$$K_{ZnY}(\alpha_{Zn^{2+}})(\alpha_{Y^{4-}}) = K_{Zn'Y'} = \frac{[ZnY^{2-}]}{[Zn'][Y']} \tag{9-8}$$

where $K_{Zn'Y'}$ is called the *conditional constant*. The word "conditional" is appropriate because $\alpha_{Zn^{2+}}$ and $\alpha_{Y^{4-}}$ are constant only under certain conditions. It is for this reason that you will not find a table of conditional formation constants. If the solution is properly buffered, the values for α will remain essentially constant throughout a given titration. The end result is as follows: If the *conditional* formation constants are used, the concentrations of metal and EDTA may be expressed as $[M']$ and $[Y']$, making the assumptions and problem-solving procedure the same as those used in acid-base calculations.

Example 9-3

Calculate pZn^{2+} in 5.00×10^{-3} M ZnY^{2-} at pH 10.0 if $\alpha_{Zn^{2+}}$ is 2.0×10^{-9} and $K_{ZnY^{2-}}$ is 3.16×10^{16}.

Solution The only source of Zn^{2+} in the solution is the dissociation of ZnY^{2-}. Since complexation equilibria are conventionally written as formations,

$$Zn^{2+} + Y^{4-} \rightleftharpoons ZnY^{2-}$$

for which

$$K_{ZnY^{2-}} = \frac{[ZnY^{2-}]}{[Zn^{2+}][Y^{4-}]}$$

Using the conditional formation constant, the expression can be written in terms of Zn' and Y':

$$K_{Zn'Y'} = K_{ZnY^{2-}}(\alpha_{Zn^{2+}})(\alpha_{Y^{4-}}) = \frac{[ZnY^{2-}]}{[Zn'][Y']}$$

Since the dissociation is responsible for all of the uncomplexed zinc and EDTA present,

$$[Zn'] = [Y']$$

Substituting in the equilibrium equation gives

$$K_{ZnY^{2-}}(\alpha_{Zn^{2+}})(\alpha_{Y^{4-}}) = \frac{[ZnY^{2-}]}{[Zn']^2}$$

or

$$(3.16 \times 10^{16})(2.0 \times 10^{-9})(0.35) = \frac{5.00 \times 10^{-3}}{[Zn']^2}$$

where $\alpha_{Y^{4-}}$ was obtained from Table 9-3. Thus

$$[Zn'] = \sqrt{(5.00 \times 10^{-3})/(2.21 \times 10^7)} = 1.50 \times 10^{-5} M$$

Finally,

$$[Zn^{2+}] = \alpha_{Zn^{2+}}[Zn'] = (2.0 \times 10^{-9})(1.50 \times 10^{-5}) = 3.0 \times 10^{-14} M$$

and

$$pZn^{2+} = -\log(3.0 \times 10^{-14}) = 13.52$$

It is possible for a formation constant to be conditional with respect to only one of the ions. Thus

$$K_{ZnY'} = K_{ZnY}\alpha_{Y^{4-}} = \frac{[ZnY^{2-}]}{[Zn^{2+}][Y']} \tag{9-9}$$

and

$$K_{Zn'Y} = K_{ZnY}\alpha_{Zn^{2+}} = \frac{[ZnY^{2-}]}{[Zn'][Y^{4-}]} \tag{9-10}$$

It is useful to remember that Equations 9-8 through 9-10 are equally valid in representing a zinc–EDTA equilibrium system. Which equation should be used in solving a numerical problem should be determined entirely by the data available, as illustrated by the following example.

Example 9-4

Calculate the concentration of Ca^{2+} in a solution containing 7.50×10^{-3} M CaY^{2-} and 1.00×10^{-2} M excess EDTA. $K_{CaY^{2-}} = 5.01 \times 10^{10}$; $\alpha_{Ca^{2+}} = 4.2 \times 10^{-3}$; $\alpha_{Y^{4-}} = 6.4 \times 10^{-1}$.

Solution The equilibrium reaction responsible for the formation of calcium is

$$Ca^{2+} + Y^{4-} \rightleftharpoons CaY^{2-}$$

By using the equilibrium constant conditional only in EDTA, the free calcium-ion concentration can be calculated directly.

$$K_{CaY'} = K_{CaY^{2-}}(\alpha_{Y^{4-}}) = \frac{[CaY^{2-}]}{[Ca^{2+}][Y']}$$

Substituting the known values gives

$$(5.01 \times 10^{10})(6.4 \times 10^{-1}) = \frac{7.5 \times 10^{-3}}{[Ca^{2+}](1.00 \times 10^{-2})}$$

or

$$[Ca^{2+}] = 2.3 \times 10^{-11} \ M$$

Note that if an equilibrium constant conditional in *both* calcium and EDTA had been used, $[Ca']$ would have been calculated, and another step would be required to obtain $[Ca^{2+}]$.

TITRATION CURVES

The course of a complexation titration can be followed by examining the metal-ion concentration as the titration progresses and plotting pM^{n+} versus the volume of titrant added. As is the case with acid-base titrations, the data for such a graph often can be measured with an ion-selective electrode and meter (Chapter 13). Also, the data can be calculated using equations derived earlier in this chapter.

Calculating Concentrations

Consider the titration of Ca^{2+} with EDTA. The titration reaction may be written as

$$Ca^{2+} + Y^{4-} \ \rightleftharpoons \ CaY^{2-} \tag{9-11}$$

The equilibrium constant for this reaction is the formation constant for the complex. If solution conditions are chosen such that the conditional formation constant is large, it will be possible to say that Reaction 9-11 goes to completion and any amount of EDTA added prior to the equivalence point will consume a stoichiometric amount of calcium. The specific equations used to calculate the calcium-ion concentration depend on the composition of the solution, which in turn depends on the stage or region of the titration. A complexation titration can be divided into three distinct regions:

Region	Major constituent	Major supplier of Ca^{2+}
1. Before the equivalence point	$Ca' + CaY^{2-}$	Ca'
2. At the equivalence point	CaY^{2-}	CaY^{2-}
3. After the equivalence point	$CaY^{2-} + Y'$	CaY^{2-}

Each region shall be considered separately, concentrating first on how to determine the composition of the solution and then on how to find the concentration of free metal ion.

Region 1: Before the equivalence point. Since the reaction stoichiometry is $1:1$ and the conditional formation constant is large (as it must be for a successful titration), the amount of untitrated calcium can be found by subtracting the amount of titrant added from the amount of calcium present initially.

$$\begin{array}{c} \text{amount Ca' initial} \\ -\left\{ \begin{array}{c} \text{amount Y' added} \\ \| \\ \text{amount Ca' consumed} \end{array} \right\} \\ \hline \text{amount Ca' remaining} \end{array}$$

and

$$[Ca'] = \frac{\text{amount Ca' remaining (mmol)}}{\text{volume (mL)}}$$

Note that the amount of calcium generated by the dissociation of the complex is negligible compared to the amount of untitrated calcium until very close to the equivalence point. The equilibrium concentration of free calcium ions can be obtained from the equivalent of Equation 9-6:

$$[Ca^{2+}] = \alpha_{Ca^{2+}}[Ca']$$

Region 2: At the equivalence point. The only uncomplexed calcium in the solution is that resulting from the slight dissociation of CaY^{2-}. As was pointed out earlier, chemists generally prefer to represent complexation equilibria using formation rather than dissociation reactions. Thus

$$Ca^{2+} + Y^{4-} \rightleftharpoons CaY^{2-}$$

for which

$$K_{CaY^{2-}} = \frac{[CaY^{2-}]}{[Ca^{2+}][Y^{4-}]}$$

If Ca^{2+} and Y^{4-} are involved in any other equilibria, we should rewrite this equation in terms of the appropriate conditional formation constant:

$$K_{Ca'Y'} = \frac{[CaY^{2-}]}{[Ca'][Y']}$$

The concentrations of Ca' and Y' are equal since they represent the amounts of calcium and EDTA that result from the dissociation of CaY^{2-}. Thus

$$K_{Ca'Y'} = \frac{[CaY^{2-}]}{[Ca']^2} \quad \text{or} \quad [Ca'] = \sqrt{[CaY^{2-}]/K_{Ca'Y'}}$$

As before, the equilibrium concentration of free calcium ions can be obtained from the equivalent of Equation 9-6:

$$[Ca^{2+}] = \alpha_{Ca^{2+}}[Ca']$$

Region 3: After the equivalence point. The CaY^{2-} complex is still the only source of Ca' in the solution, but it is not the major source of Y'. The amounts of complex and excess EDTA can be found by subtracting the amount of calcium present initially from the amount of EDTA added:

amount Y' added

$$- \left\{ \begin{array}{c} \text{amount Ca' initial} \\ \| \\ \text{amount Y' consumed} \end{array} \right\}$$

amount Y' remaining

amount CaY^{2-} formed = amount Ca' initial

and

$$[CaY^{2-}] = \frac{\text{amount } CaY^{2-} \text{ (mmol)}}{\text{volume solution (mL)}}$$

$$[Y'] = \frac{\text{amount } Y' \text{ (mmol)}}{\text{volume solution (mL)}}$$

These concentrations can be used in the equivalent of Equation 9-9 to calculate the concentration of Ca^{2+}:

$$K_{CaY'} = K_{CaY}(\alpha_{Y^{4-}}) = \frac{[CaY^{2-}]}{[Ca^{2+}][Y']}$$

Example 9-5

A 25.0-mL sample containing 2.50×10^{-2} M $Ca(NO_3)_2$ was titrated with 2.00×10^{-2} M EDTA at pH 10.0. Calculate the pCa^{2+} after addition of 20.00 mL, 31.25 mL, and 45.00 mL of EDTA. The formation constant for CaY^{2-} is 5.01×10^{10} and the values of $\alpha_{Ca^{2+}}$ and $\alpha_{Y^{4-}}$ are 0.140 and 0.350, respectively.

Solution

After 20.00 mL of EDTA *added.* The composition of the solution is determined by comparing the amount of calcium present with the amount of EDTA added.

$$
\left.
\begin{array}{l}
\text{amount Ca' initial} \\
-\left\{
\begin{array}{l}
\text{amount Y' added} \\
\quad\| \\
\text{amount Ca' consumed}
\end{array}
\right. \\
\hline
\text{amount Ca' remaining}
\end{array}
\right.
\begin{array}{l}
= 25.00 \text{ mL} \times 2.50 \times 10^{-2} \ M = 0.625 \text{ mmol} \\
\\
= 20.00 \text{ mL} \times 2.00 \times 10^{-2} \ M = 0.400 \text{ mmol} \\
\\
\hline
= 0.225 \text{ mmol}
\end{array}
$$

The concentration is obtained by dividing by the new volume,

$$[Ca'] = \frac{\text{amount Ca' (mmol)}}{\text{volume (mL)}} = \frac{0.225 \text{ mmol}}{45.00 \text{ mL}} = 5.00 \times 10^{-3} \ M$$

Finally,

$$[Ca^{2+}] = \alpha_{Ca^{2+}}[Ca'] = (0.140)(5.00 \times 10^{-3}) = 7.00 \times 10^{-4} \ M$$

and

$$pCa^{2+} = -\log(7.00 \times 10^{-4}) = 3.155$$

After 31.25 mL of EDTA *added.* Again, determining the composition, we have

$$
\left.
\begin{array}{l}
\text{amount Ca' initial} \\
-\left\{
\begin{array}{l}
\text{amount Y' added} \\
\quad\| \\
\text{amount Ca' consumed}
\end{array}
\right. \\
\hline
\text{amount Ca' or Y' remaining}
\end{array}
\right.
\begin{array}{l}
= 25.00 \text{ mL} \times 2.50 \times 10^{-2} \ M = 0.625 \text{ mmol} \\
\\
= 31.25 \text{ mL} \times 2.00 \times 10^{-2} \ M = 0.625 \text{ mmol} \\
\\
\hline
= \quad 0 \text{ mmol}
\end{array}
$$

Therefore, the 0.625 millimoles of Ca' present initially is now present as CaY^{2-}. The only source of Ca^{2+} is the dissociation of CaY^{2-}, whose equilibrium is represented by

$$Ca^{2+} + Y^{4-} \rightleftharpoons CaY^{2-}$$

Titration Curves

for which

$$K_{CaY^{2-}} = \frac{[CaY^{2-}]}{[Ca^{2+}][Y^{4-}]}$$

Since both calcium and EDTA are involved in other, competing equilibria, we must use the conditional formation constant,

$$K_{Ca'Y'} = K_{CaY}(\alpha_{Ca^{2+}})(\alpha_{Y^{4-}}) = \frac{[CaY^{2-}]}{[Ca'][Y']}$$

But

$$[Ca'] = [Y']$$

Therefore,

$$K_{CaY}(\alpha_{Ca^{2+}})(\alpha_{Y^{4-}}) = \frac{[CaY^{2-}]}{[Ca']^2}$$

or

$$[Ca'] = \sqrt{\frac{[CaY^{2-}]}{K_{CaY}(\alpha_{Ca^{2+}})(\alpha_{Y^{4-}})}}$$

$$= \sqrt{\frac{(0.625 \text{ mmol}/56.25 \text{ mL})}{(5.01 \times 10^{10})(0.140)(0.350)}} = 2.13 \times 10^{-6} \ M$$

Finally,

$$[Ca^{2+}] = \alpha_{Ca^{2+}}[Ca'] = (0.140)(2.13 \times 10^{-6}) = 2.98 \times 10^{-7} \ M$$

and

$$pCa^{2+} = -\log (2.98 \times 10^{-7}) = 6.526$$

After 45.00 mL of EDTA added. As before,

$$-\left\{ \begin{array}{c} \text{amount Ca' initial} \\ \| \\ \text{amount Y' consumed} \end{array} \right\} = 25.00 \text{ mL} \times 2.50 \times 10^{-2} \ M = 0.625 \text{ mmol}$$

$$\frac{\text{amount Y' added}}{\text{amount Y' remaining}} = 45.00 \text{ mL} \times 2.00 \times 10^{-2} \ M = \frac{0.900 \text{ mmol}}{= 0.275 \text{ mmol}}$$

amount CaY^{2-} formed = amount Ca' initial = 0.625 mmol

Again, the only source of Ca^{2+} is the dissociation of the complex whose equilibrium is represented by

$$Ca^{2+} + Y^{4-} \rightleftharpoons CaY^{2-}$$

Using the equilibrium constant conditional only in EDTA enables the concentration of Ca^{2+} to be calculated directly:

$$K_{CaY'} = K_{CaY}\alpha_{Y^{4-}} = \frac{[CaY^{2-}]}{[Ca^{2+}][Y']}$$

where

$$[CaY^{2-}] = 0.625 \text{ mmol}/70.00 \text{ mL} = 8.93 \times 10^{-3} \ M$$

and

$$[Y'] = 0.275 \text{ mmol}/70.00 \text{ mL} = 3.93 \times 10^{-3} M$$

Substitution yields

$$(5.01 \times 10^{10})(0.350) = \frac{(8.93 \times 10^{-3})}{[Ca^{2+}](3.93 \times 10^{-3})}$$

or

$$[Ca^{2+}] = 1.29 \times 10^{-10} M$$

and

$$pCa^{2+} = -\log (1.29 \times 10^{-10}) = 9.889$$

Factors Affecting the Shape of Titration Curves

It was pointed out in Chapter 7 that the analyte–titrant concentrations and the extent of the titration reaction are the primary factors that influence the shape of an acid-base titration curve. These same two factors affect the shape of a complexation titration curve.

Concentration

Figure 9-3 shows a family of curves for the titrations of different concentrations of calcium ion with EDTA. It is clear that as the concentration of either or both analyte

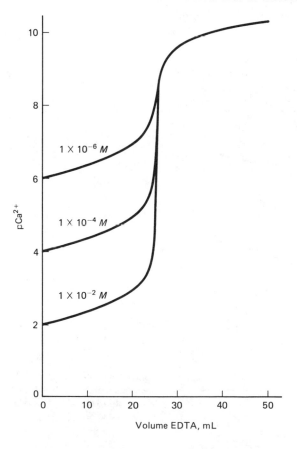

Figure 9-3 Curves for the titration of 25.0 mL of different concentrations of Ca^{2+} with an equal concentration of EDTA at pH 10.0.

Titration Curves

and titrant decrease, the magnitude and sharpness of the break at the equivalence point decrease. This is exactly the same effect that was observed with strong acid–strong base titrations, as illustrated in Figure 7-1. As a general rule, metal-ion concentrations less than about 0.001 M cannot be titrated with EDTA or similar ligands without unacceptably large errors.

Completeness of reaction

The completeness or extent of a complexation reaction is described by its conditional formation constant, which, in turn, depends partly on the values of α for the metal and the ligand. The effect of pH on the value of $\alpha_{Y^{4-}}$ has already been shown in Figure 9-2. Knowing that the equivalence-point break is less distinct for reactions that are less complete, there are several ways to predict the effect of pH on the shape of a complexation titration curve. The effect can be predicted qualitatively simply by looking at the titration reaction and using the principle of equilibrium. Consider the titration of Mg^{2+} with EDTA,

$$Mg^{2+} + Y^{4-} \rightleftharpoons MgY^{2-}$$

As the solution acidity increases, more Y^{4-} becomes protonated and the decreased concentration of Y^{4-} shifts the equilibrium to the left. The quantitative extent of this effect can be calculated using Equations 9-4 and 9-9. Figure 9-4 shows the curves for the titration of Mg^{2+} with EDTA at different pH values.

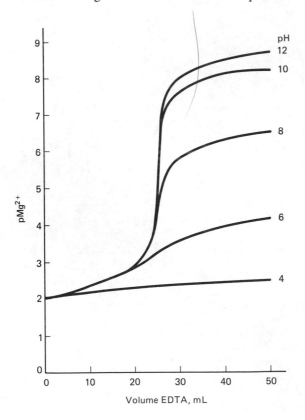

Figure 9-4 Effect of pH on the curve for the titration of 25.0 mL of 0.0100 M Mg^{2+} with 0.0100 M EDTA. The value of $\alpha_{Mg^{2+}}$ is assumed to be 1.0.

Chapter 9   Complexation Equilibria and Titrations

As a general rule, the conditional formation constant multiplied by the concentration of analyte must equal or exceed about 10^6 for a successful titration. Obviously, this minimum will not occur at the same pH for every metal because the values of K_{MY} are different. Figure 9-5 shows the minimum pH values at which various metal ions (at a concentration of 0.1 M) can be titrated assuming that they form no auxiliary complexes ($\alpha_{M^{n+}} = 1.0$) in solution. The data in this figure are useful in helping chemists to determine the solution conditions that will permit the selective titration of one metal ion in a mixture. For example, Mg^{2+} forms a relatively weak complex with EDTA, as evidenced by the fact that the pH must be at least 8.5 to have a successful titration. Zinc, on the other hand, forms a much stronger complex and can be titrated at a pH as low as

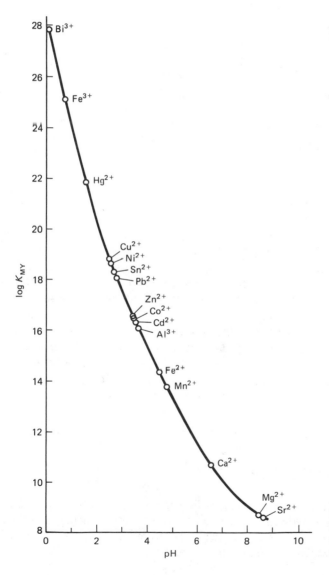

Figure 9-5 Minimum pH needed to make the conditional formation constant large enough to yield a successful titration. $C_M = 0.10\ M$; $\alpha_{M^{n+}} = 1.0$.

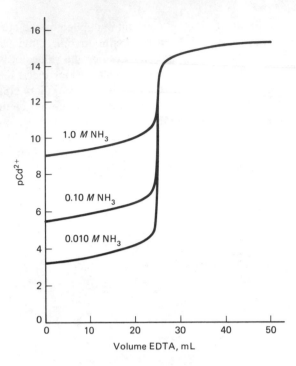

Figure 9-6 Effect of ammonia concentration on the curve for the titration of 25.0 mL of 1.00×10^{-2} M Cd^{2+} with 1.00×10^{-2} M EDTA at pH 9.00.

3.5. If a solution containing both ions at pH 7 was titrated with EDTA, only the zinc would yield a detectable end point.

The data in Figure 9-5 are based on the assumption that $\alpha_{M^{n+}}$ is 1.0; that is, the metal ions do not react with any substance in the solution other than EDTA. It was pointed out earlier that this assumption is seldom valid because many metal cations require the presence of auxiliary complexing agents to prevent the formation of insoluble hydroxides or basic oxides. As a result, the extent of a metal–EDTA titration reaction may be affected by the concentration of the auxiliary complexing agent, through its influence on $\alpha_{M^{n+}}$, in much the same way that it is affected by the hydronium-ion concentration through its influence on $\alpha_{Y^{4-}}$. Figure 9-6 shows the effect of ammonia concentration on the shape of the curve for the titration of Cd^{2+} at pH 9.0 with EDTA. This situation can be somewhat complicated for the analyst charged with finding appropriate conditions for the titration of a particular metal ion, because the concentration of the auxiliary complexing agent and the pH usually cannot be dealt with independently.

DETERMINING THE EQUIVALENCE POINT

Equivalence points in complexation titrations traditionally have been determined using the color changes of chemical indicators. In recent years, the development of ion-selective electrodes (Chapter 13) for a variety of metal ions has made the graphic method of equivalence-point determination a suitable alternative.

Theory of Complexation Indicators

Complexation indicators function in much the same way as acid-base indicators. These indicators are weak complexing agents that exhibit different colors in their complexed and uncomplexed forms. When such an indicator is added to the solution to be titrated, it forms a colored complex with the analyte:

$$M + In \rightleftharpoons MIn$$

where M and In refer to metal ion and indicator, respectively. The charges have been omitted for the sake of clarity. The titration flask now contains M and MIn. When titrant is added, it reacts with the free metal until essentially none is left, at which point it takes the metal away from MIn (charges omitted):

$$\underset{(\text{color A})}{MIn} + Y \rightleftharpoons \underset{(\text{colorless})}{MY} + \underset{(\text{color B})}{In} \qquad (9\text{-}12)$$

This constitutes the end-point reaction and is responsible for the color change that signals the analyst to stop the titration. Clearly, if Reaction 9-12 is to occur, the metal–titrant complex must be more stable than the metal–indicator complex.

Most indicators are affected by pH in the same manner as EDTA. That is, they form different protonated species that may exhibit different reactivities toward metal ions. In addition, these species often have different colors. For example, the indicator Eriochrome Black T (EBT) is a triprotic acid that may be represented as H_3In. In aqueous solution one proton is always completely dissociated. The acid-base properties of the indicator are summarized by the following ionization reactions:

$$\underset{(\text{red})}{H_2In^-} + H_2O \rightleftharpoons H_3O^+ + \underset{(\text{blue})}{HIn^{2-}} \qquad \begin{aligned} K_{a_2} &= 5 \times 10^{-7} \\ (pK_{a_2} &= 6.3) \end{aligned}$$

$$\underset{(\text{blue})}{HIn^{2-}} + H_2O \rightleftharpoons H_3O^+ + \underset{(\text{orange})}{In^{3-}} \qquad \begin{aligned} K_{a_3} &= 3 \times 10^{-12} \\ (pK_{a_3} &= 11.5) \end{aligned}$$

Metal complexes of EBT are generally red. Thus, if a color change is to be observed with this indicator, the pH of the solution must be between about 7 and 11, so that the blue form of the indicator dominates when the titrant breaks up the red metal–EBT complex at the end point. At a pH of 10, the end-point reaction is

$$\underset{(\text{red})}{MIn^-} + Y^{4-} + H^+ \rightleftharpoons MY^{2-} + \underset{(\text{blue})}{HIn^{2-}}$$

Common Indicators

Several hundred organic compounds have been suggested as potential indicators for various metal-ion titrations with EDTA. A few of those used most commonly are listed in Table 9-4.

Eriochrome Black T

This is one of the oldest and most widely used complexation indicators. It is used exclusively in the pH range 7 to 11, where the blue form of the indicator predominates

TABLE 9-4 COMMON INDICATORS FOR COMPLEXATION TITRATIONS

Name	Structure	Important pH equilibria
Eriochrome Black T		H_2In^- (red) $\xrightleftharpoons{pK_{a_2} = 6.3}$ HIn^{2-} (blue) HIn^{2-} (blue) $\xrightleftharpoons{pK_{a_3} = 11.6}$ In^{3-} (orange)
Calmagite		H_2In^- (red) $\xrightleftharpoons{pK_{a_2} = 8.1}$ HIn^{2-} (blue) HIn^{2-} (blue) $\xrightleftharpoons{pK_{a_3} = 12.4}$ In^{3-} (orange)
Arsenazo I		H_2In^{2-} $\xrightleftharpoons{pK_{a_3} = 8.33}$ HIn^{3-} HIn^{3-} $\xrightleftharpoons{pK_{a_4} = 11.76}$ In^{4-}
Xylenol Orange		H_5In^- (yellow) $\xrightleftharpoons{pK_{a_2} = 2.32}$ H_4In^{2-} (yellow) H_4In^{2-} (yellow) $\xrightleftharpoons{pK_{a_3} = 2.85}$ H_3In^{3-} (yellow) H_3In^{3-} (yellow) $\xrightleftharpoons{pK_{a_4} = 6.70}$ H_2In^{4-} (violet) H_2In^{4-} (violet) $\xrightleftharpoons{pK_{a_5} = 10.47}$ HIn^{5-} (violet)
Murexide		H_4In^- (red-violet) $\xrightleftharpoons{pK_{a_2} = 9.2}$ H_3In^{2-} (violet) H_3In^{2-} (violet) $\xrightleftharpoons{pK_{a_3} = 10.9}$ H_2In^{3-} (blue)

in the absence of metal ions. Although EBT forms red complexes with about 30 metals, only a few of the complexes have the necessary stability to permit a proper end-point color change in a direct titration with EDTA. The indicator is used most often in the direct titration of Mg^{2+}, Ca^{2+}, Cd^{2+}, Zn^{2+}, and Pb^{2+}. Common ions such as Al^{3+}, Cu^{2+}, Fe^{3+}, and Ni^{2+} form such stable complexes with EBT that the end-point reaction (Reaction 9-12) will not occur. These ions are said to "block" the indicator and must be absent or chemically masked when using EBT indicator. In solution Eriochrome Black T is oxidized slowly by dissolved oxygen, and a reducing agent such as ascorbic acid is sometimes added to retard this reaction. Another way to avoid oxidation is to add *solid* EBT directly to the sample solution just prior to the start of the titration.

Calmagite

The structure of this indicator is very similar to that of EBT, so it should not be a surprise to learn that it exhibits almost identical indicating properties. Calmagite is more stable than EBT in aqueous solution, which accounts for its growing popularity as a replacement for EBT.

Arsenazo I

This is an excellent indicator for EDTA titrations of the rare earths. It also finds some utility in calcium and magnesium titrations because, unlike EBT and Calmagite, it is not blocked by small amounts of copper or iron(III).

Xylenol Orange

This is one of the few indicators that can be used in acidic solutions. The free indicator is yellow below pH 6, while its metal complexes are red or violet. It is used most commonly in the direct titration of bismuth(III) and thorium(IV) and in back titrations where the excess EDTA is titrated with bismuth(III).

TITRANTS

EDTA is, by far, the most commonly used titrant for complexation reactions. It has most of the important properties sought in a titrant: It forms very stable, 1:1 complexes with most metal ions; its dilute solutions are very stable; and it is sensitive to solution conditions, particularly pH, that can make it somewhat selective in its reactivity with different metal ions.

Preparing EDTA Titrants

The forms of EDTA most commonly available are the free acid, H_4Y, and the disodium salt, Na_2H_2Y. The free acid is not very soluble in water and must be dissolved in dilute sodium hydroxide, where it is partially neutralized to Na_2H_2Y or Na_3HY. Neither form of EDTA is generally available in primary standard-grade purity, and therefore, solutions of approximate concentration are prepared and then standardized.

Dilute solutions of EDTA are quite stable and can be kept for long periods without undergoing changes in concentration. The surface of glass and even plastic containers is often contaminated with minute quantities of metal ions that are not removed during normal washing procedures. When EDTA solutions are stored in such containers, they can slowly leach these ions off the walls, converting them to soluble complexes. As a result, the concentration of the EDTA solution is decreased. Fortunately, the quantity of metal ions is usually very small, and their effect is not a problem unless the concentration of the EDTA is very low.

Standardizing EDTA Titrants

EDTA is usually standardized against a solution of calcium ions prepared by dissolving pure, dried calcium carbonate in hydrochloric acid and boiling to remove the evolved carbon dioxide. The titration is straightforward when Arsenazo I is used as the indicator, but when the still popular EBT (or Calmagite) is used, a minor difficulty is encountered. EBT is actually not a suitable indicator for the titration of calcium because it forms too weak a complex. In the early stages of the titration, the calcium-indicator complex will not dissociate appreciably, due to the presence of a large excess of untitrated calcium ions. That is, the position of equilibrium lies to the right for the reaction

$$Ca^{2+} + HIn^{2-} \rightleftharpoons CaIn^- + H^+$$
$$\text{(blue)} \qquad\qquad \text{(red)}$$

and the solution remains red. As the titration proceeds and more Ca^{2+} is complexed with the titrant, the position of equilibrium shifts to the left, causing a gradual color change to occur. Chemists have devised a solution to this problem based on three facts:

1. At pH 10, $K_{CaY'}$ is 1.8×10^{10} and $K_{MgY'}$ is 1.8×10^8, which is to say, CaY^{2-} is more stable than MgY^{2-}.
2. At pH 10, K_{MgIn^-} is 1.3×10^7 and K_{CaIn^-} is 1.3×10^5, which is to say, $MgIn^-$ is more stable than $CaIn^-$.
3. $MgIn^-$ is sufficiently stable that it will not dissociate appreciably prior to the equivalence point.

If a small amount of magnesium ion is added to an unstandardized EDTA solution, MgY^{2-} will form. Although this changes slightly the EDTA concentration, it is of no consequence since the titrant has not yet been standardized. When this titrant is added to the calcium-ion standard, an exchange reaction with calcium ions occurs:

$$Ca^{2+} + MgY^{2-} \longrightarrow CaY^{2-} + Mg^{2+}$$

The liberated Mg^{2+} is now free to displace calcium ion from its indicator complex:

$$Mg^{2+} + CaIn^- \longrightarrow MgIn^- + Ca^{2+}$$

The $MgIn^-$ produced, being more stable than its calcium analog, will remain essentially undissociated until the equivalence point is reached, and the problem of a gradual color change is avoided. This situation is illustrated in Figure 9-7, which shows the color transition for calcium-EBT occurring slightly before the equivalence point of the calcium-EDTA titration. The magnesium-EBT color transition matches the equivalence point break almost perfectly.

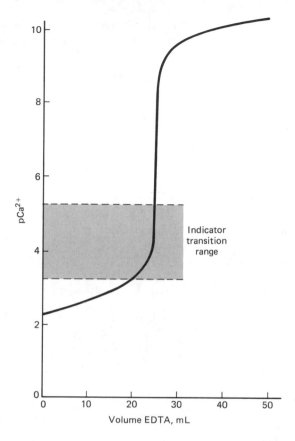

Figure 9-7 Curve for the titration of 25.0 mL of $5.00 \times 10^{-3}\ M\ Ca^{2+}$ at pH 10.0 with $5.00 \times 10^{-3}\ M$ EDTA.

APPLICATIONS

The fact that EDTA forms complexes with virtually all cations might lead one to conclude that it is completely lacking in selectivity and therefore is quite useless in the world of "real" samples. This is not the case, however. We are able to achieve considerable control over the reactivity of both EDTA and metal ion by regulating the pH and concentration of auxiliary complexing agents. Specific procedures and conditions are too numerous and varied to describe here. Instead, a brief description is given of the different types of EDTA titrations that are commonly found in the literature.

Direct Titrations

At least three dozen metals have been determined by direct titration with EDTA. Those forming "weak" complexes such as Ca^{2+} and Mg^{2+} must be titrated in basic solution where EBT, Calmagite, and Arsenazo I are the preferred indicators. Metal ions forming very stable complexes may be titrated in either acidic or basic solution. The "hardness" of water, which is due to the presence of calcium and magnesium salts, is commonly determined by a direct EDTA titration. When the titration is carried out at pH 10, both cations react together, allowing the sum to be determined. The problem of premature metal-indicator dissociation encountered in the standardization of EDTA with calcium does not occur here because the magnesium needed to make the indicator function properly is present in the sample.

Example 9-6

A 100.0-mL aliquot of city drinking water was treated with a small amount of an ammonia-ammonium chloride buffer to bring the pH to 10. After the addition of Calmagite indicator, the solution required 21.46 mL of 5.140×10^{-3} M EDTA for titration. Calculate the water hardness in terms of parts per million calcium carbonate.

Solution If we assume the density of the water sample to be 1.000 g/mL, then

$$\text{ppm CaCO}_3 = \frac{\text{wt CaCO}_3 \text{ (mg)}}{\text{vol sample (L)}}$$

The weight of $CaCO_3$ can be obtained from the titration data.

$$\text{amount Ca}^{2+} = \text{amount EDTA} \times \frac{1\text{Ca}^{2+}}{1\text{Y}^{4-}}$$

$$= (21.46 \text{ mL})(5.140 \times 10^{-3} \text{ } M)(1/1) = 0.1103 \text{ mmol}$$

$$\text{amount CaCO}_3 = \text{amount Ca}^{2+} \times \frac{1\text{CaCO}_3}{1\text{Ca}}$$

$$= (0.1103 \text{ mmol})(1/1) = 0.1103 \text{ mmol}$$

Converting amount to weight,

$$\text{wt CaCO}_3 = \text{amount CaCO}_3 \times \text{MW}$$

$$= (0.1103 \text{ mmol})(100.1 \text{ mg/mmol}) = 11.04 \text{ mg}$$

Finally,

$$\text{ppm CaCO}_3 = \frac{11.04 \text{ mg}}{0.1000 \text{ L}} = 110.4$$

The three things that most often make direct titrations unsuccessful are: a slow analyte-EDTA reaction, inability to keep the analyte soluble at the conditions necessary for a direct titration, and lack of a suitable indicator. Some of these problems may be overcome by using another type of titration.

Back Titrations

Cations that cannot be titrated directly but form very stable EDTA complexes are good candidates for a back titration. A measured amount of EDTA is added to the analyte solution and the excess is back titrated with a standard magnesium-ion solution. A "weak" metal such as magnesium is used as the titrant for the back titration because it must not displace the analyte from its EDTA complex.

The minimum pH for a quantitative reaction between aluminum and EDTA is about 4. Even at this acidic pH, $Al(OH)_3$ can precipitate. Furthermore, Al^{3+} blocks common indicators such as EBT and Calmagite. In the back titration of aluminum, the acidic sample is treated with EDTA, adjusted to about pH 7 with a buffer, and boiled to ensure complete formation of the complex. The AlY^- complex is quite stable at this pH and Calmagite can safely be added for the back titration of the excess EDTA with a standard magnesium-ion solution.

Example 9-7

A 25.00-mL aliquot of a solution containing Hg^{2+} in dilute nitric acid was treated with 10.00 mL of 0.04882 M EDTA and the pH was adjusted to 10.0 with an ammonia buffer. Two drops of a freshly prepared EBT indicator solution were added and the excess EDTA back titrated with 0.01137 M Mg^{2+}, requiring 24.66 mL to reach the end point. What is the molarity of Hg^{2+} in the sample?

Solution Both Hg^{2+} and Mg^{2+} form 1:1 complexes with EDTA, so we can determine the amounts easily.

$$
\begin{aligned}
\text{amount EDTA added} &= 10.00 \text{ mL} \times 0.04882 \ M = 0.4882 \text{ mmol} \\
-\left\{
\begin{array}{l}
\text{amount } Mg^{2+} \text{ back titration} \\
\qquad \parallel \\
\text{amount EDTA remaining}
\end{array}
\right\} &= 24.66 \text{ mL} \times 0.01137 \ M = 0.2804 \text{ mmol} \\
\overline{\text{amount EDTA used} = \text{amount } Hg^{2+} \text{ present}} &\phantom{= 24.66 \text{ mL} \times 0.01137 \ M} = \overline{0.2078 \text{ mmol}}
\end{aligned}
$$

Then

$$
[Hg^{2+}] = \frac{0.2078 \text{ mmol } Hg^{2+}}{25.00 \text{ mL}} = 8.312 \times 10^{-3} \ M
$$

Replacement Titrations

The unavailability of a suitable indicator may also be overcome by replacing the analyte with an acceptable substitute. For example, Hg^{2+} forms a strong complex with EDTA, but it cannot be titrated using EBT or Calmagite indicator. A sample containing Hg^{2+}, when treated with an excess of MgY^{2-}, undergoes a replacement reaction:

$$
\underset{\text{less stable}}{Hg^{2+} + MgY^{2-}} \rightleftharpoons \underset{\text{more stable}}{HgY^{2-} + Mg^{2+}}
$$

This reaction can be carried out in weakly acidic solution to prevent the formation of insoluble hydrated mercuric oxide [mercury(II) does not form a true hydroxide]. Once the reaction is complete, the solution can be made basic, EBT or Calmagite added, and the liberated Mg^{2+} titrated with EDTA. For this procedure to be successful, the formation constant for the analyte–EDTA complex must be greater than that for the metal–EDTA complex to be replaced.

PROBLEMS

9-1. Calculate the formation constants conditional in both metal and ligand for CdY^{2-} at the following pH values. Assume that $\alpha_{Cd^{2+}}$ is 4.5×10^{-4} in all cases.

    (a) 5.00   **(b)** 8.00   **(c)** 11.00

9-2. What is the equilibrium concentration of Cu^{2+} in $7.50 \times 10^{-3} \ M$ CuY^{2-} at pH 9.00 if $\alpha_{Cu^{2+}}$ is 5.20×10^{-5}?

9-3. A solution containing $1.25 \times 10^{-2} \ M$ ZnY^{2-} at pH 6.00 has a concentration of Zn^{2+} of $8.44 \times 10^{-9} \ M$. What is the value of $\alpha_{Zn^{2+}}$?

9-4. A 25.00-mL aliquot of $3.17 \times 10^{-2} \ M$ $Pb(NO_3)_2$ was titrated at pH 10.00 with $4.83 \times$

10^{-2} M EDTA. If $\alpha_{Pb^{2+}}$ is 0.0147, calculate the value of pPb^{2+} at the following points in the titration.

(a) before the addition of any titrant
(b) after the addition of 10.00 mL of titrant
(c) at the equivalence point
(d) after the addition of 35.00 mL of titrant

9-5. A solution containing 100.0 mL of 5.04×10^{-3} M EDTA was titrated at pH 3.00 with 2.19×10^{-2} M $Fe(NO_3)_3$. If $\alpha_{Fe^{3+}}$ was 0.680, calculate the value of pFe^{3+} at the following points in the titration.

(a) before the addition of any titrant
(b) after the addition of 15.00 mL of titrant
(c) at the equivalence point
(d) at 15.00 mL past the equivalence point

9-6. If $\alpha_{M^{n+}} = 5.0 \times 10^{-3}$, calculate the minimum pH, rounded to the nearest whole number, at which 0.0200 M solutions of each of the following cations can be titrated successfully with EDTA. You may use Table 9-3.

(a) Fe^{2+}	(c) Ni^{2+}	(e) Ca^{2+}
(b) Zn^{2+}	(d) Mg^{2+}	

9-7. Determine which of the following metal ions (at 0.010 M) can be titrated successfully at pH 7.0 with EDTA.

(a) Ba^{2+}; $\alpha_{Ba^{2+}} = 1.0$	(d) Cu^{2+}; $\alpha_{Cu^{2+}} = 0.40$
(b) Mn^{2+}; $\alpha_{Mn^{2+}} = 3.3 \times 10^{-4}$	(e) Sn^{2+}; $\alpha_{Sn^{2+}} = 8.5 \times 10^{-8}$
(c) Al^{3+}; $\alpha_{Al^{3+}} = 2.0 \times 10^{-5}$	

9-8. Calculate the smallest value of $\alpha_{M^{n+}}$ that will still permit each of the following to be titrated successfully with EDTA.

(a) 0.010 M Bi^{3+} at pH 5.0	(d) 0.0010 M Cd^{2+} at pH 10.0
(b) 0.010 M Hg^{2+} at pH 5.0	(e) 0.50 M Mg^{2+} at pH 12.0
(c) 0.10 M Ni^{2+} at pH 4.0	

9-9. Determine the molar concentration of an EDTA solution of which 26.44 mL were required to titrate a sample containing the Ca^{2+} from 287.4 mg of primary standard $CaCO_3$.

9-10. What weight of calcium carbonate must be dissolved in 250.0 mL of solution if a 50.00-mL aliquot is to require about 30 mL of 2.0×10^{-2} M EDTA for titration?

9-11. A 100.0-mL sample of drinking water was buffered at pH 10.0 and, after addition of Calmagite indicator, required 38.41 mL of 4.652×10^{-3} M EDTA for titration. Calculate the total hardness of the water as indicated.

(a) ppm $CaCO_3$ (b) ppm CaO

9-12. A sample of powdered milk weighing 1.450 g was mixed with an aqueous buffer of pH 11.5. A few drops of Calmagite indicator were added and the solution required 31.62 mL of 1.538×10^{-2} M EDTA to reach the end-point color change. Calculate the percentage of calcium in the milk.

9-13. The sulfate in a 247.1-mg sample was precipitated as $BaSO_4$ by addition of 25.00 mL of 0.03992 M $BaCl_2$. The precipitate was removed by filtration and the remaining $BaCl_2$ consumed 36.09 mL of 0.02017 M EDTA for titration to the Calmagite end point. Calculate the % SO_3 in the sample.

9-14. A 10.00-mL aliquot of a commercial algaecide containing an organomercury compound was treated with concentrated nitric acid and evaporated just to dryness. The residue of mercuric nitrate was dissolved in dilute nitric acid and diluted to 250.0 mL with water. A 50.00-mL aliquot of this solution was treated with 20.00 mL of 0.04966 M EDTA and mixed thoroughly for 10 min. After adjusting the pH to 10, the excess EDTA required

18.04 mL of 0.04711 M $MgCl_2$ for titration to the EBT end point. Calculate the concentration of mercury in the original sample in units of mg Hg/mL.

9-15. The bismuth in a 0.4840-g sample of Wood's metal (a low-melting alloy used in emergency water-sprinkling systems) was separated from the other metals and dissolved in nitric acid. This solution was treated with an excess of MgY^{2-} and, after adjusting the pH to 10, the liberated Mg^{2+} required 22.61 mL of 0.05215 M EDTA for titration to the EBT end point. Calculate the percentage of bismuth in the Wood's metal.

9-16. An antacid tablet weighing 15.476 g was dissolved in acid and diluted to 500.0 mL. A 25.00-mL aliquot of the solution was made sufficiently basic to precipitate the aluminum as $Al(OH)_3$. The remaining magnesium required 16.49 mL of 1.043×10^{-2} M EDTA for titration. A second 25.00-mL aliquot was withdrawn from the 500-mL flask and treated with 50.00 mL of the EDTA. This solution was made basic and the excess EDTA back titrated with 11.73 mL of 5.594×10^{-3} M $MgCl_2$. Calculate the percentage of both magnesium and aluminum in the sample.

9-17. A second 100-mL aliquot of the sample described in Problem 9-11 was treated with excess sodium oxalate. The precipitated calcium oxalate was removed and discarded. The remaining solution required 16.1 mL of 4.65×10^{-3} M EDTA for titration. Calculate the concentration of both calcium and magnesium ions in parts per million.

9-18. The cadmium and lead ions in a 50.00-mL sample required 40.09 mL of 0.001870 M EDTA for titration. A 75.00-mL portion of the same sample was made basic and treated with excess KCN, masking the cadmium as $Cd(CN)_4^{2-}$. This solution required 31.44 mL of the EDTA for titration. Calculate the concentration of Cd^{2+} and Pb^{2+} in the sample in parts per million.

9-19. A 1.3174-g sample containing the chloride salts of magnesium, mercury(II), and zinc was dissolved in 250.0 mL of solution. A 50.00-mL aliquot was treated with 10 mL of an NH_3/NH_4Cl buffer at pH 10 followed by 25.00 mL of 0.05331 M EDTA. After a few minutes of mixing, the excess EDTA was back titrated with 11.43 mL of 0.01816 M $MgCl_2$. A second 50.00-mL aliquot was made basic and treated with excess NaCN, complexing both the mercury and zinc. The magnesium in this sample required 16.83 mL of 0.005583 M EDTA for titration. The solution remaining at the end point of this titration was treated with excess formaldehyde, which reacts with the free CN^- and with $Zn(CN)_4^{2-}$:

$$CN^- + HCHO + H_2O \longrightarrow H_2C \overset{\nearrow \; OH}{\underset{\searrow \; CN}{}} + OH^-$$

The liberated Zn^{2+} required 28.47 mL of the EDTA for titration. Calculate the percentage of each metal in the sample.

10

OXIDATION-REDUCTION EQUILIBRIA

Modern analytical chemistry relies heavily on the measurements of certain electrical properties of materials such as potential and current that are based on the ability of substances to react by accepting or donating electrons. The fundamental aspects of electron transfer processes as they relate to the measurement and use of electrode potentials are the foundation on which the later treatment of redox titrations, potentiometry, and advanced electrochemical techniques lay. An ability to balance redox reactions by the method of half-reactions is essential to the understanding of much of the material presented in this chapter. Students are urged to read Appendix B, which reviews the methodology of balancing redox reactions and to practice their balancing skills.

THE OXIDATION-REDUCTION PROCESS

An oxidation-reduction (or redox) reaction is said to occur if the reactants undergo a change in oxidation number. *Oxidation* describes an increase in oxidation number, a process resulting from a loss of electrons. *Reduction* describes a decrease in oxidation number, a process resulting from a gain of electrons. Since free electrons are not thought to exist in solution, it should be clear that any oxidation (electron loss) must be accompanied by a corresponding reduction (electron gain). The substance causing the oxidation is called the *oxidizing agent* or *oxidant*, while the substance causing the reduction is called the *reducing agent* or *reductant*. To cause an oxidation, the oxidizing agent must take one or more electrons from the substance being oxidized. As a result, the oxidizing agent is reduced. Conversely, a reducing agent is oxidized in a chemical reaction. Applying this terminology to the reaction

$$2Fe^{3+} + Sn^{2+} \rightleftharpoons Sn^{4+} + 2Fe^{2+} \qquad (10\text{-}1)$$

$$\underset{\substack{\text{oxidizing} \\ \text{agent}}}{2Fe^{3+}} \quad \underset{\substack{\text{reducing} \\ \text{agent}}}{Sn^{2+}}$$

we would say that Sn^{2+} is oxidized by the oxidizing agent, Fe^{3+}, or that Fe^{3+} is reduced by the reducing agent, Sn^{2+}.

Chemists find it convenient to view a redox reaction as occurring in two steps, called half-reactions. A half-reaction is simply a representation of the oxidation or reduction process alone. Accordingly, Reaction 10-1 can be viewed as consisting of the two half-reactions

$$\text{Reduction: } Fe^{3+} + e^- \rightleftharpoons Fe^{2+} \qquad (10\text{-}2)$$
$$\text{Oxidation: } Sn^{2+} \rightleftharpoons Sn^{4+} + 2e^- \qquad (10\text{-}3)$$

One of the more popular methods of balancing redox reactions uses this concept of half-reactions on the premise that it is easier to balance half the reaction at a time than the entire reaction at once. When the halves are balanced, they may be combined to get the whole reaction. If we were given the two half-reactions in Reactions 10-2 and 10-3, the complete reaction could be obtained by multiplying the iron half-reaction by 2, to make the electrons equal in both halves, and then adding:

$$2Fe^{3+} + 2e^- \rightleftharpoons 2Fe^{2+}$$
$$Sn^{2+} \rightleftharpoons Sn^{4+} + 2e^-$$
$$\overline{2Fe^{3+} + 2e^- + Sn^{2+} \rightleftharpoons 2Fe^{2+} + Sn^{4+} + 2e^-}$$

There is another way that two half-reactions can be combined to yield a whole reaction: The two half-reactions can be written as reductions and subtracted from each other.

$$2Fe^{3+} + 2e^- \rightleftharpoons 2Fe^{2+}$$
$$Sn^{4+} + 2e^- \rightleftharpoons Sn^{2+}$$
$$\overline{2Fe^{3+} + 2e^- - Sn^{4+} - 2e^- \rightleftharpoons 2Fe^{2+} - Sn^{2+}}$$

or

$$2Fe^{3+} + Sn^{2+} \rightleftharpoons 2Fe^{2+} + Sn^{4+}$$

Although subtracting is perhaps slightly more complicated than adding, it has the advantage of familiarizing us with reduction half-reactions, which, as we shall see shortly, are used in conjunction with the equation for calculating electrode potentials.

Galvanic Cells

It is easy to demonstrate in the laboratory that the following reaction takes place spontaneously:

$$Zn(s) + Cu^{2+} \rightleftharpoons Cu(s) + Zn^{2+}$$

The Oxidation-Reduction Process

In reacting, each zinc atom gives two electrons to a copper(II) ion. If we construct a physical arrangement such that Zn(s) and Cu^{2+} do not come into direct contact with each other, the transferred electrons can be made to flow through an external circuit; that is, we create a source of usable electricity. Such devices are popularly referred to as batteries, but chemists prefer a slightly more restrictive name—*galvanic cells*. When a battery is discharging, it is behaving as a galvanic cell, but when it is charging, it is called an *electrolytic* cell. A typical galvanic cell is shown in Figure 10-1. The strips of zinc and copper metal are called *electrodes* and the inverted U-shaped tube connecting the solutions is referred to as a *salt bridge*. The reactions occurring at the two electrodes are:

$$\text{At Zn electrode:} \quad Zn(s) \;\rightleftharpoons\; Zn^{2+} + 2e^-$$
$$\text{At Cu electrode:} \quad Cu^{2+} + 2e^- \;\rightleftharpoons\; Cu(s)$$

When a zinc atom leaves the electrode and enters the solution as a zinc ion, two electrons are left on the electrode, making it electron-rich or negative. At the copper electrode, the reverse process causes an electron-deficient condition, resulting in a positive electrode. Electrons flow through the wire (external circuit) from where they are formed (the negative electrode) to where they are consumed (the positive electrode).

Anode and cathode

Each electrode in a cell is defined in terms of the reaction that occurs at its surface and not, as is sometimes mistakenly taught, in terms of its electrical charge. The *anode* is the electrode at which oxidation occurs and the *cathode* is the electrode at which reduction occurs. These definitions apply to both galvanic and electrolytic cells.

The zinc and copper electrodes in the cell depicted in Figure 10-1 are not only metallic conductors of electricity but also are reactants and products of the cell reaction. It is not uncommon for cells to have electrodes that are *only* metallic conductors and not

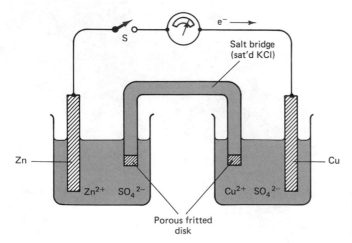

Figure 10-1 Galvanic cell with liquid salt bridge. Arrow shows direction of electron flow when switch S is closed.

Chapter 10   Oxidation-Reduction Equilibria

directly involved in the cell reaction. These *inert electrodes* simply act like an electron bank, dispensing and receiving electrons to and from the substances in the solution. Such a cell is shown in Figure 10-2. The cell reaction is

$$2Fe^{3+} + Sn^{2+} \rightleftharpoons 2Fe^{2+} + Sn^{4+}$$

In the breaker on the left, stannous ions come to the platinum electrode to leave two electrons as they become oxidized to stannic ions, thereby making the electrode negative. At the platinum electrode in the right-hand beaker, ferric ions are being reduced to ferrous ions by acquiring an electron.

Salt bridge

If the electrode reactions are to occur to any measurable extent, there must be a way for *ions* to move between the solutions in order to maintain a condition of electroneutrality. When a zinc atom is oxidized and goes into solution as the ion, the solution acquires a net positive charge of 2 units. If this charge is not lowered, it will inhibit the formation of additional zinc ions that would increase the net positive charge. The same argument applies to the copper solution except that a net negative charge occurs. The salt bridge allows ions to flow in and out of the solutions as necessary to maintain electroneutrality.

It is now possible to trace the complete flow of current in a cell. Electroneutrality can be maintained by the left-to-right flow of positive ions (Figure 10-3a) or the right-to-left flow of negative ions (Figure 10-3b) through the solutions. Examining Figure 10-3a, we observe that for each Zn^{2+} entering the solution from the electrode, another Zn^{2+} leaves by moving into the salt bridge. In turn, the salt bridge maintains its electroneutrality by moving two potassium ions into the copper solution. These two positive ions are exactly what is needed to replace the Cu^{2+} that is reduced at the electrode. This process represents a flow of positive electricity from left to right through the solution. An alternative mechanism for satisfying electroneutrality is illustrated in Figure 10-3b, where negative electricity flows from right to left through the solution. The two pro-

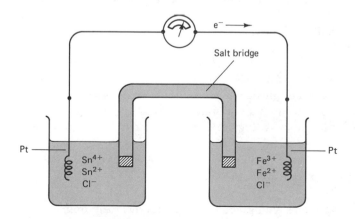

Figure 10-2 Galvanic cell with inert metal electrodes.

The Oxidation-Reduction Process

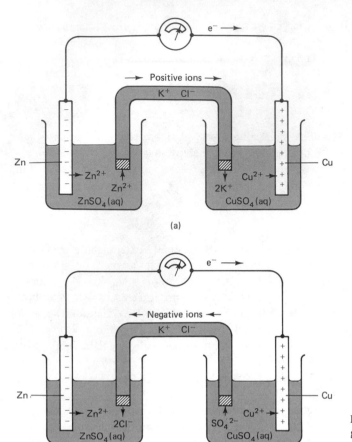

Figure 10-3 Movement of charge in a galvanic cell: (a) left-to-right flow of positive ions; (b) right-to-left flow of negative ions.

cesses depicted in Figure 10-3 are not mutually exclusive; that is, they both can (and usually do) occur simultaneously.

For reasons that are described later, it is desirable that the cation and anion comprising the electrolyte in the salt bridge have about the same mobility (rates of diffusion and migration) in the solution. Potassium chloride fulfills this requirement quite well and is, by far, the electrolyte used most commonly in salt bridges.

Cells without salt bridges

Sometimes useful cells can be constructed in which the electrodes share a common electrolyte, thereby eliminating the need for a salt bridge. The lead storage battery used in automobiles is a collection of such cells (Figure 10-4). The electrode reactions are:

At the cathode: $PbO_2(s) + 4H^+ + SO_4^{2-} + 2e^- \rightleftharpoons PbSO_4(s) + 2H_2O$
At the anode: $Pb(s) + SO_4^{2-} \rightleftharpoons PbSO_4(s) + 2e^-$

Both the oxidizing and reducing agents are insoluble, which prevents them from mixing in the solution and reacting directly with each other.

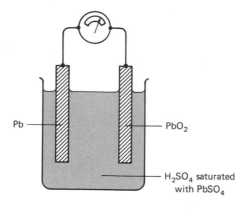

Figure 10-4   Galvanic cell containing no salt bridge.

Schematic representation of cells

Chemists have developed a shorthand notation for representing electrochemical cells. By convention, the chemical constituents comprising the cell are listed in the same order in which they would be encountered if you began at the anode and traveled through the cell solutions to the cathode. This corresponds to placing the anode on the left, listing the substances in the order of an oxidation, and the cathode on the right, listing the substances in the order of a reduction. A phase boundary across which a potential difference exists is represented by a single vertical line. A salt bridge is represented by double vertical lines. Accordingly, the cells shown in Figures 10-1 and 10-2 are represented by

$$Zn(s) \,\vert\, ZnSO_4(conc) \,\Vert\, CuSO_4(conc) \,\vert\, Cu(s)$$

and

$$Pt \,\vert\, SnCl_2(conc),\ SnCl_4(conc) \,\Vert\, FeCl_3(conc),\ FeCl_2(conc) \,\vert\, Pt$$

The contents of the salt bridge generally are not specified in the shorthand notation. When they are known, the concentrations of the various solutes are shown in parentheses.

STRENGTH OF OXIDANTS AND REDUCTANTS

It is of considerable importance to know the relative strengths of oxidants and reductants because this enables us to predict both the direction and extent of redox reactions. You may recall that the strength of an acid is represented by its ability to donate a proton. Therefore, it seems logical that the strength of a reductant can be represented by its ability to donate electrons. The manner in which this ability is manifested is examined in the following section.

Electrode Potentials

When a metal electrode is placed in a solution containing the metal's ions, it has a tendency to acquire a charge as the net result of two opposing processes. Consider the situation shown in Figure 10-5:

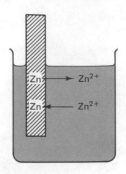

Figure 10-5 Zinc–zinc ion half-cell.

1. Zinc atoms on the electrode oxidize, going into solution as Zn^{2+} and leaving behind two electrons, thereby causing the electrode to become negative.

$$Zn(s) \longrightarrow Zn^{2+} + 2e^- \qquad (10\text{-}4)$$

2. Zinc ions in the solution reduce, acquiring two electrons from the electrode and depositing as zinc atoms, thereby causing the electrode to become positive.

$$Zn^{2+} + 2e^- \longrightarrow Zn(s) \qquad (10\text{-}5)$$

Such a system is referred to as a half-cell (one-half of a galvanic cell) and the two opposing processes are represented in the redox half-reaction:

$$Zn^{2+} + 2e^- \rightleftharpoons Zn(s)$$

Whether the zinc electrode becomes negative or positive depends on which tendency is greater. The net charge buildup is called an *electrode potential* (E), and if its sign and magnitude could be determined, we would have a quantitative measure of the relative desire for Zn to become oxidized or Zn^{2+} to become reduced.

Standard conditions

It should be apparent that the concentration (activity) of zinc ions in solution will affect the potential acquired by the zinc electrode. Increasing the concentration of Zn^{2+} will *inhibit* Reaction 10-4 and *encourage* Reaction 10-5, thereby causing the electrode to become more positive. A decrease in concentration would, of course, have the opposite effect. The effective use of electrode potentials is limited unless we know the concentrations at which they were measured. When all substances involved in the half-reaction are present in their standard-state concentrations, an electrode has its *standard potential, $E°$.*

It was noted in Chapter 2 that the activity of a *solute* is approximately equal to its concentration, especially in dilute solution; the activity of a pure substance (solid or liquid) is defined as unity; and the activity of a gas is the same as its partial pressure in atmospheres. As was done in earlier chapters, concentrations will be taken as acceptable estimates of the activity of a solute. The quantitative effects of concentration on electrode potential will be dealt with in a later section.

Measuring Cell Voltages

Unfortunately, it is not possible to determine the absolute value of the potential of a single electrode. However, the *difference* between two potentials, which is called *volt-*

age, can be measured easily and accurately. The required experimental setup is that of a galvanic cell. The measured difference in potential, ΔE, is given by

$$\Delta E_{cell} = V_{cell} = E_{cathode} - E_{anode} \qquad (10\text{-}6)$$

If we constructed several cells, each with a different cathode but the same anode, the difference in the voltages of these cells would be due entirely to the potentials of the cathodes. To facilitate comparisons, chemists have agreed to use one electrode system as an ultimate reference against which all electrode potentials are measured. In this way, a list of relative electrode potentials can be compiled.

The standard hydrogen electrode

The universally accepted reference electrode is the *standard hydrogen electrode* (SHE), shown in Figure 10-6. The metal electrode itself consists of a small piece of platinum foil coated with a layer of finely divided platinum (called platinum black because of its dark appearance). The coated foil is immersed in an acidic solution whose hydrogen-ion activity is 1.0 and through which hydrogen gas at a partial pressure of 1.0 atm (unit activity) is being passed. The platinum black has a very large surface area that enables it to adsorb a substantial quantity of hydrogen gas, bringing it into contact with aqueous hydrogen ions at the electrode surface. The platinum electrode acquires a potential that is determined by the relative tendencies of H^+ to reduce and $H_2(g)$ to oxidize, although we have no way of measuring its value. By convention, the potential of this electrode is *assigned* a value of exactly zero at all temperatures.

Measuring standard electrode potentials

To measure a standard electrode potential, a galvanic cell of the type shown in Figure 10-7 must be constructed. The voltage of such a cell is the difference between the electrode potentials,

$$\Delta E^{\circ}_{cell} = E^{\circ}_M - E^{\circ}_{SHE}$$

Since the potential of the standard hydrogen electrode is defined as zero,

$$\Delta E^{\circ}_{cell} = E^{\circ}_M - 0 = E^{\circ}_M$$

the voltage of this particular cell is the standard electrode potential of the metal, M. It is the *only* cell voltage that may be referred to as an electrode potential and it is the only

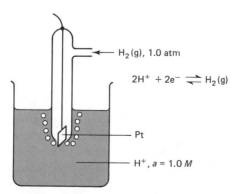

$H_2(g)$, 1.0 atm

$2H^+ + 2e^- \rightleftharpoons H_2(g)$

Pt

H^+, $a = 1.0\ M$

Figure 10-6 The standard hydrogen electrode.

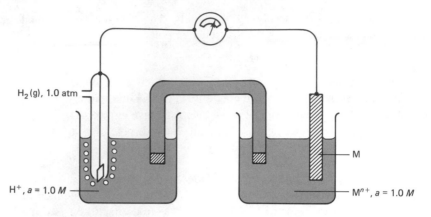

H$_2$(g), 1.0 atm

H$^+$, a = 1.0 M

M

M^{n+}, a = 1.0 M

Figure 10-7 Galvanic cell for measuring a standard electrode potential.

cell voltage that will be represented with the symbol $E°$ or E. The symbol ΔE or V will be used to designate other cell voltages.

The *sign* of the electrode potential is arbitrary, depending only on which way the potentials of the electrodes in Figure 10-7 are subtracted from each other. During the early years of modern electrochemistry, chemists did not always agree on a single sign convention. In 1953, the International Union of Pure and Applied Chemistry (IUPAC) proposed that the sign of the electrode potential be made the same as the charge on the electrode when it is in a galvanic cell with the standard hydrogen electrode. This sign convention has been accepted by virtually all major chemical societies and it is the convention used in this book. The charge on the metal electrode in the cell of Figure 10-7 can be determined by observing either the direction of electron flow in the external circuit or the reaction taking place at the metal electrode. When the metal electrode is zinc, the voltage is 0.762 and electrons are observed to flow from the zinc to the platinum electrode, which means that zinc is the negative electrode. We reach the same conclusion when the reaction at the zinc electrode is observed to be

$$Zn(s) \longrightarrow Zn^{2+} + 2e^-$$

which says that zinc atoms are going into solution and leaving their electrons on the electrode. Thus, according to the accepted convention, the standard electrode potential for the zinc half-cell is -0.762 V. Standard electrode potentials for numerous other half-cells have been measured in a similar manner or, in some cases, calculated from certain equilibrium or thermodynamic data. The $E°$ values for half-cells are usually arranged in numerically descending order as illustrated in Table 10-1. A larger list of standard electrode potentials appears in Appendix K.

Interpreting Electrode Potentials

There is a very important consequence of the IUPAC electrode potential sign convention that must be remembered: *As the tendency for a half-cell reaction to proceed in the direction of a reduction increases, the electrode potential of the half-cell also increases (becomes more positive).* We can determine which of two substances is the better oxidizing agent (more willing to become reduced) simply by comparing their standard elec-

TABLE 10-1 ABBREVIATED LIST OF ELECTRODE POTENTIALS[a]

Oxidizing agent		Reducing agent	$E°$ (V)
$MnO_4^- + 8H^+ + 5e^-$	$\rightleftharpoons$	$Mn^{2+} + 4H_2O$	1.51
$O_2(g) + 4H^+ + 4e^-$	$\rightleftharpoons$	$2H_2O$	1.229
$Fe^{3+} + e^-$	$\rightleftharpoons$	Fe^{2+}	0.771
$VO^{2+} + 2H^+ + e^-$	$\rightleftharpoons$	$V^{3+} + H_2O$	0.337
$Sn^{4+} + 2e^-$	$\rightleftharpoons$	Sn^{2+}	0.151
$2H^+ + 2e^-$	$\rightleftharpoons$	$H_2(g)$	0.000
$AgI(s) + e^-$	$\rightleftharpoons$	$Ag(s) + I^-$	−0.152
$Cd^{2+} + 2e^-$	$\rightleftharpoons$	$Cd(s)$	−0.403
$Zn^{2+} + 2e^-$	$\rightleftharpoons$	$Zn(s)$	−0.762
$Al^{3+} + 3e^-$	$\rightleftharpoons$	$Al(s)$	−1.662

[a]See Appendix K for a more comprehensive list.

trode potentials. For example, Cu^{2+} is a stronger oxidizing agent than Zn^{2+} because the standard electrode potential for the Cu^{2+}/Cu half-cell is more positive:

$$Cu^{2+} + 2e^- \rightleftharpoons Cu(s) \qquad E° = 0.521 \text{ V}$$

$$Zn^{2+} + 2e^- \rightleftharpoons Zn(s) \qquad E° = -0.762 \text{ V}$$

In organizing information about the reaction properties of oxidizing and reducing agents, we find it helpful to use the idea of conjugates in much the same way that it was used in acid-base equilibria. If we consider the oxidized and reduced forms of a substance to be an oxidizing agent and conjugate reducing agent, it is easy to see that their relative strengths are inversely related. Thus if MnO_4^- is a very good oxidizing agent because it has a very positive $E°$ (Table 10-1), its conjugate, Mn^{2+}, must be considered a poor reducing agent. Conversely, Al^{3+} is a poor oxidizing agent because its standard electrode potential is −1.66 V, which means that $Al(s)$ is a very good reducing agent.

Effect of Concentration on Electrode Potential

Earlier it was pointed out that the concentration of zinc ions in solution affects the relative tendency of zinc metal to oxidize and zinc ion to reduce. Since the electrode potential is a measure of this relative tendency, it also depends on the concentration of zinc ions. The quantitative relationship between the concentration of substances comprising a redox half-cell and the electrode potential of the half-cell was first described by the German chemist Nernst, and the equation bears his name. The IUPAC electrode potential convention requires that we represent a half-cell with a half-reaction written *as a reduction*. Thus, for the general half-reaction

$$aA + bB + ne^- \rightleftharpoons cC + dD$$

the Nernst equation is

$$E = E° - \frac{RT}{nF} \ln \frac{[C]^c[D]^d}{[A]^a[B]^b}$$

Strength of Oxidants and Reductants

where R is the molar gas constant, $8.314 \ \text{J} \cdot \text{mol}^{-1} \cdot \text{K}^{-1}$, T the temperature in Kelvin, n the number of electrons in the balanced half-reaction, $\text{eq} \cdot \text{mol}^{-1}$, and F the Faraday constant, $96{,}485 \ \text{C} \cdot \text{eq}^{-1}$. Inserting these numerical values with a temperature of 298 K (25°C) and converting the natural logarithm to base 10 gives the commonly used form of the Nernst equation,

$$E = E° - \frac{0.0592}{n} \log \frac{[C]^c [D]^d}{[A]^a [B]^b} \tag{10-7}$$

where $0.0592/n$ has units of volts. There are three important things to remember about the Nernst equation. First, the terms $[A]^a$, $[B]^b$, $[C]^c$, and $[D]^d$ represent *all* the reactants and products of the net, *ionic* half-reaction, not just the substances changing oxidation state. Second, the log term is of the same form as an equilibrium constant, being the ratio of the concentration of products to the concentration of reactants *all raised to the power equal to their coefficient in the balanced half-reaction*. Third, like equilibrium expressions, the exact form of the equation is written in terms of activities rather than concentrations and the activity of a pure substance is unity. As was done before, concentration will be taken as an acceptable estimate of the activity of a solute.

Nernst equations for several different types of half-reactions are given below.

$$\text{Ni}^{2+} + 2e^- \rightleftharpoons \text{Ni(s)} \qquad\qquad E = E° - \frac{0.0592}{2} \log \frac{1}{[\text{Ni}^{2+}]}$$

$$\text{Sn}^{4+} + 2e^- \rightleftharpoons \text{Sn}^{2+} \qquad\qquad E = E° - \frac{0.0592}{2} \log \frac{[\text{Sn}^{2+}]}{[\text{Sn}^{4+}]}$$

$$\text{Cr}_2\text{O}_7^{2-} + 14\text{H}^+ + 6e^- \rightleftharpoons 2\text{Cr}^{3+} + 7\text{H}_2\text{O} \quad E = E° - \frac{0.0592}{6} \log \frac{[\text{Cr}^{3+}]^2}{[\text{Cr}_2\text{O}_7^{2-}][\text{H}^+]^{14}}$$

$$\text{O}_2(\text{g}) + 4\text{H}^+ + 4e^- \rightleftharpoons 2\text{H}_2\text{O} \qquad\qquad E = E° - \frac{0.0592}{4} \log \frac{1}{p_{\text{O}_2}[\text{H}^+]^4}$$

Example 10-1

Calculate the electrode potential of a half-cell containing $0.100 \ M$ KMnO$_4$ and $0.0500 \ M$ MnCl$_2$ in a solution whose pH is 1.000.

Solution The balanced half-reaction, written as a reduction, for the half-cell is

$$\text{MnO}_4^- + 8\text{H}^+ + 5e^- \rightleftharpoons \text{Mn}^{2+} + 4\text{H}_2\text{O}$$

for which the Nernst equation is

$$E = E°_{\text{MnO}_4^-/\text{Mn}^{2+}} - \frac{0.0592}{n} \log \frac{[\text{Mn}^{2+}]}{[\text{MnO}_4^-][\text{H}^+]^8}$$

At pH = 1.000,

$$[\text{H}_3\text{O}^+] = 1.00 \times 10^{-1} \ M$$

Substituting this and the other concentrations in the Nernst equation gives

$$E = 1.51 - \frac{0.0592}{5} \log \frac{0.0500}{(0.100)(1.00 \times 10^{-1})^8} = 1.42 \text{ V}$$

The value for $E^\circ_{MnO_4^-/Mn^{2+}}$ was obtained from Appendix K.

Example 10-2

What is the electrode potential of a half-cell consisting of a platinum electrode immersed in a 0.100 M hydrochloric acid solution through which chlorine gas is being passed at a partial pressure of 1.15 atm?

Solution The balanced half-reaction occurring at the platinum electrode is

$$Cl_2(g) + 2e^- \rightleftharpoons 2Cl^- \qquad E^\circ_{Cl_2/Cl^-} = 1.36 \text{ V}$$

The Nernst equation for this half-reaction is

$$E = E^\circ_{Cl_2/Cl^-} - \frac{0.0592}{2} \log \frac{[Cl^-]^2}{p_{Cl_2}}$$

Subsituting the known values yields

$$E = 1.36 - \frac{0.0592}{2} \log \frac{(0.100)^2}{1.15} = 1.42 \text{ V}$$

Example 10-3

A beaker containing 50 mL of 0.150 M H_3AsO_3 and 0.0610 M H_3AsO_4 in acidic solution has an electrode potential of 0.494 V. What is the pH of the solution?

Solution The two compounds contain arsenic in different oxidation states and constitute a half-cell with the half-reaction:

$$H_3AsO_4 + 2H^+ + 2e^- \rightleftharpoons H_3AsO_3 + H_2O \qquad E^\circ_{H_3AsO_4/H_3AsO_3} = 0.559 \text{ V}$$

Thus

$$E = E^\circ - \frac{0.0592}{2} \log \frac{[H_3AsO_3]}{[H_3AsO_4][H^+]^2}$$

The hydrogen-ion concentration is the only unknown in the equation. Substituting the known values gives us

$$0.494 = 0.559 - \frac{0.0592}{2} \log \frac{0.150}{(0.0610)(H^+)^2}$$

or

$$\log \frac{0.150}{(0.0610)[H^+]^2} = \frac{2(0.559 - 0.494)}{0.0592} = 2.20$$

Taking the antilog of both sides and solving for the concentration of H^+, we have

$$\frac{0.150}{(0.0610)\,[H^+]^2} = 1.6 \times 10^2$$

$$[H^+] = \sqrt{\frac{0.150}{(0.0610)\,(1.6 \times 10^2)}} = 0.12\ M$$

Finally,

$$pH = -\log 0.12 = 0.92$$

USES OF ELECTRODE POTENTIALS

We have already seen how standard electrode potentials may be used to predict the direction of a redox reaction. With the Nernst equation, chemists can calculate the voltages of cells in which the participating substances are not in their standard-state concentrations. Such a voltage is a quantitative measure of the driving force behind a reaction. In addition, electrode potentials can be used to calculate equilibrium constants for reactions. These applications are discussed in this section.

Calculating Cell Voltage

Earlier it was stated (Equation 10-6) that the voltage of a galvanic cell can be obtained by subtracting the two half-cell electrode potentials as follows:

$$\Delta E_{cell} = E_{cathode} - E_{anode} \tag{10-8}$$

Since the cathode is placed on the right and the anode on the left of the salt bridge in the shorthand notation for describing cells, Equation 10-8 may be written as

$$\Delta E_{cell} = E_{right} - E_{left} \tag{10-9}$$

Because the potential of either electrode may depend on the concentrations of the substances constituting the half-cell, the voltage of the cell is concentration dependent. To calculate the voltage, the individual electrode potentials are determined and subtracted.

Example 10-4

Calculate the voltage of the following cell:

$$Zn(s)\,\big|\,ZnCl_2(0.120\ M)\,\big|\big|\,Cl_2(g)\,(1.15\ atm),\ KCl(0.105\ M)\,\big|\,Pt$$

Solution   The balanced half-reaction for the right-hand half-cell is

$$Cl_2(g) + 2e^- \rightleftharpoons 2Cl^- \qquad E^{\circ}_{Cl_2/Cl^-} = 1.36\ V$$

and

$$E_r = E^{\circ} - \frac{0.0592}{n} \log \frac{[Cl^-]^2}{p_{Cl_2}} = 1.36 - \frac{0.0592}{2} \log \frac{(0.105)^2}{1.15} = 1.42\ V$$

Even though it is assumed that zinc is being oxidized at the left-hand electrode, the half-reaction is written as a reduction for the purpose of solving the Nernst equation:

$$Zn^{2+} + 2e^- \rightleftharpoons Zn(s) \qquad E^{\circ}_{Zn^{2+}/Zn} = -0.762\ V$$

and

$$E_l = E° - \frac{0.0592}{n} \log \frac{1}{[Zn^{2+}]} = -0.762 - \frac{0.0592}{2} \log \frac{1}{0.120} = -0.789 \text{ V}$$

Finally, the two electrode potentials are subtracted to obtain the cell voltage:

$$\Delta E_{cell} = E_r - E_l = 1.42 - (-0.789) = 2.21 \text{ V}$$

The reaction that occurs spontaneously in a cell can be determined by subtracting the half-reactions (both written as reductions) in the same manner as the electrode potentials. That is, the half-reaction for the anode or left-hand electrode is *subtracted from* the half-reaction for the cathode or right-hand electrode. Thus the spontaneous reaction for the cell described in Example 10-4 is

$$\begin{array}{ll} Cl_2(g) + 2e^- \rightleftharpoons 2Cl^- & E = 1.42 \text{ V} \\ -[Zn^{2+} + 2e^- \rightleftharpoons Zn(s)] & -(E = -0.789 \text{ V}) \\ \hline Cl_2(g) + Zn(s) \rightleftharpoons Zn^{2+} + 2Cl^- & \Delta E = 2.21 \text{ V} \end{array}$$

Before subtracting half-reactions containing different numbers of electrons, multiply one or both of them with integers that will make the number of electrons the same. This is, of course, the same procedure used in the half-reaction method of balancing redox reactions. Note that multiplying both sides of a half-reaction with a constant does *not* change the electrode potential. The Nernst equation for each of the following half-reactions will produce the same value for E:

$$Fe^{3+} + e^- \rightleftharpoons Fe^{2+} \qquad E = E° - \frac{0.0592}{1} \log \frac{[Fe^{2+}]}{[Fe^{3+}]}$$

$$2Fe^{3+} + 2e^- \rightleftharpoons 2Fe^{2+} \qquad E = E° - \frac{0.0592}{2} \log \frac{[Fe^{2+}]^2}{[Fe^{3+}]^2}$$

Sometimes it is not known in advance which electrode in a cell is the cathode or right-hand electrode. In such cases, one electrode is assumed to be the cathode and the cell voltage is calculated accordingly. If the assumption was correct, the calculated voltage will be positive and the reaction spontaneous. If the assumption was incorrect, the calculated cell voltage will be negative and the spontaneous cell reaction will be opposite to the calculated reaction.

Example 10-5

Suppose it was thought that zinc was the cathode in the cell described in Example 10-4. Accordingly, it would be made the right-hand electrode and the cell would be written

$$Pt \,|\, KCl(0.105 \text{ M}), Cl_2(g)(1.15 \text{ atm}) \,||\, ZnCl_2(0.120 \text{ M}) \,|\, Zn$$

Calculate the voltage and spontaneous reaction of the cell.

Solution According to the calculations in Example 10-4, $E_{Zn} = -0.789$ V and $E_{Pt} = 1.42$ V. Since zinc is the right-hand electrode,

$$\Delta E_{cell} = E_r - E_l = -0.789 - 1.42 = -2.21 \text{ V}$$

Uses of Electrode Potentials

The implied cell reaction is

$$Zn^{2+} + 2e^- \rightleftharpoons Zn(s)$$
$$-[Cl_2(g) + 2e- \rightleftharpoons 2Cl^-]$$
$$\overline{\qquad\qquad\qquad\qquad\qquad\qquad\qquad}$$
$$Zn^{2+} + 2Cl^- \rightleftharpoons Zn(s) + Cl_2(g)$$

Since the calculated voltage is negative, the cell reaction is not spontaneous in the direction written, but rather in the reverse direction.

Since every redox reaction is comprised of two half-reactions, each representing one half of a galvanic cell, the difference in the electrode potential of the half-cells can be taken as a measure of the driving force behind the reaction. To establish the direction of a reaction, we need only to determine which half-cell has the more positive electrode potential. The reaction for that half-cell will be a reduction and the reaction for the other half-cell must be an oxidation. In the event that the electrode potentials of both half-cells are the same, there is no net driving force and no net reaction takes place. That is, the system is at equilibrium.

Example 10-6

Ten milliliters of a solution containing $0.0835\ M\ Cr^{3+}$ and $0.119\ M\ Cr^{2+}$ is added to 25.0 mL of a solution containing $0.0361\ M\ V^{3+}$ and $0.0904\ M\ V^{2+}$. Does a reaction occur and, if so, what reaction?

Solution The two solutions can be viewed as being half-cells of a galvanic cell. The Nernst equation is used to calculate the electrode potential for each half-cell:

$$Cr^{3+} + e^- \rightleftharpoons Cr^{2+} \qquad E_{Cr} = E° - \frac{0.0592}{n} \log \frac{[Cr^{2+}]}{[Cr^{3+}]}$$

$$= -0.407 - \frac{0.0592}{1} \log \frac{0.119}{0.0835} = -0.416\ V$$

$$V^{3+} + e^- \rightleftharpoons V^{2+} \qquad E_V = E° - \frac{0.0592}{n} \log \frac{[V^{2+}]}{[V^{3+}]}$$

$$= -0.255 - \frac{0.0592}{1} \log \frac{0.0904}{0.0361} = -0.279\ V$$

Since the vanadium half-cell has the more positive potential, V^{3+} is reduced and Cr^{2+} must be oxidized. The reaction can be obtained by subtracting the chromium half-reaction from the vanadium half-reaction to give

$$V^{3+} + Cr^{2+} \rightleftharpoons V^{2+} + Cr^{3+}$$

Extent of a Reaction: The Equilibrium Constant

It was noted earlier that the magnitude of a cell voltage is a measure of the driving force of the cell reaction. A large driving force means a more complete reaction. You are already familiar with another term that measures the completeness of a reaction—the equilibrium constant. Since both ΔE_{cell} and K_{eq} are related to the same thing, there must be some relationship between them. Consider the general redox reaction,

$$aA_{ox} + bB_{red} \rightleftharpoons cA_{red} + dB_{ox}$$

The voltage of a galvanic cell that would yield this reaction is given by

$$\Delta E_{cell} = E_A - E_B$$

At equilibrium, the cell voltage is zero and

$$E_A = E_B$$

Using the Nernst equation to express the values of E_A and E_B, we have

$$E_A^\circ - \frac{0.0592}{n} \log \frac{[A_{red}]^c}{[A_{ox}]^a} = E_B^\circ - \frac{0.0592}{n} \log \frac{[B_{red}]^b}{[B_{ox}]^d}$$

If the number of electrons in the two half-reactions are different, they must be normalized before writing the Nernst equation (see Example 10-7). Rearranging the equation and combining the log terms gives

$$E_A^\circ - E_B^\circ = \frac{0.0592}{n} \log \frac{[A_{red}]^c [B_{ox}]^d}{[A_{ox}]^a [B_{red}]^b}$$

But the log term is the equilibrium constant for the reaction. Thus

$$E_A^\circ - E_B^\circ = \frac{0.0592}{n} \log K_{eq}$$

or

$$\log K_{eq} = \frac{n(E_A^\circ - E_B^\circ)}{0.0592} \qquad (10\text{-}10)$$

Remember that n is the *normalized* number of electrons for the two half-reactions and E_A° is the standard electrode potential for the *substance being reduced* in the reaction.

Example 10-7

Calculate the equilibrium constant for the reaction

$$2Fe^{3+} + H_3AsO_3 + H_2O \rightleftharpoons 2Fe^{2+} + H_3AsO_4 + 2H^+$$

Solution   The reaction can be divided into the two half reactions

$$2Fe^{3+} + 2e^- \rightleftharpoons 2Fe^{2+} \qquad\qquad E^\circ = 0.771 \text{ V}$$

and

$$H_3AsO_4 + 2H^+ + 2e^- \rightleftharpoons H_3AsO_3 + H_2O \qquad E^\circ = 0.559 \text{ V}$$

The iron half-reaction has been multiplied through by 2 to normalize the number of electrons in the half-reactions. Note that this does *not* affect the value of E°. Iron is being reduced, so it is represented by A in Equation 10-10.

$$\log K_{eq} = \frac{2(0.771 - 0.559)}{0.0592} = 7.16$$

or

$$K_{eq} = 10^{7.16} = 1.4 \times 10^7$$

Example 10-8

What is the equilibrium constant for the following reaction?

$$Cu^{2+} + 2Ag(s) \rightleftharpoons Cu(s) + 2Ag^+$$

Solution The two half-reactions making up the reaction are

$$Cu^{2+} + 2e^- \rightleftharpoons Cu(s) \qquad E° = 0.337 \text{ V}$$

and

$$2Ag^+ + 2e^- \rightleftharpoons 2Ag(s) \qquad E° = 0.800 \text{ V}$$

where the latter has been multiplied through by 2 to normalize the electrons. Since copper is being reduced in the reaction,

$$\log K_{eq} = \frac{n(E°_{Cu} - E°_{Ag})}{0.0592} = \frac{2(0.337 - 0.800)}{0.0592} = -15.6$$

or

$$K_{eq} = 10^{-15.6} = 3 \times 10^{-16}$$

Determining Equilibrium Constants of Nonredox Reactions

The measurement of electrode potentials is an especially convenient way to determine the concentration of certain solutes. Not only can such measurements be made easily and accurately, they also can be made without disturbing the chemical equilibrium of a system. This allows us to get the data needed to calculate equilibrium constants for a variety of chemical systems. In general, a solution is prepared in which the concentrations of all the substances involved in the equilibrium process except one are known. This solution, then, is made a part of a galvanic cell and the unknown concentration is calculated from the measured voltage. The following two examples serve to illustrate how different types of equilibrium constants can be determined.

Example 10-9

Silver forms a complex ion with ammonia

$$Ag^+ + 2NH_3 \rightleftharpoons Ag(NH_3)_2^+$$

A solution was prepared containing $0.0410 \ M$ $Ag(NH_3)_2^+$ and $0.0115 \ M$ NH_3. A silver electrode was placed in the solution and connected via a salt bridge to a standard hydrogen electrode. The voltage of the cell was 0.529 V, with the silver electrode behaving as the cathode (right-hand electrode). Calculate the overall formation constant for the complex ion.

Solution The formation constant is given by

$$K_f = \frac{[Ag(NH_3)_2^+]}{[Ag^+][NH_3]^2}$$

The concentrations of the complex and ammonia are known and the silver-ion concentration can be calculated from the Nernst equation.

$$\Delta E_{cell} = E_{cathode} - E_{anode} = E_{Ag} - E_{SHE}$$

or

$$0.529 = E_{Ag} - 0$$

$$E_{Ag} = 0.529 \text{ V}$$

The half-reaction and Nernst equation for the silver half-cell are

$$Ag^+ + e^- \rightleftharpoons Ag(s)$$

$$E_{Ag} = E_{Ag}^\circ - \frac{0.0592}{n} \log \frac{1}{[Ag^+]}$$

Substituting gives us

$$0.529 = 0.800 - \frac{0.0592}{1} \log \frac{1}{[Ag^+]}$$

$$\log \frac{1}{[Ag^+]} = \frac{0.800 - 0.529}{0.0592} = 4.58$$

$$\frac{1}{[Ag^+]} = 10^{4.58}$$

or

$$[Ag^+] = 10^{-4.58} = 2.6 \times 10^{-5} \, M$$

Substituting in the K_f expression, we obtain

$$K_f = \frac{0.0410}{(2.6 \times 10^{-5})(0.0115)^2} = 1.2 \times 10^7$$

Example 10-10

A small amount of solid AgBr was added to a 0.0100 M solution of NaBr, causing it to become saturated with AgBr. A silver electrode was placed in the solution along with a salt bridge, the other end of which was connected to a reference half-cell whose electrode potential was 0.246 V. The voltage of the cell was 0.055 V with silver acting as the anode. Calculate the K_{sp} for AgBr.

Solution   The solubility product constant is

$$K_{sp} = [Ag^+][Br^-]$$

The concentration of Br^- is known and the concentration of Ag^+ can be calculated from the Nernst equation if the electrode potential of the silver half-cell is obtained first:

$$\Delta E_{cell} = E_{cathode} - E_{anode} = E_{ref} - E_{Ag}$$

$$0.055 = 0.246 - E_{Ag}$$

$$E_{Ag} = 0.246 - 0.055 = 0.191 \text{ V}$$

The half-reaction for the silver half-cell is

$$Ag^+ + e^- \rightleftharpoons Ag(s)$$

for which

$$E_{Ag} = E_{Ag^+/Ag}^\circ - \frac{0.0592}{n} \log \frac{1}{[Ag^+]}$$

Uses of Electrode Potentials   **253**

Substituting the known values and solving for the concentration of Ag^+:

$$0.191 = 0.800 - \frac{0.0592}{1} \log \frac{1}{[Ag^+]}$$

$$\log \frac{1}{[Ag^+]} = \frac{0.800 - 0.191}{0.0592} = 10.3$$

$$\frac{1}{[Ag^+]} = 10^{10.3}$$

$$[Ag^+] = 10^{-10.3} = 5 \times 10^{-11}\ M$$

Substituting this value, along with the bromide-ion concentration in the K_{sp} expression, we obtain

$$K_{sp} = [Ag^+][Br^-] = (5 \times 10^{-11})(0.0100) = 5 \times 10^{-13}$$

LIMITATIONS TO THE USE OF ELECTRODE POTENTIALS

It should be evident by now that electrode potentials are enormously important to our understanding of electroanalytical processes. There are, however, certain properties of chemical systems that limit the accuracy with which we can measure electrode potentials and our ability to use electrode potentials to predict chemical behavior. Many of these limiting effects are difficult to quantify and a detailed discussion of them is beyond the scope of this book. It is important, however, to be aware of the limitations and of the chemical properties that are responsible for them.

Calculated Versus Measured Values

Often, one of the first discoveries made by students studying electrochemistry is that the voltages measured in the laboratory are sometimes different from calculated values and the differences are too large to ascribe to simple experimental errors. Three important reasons for such differences are discussed below.

Liquid junction potentials

Most chemical cells cannot be constructed without having electrolyte solutions of different composition in contact with each other (i.e., without salt bridges). When such liquid junctions exist, a potential develops at the interface, caused by differences in rates of diffusion of ions between the solutions. To see how this junction potential arises, consider the simple example of a hydrochloric acid solution in contact with pure water. Since the natural tendency is for ions to diffuse from regions of high to low concentration, both H^+ and Cl^- will diffuse from the hydrochloric acid solution into the pure water. The driving force for this movement is the concentration difference in the two liquids. The *rate* at which the ions diffuse depends, among other things, on their size, shape, and weight. The small, light H^+ is more mobile than the much larger and heavier Cl^-. As a result, it tends to move ahead in the diffusion process, as illustrated in Figure 10-8. The pure water side of the boundary region acquires a net positive charge due to the excess of hydrogen ions, while the HCl side acquires a negative charge due to the

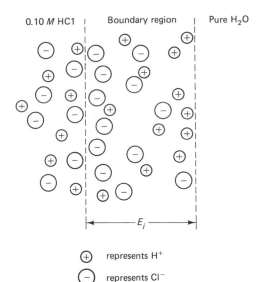

0.10 *M* HC1 | Boundary region | Pure H$_2$O

$\longmapsto E_j \longrightarrow$

⊕ represents H$^+$

⊖ represents Cl$^-$

Figure 10-8 Schematic representation of a liquid-liquid junction showing the source of the junction potential.

excess of chloride ion. This potential gradient is referred to as a liquid-junction potential. The magnitude of the potential depends on the relative mobilities of the ions and the magnitude of the concentration gradient. In the worst cases, it can be as large as 30 mV.

The voltage of a galvanic cell containing one or more liquid junctions is due to the potential difference developed at *each* of the electrode–solution and liquid–liquid interfaces. As it is usually difficult or impossible to either eliminate liquid junctions in a cell or calculate their potentials, chemists try to *minimize* their effect on the cell voltage by using a high concentration of an electrolyte in the salt bridge whose anion and cation mobilities are nearly identical. Potassium chloride is a common choice for salt bridge solutions because it is very soluble (saturated solution is greater than 4 *M*) and the mobility of K$^+$ and Cl$^-$ in water differ by only about 4%. Saturated potassium chloride salt bridges generally produce junction potentials of less than 5 mV, which is usually a negligible quantity.

Reversibility of the electrode reaction

For the potential of an electrode to be described properly by the Nernst equation, the electrode reaction must be *reversible*; that is, the oxidation and reduction reactions must be the exact opposite of each other and each must occur rapidly. Many half-cells do not behave in a truly reversible manner. The electrode potential for the CO$_2$/H$_2$C$_2$O$_4$ half-cell,

$$2CO_2(g) + 2H^+ + 2e^- \rightleftharpoons H_2C_2O_4$$

does not obey the Nernst equation because the rate at which carbon dioxide combines to produce oxalic acid is extremely slow. As a result, changes in the concentration or partial pressure of CO$_2$ have little effect on the measured electrode potential, in contradiction to what we expect from the Nernst equation.

The electrode potential for the MnO$_4$$^-$/Mn^{2+} half-cell

$$MnO_4^- + 8H^+ + 5e^- \rightleftharpoons Mn^{2+} + 4H_2O$$

Limitations to the Use of Electrode Potentials

does not obey the Nernst equation because, at the electrode surface, the reduction and oxidation reactions are not the opposite of each other. Studies have shown that MnO_4^- is reduced to $MnO_2(s)$ and Mn^{2+} is oxidized to $MnO_2(s)$ at the electrode. Consequently, changing the concentration of MnO_4^- or Mn^{2+} does not have the direct effect on the electrode potential that is expected from the equation.

Students are cautioned not to misunderstand the foregoing discussion and assume that an electrode immersed in a solution containing MnO_4^- and Mn^{2+} will have *no* potential. It *will* have a potential, and that potential will vary with the concentration of MnO_4^- and Mn^{2+}, but not exactly as described by a simple Nernst equation.

Current flow during measurement

In the process of measuring an electrode potential using the cell in Figure 10-7, some current must flow through the cell. But as current flows, reactions occur at the electrodes, causing changes in the concentrations of the substances responsible for the potentials. That is, we perturb the system in the act of making a measurement on it. Fortunately, modern measurement devices draw very small currents and do not usually perturb the system appreciably.

Predictions Using Electrode Potentials

Electrode potentials represent thermodynamic properties of redox substances. That is, they may be used to predict whether a given reaction *can* occur, but offer no direct information about the rate of a possible reaction. For example, comparing the standard electrode potentials for the $Cr_2O_7^{2-}/Cr^{3+}$ and O_2/H_2O half-cells, one concludes that dichromate ion is capable of oxidizing water at the standard-state concentrations. Yet every experienced chemist knows that aqueous potassium dichromate solutions are exceedingly stable. It turns out simply that the reaction between dichromate and water is so slow as to be essentially nonexistent. To understand the rates at which various reactions occur, we need information about the detailed mechanism by which the reaction proceeds. This is an area of great importance but beyond the scope of this book.

PROBLEMS

10-1. Write balanced half-reactions (as reductions) for each of the following redox conjugate pairs.

(a) Fe^{3+}/Fe^{2+} (e) MnO_4^-/Mn^{2+} (i) VO_2^+/VO^{2+}

(b) Sn^{4+}/Sn^{2+} (f) O_2/H_2O_2 (j) $CO_2/H_2C_2O_4$

(c) I_2/I^- (g) $S_4O_6^{2-}/S_2O_3^{2-}$

(d) $Cr_2O_7^{2-}/Cr^{3+}$ (h) H_3AsO_4/H_3AsO_3

10-2. Represent the following cells with the schematic (shorthand) notation.

	Beaker 1	*Beaker 2*
(a)	$ZnCl_2(0.10\ M)$, Zn	$SnCl_2(0.25\ M)$, $SnCl_4(0.10\ M)$, Pt
(b)	$Hg_2Cl_2(s)$, $Hg(l)$, KCl(sat'd)	$Cu(NO_3)_2(0.20\ M)$, Cu

(c) $HCl(0.010\ M)$, $H_2(g)(1.1\ atm)$, Pt $I_2(s)$, $KI(2.0\ M)$, Pt
(d) $U(SO_4)_2(0.020\ M)$, $UO_2SO_4(0.15\ M)$, $Ce(SO_4)_2(0.10\ M)$, $Ce_2(SO_4)_3$
   $H_2SO_4(1\ M)$, Pt $(0.10\ M)$, Pt
(e) $AgCl(s)$, $Ag(s)$, $KCl(1.0\ M)$ $AgNO_3(0.050\ M)$, Ag

10-3. Write the equivalent to Equation 10-7 that would apply at 60°C.

10-4. Calculate the electrode potential (voltage versus SHE) of a platinum electrode immersed in each of the following solutions.
   (a) $0.140\ M\ FeCl_3 + 0.040\ M\ FeCl_2$
   (b) $0.100\ M\ KI$ saturated with I_2
   (c) $0.350\ M\ CrCl_3 + 0.100\ M\ K_2Cr_2O_7$ at pH 2.00
   (d) $0.0500\ M\ H_2SO_3 + 0.200\ HCl$ saturated with S
   (e) $0.0150\ M\ KIO_3$ saturated with I_2 at pH 6.00
   (f) $1.00\ M\ (VO_2)_2SO_4 + 0.0100\ M\ VOSO_4$ at pH 1.00
   (g) $0.0450\ M\ MnCl_2 + 2.50 \times 10^{-3}\ M\ HCl + 0.100\ M\ KMnO_4$
   (h) $0.250\ M\ H_2C_2O_4 + 6.00\ M\ HClO_4 + CO_2(g)$ at 0.25 atm partial pressure

10-5. Calculate the electrode potential (voltage versus SHE) of a silver electrode immersed in each of the following solutions.
   (a) $0.0200\ M\ KCl$ saturated with AgCl
   (b) $2.50\ M\ KCN + 1.3 \times 10^{-3}\ M\ NaAg(CN)_2$
   (c) $4.00 \times 10^{-3}\ M\ AgNO_3$ at pH 4.00
   (d) water saturated with AgBr

10-6. Calculate the electrode potential of a half-cell comprised of $Fe(NO_3)_2$ at a concentration of 1.00 M and $Fe(NO_3)_3$ at the following concentrations.
   (a) 1.00 M   (c) $1.00 \times 10^{2}\ M$ (d) $1.00 \times 10^{-3}\ M$
   (b) $1.00 \times 10^{-1}\ M$

10-7. Calculate the electrode potential of a half-cell containing $0.100\ M\ KMnO_4$ and $0.100\ M\ MnCl_2$ at each given pH.
   (a) 1.00   (c) 3.00   (d) 4.00
   (b) 2.00

10-8. An aqueous solution saturated in Hg_2Cl_2 and Hg also contains some KCl. An inert electrode placed in this solution had a potential of 0.272 V versus the standard hydrogen electrode. What is the molar analytical concentration of KCl in the solution?

10-9. To what pH must a mixture of $1.53 \times 10^{-2}\ M\ CrCl_3$ and $6.88 \times 10^{-3}\ M\ K_2Cr_2O_7$ be adjusted so that a platinum electrode immersed in the solution will have a potential of 800 mV versus the SHE?

10-10. At what pH would there be no net reaction in the following cell?

$Pt\,|\,NO(1.00\ atm),\ HNO_2(0.0200\ M),\ HCl(pH?)\,||$

$Fe(CN)_6^{3-}(0.0100\ M),\ Fe(CN)_6^{4-}(0.0100\ M)\,|\,Pt$

10-11. Arrange the following substances in decreasing order of strength as oxidizing agents at the standard-state conditions of concentration.

$$H_3AsO_4,\ Cr^{2+},\ O_3,\ V^{3+},\ S_4O_6^{2-}$$

10-12. Arrange the following substances in decreasing order of strength as reducing agents at the standard-state conditions of concentration.

$$Al,\ H_2C_2O_4,\ Fe(CN)_6^{4-},\ Se,\ Zn$$

10-13. Theoretically, what will happen to the oxidizing power of each of the following substances if the solution acidity is increased? Substantiate your answer with a Nernst equation.
(a) $K_2Cr_2O_7$ (c) Hg_2Cl_2 (e) $NaAl(OH)_4$
(b) HNO_2 (d) Co^{3+}

10-14. For each of the following cells; (1) calculate the "implied" theoretical voltage, (2) write the spontaneous cell reaction, (3) identify the cathode, (4) identify the negative electrode, and (5) indicate the direction of electron flow in the external circuit.
(a) $Pt \mid SnCl_2(0.0250\ M),\ SnCl_4(0.100\ M) \mid\mid CuSO_4(0.200\ M) \mid Cu$
(b) $Sb \mid Sb_2O_3(sat'd),\ HCl(1.00 \times 10^{-3}\ M) \mid\mid K_2PtCl_6(0.120\ M),\ K_2PtCl_4(0.550\ M),$
     $KCl(0.0100\ M) \mid Pt$
(c) $Pt \mid TiCl_3(0.0200\ M),\ TiOCl_2(0.500\ M),\ HCl(pH\ 3.00) \mid\mid Tl(NO_3)_3(0.100\ M),$
     $TlNO_3(0.400\ M) \mid Pt$
(d) $Pt,\ H_2(1.00\ atm) \mid HCl(pH\ 4.00) \mid\mid Cd(NO_3)_2(0.0100\ M) \mid Cd$
(e) $Pt \mid NaI(1.75 \times 10^{-3}\ M),\ I_3^-(2.00 \times 10^{-4}\ M),\ AgI(sat'd) \mid Ag$
(f) $Pt \mid VSO_4(0.140\ M),\ V_2(SO_4)_3(0.550\ M),\ \mid\mid V_2(SO_4)_3(0.550\ M),\ H_2SO_4(pH\ 1.00),$
     $VOSO_4(0.220\ M) \mid Pt$
(g) $Pt \mid KI(0.100\ M),\ KIO_3(0.0100\ M),\ KOH(0.0200\ M) \mid\mid Hg_2Cl_2(sat'd),\ NaCl(3.00$
     $M) \mid Hg(l)$
(h) $Ni \mid NiSO_4(1.00 \times 10^{-4}\ M) \mid\mid NiSO_4(1.00 \times 10^{-1}\ M) \mid Ni$

10-15. At what pH will the following cell have a theoretical voltage of 0.500 V?

$$Ag \mid AgNO_3(0.100\ M) \mid\mid Pb(NO_3)_2(0.500\ M),\ HNO_3(?M) \mid PbO_2$$

10-16. At what concentration of $CrCl_3$ will the following cell have a theoretical voltage of 0.28 V?

$$Pt \mid H_2O_2(0.010\ M),\ HCl(0.20\ M),\ O_2(1.2\ atm) \mid\mid$$
$$K_2Cr_2O_7(0.020\ M),\ CrCl_3(?M),\ HCl(pH\ 2.70) \mid Pt$$

10-17. In what direction will the following reactions proceed if all substances are present initially in their standard-state concentrations?
(a) $H_2SO_3 + 2Zn(s) + 4HCl \rightleftharpoons S(s) + 2ZnCl_2 + 3H_2O$
(b) $SnCl_4 + 2FeCl_2 \rightleftharpoons SnCl_2 + 2FeCl_3$
(c) $2VCl_3 + Ni(s) \rightleftharpoons 2VCl_2 + NiCl_2$
(d) $HNO_2 + VOCl_2 \rightleftharpoons NO(g) + VO_2Cl + HCl$
(e) $2KMnO_4 + 10Na_2SO_4 + 16H_2SO_4 \rightleftharpoons 2MnSO_4 + 5Na_2S_2O_8 + 8H_2O$
(f) $Br_2(l) + 2I^- \rightleftharpoons 2Br^- + I_2(s)$

10-18. Calculate the equilibrium constants for the reactions in Problem 10-17 (in the direction they are written).

10-19. Calculate the equilibrium constant for the reaction

$$2H_2(g) + O_2(g) \rightleftharpoons 2H_2O(l)$$

Why can mixtures of H_2 and O_2 be prepared?

10-20. Calculate the minimum difference between the standard electrode potentials of the analyte and titrant necessary for each of the following reactions to be 99.9% complete. *Hint*: Can you calculate the equilibrium constant needed to achieve 99.9% completion?
(a) $A^{2+} + T^{3+} \rightleftharpoons A^{3+} + T^{2+}$ (c) $A^{2+} + 2T^{3+} \rightleftharpoons A^{4+} + 2T^{2+}$
(b) $A^{2+} + T^{3+} \rightleftharpoons A^{4+} + T^+$ (d) $3A^{2+} + T^{4+} \rightleftharpoons 3A^{3+} + T^+$

10-21. Consider the following cell:

$$Si \mid SiO_2(sat'd),\ HCl(1\ M),\ Sb_2O_3(sat'd) \mid Sb$$

(a) Calculate the initial voltage of the cell.
(b) Write the spontaneous cell reaction.

(c) Calculate the cell voltage when 40% of the solid reactants have been converted to products. Assume that the solutions remain saturated and that the concentration of H^+ is unchanged.

(d) Calculate the cell voltage when 80% of the reactants have been converted to products. The assumptions of part (c) still apply.

(e) Explain why batteries produce a constant voltage during their lifetime and then "go dead" very quickly.

10-22. If the electrode potential of the following half-cell is -0.481 V, calculate the molar solubility of TlCl.

$$Tl\,|\,TlCl(sat'd),\ KCl(0.0500\ M)$$

10-23. A cadmium electrode immersed in a solution buffered at pH 10.50 and saturated with cadmium hydroxide had a potential of -0.620 V versus the SHE. Calculate the K_{sp} for $Cd(OH)_2$.

10-24. A small amount of solid Hg_2Cl_2 and 25 mL of metallic mercury were added to a solution of 5.03×10^{-2} M NaCl. A platinum electrode immersed in the layer of mercury was connected to a reference half-cell whose electrode potential was 0.222 V. The voltage of the cell was 0.121 V with the reference electrode acting as the anode. Calculate the K_{sp} for Hg_2Cl_2.

10-25. The cell

$$Ag\,|\,AgCl(sat'd),\ KCl(1.00\ M)\,||\,NiL_2(0.0250\ M),\ Na_2L(0.150\ M)\,|\,Ni$$

has a voltage of -0.767 V. What is the overall formation constant of NiL_2?

10-26. A lead electrode, placed in 100.0 mL of a solution containing 3.144 g of $Na_2Pb(C_2O_4)_2$ and 1.226 g and $Na_2C_2O_4$, had an electrode potential of -0.261 V. Calculate the overall formation constant for the reaction

$$Pb^{2+} + 2C_2O_4^{2-} \rightleftharpoons Pb(C_2O_4)_2^{2-}$$

10-27. A solution is prepared by mixing 25.00 mL of 0.1103 M $AgNO_3$ with 25.00 mL of 0.2558 M KCN. A silver electrode placed in this solution had an electrode potential of -0.317 V. What is the overall formation constant of $Ag(CN)_2^-$?

10-28. The following cell has a voltage of 0.500 V.

$$Pt,\ H_2(1.00\ atm)\,|\,HX(1.50 \times 10^{-2}\ M),\ CaX_2(4.50 \times 10^{-3}\ M)\,||$$

$$KCl(4.00\ M),\ Hg_2Cl_2(sat'd)\,|\,Hg$$

What is the dissociation constant of the acid HX? Assume that CaX_2 is a strong electrolyte.

OXIDATION-REDUCTION TITRATIONS

Oxidation-reduction titrations are applicable to a variety of both inorganic and organic substances and their popularity may exceed that of acid-base titrations. Probably the single most significant difference between redox and acid-base titrations is the availability of many titrants and standards, each with properties that make it especially suitable for specific applications. The discussion in this chapter will, as much as possible, parallel that in Chapter 7, so that comparisons between the theory and applications of each technique can be made easily.

TITRATION CURVES

The ordinate (vertical axis) of a titration curve generally is a log function of the concentration of the substance to which the indicator responds. Thus, in the titration of HOAc with NaOH, $-\log [H_3O^+]$ is plotted rather than $-\log [HOAc]$ because the indicator is sensitive to the hydrogen-ion concentration rather than the acetic acid concentration. Most of the indicators used in redox titrations are sensitive to changes in the electrode potential of the titration solution rather than the concentration of a specific reactant or product. Furthermore, the electrode potential is a log function of the concentration of the titration reactants and products. For these reasons it is customary to plot electrode potential versus volume of titrant added in constructing redox titration curves. Although chemists often refer to the "electrode potential of a solution," what they really mean is the potential of an electrode immersed in the solution relative to the standard hydrogen electrode.

The electrode potential of a solution being titrated is determined by the composition of the solution. Once the titration has begun, the solution contains both products and reactants of the titration reaction. For example, when iron(II) is titrated with cerium(IV), the titration reaction may be written

$$Fe^{2+} + Ce^{4+} \rightleftharpoons Fe^{3+} + Ce^{3+} \qquad (11\text{-}1)$$

and the solution contains all four substances at concentrations dictated by the amounts of reactants added and the equilibrium constant for the reaction. It is easy to become confused at this point and think that the solution has two electrode potentials: one determined by the Fe^{3+}/Fe^{2+} half-reaction and the other by the Ce^{4+}/Ce^{3+} half-reaction. Such is not the case. We must realize that the titration solution reaches equilibrium *after each addition* of titrant and the electrode potentials determined by the two half-reactions are *equal*. That is, the solution has only one potential:

$$E_{Ce} = E_{Fe} = E_{solution} \qquad (11\text{-}2)$$

The electrode potential of the solution can be determined experimentally by measuring the voltage of a galvanic cell in which one half-cell is a reference and the other half-cell is the solution being titrated. Such a cell for the iron(II) titration can be represented as

$$\text{reference} \, || \, Ce^{4+}, \, Ce^{3+}, \, Fe^{3+}, \, Fe^{2+} \, | \, Pt$$

and the cell voltage given by

$$\Delta E_{cell} = E_{solution} - E_{ref}$$

If the reference electrode is the standard hydrogen electrode, the voltage of the cell is the electrode potential of the solution. If a different reference electrode is used (see Chapter 13), its potential must be added to the cell voltage to obtain the electrode potential of the solution,

$$E_{solution} = \Delta E_{cell} + E_{ref}$$

Calculating Electrode Potentials

The titration of Fe^{2+} with Ce^{4+} (Reaction 11-1) will be used to illustrate how the data necessary to construct a theoretical titration curve can be calculated. The equilibrium constant for this reaction can be determined from the E° values using Equation 10-10:

$$\log K_{eq} = \frac{n \Delta E^{\circ}_{cell}}{0.0592} = \frac{(1)(1.61 - 0.77)}{0.0592} = 14.2$$

$$K_{eq} = 10^{14.2} = 2 \times 10^{14}$$

The large value of this equilibrium constant means that the reaction goes to completion and virtually any amount of Ce^{4+} added prior to the equivalence point will consume a stoichiometric amount of Fe^{2+}. The specific equation used to calculate the electrode potential depends on the composition of the solution, which changes as the titration progresses. Redox titrations can be divided into three regions:

Region	Major constituents	Comment
1. Before the equivalence point	Fe^{2+}, Fe^{3+}, Ce^{3+}	Use Fe half-reaction
2. At the equivalence point	Fe^{3+}, Ce^{3+}	Use e.p. equation
3. After the equivalence point	Fe^{3+}, Ce^{3+}, Ce^{4+}	Use Ce half-reaction

Each region will be considered separately, concentrating first on how to determine the composition of the solution and then on how to calculate the electrode potential.

Region 1: Before the equivalence point. Since the reaction stoichiometry is $1:1$ and the equilibrium constant is large, the addition of any Ce^{4+} reacts with an equal molar amount of Fe^{2+}:

$$\begin{array}{c} \text{amount } Fe^{2+} \text{ initial} \\ -\left\{\begin{array}{c} \text{amount } Ce^{4+} \text{ added} \\ \parallel \\ \text{amount } Fe^{2+} \text{ used} \end{array}\right\} \\ \hline \text{amount } Fe^{2+} \text{ remaining} \end{array}$$

$$\text{amount } Fe^{3+} \text{ produced} = \text{amount } Fe^{2+} \text{ used}$$

We have the option of using the Nernst equation for *either* half-reaction to calculate the electrode potential. We choose the equation for the analyte half-reaction because the necessary concentration data are readily available.

$$Fe^{3+} + e^- \; \rightleftharpoons \; Fe^{2+}$$

$$E_{Fe} = E^\circ_{Fe^{3+}/Fe^{2+}} - \frac{0.0592}{1} \log \frac{[Fe^{2+}]}{[Fe^{3+}]}$$

The analytical concentration of Fe^{2+} and Fe^{3+} can be found by dividing the amount of each ion by the volume of the solution:

$$C_{Fe2+} = \frac{\text{amount } Fe^{2+} \text{ (mmol)}}{\text{vol (mL)}}$$

$$C_{Fe3+} = \frac{\text{amount } Fe^{3+} \text{ (mmol)}}{\text{vol (mL)}}$$

Since the equilibrium constant for the reaction is large, the extent of the reverse reaction is very small and has no appreciable effect on the concentrations of Fe^{2+} and Fe^{3+}. Thus we can make the usual assumption that the equilibrium and analytical concentrations are the same:

$$[Fe^{2+}] \simeq C_{Fe2+} \quad \text{and} \quad [Fe^{3+}] \simeq C_{Fe3+}$$

It is interesting to note that we cannot calculate the potential of the solution at the beginning of the titration, that is, before any titrant is added. Ostensibly, such a solution contains only the analyte, Fe^{2+}. If the concentration of Fe^{3+} is assumed to be zero, the Nernst equation cannot be solved:

$$E_{Fe} = E^\circ_{Fe^{3+}/Fe^{2+}} - \frac{0.0592}{1} \log \frac{[Fe^{2+}]}{0}$$

Clearly, as the concentration of Fe^{3+} approaches zero, the electrode potential approaches negative infinity. A solution with *no* Fe^{3+} would have incredible reducing power, more than sufficient to reduce water, components in dust, or dissolved oxygen. Equilibrium theory dictates that a small amount of Fe^{3+} must be present at all times. We assume that it is formed via the oxidation of some Fe^{2+}. Exactly how this occurs and which oxidizing agent is responsible are not of any great interest. The important thing to realize is that we have no way of calculating the concentration of Fe^{3+} or the electrode potential. This does not mean that an electrode in this solution has no potential. It is a routine matter to *measure* the electrode potential of such a solution.

Region 2: At the equivalence point. A stoichiometric amount of titrant has been added and the major constituents of the solution are the products of the titration reaction, Fe^{3+} and Ce^{3+}. From the stoichiometry of Reaction 11-1 we know that

$$[Fe^{3+}] = [Ce^{3+}] \tag{11-3}$$

The only source of Fe^{2+} and Ce^{4+} is the small back reaction. Although the concentrations of these ions are not readily known, the stoichiometry of the back reaction leading to their formation tells us they must be equal. Thus

$$[Fe^{2+}] = [Ce^{4+}] \tag{11-4}$$

The data are not immediately available to solve the Nernst equation for *either* half-reaction independently, but simultaneous equations can be solved to obtain the electrode potential. The Nernst equations for the analyte and titrant half-reactions are

$$E_{Fe} = E^{\circ}_{Fe^{3+}/Fe^{2+}} - \frac{0.0592}{n_{Fe}} \log \frac{[Fe^{2+}]}{[Fe^{3+}]} \tag{11-5}$$

or

$$n_{Fe} E_{Fe} = n_{Fe} E^{\circ}_{Fe^{3+}/Fe^{2+}} - 0.0592 \log \frac{[Fe^{2+}]}{[Fe^{3+}]} \tag{11-6}$$

and

$$E_{Ce} = E^{\circ}_{Ce^{4+}/Ce^{3+}} - \frac{0.0592}{n_{Ce}} \log \frac{[Ce^{3+}]}{[Ce^{4+}]} \tag{11-7}$$

or

$$n_{Ce} E_{Ce} = n_{Ce} E^{\circ}_{Ce^{4+}/Ce^{3+}} - 0.0592 \log \frac{[Ce^{3+}]}{[Ce^{4+}]} \tag{11-8}$$

Adding Equations 11-6 and 11-8 gives

$$n_{Fe} E_{Fe} + n_{Ce} E_{Ce} = n_{Fe} E^{\circ}_{Fe} + n_{Ce} E^{\circ}_{Ce} - 0.0592 \left(\log \frac{[Fe^{2+}]}{[Fe^{3+}]} + \log \frac{[Ce^{3+}]}{[Ce^{4+}]} \right)$$

Combining the log terms, we obtain

$$n_{Fe} E_{Fe} + n_{Ce} E_{Ce} = n_{Fe} E^{\circ}_{Fe} + n_{Ce} E^{\circ}_{Ce} - 0.0592 \log \frac{[Fe^{2+}][Ce^{3+}]}{[Fe^{3+}][Ce^{4+}]}$$

At the equivalence point, Equations 11-3 and 11-4 are valid, making the log term unity. We also know from Equation 11-2 that

$$E_{Fe} = E_{Ce} = E_{solution} \quad \text{or} \quad E_{ep}$$

Thus

$$n_{Fe}E_{ep} + n_{Ce}E_{ep} = n_{Fe}E_{Fe}^\circ + n_{Ce}E_{Ce}^\circ - 0.0592 \log \frac{[\cancel{Fe^{2+}}][\cancel{Ce^{3+}}]}{[\cancel{Fe^{3+}}][\cancel{Ce^{4+}}]} \quad (11\text{-}9)$$

and

$$(n_{Fe} + n_{Ce})\,E_{ep} = n_{Fe}E_{Fe}^\circ + n_{Ce}E_{Ce}^\circ$$

or

$$E_{ep} = \frac{n_{Fe}E_{Fe}^\circ + n_{Ce}E_{Ce}^\circ}{n_{Fe} + n_{Ce}} \quad (11\text{-}10)$$

The general form of Equation 11-10 is simply

$$E_{ep} = \frac{n_1 E_1^\circ + n_2 E_2^\circ}{n_1 + n_2} \quad (11\text{-}11)$$

Where 1 and 2 refer to analyte and titrant.

Equation 11-11 is not valid for every redox titration because its derivation depends on the cancellation of all the concentration terms in Equation 11-9. There are two situations when this may not be the case:

1. The redox reaction includes substances that do not undergo changes in oxidation state *and* the concentrations of these substances are not unity.
2. One of the redox half-reactions is stoichiometrically unsymmetrical; that is, the coefficients of the oxidized and reduced forms in the half-reaction are not equal.

The first of these situations is encountered when a titrant such as MnO_4^- is used to titrate Fe^{2+}. The titration reaction is

$$5Fe^{2+} + MnO_4^- + 8H^+ \;\rightleftharpoons\; 5Fe^{3+} + Mn^{2+} + 4H_2O$$

The Nernst equations for the half reactions are

$$n_{Fe}E_{Fe} = n_{Fe}E_{Fe^{3+}/Fe^{2+}}^\circ - 0.0592 \log \frac{[Fe^{2+}]}{[Fe^{3+}]}$$

and

$$n_{Mn}E_{Mn} = n_{Mn}E_{MnO_4^-/Mn^{2+}}^\circ - 0.0592 \log \frac{[Mn^{2+}]}{[MnO_4^-][H^+]^8}$$

Adding these two equations in the same manner used to derive Equation 11-9 gives

$$(n_{Fe} + n_{Mn})\,E_{ep} = n_{Fe}E_{Fe}^\circ + n_{Mn}E_{Mn}^\circ - 0.0592 \log \frac{[Fe^{2+}][Mn^{2+}]}{[Fe^{3+}][MnO_4^-][H^+]^8}$$

$$(11\text{-}12)$$

According to the stoichiometry of the reaction,

$$[Fe^{3+}] = 5[Mn^{2+}]$$

and

$$[Fe^{2+}] = 5[MnO_4^-]$$

at the equivalence point. Substituting these equalities in Equation 11-12 yields

$$(n_{Fe} + n_{Mn}) E_{ep} = n_{Fe} E_{Fe}^\circ + n_{Mn} E_{Mn}^\circ - 0.0592 \log \frac{\cancel{5}[\cancel{MnO_4^-}][\cancel{Mn^{2+}}]}{\cancel{5}[\cancel{Mn^{2+}}][\cancel{MnO_4^-}][H^+]^8}$$

Upon rearranging, we obtain

$$E_{ep} = \frac{n_{Fe} E_{Fe}^\circ + n_{Mn} E_{Mn}^\circ}{n_{Fe} + n_{Mn}} - \frac{0.0592}{n_{Fe} + n_{Mn}} \log \frac{1}{[H^+]^8} \qquad (11\text{-}13)$$

This tells us that the equivalence-point potential depends on the pH of the solution. Equation 11-13 reduces to the general form, Equation 11-11, only when the concentration of hydrogen ion is 1.0 M.

The second situation for which the general equation is not valid is encountered when a substance such as $Cr_2O_7^{2-}$ is used to titrate Fe^{2+}. The titration reaction is

$$6Fe^{2+} + Cr_2O_7^{2-} + 14H^+ \rightleftharpoons 6Fe^{3+} + 2Cr^{3+} + 7H_2O$$

Adding the Nernst equations for the two half reactions gives

$$(n_{Fe} + n_{Cr}) E_{ep} = n_{Fe} E_{Fe}^\circ + n_{Cr} E_{Cr}^\circ - 0.0592 \log \frac{[Fe^{2+}][Cr^{3+}]^2}{[Fe^{3+}][Cr_2O_7^{2-}][H^+]^{14}}$$

At the equivalence point

$$[Fe^{3+}] = 3[Cr^{3+}]$$

and

$$[Fe^{2+}] = 6[Cr_2O_7^{2-}]$$

Substituting these equalities in the equation above gives

$$E_{ep} = \frac{n_{Fe} E_{Fe}^\circ + n_{Cr} E_{Cr}^\circ}{n_{Fe} + n_{Cr}} - \frac{0.0592}{n_{Fe} + n_{Cr}} \log \frac{6[\cancel{Cr_2O_7^{2-}}][\cancel{Cr^{3+}}]^2}{3[\cancel{Cr^{3+}}][\cancel{Cr_2O_7^{2-}}][H^+]^{14}}$$

which reduces to

$$E_{ep} = \frac{n_{Fe} E_{Fe}^\circ + n_{Cr} E_{Cr}^\circ}{n_{Fe} + n_{Cr}} - \frac{0.0592}{n_{Fe} + n_{Cr}} \log \frac{2[Cr^{3+}]}{[H^+]^{14}}$$

In this case, the equivalence-point potential depends on the concentration of H^+ *and* Cr^{3+}. Even when the hydrogen-ion concentration is 1.0 M, this equation will not reduce to the general form (Equation 11-11).

Region 3: After the equivalence point. The solution contains both products plus excess titrant. The data needed to solve the Nernst equation for the *titrant* half-reaction are readily available.

$$-\left\{ \begin{array}{c} \text{amount Fe}^{2+} \text{ initial} \\ \parallel \\ \text{amount Ce}^{4+} \text{ used} \end{array} \right\}$$

$$\dfrac{\text{amount Ce}^{4+} \text{ added}}{\text{amount Ce}^{4+} \text{ remaining}}$$

$$\text{amount Ce}^{3+} \text{ formed} = \text{amount Ce}^{4+} \text{ used}$$

Since the concentrations of Ce^{4+} and Ce^{3+} are easily calculated, we choose to solve the Nernst equation for the titrant half-reaction to obtain the electrode potential.

$$Ce^{4+} + e^- \rightleftharpoons Ce^{3+}$$

$$E_{Ce} = E^\circ_{Ce^{4+}/Ce^{3+}} - \frac{0.0592}{1} \log \frac{[Ce^{3+}]}{[Ce^{4+}]}$$

where

$$[Ce^{4+}] \simeq C_{Ce^{4+}} = \frac{\text{amount Ce}^{4+} \text{ (mmol)}}{\text{volume (mL)}}$$

$$[Ce^{3+}] \simeq C_{Ce^{3+}} = \frac{\text{amount Ce}^{3+} \text{ (mmol)}}{\text{volume (mL)}}$$

We can summarize the conclusions of this section as follows:

Before the titration begins. No calculation possible.
Before the equivalence point. Use the Nernst equation for the analyte half-reaction.
At the equivalence point. Use Equation 11-11, if possible, or derive an appropriate equation.
After the equivalence point. Use the Nernst equation for the titrant half-reaction.

The calculated curve for the titration of Fe^{2+} with Ce^{4+} is shown in Figure 11-1. The portion of the curve prior to the equivalence point is calculated from Equation 11-5. When the titration is 50% complete, one-half the original amount of Fe^{2+} will be converted to Fe^{3+}, making the concentrations of the two ions equal. At this point in the titration the potential of the solution is equal to the standard electrode potential for iron:

$$E_{Fe} = E^\circ_{Fe^{3+}/Fe^{2+}} - \frac{0.0592}{1} \log \frac{[Fe^{2+}]}{[Fe^{3+}]}$$

$$= 0.771 - 0.0592 \log 1$$

$$= 0.771 \text{ V}$$

This point is shown by the first dashed line in Figure 11-1. The portion of the curve past the equivalence point is calculated from Equation 11-7. At 200% titrated, the concentration of excess Ce^{4+} equals the concentration of Ce^{3+} produced in the titration, and E is equal to $E^\circ_{Ce^{4+}/Ce^{3+}}$. This point is shown by the second dashed line in Figure 11-1.

Chapter 11   Oxidation-Reduction Titrations

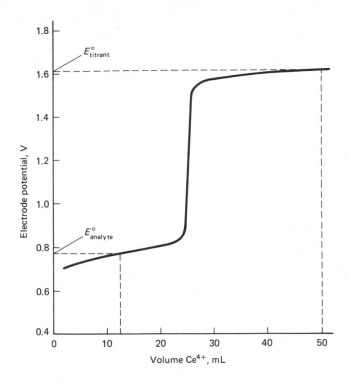

Figure 11-1 Curve for titration of 25.0 mL of 0.100 M Fe^{2+} with 0.100 M Ce^{4+}.

Example 11-1

A 25.0-mL aliquot of 0.112 M Fe^{2+} is titrated with 0.0258 M $KMnO_4$. Assuming that the hydrogen-ion concentration is 1.0 M throughout the titration, calculate the electrode potential of the solution after addition of 10.0 mL of titrant, at the equivalence point, and after addition of 40.0 mL of titrant.

$$E^\circ_{Fe^{3+}/Fe^{2+}} = 0.771 \text{ V} \qquad E^\circ_{MnO_4^-/Mn^{2+}} = 1.51 \text{ V}$$

Solution

After the addition of 10.0 mL of titrant. The titration reaction is

$$5Fe^{2+} + MnO_4^- + 8H^+ \rightleftharpoons 5Fe^{3+} + Mn^{2+} + 4H_2O$$

The composition of the solution is determined by comparing the amounts of each reactant:

$$
\begin{array}{lll}
\text{amount } Fe^{2+} \text{ initial} & = 25.0 \text{ mL} \times 0.112 \ M & = 2.80 \text{ mmol} \\
\left\{ \begin{array}{l} \text{amount } Fe^{2+} \text{ reacted} \\ \quad \| \\ \text{amount } MnO_4^- \text{ added} \times RR \end{array} \right. & = 10.0 \text{ mL} \times 0.0258 \ M \times \dfrac{5Fe^{2+}}{1MnO_4^-} & = 1.29 \text{ mmol} \\
\overline{\text{amount } Fe^{2+} \text{ remaining}} & & = \overline{1.51 \text{ mmol}}
\end{array}
$$

amount Fe^{2+} reacted = amount Fe^{3+} formed

where RR is the reacting ratio. Converting amounts to concentrations gives us

$$[Fe^{2+}] = \frac{1.51 \text{ mmol}}{35.0 \text{ mL}} = 0.0431 \ M$$

$$[Fe^{3+}] = \frac{1.29 \text{ mmol}}{35.0 \text{ mL}} = 0.0369 \ M$$

Titration Curves

Substituting in the Nernst equation for iron yields

$$E_{solution} = E_{Fe} = E^{\circ}_{Fe^{3+}/Fe^{2+}} - \frac{0.0592}{1} \log \frac{[Fe^{2+}]}{[Fe^{3+}]}$$

$$= 0.771 - 0.0592 \log \frac{0.0431}{0.0369}$$

$$= 0.767 \text{ V}$$

At the equivalence point. Equation 11-13 can be solved to obtain the equivalence-point potential.

$$E_{ep} = \frac{(1)(0.771) + (5)(1.51)}{1+5} - \frac{0.0592}{1+5} \log \frac{1}{(1)^8}$$

$$= \frac{0.771 + 7.55}{6} - 0$$

$$= 1.39 \text{ V}$$

After the addition of 40.0 mL of titrant. The composition of the solution is obtained by comparing the amounts of the reactants in the usual way:

amount MnO_4^- added $= 40.0 \text{ mL} \times 0.0258 \; M$ $= 1.03 \text{ mmol}$

$-\left\{ \begin{array}{l} \text{amount } MnO_4^- \text{ used} \\ \quad\quad\parallel \\ \text{amount } Fe^{2+} \text{ initial} \times RR \end{array} \right\} = 20.0 \text{ mL} \times 0.112 \; M \times \frac{1MnO_4^-}{5Fe^{2+}} = 0.45 \text{ mmol}$

amount MnO_4^- remaining $= 0.58 \text{ mmol}$

amount Mn^{2+} formed = amount MnO_4^- used

Converting amounts to concentrations, we obtain

$$[Mn^{2+}] = \frac{0.45 \text{ mmol}}{60.0 \text{ mL}} = 7.5 \times 10^{-3} \; M$$

$$[MnO_4^-] = \frac{0.58 \text{ mmol}}{60.0 \text{ mL}} = 9.7 \times 10^{-3} \; M$$

Using the Nernst equation for the permanganate half-reaction gives us

$$E_{solution} = E_{Mn} = E^{\circ}_{MnO_4^-/Mn^{2+}} - \frac{0.0592}{5} \log \frac{[Mn^{2+}]}{[MnO_4^-][H^+]^8}$$

$$= 1.51 - \frac{0.0592}{5} \log \frac{7.5 \times 10^{-3}}{(9.7 \times 10^{-3})(1.0)^8}$$

$$= 1.51 \text{ V}$$

Factors Affecting the Shape of Titration Curves

The concentration of the reactants and the completeness of the reaction have been the primary factors influencing the shape of acid-base and complexation titration curves. This section is devoted to an examination of how these factors influence the shape of redox titration curves.

Concentration

The electrode potentials used to construct the titration curve in Figure 11-1 were calculated using Equations 11-5 and 11-7. In both equations, the electrode potential is a function of the *ratio* of concentrations and is therefore *independent* of dilution. If the ferrous and ceric ion concentrations used for the titration were 0.001 M rather than 0.1 M, the curve would not change. This raises an interesting question: If the end-point break of the titration curve does not get smaller as the analyte and titrant concentrations decrease, what limits the smallest concentrations that can be titrated? Somehow we "know" that a 10^{-10} M Fe^{2+} solution cannot be titrated successfully with 10^{-10} M Ce^{4+}. The answer lies in the realization that the analyte solution will always contain small amounts of "other" reactants. A chemical indicator is one such reactant. Ordinarily, the concentrations of these other reactants are very small compared to the concentration of the analyte and, therefore, consume a negligible amount of titrant. However, if the analyte concentration is very small, the relative concentration of these "other" reactants may be large and they may consume a significant portion of the titrant. Instead of titrating the analyte, we may find ourselves titrating impurities or the indicator. In practice, the lower concentration limit for most redox titrations is about 1 millimolar.

Completeness of reaction

The change in electrode potential in the vicinity of the equivalence point of a redox titration is directly related to the completeness of the titration reaction. This is the same behavior encountered with other types of reactions. The completeness of a redox reaction is determined most conveniently from the difference in the standard electrode potentials of the two reacting substances (Equation 10-10). Figure 11-2 shows a family of curves in which Sn^{2+} ($E° = 0.15$ V) is titrated with a series of hypothetical oxidants whose standard electrode potentials vary from 0.25 to 1.75 V. Since most redox indicators need a change of about 0.12 V to undergo a good color transition, curve D of Figure 11-2 has about the smallest break that could still produce an acceptable indicator color change. The curves in Figure 11-2 are for reactions in which both analyte and titrant undergo a two-electron change. The curves for titrations involving one-electron changes exhibit slightly smaller equivalence-point breaks.

Hydrogen ion is a reactant or product of many redox reactions, and therefore its concentration affects the completeness of the reaction. The direction of the effect is determined easily from our knowledge of equilibrium. When hydrogen ion is a reactant, increasing its concentration improves the completeness of the titration reaction and increases the magnitude of the equivalence-point break. When hydrogen ion is a product, the opposite occurs. For example, the equivalence-point break for the titration of Fe^{2+} with MnO_4^- improves with increasing hydrogen-ion concentration:

$$5Fe^{2+} + MnO_4^- + 8H^+ \rightleftharpoons 5Fe^{3+} + Mn^{2+} + 4H_2O$$

Calculated curves for this titration at different hydrogen-ion concentrations are shown in Figure 11-3.

Mixtures

The successive titration of two or more oxidants or reductants in a sample is possible if each constituent reacts stepwise and if each reaction is complete enough to yield an observable end point. This is comparable to the stepwise titration of two acids having

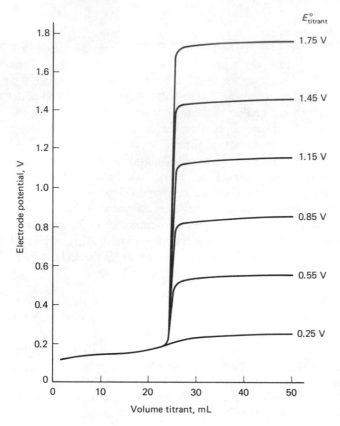

Figure 11-2 Curves for the titration of 25.0 mL of 0.100 M Sn^{2+} with different hypothetical titrants. $E^{\circ}_{Sn^{4+}/Sn^{2+}} = 0.15$ V. Titrant is assumed to undergo a two-electron reduction.

sufficiently different ionization constants. Generally speaking, the standard electrode potentials for two analytes must differ by at least 0.2 V for the reaction to be stepwise. Substances exhibiting three or more common oxidation states are analogous to polyprotic acids. Vanadium is one such substance as illustrated by the two half-reactions:

$$VO_2^+ + 2H^+ + e^- \rightleftharpoons VO^{2+} + H_2O \qquad E^{\circ} = 1.00 \text{ V}$$

$$VO^{2+} + 2H^+ + e^- \rightleftharpoons V^{3+} + H_2O \qquad E^{\circ} = 0.337 \text{ V}$$

The curve for the titration of V^{3+} with a strong oxidant will have two equivalence-point breaks; the first corresponding to the oxidation of V^{3+} to VO^{2+}, and the second to the oxidation of VO^{2+} to VO_2^+.

The titration curves for mixtures and substances with multiple oxidation states can be derived quite easily. Consider, for example, a mixture of Sn^{2+} and Fe^{2+} being titrated with potassium permanganate. The half reactions for the two analytes are:

$$Sn^{4+} + 2e^- \rightleftharpoons Sn^{2+} \qquad E^{\circ} = 0.15 \text{ V}$$

$$Fe^{3+} + e^- \rightleftharpoons Fe^{2+} \qquad E^{\circ} = 0.77 \text{ V}$$

Tin(II) is the stronger *reducing agent* as indicated by the fact that the E° for its half-reaction is *smallest*. Consequently, when the permanganate titrant is added, it reacts first

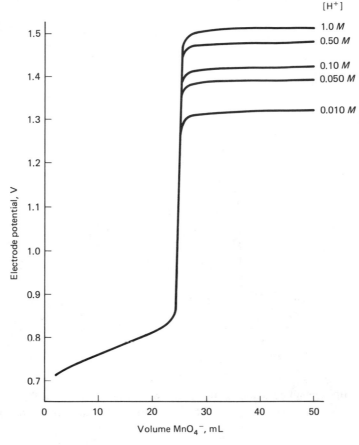

Figure 11-3 Curves for the titration of 25.0 mL of 0.100 M Fe^{2+} with 0.0200 M MnO$_4^-$ at different hydrogen-ion concentrations.

with Sn^{2+}, then with Fe^{2+}. Before the first equivalence point the major constituents in the solution are Sn^{4+}, Sn^{2+}, Fe^{2+}, and Mn^{2+} (from the reduction of the titrant). The solution potential is determined by the Sn^{4+}/Sn^{2+} redox couple and can be calculated from the Nernst equation:

$$E = E^\circ_{Sn^{4+}/Sn^{2+}} - \frac{0.0592}{2} \log \frac{[Sn^{2+}]}{[Sn^{4+}]}$$

Beyond the first equivalence point the major constituents of the solution are Sn^{4+}, Fe^{3+}, Fe^{2+}, and Mn^{2+}, and the potential is calculated most easily from the Nernst equation for the Fe^{3+}/Fe^{2+} couple:

$$E = E^\circ_{Fe^{3+}/Fe^{2+}} - \frac{0.0592}{1} \log \frac{[Fe^{2+}]}{[Fe^{3+}]}$$

Titration Curves

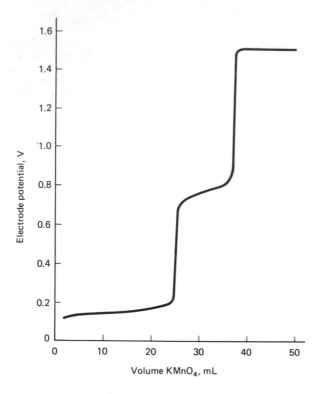

Figure 11-4 Curve for the titration of 25.0 mL of a solution containing 0.100 M Sn^{2+} and 0.100 M Fe^{2+} with 0.0400 M $KMnO_4$. The hydrogen-ion concentration is constant at 1.00 M.

After the second equivalence point, the potential is calculated from the Nernst equation for the MnO_4^-/Mn^{2+} couple. The curve for such a titration is shown in Figure 11-4.

DETERMINING THE EQUIVALENCE POINT

The techniques used to determine the equivalence point of redox titrations are the same as those for acid-base and complexation titrations: namely, visual and graphic. As pointed out previously, titration graphs are easily constructed by plotting the measured electrode potential versus the volume of titrant added.

Visual Indicators

There are two types of visual indicators used in redox titrations. *Nonspecific* or true equilibrium indicators respond only to the electrode potential of the solution, whereas *specific* indicators respond to the concentration of a particular substance in the solution. Both types are discussed in the following sections.

Nonspecific redox indicators

This type of indicator is a redox reagent whose oxidized and reduced forms are different colors. The indicator acts as a second oxidant or reductant in the solution and therefore must be *weaker* than the analyte to ensure that it reacts last with the titrant.

The observed color of the indicator depends on the ratio of its two redox forms. The half-reaction relating the two forms can be written as

$$In_{ox} + ne^- \rightleftharpoons In_{red}$$
$$\text{color A} \qquad\qquad \text{color B}$$

and the concentration ratio of the oxidized and reduced forms is a function of the electrode potential of the solution as described by the Nernst equation:

$$E = E_{In}^\circ - \frac{0.0592}{n} \log \frac{[In_{red}]}{[In_{ox}]} \tag{11-14}$$

Since the indicator is present in low concentration, the electrode potential of the solution is determined essentially by either the analyte or the titrant redox system. Thus the concentration of In_{red} and In_{ox} must be consistent with Equation 11-14 even though they are not primarily responsible for the value of E. We are interested in knowing the *change* in electrode potential that must occur in order to bring about a distinct, easily recognized change in color. Assuming that only the color of pure In_{ox} (color A) is observed when $[In_{red}]/[In_{ox}] < 1/10$ and only the color of pure In_{red} (color B) is observed when $[In_{red}]/[In_{ox}] > 10/1$, Equation 11-14 can be used to calculate the range of electrode potential over which the indicator will change color:

$$E_{color\ A} = E_{In}^\circ - \frac{0.0592}{n} \log \frac{1}{10} = E_{In}^\circ + \frac{0.0592}{n}$$

$$E_{color\ B} = E_{In}^\circ - \frac{0.0592}{n} \log \frac{10}{1} = E_{In}^\circ - \frac{0.0592}{n}$$

or

$$\Delta E_{color\ change} = \left(E_{In}^\circ + \frac{0.0592}{n} \right) - \left(E_{In}^\circ - \frac{0.0592}{n} \right) = \frac{2(0.0592)}{n}$$

This equation tells us that a change in electrode potential of at least 0.0592 V is required to bring about the desirable color change for a two-electron redox indicator. The potential range for a one-electron indicator is twice as large, or 0.118 V.

In theory, the indicator potential range is centered about E° for the indicator. That is,

$$\Delta E_{color\ change} = E_{In}^\circ \pm \frac{0.0592}{n} \tag{11-15}$$

This equation assumes that the two indicator forms have the same color intensity and that the human eye exhibits equal sensitivity to both colors. In fact, this is seldom the case and actual indicator ranges are somewhat different than those calculated from Equation 11-15. Furthermore, when hydrogen ion is part of the indicator half-reaction, the indicator range will be influenced by the pH of the solution. A selective list of common

TABLE 11-1 SELECTIVE LIST OF REDOX INDICATORS

Indicator	Color		$E°$ (V)
	Oxidized	Reduced	
Tris (2,2'-bipyridine) ruthenium (II)	Pale blue	Yellow	1.29
Tris (1,10-phenanthroline) iron (II)	Pale blue	Red	1.11
Diphenylaminesulfonic acid	Red-violet	Colorless	0.85
Diphenylamine	Violet	Colorless	0.76
Methylene blue	Blue	Colorless	0.53
Indigotetrasulfonate	Blue	Colorless	0.36
Phenosafranine	Red	Colorless	0.28

redox indicators is given in Table 11-1. Two common indicators, tris (1,10-phenanthroline) iron (II) and diphenylaminesulfonic acid, are discussed below.

Tris (1,10-phenanthroline) iron (II).   This complex, often called *ferroin*, is an iron (II) chelate. Three, bidentate 1,10-phenanthroline ligands coordinate to a ferrous ion, giving

The complex ion undergoes a reversible redox reaction represented by

$$\underset{\text{ferriin (pale blue)}}{Fe(phen)_3^{3+}} + e^- \rightleftharpoons \underset{\text{ferroin (red)}}{Fe(phen)_3^{2+}} \qquad E° = 1.06 \text{ V}$$

Because the color of the iron (III) complex is quite pale, only about 10% of the indicator needs to be in the iron (II) form for the solution to appear completely red. As a result, an end point based on the formation of a red color occurs at a potential of about 1.11 V in 1 M H_2SO_4.

Ferroin is probably the most nearly ideal nonspecific redox indicator. Its color change is very sharp and noticeable, its solutions are easily prepared and quite stable, and the indicator reaction is usually fast and reversible. It is not surprising, therefore, that a number of substituted phenanthrolines and similar bipyridines have been investigated as potential redox indicators.

Diphenylaminesulfonic Acid.   The redox behavior of this indicator is somewhat complicated, being thought to undergo a two-step oxidation:

diphenylaminesulfonate
(colorless)

irreversible

diphenylbenzidinesulfonate
(colorless)

reversible

diphenylbenzidine violet
(violet)

The second step, which is the actual indicator reaction, has an $E°$ of 0.85 V. It is somewhat surprising that the potential at which the color change occurs does not vary significantly with acidity, as hydrogen ion is a product of the oxidation. The reason for this behavior is not understood.

Specific indicators

The functioning of a specific indicator depends on the concentration of a particular analyte or titrant in the solution rather than the electrode potential. Specific indicators, by their nature, are used with a specific titrant or analyte.

Starch. Starch forms a deep blue complex with iodine but not with iodide. It is used in direct titrations, where iodine is the titrant, and in replacement titrations, where the analyte is replaced with iodine. Starches are polymeric substances consisting of two major fractions, amylose and amylopectin, the proportions of which vary depending on the source of the starch. The active fraction, amylose, is a polymer of the sugar α-D-glucose. It has the shape of a coiled helix into which long chains of I_2 combined with I^- can fit. Chemists are not presently certain about the exact nature of the polyiodine chains; both I_5^- ($2I_2 + I^-$) and I_{11}^{3-} ($4I_2 + 3I^-$) have been suggested as probable compositions. In any event, when the normally colorless starch reacts with a polyiodine chain, the product has an intense blue color.

Determining the Equivalence Point

Starch is not very stable in solution, being degraded by various microorganisms. One of the degradation products, glucose, is a reducing agent. Thus an "old" starch solution containing some glucose can cause an appreciable titration error. Such errors are easily avoided by always using fresh starch solutions.

Permanganate ion. When $KMnO_4$ is used as a titrant in strongly acidic solution, it can serve as its own indicator. The permanganate ion is very highly colored (purple), while its reduction product, Mn^{2+}, is almost colorless (pale pink) in dilute solutions. Consequently, the small excess of potassium permanganate titrant that occurs just past the equivalence point produces the purple color that signals the end of the titration.

Selecting the Proper Indicator

Specific indicators such as starch are chosen on the basis of their chemical behavior with a specific analyte or titrant. The major criterion in selecting a *nonspecific* indicator is to have the transition range of the indicator overlap the steepest part of the titration curve (the equivalence point). As a general rule, we try to select an indicator whose $E°$ is close to the potential of the solution at the equivalence point (E_{ep}). We can select from among many indicators if the break at the equivalence point is large, but the choice is much more limited when the break is small. Thus, referring to Figure 11-2, all but the last indicator in Table 11-1 could be used when the titrant whose $E°$ is 1.75 V is used, but only indigotetrasulfonate can be used when the titrant whose $E°$ is 0.55 V is used.

SAMPLE PREPARATION

It is not uncommon for the analyte in a sample to be present in two different oxidation states. The chemist must decide what is to be determined before selecting the method to be used. Sometimes only the concentration in one oxidation state is desired, while other times the total concentration is needed. In the latter case, some pretreatment of the sample is necessary to convert all of the analyte to the desired oxidation state prior to the titration. For example, if the total iron content of an ore is to be determined by titration with potassium permanganate, the sample must be prepared in such a way as to ensure that *all* of the iron present exists as Fe^{2+} before the titration begins. This might be accomplished quite simply by adding a sufficient amount of some reducing agent to the dissolved sample. However, any excess reducing agent will be capable of reducing the potassium permanganate titrant and must be removed or destroyed before the titration is begun. The necessity of this step limits the number of reagents suitable for prereduction and preoxidation. A few of the common reagents used for this purpose are discussed in this section.

Prereduction

Sodium sulfite, Na_2SO_3, and *sodium azide*, NaN_3, are good reductants in neutral or basic solution. Sulfite is oxidized to sulfate and azide is oxidized to nitrogen. The excess of either reagent can be removed by first acidifying and then boiling the solution. Sulfite is converted to sulfurous acid and expelled as gaseous sulfur dioxide:

$$Na_2SO_3 + 2HCl \longrightarrow H_2SO_3 + 2NaCl$$

$$H_2SO_3 \xrightarrow{\Delta} H_2O + SO_2(g)$$

Azide is converted to hydrazoic acid, which breaks down into nitrogen and ammonia:

$$NaN_3 + HCl \longrightarrow HN_3 + NaCl$$

$$3HN_3 \xrightarrow{\Delta} 4N_2(g) + NH_3$$

Azide ion forms insoluble salts with certain heavy metals such as lead(II), bismuth(III), and tin(II) that are known to detonate unexpectedly with explosive force. Waste azide solutions should *never* be poured down drains, where they may come in contact with lead or tin in solder joints.

Gaseous *sulfur dioxide* and *hydrogen sulfide* are mild reducing agents that can be passed directly through the sample solution. Their reactions with oxidants are generally slow, requiring as much as an hour to prereduce an analyte quantitatively. Excesses are easily removed by boiling the acidified solutions. Both gases are toxic and quite noxious even at low concentrations, leading chemists to choose other prereductants whenever possible.

Metallic reductors

Many metals are sufficiently good reductants to make them useful in prereducing analytes. Ordinarily, small granules of the metal are packed into a column, through which the sample is passed. Once prepared, these metallic reductors are very convenient to use. Active metals such as zinc and cadmium are excellent reductors but cannot be used without modification because they rapidly reduce hydrogen ion; an undesirable reaction because it uses up the metal, contaminates the analyte solution with large amounts of metal ion, and produces hydrogen gas. The *Jones reductor* (Figure 11-5) avoids these problems by using a zinc–mercury amalgam as the reducing agent. The amalgam is made by soaking zinc granules in a dilute solution of mercuric chloride [about 2% (w/w)] for 10 minutes.

$$Zn(s) + xHgCl_2 \longrightarrow ZnCl_2 + Zn(Hg)_x(s)$$
$$\text{(zinc amalgam)}$$

The amalgam reduces hydrogen ions very slowly but otherwise acts like metallic zinc. As shown in Figure 11-5, the analyte solution, acidified with H_2SO_4, is added to the top, percolates through the amalgam where the reduction occurs, and is collected as it exits the column. The reduction may be written in general as

$$\underset{\text{(analyte)}}{Ox} + Zn(Hg)(s) \xrightarrow{H_2SO_4} Red + Zn^{2+} + Hg(1)$$

The mercury formed in the reaction will amalgamate more zinc so that only Zn^{2+} and the reduced form of the analyte exit the column. Table 11-2 lists some of the substances that can be reduced with a Jones reductor.

The *Walden reductor* uses metallic silver as the reducing agent. Ordinarily, elemental silver is not very anxious to become oxidized and is considered a very poor reductant, as evidenced by the large $E°$ for the half-reaction

$$Ag^+ + e^- \rightleftharpoons Ag(s) \qquad E° = 0.80 \text{ V}$$

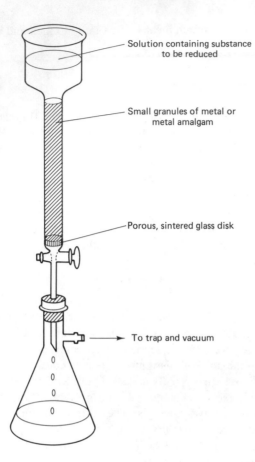

Solution containing substance to be reduced

Small granules of metal or metal amalgam

Porous, sintered glass disk

To trap and vacuum

Figure 11-5 Metallic reductor.

The presence of HCl greatly encourages the oxidation of silver because it results in the formation of stable, covalent silver chloride. The small $E°$ for the half-reaction

$$AgCl(s) + e^- \rightleftharpoons Ag(s) + Cl^- \qquad E° = 0.22 \text{ V}$$

is evidence of the increased reducing power. The reduction of an analyte on a Walden reductor may be written in general as

$$\underset{\text{(analyte)}}{Ox} + Ag(s) \xrightarrow{HCl} Red + AgCl(s)$$

This reductor is weaker and therefore more selective in its reactions than the Jones reductor, as illustrated by the data in Table 11-2.

Preoxidation

Potassium persulfate, $K_2S_2O_8$, in acidic solution is a very powerful oxidizing agent:

$$S_2O_8^{2-} + 2e^- \rightleftharpoons 2SO_4^{2-} \qquad E° = 2.01 \text{ V}$$

TABLE 11-2 REACTIONS OCCURRING
WITH METALLIC REDUCTORS

Zinc amalgam (Jones)	Silver (Walden)
$Ce^{4+} \rightarrow Ce^{3+}$	$Ce^{4+} \rightarrow Ce^{3+}$
$Cr_2O_7^{2-} \rightarrow Cr^{2+}$	$Cr_2O_7^{2-} \rightarrow Cr^{3+}$
$Cr^{3+} \rightarrow Cr^{2+}$	Cr^{3+} not reduced
$HMo_2O_6^{+} \rightarrow Mo^{3+}$	$HMo_2O_6^{+} \rightarrow MoO_2^{+}$
$MnO_4^{-} \rightarrow Mn^{2+}$	$MnO_4^{-} \rightarrow Mn^{2+}$
$Fe^{3+} \rightarrow Fe^{2+}$	$Fe^{3+} \rightarrow Fe^{2+}$
$TiO_2^{+} \rightarrow Ti^{3+}$	TiO_2^{+} not reduced
$VO_2^{+} \rightarrow V^{2+}$	$VO_2^{+} \rightarrow VO^{2+}$
$VO^{2+} \rightarrow V^{2+}$	VO^{2+} not reduced

This reagent is capable of oxidizing Mn^{2+} to MnO_4^{-}, Cr^{3+} to $Cr_2O_7^{2-}$, and Ce^{3+} to Ce^{4+}. A trace amount of silver ion often is necessary to catalyze the reactions. Excess reagent is destroyed easily and quickly by boiling the solution for a few minutes:

$$2S_2O_8^{2-} + 2H_2O \xrightarrow{\Delta} 4SO_4^{2-} + O_2(g) + 4H^{+}$$

Hydrogen peroxide, H_2O_2, is a strong oxidizing agent in acidic solution. The half-reaction is

$$H_2O_2 + 2H^{+} + 2e^{-} \rightleftharpoons 2H_2O \qquad E^{\circ} = 1.776$$

It is not powerful enough to completely oxidize Cr^{3+} to $Cr_2O_7^{2-}$ or Mn^{2+} to MnO_4^{-} but it can quantitatively convert Co^{2+} to Co^{3+} and Fe^{2+} to Fe^{3+}. After preoxidation is complete, boiling the solution will destroy the excess:

$$2H_2O_2 \xrightarrow{\Delta} 2H_2O + O_2(g)$$

Saturated solutions of *chlorine* and *bromine* (called chlorine water and bromine water) are used as preoxidants. Excess reagent is removed by boiling the acidified solution. Gaseous halogens are toxic, noxious, and very corrosive and therefore require care in handling. They should never be used outside a fume hood.

Hot, concentrated *perchloric acid*, $HClO_4$, is an extremely powerful oxidant, used mainly to destroy organic matter in a sample (oxidizing it to CO_2). In the process, the inorganic constituents are raised to their highest oxidation states. The procedure, called *wet ashing*, can result in violent explosions if not done properly. As a result, perchloric acid should be considered an *oxidant of last resort* and used only after taking all appropriate precautions. Pretreating the sample with boiling nitric acid to destroy most of the easily oxidized organic matter lowers the risk of explosion when the perchloric acid is added.

The oxidizing power of perchloric acid depends on it being *both* hot and concentrated. Cold, concentrated perchloric acid and hot, dilute perchloric acid solutions show no appreciable oxidizing power. They are still strong acid solutions, however, and should be handled with appropriate care. The oxidizing power of any excess perchloric acid is destroyed simply by cooling or diluting the solution.

It is clear from Figure 11-2 that the magnitude of the break around the equivalence point increases with the strength of the titrant. Oxidizing or reducing strength, however, is not the only consideration in choosing a titrant. The rate of the reaction between titrant and analyte, the stability and selectivity of the titrant, and the availability of a satisfactory indicator are other important considerations.

Oxidants

This section is devoted to a discussion of the chemistry of a few common oxidants used as redox titrants. The same format is used in discussing each titrant to facilitate and encourage critical comparisons between them.

Potassium permanganate

Potassium permanganate is among the oldest titrimetric oxidants. It is a powerful oxidant that is reduced to Mn^{2+} in strongly acidic solution and $MnO_2(s)$ in weakly acidic, neutral, or weakly alkaline solution:

$$pH < 2: \quad MnO_4^- + 8H^+ + 5e^- \rightleftharpoons Mn^{2+} + 4H_2O \quad\quad E° = 1.51 \text{ V}$$

$$5 < pH < 9: \quad MnO_4^- + 4H^+ + 3e^- \rightleftharpoons MnO_2(s) + 2H_2O \quad\quad E° = 1.68 \text{ V}$$

Titrations in both pH ranges are possible. Most analytes are titrated best in strongly acidic solution, but a few, including cyanide, sulfide, thiosulfate, and manganous ion, are titrated in solution at a pH between 5 and 9. Under these conditions, cyanide is oxidized to cyanate, sulfide and thiosulfate are oxidized to sulfate, and manganous ion is oxidized to manganese dioxide.

Stability. Permanganate is such a strong oxidant that it can oxidize water:

$$4MnO_4^- + 2H_2O \rightleftharpoons 4MnO_2(s) + 3O_2(g) + 4OH^- \quad\quad (11\text{-}16)$$

Very pure permanganate solutions are quite stable, however, because the reaction is slow in the absence of any catalysts. One of the substances that catalyzes Reaction 11-16 is MnO_2, making the reaction *autocatalytic*. As the reaction proceeds, more catalyst is produced, which acts to increase further the rate of the decomposition. Reaction 11-16 also is catalyzed by certain wavelengths of light, a process called photolytic decomposition.

Interestingly, permanganate is unstable in the presence of Mn^{2+}, its normal reduction product in strongly acidic solution:

$$2MnO_4^- + 3Mn^{2+} + 2H_2O \rightleftharpoons 5MnO_2(s) + 4H^+$$

Fortunately, this reaction is slow, and although it may alter significantly the concentration of a permanganate solution over a period of several days, it does not compete with fast reactions involving permanganate as a titrant.

Preparation and storage. Solutions of permanganate are prepared from the potassium salt, which is not available in a primary standard grade of purity. Certain precautions must be taken to minimize the undesirable catalytic influence of manganese dioxide and of sunlight. Not only is manganese dioxide a common contaminant in solid

$KMnO_4$ but it may be produced when permanganate reacts with trace amounts of organic matter or other reducing agents in the water used to prepare the solution. Most procedures prescribe boiling the freshly prepared solution to hasten such reactions, followed by filtration through sintered glass to remove the insoluble manganese dioxide. The filtered solution is stored in a dark brown bottle to prevent short wavelengths of light from reaching the solution and causing photolytic decomposition.

Applications. Probably the most often reported use of permanganate as a titrant is for the determination of iron. After the sample is dissolved, the iron is prereduced to Fe^{2+}, either by passage through a metallic reductor or by treatment with stannous chloride:

$$2FeCl_3 + SnCl_2 \rightleftharpoons 2FeCl_2 + SnCl_4$$

The excess stannous chloride must be destroyed; otherwise, it will compete with Fe^{2+} in reducing the permanganate titrant. This is accomplished by treating the solution with excess mercuric chloride:

$$SnCl_2 + 2HgCl_2 \rightleftharpoons SnCl_4 + Hg_2Cl_2(s)$$

At first, one may not appreciate how this can help, since we seem to be replacing one reducing agent, stannous chloride, with another, mercurous chloride. However, mercurous chloride, being insoluble, reacts too slowly with permanganate to interfere in the following titration reaction:

$$5Fe^{2+} + MnO_4^- + 8H^+ \rightleftharpoons 5Fe^{3+} + Mn^{2+} + 4H_2O \qquad (11\text{-}17)$$

Example 11-2

A sample of iron ore weighing 800 mg was treated with nitric acid, boiled to dryness, and redissolved in dilute hydrochloric acid. After filtering to remove any undissolved silica, the liquid was passed through a Walden silver reductor. The collected sample then was titrated with 0.0210 M $KMnO_4$, requiring 12.6 mL to reach the end point. Calculate the % Fe_2O_3 in the ore.

Solution The % Fe_2O_3 is calculated from the equation

$$\% \ Fe_2O_3 = \frac{\text{wt } Fe_2O_3}{\text{wt sample}} \times 100$$

The weight of Fe_2O_3 can be found from the amount (mmol) of Fe_2O_3 which is determined by the titration. According to Reaction 11-17,

$$\text{amount } Fe^{2+} = \text{amount } MnO_4^- \times \frac{5Fe^{2+}}{1MnO_4^-}$$

$$= (12.6 \ \text{mL})(0.0210 \ M) \times \frac{5}{1} = 1.32 \ \text{mmol}$$

Since each Fe_2O_3 produces $2Fe^{2+}$,

$$\text{amount } Fe_2O_3 = \text{amount } Fe^{2+} \times \frac{1Fe_2O_3}{2Fe^{2+}}$$

$$= 1.32 \ \text{mmol} \times \frac{1}{2} = 0.660 \ \text{mmol}$$

Titrants

Converting the amount in millimoles to weight, we obtain

$$wt \ Fe_2O_3 = 0.660 \ mmol \times 159.7 \ mg \ Fe_2O_3/mmol = 105 \ mg$$

Finally,

$$\% \ Fe_2O_3 = \frac{105 \ mg \ Fe_2O_3}{800 \ mg \ sample} \times 100 = 13.1$$

There are a considerable number of applications based on an indirect titration employing permanganate as the titrant. Calcium, for example, has been determined in this manner. The dissolved sample is treated with oxalic acid and made basic, leading to the precipitation of calcium oxalate, CaC_2O_4. The precipitate is collected by filtration and dissolved in dilute sulfuric acid, liberating calcium ions and oxalic acid. The determination is completed by titrating the oxalic acid with permanganate. The reactions are

$$CaC_2O_4(s) + 2H^+ \longrightarrow Ca^{2+} + H_2C_2O_4 \tag{11-18}$$

$$5H_2C_2O_4 + 2MnO_4^- + 6H^+ \longrightarrow 10CO_2(g) + 8H_2O + 2Mn^{2+} \tag{11-19}$$

Example 11-3

A limestone sample weighing 400 mg was dissolved in acid and treated with excess sodium oxalate. The solution was made basic and the resulting calcium oxalate precipitate was filtered, washed, and redissolved in dilute acid. This solution required 14.1 mL of 0.00865 M $KMnO_4$ for titration. What is the calcium content of the limestone?

Solution The percentage of calcium is calculated from the equation

$$\% \ Ca = \frac{wt \ Ca}{wt \ sample} \times 100$$

Oxalate is the connection between what is wanted (amount of Ca) and what is known (amount of MnO_4^-). According to the stoichiometry of Reaction 11-19,

$$amount \ H_2C_2O_4 = amount \ MnO_4^- \times \frac{5H_2C_2O_4}{2MnO_4^-}$$

$$= (14.1 \ mL)(0.00865 \ M) \times \frac{5}{2} = 0.305 \ mmol$$

From the stoichiometry of Reaction 11-18 and the formula for calcium oxalate,

$$amount \ Ca = amount \ H_2C_2O_4 \times \frac{1CaC_2O_4}{1H_2C_2O_4} \times \frac{1Ca}{1CaC_2O_4}$$

$$= 0.305 \ mmol \times \frac{1}{1} \times \frac{1}{1} = 0.305 \ mmol$$

Then

$$weight \ Ca = 0.305 \ mmol \times 40.1 \ mg/mmol = 12.2 \ mg$$

and

$$\% \ Ca = \frac{12.2 \ mg}{400 \ mg} \times 100 = 3.06$$

Selected methods using permanganate are summarized in Table 11-3. The examples cited include direct, replacement, and back titrations.

Cerium(IV)

In perchloric acid solution, cerium(IV) is the strongest of the common oxidizing titrants. Although the half-reaction is commonly written as

$$Ce^{4+} + e^- \rightleftharpoons Ce^{3+}$$

it is unlikely that Ce^{4+} [or more correctly, $Ce(H_2O)_n^{4+}$] exists in solution to any appreciable extent. The large charge density (charge per unit surface area) of this ion makes it especially attractive to negative ions with which it forms covalent bonds. This has the effect of stabilizing cerium(IV) and making its reduction more difficult. As a result of complexation with anions, the measured electrode potential of a half-cell consisting of cerium(IV) and cerium(III), both at an analytical concentration of 1.0 M, depends on

TABLE 11-3 SELECTED APPLICATIONS OF PERMANGANATE TITRATIONS

Analyte	Oxidation half-reaction	Comments
Fe	$Fe^{2+} \rightleftharpoons Fe^{3+} + e^-$	Prereduce with $SnCl_2$ or metal reductor; direct titration in H_2SO_4 or HCl
Mo	$2Mo^{3+} + 6H_2O \rightleftharpoons HMo_2O_6^+ + 11H^+ + 6e^-$	Prereduce with Jones reductor, collecting in excess Fe^{3+}; titrate Fe^{2+} formed
Ti	$Ti^{3+} + H_2O \rightleftharpoons TiO^{2+} + 2H^+ + e^-$	Prereduce with Jones reductor, collecting in excess Fe^{3+}; titrate Fe^{2+} formed
U	$U^{4+} + 2H_2O \rightleftharpoons UO_2^{2+} + 4H^+ + 2e^-$	Prereduce with Jones reductor. Air oxidize to U^{4+}; titrate in 1 M H_2SO_4
V	$VO^{2+} + H_2O \rightleftharpoons VO_2^+ + 2H^+ + e^-$	Prereduce with Bi reductor or $SO_2(g)$; direct titration
As	$H_3AsO_3 + H_2O \rightleftharpoons H_3AsO_4 + 2H^+ + 2e^-$	Direct titration 1 M HCl; ICl catalyst
Sb	$H_3SbO_3 + H_2O \rightleftharpoons H_3SbO_4 + 2H^+ + 2e^-$	Direct titration 2.5 M HCl
$H_2C_2O_4$	$H_2C_2O_4 \rightleftharpoons 2H^+ + 2CO_2(g) + 2e^-$	Direct titration 2 M H_2SO_4
Mg, Ca, Sr, Ba, Zn, Co	$H_2C_2O_4 \rightleftharpoons 2H^+ + 2CO_2(g) + 2e^-$	Precipitate metal oxalate, collect, and dissolve in H_2SO_4 for titration
HNO_2	$HNO_2 + H_2O \rightleftharpoons NO_3^- + 3H^+ + 2e^-$	Add excess $KMnO_4$ and back titrate after 15 min
H_2O_2	$H_2O_2 \rightleftharpoons O_2(g) + 2H^+ + 2e^-$	Direct titration in 1 M H_2SO_4 or KCl
H_2S	$S^{2-} + 8OH^- \rightleftharpoons SO_4^{2-} + 4H_2O + 8e^-$	Add excess alkaline $KMnO_4$; back titrate after boiling 5 min
Phenol, aniline, methanol, salicylic acid, ethylene glycol	Oxidation products vary	Add excess alkaline $KMnO_4$; back titrate after 24 h

Acid	$E°$ (V)^a
1.0 M HClO$_4$	1.70
1.0 M HNO$_3$	1.61
1.0 M H$_2$SO$_4$	1.44
1.0 M HCl	1.28

^{a}Analytical concentrations of Ce^{4+} and
Ce^{3+} are 1.0 M.

the particular acid present and its concentration. Values for the common strong acids are shown in Table 11-4. The electrode potential is largest in perchloric acid because ClO_4^- is considered the poorest complexing anion (of the four listed) and therefore exerts the least stabilizing effect on cerium(IV). To be consistent with the representation of other cations, Ce^{4+} will be used to represent Ce(IV) in solution.

Stability. Sulfuric acid solutions of ceric ion are stable for years when properly prepared and stored. Perchloric and nitric acid solutions of ceric salts decrease in concentration a few hundredths of a percent each day. Ceric ion is much less stable in hydrochloric acid, where it oxidizes chloride ion. The reaction is not so fast as to preclude the use of ceric ion in titrations where chloride ion is present, but it does prevent storing the oxidant in a chloride-containing solution.

Preparation and storage. Several cerium(IV) salts are available commercially, including ceric hydroxide, $Ce(OH)_4$; ammonium tetrasulfatocerate, $(NH_4)_4Ce(SO_4)_4 \cdot 2H_2O$; and ammonium hexanitratocerate, $(NH_4)_2Ce(NO_3)_6$. Only the latter is available in pure enough form to be considered a primary standard. The salts are dissolved in fairly concentrated sulfuric acid solution ($>0.2\ M\ H^+$) to avoid reactions with hydroxide ion that lead to the formation of insoluble basic salts such as $Ce(OH)_2SO_4$.

One regrettable disadvantage to the use of ceric ion titrants is their high cost. On a per mole basis, they are about 10 times as expensive as other common oxidants.

Applications. Most of the reported applications of ceric ion parallel those of permanganate, which are summarized in Table 11-3. Ceric ion is an excellent oxidant for alcohols, aldehydes, ketones, and carboxylic acids (usually converting them to CO_2) and is used in back-titration methods for determining these substances. The sample is treated with a known excess of Ce^{4+}, heated to assist the oxidation process, and the excess cerium determined by titration with a standard reductant such as Fe^{2+}.

Example 11-4

A 20.0-mL aliquot of malonic acid solution was treated with 10.0 mL of 0.250 M Ce^{4+}, leading to the reaction

$$CH_2(CO_2H)_2 + 6Ce^{4+} + 2H_2O \longrightarrow HCO_2H + 2CO_2(g) + 6Ce^{3+} + 6H^+$$

After standing for 10 minutes at 60°C, the solution was cooled and the excess Ce^{4+} was titrated with 0.100 M Fe^{2+}, requiring 14.4 mL to reach the ferroin end point. Calculate the molarity of the malonic acid in the sample.

Solution The reaction for the back titration is

$$Ce^{4+} + Fe^{2+} \rightleftharpoons Ce^{3+} + Fe^{3+}$$

The difference between the amount of Ce^{4+} added and the amount remaining is related to the amount of malonic acid present:

$$\begin{aligned}
\text{amount } Ce^{4+} \text{ added} &= 10.0 \text{ mL} \times 0.250 \, M = 2.50 \text{ mmol}\\
-\left\{\begin{array}{l}\text{amount } Ce^{4+} \text{ remaining}\\[4pt] \parallel \\ \text{amount } Fe^{2+} \times \dfrac{1Ce^{4+}}{1Fe^{2+}}\end{array}\right\} &= 14.4 \text{ mL} \times 0.100 \, M = 1.44 \text{ mmol}\\[6pt]
\overline{\text{amount } Ce^{4+} \text{ used}} & \overline{= 1.06 \text{ mmol}}
\end{aligned}$$

From the reaction between Ce^{4+} and malonic acid,

$$\text{amount } CH_2(CO_2H)_2 = \text{amount } Ce^{4+} \text{ used} \times \frac{1CH_2(CO_2H)_2}{6Ce^{4+}}$$

$$= 1.06 \text{ mmol} \times \frac{1}{6} = 0.177 \text{ mmol}$$

Finally,

$$\text{concentration} = \frac{0.177 \text{ mmol}}{20.0 \text{ mL}} = 8.83 \times 10^{-3} \, M$$

Potassium dichromate

Dichromate ion is not as strong an oxidant as permanganate or ceric ion and it reacts more slowly with some reductants. Despite these drawbacks, it is a desirable titrant because of its exceptional stability, low cost, and availability in primary standard-grade purity. In strongly acidic solution, chromium(VI) exists as $Cr_2O_7{}^{2-}$ and is reduced according to the following half-reaction:

$$Cr_2O_7{}^{2-} + 14H^+ + 6e^- \rightleftharpoons 2Cr^{3+} + 7H_2O \qquad E° = 1.33 \text{ V}$$

Like cerium, the measured electrode potential varies somewhat with the nature and concentration of the acid in solution but not to as great an extent. Dichromate is converted to chromate in basic solution,

$$Cr_2O_7{}^{2-} + 2OH^- \rightleftharpoons 2CrO_4{}^{2-} + H_2O$$

and in the process loses virtually all of its oxidizing power. The electrode potential for the reduction of chromate ion is -0.13 V. The indicators used in dichromate titrations must display very distinct color changes in order not to be hidden by the lightly colored dichromate (orange) and chromic (green) ions. Diphenylaminesulfonic acid probably is the indicator used most often with potassium dichromate.

Stability. Acidic solutions of potassium dichromate are exceptionally stable. Even boiling for long periods of time does not measurably alter their concentration (assuming the evaporated water is replaced).

Preparation and storage. Potassium dichromate is available in primary-standard-grade purity and dissolves readily in water and dilute acid. No special precautions need to be observed in storing solutions of this titrant.

Applications. The determination of Fe^{2+}, either directly or indirectly, constitutes the most important application of dichromate as a redox titrant. In terms of accuracy

Titrants

it is the equal of permanganate for the titration of Fe^{2+}. Some reductants react too slowly with dichromate to permit their direct titration. In many such cases, excess dichromate can be added and back titrated with a standard ferrous ion solution.

The chemical oxygen demand (COD) of untreated water commonly is determined by such a back titration. COD is a measure of the amount of oxygen that can be consumed by various substances such as aerobic organisms, sulfide, ammonia, and iron(II) that occur in untreated natural water and in wastewater. In this procedure, the oxidizable material is destroyed by treating the water sample with a potassium dichromate–sulfuric acid mixture containing silver sulfate as a catalyst and boiling for several hours. The excess dichromate is back titrated with a standard ferrous-ion solution and the amount of oxidizable material can be calculated from the amount of dichromate consumed.

Iodine

The half-reaction for the reduction of iodine is

$$I_2(s) + 2e^- \;\rightleftharpoons\; 2I^- \qquad E° = 0.536 \text{ V} \qquad (11\text{-}20)$$

Despite the fact that the oxidizing power of iodine is considerably less than that of ceric, permanganate, or dichromate ions, it enjoys widespread use as a redox titrant because it reacts *rapidly* with a variety of strong reductants and because there is an excellent indicator available (starch) for its titrations. These two advantages outweigh iodine's inability to react completely with weak reductants and its limited stability in solution.

Iodine is not very soluble in water—a saturated solution is only about 1.3×10^{-3} M at 20°C—but dissolves readily in an iodide solution, owing to the formation of the triiodide ion, I_3^-.

$$I_2 + I^- \;\rightleftharpoons\; I_3^- \qquad K_{eq} = 7.10 \times 10^2$$

Larger aggregates such as I_5^- are also formed but in much smaller amounts. Because most of the iodine in such solutions is present as I_3^-, it is proper to refer to them as *triiodide solutions*. The half-reaction for the reduction of I_3^- is

$$I_3^- + 2e^- \;\rightleftharpoons\; 3I^- \qquad E° = 0.536 \text{ V} \qquad (11\text{-}21)$$

There is little difference between the redox chemistry of I_2 and I_3^-, and regardless of whether we speak of iodine or triiodide ion, we are dealing with a reagent that undergoes a two-electron reduction to iodide ions (Reactions 11-20 and 11-21). Many chemists prefer the simplicity of representing the reagent as I_2 and calling it an *iodine solution*, a practice that is followed in this book.

Stability. Iodine solutions are not very stable. In acidic medium, iodide is oxidized by oxygen, thereby increasing the concentration of iodine:

$$4I^- + O_2 + 4H^+ \;\longrightarrow\; 2I_2 + 2H_2O \qquad (11\text{-}22)$$

This reaction is very slow in neutral solution but becomes more rapid as the acidity increases. In addition, the reaction is catalyzed by certain metal ions and short-wavelength light.

In alkaline solution, iodine reacts with hydroxide ion, undergoing a redox disproportionation. Between pH 7 and 9, hypoiodite and iodide are the major products, but above pH 9 the disproportionation leads to the formation of iodate:

$$\text{pH 7-9:} \quad I_2 + 2OH^- \rightleftharpoons \underset{\text{(hypoiodite)}}{IO^-} + I^- + H_2O \quad (11\text{-}23)$$

$$\text{pH} > 9: \quad 3I_2 + 6OH^- \rightleftharpoons \underset{\text{(iodate)}}{IO_3^-} + 5I^- + 3H_2O \quad (11\text{-}24)$$

The conversion of iodine into hypoiodite does not change the oxidizing capacity of the solution, as both substances are good two-electron oxidants:

$$I_2 + 2e^- \rightleftharpoons 2I^-$$

$$IO^- + H_2O + 2e^- \rightleftharpoons I^- + 2OH^-$$

Furthermore, an equilimolar mixture of hypoiodite and iodide reacts rapidly and quantitatively with many reducing agents. There are a few important reactions with iodine that are carried out in slightly alkaline solution, where much, if not all, of the iodine is present as the hypoiodite ion. Iodate ion, on the other hand, does not react rapidly with many reducing agents and therefore is not an acceptable substitute for iodine.

Preparation and storage. Iodine solutions up to about 1 *M* can be prepared quite easily by dissolving solid I_2 in excess aqueous KI. The rate of dissolution is quite slow in dilute KI solutions, so procedures usually call for dissolving the iodine in a small volume of concentrated KI solution and then diluting with water. To minimize the formation of additional iodine via the oxidation of iodide (Reaction 11-22), the solutions are not acidified. Since Reaction 11-22 is catalyzed photolytically by short-wavelength radiation, the solution is best stored in a dark brown bottle that is able to filter such radiation and prevent it from reaching the iodine.

Applications. Iodine has been used to titrate a very large and diverse group of reductants. A selected list of such substances is given in Table 11-5. Ascorbic acid (vitamin C) is typical of many organic compounds that are capable of being titrated directly with iodine:

TABLE 11-5 SELECTED APPLICATIONS OF DIRECT AND BACK TITRATIONS WITH IODINE

Analyte	Oxidation half-reaction	Comments
As	$H_3AsO_3 + H_2O \rightleftharpoons H_3AsO_4 + 2H^+ + 2e^-$	Direct titration at pH 7 to 9
Sb	$H_3SbO_3 + H_2O \rightleftharpoons H_3SbO_4 + 2H^+ + 2e^-$	Direct titration at pH 7 to 9
HCN	$2CN^- + I_2 \rightleftharpoons 2ICN + 2e^-$	Direction titration at pH 9
H_2CO	$H_2CO + 3OH^- \rightleftharpoons HCO_2^- + 2H_2O + 2e^-$	Add excess I_2 plus NaOH; back titrate I_2 after 5 min
H_2S	$H_2S \rightleftharpoons S(s) + 2H^+ + 2e^-$	Add to excess I_2 in 1 *M* HCl and back titrate I_2
Zn, Cd, Hg, Pb	$H_2S \rightleftharpoons S(s) + 2H^+ + 2e^-$	Precipitate metal sulfide, collect, and dissolve in 3 *M* HCl with excess I_2; back titrate I_2
SO_2	$H_2SO_3 + H_2O \rightleftharpoons SO_4^{2-} + 4H^+ + 2e^-$	Add to excess I_2 in acid; back titrate I_2
Ascorbic acid	See the text	Direct titration in acid

$$\text{ascorbic acid} + I_2 \rightleftharpoons \text{dehydroxyascorbic acid} + 2H^+ + 2I^-$$

ascorbic acid   dehydroxyascorbic acid

The end point is marked by the appearance of the blue starch-iodine complex.

Numerous sulfur compounds have been determined by prereducing the sulfur to hydrogen sulfide, which is then titrated with iodine. The reaction

$$H_2S + I_2 \rightleftharpoons S(s) + 2H^+ + 2I^-$$

is quantitative and fast enough to allow a direct titration, but a back titration is often preferred in order to avoid the potential loss of some H_2S by volatilization. The sulfide-containing sample is placed in a closed container with an excess of iodine and then acidified. The unreacted iodine is titrated with a standard reductant.

In 1935, a German chemist, Karl Fischer, described a direct titration procedure for determining water using an iodine titrant. The method, still very much in use, is applicable to a large variety of sample types. The water-containing sample is dissolved in anhydrous methanol and titrated with a solution consisting of iodine, sulfur dioxide, and pyridine dissolved in methanol. The titration is based on a reaction between iodine and sulfur dioxide that occurs only if water is present:

$$C_5H_5N \cdot I_2 + C_5H_5N \cdot SO_2 + C_5H_5N + H_2O \longrightarrow 2C_5H_5NH^+I^- + C_5H_5N \cdot SO_3$$

$$(11\text{-}25)$$

The pyridine acts both as a coordinating solvent for the iodine and sulfur dioxide and as a base to accept protons from the water. In a second step, the pyridine sulfur trioxide goes on to react with methanol, forming pyridinium methyl sulfate:

$$C_5H_5N \cdot SO_3 + CH_3OH \longrightarrow C_5H_5NH^+CH_3OSO_3^-$$

A simplified version of Reaction 11-25, useful for the purpose of solving titration problems, can be written as

$$I_2 + SO_2 + 3Py + H_2O \longrightarrow 2PyH^+I^- + Py \cdot SO_3 \qquad (11\text{-}26)$$

where Py is pyridine.

The titrant is somewhat troublesome to prepare but fortunately is available from commercial sources. Frequent standardization is necessary, as the titrant is not particularly stable. All glassware and solvents used in the determination must be thoroughly dried and protected from contamination by atmospheric moisture. Starch behaves differently in methanol than water and cannot be used as an indicator for this titration. Instead, the end point is detected by the appearance of the first excess of iodine or by one of several types of electrochemical measurements.

Example 11-5

A 10.0-mL aliquot (density = 1.50 g/mL) of a fluorohydrocarbon solvent, thought to be contaminated with water, was dissolved in 50 mL of anhydrous methanol and titrated with a Karl Fischer iodine reagent, requiring 22.4 mL to reach the end point. A 10.0-mL aliquot of a sample prepared by dissolving 1.00 mL of distilled water in 500 mL of anhydrous methanol required 26.7 mL of the same titrant to reach the end point. Calculate the % H_2O in the fluorohydrocarbon solvent.

Solution The second titration represents the standardization of the titrant. According to Reaction 11-26,

$$\text{amount } I_2 = \text{amount } H_2O$$

The volume of the water standard taken can be converted to weight using the density of water, and then to millimoles using the molecular weight:

$$\text{wt } H_2O = 1.00 \text{ mL} \times \frac{10.0 \text{ mL}}{500 \text{ mL}} \times 1.00 \underset{\text{(density)}}{\frac{\text{g } H_2O}{\text{mL } H_2O}} = 0.0200 \text{ g}$$

$$\text{amount } H_2O = \frac{0.0200 \text{ g} \times 1000 \text{ mg/g}}{18.0 \text{ mg/mmol}} = 1.11 \text{ mmol}$$

Substituting gives us

$$\text{amount } I_2 = \text{vol} \times \text{conc} = \text{amount } H_2O$$

$$= 26.7 \text{ mL} \times \text{conc} = 1.11 \text{ mmol}$$

or

$$\text{concentration} = \frac{1.11 \text{ mmol}}{26.7 \text{ mL}} = 0.0416 \ M$$

Now that the concentration of the titrant is known, the amount of the unknown can be determined:

$$\text{amount } H_2O = \text{amount } I_2$$

$$= (22.4 \text{ mL})(0.0416 \ M) = 0.932 \text{ mmol}$$

$$\text{wt } H_2O = \text{amount (mmol)} \times \text{MW}$$

$$= (0.932 \text{ mmol})(18.0 \text{ mg/mmol}) = 16.8 \text{ mg or } 0.0168 \text{ g}$$

Finally,

$$\% \ H_2O = \frac{\text{wt } H_2O}{\text{wt sample}} \times 100 = \frac{\text{wt } H_2O}{\text{vol sample} \times \text{density}} \times 100$$

$$= \frac{0.0168 \text{ g}}{(10.0 \text{ mL})(1.50 \text{ g/mL})} \times 100 = 0.112$$

Reductants

Reductants are used less often than oxidants as titrants because they are generally less stable, being subject to oxidation by atmospheric oyxgen. The difficulty of working in an oxygen-free environment is considerable and is avoided whenever possible. The two reductants discussed in this section are useful in certain applications partly because they react quite slowly with oxygen.

Iron(II)

Ferrous ion is one of the few reducing titrants that can be handled without special provisions for excluding atmospheric oxygen. In titrations, it undergoes a simple one-electron oxidation to ferric ion. The fairly large $E°$ for the half-reaction,

$$Fe^{3+} + e^- \rightleftharpoons Fe^{2+} \qquad E° = 0.77 \text{ V}$$

means that Fe^{2+} is a weak reducing agent.

Stability. The atmospheric oxidation of ferrous ion is quite rapid in neutral solution but slows considerably as the acidity is increased. Solutions that are $0.5 M$ in H_2SO_4 are stable enough to be used for several hours without restandardization. Ferrous ion hydrolyzes in weakly acidic, neutral, and alkaline solutions, forming insoluble ferrous hydroxide.

Preparation and storage. Solutions of iron(II) are almost always prepared from *Mohr's salt*, $Fe(NH_4)_2(SO_4)_2 \cdot 6H_2O$, or *Oesper's salt*, $Fe(enH_2)(SO_4)_2 \cdot 4H_2O$, where enH_2 = diprotonated ethylenediamine. Both reagents are available in primary standard-grade purity, and dissolve readily in acidic solution to yield Fe^{2+}. Neither salt should be oven dried, as they decompose on heating.

Applications. Iron(II) is not used much in direct titrations. Instead, it serves as a titrant for determining excess oxidants, such as MnO_4^-, Ce^{4+}, and $Cr_2O_7^{2-}$, in back titrations.

Iodide

Although iodide ion is a slightly stronger reductant than ferrous ion, it is not used as a titrant. An attempted titration with iodide produces iodine, whose rather intense color obscures any possible indicator color change. Far from being an unimportant reductant, however, iodide is used extensively in replacement or, as they are often called, *iodometric* titrations. Ferric ion is commonly determined via an iodometric titration according to the following reactions:

Replacement reaction:
$$2Fe^{3+} + 2I^- \rightleftharpoons 2Fe^{2+} + I_2$$
$$\text{(excess)}$$

Titration reaction:
$$I_2 + \text{titrant(Red)} \rightleftharpoons I^- + \text{titrant(Ox)}$$

There is no problem with the indicator color change being obscured, since the iodine is being consumed rather than formed in the titration. Although it is feasible to use the disappearance of the iodine color as the end point, it is not recommended. Iodine is not as intensely colored as a regular indicator, and it becomes too faint to see somewhat before the actual equivalence point is reached. The problem is easily overcome by adding a sensitive indicator such as starch just before the iodine color disappears completely.

Titrant for iodine. The reagent used most often to titrate iodine is sodium thiosulfate, $Na_2S_2O_3$. In reacting with iodine, thiosulfate is oxidized to tetrathionate, $S_4O_6^{2-}$:

$$I_2 + 2S_2O_3^{2-} \rightleftharpoons 2I^- + S_4O_6^{2-} \qquad (11\text{-}27)$$

or

$$\text{I}_2 + 2\ {}^{-}\text{S}-\overset{\overset{\displaystyle O}{\|}}{\underset{\underset{\displaystyle O}{\|}}{\text{S}}}-\text{O}^{-} \;\rightleftharpoons\; 2\text{I}^{-} + {}^{-}\text{O}-\overset{\overset{\displaystyle O}{\|}}{\underset{\underset{\displaystyle O}{\|}}{\text{S}}}-\text{S}-\text{S}-\overset{\overset{\displaystyle O}{\|}}{\underset{\underset{\displaystyle O}{\|}}{\text{S}}}-\text{O}^{-}$$

thiosulfate   tetrathionate

Thiosulfate ion is actually quite a strong reductant, as indicated by the small $E°$ for its half-reaction:

$$S_4O_6{}^{2-} + 2e^- \;\rightleftharpoons\; 2S_2O_3{}^{2-} \qquad E° = 0.08 \text{ V}$$

Sodium thiosulfate is readily soluble in water and is stable in both neutral and basic solutions. It undergoes a *slow* decomposition in acidic solutions with a pH less than about 5:

$$S_2O_3{}^{2-} + 2H^+ \;\xrightarrow{\text{slow}}\; H_2SO_3 + S(s) \tag{11-28}$$

A thin layer of yellow sulfur adhering to the walls of the container is visual evidence that appreciable decomposition has occurred. In addition, certain bacteria are known to metabolize thiosulfate, converting it to sulfite, sulfate, and elemental sulfur. This source of instability is rather easily overcome by using sterile conditions during preparation of the solution. Most procedures recommend using cool, freshly boiled water to dissolve the solid sodium thiosulfate and boiling water to rinse out the storage container. Once the solution has been prepared, a small amount of thymol, chloroform, or sodium benzoate is added to inhibit bacterial growth.

Although it is clear from Reaction 11-28 that thiosulfate may not be *stored* in acidic solutions, it can be used to titrate iodine in acidic solutions because its reaction with iodine (Reaction 11-27) is much faster than with hydrogen ion (Reaction 11-28). This is important because most iodometric replacement reactions must be carried out in fairly acidic solution.

Applications. Iodometric methods have been applied to the determination of a large variety of both organic and inorganic substances. Selected applications are summarized in Table 11-6. A few of these are described in a little detail in this section.

Metal ions that are easily reduced, such as Cu^{2+} and Fe^{3+}, can be determined using the iodometric technique. Iodide reduces Cu^{2+} to cuprous iodide:

$$2Cu^{2+} + 4I^- \;\rightleftharpoons\; 2CuI(s) + I_2 \tag{11-29}$$

Most reducing agents reduce Cu^{2+} to the metal, but iodide forms a strong, covalent bond with Cu^+, which stabilizes the $+1$ oxidation state. Even though hydrogen ions are not a part of Reaction 11-29, pH is important to the success of the titration. Above pH 4, insoluble cupric hydroxide forms, which reacts slowly with iodide and causes a premature end-point color change. At pH values less than 1, air oxidation of iodide becomes significant and leads to high results. Most procedures call for buffering the solution around pH 3.2 with ammonium hydrogen fluoride, NH_4HF_2. This reagent acts as an equimolar buffer of HF and F^- ($pK_a \simeq 3.2$). The presence of F^- also serves to eliminate the interference of Fe^{3+} by forming a stable complex, $FeF_6{}^{3-}$, that is not reduced by iodide.

Titrants

TABLE 11-6 SELECTED APPLICATIONS OF IODOMETRIC TITRATIONS

Analyte	Replacement reaction	Comments
Ce	$2Ce^{4+} + 2I^- \rightleftharpoons 2Ce^{3+} + I_2$	Preoxidize to Ce^{4+}; titrate in 1 M H_2SO_4
Mn	$2MnO_4^- + 10I^- + 8H^+ \rightleftharpoons 2Mn^{2+} + 5I_2 + 8H_2O$	Preoxidize with persulfate; titrate in 0.1 M HCl
Cr	$Cr_2O_7^{2-} + 6I^- + 14H^+ \rightleftharpoons 2Cr^{3+} + 3I_2 + 7H_2O$	Preoxidize with $HClO_4$; titrate after 5 min
IO_3^-	$IO_3^- + 5I^- + 6H^+ \rightleftharpoons 3I_2 + 3H_2O$	Titrate in 0.5 M HCl
Fe	$2Fe^{3+} + 2I^- \rightleftharpoons 2Fe^{2+} + I_2$	Titrate in acid
Cu	$2Cu^{2+} + 4I^- \rightleftharpoons 2CuI(s) + I_2$	Titrate in NH_4HF buffer at pH 3
Cl_2	$Cl_2 + 2I^- \rightleftharpoons 2Cl^- + I_2$	Titrate in dilute acid
Br_2	$Br_2 + 2I^- \rightleftharpoons 2Br^- + I_2$	Titrate in dilute acid
HOCl	$HOCl + 2I^- + H^+ \rightleftharpoons Cl^- + I_2 + H_2O$	Titrate in 0.5 M H_2SO_4
H_2O_2	$H_2O_2 + 2I^- + 2H^+ \rightleftharpoons I_2 + 2H_2O$	Titrate in 1 M H_2SO_4 with Na_2MoO_4 catalyst
HNO_2	$2HNO_2 + 2I^- + 2H^+ \rightleftharpoons 2NO + I_2 + 2H_2O$	Remove NO by bubbling CO_2 through solution prior to titration
O_3	$O_3 + 2I^- + 2H^+ \rightleftharpoons O_2 + I_2 + H_2O$	Pass O_3 through neutral KI; acidify with H_2SO_4

Example 11-6

A piece of brass weighing 220 mg was dissolved and prepared for an iodometric titration. Excess KI was added and the liberated iodine required 26.9 mL of 0.0847 M sodium thiosulfate to reach the starch end point. Calculate the % Cu in the brass.

Solution The balanced reactions for this determination are

$$2Cu^{2+} + 4I^- \rightleftharpoons 2CuI(s) + I_2$$
$$\text{(excess)}$$

$$I_2 + 2S_2O_3^{2-} \rightleftharpoons 2I^- + S_4O_6^{2-}$$

Accordingly,

$$\text{amount } Cu^{2+} = \text{amount } I_2 \times \frac{2Cu^{2+}}{1I_2}$$

and

$$\text{amount } I_2 = \text{amount } S_2O_3^{2-} \times \frac{1I_2}{2S_2O_3^{2-}}$$

or combining,

$$\text{amount } Cu^{2+} = \text{amount } S_2O_3^{2-} \times \frac{1I_2}{2S_2O_3^{2-}} \times \frac{2Cu^{2+}}{1I_2}$$

$$= (26.9 \text{ mL})(0.0847 \text{ } M)(1/2)(2/1)$$

$$= 2.28 \text{ mmol}$$

Converting to weight, we obtain

$$\text{wt Cu} = \text{amount Cu(mmol)} \times \text{AW}$$

$$= (2.28 \text{ mmol})(63.5 \text{ mg/mmol}) = 145 \text{ mg}$$

Then

$$\% \text{ Cu} = \frac{\text{wt Cu}}{\text{wt sample}} \times 100 = \frac{145 \text{ mg}}{220 \text{ mg}} \times 100 = 65.9$$

Hydrogen peroxide can be determined iodometrically based on the reaction

$$H_2O_2 + 2I^- + 2H^+ \longrightarrow 2H_2O + I_2$$

which occurs quantitatively in acidic solution. A trace of sodium molybdate is usually added to catalyze the reaction. Many organic peroxides react similarly and can be determined using this procedure.

Ozone has long been determined iodometrically, based on the reaction

$$O_3 + 2I^- + H_2O \longrightarrow O_2 + 2OH^- + I_2$$

The reaction is carried out by passing the gas sample through a neutral, unbuffered solution of potassium iodide. The resulting iodine solution is acidified prior to titration with sodium thiosulfate.

STANDARDS

As is usually the case, not all of the reagents used as titrants in redox reactions are available in sufficient purity to enable their concentrations to be determined by direct weighing. This section discusses some of the primary standards available for standardizing such titrants.

For Oxidants

Permanganate, iodine, and some ceric solutions must be standardized to determine their exact concentrations. The following two standards are used almost exclusively for such standardizations.

Arsenious oxide, As$_2$O$_3$

This reagent, also called arsenic trioxide or arsenic(III) oxide, is not soluble in acidic or neutral solution but dissolves readily in 1 M sodium hydroxide, yielding the oxyanion, arsenite:

$$As_2O_3 + 6NaOH \longrightarrow 2Na_3AsO_3 + 3H_2O$$

Once the oxide is dissolved, the solution may be acidified to yield stable, arsenious acid, H_3AsO_3. In fact, the acidification should take place as soon as possible to avoid oxidation of the arsenite to arsenate, AsO_4^{3-}, by atmospheric oxygen. Acidic solutions of arsenite are stable indefinitely. The reaction between an oxidant and arsenious acid may be written

$$\text{Ox} + H_3AsO_3 \ \rightleftharpoons \ \text{Red} + H_3AsO_4$$

When Ox is ceric or permanganate ion, the reaction is carried out in acidic solution. Both reactions are too slow to be useful without the addition of a catalyst. Osmium tetroxide, OsO_4, or iodine monochloride, ICl, is used in the cerium reaction and iodine monochloride or iodate ion is used in the permanganate reaction.

When H_3AsO_3 is used to standardize an iodine solution, the titration reaction must be carried out in basic solution. The equilibrium constant for the reaction

$$I_2 + H_3AsO_3 + H_2O \ \rightleftharpoons \ 2I^- + H_3AsO_4 + 2H^+$$

is only 0.55 at room temperature. By buffering the solution at a pH between 7 and 9, the hydrogen-ion concentration is kept low, forcing the reaction to completion. Below pH 7, the reaction is too slow and incomplete to yield an accurate end point; above pH 9, iodine undergoes an undesirable redox disproportionation reaction (Reaction 11-24).

Sodium oxalate, $Na_2C_2O_4$

Sodium oxalate is used to standardize ceric and permanganate titrants. It dissolves readily in acidic solutions, forming oxalic acid, $H_2C_2O_4$, which is quite stable. In reacting with an oxidant, oxalic acid is converted to water and carbon dioxide:

$$\text{Ox} + H_2C_2O_4 \ \longrightarrow \ \text{Red} + H_2O + CO_2(g)$$

When the oxidant is Ce^{4+}, a catalyst such as OsO_4 or ClO_4^- is needed to increase the rate of the reaction. Some procedures call for the use of perchloric acid solution to dissolve the primary standard sodium oxalate, thereby ensuring that a sufficient amount of catalyst will be present.

The reaction between permanganate ion and oxalic acid is complicated and proceeds slowly, even at 90°C. Interestingly, it is catalyzed by manganous ion, a product of the reaction. Thus, once started, the Mn^{2+} formed acts to increase the rate of the remaining reaction. Most procedures suggest heating the solution to about 60°C during the early part of the titration in order to help get the reaction started.

For Reductants

Thiosulfate and sometimes ferrous solutions must be standardized prior to their use as titrants. The following standards are commonly used for this purpose.

Ammonium hexanitratocerate, $(NH_4)_2Ce(NO_3)_6$

This reagent, which has been discussed already in the section on titrants, can be used to standardize ferrous solutions. The reaction is fast in strongly acidic solution, and despite the fact that much of the tetravalent cerium may be complexed with anions, the reaction is usually represented as follows:

$$Fe^{2+} + Ce^{4+} \ \rightleftharpoons \ Fe^{3+} + Ce^{3+}$$

Probably the most significant drawback to using this standard is its high cost.

Potassium dichromate, $K_2Cr_2O_7$

The low cost and excellent stability of potassium dichromate makes it a preferred standard for determining the concentration of ferrous solutions. It is also commonly used to standardize thiosulfate solutions via the iodometric technique. In this case, a known quantity of potassium dichromate is treated with excess potassium iodide and the liberated iodine is titrated with the thiosulfate solution:

$$Cr_2O_7^{2-} + 6I^- + 14H^+ \rightleftharpoons 2Cr^{3+} + 7H_2O + 3I_2$$

$$I_2 + 2S_2O_3^{2-} \rightleftharpoons 2I^- + S_4O_6^{2-}$$

Copper and iron

It is always desirable to standardize a titrant with the same substance that it will be used to titrate in the unknown. The reason for this is simple: Some of the errors associated with the analyte titration reaction will be canceled. For example, if a particular indicator color change occurs 0.1% past the equivalence point, the error in the analyte titration will be, at least partly, compensated for by a similar error in the standardization titration.

Both copper and iron wire are sufficiently pure to be used as standards for titrants such as sodium thiosulfate that are being prepared for an iodometric titration of copper or iron. The metal wire usually contains a surface layer of oxide or metal salt that should be removed before weighing. This is easily accomplished by briefly dipping the wire in a dilute acid solution.

PROBLEMS

11-1. What is the oxidation state of the underlined element in each of the following?
(a) $\underline{Cr}_2(SO_4)_3$
(b) $Na\underline{N}O_2$
(c) $H_2\underline{Se}O_3$
(d) $H_2\underline{O}_2$
(e) $Al\underline{F}_6^{3-}$
(f) $\underline{Mn}O_4^-$

11-2. Complete and balance the following reactions using the method of half-reactions. The reactions take place in dilute sulfuric acid unless otherwise specified.
(a) $KIO_3 + KI \longrightarrow$
(b) $Cu(NO_3)_2 + KI \longrightarrow$
(c) $KMnO_4 + FeSO_4 \xrightarrow{pH\ 5}$
(d) $Cr_2(SO_4)_3 + K_2S_2O_8 \longrightarrow$
(e) $FeCl_3 + Ag(s) \xrightarrow{HCl}$
(f) $Na_2S_2O_3 + I_2 \longrightarrow$
(g) $H_2C_2O_4 + KMnO_4 \longrightarrow$
(h) $I_2 + NaOH \xrightarrow{pH\ 10}$
(i) $H_3AsO_4 + KI \xrightarrow{2\ M\ H_2SO_4}$
(j) $Ti_2(SO_4)_3 + O_3 \longrightarrow$

11-3. For the titration of 25.00 mL of 0.1146 M FeSO₄ at pH 1.00, calculate the electrode potential after addition of the following volumes of 0.02292 M KMnO₄.
(a) 20.00 mL
(b) 22.50 mL
(c) 25.00 mL
(d) 27.50 mL
(e) 30.00 mL

11-4. Repeat the calculations for Problem 11-3 if the titration was performed at pH 3.00. Which pH yields the largest equivalence-point break?

11-5. For the titration of 50.00 mL of 5.00×10^{-2} M Fe²⁺ with 1.000×10^{-1} M Ce⁴⁺, calculate pFe²⁺ and E versus SHE after addition of the following volumes of titrant.
(a) 5.00 mL
(b) 15.00 mL
(c) 25.00 mL
(d) 35.00 mL
(e) 45.00 mL

Plot the titration curve in terms of pFe²⁺ and E.

11-6. Calculate the electrode potential of the half-cell solution when the following volumes of 0.08210 M KI have been added to 20.00 mL of 0.01350 M KIO₃ buffered at pH 2.20. V_{eq} is the volume required to reach the equivalence point.
(a) $0.25V_{eq}$
(b) $0.50V_{eq}$
(c) V_{eq}
(d) $1.50V_{eq}$
(e) $2.00V_{eq}$

11-7. What volume of 5.03×10^{-3} M Ce^{4+} must be added to 20.00 mL of 3.21×10^{-3} M Sn^{2+} in 1.00 M $HClO_4$ for the half-cell solution to have the following electrode potentials?
  (a) 0.130 V
  (b) 0.170 V
  (c) 1.65 V
  (d) 1.75 V

11-8. Vanadous ion, V^{2+}, can be oxidized in three discrete steps to VO_2^+ in 1.00 M H^+ by Ce^{4+}. If 20.00 mL of 0.01142 M VSO_4 is being titrated with 0.03891 M Ce^{4+}, calculate the electrode potential of the half-cell solution at the following points.
  (a) after addition of 5.00 mL
  (b) at first equivalence point
  (c) after addition of 7.50 mL
  (d) at second equivalence point
  (e) after addition of 15.00 mL
  (f) at third equivalence point
  (g) after addition of 20.00 mL
  Select an indicator that could be used at each equivalence point.

11-9. A 50.00-mL aliquot of a mixture of 0.03140 M $FeCl_2$ and 0.1108 M $TiCl_3$ buffered at pH 2.00 was titrated with 0.03442 M $K_2Cr_2O_7$. Calculate the electrode potential of the solution after the addition of the following volumes of titrants.
  (a) 10.00 mL
  (b) 20.00 mL
  (c) 30.00 mL
  (d) 40.00 mL

11-10. Is there any chance that $K_2Cr_2O_7$ could be used to titrate $CrCl_2$? If so, specify the general conditions and write a balanced reaction for the titration.

11-11. A sample of pure ascorbic acid (vitamin C) weighing 283.4 mg was dissolved in 20.00 mL of 1.00 M HCl and treated with 25.00 mL of 0.06441 M $Fe(NO_3)_3$.

ascorbic acid $+$ Fe^{3+} $\longrightarrow$ dehydroascorbic acid $+$ Fe^{2+} (unbalanced)

  The electrode potential of the half-cell solution was 0.390 V.
  (a) Calculate the standard electrode potential for the ascorbic acid half-cell.
  (b) What would the electrode potential have been if the HCl solution was 0.100 M?

11-12. The electrode potential of a $TiO^{2+}\,|\,Ti^{3+}$ half-cell depends on pH.
  (a) Does this mean that TiO^{2+} or Ti^{3+} could be determined by titration with an acid? Explain your answer.
  (b) Calculate the electrode potential of an equimolar mixture of TiO^{2+} and Ti^{3+} at pH values ranging from 1 to 7 in 1-unit increments.
  (c) Plot the electrode potential versus pH and determine the slope of the line.

11-13. The difference in electrode potential at 95% and 105% titrated can be taken as a rough measure of the size of the equivalence-point break in a titration curve. Calculate this difference for the titration of 25.00 mL of a solution containing Fe^{2+} at the following concentrations with Ce^{4+} at the same concentration as the Fe^{2+}.
  (a) 1.00×10^{-1} M
  (b) 1.00×10^{-3} M
  (c) 1.00×10^{-5} M

11-14. An indicator appears to have the color of its pure oxidized form when at least 78% of it is in the oxidized form. It appears to have the color of its pure reduced form when at least 92% is in the reduced form. If the indicator undergoes a one-electron change, calculate its color transition range in millivolts.

11-15. Write balanced reactions (not half-reactions) for the reduction of the following substances in a Walden (silver) reductor.
    (a) $FeCl_3$ **(c)** $K_2Cr_2O_7$ **(e)** $MnCl_3$
    (b) $CoCl_3$ **(d)** VO_2Cl

11-16. Write balanced reactions (not half-reactions) for the reduction of the following substances in a Jones (zinc-amalgam) reductor.
    (a) $K_2Cr_2O_7$ **(c)** $KMnO_4$ **(e)** $Fe_2(SO_4)_3$
    (b) $(VO_2)_2SO_4$ **(d)** $Cr_2(SO_4)_3$

11-17. A solution of potassium permanganate was standardized by titration with a standard Fe^{2+} solution at pH 2.0. The iron solution was prepared from 0.7417 g of pure iron wire and required 30.16 mL of the permanganate solution for titration. Calculate the molar concentration of the potassium permanganate solution.

11-18. A 416.4-mg sample of primary standard sodium oxalate was dissolved in 0.1 M H_2SO_4 and titrated with a solution of potassium permanganate. If 26.73 mL were required to reach the end point, what is the molar concentration of the potassium permanganate solution?

11-19. A sample of vanadium ore weighing 6.317 g was dissolved in acid and passed through a Walden (silver) reductor (see Table 11-2). The resulting solution was collected in a 100.0-mL volumetric flask and diluted to the mark with 0.1 M HCl. A 20.00-mL aliquot of the prereduced solution required 18.74 mL of 0.01146 M $KMnO_4$ to reach the end point. Calculate the percentage of V_2O_5 in the ore.

11-20. A 0.3017-g sample of uranium ore was dissolved and prereduced to convert all of the uranium to U^{4+}. This solution required 28.37 mL of 0.01374 M $KMnO_4$ for titration to $UO_2{}^{2+}$ in 0.1 M H_2SO_4. Calculate the percentage of U_3O_8 in the sample.

11-21. The zinc carbonate in a 3.0591-g sample of smithsonite ore was dissolved in acid and separated from the remaining insoluble substances. This solution was treated with an excess of oxalic acid and then made basic to precipitate ZnC_2O_4. The precipitate was collected, dissolved in 0.5 M H_2SO_4, and the liberated oxalic acid required 26.14 mL of 0.01875 M $KMnO_4$ for titration. Calculate the % $ZnCO_3$ in the ore.

11-22. Thioacetamide is hydrolyzed in water, forming hydrogen sulfide:

$$\underset{\begin{matrix} \\ \end{matrix}}{CH_3\overset{\displaystyle \overset{S}{\|}}{C}NH_2} + 2H_2O \xrightarrow{\Delta} CH_3\overset{\displaystyle \overset{O}{\|}}{C}ONH_4 + H_2S(g)$$

A 50.00-mL volume of aqueous thioacetamide solution was boiled for 10 minutes and the evolved gases collected in dilute sodium hydroxide. This solution required 31.84 mL of 0.06155 M $KMnO_4$ for titration.

$$S^{2-} + MnO_4{}^- \xrightarrow{OH^-} SO_4{}^{2-} + MnO_2(s) \quad \text{(not balanced)}$$

Calculate the molar concentration of thioacetamide in the original sample.

11-23. A 15.46-g sample of bacon was pureed in a blender with 100 mL of water. The suspension was filtered and the clear solution containing dissolved sodium nitrite was adjusted to pH 2. This solution was treated with 25.00 mL of 0.01548 M $KMnO_4$ to oxidize the nitrite to nitrate and, after standing for 10 minutes, the excess $KMnO_4$ was back titrated with 14.97 mL of 0.08183 M $FeSO_4$. Calculate the concentration of nitrite in the bacon as parts per million.

11-24. A ceric ion solution was standardized by titration with a solution of H_3AsO_3 prepared from 206.4 mg of primary standard As_2O_3. If the end point was reached after the addition of 31.44 mL of the cerium solution, calculate its molar concentration.

11-25. The hydrogen peroxide in a 2.00-g sample of antiseptic was titrated in 1.0 M H_2SO_4 with 0.1036 M Ce^{4+}, requiring 37.06 mL to reach the ferroin end point. Calculate the percentage of H_2O_2 in the antiseptic.

11-26. Ten milliliters of an intravenous glucose solution weighing 12.42 g was diluted to 250.0 mL with water. A 10.00-mL aliquot of this solution was treated with 25.00 mL of 0.1537 M Ce^{4+} in 5 M $HClO_4$, which converted the glucose to formic acid:

$$C_6H_{12}O_6 + Ce^{4+} \longrightarrow HCO_2H + Ce^{3+} \quad \text{(not balanced)}$$

After heating at 45°C for about 10 minutes, the excess Ce^{4+} was back titrated with 21.06 mL of 0.1146 M $FeSO_4$. Calculate the percentage of glucose in the intravenous solution.

11-27. A 961.4-mg sample containing titanium was dissolved and passed through a Jones (zinc-amalgam) reductor. The solution from the reductor, containing Ti^{3+}, was collected in 75 mL of 0.1 M $Fe_2(SO_4)_3$. The Fe^{2+} formed required 26.73 mL of 0.05477 M Ce^{4+} for titration. Calculate the % TiO_2 in the sample.

11-28. A sample of iron ore weighing 0.2146 g was dissolved in acid and passed through a Walden (silver) reductor. The resulting solution required 21.27 mL of 0.01436 M $K_2Cr_2O_7$ for titration to the diphenylamine sulfonic acid end point. Calculate the % Fe_3O_4 in the ore sample.

11-29. What sample weight must be taken so that the volume in milliliters of 0.01500 M $K_2Cr_2O_7$ titrant will equal the percentage of As_2O_3 when titrated as H_3AsO_3?

11-30. What molar concentration of $K_2Cr_2O_7$ must be used in the titration of H_3SbO_3 to H_3SbO_4 in order that the volume in milliliters divided by 10 will equal the % Sb in 1.5000-g samples?

11-31. What is the molar concentration of iodine in a solution if 38.04 mL is required to titrate the H_3AsO_3 prepared from 106.8 mg of primary standard As_2O_3?

11-32. What is the molar concentration of an iodine solution prepared by treating 1.607 g of primary standard KIO_3 with excess KI in acidic solution and diluting to 500.0 mL?

11-33. What is the molar concentration of cyanide in an electroplating solution if 5.00 mL of the solution buffered at pH 9 required 16.94 mL of 0.2874 M I_2 for titration to the starch end point?

11-34. The sulfur dioxide in 25.00 L of air was collected by drawing the sample through 200.0 mL of 0.02017 M I_2 at pH 1.0. The iodine remaining was back titrated with 42.07 mL of 0.1041 M $Na_2S_2O_3$. Calculate the concentration of SO_2 in units of mg SO_2/L air.

11-35. A 1.407-g sample of paint from an old building was dissolved and, after appropriate chemical treatment to remove potential interferences, treated with H_2S to precipitate lead sulfide. The precipitate was collected, rinsed, and dissolved in 25.00 mL of 3 M HCl containing 0.1200 M I_2. The H_2S liberated by the dissolution in HCl was oxidized to elemental sulfur. The remaining I_2 required 28.41 mL of 0.04136 M $Na_2S_2O_3$ for titration to the starch end point. Calculate the % Pb in the paint.

11-36. A sample of hydrated calcium sulfate weighing 0.2040 g was shaken with 50 mL of anhydrous methanol and titrated with a 0.1427 M Karl Fischer iodine solution, requiring 20.39 mL to reach the end point.
(a) Calculate the % H_2O in the sample.
(b) What is the average number of molecules of hydrated water per molecule of $CaSO_4$?

11-37. The iodine liberated by the action of excess KI on a sample of Cu^{2+} prepared from 0.2907 g of very pure copper wire consumed 28.37 mL of a sodium thiosulfate solution in a titration to the starch end point. Calculate the molar concentration of the sodium thiosulfate solution.

11-38. A 0.2500-g sample of swimming pool "oxidizer" containing calcium hypochlorite as the active ingredient was dissolved in 0.50 M H_2SO_4 containing excess KI. The liberated iodine required 38.17 mL of sodium thiosulfate for titration to the starch end point. The iodine liberated by the action of excess KI on 101.3 mg of primary standard $K_2Cr_2O_7$ dissolved in acid required 17.41 mL of the sodium thiosulfate solution for titration. Calculate the % $Ca(OCl)_2$ in the sample.

11-39. A sample weighing 223.9 mg contained potassium iodate and inert material. When dissolved in acid and treated with excess KI, the liberated iodine required 35.04 mL of 0.1114 M sodium thiosulfate for titration to the starch end point. Calculate the % KIO_3 in the sample.

11-40. A volume of air measuring 50.00 L was drawn through 250 mL of 0.1 M KI at pH 7.0. The solution was acidified and the iodine present consumed 39.38 mL of 0.2229 M $Na_2S_2O_3$ in a titration to the starch end point. If the air had a density of 1.20×10^{-3} g/mL, calculate the % O_3 in the air.

PRECIPITATION EQUILIBRIA AND TITRATIONS

Titrations based on reactions that produce sparingly soluble substances are referred to as precipitation titrations. They are among the oldest titrations known but are very limited in scope because so many precipitation reactions fail to meet either the stoichiometry or speed requirements for a successful titration. Coprecipitation of the analyte or titrant very often leads to nonstoichiometric reactions. The techniques of digestion and aging used to minimize coprecipitation in gravimetry cannot be applied to direct titrations because they require considerable time to become effective. Reactions must take place rapidly if direct titrations are to be successful, but the rates of formation of some precipitates are often quite slow, especially in the dilute solutions existing near the equivalence point. Only procedures using silver ion as the titrant or analyte have withstood the test of time and remained competitive with newer analytical methods.

SOLUBILITY EQUILIBRIA

An equilibrium exists between an undissolved solute and its saturated solution when the rate of precipitation equals the rate of dissolution. For an ionic compound, the reaction can be written generally as

$$M_x A_y(s) \;\rightleftharpoons\; xM + yA$$

where M represents a metal cation and A an anion. The charges on M and A are omitted for clarity. The exact mathematical description of this equilibrium is given by

$$K_{eq} = \frac{(a_M)^x (a_A)^y}{a_{M_x A_y(s)}}$$

Recalling from Chapter 2 that the activity of a pure substance is defined as unity and that the activity of a dilute solute is approximately the same as its concentration, the equilibrium expression can be rewritten as

$$K_{eq} = K_{sp} = [M]^x [A]^y$$

The equilibrium constant expressed in this way is referred to as the *solubility product constant* and is given the symbol, K_{sp}. Like all equilibrium constants, it is a measure of the completeness of a reaction and is therefore crucial in determining equilibrium concentrations from analytical concentrations. It is useful to note that when an ionic precipitate dissolves, the concentration of ions produced does not depend on the amount of precipitate present, as long as there is enough to establish an equilibrium between the solid and the solution.

Calculating Solubility

Solubility is defined as the concentration of a dissolved solute at equilibrium with its undissolved form. When a slightly soluble ionic salt dissolves, it produces cations and anions in the solution. If these ions do not undergo any reaction other than combining to re-form the salt, the solubility can be calculated easily from the solubility product expression.

Example 12-1

Calculate the molar solubility of silver bromide in water at 20°C. The K_{sp} for AgBr at this temperature is 5.0×10^{-13}.

Solution Silver bromide dissolves and dissociates according to the reaction

$$AgBr(s) \rightleftharpoons Ag^+ + Br^-$$

for which

$$K_{sp} = [Ag^+][Br^-]$$

The solubility, or concentration of dissolved silver bromide, is equal to the concentration of free silver ion:

$$solubility = [AgBr]_{dissolved} = [Ag^+]$$

According to the reaction stoichiometry, silver ion and bromide ion are formed in equal quantities. Since there is no other source of these ions and since they do not undergo reactions with any other substances, their concentrations are equal.

$$[Br^-] = [Ag^+]$$

Substituting in the equilibrium expression gives

$$K_{sp} = [Ag^+]^2$$

$$[Ag^+] = solubility = \sqrt{K_{sp}} = \sqrt{5.0 \times 10^{-13}} = 7.1 \times 10^{-7} \ M$$

Example 12-2

Calculate the molar solubility, in g/100 mL, of calcium iodate in water at 20°C.

Solution Calcium iodate dissolves and dissociates according to the reaction

$$Ca(IO_3)_2(s) \rightleftharpoons Ca^{2+} + 2IO_3^-$$

for which

$$K_{sp} = [Ca^{2+}][IO_3^-]^2$$

The solubility, or concentration of dissolved calcium iodate, is equal to the concentration of calcium ion:

$$\text{solubility} = [Ca(IO_3)_2]_{\text{dissolved}} = [Ca^{2+}]$$

According to the reaction stoichiometry, twice as much iodate as calcium is formed on dissolution. Thus

$$[IO_3^-] = 2[Ca^{2+}]$$

Substituting in the equilibrium expression gives

$$K_{sp} = [Ca^{2+}](2[Ca^{2+}])^2 = 4[Ca^{2+}]^3$$

$$[Ca^{2+}] = \text{solubility} = \sqrt[3]{\frac{K_{sp}}{4}} = \sqrt[3]{\frac{7.1 \times 10^{-7}}{4}} = 5.6 \times 10^{-3}\ M$$

Converting to the desired concentration units, we obtain

$$\text{solubility} = (5.6 \times 10^{-3}\ \text{mol/L})(389.9\ \text{g/mol}) = 2.2\ \text{g/L}$$

$$= 0.22\ \text{g/100 mL}$$

Effect of Common Ions

In the absence of any competing equilibria, a precipitate is less soluble in a solution containing an excess of one of the ions common to the precipitate than it is in pure water. This simple observation is a direct consequence of the principle of equilibrium. For example, the dissolution–precipitation equilibrium for silver chromate is given by the reaction

$$Ag_2CrO_4(s) \rightleftharpoons 2Ag^+ + CrO_4^{2-} \qquad (12\text{-}1)$$

for which

$$K_{sp} = [Ag^+]^2[CrO_4^{2-}] \qquad (12\text{-}2)$$

Upon the addition of excess Ag^+, the position of the equilibrium will shift to the left, causing some of the dissolved silver chromate to precipitate from the solution. According to Equation 12-2, the concentration of chromate ion must decrease in order for the concentration product to remain constant.

Example 12-3

Calculate the molar solubility of lead iodide (a) in water and (b) in 0.200 M sodium iodide solution. The K_{sp} for PbI_2 is 7.9×10^{-9}.

Solution

(a) The dissolution-precipitation equilibrium is given by

$$PbI_2(s) \rightleftharpoons Pb^{2+} + 2I^-$$

for which

$$K_{sp} = [Pb^{2+}][I^-]^2$$

The solubility, or concentration of dissolved PbI_2, is equal to the concentration of Pb^{2+} (or one-half the concentration of I^-):

$$\text{solubility} = [PbI_2]_{\text{dissolved}} = [Pb^{2+}]$$

From the reaction stoichiometry,

$$[I^-] = 2[Pb^{2+}]$$

Substituting in the equilibrium expression gives

$$K_{sp} = [Pb^{2+}](2[Pb^{2+}])^2 = 4[Pb^{2+}]^3$$

$$[Pb^{2+}] = \text{solubility} = \sqrt[3]{\frac{K_{sp}}{4}} = \sqrt[3]{\frac{7.9 \times 10^{-9}}{4}} = 1.3 \times 10^{-3}\ M$$

(b) The same equilibrium expression holds:

$$K_{sp} = [Pb^{2+}][I^-]^2$$

There are now two sources of iodide: the NaI and the PbI_2. The amount of iodide coming from the PbI_2 is small compared to that from the NaI. Thus

$$[I^-] = C_{\text{NaI}} + 2[Pb^{2+}] \simeq C_{\text{NaI}} = 0.200\ M$$

Then

$$[Pb^{2+}] = \text{solubility} = \frac{K_{sp}}{[I^-]^2} = \frac{7.9 \times 10^{-9}}{(0.200)^2} = 2.0 \times 10^{-7}\ M$$

Note that the solubility has decreased markedly (four orders of magnitude) upon the addition of an excess of I^-.

Effect of pH

If the cation of a substance is a weak acid or the anion is a weak base, the solubility of the substance will be affected by the pH of the solution. Consider a metal salt, MA, in which the anion is a weak base. In water, two chemical equilibria are established simultaneously:

$$MA(s) \rightleftharpoons M^+ + A^- \tag{12-3}$$

$$A^- + H_3O^+ \rightleftharpoons HA + H_2O \tag{12-4}$$

As the acidity is increased (pH lowered), the acid-base equilibrium (Reaction 12-4) is shifted to the right and the resulting decrease in the concentration of A^- causes more MA to dissolve in order to maintain the dissolution equilibrium (Reaction 12-3). If the cation is a weak acid, the two equilibria are

$$MA(s) \rightleftharpoons M^+ + A^- \tag{12-5}$$

$$M^+ + OH^- \rightleftharpoons MOH \tag{12-6}$$

An increase in acidity will lead to a decrease in the hydroxide-ion concentration, thereby causing the acid-base equilibrium (Reaction 12-6) to shift to the left. The resulting increase in the concentration of M^+ will cause more MA to precipitate from the equilibrium solution.

The calculation of the solubility of salts in which the anion and/or cation are in-

Solubility Equilibria

volved in acid-base equilibria can be done in a routine manner if *conditional* solubility product constants are used. Consider the situation described by Equations 12-3 and 12-4. The dissolution equilibrium is given by

$$K_{sp} = [M^+][A^-] \tag{12-7}$$

but the equation is not directly solvable for $[M^+]$ (the solubility) because $[A^-]$ is not known. However, the *total* concentration of dissolved A is the same as the equilibrium concentration of M^+:

$$[M^+] = [A'] = \text{solubility}$$

where

$$[A'] = [HA] + [A^-]$$

You may recall from Chapter 6 that

$$\alpha_{A^-} = \frac{[A^-]}{[A']} \quad \text{or} \quad [A^-] = [A']\alpha_{A^-} \tag{12-8}$$

Substituting this expression for $[A^-]$ in Equation 12-7 gives

$$K_{sp} = [M^+][A']\alpha_{A^-} = [M^+]^2\alpha_{A^-}$$

or

$$\frac{K_{sp}}{\alpha_{A^-}} = K'_{sp} = [M^+]^2 = (\text{solubility})^2$$

The value of α_{A^-} can be calculated from the appropriate form of Equation 6-51.

Example 12-4

Calculate the molar solubility of magnesium fluoride at (a) pH 7.00 and (b) pH 2.00.

Solution   The dissolution equilibrium is

$$MgF_2(s) \;\rightleftharpoons\; Mg^{2+} + 2F^-$$

Since F^- is a weak base and can combine with hydrogen ions, the solubility should be calculated using the conditional solubility product constant:

$$K'_{sp} = \frac{K_{sp}}{(\alpha_{F^-})^2} = [Mg^{2+}][F']^2$$

where

$$[F'] = 2[Mg^{2+}] = 2(\text{solubility})$$

Substituting gives us

$$\frac{K_{sp}}{(\alpha_{F^-})^2} = [Mg^{2+}](2[Mg^{2+}])^2 = 4[Mg^{2+}]^3$$

$$[Mg^{2+}] = \text{solubility} = \sqrt[3]{\frac{K_{sp}}{4(\alpha_{F^-})^2}} = \sqrt[3]{\frac{K'_{sp}}{4}}$$

The fraction of fluoride existing as F^- is calculated from Equation 6-51:

$$\alpha_{F^-} = \frac{K_a}{[H_3O^+] + K_a}$$

(a) At pH 7.00,

$$\alpha_{F^-} = \frac{6.8 \times 10^{-4}}{(1.00 \times 10^{-7}) + (6.8 \times 10^{-4})} = 1.0$$

and

$$\text{solubility} = \sqrt[3]{\frac{6.6 \times 10^{-9}}{(4)(1.0)^2}} = 1.2 \times 10^{-3} \, M$$

(b) At pH 2.00,

$$\alpha_{F^-} = \frac{6.8 \times 10^{-4}}{(1.00 \times 10^{-2}) + (6.8 \times 10^{-4})} = 6.4 \times 10^{-2}$$

and

$$\text{solubility} = \sqrt[3]{\frac{6.6 \times 10^{-9}}{(4)(6.4 \times 10^{-2})^2}} = 7.4 \times 10^{-3} \, M$$

The same approach may be used when the cation is a weak acid, as illustrated by Reactions 12-5 and 12-6. The dissolution equilibrium is given by Equation 12-7, but again, the equation is not directly solvable for $[A^-]$ because $[M^+]$ is not known. Using the conditional solubility product constant yields

$$K'_{sp} = \frac{K_{sp}}{\alpha_{M^+}} = [M'][A^-] = [A^-]^2 = (\text{solubility})^2 \qquad (12\text{-}9)$$

Since M^+ is a monoprotic acid (the conjugate acid of MOH), α_{M^+} can be calculated from the appropriate form of Equation 6-48:

$$\alpha_{M^+} = \frac{[H_3O^+]}{[H_3O^+] + K_a}$$

where

$$K_a \, (\text{for } M^+) = \frac{K_w}{K_b \, (\text{for MOH})}$$

It is, of course, entirely possible that *both* the cation and the anion of a slightly soluble substance will be involved in simultaneous acid-base equilibria. Such a situation requires only that the solubility product constant used for the calculation is conditional in both M^+ and A^-. In this way, the general approach used to calculate the solubility is not altered. Thus

$$K''_{sp} = \frac{K_{sp}}{(\alpha_{M^+})(\alpha_{A^-})} = [M'][A'] = [M']^2 = (\text{solubility})^2 \qquad (12\text{-}10)$$

Solubility Equilibria

Effect of Complex-Ion Formation

The presence of complexing agents that can combine with either the cation or anion of a slightly soluble substance will lead to an increase in its solubility. Thus silver chloride is quite insoluble in water but dissolves appreciably in dilute, aqueous ammonia:

$$AgCl(s) \rightleftharpoons Ag^+ + Cl^- \tag{12-11}$$

$$Ag^+ + NH_3 \rightleftharpoons AgNH_3^+ \tag{12-12}$$

$$AgNH_3^+ + NH_3 \rightleftharpoons Ag(NH_3)_2^+ \tag{12-13}$$

The combined equilibrium constant for Reactions 12-12 and 12-13 is much larger than 1 and the position of the equilibrium lies far to the right. As silver ions react with the ammonia, more silver chloride dissolves to maintain the dissolution equilibrium.

Solubilities are calculated most conveniently using the same type of conditional equilibrium constants described in the preceding section. Thus, for Reaction 12-11,

$$K_{sp} = [Ag^+][Cl^-] \tag{12-14}$$

where

$$[Ag^+] = (\alpha_{Ag^+})[Ag'] \tag{12-15}$$

Substituting for $[Ag^+]$ in Equation 12-14 gives

$$K_{sp} = (\alpha_{Ag^+})[Ag'][Cl^-]$$

or

$$\frac{K_{sp}}{\alpha_{Ag^+}} = K'_{sp} = [Ag'][Cl^-]$$

where

$$[Ag'] = [Cl^-] = \text{solubility}$$

The value of α_{Ag^+} may be calculated from an equation similar to Equation 9-7 if the successive formation constants and the ammonia concentration are known.

Example 12-5

Calculate the solubility of silver thiocyanate (a) in water and (b) in a solution whose free ammonia concentration is 0.0150 M. Neglect the effects of any acid-base properties of Ag^+ and SCN^-.

Solution The dissolution equilibrium is

$$AgSCN(s) \rightleftharpoons Ag^+ + SCN^-$$

for which

$$K_{sp} = [Ag^+][SCN^-]$$

(a) Since neither Ag^+ nor SCN^- is involved in other equilibria,

$$[Ag^+] = [SCN^-] = \text{solubility}$$

and

$$K_{sp} = [Ag^+]^2 = (\text{solubility})^2$$

or

$$\text{solubility} = \sqrt{K_{sp}} = \sqrt{1.1 \times 10^{-12}} = 1.0 \times 10^{-6} \ M$$

(b) Dissolved silver ion reacts with ammonia to form complex ions:

$$Ag^+ + NH_3 \rightleftharpoons AgNH_3^+$$

$$AgNH_3^+ + NH_3 \rightleftharpoons Ag(NH_3)_2^+$$

Thus

$$[SCN^-] = [Ag'] = \text{solubility}$$

where

$$[Ag'] = \frac{[Ag^+]}{\alpha_{Ag^+}}$$

Substituting in the solubility product expression, we have

$$K_{sp} = (\alpha_{Ag^+})[Ag'][SCN^-] = (\alpha_{Ag^+})(\text{solubility})^2$$

The value of α_{Ag^+} may be calculated from a version of Equation 9-7, appropriate for a metal ion forming two ammonia complexes:

$$\alpha_{Ag^+} = \frac{1}{1 + K_{f_1}[NH_3] + K_{f_1}K_{f_2}[NH_3]^2}$$

$$= \frac{1}{1 + (2.04 \times 10^3)(0.0150) + (2.04 \times 10^3)(8.13 \times 10^3)(0.0150)^2}$$

$$= 2.66 \times 10^{-4}$$

Substituting in the equation for the solubility gives us

$$1.1 \times 10^{-12} = (2.66 \times 10^{-4})(\text{solubility})^2$$

or

$$\text{solubility} = \sqrt{\frac{1.1 \times 10^{-12}}{2.66 \times 10^{-4}}} = 6.4 \times 10^{-5} \ M$$

It is not uncommon to encounter situations where an anion used to precipitate a metal cation is also capable of complexing the cation, usually forming a soluble substance in the process, as illustrated by the following examples:

$$Al^{3+} + 3OH^- \rightleftharpoons Al(OH)_3(s)$$

$$Al(OH)_3(s) + OH^- \rightleftharpoons Al(OH)_4^-$$

$$Zn^{2+} + 2CN^- \rightleftharpoons Zn(CN)_2(s)$$

$$Zn(CN)_2(s) + CN^- \rightleftharpoons Zn(CN)_3^-$$

$$Zn(CN)_3^- + CN^- \rightleftharpoons Zn(CN)_4^{2-}$$

Solubility Equilibria

$$Ni^{2+} + C_2O_4^{2-} \rightleftharpoons NiC_2O_4(s)$$

$$NiC_2O_4(s) + C_2O_4^{2-} \rightleftharpoons Ni(C_2O_4)_2^{2-}$$

In the absence of complex-ion formation, continued addition of excess precipitating agent causes the solubility of the precipitate to decrease via the common-ion effect. When soluble complex ions are formed, there may be an initial decrease in the solubility due to the common-ion effect, but that will be followed by an increase as the complex ion is formed. The degree to which the solubility increases depends on the excess concentration of anion and on the equilibrium constants for the formation of the precipitate and complex ions.

Complex-ion formation is a very common occurrence, and its neglect in solubility calculations may lead to highly erroneous answers. It has been noted, for example, that the solubility of mercuric sulfide in 0.10 M sodium sulfide is calculated to be 2.0 $\times$ 10^{-52} M when neglecting all competing equilibria:

$$HgS(s) \rightleftharpoons Hg^{2+} + S^{2-}$$

$$K_{sp} = [Hg^{2+}][S^{2-}]$$

$$2.0 \times 10^{-53} = [Hg^{2+}](0.10)$$

$$[Hg^{2+}] = solubility = 2.0 \times 10^{-52} \ M$$

At this concentration, the volume of water required to contain one dissolved mercuric ion is

$$(2.0 \times 10^{-52} \ mol/L)(6.02 \times 10^{23} \ ions/mol) = 1.2 \times 10^{-28} \ ions/L$$

or

$$\frac{1}{1.2 \times 10^{-28} \ ions/L} = 8.3 \times 10^{27} \ L/ion$$

This is a greater quantity of water than exists on the entire planet! Research has established that soluble mercuric sulfide complexes are present in solution, making the experimental solubility on the order of 1×10^{-3} M. Solubility calculations taking into account the formation of complex ions formed by one of the common ions of the precipitate are often quite complicated and will not be examined in this book.

TITRATION CURVES

Theoretically generated titration curves are useful for establishing the criteria for a successful titration, the necessary properties of an indicator, and the likely titration error. The data for precipitation titration curves can be calculated from concentrations of the reagents and the solubility product constant.

Calculating Concentration

To illustrate the necessary calculations, consider an analyte of bromide ion being titrated with silver ions. The titration reaction is

$$AgNO_3 + NaBr \rightleftharpoons AgBr(s) + NaNO_3$$

or in terms of the participating ions,

$$Ag^+ + Br^- \rightleftharpoons AgBr(s) \tag{12-16}$$

Reaction 12-16 is the reverse of the dissolution reaction and therefore has an equilibrium constant of $1/K_{sp}$. Since the K_{sp} is small, its reciprocal is large, which means that Reaction 12-16 goes to completion and any amount of silver ion added prior to the equivalence point will consume a stoichiometric amount of bromide.

The specific mathematical treatment depends on the composition of the solution which changes as the titration proceeds. Precipitation titrations such as the example discussed above can be divided into three distinct regions based on composition:

Region	Major constituents
1. Before the equivalence point	$AgBr(s) + NaNO_3 + NaBr$
2. At the equivalence point	$AgBr(s) + NaNO_3$
3. After the equivalence point	$AgBr(s) + NaNO_3 + AgNO_3$

Each type of solution will now be considered separately.

Region 1: Before the equivalence point. Since the reaction stoichiometry is $1:1$,

$$\frac{\text{amount NaBr initial}\ -\ \left\{\begin{matrix}\text{amount AgNO}_3\text{ added}\\ \parallel\\ \text{amount NaBr used}\end{matrix}\right\}}{\text{amount NaBr remaining}}$$

The analytical concentration of sodium bromide remaining can be obtained by dividing its amount in millimoles by the new volume. There are actually two sources of bromide ion: the remaining sodium bromide and the precipitate, silver bromide. Since silver bromide has a small K_{sp}, it supplies a negligible amount of bromide ion to the solution and

$$[Br^-] \simeq C_{NaBr}$$

Silver ion is formed from the slight dissolution of the precipitate:

$$AgBr(s) \rightleftharpoons Ag^+ + Br^-$$

and its concentration can be calculated from the equilibrium expression,

$$[Ag^+] = \frac{K_{sp}}{[Br^-]} = \frac{K_{sp}}{C_{NaBr}}$$

Region 2: At the equivalence point. The only source of bromide ion is the precipitate. It is not necessary to determine how much solid silver bromide is present, although this is not hard to do, because it is not a part of the equilibrium expression that must be solved. Since silver ion and bromide ion are formed simultaneously when the precipitate dissolves and there is no other source of either ion,

$$[Ag^+] = [Br^-]$$

Substituting in the K_{sp} expression gives

$$K_{sp} = [Ag^+]^2$$

$$[Ag^+] = \sqrt{K_{sp}}$$

Region 3: After the equivalence point. The excess silver nitrate is the major supplier of silver ion. The amount of excess silver nitrate is found by taking the difference between the amount of silver nitrate added and the amount of silver nitrate consumed, which equals the amount of sodium bromide originally present:

$$\dfrac{ - \left\{ \begin{array}{c} \text{amount NaBr initial} \\ \| \\ \text{amount AgNO}_3 \text{ used} \end{array} \right\} }{ \begin{array}{c} \text{amount AgNO}_3 \text{ added} \\ \hline \text{amount AgNO}_3 \text{ remaining} \end{array} }$$

and

$$[Ag^+] \simeq C_{AgNO_3} \text{ remaining} = \frac{\text{amount AgNO}_3 \text{ remaining}}{\text{volume of solution}}$$

As pointed out before, if the stoichiometry of the reaction is not $1:1$, this must be taken into account when determining the composition of the solution.

Example 12-6

For the titration of 25.00 mL of 0.100 M KBr, calculate the pAg after the addition of 5.00, 12.50, and 15.00 mL of a 0.200 M AgNO$_3$ titrant. The K_{sp} for AgBr is 5.00×10^{-13}.

Solution

After the addition of 5.00 mL. Since the reaction stoichiometry is $1:1$:

$$\begin{array}{ll} \text{amount KBr initial} & = 25.00 \text{ mL} \times 0.100 \ M = 2.50 \text{ mmol} \\ - \left\{ \begin{array}{c} \text{amount AgNO}_3 \text{ added} \\ \| \\ \text{amount KBr used} \end{array} \right\} & = 5.00 \text{ mL} \times 0.200 \ M = 1.00 \text{ mmol} \\ \hline \text{amount KBr remaining} & = 1.50 \text{ mmol} \end{array}$$

This excess KBr is the major source of Br$^-$; thus

$$[Br^-] \simeq C_{KBr} = \frac{1.50 \text{ mmol}}{30.00 \text{ mL}} = 0.0500 \ M$$

The silver ion comes from the dissolution of the precipitate, which is described by the K_{sp} expresssion:

$$K_{sp} = [Ag^+][Br^-]$$

Substituting for the bromide concentration, we have

$$5.00 \times 10^{-13} = [Ag^+](0.0500)$$

$$[Ag^+] = 1.00 \times 10^{-11} \ M$$

and

$$pAg = 11.00$$

After the addition of 12.50 mL. Comparing the amounts of each reactant, we get

$$amount\ KBr\ initial = 25.00\ mL \times 0.100\ M = 2.50\ mmol$$

$$amount\ AgNO_3\ added = 12.50\ mL \times 0.200\ M = 2.50\ mmol$$

We see that neither analyte nor titrant is in excess. Thus the only source of Ag^+ and Br^- is the precipitate. Using the K_{sp} expression yields

$$K_{sp} = [Ag^+][Br^-] = [Ag^+]^2$$

$$[Ag^+] = \sqrt{K_{sp}} = \sqrt{5.00 \times 10^{-13}} = 7.07 \times 10^{-7}\ M$$

and

$$pAg = 6.15$$

After the addition of 15.00 mL. Again, comparing amounts, we have

$$-\left\{\begin{array}{c} amount\ KBr\ initial \\ \| \\ amount\ AgNO_3\ used \end{array}\right\} = 25.00\ mL \times 0.100\ M = 2.50\ mmol$$

$$\frac{amount\ AgNO_3\ added}{amount\ AgNO_3\ remaining} \quad \begin{array}{l} = 15.00\ mL \times 0.200\ M = 3.00\ mmol \\ = 0.50\ mmol \end{array}$$

The excess $AgNO_3$ is the major source of Ag^+ and, because it is a strong electrolyte,

$$[Ag^+] = C_{AgNO_3} = \frac{0.50\ mmol}{40.0\ mL} = 0.013\ M$$

and

$$pAg = 1.89$$

Factors Affecting Shape

Both the magnitude and the sharpness of the end-point break influence how well the equivalence point can be estimated. Therefore, it is important for us to examine the variables that can affect the shape of the titration curve. The properties of indicators also influence how well equivalence points can be estimated, but they are of a specific nature and will be discussed together with each indicator.

Titrant and analyte concentrations

It is easy to see from Example 12-6 that the silver ion concentration before the equivalence point depends partly on the original concentration of bromide ion, while after the equivalence point it depends partly on its own original concentration. As a result, high concentrations of analyte and titrant produce larger and sharper end-point breaks than do low concentrations. This effect is illustrated by the family of curves in Figure 12-1. Had pBr been plotted rather than pAg, the curves would simply be inverted. Most indicators require a sharp, twofold or greater change in the p-value in order to function well. Figure 12-1 suggests that 0.1 M bromide can be titrated with minimal

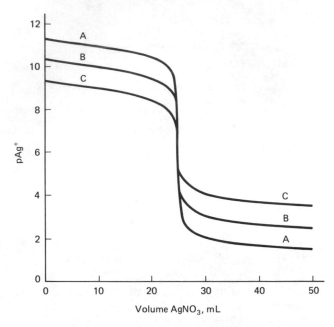

Figure 12-1 Effect of titrant and analyte concentration on the shape of the titration curve:
A. 25.0 mL of 0.100 M KBr with 0.100 M AgNO$_3$;
B: 25.0 mL of 0.0100 M KBr with 0.0100 M AgNO$_3$;
C: 25.0 mL of 0.00100 M KBr with 0.00100 M AgNO$_3$.

error but when the concentration is 0.001 M, the break is less than two pAg units, and although the titration may still be possible, the error will be large.

Completeness of the reaction

The equilibrium constant for a reaction in which a precipitate is formed usually is just the reciprocal of the solubility product constant for the precipitate. For example, when silver nitrate is titrated with potassium thiocyanate, the net ionic reaction is

$$Ag^+ + SCN^- \rightleftharpoons AgSCN(s)$$

for which the equilibrium constant is

$$K_{eq} = \frac{1}{[Ag^+][SCN^-]} = \frac{1}{K_{sp}}$$

A product that is very insoluble (small K_{sp}) will lead to a titration reaction that is very complete. Figure 12-2 shows the effect of solubility on the shape of the titration curve. The largest change in pAg occurs in the titration of iodide ion, which, among the ions considered, forms the most insoluble silver salt. The smallest change in pAg occurs with chloride ion, which forms the least insoluble silver salt.

TITRANTS AND STANDARDS

Silver nitrate is the titrant used for determining halide and thiocyanate ions. It is available in primary-standard-grade purity, but it is expensive. Titrants prepared from less pure silver nitrate can be standardized with primary-standard potassium chloride. In situations

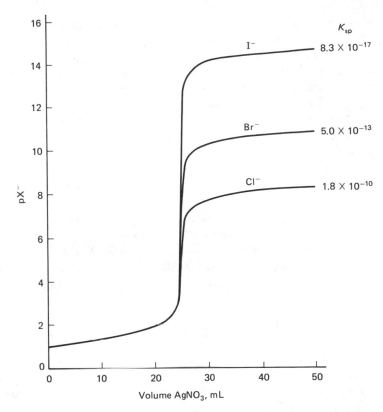

Figure 12-2 Effect of solubility on the shape of the titration curve. Each curve is for 25.0 mL of 0.100 M anion titrated with 0.100 M AgNO$_3$.

where the amount of silver ion must be determined, potassium chloride is usually the titrant.

Certain precautions must be taken with the preparation and storage of silver nitrate solutions. The stability of atomic silver makes silver ion a good reducing agent. Certain proteins in skin can reduce silver ion to atomic silver, leaving a dark brown stain that is unsightly and will remain until replaced by new skin. One should therefore avoid unnecessary skin contact and clean up spills immediately.

CHEMICAL INDICATORS

The indicators used in precipitation reactions are usually specific compound formers; that is, they react selectively with the titrant to form a colored substance. Since both the analyte A and the indicator In can react with the titrant T, they may be viewed as competitors:

Titration reaction: $A + T \rightleftharpoons AT(s)$
Indicator reaction: $In + T \rightleftharpoons InT$

Because the indicator reaction is responsible for the color change that signals the end point of the titration, it must not occur until virtually all of the analyte has reacted. The extent to which the analyte reacts in preference to the indicator is governed largely by the difference in the equilibrium constants of the two reactions. The larger the equilibrium constant for the titration reaction relative to that for the indicator reaction, the greater the preference of the titrant for the analyte.

The indicator reaction must produce a significant, observable color change upon consumption of a negligible amount of titrant if the titration error is to be small. Two conditions are necessary for this to happen: (1) the indicator reaction must proceed appreciably to the right even in the presence of a low concentration of titrant, and (2) the product of the indicator reaction must be intensely colored so that it can be seen at low concentrations. The first condition is favored by a large equilibrium constant, yet in the preceding paragraph, it was pointed out that the equilibrium constant should be small relative to that for the titration reaction. Obviously, a trade-off is necessary. It is frequently the analytical chemist's job to determine to what extent each goal must be met and to establish the range of values for the equilibrium constants that will allow a method to work properly.

Types of Indicators

Most precipitation titrations use one of three different indicators: potassium chromate, dichlorofluorescein, or ferric nitrate. The first two are used with silver nitrate titrants, while the latter is used with potassium thiocyanate titrants. Since so few indicators are used, specific properties and modes of action of each of these are described in the following section.

APPLICATIONS

Throughout the history of science, people making important discoveries often were honored by having their names attached to the discoveries. Although this practice is not as common as it was once, old methods such as the ones described below are still referred to by the name of the chemist responsible for its development.

Mohr's Method

This method was developed for the determination of chloride, bromide and cyanide ions. It uses silver nitrate as the titrant and sodium chromate as the indicator.

Titration reaction: $Ag^+ + Cl^- \rightleftharpoons AgCl(s)$
white

Indicator reaction: $2Ag^+ + CrO_4^{2-} \rightleftharpoons Ag_2CrO_4(s)$
orange-red

The molar solubility of silver chromate is nearly five times that of silver chloride. As a result, silver chloride precipitates first in the titration flask. Just after the equivalence point, the silver-ion concentration becomes large enough to initiate the precipitation of orange-red silver chromate, which signals the end of the titration.

Since the silver-ion concentration at the equivalence point is known, the chromate-ion concentration required to initiate precipitation of silver chromate can be calculated. At the equivalence point in the titration of chloride ion we can write

$$[Ag^+] = [Cl^-] = \sqrt{K_{sp, AgCl}}$$

$$[Ag^+] = \sqrt{1.8 \times 10^{-10}} = 1.3 \times 10^{-5} \, M$$

The chromate-ion concentration needed to precipitate silver chromate can be calculated from the solubility product expression:

$$K_{sp, Ag_2CrO_4} = [Ag^+]^2 [CrO_4^{2-}]$$

$$1.2 \times 10^{-12} = (1.3 \times 10^{-5})^2 [CrO_4^{2-}]$$

$$[CrO_4^{2-}] = 7.1 \times 10^{-3} \, M$$

Actually, a somewhat smaller concentration, usually about $2.5 \times 10^{-3} \, M$, is commonly recommended because, at higher concentrations, the pale yellow color of CrO_4^{2-} masks the first appearance of the orange-red silver chromate. This lower chromate-ion concentration requires a larger silver-ion concentration (more titrant) to initiate precipitation. Furthermore, a certain minimum amount of silver chromate must be formed before its color will be intense enough to be seen by the human eye. Both of these factors cause more silver nitrate titrant to be added than is necessary to react with the chloride, thereby resulting in a positive titration error. The error increases as the concentration of chloride ion being titrated decreases. At a chloride concentration of 0.1 M the titration error is negligible, but at much lower concentrations the error is usually too large to ignore. In such cases, a correction may be made by determining an indicator blank—that is, determining the amount of silver nitrate needed to "titrate" the indicator. Chloride-free calcium carbonate is usually added to provide a white suspension similar to the one that is present when the sample is titrated. As an alternative to determining an indicator blank, the silver nitrate titrant can be standardized with primary-standard sodium chloride using the Mohr method. If the volumes of silver nitrate needed to titrate the standard and the unknown are close to the same, the indicator errors for the two titrations will largely cancel each other.

Example 12-7

Calculate the relative titration error when 50.00 mL of 0.1000 M NaCl is titrated with 0.1000 M AgNO$_3$. The initial concentration of K$_2$CrO$_4$ indicator is $2.000 \times 10^{-3} \, M$. Assume that the titration is stopped at the very onset of silver chromate formation.

Solution The titration error is the difference between the actual volume of AgNO$_3$ used and the stoichiometric volume (volume to reach the *equivalence point*). The stoichiometric volume is calculated from the amount of Cl$^-$ present.

$$\text{amount AgNO}_3 \text{ required} = \text{amount NaCl} \times \frac{1 \text{AgNO}_3}{1 \text{NaCl}}$$

$$(V_{AgNO_3})(0.1000\ M) = (50.00\ \text{mL})(0.1000\ M)(1/1)$$

$$V_{AgNO_3} = 50.00\ \text{mL}$$

The concentration of Ag^+ at the equivalence point is

$$[Ag^+] = \sqrt{K_{sp}}$$

$$[Ag^+] = \sqrt{1.8 \times 10^{-10}} = 1.3 \times 10^{-5}\ M$$

The actual volume of $AgNO_3$ used will be larger by the amount necessary to initiate Ag_2CrO_4 precipitation. At the equivalence point, the concentration of CrO_4^{2-} is

$$[CrO_4^{2-}] = \frac{(50.00\ \text{mL})(2.000 \times 10^{-3}\ M)}{100.0\ \text{mL}} = 1.000 \times 10^{-3}\ M$$

The concentration of Ag^+ necessary to initiate precipitation can be calculated from the K_{sp} expression:

$$K_{sp} = [Ag^+]^2[CrO_4^{2-}]$$

$$1.2 \times 10^{-12} = [Ag^+]^2(1.000 \times 10^{-3})$$

$$[Ag^+] = 3.5 \times 10^{-5}\ M$$

The *amount* of $AgNO_3$ added to increase the concentration of Ag^+ from its equivalence-point value to this value is

amount $AgNO_3$ extra = amount $AgNO_3$ to ppt Ag_2CrO_4 − amount $AgNO_3$ to e.p.

$$= (100.0\ \text{mL})(3.5 \times 10^{-5}M) - (100.0\ \text{mL})(1.3 \times 10^{-5}M)$$

$$= 2.2 \times 10^{-3}\ \text{mmol}$$

Finally,

$$\text{amount } AgNO_3 \text{ extra} = (\text{vol}_{AgNO_3})(C_{AgNO_3})$$

$$2.2 \times 10^{-3}\ \text{mmol} = (\text{vol}_{AgNO_3})(0.1000\ M)$$

$$\text{vol}_{AgNO_3} = 0.022\ \text{mL}$$

and the relative titration error is

$$\text{rel. error} = \frac{0.022\ \text{mL}}{50.00\ \text{mL}} = 4.4 \times 10^{-4}$$

The most serious limitation of the Mohr method is the need for careful control of the pH, which must be between 6.5 and 10.3. Below pH 6.5 silver chromate becomes excessively soluble due to the reaction

$$2CrO_4^{2-} + 2H^+ \longrightarrow Cr_2O_7^{2-} + H_2O$$

Silver dichromate is considerably more soluble than silver chromate and the higher solubility increases the indicator error. Above pH 10.3 silver ion may react with hydroxide instead of chloride ion, forming insoluble silver hydroxide or silver oxide:

$$2Ag^+ + 2OH^- \rightleftharpoons 2AgOH(s) \rightleftharpoons Ag_2O(s) + H_2O$$

Transition metal cations generally interfere because they form insoluble hydroxides or basic salts in neutral or alkaline solution that have a tendency to coprecipitate chloride or bromide ions. In addition, some of the hydroxides, such as $Fe(OH)_3$, are highly colored and will mask the indicator color change. Anions such as phosphate, aresenate, and carbonate interfere by forming insoluble silver salts in neutral or alkaline solution. If they are present in appreciable amounts, a separation is required or an alternative method must be used.

Fajan's Method

Fajan's method uses an indicator that is adsorbed by the electrostatically charged colloidal precipitate immediately after the equivalence point. Such adsorption indicators have different colors in the free and adsorbed states. The indicators used most commonly in the titration of halides are dichlorofluorescein and tetrabromofluorescein, whose common name is eosin.

dichlorofluorescein (HDCF)   Tetrabromofluorescein or Eosin (HEs)

The dichlorofluorescein anion (DCF^-) has an irridescent greenish-yellow color in solution that turns pink when it is adsorbed on silver chloride. The change in color is thought to be due to a deformation of the indicator anion resulting from the adsorption process.

To illustrate how an adsorption indicator functions, we need to review the electrical behavior of colloidal silver halides. Colloidal silver chloride in the presence of excess chloride ion will adsorb some of these ions and acquire a negative charge. This negatively charged particle then attracts positive ions to a loosely bound counter-ion layer. In the presence of excess silver ion, the colloidal silver chloride will acquire a positive charge, which in turn will attract negative ions to the counter-ion layer (see Figure 4-1). During the titration of chloride ion with silver ion, an abrupt change in the charge on the silver chloride precipitate occurs in the vicinity of the equivalence point where excess chloride ion is replaced by excess silver ion. When the primary adsorption layer becomes positive, the indicator anion is attracted to the counter-ion layer and the color change occurs. The indicator reaction can be written

$$AgCl \cdot Ag^+ + DCF^- \rightleftharpoons AgCl \cdot Ag^+ : DCF^- \qquad (12\text{-}17)$$
$$\underset{\substack{\text{greenish-}\\\text{yellow}}}{} \qquad\qquad\qquad \underset{\text{pink}}{}$$

Applications   **317**

Dyes such as dichlorofluorescein and eosin that are strongly attracted to the counter-ion layer after the equivalence point are also strongly attracted to the primary adsorption layer before the equivalence point. If the indicator is adsorbed more strongly than the analyte ion, it cannot be used because the color change will occur at the beginning of the titration. Dichlorofluorescein is adsorbed less strongly than Cl^-, Br^-, I^-, or SCN^- and can be used in the titration of any of these ions. Eosin is adsorbed more strongly than Cl^- but less strongly than Br^-, I^-, or SCN^- and cannot be used in the titration of chloride ion.

For the titration of Br^-, I^-, or SCN^-, where a choice of indicators is possible, eosin is preferred because it can be used at a lower pH, thereby avoiding a number of potential interferences. In order for Reaction 12-17 to take place, the anion form of the indicator must be present in appreciable concentration. Dichlorofluorescein cannot be used below about pH 6.5 because it is a very weak acid and too little DCF^- exists at low pH. Eosin is a stronger acid than dichlorofluorescein and therefore more highly dissociated at low pH. It can be used at pH values as low as 2.0. Most of the ions that interfere in the Mohr method also interfere in Fajan's method for chloride ion, and for the same reasons. When Br^-, I^-, or SCN^- are being titrated using eosin as the indicator, the lower allowable pH eliminates the formation of some insoluble silver salts and transition metal hydroxides.

For adsorption indicators to function properly, at least a portion of the silver halide precipitate must remain colloidal in the vicinity of the equivalence point. Unfortunately, a significant fraction of the precipitate tends to coagulate just before the equivalence point, especially when high concentrations are being titrated. If the titration is continued slowly with vigorous shaking, a good end point can be obtained. Some procedures for determining chloride call for the addition of dextrin as a protective colloid to retard coagulation.

Silver halide precipitates are somewhat light sensitive. Unfortunately, many adsorption indicators greatly increase this light sensitivity. As a result, titrations should be performed rapidly and with the minimum necessary exposure to light.

Volhard's Method

This method involves the addition of an excess of silver ion to a halide solution, followed by a back titration of the excess with thiocyanate ion. Ferric ion is used as the indicator, the end point being marked by the appearance of red $FeSCN^{2+}$.

$$\text{Analyte reaction:} \quad Cl^- + Ag^+(xs) \; \rightleftharpoons \; AgCl(s)$$

$$\text{Titration reaction:} \quad Ag^+ + SCN^- \; \rightleftharpoons \; \underset{\text{white}}{AgSCN(s)}$$

$$\text{Indicator reaction:} \quad Fe^{3+} + SCN^- \; \rightleftharpoons \; \underset{\text{red}}{FeSCN^{2+}}$$

Calculations have shown that the theoretical indicator error varies only slightly as long as the concentration of ferric ion is between 0.005 and 1.5 M. In practice, concentrations greater than about 0.2 M must be avoided because the ferric ion imparts a yellow color to the solution that masks the indicator color change.

The main advantage of the Volhard method lies in the fact that it can be carried out in quite strongly acidic solutions. This is a significant advantage because it eliminates many of the interferences that plague both the Mohr and Fajan methods. The cost of this advantage is the extra time and solutions required to perform a back titration instead of a direct titration.

A special problem is encountered when the Volhard method is used for the determination of chloride. Silver chloride, unlike silver bromide and silver iodide, is more soluble than silver thiocyanate, and the following reaction can occur during the back titration:

$$AgCl(s) + SCN^- \rightleftharpoons AgSCN(s) + Cl^- \qquad (12\text{-}18)$$

If Ag^+ is added to both sides of this chemical equation, it becomes clear that the equilibrium constant for the reaction is simply the ratio of the two solubility product constants.

$$AgCl(s) + Ag^+ + SCN^- \rightleftharpoons AgSCN(s) + Ag^+ + Cl^-$$

$$K_{eq} = \frac{[Ag^+][Cl^-]}{[Ag^+][SCN^-]} = \frac{K_{sp,AgCl}}{K_{sp,AgSCN}} = \frac{1.8 \times 10^{-10}}{1.1 \times 10^{-12}} = 1.6 \times 10^2$$

Reaction 12-18 causes more thiocyanate titrant to be added than is necessary to back titrate the excess silver ion. This problem is avoided in one of two ways. After addition of the excess silver nitrate, the silver chloride precipitate is digested, causing it to coagulate and then is removed by filtration. The tedious nature of filtration makes this an unpopular procedure. The most widely used procedure consists of adding nitrobenzene to the precipitated silver chloride. Nitrobenzene coats the precipitate particles and prevents them from coming in contact with the thiocyanate titrant. Special attention should be observed with the use of nitrobenzene, as it is a known carcinogen.

Example 12-8

A 500.0-mg sample of butter was warmed and shaken vigorously with water. The undissolved material was removed by filtering and the aqueous portion was made 1.0 M in HNO_3 and 0.025 M in $Fe(NO_3)_3$. This acidified solution was treated with 10.00 mL of 0.1755 M $AgNO_3$ to precipitate the chloride ion and, after the addition of a small amount of nitrobenzene, 14.22 mL of 0.1006 M KSCN was required to back titrate the excess Ag^+. Calculate the % NaCl in the butter.

Solution The reaction stoichiometries are all 1:1; thus

$$\text{amount } Ag^+ \text{ added} = (10.00 \text{ mL})(0.1755 \ M) = 1.755 \text{ mmol}$$

$$-\left\{ \begin{array}{c} \text{amount } SCN^- \text{ required} \\ \| \\ \text{amount } Ag^+ \text{ in excess} \end{array} \right\} = (14.22 \text{ mL})(0.1006 \ M) = 1.431 \text{ mmol}$$

$$\text{amount } Ag^+ \text{ used} = \text{amount NaCl present} \qquad = 0.324 \text{ mmol}$$

Then

$$\% \text{ NaCl} = \frac{0.324 \text{ mmol} \times 58.5 \text{ mg/mmol}}{500.0 \text{ mg}} \times 100 = 3.79$$

Most procedures call for the indicator to be added *after* the excess silver nitrate. In the

Applications

determination of iodide, this particular order of reagent addition is crucial because iodide ion can reduce the ferric-ion indicator:

$$2Fe^{3+} + 2I^- \rightleftharpoons 2Fe^{2+} + I_2$$

The resulting ferrous ion does not form a colored complex with thiocyanate ion, and no end point will be detected.

PROBLEMS

12-1. Calculate the molar solubility of each of the following substances in water. Neglect the effects of any simultaneous acid-base or complexation equilibria.
- **(a)** TlCl
- **(b)** $BaCrO_4$
- **(c)** $Zn(CN)_2$
- **(d)** $Mg(OH)_2$
- **(e)** ThF_4

12-2. Following each substance below is its solubility in water. Calculate the solubility product constant for each substance.
- **(a)** AgSCN, 0.174 mg/L
- **(b)** $Al(OH)_3$, 1.37×10^{-4} mg/L
- **(c)** $Ba(IO_3)_2$, 7.21×10^{-4} mol/L
- **(d)** Tl_2CrO_4, 3.28 mg/100 mL
- **(e)** $Th(IO_3)_4$, 3.66×10^{-2} g/100 mL
- **(f)** $La_2(CO_3)_3$, 8.20×10^{-6} mmol/100 mL

12-3. For each of the three silver salts, $AgIO_3$, Ag_2CrO_4, and Ag_3PO_4, calculate the following.
- **(a)** equilibrium concentration of Ag^+ in water
- **(b)** molar solubility in water
- **(c)** molar solubility in 0.150 M $AgNO_3$

12-4. What concentration of NaI is required to give PbI_2 a molar solubility of 5.3×10^{-5} M?

12-5. What percentage of Ca^{2+} remains unprecipitated when 50.0 mL of 0.100 M $CaCl_2$ is mixed with 25.0 mL of 0.400 M Na_2SO_4?

12-6. Calculate the concentration of Pb^{2+} in a solution saturated with PbI_2 and containing the following.
- **(a)** 0.0100 M KI
- **(b)** 0.100 M KI
- **(c)** 1.00 M KI

12-7. Determine if a precipitate will form when the following solutions are mixed.
- **(a)** 10.0 mL of 4.00×10^{-3} M $Ca(NO_3)_2$ with 40.0 mL of a hydroxide solution buffered at pH 12.00
- **(b)** 25.0 mL of 3.75×10^{-3} M $BaCl_2$ with 25.0 mL of 0.100 M KIO_3; assume there are no pH effects
- **(c)** 20.0 mL of 0.120 M Na_2CO_3 and 10.0 mL of pH 7.00 buffer with 25.0 mL of 0.200 M $MgCl_2$
- **(d)** 15.0 mL of 0.0250 M $AgNO_3$ with 10.0 mL of 0.100 M KCl in 2.00 M NH_3

12-8. Calculate the molar solubility of thorium fluoride at pH 2.00. Assume that $\alpha_{Th^{4+}} = 1.00$.

12-9. What is the largest concentration of Sr^{2+} that can exist in a solution containing 0.500 M $Na_2C_2O_4$ buffered at pH 4.00?

12-10. What is the smallest value of $\alpha_{PO_4^{3-}}$ that will permit precipitation of calcium phosphate from a solution containing 0.0350 M $CaCl_2$ and a total phosphate concentration of 0.200 M?

12-11. A 25.00-mL sample of 0.1306 M $AgNO_3$ is being titrated with 0.1194 M KSCN. Calculate the pAg^+ at the following points.

(a) before addition of any KSCN
(b) after addition of 10.00 mL of KSCN
(c) after addition of 15.00 mL of KSCN

(d) at the equivalence point
(e) after addition of 40.00 mL of KSCN

12-12. Calculate pF^- at the equivalence point of a titration of 20.00 mL of 0.01039 M La(NO$_3$)$_3$ with 0.006338 M NaF.

12-13. What volume of 0.04211 M Na$_2$MoO$_4$ must be added to 25.00 mL of 0.1244 M Pb(NO$_3$)$_2$ to reach the equivalence point of the titration?

12-14. What weight of KSCN must be added to 50.00 mL of 0.1000 M AgNO$_3$ to make the concentration of silver ion 1.00×10^{-11} M?

12-15. Calculate the molar equilibrium concentration of chloride ion at the point where insoluble silver chromate just begins to form in the titration of 50.00 mL of 0.1003 M NaCl with 0.1014 M AgNO$_3$. Assume that the initial concentration of potassium chromate indicator is 3.00×10^{-3} M. Is this point prior to, at, or after the equivalence point?

12-16. What is the molar concentration of a silver nitrate solution of which 28.60 mL was required to reach the end point in a titration with a solution containing 0.2116 g of primary standard KCl?

12-17. A standard solution of silver nitrate was prepared by dissolving 6.503 g of pure AgNO$_3$ in sufficient water to make 250.0 mL of solution. A 0.6319-g sample containing MgCl$_2$ required 24.33 mL of the silver nitrate for titration. Calculate the percentage of MgCl$_2$ in the sample.

12-18. A sample of relatively pure zinc oxide weighing 0.3417 g was dissolved and titrated with 0.1103 M potassium ferrocyanide, requiring 24.79 mL to reach the end point:

$$2Fe(CN)_6^{4-} + 3Zn^{2+} + 2K^+ \longrightarrow K_2Zn_3[Fe(CN)_6]_2(s)$$

What is the % ZnO in the sample?

12-19. A 25.00-mL aliquot of saline solution was acidified with dilute nitric acid and treated with 25.00 mL of 0.2149 M AgNO$_3$. After addition of a few milliliters of nitrobenzene, the solution required 26.40 mL of 0.1322 M KSCN for titration. Calculate the molar concentration of NaCl in the saline solution.

12-20. The bismuth in 0.7405 g of an alloy was precipitated as BiOCl and separated from the solution by filtration. The washed precipitate was dissolved in nitric acid and treated with 10.00 mL of 0.1498 M AgNO$_3$, causing the precipitation of AgCl. The excess AgNO$_3$ required 12.92 mL of 0.1008 M KSCN for titration. Calculate the % Bi in the sample.

12-21. The phosphate in a 3.000-g sample of industrial detergent was precipitated by the addition of 1.000 g of AgNO$_3$. The solution was filtered and the filtrate required 18.23 mL of 0.1377 M KSCN for titration to the FeSCN^{2+} end point. Calculate the percentage of phosphate in the detergent.

12-22. The director of a chemical analysis laboratory is told to expect a large number of samples for which the bromide content must be determined. To avoid time spent on calculating the final results, she wants the volume of AgNO$_3$ used in a Fajan's titration to equal the % Br in the sample. If each sample will weigh 500.0 mg, what molar concentration of AgNO$_3$ must be used?

12-23. What weight of sample must be used in the Mohr method for chloride so that two times the volume of 0.1016 M AgNO$_3$ used will equal the % KCl in the sample?

12-24. A 1.7483-g sample containing Al(NO$_3$)$_3$, AlCl$_3$, and inert materials was dissolved in acid and divided into two equal portions. One portion was treated with 5.00 mmol of AgNO$_3$. The excess Ag$^+$ required 28.89 mL of 0.1002 M KSCN for titration. The other portion required 26.02 mL of 0.1193 M NaOH for titration:

$$Al^{3+} + 3OH^- \longrightarrow Al(OH)_3(s)$$

Calculate the % $Al(NO_3)_3$ and % $AlCl_3$ in the sample.

12-25. Silver nitrate and potassium chloride solutions are made by dissolving 4.2475 g and 1.8640 g of primary standard $AgNO_3$ and KCl, respectively, in sufficient distilled water to make 250.0 mL of each solution. A 25.00-mL aliquot of the KCl solution required 24.93 mL of $AgNO_3$ for titration by the Fajan method. What is the absolute and relative titration error?

POTENTIOMETRY

Potentiometry is the term used to describe the measurement of potentials or voltages of electrochemical cells. You learned in Chapter 10 that a metal electrode immersed in a solution of its ions develops a potential that is determined by the concentration of the ions. In this chapter we review those electrodes, discuss a type of electrode that responds to concentration by a somewhat different mechanism, and examine various ways in which measured electrode potentials can be used in quantitative determinations.

In making a potentiometric measurement, both an indicator and a reference electrode are required. Each type of electrode is discussed in this chapter.

REFERENCE ELECTRODES

A reference electrode must have a potential that is independent of the solution in which it is immersed and that does not change significantly when a small amount of current is passed through it. The large majority of electrochemical measurements are made with one of three different reference electrodes.

Standard Hydrogen Electrode

The construction and operation of the standard hydrogen electrode (SHE) was described in Chapter 10 and you may recall that it is the ultimate reference against which the potentials of all other electrodes are measured, either directly or indirectly. As might be expected, the electrode is extremely reproducible and it develops a potential that is very close to that predicted theoretically. Unfortunately, it is not a *convenient* electrode to use for routine measurements. The actual electrode surface is quite fragile and should not be touched, scraped, or allowed to dry out. Furthermore, hydrogen gas and associated pressure and flow regulators are required. Most routine measurements are made using one of the following two electrodes.

Saturated Calomel Electrode

A schematic drawing of a commercial saturated calomel electrode (SCE) is shown in Figure 13-1a. The working part of the electrode, consisting of a platinum wire immersed in a slurry of solid mercurous chloride (whose common name is *calomel*), liquid mercury, and aqueous saturated potassium chloride, is contained in the inner tube. The outer tube is merely a saturated potassium chloride salt bridge that permits the entire assembly to be placed directly in the solution to be measured. The salt bridge tube has a small porous asbestos thread or ceramic fiber at the bottom that permits electrical contact to be made between one side of the salt bridge and the test solution without allowing any appreciable flow of liquid between the two solutions. The opening near the top of the outside tube is for addition or replacement of the potassium chloride solution, should it become necessary.

The calomel electrode half-cell may be represented as

$$\text{Hg} \,\big|\, \text{Hg}_2\text{Cl}_2\,[\,\text{sat'd}\,],\, \text{KCl}\,[\,\text{sat'd}\,]\,\big|\big|$$

Mercurous chloride is reduced and elemental mercury is oxidized in the reversible electrode half-reaction:

$$\text{Hg}_2\text{Cl}_2\,(\text{s}) + 2e^- \;\rightleftharpoons\; 2\text{Hg}\,(l) + 2\text{Cl}^-$$

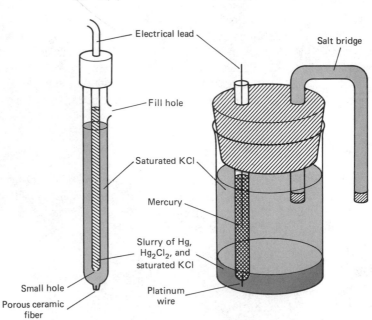

(a) (b)

Figure 13-1 Saturated calomel reference electrodes: (a) commercial; (b) laboratory prepared.

Since the activities of the liquid Hg and solid Hg_2Cl_2 are both unity, the potential of the electrode is described by the Nernst equation:

$$E = E^{\circ}_{Hg_2Cl_2/Hg} - \frac{0.0592}{2} \log \frac{(1)[Cl^-]^2}{1}$$

The concentration of Cl^- is fixed (saturated KCl is about 4.2 M), so the electrode potential remains constant as long as the salt bridge solution is not permitted to mix with the test solution.

Not all calomel electrodes are prepared with saturated potassium chloride solution; some use 1.0 M or 0.1 M KCl. The advantage of using a saturated solution lies in the fact that the concentration of Cl^- does not change if some of the solvent evaporates. It also results in a relatively small junction potential at the two salt bridge–solution interfaces. On the other hand, the solubility of KCl, and therefore the concentration of a saturated solution, depends on the temperature. Thus a small change in temperature causes a substantial change in the potential of the electrode. The temperature effect is much smaller in electrodes using unsaturated potassium chloride solutions, as shown by the data in Table 13-1.

Silver–Silver Chloride Electrode

This electrode, shown in Figure 13-2, consists of a silver wire coated with silver chloride that is immersed in a potassium chloride solution saturated with silver chloride. The half-cell is represented as

$$Ag(s)\,|\,AgCl[sat'd],\,KCl[x\,M]\,||$$

for which the half-reaction is

$$AgCl(s) + e^- \; \rightleftharpoons \; Ag(s) + Cl^- \tag{13-1}$$

According to the Nernst equation, the potential of the electrode depends only on the concentration of Cl^-:

$$E = E^{\circ}_{AgCl/Ag^+} - \frac{0.0592}{1} \log [Cl^-] \tag{13-2}$$

TABLE 13-1 POTENTIALS OF REFERENCE ELECTRODES AT DIFFERENT TEMPERATURES

Temperature (°C)	Potential (V)			
	Calomel (0.1 M KCl)	Calomel (1.0 M KCl)	Calomel (sat'd KCl)	Ag/AgCl (1.0 M KCl)
10	0.3362		0.2539	0.2314
15	0.3361		0.2511	0.2286
20	0.3358		0.2478	0.2256
25	0.3356	0.2810	0.2445	0.2223
30	0.3354		0.2412	0.2190
35	0.3351		0.2376	0.2157
40	0.3345		0.2345	0.2121

Reference Electrodes

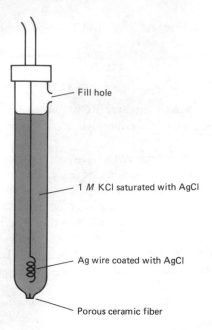

Figure 13-2 Commercial silver–silver chloride reference electrode.

This electrode, like its cousin the calomel electrode, is normally prepared using saturated potassium chloride solution. The advantages of this electrode over the popular calomel electrode are that it can be made very small and it can be used at somewhat higher temperatures.

METAL INDICATOR ELECTRODES

Indicator electrodes for potentiometric measurements are classified according to the mechanism by which the electrode potential is produced. *Metal* indicator electrodes develop a potential that is determined by the equilibrium position of a redox half-reaction at the electrode surface. *Membrane* indicator electrodes develop a potential determined by the difference in concentration of a particular ion on two sides of a special membrane. The remainder of this section is devoted to metal indicator electrodes. Membrane electrodes are discussed in the next section.

First-Order Electrodes

Electrodes of the first order (or first kind) are comprised of a metal immersed in a solution of its ions, such as a silver wire dipping into a silver nitrate solution. For the reversible half-reaction

$$\mathrm{Ag^+ + e^- \rightleftharpoons Ag(s)} \qquad E° = 0.800 \text{ V}$$

the Nernst equation is

$$E = 0.800 - \frac{0.0592}{1} \log \frac{1}{[Ag^+]} \qquad (13\text{-}3)$$

Only a few metals such as silver, mercury, copper, cadmium, zinc, bismuth, lead, and tin exhibit reversible half-reactions with their ions and are suitable for use as first-order electrodes. Other metals, including iron, nickel, cobalt, chromium, and tungsten, develop nonreproducible potentials that are influenced by impurities and crystal irregularities in the solid and by oxide coatings on their surfaces. This nonreproducible behavior makes them unsatisfactory as first-order electrodes.

Second-Order Electrodes

A metal electrode can sometimes be made responsive to the concentration of an anion that forms a precipitate or complex ion with cations of the metal. These are called second-order or second-kind electrodes because they respond to an ion not directly involved in the electron-transfer process. The silver–silver chloride reference electrode discussed in the preceding section is an example of a second-order electrode. When the coated silver wire is immersed in a solution, enough silver chloride dissolves to saturate the layer of solution in contact with the electrode surface. The silver-ion concentration in this solution layer is determined by the position of the solubility product equilibrium:

$$AgCl(s) \rightleftharpoons Ag^+ + Cl^- \qquad K_{sp} = [Ag^+][Cl^-] \qquad (13\text{-}4)$$

The potential of the electrode actually is governed by the Ag^+/Ag half-reaction for which Equation 13-3 applies. Solving the K_{sp} equilibrium expression for the silver-ion concentration and substituting in Equation 13-3 gives

$$E = E^\circ_{Ag^+/Ag} - \frac{0.0592}{1} \log \frac{[Cl^-]}{K_{sp}}$$

or

$$E = E^\circ_{Ag^+/Ag} + \frac{0.0592}{1} \log K_{sp} - \frac{0.0592}{1} \log [Cl^-]$$

$$= 0.800 + \frac{0.0592}{1} \log 1.82 \times 10^{-10} - \frac{0.0592}{1} \log [Cl^-]$$

$$= 0.223 - \frac{0.0592}{1} \log [Cl^-] \qquad (13\text{-}5)$$

which is identical to Equation 13-2.

The limitations of these electrodes are quite severe. They can be used only over a range of anion concentrations such that the solution remains saturated with the substance coating the metal. For the Ag/AgCl electrode, too low a chloride-ion concentration causes the silver chloride coating to dissolve completely. Too high a chloride-ion concentration has the same effect by forming soluble complex ions:

$$AgCl(s) + Cl^- \rightleftharpoons AgCl_2^-$$

$$AgCl_2^- + Cl^- \rightleftharpoons AgCl_3^{2-}$$

Other anions may interfere if they form salts with silver ion that are less soluble than silver chloride. Thus Br^-, I^-, SCN^-, CN^-, and S^{2-} interfere when using a silver–silver chloride electrode to determine chloride-ion concentrations.

Inert Electrodes

Chemically inert conductors such as gold, platinum, or carbon that do not participate, directly or indirectly, in the redox process are called *inert* electrodes. The potential developed at an inert electrode depends on the nature and concentration of the various redox reagents in the solution. For example, a platinum electrode immersed in a solution containing ferrous and ferric ions develops a potential that is described by the Nernst equation for iron:

$$E = E^{\circ}_{Fe^{3+}/Fe^{2+}} - \frac{0.0592}{1} \log \frac{[Fe^{2+}]}{[Fe^{3+}]}$$

Inert electrodes respond to *any* reversible redox system and show no chemical selectivity whatsoever.

MEMBRANE INDICATOR ELECTRODES

The potential developed at this type of electrode results from an unequal charge buildup at opposing surfaces of a special membrane. The charge at each surface is governed by the position of an equilibrium involving analyte ions, which, in turn, depends on the concentration of those ions in the solution. The immense value of this type of electrode stems from our ability to design and manufacture membranes that exhibit considerable selectivity toward a specific ion and that produce potentials described by a Nernst-type equation. The excellent selectivity of these electrodes have led chemists to call them *ion-selective electrodes*.

The electrodes are categorized according to the type of membrane they employ: glass, polymer, or crystalline. A fourth type uses a glass membrane in conjunction with a gas-permeable membrane to determine the concentration of gases. The construction, behavior, and properties of each type of electrode are discussed in the following sections.

Glass Membrane Electrodes

The construction of glass membrane electrodes is illustrated by the pH electrode shown in Figure 13-3. The internal element consists of a silver–silver chloride electrode immersed in a pH 7 buffer saturated with silver chloride. The thin, ion-selective glass membrane is fused to the bottom of a sturdy, nonresponsive glass tube so that the entire membrane can be submerged during measurements. When placed in a solution containing hydrogen ions, this electrode can be represented by the half-cell:

$$Ag(s)\big|AgCl[sat'd], Cl^-(inside), H^+(inside)\big|glass\ membrane\big|H^+(outside)$$

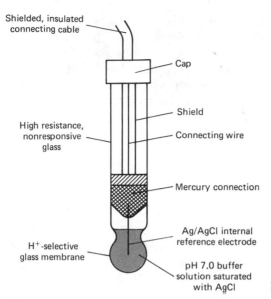

Figure 13-3 Glass-membrane pH electrode.

whose potential is given by

$$E = E^\circ_{AgCl/Ag} - \frac{0.0592}{1} \log [Cl^-] + \frac{0.0592}{1} \log \frac{[H^+]_{outside}}{[H^+]_{inside}} \qquad (13\text{-}6)$$

Separating the ratio of hydrogen-ion concentrations into two log terms gives

$$E = E^\circ_{AgCl/Ag} - 0.0592 \log [Cl^-] + 0.0592 \log \frac{1}{[H^+]_{inside}} + 0.0592 \log [H^+]_{outside}$$

Since the concentrations (activities) of Cl^- and H^+ in the internal electrolyte solution are constant, the first three terms to the right of the equal sign may be combined into a single constant, Q, and the equation rewritten as

$$E = Q + 0.0592 \log [H^+]_{outside} \qquad (13\text{-}7)$$

Composition of glass membranes

Despite a significant effort by many research groups, the relationship between the chemical composition and structure of a membrane and its response selectivity is poorly understood. Although chemists have had some success in making glass membranes with different selectivities, only those for H^+ and Na^+ are good enough to have appreciable value. Both glasses consist mainly of SiO_2 (about 70%). In hydrogen-selective glass, the remaining 30% is a mixture of CaO, BaO, Li_2O, and Na_2O, while in sodium-selective glass it is Al_2O_3 and Na_2O.

Membrane behavior

Structurally, glass membranes are an irregular network of tetrahedral SiO_4 units connected through shared oxygen atoms. Such a network contains many negatively charged oxygens which bind cations such as Li^+ and Ca^{2+}. When the membrane comes in contact with an aqueous solution, the outer surfaces absorb considerable amounts of water, forming hydrated, gel-like layers. The thickness of these layers depends on the composition of the glass but is usually in the range 10^{-5} to 10^{-4} mm. These layers act as cation-exchange membranes with a *high selectivity toward a specific cation*. In the case of the glass pH electrode, hydrogen ions diffuse from the solution into the hydrated layer and displace lithium ions from the negative binding sites. This ion-exchange process is depicted in Figure 13-4. Chemists do not fully understand the mechanism by which the membrane acquires a potential difference between its inside and outside surfaces. Although the following explanation may not be entirely accurate, it is consistent with most of our experimental observations. If the glass in the hydrated layer is considered to be like a weak acid, it would be in equilibrium with its ions:

$$\text{HGl} \quad \rightleftharpoons \quad \text{H}^+ \quad + \quad \text{Gl}^- \qquad (13\text{-}8)$$

$$\text{(membrane)} \qquad \quad \text{(solution)} \quad \text{(membrane)}$$

where Gl^- represents the cation binding site in the glass membrane. The hydrogen-ion concentration in the solution affects the position of this equilibrium, which, in turn, determines the number of free, negative binding sites. Thus each side of the glass membrane acquires a negative charge, the magnitude of which is determined by the hydrogen-ion concentration of the solution with which it is in contact. This is the potential difference responsible for the last term in Equation 13-6.

Selectivity

The selectivity of an ion-selective electrode is determined by the exclusivity of the reaction that brings the analyte ions into the hydrated layer. Undoubtedly, the composition and structure of the glass are largely responsible for this selectivity. Unfortunately, no electrode responds exclusively to a single type of ion. The relative response of an electrode to one type of ion, B, compared to that for the ion of interest, A, at the same concentration is called its *selectivity coefficient*, $k_{B/A}$:

$$k_{B/A} = \frac{\text{response to B}}{\text{response to A}}$$

Small selectivity coefficients are desired. When $k_{B/A}$ is 0.001, the concentration of B must be 1000 times greater than the concentration of A (assuming both have the same ionic charge) in order to produce the same electrode response. Put another way, in a solution with equal concentrations of A and B, only 0.1% (0.001 of the total) of the electrode potential is due to ion B. It is common practice to talk about the selectivity of an electrode for "the analyte over another constituent." It is important to remember,

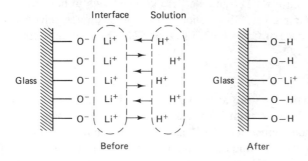

Before

After

Figure 13-4 Ion-exchange process at glass membrane–solution interface.

Chapter 13 Potentiometry

however, that the selectivity *coefficient* is defined as the response of the other constituent over the response of the analyte.

The potential of most ion-selective electrodes is described by the general equation

$$E = Q + \frac{0.0592}{z_A} \log \left([A] + k_{B/A}[B]^{z_A/z_B} + k_{C/A}[C]^{z_A/z_C} + \cdots \right)$$

or

$$E = Q + \frac{0.0592}{z_A} \log \left([A] + \sum k_{X/A}[X]^{z_A/z_X} \right) \tag{13-9}$$

where A represents the analyte ion, X represents the interfering ion(s), and z_A and z_X are the charges on those ions. It is important to remember that the electrode response is actually related to the *activity* of the various ions and that the use of concentrations in Equation 13-9 represents an approximation made for convenience.

The hydrogen-ion or pH electrode exhibits remarkably high selectivity toward H^+ over all other ions. Figure 13-5 shows how a glass pH electrode responds to pH changes in solutions that are 0.1 M in various alkali metal cations. The response is linear over

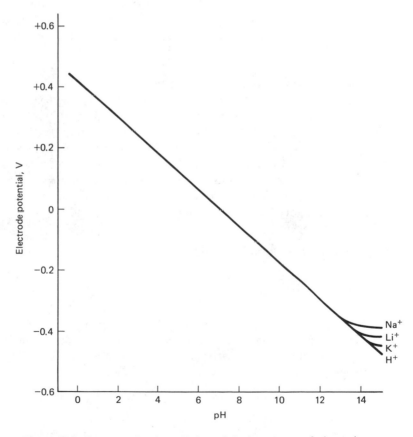

Figure 13-5 Response of a glass pH electrode in the presence of other cations (0.1 M in each case).

Membrane Indicator Electrodes

12 orders of magnitude of hydrogen-ion concentration. The selectivity coefficients are so small that the alkali metals do not interfere until their concentrations are about 100 billion (10^{11}) times greater than the hydrogen-ion concentration! No other membrane electrode can match the glass pH electrode in terms of selectivity. The so-called sodium-ion electrode actually is more responsive toward H^+ than Na^+, but this is not a debilitating limitation as it still can be used to measure as little as 10^{-5} M sodium ion in neutral or weakly alkaline solution where the hydrogen-ion concentration is very low.

Asymmetry potential

If identical solutions and identical reference electrodes are placed on both sides of a glass membrane, the measured potential difference is expected to be zero. In fact, a small potential difference, called the *asymmetry potential*, almost always exists under these conditions. Moreover, this potential is neither constant with time nor reproducible between electrodes.

Clearly, the asymmetry potential results from the fact that the two surfaces of the membrane are not identical on a microscopic scale. The responsible factors are not well established but probably include surface imperfections and mechanical strains that occur during manufacture along with chemical and mechanical attack and contamination of the outer surface that occur during routine use. Fortunately, the asymmetry potential does not change rapidly and its effect on the desired measurement can be corrected by periodic calibration of the electrode with a standard pH buffer (see the section "Electrode Calibration").

Types of glass membrane electrodes

Glass electrodes are fabricated in a variety of sizes and shapes. Semimicro electrodes will fit inside small test tubes and require as little as 0.2 mL of sample solution. Spear-tip electrodes are designed to penetrate and measure the pH of various semisoft materials such as bread, cheese, and fruit. Most of the pH electrodes are also available physically combined with an external silver–silver chloride reference electrode, all encased in a single tube (Figure 13-6). These *combination electrodes* have the advantage of being more compact than a separate, two-electrode system. They are easily recognizable in the laboratory by their two-pronged connecting lead.

Polymer (Liquid) Membrane Electrodes

A polymer membrane electrode selective for calcium ion is shown in Figure 13-7. Schematically, it is similar to the glass pH electrode in that it contains an internal silver–silver chloride electrode and an internal reference solution of fixed composition. The membrane consists of liquid calcium di(n-decyl)phosphate, $[(CH_3(CH_2)_8CH_2O)_2 PO_2]_2Ca$, immobilized in a thin disk of polyvinyl chloride that cannot be penetrated by water. At each membrane surface, the calcium compound establishes an equilibrium with its ions:

$$[(RO)_2PO_2]_2Ca \rightleftharpoons 2(RO)_2PO_2^- + Ca^{2+} \qquad (13\text{-}10)$$

$$\text{(membrane)} \qquad\qquad \text{(membrane)} \qquad \text{(aqueous)}$$

where R is the n-decyl hydrocarbon chain. Note the similarity of this equation to Equa-

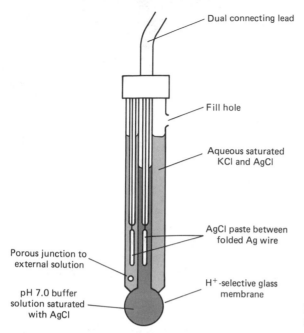

Figure 13-6 Combination glass pH and silver–silver chloride reference electrode.

tion 13-8 for the glass pH electrode. It is important to recognize that the didecylphosphate anion is a fixed part of the nonaqueous liquid membrane. Since the concentration of calcium ions in the solutions on each side of the membrane may be different, the concentration of anions at each membrane surface may be different also, giving rise to a potential described by the equation

$$E = E^{\circ}_{AgCl/Ag^+} - \frac{0.0592}{1} \log [Cl^-] + \frac{0.0592}{2} \log \frac{[Ca^{2+}]_{outside}}{[Ca^{2+}]_{inside}}$$

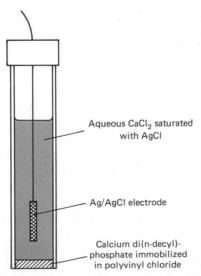

Figure 13-7 Calcium-ion liquid-membrane electrode.

TABLE 13-2 CHARACTERISTICS OF SELECTED LIQUID-MEMBRANE ELECTRODES

Analyte ion	Active membrane material	Concentration range (M)	Recommended pH range	Selectivity coefficients
NH_4^+	Nonactin/monactin	10^{-1}–10^{-5}	5–8	$K^+ = 0.12$; $Na^+ = 0.002$; $Li^+ = 0.0042$; $Mg^{2+} = 2 \times 10^{-4}$
Ca^{+2}	Calcium di(n-decyl)-phosphate	10^0–10^{-6}	6–10	$Zn^{2+} = 3.2$; $Fe^{2+} = 0.8$; $Mg^{2+} = 0.14$; $Na^+ = 0.003$
$Ca^{2+} + Mg^{2+}$ (divalent cation)	Similar to that for Ca^{2+} electrode	10^{-2}–10^{-5}	6–10	$Zn^{2+} = 3.5$; $Fe^{2+} = 3.5$; $Cu^{2+} = 3.1$; $Na^+ = 0.015$
K^+	Valinomycin	10^0–10^{-6}	3–10	$NH_4^+ = 0.013$; $H^+ = 0.01$; $Na^+ = 6 \times 10^{-5}$; $Li^+ = 2 \times 10^{-4}$; $Ca^{2+} = 2 \times 10^{-4}$
Cl^-	Dimethyldioctadecyl-ammonium chloride	10^0–10^{-5}	3–10	$I^- = 17$; $NO_3^- = 4.2$; $Br^- = 1.6$; $HCO_3^- = 0.19$; $F^- = 0.10$; $SO_4^{2-} = 0.14$
ClO_4^-	Tris(substituted 1,10-phenan-throline)iron(II) perchlorate	10^0–10^{-5}	4–10	$I^- = 0.012$; $NO_3^- = 0.0015$; $Br^- = 0.00056$; $Cl^- = 0.00022$; $OH^- = 1$
NO_3^-	Tridodecylhexadecyl-ammonium nitrate	10^0–10^{-5}	3–8	$I^- = 17$; $Br^- = 0.1$; $NO_2^- = 0.07$; $Cl^- = 0.005$; $SO_4^{2-} < 10^{-5}$; $ClO_4^- = 800$

Collecting the constants into a single term as was done with Equation 13-6 gives

$$E = Q + \frac{0.0592}{2} \log [Ca^{2+}]_{\text{outside}} \qquad (13\text{-}11)$$

The selectivity of this electrode is determined by the ability of the di(n-decyl)phosphate anion in the membrane to combine exclusively with calcium ions. Interference from hydrogen ion is substantial below a pH of about 5, due to the replacement of calcium with hydrogen ions on the di(n-decyl)phosphate:

$$[(RO)_2PO_2]_2Ca + 2H^+ \rightleftharpoons 2(RO)_2PO_2H + Ca^{2+}$$

As this occurs, the electrode becomes sensitive to hydrogen ions as well as calcium ions. Selectivity coefficients for several liquid membrane electrodes are given in Table 13-2. When the selectivity coefficients are known, Equation 13-9 can be used to calculate the effect of diverse ions on the potential of these electrodes.

The lower limit of detection is determined primarily by the solubility of the immobilized ion exchanger in water. For a calcium-ion electrode in the absence of interfering ions, Equation 13-11 remains valid until the concentration of calcium ion in the test solution is about 100 times greater than the solubility of calcium di(n-decyl)phosphate. At this point the electrode response begins to level off at a constant value reflecting the calcium-ion concentration in the saturated aqueous solution.

Example 13-1

A nitrate-ion electrode in 1.64×10^{-4} M KNO_3 has an electrode potential of 0.017 V. Enough potassium nitrite is added to the solution to make its concentration 4.76×10^{-2}

Chapter 13   Potentiometry

M, without changing the volume. The new electrode potential is -0.049 V. Calculate the selectivity of the electrode for nitrate over nitrite ions.

Solution   According to Equation 13-9, the electrode potential of the solution containing both nitrate and nitrite is

$$E = Q + \frac{0.0592}{-1} \log \left([NO_3^-] + k_{NO_2^-/NO_3^-} [NO_2^-]^{-1/-1} \right)$$

The value of Q can be calculated from this equation using the electrode potential measured with no nitrite present:

$$0.017 = Q + \frac{0.0592}{-1} \log \left[1.64 \times 10^{-4} + k(0) \right] = Q + 0.224$$

or

$$Q = 0.017 - 0.224 = -0.207 \text{ V}$$

Using the data for the mixture, we obtain

$$-0.049 = -0.207 + \frac{0.0592}{-1} \log \left[(1.64 \times 10^{-4}) + k_{NO_2^-/NO_3^-} (4.76 \times 10^{-2})^{-1/-1} \right]$$

Rearranging yields

$$\log \left[(1.64 \times 10^{-4}) + k_{NO_2^-/NO_3^-} (4.76 \times 10^{-2}) \right] = \frac{-0.049 + 0.207}{-0.0592} = -2.67$$

Taking the antilog of both sides gives us

$$1.64 \times 10^{-4} + k_{NO_2^-/NO_3^-} (4.76 \times 10^{-2}) = 10^{-2.67} = 2.1 \times 10^{-3}$$

and

$$k_{NO_2^-/NO_3^-} = \frac{2.1 \times 10^{-3} - 1.64 \times 10^{-4}}{4.76 \times 10^{-2}} = 0.041$$

Crystalline Membrane Electrodes

These electrodes differ from glass electrodes only in that they are constructed with a crystalline rather than glass membrane. Interest in such electrodes stems partly from the opportunity to devise electrodes responsive to anions by employing a membrane containing selective anionic sites. The fluoride-ion electrode was the first such electrode developed. The membrane consists of a single crystal of lanthanum fluoride, doped with a trace of europium(II) to generate the crystal defects necessary for its electrical conductivity. The potential developed at each surface of the membrane is determined by the position of the equilibrium,

$$\begin{array}{cccc} LaF_3 & \rightleftharpoons & La^{3+} & + & 3F^- \\ \text{(membrane)} & & \text{(membrane)} & & \text{(aqueous)} \end{array}$$

and is described by the equation

$$E = Q + \frac{0.0592}{-1} \log [F^-]_{outside} = Q - 0.0592 \log [F^-]_{outside}$$

Membrane Indicator Electrodes

The electrode is at least 1000 times more selective for fluoride ion than other common anions, with the exception of hydroxide. It is generally reported that the maximum concentration of hydroxide ion that can be tolerated is one-tenth of the fluoride-ion concentration. At low pH, fluoride ion is converted to the weak acid HF ($pK_a = 3.17$), to which the electrode is insensitive.

Although many insoluble crystalline salts exhibit good selectivity toward both cation and anion exchange, generally they are not sufficiently conductive to be useful membranes in ion-selective electrodes. It has been found, however, that silver salts or sulfide salts, mixed in an approximately $1:1$ molar ratio with crystalline silver sulfide and pressed into a thin disk, become good electrical conductors owing to the mobility of the silver ion in the sulfide matrix. Thus membranes of Ag_2S with $AgBr$ or CuS are used to make a bromide-ion or cupric-ion electrode. The properties of selected crystalline membrane electrodes are summarized in Table 13-3.

Gas-Sensing Electrodes

These electrodes are used to measure the concentration of various gases either in solution or in the gas phase. A regular membrane electrode is used as the actual sensing device. Selectivity is obtained by encasing the sensing membrane with a thin, replaceable, gas-permeable membrane. The construction of a gas-sensing electrode is illustrated by the carbon dioxide electrode shown in Figure 13-8. A small volume of an aqueous sodium bicarbonate solution is encased around the working end of a glass pH–reference com-

TABLE 13-3 CHARACTERISTICS OF SELECTED CRYSTALLINE MEMBRANE ELECTRODES

Analyte ion	Membrane composition	Concentration range (M)	Recommended pH range	Selectivity coefficients
F^-	LaF_3	10^0–10^{-6}	5–8	OH^- (see text)
Cl^-	$AgCl/Ag_2S$	10^0–10^{-4}	2–11	$Br^- = 300$; $I^- = 2 \times 10^6$; $OH^- = 0.01$; $CN^- \gg 1$; $S^{2-} \gg 1$
Br^-	$AgBr/Ag_2S$	10^0–10^{-5}	2–12	$Cl^- = 0.003$; $OH^- = 3 \times 10^{-5}$; $I^- = 5000$; $CN^- \gg 1$; $S^{2-} \gg 1$
I^-	AgI/Ag_2S	10^0–10^{-7}	3–12	$Cl^- = 4 \times 10^{-7}$; $Br^- = 2 \times 10^{-4}$; $CN^- = 0.5$; $OH^- = 1 \times 10^{-8}$; $S^{2-} \gg 1$
CN^-	$AgCN/Ag_2S$	10^0–10^{-6}	11–13	$Cl^- = 1 \times 10^{-6}$; $Br^- = 2 \times 10^{-4}$; $I^- = 1.5$
S^{2-}	Ag_2S	10^0–10^{-7}	13–14	—
Ag^+	Ag_2S	10^0–10^{-7}	2–9	$Hg^{2+} \gg 1$
Cu^{2+}	CuS/Ag_2S	10^0–10^{-8}	3–7	$Pb^{2+} = 0.002$; $Cd^{2+} = 3 \times 10^{-4}$; $Zn^{2+} = 2 \times 10^{-4}$; $Ni^{2+} = 2 \times 10^{-4}$
Cd^{2+}	CdS/Ag_2S	10^{-1}–10^{-7}	3–7	$Pb^{2+} = 0.5$; $Zn^{2+} = 1 \times 10^{-4}$; $Ni^{2+} = 5 \times 10^{-6}$
Pb^{2+}	PbS/Ag_2S	10^{-1}–10^{-6}	3–7	$Cd^{2+} = 0.3$; $Zn^{2+} = 2 \times 10^{-4}$; $Fe^{2+} = 0.05$

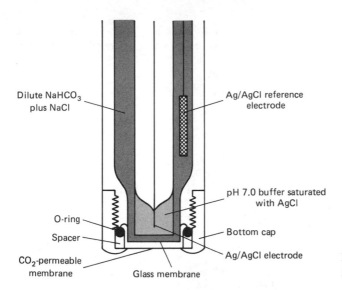

Dilute NaHCO₃ plus NaCl

Ag/AgCl reference electrode

pH 7.0 buffer saturated with AgCl

O-ring

Spacer

Bottom cap

CO_2-permeable membrane

Ag/AgCl electrode

Glass membrane

Figure 13-8 Schematic diagram of a CO_2 gas-sensing electrode.

bination electrode by a gas-permeable membrane made of a microporous, hydrophobic polymer through which water and electrolytes cannot pass. When the electrode is immersed in a solution containing dissolved CO_2, the pores of the membrane transport the molecules to the internal solution. Because there are many pores in the membrane and because the volume of the internal solution is very small, equilibration between the external and internal solutions is quite fast (a few seconds to a few minutes). As carbon dioxide enters the internal solution, it changes the composition via the reaction

$$CO_2(aq) + H_2O \rightleftharpoons H_3O^+ + HCO_3^-$$
(internal solution)

The resulting change in pH is detected by the glass–reference combination electrode.

The overall reaction for the process may be obtained by adding three reactions:

$$CO_2(aq) \rightleftharpoons CO_2(g)$$
external   membrane
solution

$$CO_2(g) \rightleftharpoons CO_2(aq)$$
membrane   internal
             solution

$$CO_2(aq) + H_2O \rightleftharpoons H_3O^+ + HCO_3^-$$
internal solution   internal solution

$$CO_2(aq) + H_2O \rightleftharpoons H_3O^+ + HCO_3^- \qquad (13\text{-}12)$$
external solution   internal solution

The equilibrium constant for this reaction is given by

$$K = \frac{[H_3O^+][HCO_3^-]}{[CO_2(aq)]}$$

Membrane Indicator Electrodes

By using a high concentration of bicarbonate ion in the internal solution, the HCO_3^- formed by Reaction 13-12 will not alter its concentration significantly. Therefore,

$$\frac{[H_3O^+]}{[CO_2(aq)]} = \frac{K}{[HCO_3^-]} = K'$$

or

$$[H_3O^+] = K'[CO_2(aq)]$$

Substituting this expression in Equation 13-6 for a glass pH electrode yields

$$E = Q + 0.0592 \log \left(K'[CO_2(aq)] \right)$$

or

$$E = Q' + 0.0592 \log [CO_2(aq)] \tag{13-13}$$

where

$$Q' = Q + 0.0592 \log K'$$

Other gases, such as SO_2 and NO_2, that dissolve to yield acidic solutions, can be detected in the same manner. To the extent that such gases can penetrate the gas-permeable membrane, they will interfere with each other in a determination. Gas-sensing electrodes are commercially available for CO_2, SO_2, NO_2, H_2S, HCN, HF, and NH_3.

MEASURING ELECTRODE POTENTIALS

The potentials of ion-selective electrodes are measured by making them, along with a suitable reference electrode, part of an electrochemical cell. Since the potential of the reference electrode is known, we can write

$$\Delta E_{cell} = E_{ind} - E_{ref}$$

or

$$E_{indicator} = \Delta E_{cell} + E_{reference}$$

A detailed discussion of the instrumentation necessary to measure cell voltages accurately is beyond the scope of this book. It is, however, important to recognize that the electrical resistance of many ion-selective electrodes is very large, reaching 100 MΩ or more. You may recall from Chapter 10 that any current allowed to flow in a galvanic cell serves to decrease the voltage of the cell. At a resistance of 100 MΩ, the current through the measuring circuit must be kept to 10^{-12} A or less in order to avoid a significant voltage drop across the measuring device. To accomplish this, the measuring device must have an internal resistance of at least 10^6 MΩ.

Potentiometry, especially with ion-selective electrodes, offers the chemist a fast, simple, and inexpensive way to determine the concentrations of a variety of simple anions, cations, and molecules. As the electrodes are nondestructive and unaffected by color or turbidity, they have applications in a wide variety of problems, from determining the fluoride-ion concentration of drinking water to determining the amount of potassium ion transported across cell membranes as a result of some biochemical process. In addition, they are used to follow concentration changes occurring during chemical reactions, leading to determination of end points in titrations and reaction-rate constants in the study of reaction kinetics.

Direct Potentiometric Measurements

Direct potentiometric measurements rely on a comparison of the potential developed by an indicator electrode immersed in the test solution with its potential when immersed in a standard solution of the analyte. The advantage of this technique lies mainly in its simplicity; if the electrode is sufficiently selective toward the ion of interest and largely free of sample matrix effects, little or no sample pretreatment is needed. In addition, direct measurements are often readily adapted to continuous monitoring of a variety of industrial and natural processes. There are several ways in which direct measurements can be performed.

Electrode calibration

This technique involves measuring the potential of an indicator electrode immersed in a solution of the unknown and using the electrode response equation to calculate the concentration. The measured voltage of the cell is given by

$$\Delta E_{meas} = E_{ind} - E_{ref} + E_j \qquad (13\text{-}14)$$

where E_j is the sum of the various junction potentials in the cell. The two most important junction potentials occur where the salt bridge contacts the analyte and reference electrode solutions. Ideally, the net junction potential will be small, but in practice this is not always the case. If Equation 13-7 describes the response of the indicating electrode for analyte A, it can be combined with Equation 13-14 to give

$$\Delta E_{meas} = Q + \frac{0.0592}{n} \log [\text{A}] - E_{ref} + E_j$$

$$= Q - \frac{0.0592}{n} p\text{A} - E_{ref} + E_j$$

Rearranging, we obtain

$$p\text{A} = \frac{(Q + E_j - E_{ref}) - \Delta E_{meas}}{0.0592/n} = \frac{K - \Delta E_{meas}}{0.0592/n} \qquad (13\text{-}15)$$

where

$$K = Q + E_j - E_{ref}$$

Since E_j cannot be evaluated from theory, the combined constant K must be determined experimentally using a solution whose analyte concentration is known. Once K has been determined, ΔE for the unknown is measured and this value along with K is used in Equation 13-15 to calculate pA.

Example 13-2

A 20.0-g sample of hot-dog meat was placed in a blender with 200 mL of water and, after several minutes of blending, the solution was filtered, rinsed, and diluted to 250 mL in a volumetric flask. A combination sodium-ion/reference electrode placed in the filtrate produced a voltage of 0.090 V. The same combination electrode in a 4.35×10^{-2} M Na^+ standard had a voltage of 0.120 V. Calculate the percentage of Na^+ in the hot dog.

Solution Equation 13-15 is used to calculate K,

$$pNa^+ = \frac{K - \Delta E_{meas}}{0.0592/n}$$

$$-\log 4.35 \times 10^{-2} = \frac{K - 0.120}{0.0592/1}$$

$$K = 0.201 \text{ V}$$

Using the value of K and ΔE for the unknown, Equation 13-15 can now be solved for pNa^+:

$$pNa^+ = \frac{0.201 - 0.090}{0.0592/1} = 1.88$$

$$C_{Na^+} = 1.3 \times 10^{-2} \text{ } M$$

Finally,

$$\text{wt } Na^+ = (1.3 \times 10^{-2} \text{ mol/L})(23.0 \text{ g/mol})(0.250 \text{ L}) = 7.5 \times 10^{-2} \text{ g}$$

and

$$\% \text{ Na} = \frac{\text{wt Na}}{\text{wt sample}} \times 100 = \frac{7.5 \times 10^{-2} \text{ g}}{20.0 \text{ g}} \times 100 = 0.38$$

This is precisely the manner in which pH is measured with a glass electrode except the pH meter does the calculation electronically. Equation 13-15, written in terms of pH, can be rearranged to give

$$\Delta E_{meas} = K - 0.0592 \text{ pH} \tag{13-16}$$

When the pH and reference electrodes are immersed in a standard solution whose pH is known accurately, the appropriate meter reading is set with a knob labeled "calibrate." In essence, the meter is being told the value of K. All pH meters are designed to change 1 pH unit for every 0.0592-V change in cell voltage at 25°C, so when the electrodes are transferred to the unknown solution, the meter senses the new voltage and displays the appropriate pH. It is not unusual to have an electrode response that is not quite "Nernstian"; that is, the change in potential resulting from a 1-unit change in pA is not exactly $0.0592/n$ V. All pH meters are designed to compensate for a non-Nernstian response. Once the system has been standardized with a buffer of known pH, the electrodes are transferred to a second standard pH buffer. If the meter reading is incorrect due to a non-Nernstian response, it can be adjusted to the correct value with a knob labeled "slope" or "temperature." Non-Nernstian responses are quite common among ion-selective electrodes.

Many substances are used as pH standards. Laboratories may purchase their standards from chemical supply companies or prepare them "in-house" using an appropriate recipe. The National Bureau of Standards has identified seven buffers that it considers to be highly appropriate for use as pH standards. The composition and certified pH of these buffers at several temperatures are given in Table 13-4.

The validity of direct potentiometric measurements depends on our ability to make the junction potential negligibly small or, failing that, to ensure that it is nearly the same in the measurement of both the standard and unknown solutions. Unfortunately, we are often unable to do either because the composition of the unknown will almost inevitably differ from that of the standard solution used for calibration of the electrode. It is equally unfortunate that this source of error is very difficult to detect.

The response of an ion-selective electrode is governed by the activity rather than concentration of the analyte. To calculate concentration from activity, the activity coefficient must be known. In most applications, it cannot be determined because the ionic strengths of the solutions are not known. As a consequence, the electrode calibration technique is best suited to those situations where activity rather than concentration is being sought; when the ionic strength is very low, making the difference between activity and concentration negligible (activity coefficient approximately unity); or when the ionic strength of the sample and standard can be made similar. This latter condition often can be achieved by adding a large amount of inert electrolyte to each solution.

Although extremely simple and fast, the electrode calibration technique suffers

TABLE 13-4 PRIMARY pH BUFFERS CERTIFIED BY THE NATIONAL BUREAU OF STANDARDS

	Temperature (°C)				
	20	25	30	35	40
Potassium hydrogen tartrate, saturated at 25°C	—	3.557	3.552	3.549	3.547
Potassium dihydrogen citrate, 0.05000 M	3.788	3.776	3.766	3.759	3.753
Potassium hydrogen phthalate, 0.05000 M	4.002	4.008	4.015	4.024	4.035
Potassium dihydrogen phosphate, 0.02500 M + disodium hydrogen phosphate, 0.02500 M	6.881	6.865	6.853	6.844	6.838
Potassium dihydrogen phosphate, 0.008695 M + disodium hydrogen phosphate, 0.03043 M	7.429	7.414	7.400	7.389	7.380
Sodium tetraborate, 0.01000 M	9.225	9.180	9.139	9.102	9.068
Sodium bicarbonate, 0.02500 M + sodium carbonate, 0.02500 M	10.062	10.012	9.966	9.925	9.889

Techniques and Applications

from the fact that a small error in the measured voltage translates to a large error in the calculated concentration. If ΔE is the correct voltage, the analyte concentration as pA is given by Equation 13-15:

$$pA = \frac{K - \Delta E}{0.0592/n}$$

where pA represents the negative logarithm of the "true" concentration of analyte. Suppose that the measured cell voltage is in error by 1 mV (0.001 V). Then the calculated concentration pA' is given by

$$pA' = \frac{K - (\Delta E - 0.001)}{0.0592/n}$$

where pA' represents the negative logarithm of the "experimental" concentration. Subtracting these equations, we obtain

$$pA - pA' = \frac{K - \Delta E}{0.0592/n} - \frac{K - (\Delta E - 0.001)}{0.0592/n} = \frac{0.001}{0.0592/n} = n(0.0169)$$

or

$$\log \frac{1}{[A]} - \log \frac{1}{[A']} = \log \frac{[A']}{[A]} = n(0.0169)$$

If n is 1,

$$\frac{[A']}{[A]} = 10^{0.0169} = 1.04$$

Rearranging and subtracting [A] from both sides gives

$$[A'] - [A] = 1.04[A] - [A]$$

where the difference between [A'], the "experimental" concentration, and [A], the "true" concentration, is the error. Thus

$$error = \Delta A = 1.04[A] - [A] = 0.04[A]$$

or, in relative terms,

$$\% \text{ error} = \frac{\Delta A}{[A]} \times 100 = \frac{0.04[A]}{[A]} \times 100 = 4$$

Thus a 1-mV error in the measured value produces a 4% error in the calculated concentration. The relative error is twice as great for a divalent ion ($n = 2$) and three times as great for a trivalent ion.

System calibration

The electrode calibration technique has two serious limitations: It assumes that Equation 13-15 (or one similar to it) is valid, and it relies on a single standard to establish the value of K in the equation. System calibration uses a series of standards to define experimentally the relationship between the signal and the concentration. The measured electrode potentials are plotted versus the negative logarithm of the concentrations (or activities) as shown in Figure 13-9. It is customary to prepare standards whose concentrations cover the range within which the concentration of the unknown is expected to fall. An unexpected large error in one measured value is easily recognized because the point falls far off the curve described by the other points. Also, the effect of small random errors is decreased by using a number of values to determine the position of the curve.

Since the electrode potential is determined by the *activity* of the analyte ion, the validity of this procedure depends on the activity coefficients for the standards and unknown being nearly the same. You may recall from Chapter 2 that the activity coefficient is determined, in part, by the ionic strength of a solution. In many cases, the ionic strength of the solutions can be "equalized" by adding a large, measured excess of an inert electrolyte to each standard and unknown.

Standard addition

This technique uses a different approach to handling the ionic strength problem. The electrode potential of the sample is measured before and after the addition of a small amount of a standard solution of the analyte. It is assumed that the added standard solution does not change the ionic strength, and therefore the activity coefficient, significantly.

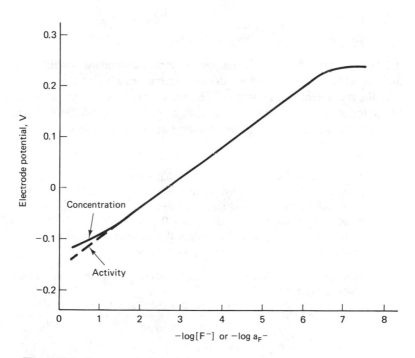

Figure 13-9 Response of a fluoride-ion electrode as a function of fluoride-ion concentration.

If Equation 13-15 applies, we can write

$$pA = -\log C_A = \frac{K - \Delta E_1}{0.0592/n}$$

or

$$\log C_A = \frac{n(\Delta E_1 - K)}{0.0592} \tag{13-17}$$

After the addition of a known volume of a standard solution, the equation can be written as

$$\log \left[\frac{C_A V_A + C_S V_S}{V_A + V_S} \right] = \frac{n(\Delta E_2 - K)}{0.0592} \tag{13-18}$$

where C_A and C_S are the concentrations of the unknown and standard solution, respectively, V_A is the volume of the original solution, V_S is the volume of standard solution added, and ΔE_2 is the new measured cell voltage. Subtracting Equation 13-18 from 13-17, we obtain

$$\log C_A - \log \left[\frac{C_A V_A + C_S V_S}{V_A + V_S} \right] = \frac{n(\Delta E_1 - K)}{0.0592} - \frac{n(\Delta E_2 - K)}{0.0592}$$

which can be solved for C_A, giving

$$C_A = \frac{C_S V_S}{(V_A + V_S)\, 10^{-n(\Delta E_1 - \Delta E_2)/0.0592} - V_A} \tag{13-19}$$

It is somewhat difficult to judge just how much standard should be added. When too little is added, the cell voltage doesn't change much and the relative error associated with the measurement of a small difference between two large values can be appreciable. When too much is added, the ionic strength changes significantly, thereby defeating one of the main reasons for using the standard addition technique.

Example 13-3

The wastewater from a chemical plant manufacturing sulfuric acid was thought to be contaminated with lead. A 50-mL aliquot of the water was adjusted to pH 5 and a lead indicating and saturated calomel reference electrode were placed in the solution. The potential difference between the electrodes was -0.118 V. After the addition of 5.00 mL of a 6.00 $\times$ 10^{-3} M Pb^{2+} standard, the potential difference was -0.109 V. Calculate the molar concentration of Pb^{2+} in the wastewater sample.

Solution The solution is quite simple, requiring only the direct substitution of the given values in Equation 13-19.

$$[Pb^{2+}] = \frac{(6.00 \times 10^{-3} \text{ mmol/mL})(5.00 \text{ mL})}{[(50.0 \text{ mL} + 5.00 \text{ mL})\, 10^{-2[-0.118-(-0.109)]/0.0592}] - 50.0 \text{ mL}}$$

$$= \frac{3.00 \times 10^{-2} \text{ mmol}}{(55.0 \text{ mL})\, 10^{0.304} - 50.0 \text{ mL}} = 4.94 \times 10^{-4} \text{ mmol/mL}$$

Potentiometric Titrations

Since the response of an ion-selective electrode is a logarithmic function of concentration, it is an excellent probe for monitoring the progress of a titration. All that is needed is to place the appropriate indicator and reference electrodes in the sample solution and record the cell voltage as increasing amounts of titrant are added. In contrast to direct potentiometric measurements, electrode selectivity is not of great importance in potentiometric titrations. Selectivity comes from the titrant, which is chosen for its ability to react *only* with the analyte. Although the electrode response may be due to both the analyte and another ion, only the analyte concentration *changes* during the titration, as illustrated in Figure 13-10. The lower curve is obtained when the electrode responds *only* to the analyte, whereas the upper curve is obtained when the electrode responds to *both* analyte and diverse ion (but only the analyte reacts with the titrant). Note that the same end point is obtained from both curves.

Finding the end point

There are several ways to graph potentiometric titration data so that the end point can be determined. These can be illustrated using the titration data in Table 13-5. The most common method, with which you are already familiar, is to plot the cell voltage versus the volume of titrant added. The end point is taken as the inflection point (point of maximum slope), as shown in Figure 13-11a. The accuracy with which this point can

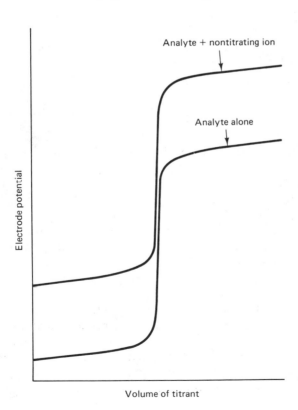

Figure 13-10 Effect of an electrode-responsive but nontitrating ion on a potentiometric titration curve.

TABLE 13-5 POTENTIOMETRIC DATA FOR THE TITRATION OF 25.0 mL OF 0.0100 M SODIUM FLUORIDE WITH 0.0100 M LANTHANUM NITRATE

Volume of La(NO$_3$)$_3$ (mL)	E vs. SCE (mV)	$\Delta E/\Delta$mL (mV/mL)	$\Delta(\Delta E/\Delta$mL$)/\Delta$mL (mV/mL^2)
2.0	−250		
		0.5	
4.0	−249		0.25
		1	
5.0	−248		0
		1	
6.0	−247		1
		2	
6.5	−246		4
		4	
			32
7.0	−244		
		20	
7.2	−240		0
		20	
7.4	−236		25
		25	
7.6	−231		50
		35	
7.8	−224		75
		50	
			200
8.0	−214		
		90	
8.1	−205		600
		150	
8.2	−190		900
		260	
8.3	−164		−700
		190	
8.4	−145		−800
		110	
			−300
8.5	−134		
		80	
8.6	−126		−225
		45	
8.8	−117		−75
		30	
9.0	−111		−50
		20	
9.2	−107		−35
		13	
			−23
9.5	−103		
		6	
10.0	−100		−6
		3	
11.0	−97		−1
		2	
12.0	−95		−0.75
		0.5	
14.0	−94		

be found depends in part on the number of data points collected in the vicinity of the end point.

A careful examination of Figure 13-11a shows an increasing slope up to the end point and a decreasing slope thereafter. This suggests another approach to finding the end point—plot the *slope* of curve (a) versus the volume of titrant added. This is called a *first derivative* curve and is shown in Figure 13-11b. The end point is the extrapolated intersection of the rising and falling portions of the curve. It is also possible to prepare a *second derivative* curve by plotting the slope of the first derivative curve versus the volume of titrant added, as shown in Figure 13-11c. The point where the line passes through zero on the vertical axis is taken as the end point.

Collecting data similar to that in Table 13-5 is slow and tedious when done manually. Instruments are available that record the first- or second-derivative signals and shut the buret off when the signal changes direction (first derivative) or passes through zero (second derivative). These *automatic titrators* do not produce results that are more accurate than those obtained manually, but they are much faster. As you might expect, they are generally found in laboratories where a large number of titrations are performed regularly.

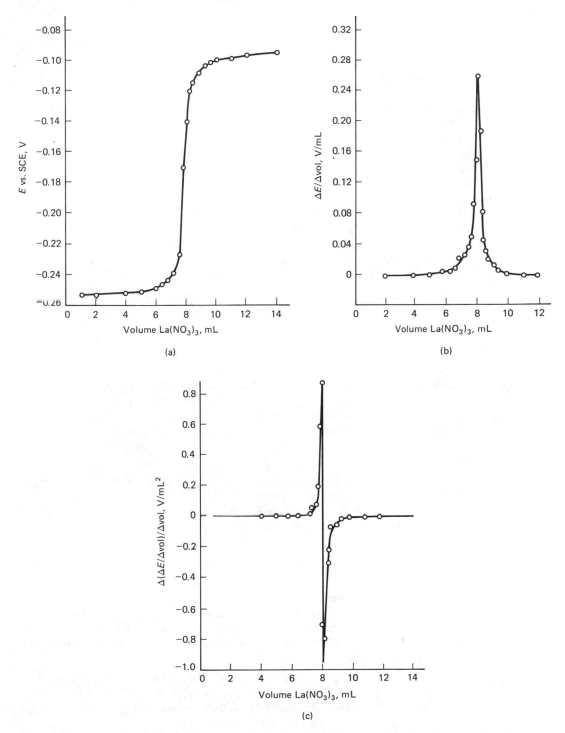

Figure 13-11 Potentiometric titration curves of 25.0 mL of 0.0100 M NaF with 0.0104 M La(NO$_3$)$_3$; (a) regular curve; (b) first-derivative curve; (c) second-derivative curve.

PROBLEMS

13-1. The electrode potential of a saturated calomel electrode is 0.2364 V at 37°C. If the molar solubility of KCl is 3.42 M at this temperature, what is the electrode potential, at the same temperature, of a calomel electrode prepared with the following concentrations of KCl?
  (a) 1.000 M KCl   (b) 0.1000 M KCl   (c) 0.01000 M KCl

13-2. Starting with Equation 13-3, calculate the standard electrode potentials for the half-cells with the following half-reactions.
  (a) $AgBr(s) + e^- \rightleftharpoons Ag(s) + Br^-$
  (b) $AgI(s) + e^- \rightleftharpoons Ag(s) + I^-$
  (c) $AgSCN(s) + e^- \rightleftharpoons Ag(s) + SCN^-$

13-3. Calculate the theoretical potential of a platinum electrode immersed in each of the following solutions.
  (a) $UO_2^{2+}(0.01000 \ M) + U^{4+}(0.1475 \ M) + HCl(0.1000 \ M)$
  (b) $Sn^{4+}(0.6000 \ M) + Sn^{2+}(1.000 \times 10^{-1} \ M)$
  (c) $TiO^{2+}(1.000 \ M) + Ti^{3+}(0.05620 \ M) + HCl(1.000 \ M)$
  (d) $HCl(2.749 \ M) + H_2(1.000 \times 10^{-5} \ atm)$

  Comment on the selectivity of the platinum electrode.

13-4. A glass-membrane pH electrode using a silver–silver chloride internal element in 1.00 M HCl saturated with AgCl is placed in a buffer of pH 4.00.
  (a) Calculate the potential of the electrode (versus SHE).
  (b) Calculate the potential of the electrode if the internal solution is 0.100 M HCl.

13-5. A glass-membrane pH electrode has a selectivity coefficient for H^+ over Na^+ of 5×10^{-14}. Calculate the error in units of millivolts and in units of pH when the electrode is used to measure a solution that is 1.00 M Na^+ combined with the following.
  (a) $1.00 \times 10^{-13} \ M \ H^+$ (b) $1.00 \times 10^{-14} \ M \ H^+$

13-6. What is the potential difference across the glass membrane of a pH electrode for the following differences in pH between the internal and external solutions?
  (a) 4.00   (c) 6.00   (d) 7.00
  (b) 5.00

13-7. A glass-membrane pH electrode and a saturated calomel reference electrode were placed in a buffer at pH 3.56 and the voltage across the electrodes was −0.111 V. When immersed in another solution of unknown pH, the voltage was −0.041 V. Calculate the pH of the unknown solution. Why was the first measurement necessary?

13-8. The voltage across the leads of a glass pH–reference combination electrode in a solution of pH 8.00 was 133 mV. Calculate the expected voltage if the pH of the solution is raised to 9.00.

13-9. The potential difference between a potassium-ion and saturated calomel electrode immersed in 100.0 mL of $1.073 \times 10^{-3} \ M$ KCl was 362 mV. When 594 mg of NH_4Cl was added to the KCl solution (no volume change), the potential difference was 386 mV. Calculate the selectivity coefficient for K^+ over NH_4^+. Assume that there is no loss of NH_4^+ due to acid-base equilibrium effects.

13-10. A calcium-ion electrode has a selectivity coefficient for Ca^{2+} over Na^+ of 0.0030. When immersed in a solution of $4.45 \times 10^{-4} \ M \ Ca^{2+}$, the electrode had a potential of 0.214 V versus the SCE. What potential (versus SCE) will this electrode have in a solution with the same concentration of Ca^{2+} but also containing $2.74 \times 10^{-1} \ M \ Na^+$?

13-11. What is the absolute error due to the Na^+ in terms of pCa^{2+} for the second measurement described in Problem 13-10?

13-12. The voltage across a lead-ion indicator electrode and silver–silver chloride reference electrode immersed in 2.50×10^{-3} M $Pb(NO_3)_2$ at pH 4.0 was 0.278 V. If $k_{Cd^{2+}/Pb^{2+}}$ is 0.31, calculate the absolute and relative error in terms of the concentration of Pb^{2+} when the electrodes are used to measure the lead-ion concentration in a solution containing 3.17×10^{-2} M $Pb(NO_3)_2$ and 8.03×10^{-3} M $Cd(NO_3)_2$.

13-13. The value of $k_{NO_2^-/NO_3^-}$ is 0.070 for a nitrate-ion electrode. What is the largest concentration of NO_2^- that can be tolerated in the measurement of 1.63×10^{-4} M NO_3^- without introducing a relative error greater than 5.0%?

13-14. A 5.00-mL aliquot of a liquid cleanser is transferred to a 250-mL volumetric flask and diluted to volume. An ammonia–reference electrode pair placed in the diluted solution produced a voltage of 0.243 V. The same electrodes in 1.022×10^{-2} M NH_3 produced a voltage of 0.268 V. Calculate the molar concentration of NH_3 in the cleanser.

13-15. A combination fluoride-ion/reference electrode was calibrated using a reference solution prepared by dissolving 391.6 mg of primary standard NaF in 1.000 L of solution. When immersed in this solution, the voltage across the electrodes was -108 mV. When placed in 50.00 mL of wastewater from a manufacturing process, the voltage was -123 mV. Calculate the concentration of fluoride in the wastewater as ppm F^-.

13-16. The following data were collected using a calcium-ion indicating electrode and calomel reference electrode:

$[Ca^{2+}]$	ΔE_{meas}
1.104×10^{-1}	0.315
1.167×10^{-2}	0.288
9.985×10^{-4}	0.258
1.208×10^{-4}	0.231
1.006×10^{-5}	0.200
Unknown	0.226

 (a) Plot the data for the standards in a manner that will lead to a straight line.
 (b) Determine the slope of the line from part (a). Is it the expected theoretical slope?
 (c) Use the graph to find the molar concentration of the unknown.
 (d) Calculate the molar concentration of the unknown from Equation 13-15 using the last two lines of data in the table.
 (e) Why are the concentrations of Ca^{2+} from parts (c) and (d) different? Which is probably closer to the correct value?

13-17. A 50.00-mL aliquot of liquid from a sewage holding tank was brought to pH 13 with sodium hydroxide and diluted to 100.0 mL. The voltage across a sulfide-ion and calomel electrode immersed in the solution was 0.337 V. After addition of 10.00 mL of 2.50×10^{-3} M Na_2S, the voltage across the electrodes was 0.324 V. Calculate the concentration of sulfide in the water as ppm S^{2-}.

13-18. A 0.9164-g sample containing soluble fluoride salts was dissolved in a phosphate buffer of pH 6.5. A fluoride-ion and calomel electrode were placed in the solution, which was then titrated with 0.01249 M $La(NO_3)_3$. Voltages were recorded after each addition of titrant and the titration graph constructed from these data showed an end point after the addition of 31.82 mL of titrant. Calculate the percentage of F^- in the sample.

13-19. A 50.00-mL sample of treated drinking water was diluted to 100.0 mL and adjusted to pH 10 with an ammonia–ammonium chloride buffer. This solution was titrated with 7.022×10^{-3} M EDTA using a total hardness (divalent cation) electrode to monitor the progress of the titration. The titration graph showed an end point after addition of 21.40 mL of the EDTA. Calculate the total hardness of the sample as ppm $CaCO_3$.

ELECTROANALYTICAL METHODS BASED ON ELECTROLYSIS

Chapters 10, 11, and 13 dealt with electrochemical properties of spontaneous chemical reactions. In this chapter we deal with techniques based on nonspontaneous redox reactions that are driven by an external source of electricity. Although there are many such electroanalytical techniques, the discussion here is limited to electrogravimetry, coulometry, and polarography. Each of these techniques has important applications in analytical chemistry. Before the methodology of these techniques can be understood, something of the underlying principles of electrolysis must be learned.

PRINCIPLES OF ELECTROLYSIS

Electrolytic cells are those in which nonspontaneous reactions are caused to occur by the application of a sufficient voltage to the electrodes. The process of causing reactions to occur in this manner is called *electrolysis*. In many respects, an electrolytic cell is the opposite of a galvanic cell. During electrolysis, a galvanic cell is created from the accumulation of the electrolysis products at the electrode surfaces. The voltage of this galvanic cell is always opposed to the applied voltage, making continued electrolysis more difficult. If the electrolysis reaction is reversible, the galvanic voltage increases until it equals the applied voltage, at which point the electrolysis stops. Removing the applied voltage leaves a galvanic cell ready to produce electricity. The lead-acid or lead-storage battery used in automobiles is an example of this type of ''rechargeable'' cell. It behaves as a galvanic cell when supplying power to start the engine and as an electrolytic cell when being charged by the alternator operating off the running engine. One cell of a lead-acid battery is shown in Figure 14-1. The standard 12-volt car battery

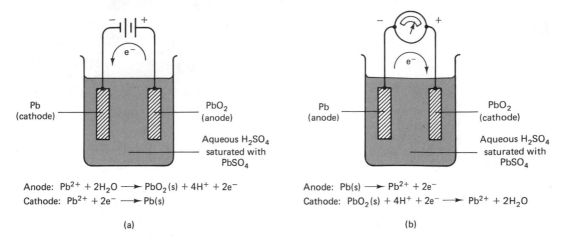

Figure 14-1 One cell of a lead-acid battery: (a) electrolytic (charging) mode, (b) galvanic (discharging) mode.

consists of six such cells connected in series. It is interesting to note that while the lead electrode is always negative, oxidation occurs there during discharging (galvanic cell) but reduction occurs during charging (electrolytic cell). It is for this reason that the definition of anode and cathode should *not* be based on electrode charge.

Applied Voltage Required for Electrolysis

There are three barriers to current flow in an electrolytic cell: the opposing voltage of the galvanic cell, the overvoltage of the electrodes, and the ohmic voltage of the solution. Thus the minimum applied voltage required to cause electrolysis is given by

$$\Delta E_{applied} = \Delta E_{galvanic} + \Delta E_{overvoltage} + \Delta E_{ohmic} \qquad (14\text{-}1)$$

Galvanic voltage

If the electrode reactions along with the concentrations of reactants and products are known, the galvanic cell voltage can be calculated from the appropriate Nernst equations. Consider the electrolysis of $0.0100\ M$ $AgNO_3$ in $0.10\ M$ HNO_3 using two identical platinum electrodes. The electrode reactions are

$$\text{Cathode:} \quad Ag^+ + e^- \rightleftharpoons Ag(s)$$
$$\text{Anode:} \quad 2H_2O \rightleftharpoons O_2(g) + 4H^+ + 4e^-$$

The electrode potential of the cathode is given by

$$E_c = E^\circ_{Ag^+/Ag} - \frac{0.0592}{1} \log \frac{1}{[Ag^+]}$$

$$= 0.800 - 0.0592 \log \frac{1}{0.0100} = 0.682\ V$$

Similarly, for the anode:

$$E_a = E°_{O_2/H_2O} - \frac{0.0592}{4} \log \frac{1}{p_{O_2}[H^+]^4}$$

Since the partial pressure of oxygen in the atmosphere is about 0.21,

$$E_a = 1.23 - 0.0148 \log \frac{1}{(0.21)(0.10)^4} = 1.16 \text{ V}$$

The spontaneous reaction of the galvanic cell produced during electrolysis is the reverse of the electrolysis reaction:

$$4Ag(s) + O_2(g) + 4H^+ \rightleftharpoons 4Ag^+ + 2H_2O$$

and the galvanic voltage of this cell is given by

$$\Delta E_{galv} = E_a - E_c = 1.16 - 0.682 = 0.48 \text{ V}$$

It is easy to get confused at this point about the designation of anode and cathode. Since the overall discussion concerns electrolysis cells, anode and cathode refer to the electrodes where oxidation and reduction occur, respectively, during the *electrolytic* operation of the cell. It is important to realize that as the electrolysis proceeds, the concentration of Ag^+ decreases and the concentration of H^+ increases. These changes act to *increase* the galvanic cell voltage. As a result, the minimum applied voltage needed to sustain the electrolysis must also increase.

Overvoltage

An electrode whose actual potential is different from its calculated value is said to be *polarized* and the amount of polarization is called the *overpotential* of the electrode. Regardless of its source, overpotential always operates to make the reaction at the electrode more difficult. The term *overvoltage* normally refers to the difference in the overpotentials of the two electrodes in a cell.

There are two types of electrode polarization. *Concentration polarization* occurs when the rate of oxidation or reduction at an electrode surface is faster than the rate at which fresh reactants can be supplied to or products removed from the surface. When this occurs, the concentration of reactants and products are different at the electrode surface from in the bulk solution. It should be clear that any increase in the current passed at an electrode will cause the overpotential to increase. However, at a constant applied potential, the current varies inversely with the electrode area. Consequently, chemists often prefer to use current density, which is the current per unit surface area, to account for both of these parameters when dealing with overpotentials. Table 14-1 lists overpotentials for the evolution of hydrogen and oxygen at several different electrodes and current densities. The degree of concentration polarization is decreased by any change that increases the rate of mixing of the solute at the interface with that in the bulk solution. Mechanical stirring is the most obvious way to reduce concentration polarization, but raising the temperature also aids the mixing process because it increases the rate of diffusion. Similarly, addition of an inert electrolyte can decrease electrostatic

TABLE 14-1 OVERPOTENTIALS (V) FOR THE EVOLUTION OF HYDROGEN
AND OXYGEN AT INERT ELECTRODES

Current density (A/cm²)	Ag		C (graphite)		Pt (bright)		Pt (platinized)	
	H_2	O_2	H_2	O_2	H_2	O_2	H_2	O_2
1.00×10^{-3}	0.475	0.580	0.600	—	0.024	0.72	0.015	0.398
1.00×10^{-2}	0.762	0.729	0.779	—	0.068	0.85	0.030	0.521
1.00×10^{-1}	0.875	0.984	0.977	—	0.288	1.28	0.041	0.638
1.00×10^{0}	1.089	1.131	1.220	—	0.676	1.49	0.048	0.766

forces between the electroactive ions and the electrode, thereby preventing a concentration gradient from forming.

Chemists have observed that even in the absence of concentration polarization, an electrode may still exhibit an overpotential. In order to have a finite current at an electrode, it is necessary for the potential to shift from its equilibrium value, a process called *kinetic polarization*. The magnitude of this potential shift is the *activation overpotential*. Kinetic polarization occurs when the current at the electrode surface is controlled by the rate of the electrode reaction rather than the rate of mass transfer as it is in concentration polarization. This type of polarization is usually quite small for the oxidation of metals and the reduction of metal ions but can be pronounced for electrode reactions involving gaseous reactants or products.

Overvoltages are very dependent on the physical nature of the electrode, generally being larger for soft metals such as copper, silver, mercury, and lead. Unfortunately, overvoltages cannot be predicted from theory with any reasonable degree of reliability. As a result, chemists must rely on previously measured values or on crude estimates calculated from empirical data.

Ohmic voltage

Every cell possesses some electrical resistance. The voltage required to overcome this resistance, called the *ohmic voltage*, is defined by Ohm's law:

$$\Delta E_{ohmic} = IR$$

where I is the current in amperes and R is the cell resistance in ohms. The ohmic voltage always acts to increase the applied voltage required to operate an electrolytic cell and to decrease the voltage obtained from a galvanic cell. In situations where a large current must be drawn through a cell, it is important to keep the cell resistance small in order to avoid a large ohmic voltage.

Predicting Electrode Reactions

Generally, there are several reactions that conceivably may occur in an electrolytic cell. The reaction that does occur is the one requiring the *smallest* applied voltage. That is, the oxidation requiring the smallest positive potential occurs at the anode and the reduction requiring the smallest negative potential occurs at the cathode. The applied potential required for each electrode reaction can be calculated if the overpotentials are known:

Principles of Electrolysis

$$E_{appl} = E_{galv} + E_{overpot} \qquad \text{(14-2)}$$

To illustrate, consider the electrolysis of $0.10\ M\ Cu^{2+}$ in $1.0\ M\ HCl$ using two identical platinum electrodes. There are two possible oxidations:

$$2H_2O \rightleftharpoons O_2(g) + 4H^+ + 4e^-$$
$$2Cl^- \rightleftharpoons Cl_2(g) + 2e^-$$

The Nernst equation may be used to calculate the galvanic potentials (relative to SHE), but first the half-reactions are rewritten as reductions. Thus

$$O_2(g) + 4H^+ + 4e^- \rightleftharpoons 2H_2O$$

$$E = E^\circ_{O_2/H_2O} - \frac{0.0592}{4} \log \frac{1}{p_{O_2}[H^+]^4}$$

The potential is indeterminate if no oxygen is present initially in the solution, but if the reaction occurs, the partial pressure of oxygen will become 1.0 atm and

$$E = 1.23 - \frac{0.0592}{4} \log \frac{1}{(1.0)(1.0)^4} = 1.23\ V$$

The overpotential for this oxidation depends on the current and the surface area of the electrode, but typically is about 1.0 V. Thus the minimum applied potential required to obtain this reaction is 1.23 V + 1.0 V or 2.2 V. Remember that overpotential always acts to make the electrolysis more difficult, which translates to an *increased* anode potential.

Similarly, for the chloride reaction,

$$Cl_2(g) + 2e^- \rightleftharpoons 2Cl^-$$

$$E = E^\circ_{Cl_2/Cl^-} - \frac{0.0592}{2} \log \frac{[Cl^-]^2}{p_{Cl_2}}$$

If this reaction occurs and the solution becomes saturated with chlorine,

$$E = 1.36 - \frac{0.0592}{2} \log \frac{(1.0)^2}{1.0} = 1.36\ V$$

The overpotential for chlorine evolution at a bright platinum electrode is negligible and the applied potential necessary to cause this reaction to occur is 1.36 V. Thus chloride is oxidized at the anode because it requires the *smallest* positive potential.

Turning our attention to the cathode, again we see two possible reactions:

$$Cu^{2+} + 2e^- \rightleftharpoons Cu(s)$$
$$2H^+ + 2e^- \rightleftharpoons H_2(g)$$

The electrode potentials can be calculated from the respective Nernst equations:

$$E_{Cu} = 0.34 - \frac{0.0592}{2} \log \frac{1}{0.10} = 0.31 \text{ V}$$

$$E_H = 0.00 - \frac{0.0592}{2} \log \frac{1.0}{(1.0)^2} = 0.00 \text{ V}$$

The overpotential for the deposition of copper on platinum is essentially zero, but for the reduction of hydrogen it is typically 0.4 V or more. Consequently, the cathode reaction is

$$Cu^{2+} + 2e^- \rightleftharpoons Cu(s)$$

because it will take place at a more positive (less negative) applied potential than the reduction of H^+:

$$E_{appl, Cu} = 0.31 - 0.0 = 0.31 \text{ V}$$

$$E_{appl, H} = 0.00 - 0.4 = -0.4 \text{ V}$$

The necessity of relying on rather crude estimates of the electrode overpotentials, and the possibility that we may not know the initial concentrations, limit the reliability of our conclusions regarding the electrode reactions. Furthermore, the concentration of the reactants and products, and therefore the galvanic voltage, changes as the electrolysis proceeds. Thus it is possible for the reaction to change during the electrolysis. This situation is discussed in the following section.

Current–Voltage Behavior During Electrolysis

Electrolytic procedures can be divided into two general types: those in which the electrode potentials are controlled and those in which the cell current is controlled. An understanding of the relationship between potential and current is fundamental to the study of any of the electrolysis techniques.

Changes in current at constant applied voltage

In an electrolysis cell, the number of electrons exchanged at the cathode and anode must be equal and the cell current is therefore the same as the current at either electrode. This is true despite the fact that one electrode may have a greater current-passing capability than the other. The cell current is always limited by the electrode that can pass the smallest current. Suppose that we place two identical electrodes in a sulfuric acid solution containing 0.1 M $CuSO_4$ and apply a constant voltage sufficient to cause the following electrode reactions:

Cathode: $Cu^{2+} + 2e^- \rightleftharpoons Cu(s)$
Anode: $2H_2O \rightleftharpoons O_2(g) + 4H^+ + 4e^-$

The current decreases with time as shown in Figure 14-2. The initial current is high because the concentration of electroactive species at each electrode surface is large. Once

Principles of Electrolysis

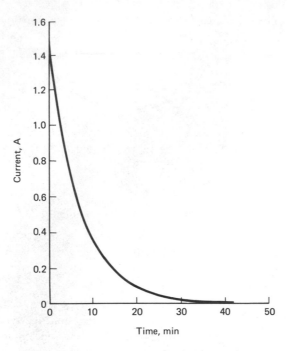

Figure 14-2 Change in current with time during electrolysis.

the electrolysis begins, these species are partially depleted at the electrode surfaces and the electrodes become polarized. At this point, the current becomes controlled by the rate of transport of electroactive species to the electrode surface, and at the cathode, this rate decreases rapidly as the concentration of Cu^{2+} decreases. The anode is not current limiting because the concentration of water does not change appreciably during the electrolysis. If the applied voltage is controlled so that only Cu^{2+} can be reduced at the cathode, a quantitative electrolysis may take a very long time.

Changes in voltage at constant current

The current and voltage of an electrolysis cell cannot be kept constant simultaneously. Consider the copper sulfate electrolysis just described. The current tends to drop as the concentration of Cu^{2+} decreases during the electrolysis. To compensate, the applied voltage must get larger in order to increase the rate of transfer of Cu^{2+} to the electrode surface. Eventually, the applied voltage becomes large enough to permit other electroactive substances to reduce at the cathode, and the cell current becomes a function of the concentration of two or more electroactive substances. Although this may shorten the time necessary to reduce all of the Cu^{2+}, it also leads to a lack of specificity in terms of the electrode reaction.

Chemists have developed electrochemical techniques that rely on both constant-current and constant-voltage electrolysis. In addition, there are numerous techniques that employ a controlled, but not constant, potential or current. A few of the more simple electrochemical techniques are discussed in the following sections of this chapter.

ELECTRODEPOSITION

As the name implies, electrodeposition refers to the process of electrolytically depositing a substance on an electrode. It is used mainly in conjunction with gravimetric procedures and as a means for separating metals in solution.

Electrogravimetry

In this technique the analyte is deposited as a solid on a preweighed electrode. The weight of the analyte is determined by reweighing the electrode when the deposition is complete. Selectivity is achieved by using an applied voltage that allows only the analyte to deposit on the electrode.

The apparatus required to perform electrogravimetry is shown in Figure 14-3. The power supply can be any stable voltage source with a relatively high current rating that is capable of supplying up to about 10 V dc. The electrodes are usually made of platinum, silver, or copper. Despite its great expense, platinum is preferred because it is very unreactive and can be ignited (heated to a very high temperature) to remove volatile organic impurities such as grease that may affect the adhesion of a deposited metal.

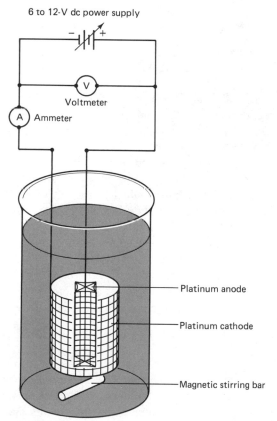

Figure 14-3 Apparatus for electrogravimetry.

Electrolysis devices that use two polarizable electrodes are simple but may require fairly frequent adjustment of the applied voltage. Although it is not obvious, the decrease in current that occurs during electrolysis is accompanied by a change in the cathode potential to more negative values. Such a change may be harmful by enabling undesirable electrode reactions to occur. The basis for the potential change lies in Equation 14-1, which states that if the applied voltage is constant, a decrease in current (or ohmic voltage) and cathode overpotential must be accompanied by a corresponding increase in the galvanic voltage. Most, if not all, of the increase must be borne by the cathode, since little polarization occurs at the anode. To avoid this negative shift in cathode potential, the operator must periodically decrease the applied voltage as the current decreases. There are instruments available called *potentiostats*, which are capable of continuously adjusting the applied voltage in order to maintain a constant cathode or anode potential, and these are discussed in the next section.

One particularly undesirable feature of allowing the cathode potential to become more negative during an electrolysis is the eventual reduction of hydrogen ions. The coformation of hydrogen gas at the electrode often prevents the analyte from adhering properly. One simple way of avoiding the evolution of hydrogen gas is to add a substance that reduces at a potential less negative than hydrogen ion. Such a substance is called a *cathode depolarizer*. Nitrate ion is commonly used for this purpose, being reduced to ammonium ion according to the reaction

$$NO_3^- + 10H^+ + 8e^- \longrightarrow NH_4^+ + 3H_2O$$

Codeposition of other metals can sometimes be avoided by adding a selective complexing agent to stabilize the diverse metal ion and shift its reduction potential to a more negative value.

Electrolytically deposited metals should be dense, smooth, and strongly adherent to the electrode, so that rinsing, drying, and weighing can be accomplished without loss or reaction of the metal. Metallic deposits that appear powdery or flaky are likely to be less pure and less adherent than those with a smooth surface and bright metallic luster. Factors that affect the physical properties of the deposit include current density, complexing agents, temperature, and stirring. The best deposits are obtained when the current density is small (less that about 0.1 A/cm^2), which is the main reason that large electrodes are preferred. Although we are not entirely sure why, many metals form better deposits when electrolyzed from solutions in which their ions are complexed. Ammonia and cyanide are often used for this purpose, as they are excellent complexing agents for a variety of metals and their solutions are basic, which helps avoid codeposition of hydrogen gas. The effect of temperature on the quality of the deposit is variable and must be determined experimentally. In any event, small variations in temperature seldom have a noticeable effect. Stirring almost always improves the quality of the deposit.

Separation at a Mercury Cathode

Mercury is not used as an electrode in electrogravimetry because it is volatile, toxic, and hard to handle. It is, however, an excellent electrode for *separating* reducible metals from solution. In many cases the metal forms an amalgam with mercury and the reduction may be written

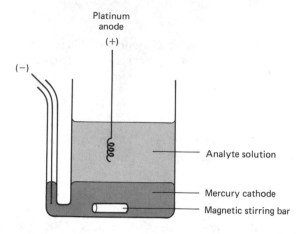

Platinum
anode

(+)

(−)

Analyte solution

Mercury cathode

Magnetic stirring bar

Figure 14-4 Mercury cathode electrolysis cell.

$$M^{n+} + Hg(l) + ne^- \longrightarrow M(Hg)(l) \qquad (14\text{-}3)$$

Selective reduction is obtained by controlling the potential of the mercury electrode.

Since many metal ions are not soluble in neutral or alkaline solution, Reaction 14-3 is normally carried out in acidic solution. Fortunately, the reduction potential for H^+ is shifted in the negative direction due to a large overpotential with the mercury electrode. This, coupled with the fact that the reduction potentials of many metal ions are shifted in the positive direction because their reduced forms are stabilized as mercury amalgams, decreases the likelihood of interfering hydrogen-ion reduction.

A simple mercury-cathode electrolysis cell is shown in Figure 14-4. The large surface area of the mercury electrode permits fairly large currents to be used, which, in turn, shortens the electrolysis times. Furthermore, the electrode surface can be refreshed constantly by stirring. Most of the first- and second-series transition metals, together with a few of the metaloids, can be removed from solution quantitatively at a mercury cathode.

COULOMETRY

Coulometry is a general term used to described various techniques for measuring the *amount* of electricity required to react, directly or indirectly, with the analyte. This amount of electricity can be related directly to the weight of analyte present using known physical constants. Thus coulometry, like gravimetry, has the advantage of ordinarily not requiring a calibration or standardization step. Coulometric methods rival gravimetric and volumetric methods in terms of accuracy and precision and usually are faster and more easily automated.

Coulometry is performed either at constant electrode potential or constant cell current. At constant potential it is similar to electrogravimetry but more versatile because the electrode reaction does not have to produce a weighable substance. This makes the method applicable to a much larger number of electrochemical reactions. In addition, much smaller quantities of electricity than those required to yield "weighable" analytes can be measured.

The Faraday Constant

The amount of electricity used in a process is commonly represented in units of *coulombs* (C), where one coulomb is the quantity of electricity transported in 1 second by a constant current of 1 ampere. If a constant current of I amperes flows for t seconds, the number of coulombs Q is given by the expression

$$Q = It \qquad (14\text{-}4)$$

If the current is not constant with time, the quantity of electricity is more difficult to determine, requiring integration of the current with respect to time:

$$Q = \int_0^t I(t)\, dt \qquad (14\text{-}5)$$

The quantity of electricity that contains 1 mole of electrons is 96,485 coulombs or 1 *Faraday*. Since 1 mole of electrons will result in one equivalent of a substance being oxidized or reduced at an electrode, the Faraday has units of coulombs per equivalent and is the proportionality constant relating the quantity of electricity passed with the amount of substance oxidized or reduced at an electrode.

Example 14-1

A solution containing 75.0 mg of copper was electrolyzed at a constant current of 0.250 A, causing metallic copper to deposit on a platinum cathode. What was the percentage of copper remaining in the solution after 15.0 min?

Solution   The quantity of electricity is determined using Equation 14-4

$$Q = 0.250 \text{ A} \times 15.0 \text{ min} \times 60 \text{ s/min} = 225 \text{ C}$$

and is converted to equivalents of copper using the Faraday constant

$$\text{amount Cu} = \frac{225 \text{ C}}{96{,}485 \text{ C/eq}} = 2.33 \times 10^{-3} \text{ eq}$$

Since the electrode reaction is

$$Cu^{2+} + 2e^- \longrightarrow Cu(s)$$

The weight of copper deposited is

$$\text{wt Cu} = 2.33 \times 10^{-3} \text{ eq} \times 63.5 \frac{\text{g Cu}}{\text{mol Cu}} \times \frac{1 \text{ mol Cu}}{2 \text{ eq Cu}} = 0.0740 \text{ g}$$

Then

$$\text{wt Cu remaining} = 75.0 \text{ mg} - 74.0 \text{ mg} = 1.0 \text{ mg}$$

and

$$\% \text{ Cu remaining} = \frac{1.0 \text{ mg}}{75.0 \text{ mg}} \times 100 = 1.3$$

Constant-Potential Coulometry

In this technique the potential of the working electrode is held constant at a value that enables only the analyte to react. We saw earlier that as an electrolysis proceeds, changes in the concentration polarization and overpotential can cause an electrode potential to change even though the applied voltage is held constant. A three-electrode cell, shown in Figure 14-5, is used to overcome this problem. An electronic device called a potentiostat controls the potential of the working electrode by comparing it to the constant potential of a saturated calomel reference electrode. To accommodate the changes in cell current that must occur during electrolysis, a counter or auxiliary electrode is introduced into the circuit. This electrode is polarizable and free to adopt any potential, relative to the working electrode, necessary to maintain the appropriate cell current. To avoid any possible interference by the product of the electrode reaction, the counter electrode usually is immersed in a separate electrolyte solution that is connected to the analyte solution via a salt bridge.

In addition to being able to hold the potential of the working electrode to a set value, most good potentiostats contain an electronic current integrator to measure the amount of electricity that passes through the cell during the electrolysis. If such an integrator is unavailable, the chemist can construct a *chemical* current integrator (or

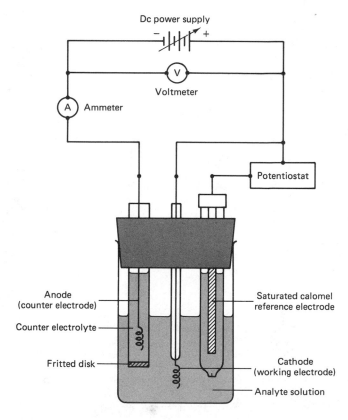

Figure 14-5 Constant-potential coulometer using a three-electrode cell.

chemical coulometer) and place it in series with the cell containing the analyte. Such an integrator is shown in Figure 14-6. As current flows through the circuit, water is electrolyzed at each electrode:

Cathode: $\quad H_2O + e^- \quad \longrightarrow \quad \frac{1}{2}H_2(g) + OH^-$

Anode: $\quad \frac{1}{2}H_2O \quad \longrightarrow \quad \frac{1}{4}O_2(g) + H^+ + e^-$

For every mole of electrons flowing through the cell, three-fourths of a mole of gas ($\frac{1}{2}$ mole of H_2 plus $\frac{1}{4}$ mole of O_2) are formed and collected by water displacement in an inverted closed buret. Using the ideal gas law, the measured volume of gas collected can be related to the number of moles of electrons passed, which, in turn, can be converted to coulombs using the Faraday constant. Another chemical current integrator uses silver or platinum electrodes in a silver nitrate solution. Current through the cell causes the deposition of silver onto the preweighed cathode. The increase in weight is related to the quantity of electricity passed.

The specificity of the electrode reaction is much greater with the three-electrode cell than with the two-electrode cell used in constant applied-voltage electrolysis. We

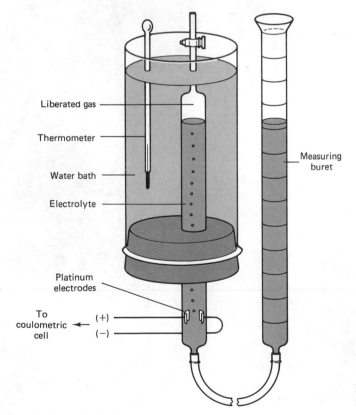

Liberated gas

Thermometer

Water bath

Electrolyte

Platinum electrodes

To coulometric cell

(+)

(−)

Measuring buret

Figure 14-6 Hydrogen-oxygen gas coulometric current integrator.

pay for this advantage with increased electrolysis time. At constant electrode potential there is an inverse logarithmic relationship between the current and the concentration of analyte. If the electrolysis is to be "quantitative," the current must go to essentially zero, which may take a very long time.

Applications

Constant-potential coulometry has been used for the determination of more than two dozen metal ions with mercury as the preferred electrode. Ions such as Ag^+, Cu^{2+}, Cd^{2+}, Bi^{3+}, and Pb^{2+} are reduced to their elemental state. Other metal ions, including those of Ti, V, and Mo, have been determined by reducing them to a soluble, lower oxidation state. Iron(II) and arsenic(III) have been determined by oxidation at a platinum anode.

Iodide, bromide, and chloride can be determined indirectly with a silver anode, depositing them as insoluble silver salts. Here the electrons are used to oxidize silver, which then reacts with the halide.

Constant-Current Coulometry (Coulometric Titrations)

Constant-current coulometry is a titration technique in which the titrant is generated, usually in situ, via an electrochemical reaction. For example, Fe^{2+} can be titrated with Ce^{4+} generated electrochemically from Ce^{3+} at a platinum anode.

$$\text{Electrode reaction:} \quad Ce^{3+} \rightleftharpoons Ce^{4+} + e^-$$
$$\text{Titration reaction:} \quad Fe^{2+} + Ce^{4+} \rightleftharpoons Fe^{3+} + Ce^{3+}$$

Concentration polarization at the anode is kept small and relatively constant by using a large excess of Ce^{3+} in the solution. As a result, the anode potential will not change much from its initial value and does not need to be controlled externally. This allows the cell to be operated at constant current, and the quantity of titrant generated to be determined simply by measuring the time of generation.

In some titrations, part of the analyte may react directly at the electrode rather than with the titrant. For example, in the titration of Fe^{2+}, the *initial* reaction at the anode is the oxidation of Fe^{2+} to Fe^{3+} because it occurs at a less positive potential than the oxidation of Ce^{3+}. As the titration proceeds, the concentration of Fe^{2+} decreases and the potential of the anode becomes more positive until it reaches a value where Ce^{3+} is oxidized rapidly. In the end, it does not matter whether the analyte reacts only with the titrant or with both titrant and the electrode, since the same amount of electricity (same number of electrons) must pass through the cell in either case.

Like its cousin the volumetric titration, the coulometric titration requires a reaction that is both fast and quantitative in addition to some means for detecting the equivalence point. Most of the end points used in volumetric titrations are equally applicable to coulometric titrations.

Most commercial coulometric titrators consist of a constant-current source and an associated electronic timer with a digital readout. The device may be capable of delivering a range of currents to accommodate the titration of both large and small amounts

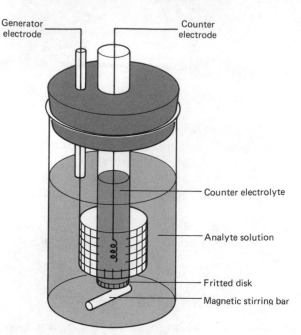

Generator
electrode

Counter
electrode

Counter electrolyte

Analyte solution

Fritted disk

Magnetic stirring bar

Figure 14-7 Coulometric titration
apparatus.

of analyte. Often the currents are selected such that the readout is directly in micro-
equivalents of electrons (or analyte). The timer is always switched on and off synchro-
nously with the current source.

A typical coulometric titration cell is shown in Figure 14-7. The electrode at which
the titrant is formed is called a *generator electrode*. Its surface area is large in order to
ensure that a sufficient amount of reactant can reach the electrode fast enough to avoid
unwanted polarization. The counter electrode is often isolated from the analyte solution
to avoid possible interferences from the products of the electrode reaction. Remember,
if something is oxidized at the generator electrode, something else is reduced at the
counter electrode and if these two electrode products mix and react with each other, the
titration stoichiometry will be destroyed.

Example 14-2

A 25.0-mL aliquot of a solution containing phenol was treated with 5 mL of 1 M KBr and
10 mL of 1.0 M HC1. A platinum generator electrode was made the anode, where Br^- was
oxidized to Br_2. A constant current of 8.00 mA was passed for 2 min 38 s to reach the end
point of the titration:

$$\text{phenol} + 3Br_2 \longrightarrow \text{tribromophenol} + 3H^+ + 3Br^-$$

Calculate the molar concentration of phenol in the original sample.

Solution As in any direct titration, the amount of analyte can be found from the amount of titrant used. In this case, the amount of Br_2 (titrant) produced was

$$\text{amount } Br_2 = \frac{\text{coulombs of electricity}}{\text{Faraday}} = \frac{(8.00 \times 10^{-3} \text{ A})(158 \text{ s})}{96,485 \text{ C/eq}}$$

$$= 1.31 \times 10^{-5} \text{ eq}$$

Since it takes 2 electrons to produce 1 Br_2:

$$\frac{1.31 \times 10^{-5} \text{ eq } Br_2}{2 \text{ eq/mol}} = 6.55 \times 10^{-6} \text{ mol } Br_2$$

From the titration reaction stoichiometry,

$$6.55 \times 10^{-6} \text{ mol } Br_2 \times \frac{1 \text{ phenol}}{3 \text{ } Br_2} = 2.18 \times 10^{-6} \text{ mol phenol}$$

Finally,

$$C_{\text{phenol}} = \frac{2.18 \times 10^{-6} \text{ mol phenol}}{0.0250 \text{ L solution}} = 8.73 \times 10^{-5} \text{ } M$$

Applications

The source of one of the major limitations of coulometric titrimetry is also the source of one of its important advantages: the electrolytic generation of the titrant. The limitation lies in the fact that only titrants capable of being generated with 100% current efficiency can be used. Relatively few of the many volumetric titrants meet this requirement (see Table 14-2). Important volumetric titrants such as MnO_4^-, $Cr_2O_7^{2-}$, and $S_2O_3^{2-}$ are not on the list because they cannot be generated with the required 100%

TABLE 14-2 SELECTED APPLICATIONS OF COULOMETRIC TITRATIONS

Reagent	Generator electrode reaction	Substance determined
H^+	$2H_2O \rightleftharpoons 4H^+ + O_2(g) + 4e^-$	Bases
OH^-	$2H_2O + 2e^- \rightleftharpoons 2OH^- + H_2(g)$	Acids
Ag^+	$Ag \rightleftharpoons Ag^+ + e^-$	Halides, CN^-, SCN^-, S^{2-}, mercaptans
HY^{3-}	$HgNH_3Y^{2-} + NH_4^+ + 2e^- \rightleftharpoons Hg(l) + 2NH_3 + HY^{3-}$	Ca^{2+}, Cu^{2+}, Zn^{2+}, Pb^{2+}
Br_2	$2Br^- \rightleftharpoons Br_2 + 2e^-$	As(III), Sb(III), U^{4+}, I^-, SCN^-, NH_3, NH_2OH, N_2H_4, phenols, anilines, 8-hydroxyquinolines, alkenes, mercaptans
I_2	$2I^- \rightleftharpoons I_2 + 2e^-$	As(III), Sb(III), H_2S, $S_2O_3^{2-}$
Ce^{4+}	$Ce^{3+} \rightleftharpoons Ce^{4+} + e^-$	Fe^{2+}, Ti^{3+}, U^{4+}, I^-, As(III)
Fe^{2+}	$Fe^{3+} + e^- \rightleftharpoons Fe^{2+}$	Ce^{4+}, MnO_4^-, $Cr_2O_7^{2-}$, VO_3^-
Ti^{3+}	$TiO^{2+} + 2H^+ + e^- \rightleftharpoons Ti^{3+} + H_2O$	Ce^{4+}, VO_3^-, UO_2^{2+}, Fe^{3+}
$CuCl_2^-$	$Cu^{2+} + 2Cl^- + e^- \rightleftharpoons CuCl_2^-$	Br_2, $Cr_2O_7^{2-}$, VO_3^-, IO_3^-

current efficiency. A significant advantage is derived from the fact that the titrant is consumed very shortly after it is produced, thereby eliminating the need for long-term stability. Thus bromine and chlorine, which are too volatile for use as volumetric titrants, are commonly used in coulometry as oxidants and halogenating agents for organic compounds. Similarly, unstable reducing agents, such as titanium(III) and copper(I), have been used to titrate a variety of oxidants.

The very versatile EDTA has been generated electrochemically via the reduction of a mercury–ammonia–EDTA complex at a mercury cathode:

$$HgNH_3Y^{2-} + NH_4^+ + 2e^- \longrightarrow Hg(l) + 2NH_3 + HY^{3-}$$

Once released, the EDTA may react with any of several metal ions such as Ca^{2+}, Cu^{2+}, Zn^{2+}, and Pb^{2+}. The mercury complex is more stable than most metal chelates and will not react directly with the metal ions.

Back titrations may be employed when direct titrations are not feasible. For example, aniline reacts too slowly with bromine to permit a good estimate of the equivalence point in a direct titration. In the back-titration technique, the sample is mixed with an excess of KBr and $CuSO_4$ in hydrochloric acid solution, and a measured excess of bromine is electrogenerated at the anode.

Electrode reaction: $2Br^- \rightleftharpoons Br_2 + 2e^-$

Analyte reaction:

After waiting a few minutes for the bromination reaction to reach completion, the electrode polarity is reversed and the excess bromine is titrated with electrogenerated $CuCl_2^-$:

Electrode reaction: $Cu^2 + 2Cl^- + e^- \rightleftharpoons CuCl_2^-$

Back-titration reaction: $Br_2 + 2CuCl_2^- \rightleftharpoons 2Br^- + 2Cu^{2+} + 4Cl^-$

Occasionally, it is not feasible to generate the desired titrant directly in the sample solution. In such cases it may be possible to use an externally generated titrant. Coulometric acid-base titrations are the best examples of the use of the external generation technique. The generation apparatus is shown in Figure 14-8. The incoming reservoir solution is split and swept past platinum electrodes sealed in the opposing arms of a T-joint. The products of each electrode reaction are swept along by the continuous flow of the reservoir solution. When an aqueous sodium sulfate solution is used, the electrode reactions are:

Cathode: $2H_2O + 2e^- \rightleftharpoons H_2(g) + 2OH^-$

Anode: $2H_2O \rightleftharpoons 4H^+ + O_2(g) + 4e^-$

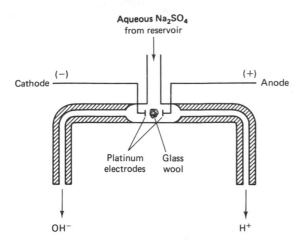

Aqueous Na$_2$SO$_4$
from reservoir

Cathode — (−)

(+) — Anode

Platinum electrodes   Glass wool

OH$^-$

H$^+$

Figure 14-8 Cell for external generation of acid and base.

Such a generator can be used to supply either H$^+$ for titrating bases or OH$^-$ for titrating acids, simply by selecting the appropriate delivery arm.

POLAROGRAPHY

Polarography is based on measuring the current of an electrolysis cell in which the potential of a working electrode is varied continuously. This technique is distinguished from ordinary electrolysis in that only a tiny, insignificant portion of the analyte is electrolyzed. Qualitative information is obtained by noting the potential at which oxidation or reduction occurs and quantitative information is contained in the magnitude of the cell current.

Apparatus

A basic polarograph is diagrammed in Figure 14-9. The three-electrode cell consists of a dropping mercury working electrode, a saturated calomel reference electrode, and a platinum counter electrode. The *dropping mercury electrode* or *dme* is created by forcing mercury through a glass capillary tube 5- to 10-cm long and about 0.05 mm in inside diameter. When the column of mercury above the tube is about 50 cm or more, a steady flow of identical mercury droplets is obtained. The characteristics of this electrode are discussed in some detail in a later section.

The potential of the dme is controlled relative to the saturated calomel electrode by a power source capable of varying the voltage continuously over a range of about ±3.0 V. Current flowing as a result of the electrolysis at the dme is passed through the counter electrode, which may or may not be physically (but not electrically) isolated from the analyte solution. This current is usually measured by converting it to a voltage that, in turn, drives the pen on a strip-chart recorder. In this way, a continuous recording of current as a function of applied voltage or electrode potential is obtained. Such a recording is called a *polarogram*.

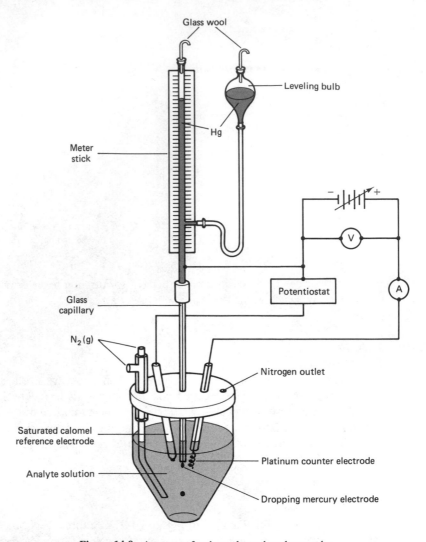

Figure 14-9 Apparatus for three-electrode polarography.

Polarograms

In recording polarograms, the potential of the dme normally is made increasingly negative relative to the reference electrode, and current flowing due to a cathodic (reduction) process is given a positive value. Polarograms of 1 M HCl alone and with 5×10^{-4} M $CdCl_2$ are shown in Figure 14-10. In the polarogram of HCl alone, the sharp increase in current at about -1.2 V is due to the onset of hydrogen-ion reduction at the cathode. This rise would occur at a much less negative potential were it not for the large overpotential involved. The small, slightly increasing current flowing even in the absence of any electroactive analyte is called the *residual current*. A separation of charge takes place at the boundary between the mercury drop and the solution, making the interface behave like a capacitor. Current is required to charge this capacitor, and because the

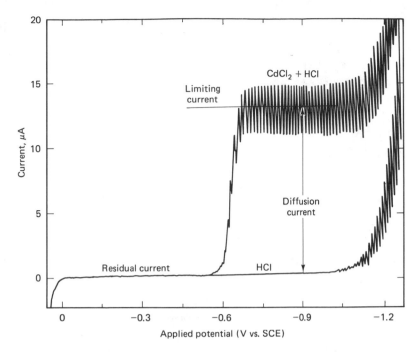

Figure 14-10 Polarograms of 1.0 *M* HCl alone and with 5.0×10^{-4} *M* CdCl$_2$. [From D. T. Sawyer and J. L. Roberts, Jr., *Experimental Electrochemistry for Chemists* (New York: John Wiley & Sons, Inc., 1974) p. 334. Copyright © 1974 by John Wiley & Sons, Inc. Reprinted by permission of John Wiley & Sons, Inc.]

electrode surface repeatedly grows, then suddenly falls to zero when the drop falls, the current fluctuates in the same manner. As the potential on the electrode increases, the current required to charge the capacitor also increases. In essence, the residual current is considered a baseline current. The steep increase in the current at about -0.6 V in the polarogram of CdCl$_2$ is due to the reduction of Cd^{2+}:

$$Cd^{2+} + Hg(l) + 2e^- \rightleftharpoons Cd(Hg)(l)$$

As the electrode is made more negative, cadmium ions are reduced as fast as they reach the surface and the electrode becomes completely polarized. At this point (about -0.7 V in Figure 14-10), the current becomes independent of the applied potential. This constant current is called the *limiting current*. The sharp increase from the residual current to the limiting current is called a *polarographic wave*. Eventually, the electrode becomes sufficiently negative to reduce H$^+$ and the current rises again.

The rate at which electroactive ions reach the electrode is controlled by three processes: convection (mechanical and thermal), electrostatic attraction (migration), and diffusion. The success of the polarographic experiment depends on our ability to make diffusion solely responsible for the transport of electroactive material to the electrode. Convection is minimized by not stirring and avoiding any temperature differentials within the solution. Electrostatic attraction is effectively eliminated by addition of a large amount of an inert electrolyte, usually called a *supporting electrolyte*. The ions of the supporting electrolyte surround the charged electrode and effectively neutralize its ability to project an electric field very far into the solution, where it may attract an oppositely charged analyte. Under these conditions, the difference between the limiting and residual currents is the *diffusion current*, I_d. The rate of diffusion is a physical property that is directly proportional to concentration, and therefore the diffusion current is of considerable importance in a quantitative determination.

Half-wave potential

The two most common types of reactions at a dropping mercury electrode are

$$M^{n+} + Hg(l) + ne^- \rightleftharpoons M(Hg)(l)$$

$$M^{z+} + ne^- \rightleftharpoons M^{(z-n)+}$$

If these half-reactions are reversible, the following relationship (called the Heyrovsky equation) can be derived:

$$E = E_{1/2} - \frac{0.0592}{n} \log \frac{I}{I_d - I} \tag{14-6}$$

where $E_{1/2}$, called the *half-wave potential*, is the potential at which the current I equals $I_d/2$. The half-wave potential is independent of concentration but characteristic of the substance responsible for the wave. The saturated calomel reference electrode is used so commonly in polarography that chemists usually report $E_{1/2}$ values relative to that electrode rather than to the standard hydrogen electrode. Equation 14-6 often is used as a test of reversibility of the electrode reaction. According to the equation, a plot of E versus $\log I/(I_d - I)$ will be a straight line having a slope of $-0.0592/n$ if the reaction responsible for the wave is reversible.

If Equation 14-6 had been derived rather than just stated, the relation between the half-wave potential and the standard electrode potential would have been seen immediately. For an $E_{1/2}$ relative to the standard hydrogen electrode, the relationship is given by

$$E_{1/2} = E^\circ - \frac{0.0592}{n} \log \frac{\sqrt{D_{ox}}}{\sqrt{D_{red}}} \tag{14-7}$$

where D_{ox} and D_{red} are the diffusion coefficients for the oxidized and reduced forms of the analyte. In many cases the diffusion coefficients are so similar that the log term is essentially zero and $E_{1/2}$ equals E°.

It is necessary to remember that Equation 14-6 is valid only for reactions that are *reversible* at the polarographic electrode. The behavior of polarographic waves for irreversible reactions is quite different, and often the half-wave potential is concentration dependent.

Effect of complex formation

The presence of a complexing agent has little effect on the shape of a polarographic wave; that is, Equation 14-6 is still valid, although the value of I_d for a given concentration of metal may be slightly different. What does change is the half-wave potential. Addition of a complexing agent can change the relative stability of the two oxidation states of a metal. If the oxidized form is preferentially stabilized, the reduction is more difficult and will occur at a more negative potential. If the reduced form is preferentially stabilized, the opposite effect is observed. For a general reduction,

$$ML_p + ne^- \rightleftharpoons ML_q + (p - q)L$$

where M is the metal ion and L is the ligand (charges have been omitted for simplicity), the half-wave potential is given by the expression

$$E_{1/2} = E^\circ - \frac{0.0592}{n} \log \frac{K_p}{K_q} - \frac{0.0592(p - q)}{n} \log [L] \qquad (14\text{-}8)$$

where K_p and K_q are the formation constants for the complexes, ML_p and ML_q. A plot of $E_{1/2}$ versus $\log [L]$ will be a straight line with the following characteristics:

$$\text{slope} = -\frac{0.0592(p - q)}{n}$$

$$\text{intercept} = E^\circ - \frac{0.0592}{n} \log \frac{K_p}{K_q}$$

If only one oxidation state forms a complex, Equation 14-8 will simplify accordingly. Equation 14-8 is also valid if the oxidized and reduced forms of the analyte are weak acids. One needs only to consider that L represents H^+. In such a case, K_p and K_q are merely the reciprocals of the acid dissociation constants.

Properties of the Dropping Mercury Electrode

When polarograms are obtained using a dropping mercury electrode, the current fluctuations are so large and fast that a well-damped detection system that records only a portion of the total current oscillation is needed. The integrated current is usually taken as the average of the recorded drop oscillation, as shown by the solid line in Figure 14-10. It should be apparent that the average current will depend not only on the rate at which electroactive substance moves toward the electrode but also on the rate at which the electrode expands into the solution. The effect of various factors on the average diffusion current is given by the Ilkovic equation:

$$I_d = 607nD^{1/2}m^{2/3}t^{1/6}C \qquad (14\text{-}9)$$

where m is the rate of mercury flow in mg/s, t is the drop time in seconds, and C is the concentration of electroactive substance in mol/L. The rate of mercury flow and the drop time are determined almost entirely by the dimensions of the capillary and the height of the mercury column above it. These values remain essentially constant throughout the recording of a polarogram. Perhaps the most notable feature of Equation 14-9 is the direct proportionality between the diffusion current and the concentration of the electroactive substance.

The advantages of the dropping mercury electrode are considerable. It has a large overpotential for the reduction of H^+, which permits examination of the reduction of many substances in acidic solution without interference from H^+. Equally important, a fresh electrode surface is produced with each new drop. Without this, the electrode surface can become contaminated with the product of the electrode reaction, thereby changing the characteristics of the electrode. Since the behavior of the electrode (current drop) is independent of its history (previous drop), very reproducible current–voltage curves are obtained.

The one significant disadvantage of the mercury electrode is the ease with which it is oxidized (about $+0.4\ V$ versus SCE). The large anodic current from this oxidation will mask the current from any other oxidizable substance. Less troublesome, but still important, is the extra care that must be taken in handling mercury due to its toxicity. It has an appreciable vapor pressure and must be covered with liquid or contained at all times. Great care should be taken to avoid spills because cleanup can be very difficult.

Current maxima

Polarographic waves are sometimes distorted from their predicted shape. The distortion usually takes the form of a peak in the current at the top of the polarographic wave, as shown in Figure 14-11. Such distortions are referred to as current maxima. Chemists have not unequivocally identified all the reasons why such maxima occur, but they have learned that the maxima are diminished in size or eliminated by the addition of very small concentrations of certain surface active agents such as gelatin, methyl red, and Triton X-100.

Dissolved oxygen

Oxygen is electroactive at a mercury cathode giving rise to two waves. Both reactions responsible for the waves are irreversible:

$$O_2 + 2H^+ + 2e^- \longrightarrow H_2O_2 \qquad E_{1/2} \simeq -0.15\ \text{V (vs. SCE)}$$

$$H_2O_2 + 2H^+ + 2e^- \longrightarrow 2H_2O \qquad E_{1/2} \simeq -0.9\ \text{V (vs. SCE)}$$

In certain instances, these waves may be used for the determination of dissolved oxygen. At other times, they may interfere with the accurate measurement of other polarographic waves. Since it is very difficult to keep oxygen away from solutions while they are being

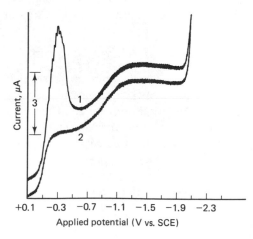

Figure 14-11 Polarogram of oxygen: Curve 1, 0.05 M KCl saturated with air; Curve 2, after addition of a trace of methyl red. [From J. J. Lingane, *Electroanalytical Chemistry*, 2nd ed. (New York: John Wiley & Sons, Inc., 1958) p. 240. Copyright © 1958 by James Lingane. Reprinted by permission of John Wiley & Sons, Inc.]

handled, it is usually not removed until the sample has been placed in the polarographic cell. Bubbling an inert gas such as nitrogen or argon through the solution for 5 to 10 minutes will reduce the oxygen concentration below the normal detection limit.

Applications

Since the half-wave potential is characteristic of the substance being reduced or oxidized at a polarographic electrode, we might expect that it could be used for the purpose of identification. Unfortunately, this is seldom possible because $E_{1/2}$ also depends on the composition of the solution. Ordinarily, the composition of the sample is not sufficiently well known to permit the preparation of matching standards.

Quantitative determinations can be carried out using Equation 14-9. The most common technique is to prepare a series of standard solutions that covers the concentration range within which the unknown is expected to fall. The composition of these standards should match that of the unknown as closely as possible. A standard curve is prepared by measuring the diffusion current for each standard and plotting it versus the concentration, and the concentration of the unknown is read from this graph. When the sample composition is unknown or difficult to duplicate in the standard solutions, the standard addition technique may be used.

A very large number of both inorganic and organic compounds can be reduced or oxidized at a polarographic electrode. Simple metallic cations are reduced at the dropping mercury electrode to form the free metal or an ion of lower oxidation state. Most metal complexes are also reducible. The most important applications of polarography have been in the area of organic determinations. Some of the different functional groups that can be reduced or oxidized are listed in Table 14-3.

On the surface it would seem that polarography is an excellent technique for determining a wide variety of substances. In practice, this potential is not fully realized because many samples are too complex and the polarographic wave of the analyte cannot be adequately separated from waves of the other sample constituents. There are several more sophisticated electrochemical techniques with many important applications in quantitative analysis, including pulse polarography, ac polarography, cyclic voltam-

TABLE 14-3 POLAROGRAPHIC BEHAVIOR OF SELECTED ORGANIC FUNCTIONAL GROUPS

Group	Reaction		
C=C	$ArCH{=}CH_2 + 2H^+ + 2e^-$	$\longrightarrow$	$ArCH_2CH_3$
C≡C	$ArC{\equiv}CH + 2H^+ + 2e^-$	$\longrightarrow$	$ArCH{=}CH_2$
C=O	$ArCOR + 2H^+ + 2e^-$	$\longrightarrow$	$ArCHOHR$
C—X	$RCH_2Br + 2H^+ + 2e^-$	$\longrightarrow$	$RCH_3 + Br^-$
N=N	$RN{=}NR' + 2H^+ + 2e^-$	$\longrightarrow$	$RNH{-}NHR'$
N=O	$RN{=}O + 2H^+ + 2e^-$	$\longrightarrow$	$RNHOH$
NO_2	$RNO_2 + 2H^+ + 2e^-$	$\longrightarrow$	$RN{=}O + H_2O$
O—O	$RO{-}OR' + 2H^+ + 2e^-$	$\longrightarrow$	$ROH + R'OH$
S—S	$RS{-}SR' + 2H^+ + 2e^-$	$\longrightarrow$	$RSH + R'SH$
S=O	$R_2S{=}O + 2H^+ + 2e^-$	$\longrightarrow$	$R_2S + H_2O$

metry, and stripping analysis, that are founded, in large part, on the principles of polar-ography. A discussion of these techniques is usually reserved for a course in instrumental analysis.

PROBLEMS

14-1. Write the expected anode and cathode reaction for the electrolysis of a 0.1 M aqueous solution of each of the following substances in 0.1 M H^+. The following values are the applicable overpotentials:

$$Zn^{2+}, Ni^{2+}, Cd^{2+}, Cl^-, Br^-, I^- = 0\ V$$

$$H^+ = 0.4\ V;\ H_2O = 1.0\ V;\ SO_4{}^{2-} = 0.5\ V$$

(a) $ZnCl_2$ (c) HBr   (d) KF

(b) $NiSO_4$

14-2. The silver content of some coins was determined by dissolving a piece of one coin weighing 7.638 g in nitric acid and electrolytically depositing the silver on a weighed silver cathode. If the silver electrode weighed 28.471 g before and 28.893 g after electrodeposition, calculate the % Ag in the coin.

14-3. How long would it take to deposit 650 mg of silver on a metal cathode at a constant cell current of 0.850 A? Assume that the reduction of silver ion is the exclusive cathode reaction.

14-4. For a cell containing $Pb(NO_3)_2$ that is operated at a constant current of 1.00 A, calculate the time required to deposit 1.00 g of the following:

(a) Pb at the cathode   (b) PbO_2 at the anode

14-5. In the electrodeposition of nickel from 200 mL of 0.0240 M $Ni(CN)_4{}^{2-}$ at a constant current of 1.35 A, what percentage of nickel remains in solution after 10 min?

14-6. How long will it take to remove 95% of the copper from 250 mL of 1.40×10^{-2} M $CuSO_4$ at a constant current of 0.500 A?

14-7. The hydrazine in a 3.104-g sample was determined coulometrically through its oxidation at constant potential:

$$N_2H_4 \longrightarrow N_2(g) + 4H^+ + 4e^-$$

A silver coulometer in series with the analyte electrolysis cell was employed to determine the amount of electricity used in the oxidation. If the weight of the silver electrode increased 0.4146 g during the electrolysis, calculate the percentage of hydrazine in the sample.

14-8. Nitrite can be oxidized to nitrate at a platinum electrode in weakly acidic solution. A cell containing 0.3550 g of a soluble food preservative, placed in series with a hydrogen/oxygen gas coulometer, was electrolyzed at a potential conducive to the oxidation of nitrite. When the current became negligible, it was determined that the combined volume of hydrogen and oxygen gas produced by the coulometer was 26.40 mL at STP. Calculate the % $NO_2{}^-$ in the sample.

14-9. A 541.7-mg sample of German silver was dissolved in acid, diluted with water, and trans-ferred to an electrolysis cell equipped with an electronic coulometer. The solution was electrolyzed at -0.30 V versus SCE to reduce Cu^{2+} to Cu^0, which required 98.66 C of electricity. The potential of the working electrode was adjusted to -0.85 V versus SCE, where 24.07 C was required to reduce Ni^{2+} to Ni^0. Finally, 41.13 C was required to reduce Zn^{2+} to Zn^0 at -1.40 V versus SCE. Calculate the % Cu, % Ni, and % Zn in the sample.

14-10. A 300.4-mg sample of ore containing stibnite, Sb_2S_3, was ground and heated in a stream of oxygen for several hours during which the following reaction took place:

$$Sb_2S_3(s) + O_2(g) \longrightarrow Sb_4O_6(s) + SO_2(g)$$

The remaining oxide was dissolved ($Sb_4O_6 \rightarrow H_3SbO_3$), diluted with about 200 mL of aqueous 0.050 M KI, and titrated with iodine generated electrolytically at a constant current of 25.00 mA. If the end point was reached after 4 min and 38 s, calculate the % Sb_2S_3 in the ore.

14-11. A 5.00-mL aliquot of liquid bleach is diluted to 25.00 mL in a volumetric flask. A 10.00-mL aliquot of this solution was placed in a coulometric titration cell containing 75 mL of 0.1 M Fe^{3+}. The hypochlorite in the bleach was titrated with electrogenerated Fe^{2+}:

$$OCl^- + 2Fe^{2+} + 2H^+ \longrightarrow Cl^- + 2Fe^{3+} + H_2O$$

At a constant current of 20.00 mA, it took 244.3 s to reach the end point. Calculate the concentration of sodium hypochlorite as mg NaOCl/mL bleach.

14-12. A 5.00-mL aliquot of a petroleum feedstock (density = 0.830 g/mL) was shaken with 25 mL of 1 M aqueous NaOH. The aqueous phase was transferred to an electrolysis cell containing a silver anode and the extracted organosulfides (mercaptans) were titrated with electrogenerated silver ion at a constant current of 5.00 mA:

$$RSNa + Ag^+ \longrightarrow RSAg(s) + Na^+$$

If the titration required 108 s to reach the end point, calculate the % S in the feedstock.

14-13. A 10.00-mL sample containing aniline was transferred to a coulometric titration cell, treated with 10 mL of 5 M KBr, 10 mL of 0.1 M $CuCl_2$, and diluted to about 100 mL with 0.2 M H_2SO_4. Bromine was generated at a large platinum anode at a constant current of 15.00 mA for exactly 6.00 min:

The polarity of the platinum electrode was reversed and the excess bromine back titrated with electrogenerated $CuCl_2^-$:

$$Br_2 + CuCl_2^- \longrightarrow Br^- + CuCl_2 \quad \text{(not balanced)}$$

The back titration took 83.5 s at a constant current of 10.00 mA. Calculate the molar concentration of aniline in the sample.

14-14. A 408.1-mg sample containing aluminum was dissolved and treated with 25.00 mL of 0.02006 M 8-hydroxyquinoline. The precipitated AlQ_3 (Q = 8-hydroxyquinolate) was removed by filtration and discarded. The 8-hydroxyquinoline remaining in the solution was titrated with electrogenerated bromine at a constant current of 35.0 mA, requiring 131 s to reach the equivalence point:

$+ 2Br_2 \longrightarrow$ $+ 2HBr$

Calculate the percentage of Al_2O_3 in the sample.

14-15. A series of standard solutions of Pb^{2+} gave the following polarographic diffusion currents with a dropping mercury electrode:

$[Pb^{2+}]$ (M)	$I_d(\mu A)$
5.00×10^{-5}	6.8
1.00×10^{-4}	13.9
2.50×10^{-4}	35.3
5.00×10^{-4}	70.9
7.50×10^{-4}	108.4
1.00×10^{-3}	142.0

A solution of unknown lead concentration had a diffusion current of 23.6 μA. Prepare a standard calibration curve and determine the molar concentration of lead from the graph.

14-16. Gallium(III) forms a complex of uncertain stoichiometry with an organic ligand L. A series of solutions were prepared, all containing the same concentration of Ga(III) but with increasing concentrations of excess L. Polarograms of the solutions were obtained and the computed values for $E_{1/2}$ were plotted against log [L]. A straight line fit to the points had a slope of -78.7 mV. The reaction at the dropping mercury electrode was

$$GaL_x + 3e^- \longrightarrow Ga(l) + xL \quad \text{(charges omitted for clarity)}$$

Calculate the number of ligands, x, bound to each gallium ion.

PRINCIPLES
AND INSTRUMENTS
OF SPECTROPHOTOMETRY

Spectrometry is a general term describing various methodologies dealing with the production, use, and measurement of electromagnetic radiant energy or radiation. The range of spectrometric methods is immense and includes the use of radiation from virtually every part of the electromagnetic spectrum. This introductory discussion will be restricted to the simpler, more widely used methods. However, many of these principles and techniques are applicable to other more sophisticated spectrometric methods. This chapter is devoted to the technique of spectrophotometry and includes a discussion of the fundamental principles of absorption and emission of ultraviolet and visible radiation by molecules and how such radiation is produced and detected. The methodology and applications of molecular absorption spectrophotometry are discussed in Chapter 16.

ELECTROMAGNETIC RADIATION

Electromagnetic radiation is a type of energy that interacts with matter in a variety of ways and is therefore of great importance in chemistry. In order to understand the nature of these interactions, some familiarity with the properties of radiant energy is necessary. It is not easily characterized, having some properties consistent with wave theory but others that demand we treat it as a particle.

Wave Properties

Electromagnetic radiation is an oscillating electric force field transmitted through space in the form of a transverse wave. At right angles to the electric field is a similar, oscillating magnetic field. The discussion here is limited to the electric field of the wave

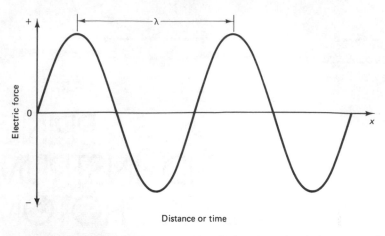

Figure 15-1 Plane-polarized electromagnetic radiation of wavelength, λ, propagated along the x-axis.

because it is responsible for phenomena that are of interest to chemists such as transmission, absorption, refraction, and reflection. Figure 15-1 is a graphic representation of the electric force field of a single radiant energy. The length of one wave, or the distance between waves, is called the *wavelength*, λ. The reciprocal of the wavelength, or the number of waves per unit length, is called the *wavenumber*, $\bar{v}$. It is possible also to characterize a wave in terms of time rather than length. The time required for one wave to pass a fixed point in space is called the *period*, p. Its reciprocal, the *frequency*, v, is the number of waves passing a fixed point per unit time. These four characteristics are summarized in Table 15-1. Wavelength and frequency are used most often by chemists in characterizing radiant energy. Although wavelength is commonly expressed with any convenient unit (e.g., nm, cm), wavenumber is always given in cm^{-1}.

A simple relationship exists between the length of a wave and the time required for it to move past a fixed point in space. This relationship says that the product of the wavelength and the frequency equals the *velocity of propagation* of the wave, v:

$$\lambda v = v$$

In a vacuum, the velocity of propagation or "speed of light," as it is sometimes called, is given the symbol c and has a value of 2.997958×10^8 or approximately 3.00×10^8 m/s.

$$(\lambda v)_{vac} = c \tag{15-1}$$

TABLE 15-1 SUMMARY OF WAYS TO CHARACTERIZE A WAVE

Name	Symbol	Definition
Wavelength	λ	Length of one wave
Wavenumber	$\bar{v}$	Number of waves/unit length
Period	p	Time for one wave to pass a fixed point in space
Frequency	v	Number of waves passing fixed point/unit time

The velocity in any other medium is always less because of interactions between the electric field of the radiation and bound electrons of the atoms or molecules comprising the matter. Frequency is determined *solely* by the source of the radiation and is not affected by the medium through which it travels; therefore, the wavelength must decrease along with the velocity of propagation as radiation passes from a vacuum to some other medium. The change is not great in passing from a vacuum to air because the velocity of propagation is only about 0.03% less in air than in a vacuum. It is appreciably different, however, in more dense media such as glass.

Radiant energy whose electric field oscillates in a single plane as shown in Figure 15-1 is said to be plane polarized. Most common sources produce unpolarized radiation in which the oscillations occur in all the planes perpendicular to the direction of propagation.

Radiant power and intensity

The energy of radiation striking a given area per unit time is called the *radiant power*, P, and is related to the square of the amplitude of the wave. Power is *not* directly related to wavelength. It is possible to have two beams of radiation each with the same wavelength but with different powers or amplitudes, as illustrated in Figure 15-2. *Intensity*, I, is the power per unit solid angle. Although it is not always correct to do so, power and intensity are often used interchangeably.

Interference

When two or more waves traverse the same space they interact with one another, a process scientists call interference. The result of the interaction may be represented as an addition of the electric field of the individual waves as shown in Figure 15-3. The combination of two identical waves that are completely in phase is called *constructive interference* and is represented as a new wave with the same wavelength or frequency but twice the amplitude of the combining waves. *Destructive interference* describes the combining of two identical waves that are completely out of phase. The result is no wave at all. Thus destructive interference leads to the canceling of radiation.

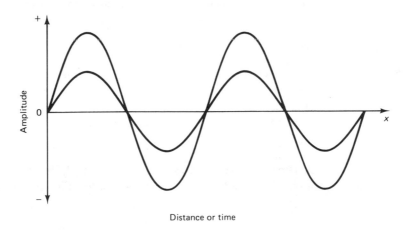

Figure 15-2 Radiation with identical wavelengths but different powers.

Electromagnetic Radiation

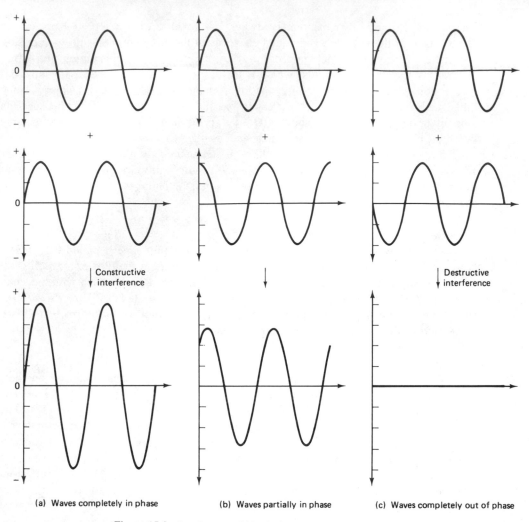

Figure 15-3   Interference of identical waves at different phases to each other.

(a) Waves completely in phase   (b) Waves partially in phase   (c) Waves completely out of phase

Particle Properties

Not all interactions between electromagnetic radiation and matter can be explained in terms of simple wave theory. An understanding of certain interactions requires radiation to be viewed as a particle or packet of energy called a *photon*. The energy of a photon is proportional to the frequency of the radiation and is given by

$$E = h\nu \qquad (15\text{-}2)$$

where h is Planck's constant and has a value of 6.63×10^{-34} J·s. Since frequency and wavelength are inversely related, photon energy is described in terms of wavelength by the equation

$$E = \frac{hc}{\lambda} \qquad (15\text{-}3)$$

In trying to rationalize simultaneous wave and particle characteristics, it is convenient to think of electromagnetic radiation as small packets of energy moving through space in a wave-like manner.

Radiant power or intensity should not be confused with photon energy. Power is a more inclusive unit, representing the energy of the individual photons and the number of those photons present in the beam. A beam containing relatively few photons of high energy may have as much power as another beam containing more photons but each with a lower energy.

The Electromagnetic Spectrum

Although the electromagnetic spectrum is continuous over all possible wavelengths, it is subdivided in name according to how the radiation interacts with matter. The portion of the electromagnetic spectrum of most interest to chemists is shown in Figure 15-4. The exact dividing line between regions is arbitrary, but chemists seldom argue about them because they are not important. The visible portion of the spectrum, those wavelengths to which the human eye produces a sensation we call color, is but a tiny part of

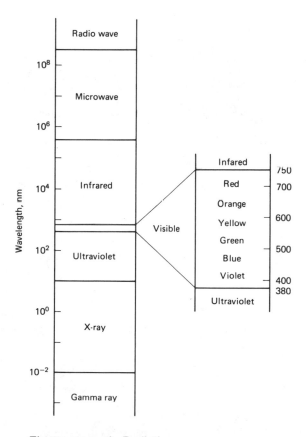

Figure 15-4 The electromagnetic spectrum.

the usable spectrum. Individuals are remarkably consistent in the maximum (750 nm) and minimum (380 nm) wavelengths that can be ''seen.''

This and subsequent chapters are concerned almost exclusively with electromagnetic radiation in the visible and ultraviolet portions of the spectrum.

REVIEW OF ATOMIC AND MOLECULAR ENERGY

You may recall that in addition to kinetic energy, atoms may have internal or potential energy if their electrons are not in the orbitals of lowest energy. Figure 15-5 illustrates how chemists represent the allowed potential energy levels for free, unbound atoms. The energy diagrams start with the orbital normally occupied by the valence electron(s) because it is generally this electron that moves from one orbital to another, giving rise to a new potential energy for the atom. There are three important observations that should be made in examining Figure 15-5: (1) atoms can exist in relatively few energy states,

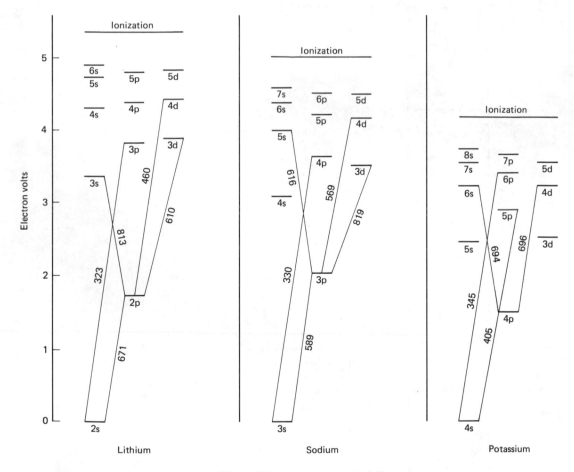

Figure 15-5 Atomic energy-level diagrams.

(2) the energy states are fairly different from one another, and (3) the energy states are characteristic of the element. The numbers on the lines connecting two energy states represent the wavelengths of radiation (in nanometers) corresponding to the energy difference between the two states.

The situation with molecules is considerably more complicated. Not only can bonding electrons move from one molecular orbital to another (called an electronic transition), but atoms and groups of atoms within molecules can undergo various types of vibrations and rotations, each requiring a discrete amount of energy to initiate or maintain. Thus each molecular energy state is comprised of an electronic, vibrational, and rotational component such that

$$E = E_{\text{electronic}} + E_{\text{vibrational}} + E_{\text{rotational}}$$

Although the specific energies of these three components vary with the level and the substance, they generally fall within the ranges given in Table 15-2. A simple molecular energy-level diagram with the vibrational and rotational components included is shown in Figure 15-6. The lowest-energy state (bottom line in Figure 15-6) is called the *ground state*. All higher-energy states are excited states. They may be electronic, vibrational, or rotational excited states, or some combination of the three. As shown in the figure, an appreciable overlap can occur in the energies of the different electronic states. That is, the lowest vibrational level of the first excited electronic state may be lower in energy than a higher vibrational level of the zeroth or ground electronic state. The significance of this fact will become apparent in the later discussion on how molecules lose absorbed energy. For the sake of clarity, no further distinction will be made between rotational and vibrational states.

Molecular energy-level diagrams are actually more complex than illustrated in Figure 15-6. Most molecules contain an even number of electrons which, in the ground state, are distributed two to each orbital. The Pauli exclusion principle states that two electrons in the same orbital must have opposing spins (spins paired). An electronic state in which all electron spins are paired is called a *singlet* state, S_x. The subscript x denotes the level of the state: 0 for the ground electronic state, 1 for the first excited state, and so forth. If one electron is promoted to an orbital of higher energy, the spin usually does not change; that is, it remains paired with the spin of the remaining ground state electron. The molecule is now said to be in the first excited singlet state, S_1. In some cases, the spin changes as the electron moves to a higher orbital, producing what is referred to as a *triplet state, T*. These states are represented in Figure 15-7. The names singlet and triplet are derived from the behavior of excited molecules in the presence of a magnetic field, which is not of any consequence to the discussion here.

TABLE 15-2 ENERGY AND WAVELENGTH RANGES FOR TYPES OF INTERNAL ENERGY

Type of internal energy	Range (eV)	Wavelength corresponding to energy
Rotational	$1.24 \times 10^{-2} - 1.24 \times 10^{-4}$	$100-10{,}000\ \mu m$
Vibrational	$0.828-0.0124$	$1.5-100\ \mu m$
Electronic	$8.28-0.828$	$150-1500\ nm$

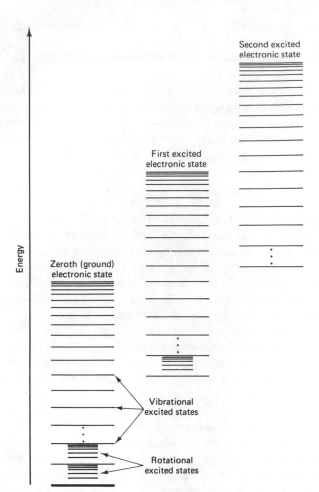

Energy

Second excited
electronic state

First excited
electronic state

Zeroth (ground)
electronic state

Vibrational
excited states

Rotational
excited states

Figure 15-6 Simple molecular energy-
level diagram.

ABSORPTION AND EMISSION
OF ELECTROMAGNETIC RADIATION

This section is devoted mainly to how *molecules* absorb radiation and what happens to
that radiation. The processes with *atoms* are basically similar, but the details of atomic
absorption and emission are discussed in Chapter 18.

Absorption

When a molecule absorbs a photon, its energy is increased by an amount exactly equal
to the energy of the photon, as depicted by line *A* in Figure 15-7. Since the molecule is
restricted to discrete potential energy states, the energy (wavelength) of the photon must
correspond exactly to the difference in energy between the initial state of the molecule

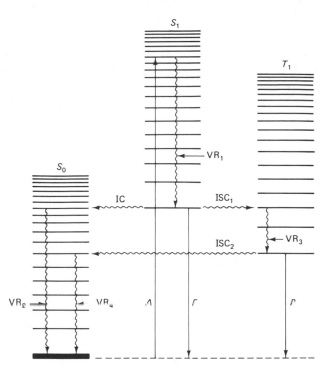

Figure 15-7 Pathways by which molecules lose absorbed energy. *A*, Absorption; *F*, fluorescence; *P*, phosphorescence; VR, vibrational relaxation; IC, internal conversion; ISC, intersystem crossing.

and the state to which it will be increased upon absorption. The process can be represented by the equation

$$M + h\nu \longrightarrow M*$$

where $M*$ represents the molecule in some excited state. A photon whose energy is not exactly the same as the difference between the two energy states will not be absorbed. Under normal experimental conditions, molecules cannot temporarily hold a photon of insufficient energy and then add to it another photon to complete the transition. Neither can a molecule divide the energy of a photon and use only that portion necessary for a particular transition.

A typical molecule has a very large number of different energy states and, in principle, is capable of absorbing many different wavelengths of radiation. However, according to quantum mechanical calculations, not all transitions are equally probable. That is, even though photons of the correct energy or wavelength are present, they may not be absorbed by the molecules. Chemists have devised a set of *selection rules* that divide transitions into two categories: allowed and forbidden. The choice of words is unfortunate, as their meaning should not be interpreted literally. *Allowed* refers to transitions that have the highest probability of occurrence, while *forbidden* refers to those with the lowest probability of occurrence. For example, transitions between singlet and triplet states (in either direction) are considered "forbidden" not because they do not

Absorption and Emission of Electromagnetic Radiation

occur, but because they are considerably less likely to occur than transitions between two singlet states, which are "allowed."

The lifetime of a molecule in an excited state is short: generally between 10^{-9} and 10^{-6} s for a singlet state (with allowed transitions) and 10^{-4} and 10^{1} s for a triplet state (with forbidden transitions). As a result, it is very difficult to keep an appreciable fraction of the available molecules in a sample in the excited state. This is somewhat akin to the difficulty a juggler faces in trying to keep a large number of balls in the the air at all times. If the source of photons is continuous, the process of absorption is continuous, with molecules absorbing, returning to the ground state, reabsorbing, and so on. We know how the juggler's ball loses its potential energy; now let's consider what happens to a molecule's excess potential energy.

Loss of Absorbed Energy

Molecules can lose energy by the emission of heat and radiation. Heat is the end product of a process called *vibrational relaxation* or collisional deactivation. In this process, an excited molecule transfers internal energy to other molecules through a series of collisions. The amount of energy transferred in a single collision is small, roughly comparable to the difference in energy of adjacent vibrational and rotational levels. The transferred energy shows up as increased kinetic energy (heat) in the colliding molecules.

The rate at which collisions occur in solution is very large, making vibrational relaxation a very efficient process. Not every collision will be exactly right for an energy transfer, but a single excited molecule will undergo several thousand collisions during an average singlet-state lifetime of 10^{-8} s.

The emission of radiation is simply the opposite of absorption. At some point an excited molecule may fall to a lower energy state by emitting a photon. The energy of the photon will equal the difference between the initial and final energy states of the molecule. Since these states are highly characteristic of a given substance, so is the energy or wavelength of the emitted photon.

Figure 15-7 illustrates some of the pathways by which molecules in an excited electronic state can return to the ground state. In this example, absorption (shown by solid arrow A) promotes the molecule from the lowest vibrational level of the singlet ground state, S_0, to an upper vibrational and rotational level of the excited electronic state S_1. The absorption is followed quickly by vibrational relaxation to the lowest vibrational level of S_1 (shown by VR_1). At this point, the molecule may do one of several things. It could transfer to an upper vibrational level of S_0 by a process called *internal conversion* (IC in Figure 15-7). The rate of internal conversion is fast when there is a molecular energy level in S_0 that closely matches the present level in S_1. Once internal conversion is completed, the molecule can continue to undergo vibrational relaxation (VR_2) to reach the ground state. This is by far the most common of the various pathways for returning to the ground state. It is a totally radiationless process in which all of the absorbed energy is converted to heat and could be described by the equation

$$M^* \longrightarrow M + \text{heat}$$

When internal conversion is not feasible, the molecule at S_1 may transfer to a triplet state by a process called *intersystem crossing* (ISC_1 in Figure 15-7). As was the case with internal conversion, the probability of intersystem crossing depend ability of a triplet-state energy level that closely matches the present ener; ing to the triplet state allows some additional vibrational relaxation to th tional level of T_1 (VR$_3$). If an appropriate energy level is present in S may undergo another intersystem crossing (ISC$_2$) followed by relaxatio state (VR$_4$). This is also a totally radiationless pathway of deactivation probable due to the lack of a suitable vibrational level in S_0, the molecule can return to the ground state by emission of a photon (P in Figure 15-7). Photon emission from a triplet state is called *phosphorescence*. You may recall that $T \rightarrow S$ transitions are "forbidden" (not very probable), which suggests that phosphorescence is a somewhat uncommon occurrence among molecules.

If neither internal conversion nor intersystem crossing are very probable, the molecule at S_1 may simply undergo an $S_1 \rightarrow S_0$ transition via photon emission (F in Figure 15-7). Photon emission from a singlet state is called *fluorescence*. Since $S_1 \rightarrow S_0$ transitions are "allowed," fluorescence from molecules is observed more frequently than phosphorescence.

It is important to realize that the pathways by which molecules lose absorbed energy are not mutually exclusive. A substance does not have to fluoresce or phosphoresce, some fraction of the excited molecules may deactivate by one pathway while the remainder use a different route to the ground state. In the large majority of cases, vibrational relaxation is the predominant pathway because it is the fastest. It is somewhat like people driving home from work—the largest number use the route that gets them home the fastest.

It is worth noting at this point that the chemistry of a molecule in an excited electronic state may be quite different from its chemistry in the ground state. Two of the most common types of electronic transitions in organic molecules are the promotion of nonbonding and pi-bonding electrons to antibonding orbitals. Populating antibonding orbitals weakens bonds in the molecule, which, in turn, may cause the molecule to react differently than it would in the ground state. Reactions with electronically excited molecules are called *photochemical reactions* and are very important in organic synthesis.

ABSORPTION SPECTRA

It is known that molecules absorb certain wavelengths of radiation but not others and that even if an absorption *can* occur, it does not have to; that is, there is a probability of absorption associated with each wavelength. A plot of the amount of radiation absorbed by a sample as a function of the wavelength is called an *absorption spectrum*. The absorption spectrum of potassium permanganate is shown in Figure 15-8. To understand why a broad band rather than a series of discrete lines is observed, the molecular energy-level diagram must be reexamined. A molecule in the ground state (lowest vibrational level of S_0) may absorb a photon of sufficient energy (wavelength) to raise itself to the first vibrational level (V_1) of the excited state S_1. This process is shown by line 1 in Figure 15-9. The molecule may also be capable of absorbing slightly more energetic (shorter wavelength) photons and ending up in the second and third vibrational levels, V_2 and V_3, of the excited state (lines 2 and 3). Although most molecules may

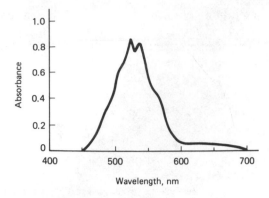

Figure 15-8 Spectrum of $1.27 \times 10^{-4}\ M\ KMnO_4$ in $0.75\ M\ H_2SO_4$. Cell path length is 1.00 cm.

reside in the lowest vibrational level of the S_0 state, a significant number may be in higher vibrational levels. These molecules may absorb also, as shown by lines 4 through 6 in Figure 15-9. Although they are not shown, remember that within each vibrational level there are many closely spaced rotational levels, and absorption can originate from and terminate at many of these levels. Thus the broad band of absorption in Figure 15-8 actually consists of a large number of lines so closely spaced on the wavelength axis that they are not independently distinguishable.

It is appropriate at this point to consider the relationship between the color of a substance and its absorption spectrum. Dilute potassium permanganate solution is distinctly purple because, according to the spectrum shown in Figure 15-8, it absorbs mostly the green and yellow wavelengths. It does *not* absorb the red or blue wavelengths (see

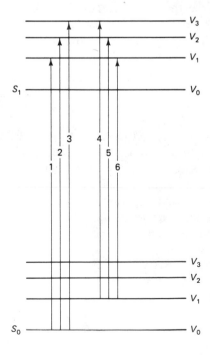

Figure 15-9 Simplified energy-level diagram showing reason for band rather than line absorption.

Figure 15-4). When we look at the solution we see the light that passes through it, not that which is absorbed. Blue and red combine to appear purple to our eyes. An opaque chalkboard appears green because it *reflects* green or both yellow and blue wavelengths of radiation. Our eyes cannot tell which, but you should be able to devise a simple experiment that could distinguish between the two possibilities.

Since the color of an object is determined by what it transmits or reflects, the nature of the source radiation can influence its color. If the light used to irradiate a potassium permanganate solution contained no red wavelengths, the solution could not transmit them and would therefore appear blue. We encounter three major sources of light every day: the sun, the incandescent lamp, and the fluorescent lamp. The spectral output of each source is different and is the reason why some objects have a slightly different color when observed in a different kind of light. Dye manufacturers rely on chemists to ensure that their products ''appear'' to have the same color regardless of the type of illumination.

RELATIONSHIP BETWEEN ABSORPTION AND CONCENTRATION

From the foregoing discussion it is clear that a beam of radiant energy is reduced in power as it passes through a solution containing an absorbing substance. The relationship between the amount of radiation absorbed or transmitted and the amount of absorbing species is called *Beer's law* or sometimes the *Beer–Lambert law*, after the scientists responsible for its derivation.

Beer's law is not difficult to derive, but it does require some knowledge of calculus. Consider a length of sample divided into infinitesimally thin layers of thickness dx, as shown in Figure 15-10. The decrease in the power of a single-wavelength beam passing through each layer is proportional to the thickness (dx), the concentration of the absorbing species (C), and the incident radiant power (P_x):

$$dP = -kP_x C\, dx \qquad (15\text{-}4)$$

where k is the proportionality constant. The minus sign indicates that the power is decreasing. Equation 15-4 can be rearranged and integrated over the entire length of the sample (b):

$$-\frac{dP}{P_x} = kC\, dx$$

$$-\int_{P_0}^{P} \frac{dP}{P_x} = kC \int_{0}^{b} dx$$

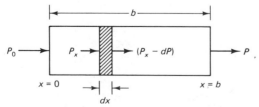

Figure 15-4).

Figure 15-10 Decrease in radiant power P_0 upon passing through a solution of length b and with a concentration of absorbing solute, C.

where the integration limits P_0 and P are the values of P_x at $x = 0$ and $x = b$, respectively. Evaluating this integral gives the expression

$$-(\ln P - \ln P_0) = kCb$$

which, when rearranged and converted to base 10 logarithms, gives the Beer's law equation

$$\log \frac{P_0}{P} = \frac{k}{2.303} Cb = \epsilon bC \qquad (15\text{-}5)$$

Nomenclature and Units

The *transmittance*, T, is the fraction of the incident radiation that passes through the sample

$$T = \frac{P}{P_0}$$

and can range in value from 0 to 1. Percent transmittance is merely $T \times 100$. Chemists also use a term called *absorbance, A*, which is defined as

$$A = \log \frac{P_0}{P} = \log \frac{1}{T} = \log \frac{100}{\% \, T} \qquad (15\text{-}6)$$

Thus Equation 15-5 may be written

$$A = \log \frac{1}{T} = \epsilon bC \qquad (15\text{-}7)$$

Absorbance represents, in a logarithmic manner, the amount of radiation absorbed. When no radiation is absorbed, $P = P_0$ and $A = 0$. When 90% of the radiation is absorbed, 10% is transmitted. Thus $P = 0.1 \times P_0$, and from Equation 15-6, $A = 1$. When 99% is absorbed, 1% is transmitted and $A = 2$. There is no finite upper limit of absorbance; as P approaches 0, A approaches infinity. Since they involve ratios of radiant power, both transmittance and absorbance are unitless values.

The constant ϵ in Equation 15-7 is called the *molar absorptivity* and has units of $M^{-1} \cdot cm^{-1}$ or $L \cdot mol^{-1} \cdot cm^{-1}$. The use of this symbol specifically requires that we express the concentration in units of molarity and the sample path length in centimeters. Spectrophotometry is an old technique and some of the early terminology, which is no

TABLE 15-3 SUMMARY OF TERMS AND SYMBOLS USED
IN ABSORPTION MEASUREMENTS

Correct name	Symbol	Definition	Old names and symbols
Absorbance	A	$\log P_0/P$	Extinction, E; optical density, OD
Transmittance	T	P/P_0	Transmittancy
Path length	b	—	l, d
Molar absorptivity[a]	ϵ	A/bC	Molar extinction coefficient
Absorptivity[b]	a	A/bC	Extinction coefficient, k

[a] C has units of M, b has units of cm.
[b] C may have any unit of concentration other than M, b has units of cm.

longer recommended, is still encountered in the literature. Table 15-3 is a summary of the common terms and symbols of absorption spectrophotometry, both past and present.

Molar absorptivity is a property characteristic of a substance. It is constant at any given wavelength and may be considered a measure of the probability of absorption occurring when the necessary photons are present. Absorbance, on the other hand, is a property characteristic of the sample. Not only does it depend on the wavelength, but also on the concentration of the absorbing substance and the pathlength of the sample.

Example 15-1

A solution of tris(1,10-phenanthroline)iron(II) has a transmittance of 0.730 in a 1.00-cm cell. What will be its transmittance in a 5.00-cm cell?

Solution The relationship between transmittance and pathlength is given by Equation 15-7. Thus

$$\log 1/T_1 = \epsilon b_1 C \quad \text{and} \quad \log 1/T_2 = \epsilon b_2 C$$

where 1 and 2 refer to the two measurements. Taking the ratio of the two equations,

$$\frac{\log 1/T_1}{\log 1/T_2} = \frac{\cancel{\epsilon} b_1 \cancel{C}}{\cancel{\epsilon} b_2 \cancel{C}}$$

and substituting the known values gives

$$\frac{\log 1/0.730}{\log 1/T_2} = \frac{1.00 \text{ cm}}{5.00 \text{ cm}}$$

or

$$\log 1/T_2 = 0.683$$

and

$$T_2 = 10^{-0.683} = 0.207$$

Example 15-2

Exactly 0.1374 g of pure potassium dichromate was dissolved in 10 mL of 2 M H_2SO_4, transferred to a 500-mL volumetric flask, and diluted to the mark with distilled water. A 25-mL aliquot of this solution was transferred to another 500-mL volumetric flask and diluted to the mark with water. This final solution had an absorbance of 0.317 in a 2.00-cm cell. What is the molar absorptivity of $K_2Cr_2O_7$?

Solution The molar absorptivity can be calculated from Beer's law if the concentration of the measured solution is known. The concentration in the first solution is given by

$$\frac{137.4 \text{ mg } K_2Cr_2O_4}{294.2 \text{ mg } K_2Cr_2O_7/\text{mmol} \times 500 \text{ mL}} = 9.34 \times 10^{-4} \text{ mmol/mL}$$

After the second dilution,

$$C_{K_2Cr_2O_7} = 9.34 \times 10^{-4} \text{ mmol/mL} \times \frac{25.0 \text{ mL}}{500 \text{ mL}} = 4.67 \times 10^{-5} \text{ } M$$

Then

$$A = \epsilon bC$$

$$0.317 = \epsilon(2.00 \text{ cm})(4.67 \times 10^{-5} \, M)$$

or

$$\epsilon = 3.39 \times 10^3 \, M^{-1} \cdot \text{cm}^{-1}$$

Deviations from Beer's Law

The most common test of conformity to Beer's law (Equation 15-7) is to plot absorbance versus concentration. A chemical system conforming to the equation will yield a straight line through the origin with a slope equal to ϵb. Deviations from Beer's law are sometimes encountered, but in many cases they are apparent rather than real. Real deviations occur only at high concentrations where the solute molecules are packed so closely that the charge distribution of a molecule becomes distorted by its neighbors, thereby changing the energy-level pattern and its absorption properties. In addition, although the molar absorptivity is independent of concentration, it does depend on the refractive index of the solution, which, in turn, varies with the concentration of solute. This effect is rarely significant, however, at concentrations less than about 0.01 M.

Apparent deviations from Equation 15-7 are more common and generally result from equilibrium processes that cause the concentrations of absorbing species to be something different from what the analyst thinks they are. For example, in a series of standard solutions of a colored metal chelate prepared by serial dilution, the percent dissociation increases as the concentration decreases, and the *actual* concentration of solute will differ from the concentration calculated solely on the basis of dilution.

Solutes that fluoresce may not appear to obey Beer's law because a portion of the fluoresced radiation will be emitted along the optical path of the transmitted radiation. Since, as will be seen shortly, the radiation detectors used are not wavelength selective, the fluoresced radiation will be detected and interpreted as transmitted radiation.

INSTRUMENTATION FOR MEASURING ABSORPTION OF RADIATION

Instruments for measuring absorption vary tremendously in their level of sophistication and cost. Nonetheless, they are all comprised of five basic components, as shown in Figure 15-11. This section is a discussion of the important aspects of these components and how they are combined in the complete instrument.

Radiation Sources

A good spectrophotometric source should have a stable, high-intensity output that covers a wide spectral range. No single source is suitable for all of the spectral regions of interest to chemists, and an instrument designed to make measurements over a wide range of wavelengths may contain two or even three different sources.

Figure 15-11 Basic components of a spectrophotometer.

Deuterium discharge lamp

This source uses an electrical discharge to dissociate deuterium molecules into atoms. The process is accompanied by the emission of continuous ultraviolet radiation in the range of about 160 to 380 nm

$$D_2(g) \xrightarrow[\text{discharge}]{\text{electrical}} D_2^*(g) \longrightarrow 2D(g) + h\nu$$

Internal energy of the excited deuterium molecules is converted partly to kinetic energy of the atoms and partly to radiant energy. Since the kinetic energies of the atoms are continuous (not quantized), the photon energies are also continuous.

The discharge, carried out in a sealed tube containing deuterium at a pressure around 1 to 5 torr, takes place between a heated filament and a metal electrode held at a potential difference of about 40 V. A quartz window must be built into the tube to allow the radiation to escape because glass absorbs strongly below 325 nm.

Tungsten filament lamp

The most common source of visible radiation is the ordinary tungsten filament lamp. It consists of a thin, coiled tungsten wire sealed in an evacuated glass bulb. Electrical energy passing through the filament is converted to heat, causing it to glow "white hot." The spectral distribution of an incandescent filament is essentially that of a blackbody radiator. The output intensity and wavelength distribution of a blackbody radiator is independent of the material, depending only on the temperature, as shown in Figure 15-12. The filament temperature of an ordinary tungsten lamp is about 2900 K and the lamp output (325 to 3000 nm) covers part of the ultraviolet, the entire visible, and a sizable portion of the infrared spectrum.

The tungsten-halogen or quartz-halogen lamp is a tungsten filament lamp that has been modified to operate at a higher temperature (about 3500 K), and therefore has a more intense output. The ordinary tungsten lamp has a very short lifetime when operated at 3500 K because the filament vaporizes and eventually breaks. By adding a small amount of iodine to the evacuated bulb, much of the vaporized tungsten reacts to form a gaseous tungsten-iodide compound of variable stoichiometry. This compound is relatively stable and remains in the bulb until it strikes the hot filament, where it decomposes into solid tungsten and iodine vapor. The process can be represented by the reactions

$$W(s) \longrightarrow W(g)$$
$$W(g) + I_2(g) \longrightarrow WI_x(g)$$
$$WI_x(g) + W(s) \longrightarrow 2W(s) + I_2(g)$$
$$\text{hot}$$
$$\text{filament}$$

Instrumentation for Measuring Absorption of Radiation

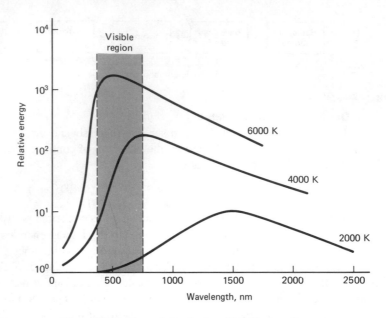

Figure 15-12 Spectral distribution of blackbody radiators.

Ordinary glass becomes soft at the higher operating temperature, so the bulb is usually made of quartz; thus the name *quartz-halogen* lamp.

Tungsten-halogen lamps have a more intense spectral output when operated at conditions that give them the same lifetime as an ordinary tungsten lamp. On the other hand, when operated to produce the same spectral output as an ordinary lamp, they have much longer lifetimes. The energy output of an incandescent lamp in the visible spectrum varies approximately as the fourth power of the applied voltage. As a result, high-quality voltage regulators are required to produce a steady light output.

Wavelength Selectors

Since the radiation sources are continuous, some means is needed of selecting the particular wavelength at which a measurement is to be made. It is physically impossible to isolate a single wavelength from a continuum, so a very narrow band of wavelengths has to suffice. Devices that isolate a narrow band of wavelengths from a continuum are called *monochromators* and almost always employ either a grating or prism as the dispersive element.

Grating monochromators

A reflection grating is comprised of a series of closely spaced, parallel grooves cut or etched into a solid material whose surface has been made reflective by polishing or by evaporating a thin film of aluminum onto it. An enlarged section of a grating surface is shown in Figure 15-13. Lines 1 and 2 are the paths followed by two photons of the same wavelength that are completely in phase (constructive interference) as they travel

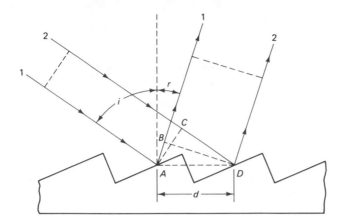

Figure 15-13 Schematic diagram of diffraction from a grating.

toward the grating. Photon 2 must travel farther before being reflected and, in order for constructive interference to occur in the reflected beams, this extra distance must be a whole-number multiple of the wavelength. That is,

$$n\lambda = (\overline{CD} - \overline{AB}) \qquad (15\text{-}8)$$

where $\overline{CD}$ is the extra distance traveled by wave 2 before reflection, $\overline{AB}$ is the extra distance traveled by wave 1 after reflection, and n is an integer called the *order*. The incident and reflected angles i and r are equal to angles CAD and BDA, respectively. Therefore,

$$d \sin i = \overline{CD}$$

and

$$-d \sin r = \overline{AB}$$

where d is the spacing of the grooves. By convention, the minus sign is used in conjunction with the relected beam. Substituting these expressions into Equation 15-8 gives

$$n\lambda = d(\sin i + \sin r) \qquad (15\text{-}9)$$

Three important observations can be made about Equation 15-9. First, if we wish to change the wavelength of radiation that appears (undergoes constructive interference) at angle r, we need only to change angle i. Second, a wavelength from each order will appear in the beam reflected at angle r. For example, if a first-order wavelength ($n = 1$) of 600 nm is found in the beam, a second-order wavelength of 300 nm, a third-order wavelength of 200 nm, and so forth, will also be present. Ordinarily, gratings are designed to concentrate the radiant power in the first order. Higher-order wavelengths are removed by filters or another monochromator. Finally, Equation 15-9 shows the relationship between the groove spacing and the first-order wavelengths. To obtain ultraviolet and visible wavelengths in the first order the grooves must be very closely spaced— about 2000 to 6000 grooves/mm. A spacing on the order of 20 grooves/mm is appropriate for first-order wavelengths in the far-infrared spectrum.

Instrumentation for Measuring Absorption of Radiation

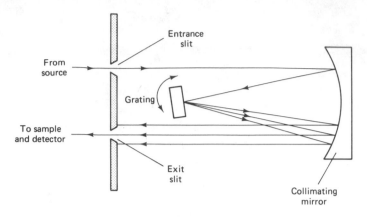

Figure 15-14 Ebert mount grating monochromator.

To fashion a monochromator from a reflection grating, a pair of slits and a mirror must be added, as shown in Figure 15-14. The three lines leaving the grating represent three different wavelengths in the diffracted beam. The wavelength falling on the exit slit can be changed by rotating the grating. This particular monochromator design is merely one of several used in modern instruments. A discussion of the relative merits of different monochromator designs is beyond the scope of this book.

The range of wavelengths isolated by the monochromator depends on the width of the exit slit and the dispersion of the grating. *Dispersion* is a measure of how well a device can separate wavelengths in space. Gratings produce a constant dispersion over a reasonably wide range of wavelengths, which means that a grating monochromator physically separates wavelengths 200 and 201 nm by the same amount that it separates 800 and 801 nm. Consequently, the range in wavelength of the radiation passing a given exit slit, called the *bandwidth*, is the same regardless of the wavelength setting of the monochromator.

The ideal monochromator would pass radiation of a single wavelength, called *monochromatic radiation*. The bandwidth of an actual monochromator can be decreased by narrowing the exit slit but only at the expense of reduced radiant power. In general, the bandwidth should be substantially less than the natural width of the spectral absorption band being recorded. There is no real difficulty in accomplishing this with ultraviolet and visible spectrophotometers because the spectral absorption bands of molecules in solution are generally quite wide.

Prism monochromators

At one time, prisms were used almost exclusively in monochromators, but recent technological advances have greatly improved the quality and reduced the cost of gratings. As a result, prisms are generally found only in older instruments. Prisms disperse radiant energy by refraction, a process in which the paths of dissimilar wavelengths of radiation are bent differently upon passing between two media of different refractive index. One of the most common prism configurations is shown in Figure 15-15. The back side of the 30° prism is mirrored to reflect the incident radiation back through the

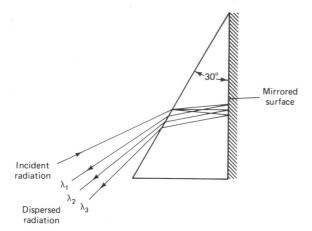

Figure 15-15 Littrow prism.

prism, thereby doubling the degree of separation. Unlike gratings, prisms do not produce a constant dispersion; they separate short wavelengths more efficiently than long ones. This means that the bandwidth of a prism monochromator varies with its wavelength setting.

Prisms have a further disadvantage in that the radiant energy to be dispersed must pass through the prism material. Consequently, the prism must be made from a material transparent to the radiation that will be employed. Flint glass is an excellent material for prisms that are to be used with visible radiation, but it absorbs ultraviolet radiation. Prisms for use in the ultraviolet generally are made of quartz.

Sample Holders

Sample containers, called *cells* or *cuvets*, come in a variety of shapes and sizes appropriate for various instruments and experiments. The simplest containers are test tubes, but generally they are not preferred because of optical distortions that result from the curved surface. A few representative cells are shown in Figure 15-16. They are characterized by having two flat, parallel windows through which the radiation passes. The rectangular cell of 1-cm path length is certainly the most widely used of the various

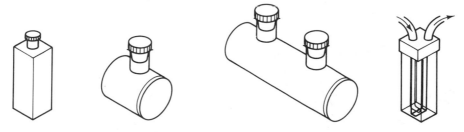

Figure 15-16 Selected sample cells.

Instrumentation for Measuring Absorption of Radiation

types. Longer-path-length cells get more analyte in the optical beam and thereby increase the measured absorbance (recall Beer's law).

Glass cells are suitable for measurements with visible radiation but cannot be used in the ultraviolet because of their strong absorption. Quartz or fused silica cells, although more expensive, can be used in either ultraviolet or visible spectrophotometry.

Instruments often are designed to accommodate two cells simultaneously. As a result, cells are sold in matched sets of two or four. The cells in a set are always made from the same batch of glass or quartz, and therefore have nearly identical optical qualities.

Radiation Detectors

Radiation detectors used in ultraviolet and visible spectrometry are devices that convert radiant energy into electrical energy. A good detector must be nonselective over a fairly wide range of wavelengths. Ideally, it should respond equally to every wavelength within the range, but this is never realized in practice. It should also have a high efficiency; which means a large electrical signal should be produced in response to a low power of radiation. Two detectors dominate the present commercial market: the phototube and the photomultiplier tube.

Phototube

The phototube, shown in Figure 15-17, is a simple device consisting of a semi-cylindrical cathode and a wire anode in an evacuated tube containing a quartz window. The surface of the cathode is coated with a photoemissive material that emits electrons upon being irradiated. By applying a voltage across the electrodes (about 100 V), the emitted electrons are caused to migrate to the wire anode, thereby generating an electrical current, the magnitude of which is directly proportional to the power of the radiation in the beam. The electrical signals generated by these tubes in normal applications are quite small and always require substantial amplification before they can be displayed.

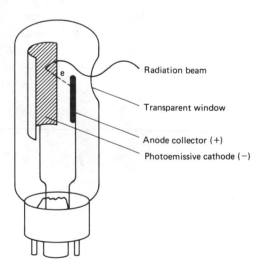

Radiation beam

Transparent window

Anode collector (+)
Photoemissive cathode (−)

Figure 15-17 Phototube detector for ultraviolet and visible radiation.

Several different photoemissive surfaces are found among commercial phototubes. Some consist of silver and one or more alkali metal mixed with their oxides. Others consist of alkali metal semiconductors such as Cs_3Sb, K_2CsSb, or Na_2KSb mixed with a trace of cesium. The different surfaces exhibit different sensitivities toward certain wavelengths, such that no single phototube is best for the entire 180 to 1000-nm wavelength range of ultraviolet-visible spectrophotometers.

Photomultiplier tube

The photomultiplier tube or PMT is really a sophisticated version of a phototube. A topside view of this device is shown in Figure 15-18. In addition to the photoemissive cathode, the tube contains a series of coated electrodes called dynodes, each held at a potential about 100 V more positive than the preceding one. The cathode is coated with a photoemissive surface like that used in a phototube. The dynodes are coated with a compound, such as BeO, CsSb, or GaP, that ejects several electrons when bombarded with a single high-energy electron. Each dynode is shaped to focus the emitted electrons toward the next dynode. Thus a photoelectron leaving the cathode is multiplied at each successive dynode. The anode may collect as many as 10^7 electrons for each photoelectron generated at the cathode. This high internal amplification means that very low radiant power can be detected without the need of a large external amplifier.

Photodiode array

This detector consists of a two-dimensional array of tiny, individual photodiode sensors, each consisting of a reversed-biased *p-n* junction formed on a silicon chip, as shown in Figure 15-19. The device is biased such as to make the depletion layer almost totally nonconducting in the absence of electromagnetic radiation in the wavelength range

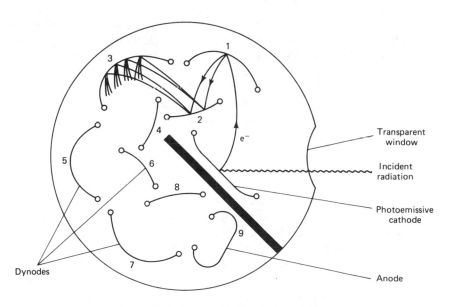

Figure 15-18 Photomultiplier tube (top view).

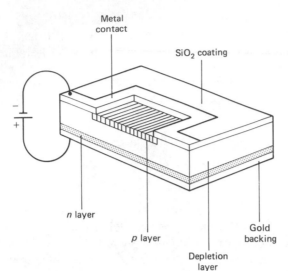

Metal contact

SiO$_2$ coating

n layer

p layer

Depletion layer

Gold backing

Figure 15-19 Schematic design of a reversed-bias *p-n* junction photodiode.

to be detected. When photons within an appropriate energy range strike the diode, electrons are promoted to a conduction band, causing the depletion layer to become conductive. The resulting current is proportional to the incident radiant power.

Like photoemissive devices, photodiodes can be made that are sensitive to wavelengths from different spectral regions. Those responding to radiation in the range from about 400 to 1000 nm have sensitivities that are intermediate between those of simple phototubes and photomultiplier tubes. Photodiodes responding to ultraviolet radiation (200 to 400 nm) are available, but are less sensitive than those responding to visible radiation.

Typical arrays contain several hundred to several thousand individual diodes, each with a small storage capacitor wired in parallel, sequentially connected to a common output through a transistor switch. A solid-state device known as a shift register momentarily closes each transistor switch in sequence, allowing the capacitor to charge to a predetermined voltage. Spectral irradiation of any of the photodiodes causes them to become conductive, which allows the associated capacitors to discharge partially. The lost charge is measured during the next cycle, when each capacitor is recharged to the predetermined value.

Readout Devices

The readout device must take the amplified signal from the detector and display it in some form useful to the operator. Obviously, it must be accurate, but it should also be fast and capable of displaying both absorbance and transmittance.

The simplest readout device is an analog meter. Although inexpensive, the readability of analog meters is poor, generally about $\pm 0.1\%$ T. Digital displays are usually more versatile than analog meters. For one thing, they can use enough digits so that readability is not a limiting factor. Often, digital displays are used in conjunction with a microprocessor. These small on-board computers can store information on standards so that when an unknown is measured the concentration can be computed and displayed. Laboratory recorders are also popular readout devices. They have the considerable advantage of being able to record continuously the absorbance or transmittance as a function of time or wavelength.

Optical Systems

Commercial instruments for routine use are designed with either one or two optical paths and are referred to as single-beam or double-beam spectrophotometers. The advantages of single-beam instruments lie in their simplicity and low cost. Generally, they are suited only for quantitative determinations involving measurements at a single wavelength. Double-beam spectrophotometers are more expensive but also much more versatile than their single-beam counterparts. They are designed to allow continuous recording of absorbance or transmittance as a function of wavelength.

Single-beam design

The optical layout of a typical single-beam instrument is shown in Figure 15-20. The output of the source is monitored by a reference phototube that serves to compensate for fluctuations in radiant power due to inadequate regulation of the power supply voltage. The readout is based on the amplified difference signal from the two phototubes.

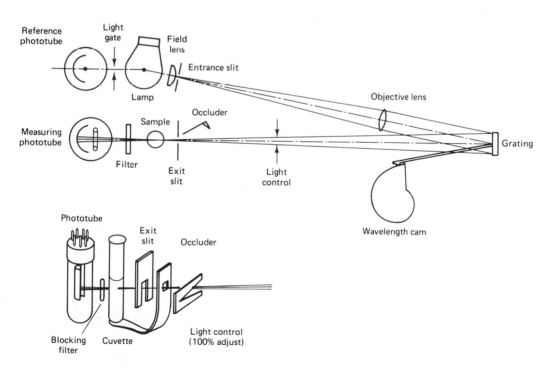

Figure 15-20 Schematic diagram of the optical layout of the Milton Roy Spectronic 20 single-beam spectrophotometer. (Courtesy of Milton Roy Co.)

Instrumentation for Measuring Absorption of Radiation **401**

The light control is a V-shaped slot that can be moved farther into or out of the beam to decrease or increase the amount of radiation reaching the detector.

According to Equation 15-5, the ratio of P_0 to P must be measured to determine the absorbance of a solution. To account for reflection off the cell windows and extraneous absorption by diverse constituents added to the unknown during sample pretreatment, P_0 is normally taken as the measured signal when the cell contains a reagent blank. The measuring procedure is as follows. With the cell compartment empty (which causes the occluder to block the beam) the detection circuit is adjusted to minimize any residual current, called the *dark current*. Then a reagent blank is placed in the cell compartment and the light control is adjusted until the readout is 100% T or 0 A. At this point, the radiant power reaching the detector is being defined as P_0. Now the blank is replaced with the sample and the absorbance or transmittance is displayed on the readout meter. If measurements at a different wavelength are desired, the entire procedure must be repeated, clearly a slow and tedious process.

Double-beam design

In this type of spectrophotometer the beam is split and passed alternately through the sample and reference cells. The optical layout for a Varian model 2300 UV-Vis spectrophotometer is shown in Figure 15-21. Radiation from the source is passed through the monochromator twice to improve the "monochromaticity" of the incident beam.

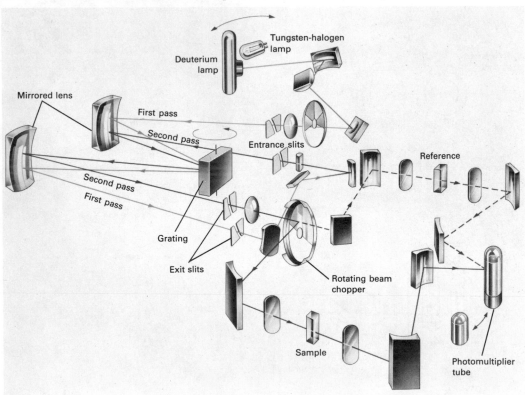

Figure 15-21   Schematic diagram of the optical layout of the Varian 2300 UV-Vis double-beam recording spectrophotometer. (Courtesy of Varian Associates, Inc.)

The beam chopper is a rotating half-mirror which divides the beam and passes it alternately through a reference and sample cell. The two paths lengths are the same, so when the beams are combined the detector sees alternating segments of radiation that have passed through the cells. A simple electronic comparison of these values allows a direct readout of absorbance. This system automatically corrects for changes in the source intensity or detector response with time, because the power emerging from both cells is compared frequently (perhaps several hundred times per second).

Double-beam spectrophotometers are nicely suited to wavelength scanning. Two synchronous motors are used: One drives a cam that continuously alters the position of the grating (and therefore the wavelength reaching the cells), while the other runs the paper feed of a laboratory chart recorder. The detector signal is used to drive a pen across the chart paper, thereby providing a permanent recording of the absorbance or transmittance as a function of wavelength.

Multichannel, diode-array design

The diode-array spectrometer may be a single- or double-beam instrument, but it contains several important design features that enable the diode array to act as a multi-

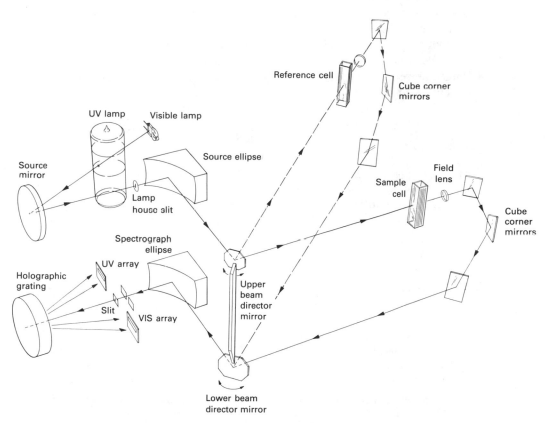

Figure 15-22 Optical schematic of the Hewlett-Packard HP 8450A diode-array spectrometer. (Courtesy of Hewlett-Packard Co.)

Instrumentation for Measuring Absorption of Radiation

channel detector. A schematic diagram of a double-beam instrument designed to measure throughout the 200- to 800-nm spectral range is shown in Figure 15-22. A careful examination of the diagram reveals several interesting features. The tungsten-halogen lamp is located directly behind a "see-through" deuterium lamp so that the optical beam carries radiation that is continuous from 200 to 800 nm, and no source switching is necessary. The sample and reference cells are located *between* the source and monochromator and there is no monochromator exit slit. In this way, radiation reflecting from the grating contains *all* of the spectral information simultaneously. Actually, the dispersive element is comprised of two gratings formed on a single substrate but tilted at a few degrees from each other so that the diffracted radiation from 200 to 400 nm is directed to one array detector while that from 400 to 800 nm is directed toward a second detector.

The individual diodes in the arrays are small, typically 10 to 50 μm wide and 400 to 500 μm long. With the highly dispersive gratings used, spectral bandwidths of 1 to 2 nm per diode are possible. The array detectors are scanned (capacitors charged) at a rate of about 10 μs per diode, so the spectral information in a 512-diode array can be read out in 5.12 ms. A computer is required not only to control the scanning of the arrays but also to store the spectral information for later display in a form understandable to the scientist.

Diode-array spectrophotometers have been available commercially for only a few years and are fairly expensive by comparison with older-style instruments, but it is clear that they are the spectrophotometer of the future. Chemists are especially excited about their applicability to the study of transient intermediates in fast reactions, the fast simultaneous determination of multiple analytes in mixtures, the rapid automated processing of large numbers of samples, and the determination of separated components being eluted from chromatography columns.

PROBLEMS

15-1. Convert the following frequencies to wavelengths in nanometers. Indicate the spectral region within which the wavelength falls.

(a) 4.283×10^{14} s^{-1} (c) 1.333×10^{15} s^{-1} (e) 2.500×10^{16} s^{-1}
(b) 1.053×10^{13} s^{-1} (d) 6.667×10^{14} s^{-1} (f) 3.371×10^{10} s^{-1}

15-2. Convert the following wavelengths to frequencies in hertz (waves/s).

(a) 536 nm (c) 24.6 cm (e) 6.42 pm
(b) 14.3 μm (d) 18.3 mm (f) 225 nm

15-3. A beam of radiation has a wavelength of 589.3 nm in a vacuum. Calculate its wavelength in nanometers in each of the following media. The values in parentheses are the velocities of propagation in that medium.

(a) glass $(2.015 \times 10^8$ m·s$^{-1})$
(b) fused quartz $(2.052 \times 10^8$ m·s$^{-1})$
(c) water $(2.251 \times 10^8$ m·s$^{-1})$
(d) ethanol $(2.205 \times 10^8$ m·s$^{-1})$

15-4. Calculate the energy in joules of the radiation described in parts (a), (b), and (c) of Problems 15-1 and 15-2. Planck's constant is 6.626×10^{-34} J·s.

15-5. What is the wavelength in nanometers and frequency in reciprocal seconds of radiation with the following energies? See Problem 15-4 for the value of Planck's constant.
   (a) 4.000×10^{-19} J   (c) 8.000×10^{-20} J   (d) 2.410×10^{-19} J
   (b) 1.000×10^{-18} J

15-6. Convert the following percent transmittances to absorbances.
   (a) 0.100   (c) 10.0   (e) 41.4
   (b) 1.00   (d) 26.3   (f) 83.7

15-7. Convert the following absorbances to percent transmittance.
   (a) 0.0040   (c) 0.314   (e) 1.000
   (b) 0.100   (d) 0.799   (f) 2.316

15-8. The percent transmittance of a solution containing 2.46×10^{-5} M Na_2MoO_4 was 22.6 at 210 nm in a 1.00-cm cell. Calculate the molar absorptivity of sodium molybdate at 210 nm.

15-9. A 25.00-ppm $KMnO_4$ solution had an absorbance of 0.674 at 515 nm in a 5.00-cm cell. Calculate the molar absorptivity of $KMnO_4$ at this wavelength. What would be the molar absorptivity in a 1.00-cm cell?

15-10. A solution was prepared by dissolving 118.9 mg of primary standard $K_2Cr_2O_7$ in dilute sulfuric acid and diluting to 1.000 L with water. A 10.00-mL aliquot was transferred to a 100.0-mL volumetric flask and diluted to the mark. This solution had absorbances of 0.206 and 0.143 at 400 and 420 nm, respectively, in a 1.00-cm cell. Calculate the molar absorptivity of $K_2Cr_2O_7$ at the two wavelengths.

15-11. A solution of $FeSCN^{2+}$ had an absorbance of 0.917 at 575 nm in a 2.00-cm cell. If the molar absorptivity of $FeSCN^{2+}$ at this wavelength is 7.00×10^3, calculate the molar concentration of $FeSCN^{2+}$.

15-12. The molar absorptivity of the complex formed between bismuth(III) and 1-(2-pyridylazo)-2-naphthol is 2.10×10^4 at 580 nm. What concentration of the complex is needed to produce a percent transmittance of 20.0 in a 1.00-cm cell?

15-13. Two blue solutions, each known to contain only one absorbing species, had the following absorbances in a 1.00-cm cell:

Solution	A at 770 nm	A at 820 nm
1	0.622	0.417
2	0.391	0.240

Do the solutions contain the same substance? Explain how you know.

15-14. What should be the absorbance of 2.63×10^{-5} M tris(1,10-phenanthroline)iron(II) at 508 nm in a 2.00-cm cell if the molar absorptivity of the complex is 1.10×10^4?

15-15. The ligand 4,4',4''-triphenyl-2,2',2''-terpyridine forms a 1:1 complex with Fe^{2+} whose molar absorptivity is 3.02×10^3 at 583 nm. If the lowest detectable absorbance is 0.002, what is the lowest concentration of Fe^{2+} that can be detected with this complex in a 10.0-cm cell?

15-16. Titanium(III) forms a purple 1:2 (Ti:L) complex with salicylfluorone at pH 3.0. A solution containing 0.250 ppm Ti and excess salicylfluorone had an absorbance of 0.958 at 530 nm in a 2.00-cm cell. Calculate the following.
   (a) molar absorptivity of the complex
   (b) percent transmittance of the solution

Problems

(c) percent transmittance of the solution in a 1.00-cm cell

(d) molar concentration of complex that would have an absorbance of 0.500 in a 1.00-cm cell

15-17. A solution containing 6.94×10^{-6} M CrQ_3 (Q = 8-hydroxyquinolate ion) had a percent transmittance of 31.4 in a 1.00-cm cell. Calculate the following.

(a) absorbance of the solution

(b) molar absorptivity of the complex

(c) absorbance in a 5.00-cm cell of a solution one-half the foregoing concentration

(d) cell path length needed to give a percent transmittance of 10.0

15-18. Suppose that the iron-phenanthroline complex in Problem 15-14 was 1.00% dissociated and neither of the dissociation products absorb.

(a) Calculate the measured absorbance of the solution.

(b) Calculate the percent relative absorbance error.

(c) How could the error caused by dissociation be decreased in a procedure for determining iron?

15-19. The molar absorptivities of ZnL_2 and L^- at 552 nm are 9.88×10^3 and 6.41×10^1, respectively. Calculate the absorbance in a 2.00-cm cell of a solution prepared by mixing 5.00 mL of 1.25×10^{-3} M $ZnCl_2$ with 10.00 mL of 0.100 M NaL and diluting to 500 mL. Assume the reaction goes to completion.

15-20. What is the maximum molar concentration of L^- that can be present in the solution described in Problem 15-19 if the measured absorbance is to be within 99% of the absorbance due only to ZnL_2?

15-21. Suppose that the iron-phenanthroline complex in Problem 15-14 was 1.00% dissociated and the free 1,10-phenanthroline had a molar absorptivity of 1.25×10^3 at 508 nm.

(a) Calculate the measured absorbance of the solution.

(b) Calculate the percent error in absorbance.

(c) What molar concentration of iron complex would be calculated from the measured absorbance if it was assumed that no dissociation occurred?

(d) What is the percent error in molar concentration of iron complex?

METHODS AND APPLICATIONS OF SPECTROPHOTOMETRY

It is difficult to overstate the importance of absorption spectrophotometry in chemical analysis. Spectrophotometric methods of determination based on ultraviolet and visible absorption probably are used more often, and for a larger variety of substances, than any other analytical method. Many organic compounds absorb strongly, especially in the ultraviolet, and may be determined by simple direct procedures. Although the majority of inorganic substances do not absorb strongly, many are determined routinely by converting them to absorbing species via specific chemical reactions. The high selectivity of many spectrophotometric methods is, in fact, due to the selective nature of these reactions.

One of the major accomplishments of the last 15 years has been the development of automated systems capable of obtaining data on multiple samples without much operator assistance. Many such systems are based on spectrophotometric measurements, mainly because the measurements can be made quickly, easily, and without disturbing the solution. Automated systems were first used in the clinical laboratory but now are found in environmental, industrial, and pharmaceutical quality control laboratories as well.

Absorption spectrophotometry generally is best suited for determining constituents in the 1 to 50 parts per million concentration range. As such, it complements rather than competes with techniques such as gravimetry and titrimetry which are applicable to larger concentrations.

CONSIDERATIONS IN CHOOSING A METHOD

The scientific literature is filled with thousands of different spectrophotometric methods. One can easily find 50 or more different methods for determining a common substance such as iron. The question naturally arises of how to choose the "best" method. There is no best method for *every* situation or type of sample. Above all, the chosen method must be capable of supplying a result that can solve or contribute to the solution of the original problem. In situations requiring the frequent analysis of samples of very similar composition, certain methods may gain sufficiently widespread acceptance to be considered "standard" methods. For example, most clinical analyses are performed using standard methods. The overall composition of blood and urine does not vary much from one patient to another and the accepted methods have been demonstrated to give accurate results with these samples. It does not follow, however, that a method used for inorganic phosphate in blood will be appropriate for determining the phosphate content of soil.

The choice between several possible methods usually depends on a number of considerations. Every method has certain intrinsic characteristics such as accuracy, sensitivity, selectivity, speed, and cost. But these are not the only factors to be considered. The number of samples to be analyzed, the time and instrumentation available, and the expertise of the analyst can also influence the final choice. For example, a particular method may give excellent results but take 3 hours to complete. That method might be an acceptable choice if only a few samples a month are to be analyzed, but would be unacceptable if the laboratory received 200 samples each day.

FUNDAMENTAL FEATURES OF SPECTROPHOTOMETRIC METHODS

Methods that are well characterized and widely accepted generally have very carefully prescribed procedures for preparing the solutions and making the measurements. Ordinarily, the chemist needs only to follow the procedure exactly to obtain a suitable result. However, more often than not, there is no accepted, standard method and the composition of samples varies so much that no single procedure will always provide satisfactory results. In such cases, the chemist must be able to select a suitable method and develop a procedure appropriate to the nature of the sample. In this section we describe some of the individual steps involved in a typical determination and provide some insight into how good procedures are developed.

Forming an Absorbing Species

Relatively few analytes absorb strongly enough to be determined directly without prior chemical treatment. In most cases, this problem can be overcome by adding a substance (or substances) that reacts with the analyte to produce an absorbing species. Such substances are called *chromophoric reagents* or just *chromophores*. The reaction between the analyte and the chromophore must be quantitative, even at the normally low concentrations of analyte, and the absorbing product should be stable long enough to allow the chemist to make the necessary measurements.

Established procedures always specify appropriate reaction conditions such as pH, solvent, order of reagent addition, and temperature, because they commonly influence the extent and rate of the reaction between the analyte and chromophoric reagent and the stability of the absorbing product. These parameters also may influence reactions between other constituents and the chromophoric reagent that are responsible for interferences in the method. As a result, the chemist *must* have information about what substances are likely to be present in the sample being analyzed and how they are affected by changes in the solution conditions.

Measuring the Sample and Standards

Most modern instruments offer the capability of measuring either absorbance or transmittance. The measurement of absorbance is generally preferred because of its direct relationship to concentration (Equation 15-7). Older, inexpensive instruments often have an analog meter readout device that is calibrated logarithmically in absorbance and linearly in transmittance. With these instruments it is often better to measure the transmittance (especially at low values) and convert it to absorbance mathematically using Equation 15-6. Modern instruments equipped with microprocessors (small on-board computers) can store the absorbances and concentrations of the standards. When the unknown is measured, the microprocessor executes a program to solve the Beer's law equation and report the concentration on a digital display.

Numerical limits of measurement

The instrumental measurement of absorbance (or transmittance) is more accurate at intermediate values than at very small or very large values. Consequently, a sample size should be selected that will produce a final solution whose absorbance falls within the optimum measurement range. This range varies somewhat with the quality of the instrument and the type of detector it uses. In general, the optimum absorbance range for an inexpensive instrument using a simple phototube detector is from 0.2 to 0.7. The range for higher-quality instruments with photomultiplier tube detectors may be from 0.2 to 1.5 A or more. You should not conclude that values outside these ranges cannot be measured; it is only that such values will have somewhat larger errors associated with them.

The lower limit of the optimum absorbance range is determined by a combination of uncertainties, including those associated with the dark current of the detector and with the positioning of the sample and reference cells in the optical path. The upper limit is determined largely by the noise level of the detector. When the absorbance is large very little radiation reaches the detector. Consequently, the background electrical noise in the detector circuit becomes an appreciable fraction of the total signal.

Selecting the measurement wavelength

Ordinarily, for an analyte to be determined spectrophotometrically, its absorbance must be measured at a wavelength where the other substances in the sample solution do not absorb. Consider the situation illustrated in Figure 16-1. Curves A and B are the spectra of an analyte and diverse substance, respectively. Curve C is the spectrum of these two substances combined in a sample. Clearly, when both substances are present, absorbance measurements for the analyte cannot be made below 560 nm without interference from the diverse substance. At 520 nm, the absorbance due to the analyte is only 0.70, but the measured value is 0.84, a 20% error. Measuring on the shoulder at a wavelength between 560 and 565 nm would be preferable because there is no interference from the diverse substance at these wavelengths. In the absence of interfering substances, absorbance measurements should be made at the wavelength corresponding to

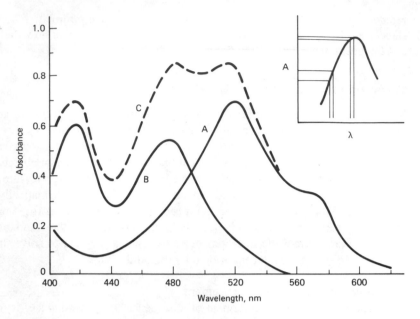

Figure 16-1 Absorption spectra of two substances (A and B) and a mixture of the two (C). Inset: Effect of wavelength error on absorbance.

the largest absorption peak because this is where the absorbance is most sensitive to changes in concentration and least sensitive to changes in wavelength. The sensitivity to changes in concentration is not difficult to rationalize. If the concentration of the substance responsible for curve A in Figure 16-1 is cut in half, the absorbance at 520 nm will change by 0.35 (0.70 − 0.35) but the change at 565 nm will be only 0.16 (0.32 − 0.16). The sensitivity of the absorbance to changes in wavelength is illustrated in the inset to Figure 16-1. A small error in setting the wavelength produces a much larger error in absorbance when the measurement is made on the slope rather than at the peak.

Establishing a Relationship between Absorbance and Concentration

According to Equation 15-7, the concentration of an unknown can be determined with a single absorbance measurement if the values of ϵ and b are known. Unfortunately, this simple approach is seldom possible because either the system does not strictly adhere to Equation 15-7 or ϵ is not known with a sufficiently high degree of accuracy. The overwhelming majority of methods rely on the use of a calibration curve constructed from the measured absorbances of a series of standard solutions. These standards are prepared in the same manner as the unknown (and preferably at the same time) and cover the concentration range within which the unknown is expected to fall. Two typical calibration curves are shown in Figure 16-2. System A conforms to Equation 15-7 while system B does not. Although nonlinear calibration curves are sometimes used, generally they are an indication that the chemical behavior of the system is not being interpreted correctly or that the instrumental parameters are not optimized. The slope of the calibration curve is a measure of the *sensitivity* or ability to distinguish between different concen-

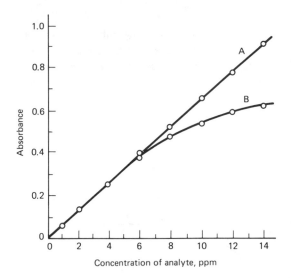

Figure 16-2 Spectrophotometric calibration curves: A, conforming to Beer's law; B, not conforming to Beer's law.

trations. The calibration curve of system B in Figure 16-2 is relatively useless at concentrations above 10 ppm because of its diminishing slope.

When the sample contains other constituents that absorb or influence the absorbance of the analyte, the values obtained from a regular calibration curve will be in error. In such situations, a *standard addition* technique may yield better results. The basis of this technique was discussed in Chapter 13. In general, known quantities of standard are added to identical aliquots of sample, and the calibration curve is prepared by plotting the measured absorbances versus the concentration of added standard. If the system conforms to Beer's law (Equation 15-7), a straight line is obtained that can be extrapolated to the concentration axis to give the unknown concentration, as shown in Figure 16-3. Since a linear extrapolation is necessary, it is absolutely essential that the chemical system conform to Beer's law.

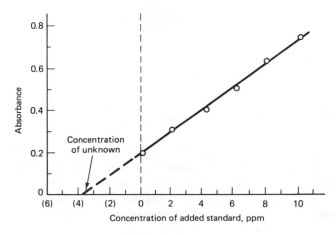

Figure 16-3 Standard addition calibration curve.

Calculations

Ordinarily, the calibration curve is prepared using the same unit of concentration that is desired for the unknown, and the only additional calculation necessary is the correction for sample dilution during pretreatment. Even this may be unnecessary if the standards and unknown are diluted to the same extent.

Example 16-1

A 25.0-mL aliquot of Mississippi River water was treated with a mild reducing agent and then excess 2,9-dimethyl-1,10-phenanthroline. This reagent reacts with copper but not iron(II). After diluting to 50.0 mL, the solution had an absorbance of 0.388 at 455 nm. From the calibration curve, this absorbance corresponded to a concentration of 3.1 ppm Cu. Calculate the copper concentration in parts per million in the original sample.

Solution   The *amount* of Cu in the original sample is the same as the amount in the measured solution. Thus

$$C_{Cu} \times 25.0 \text{ mL} = 3.1 \text{ ppm} \times 50.0 \text{ mL}$$

and

$$C_{Cu} = 3.1 \text{ ppm} \times \frac{50.0 \text{ mL}}{25.0 \text{ mL}} = 6.2 \text{ ppm}$$

Example 16-2

If the density of the water sample in Example 16-1 was 1.00 g/mL, calculate the percent copper in the water.

Solution   If the concentration of Cu was 6.2 ppm, the weight of Cu in the 25.0-mL sample is given by

$$\text{wt Cu} = 6.2 \text{ mg Cu/L} \times 0.0250 \text{ L} = 0.155 \text{ mg}$$

Then

$$\% \text{ Cu} = \frac{\text{wt Cu}}{\text{wt sample}} \times 100$$

$$= \frac{0.155 \text{ mg}}{25.0 \text{ mL} \times 1.00 \text{ g/mL} \times 1000 \text{ mg/g}} \times 100$$

$$= 6.2 \times 10^{-4}$$

TECHNIQUES OF SPECTROPHOTOMETRIC DETERMINATIONS

Spectrophotometric measurements have been applied to quantitative determinations in a variety of ways. Substances that absorb or can be made to absorb can be determined either singly or simultaneously. The reliability, speed, and ease with which absorbance can be measured make spectrophotometric methods almost ideal for automated systems designed to handle large numbers of samples with a minimum of operator assistance. In addition, absorbance may be used as a concentration monitor to follow the progress of a titration.

Direct Determinations

Procedures that call for measuring the absorbance of the analyte in some suitable form are classified as *direct determinations*. They are very common in spectrophotometry, mainly because of their relative simplicity and speed.

Organic compounds

Many organic compounds absorb quite strongly, especially in the ultraviolet. Their absorption is due largely to the presence of a particular collection of atoms in the molecule, called a chromophoric group. Table 16-1 lists some of the common chromophoric groups along with their approximate molar absorptivity and wavelength of maximum absorption. You may note that all of these chromophores contain multiple bonds or atoms with unshared bonding electrons. The shared electrons in single bonds are held so tightly that the energies required to excite them correspond to wavelengths below 180 nm, in a region called the vacuum ultraviolet. The region is so named because high vacuum is required in the instruments to avoid absorption of the radiation by CO_2 and O_2 in air. The data in Table 16-1 show clearly that increasing the degree of conjugation in an organic compound shifts the wavelength of maximum absorbance (λ_{max}) to larger values and increases the molar absorptivity.

Inorganic compounds

As pointed out earlier, few inorganic substances absorb strongly enough to be determined directly without modification. The normal practice is to add a complexing agent that will react selectivity with the inorganic ion, converting it to a highly absorbing substance. Carrying out a quantitative reaction at the trace concentration level may be inconvenient, but it provides a means of obtaining selectivity for the analyte.

Although a few inorganic ligands such as iodide, cyanide, and thiocyanate have been used in the determination of metals, organic chelating agents generally are preferred because they can be "tailor-made" for a particular metal ion. Most chelating agents contain strongly absorbing chromophoric groups that participate in bonding to the metal ion. As a result, the absorption spectra of the metal chelate and the unbound chelating agent usually are sufficiently different to permit the absorbance of the former to be measured without interference from the latter. A selected list of organic reagents is given in Table 16-2. The analyte listed is the one for which the reagent is used most commonly. Often, a single reagent can be used to determine several substances, depending on the reaction conditions and measurement wavelength employed.

Indirect Determination

Occasionally, a situation is encountered where the analyte does not absorb and cannot be converted conveniently to an absorbing species. In such cases it may be possible to determine the analyte indirectly by its ability to decrease the absorbance of a substance with which it reacts. Fluoride ion commonly is determined by such a technique. Highly colored zirconium eriochrome cyanine R is rapidly decolorized by fluoride, which re-

TABLE 16-1 ABSORPTION CHARACTERISTICS OF COMMON CHROMOPHORIC GROUPS

Group	Example	λ_{max} (nm)	ϵ_{max}
Hydrocarbon Alkene	$H_2C{=}CH_2$	170	16,000
	$H_2C{=}CH{-}CH{=}CH_2$	220	21,000
	$H_2C{=}CH{-}CH{=}CH{-}CH{=}CH_2$	260	35,000
Aromatic	(benzene ring)	184	60,000[a]
	(naphthalene)	220	100,000[a]
	(anthracene)	253	200,000[a]
Sulfide	$C_2H_5{-}S{-}C_2H_5$	194	4,600
		215	1,600
Disulfide	$C_2H_5{-}S{-}S{-}C_2H_5$	194	5,500
Amine Aliphatic	$(C_2H_5)_2NH$	193	2,500
	$(C_2H_5)_3N$	199	3,950
Aryl	(phenyl)$-NH_2$	230	8,600
		280	1,430
Aromatic	(pyridine)	195	7,500
		251	2,000
	(quinoline)	226	35,500[a]
Nitro	(phenyl)$-NO_2$	269	7,800
Nitrite	$C_4H_9{-}O{-}N{=}O$	218	1,050
Isothiocyanate	$C_2H_5{-}N{=}C{=}S$	250	1,200
Alcohol, Aryl	(phenyl)$-OH$	211	6,200
		270	1,450

TABLE 16-1 ABSORPTION CHARACTERISTICS OF COMMON CHROMOPHORIC GROUPS (*cont.*)

Group	Example	λ_{max} (nm)	ϵ_{max}
Aldehyde, Aryl	⬡—CHO	242	14,000[a]
Acid, Aryl	⬡—CO$_2$H	202 228	8,000 10,000
Quinone	O=⬡=O	242	24,000

[a]Largest of several peaks.

moves zirconium from the complex, forming stable $ZrOF_2$. A calibration curve is prepared from the absorbances of a series of solutions, each containing the same amount of zirconium complex but different amounts of fluoride.

Simultaneous Multicomponent Determinations

Sometimes, the absorption spectra of two substances overlap to such an extent that no wavelength exists where one substance absorbs without interference from the other. Such a situation is illustrated in Figure 16-4. If their absorbances are additive and adhere to Equation 15-7, it may be possible to determine both components by making measurements at two wavelengths and solving simultaneous equations. The absorbances of the mixture at wavelengths 1 and 2 are given by the expressions

$$A_{\lambda_1} = A_{x,\lambda_1} + A_{y,\lambda_1} = \epsilon_{x,\lambda_1} b C_x + \epsilon_{y,\lambda_1} b C_y$$

$$A_{\lambda_2} = A_{x,\lambda_2} + A_{y,\lambda_2} = \epsilon_{x,\lambda_2} b C_x + \epsilon_{y,\lambda_2} b C_y$$

If the molar absorptivities of both components are known at the selected wavelengths, these equations can be solved simultaneously to give

$$C_x = \frac{A_{\lambda_1}\epsilon_{y,\lambda_2} - A_{\lambda_2}\epsilon_{y,\lambda_1}}{b(\epsilon_{x,\lambda_1}\epsilon_{y,\lambda_2} - \epsilon_{y,\lambda_1}\epsilon_{x,\lambda_2})} \qquad (16\text{-}1)$$

$$C_y = \frac{A_{\lambda_2} - \epsilon_{x,\lambda_2} b C_x}{\epsilon_{y,\lambda_2} b} \qquad (16\text{-}2)$$

If the molar absorptivities are not known, they can be determined experimentally from the measured absorbances of known concentrations of each component.

Selecting the appropriate wavelengths is not a routine task because the absorbances may not be additive and Beer's law may not be obeyed at all wavelengths. Furthermore, there are mathematical constraints on the relative values of the molar absorptivities if the errors in the calculated concentrations are to be kept at an acceptable level. For

Techniques of Spectrophotometric Determinations

TABLE 16-2 SELECTED ORGANIC REAGENTS USED
IN THE SPECTROPHOTOMETRIC DETERMINATION OF METALS

Name	Structure	Metal	λ_{max}(nm)	ϵ_{max}
Alizarin (1,2-dihydroxy-anthraquinone)		Zr	525	5.3×10^3
Arsenazo III (2,2'-[1,8-dihydroxy-3,6-disulfo-2,7-naphthylene(azo)] dibenzenearsonic acid		Th	655	1.2×10^5
Chromotropic acid (1,8-dihydroxynaphthalene-3,6-disulfonic acid)		Ti	460	1.7×10^4
Diphenylthiocarbazone (3-mercapto-1,5-diphenyl formazan)		Pb	520	6.6×10^4
8-Hydroxyquinoline		Al	386	6.6×10^3
Nitroso-R salt (1-nitroso-2-hydroxy-3,6-naphthalene disodium sulfonate)		Co	500	1.5×10^4

416

TABLE 16-2 SELECTED ORGANIC REAGENTS USED IN THE
SPECTROPHOTOMETRIC DETERMINATION OF METALS (*cont.*)

Name	Structure	Metal	λ_{max} (nm)	ϵ_{max}
1,10-Phenanthroline		Fe	508	1.1×10^4
N-Phenylbenzohydroxamic acid		V	525	5.1×10^3
Potassium thiocyanate	KSCN	Mo	470	2.0×10^4
1-(2-Pyridylazo)-2-naphthol		Zn	515	2.3×10^4
Sodium diethyldithiocarbamate		Cu	436	1.3×10^5

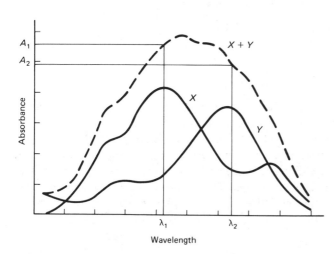

Figure 16-4 Overlapping spectra of two absorbing substances.

example, if the molar absorptivities are such that the denominator in Equation 16-1 is nearly zero, a small error in one of the molar absorptivities or absorbances will cause a large error in the calculated concentration. Simultaneous determinations of more than three components are rare due to the difficulty of finding wavelengths that do not lead to unacceptably large errors in the calculated concentrations.

Example 16-3

The ultraviolet spectra of o- and p-cresol overlap extensively. A 5.00-mL sample containing the two compounds was dissolved and diluted to 50.0 mL with isooctane. This solution had absorbances of 0.716 at 264 nm and 0.318 at 272 nm. A solution of 4.63×10^{-4} M o-cresol had absorbances of 0.730 at 264 nm and 0.178 at 272 nm. A solution of the para isomer of the same concentration had absorbances of 0.548 at 264 nm and 0.433 at 272 nm. If all measurements were made in 1-cm cells, calculate the molar concentration of each isomer in the sample.

Solution The molar absorptivities of each isomer at the two wavelengths can be calculated using the Beer's law equation. For the ortho isomer:

At 264 nm: $0.730 = \epsilon_{o,264}(1.00 \text{ cm})(4.63 \times 10^{-4} M)$

$$\epsilon_{o,264} = 1577 \; M^{-1} \cdot cm^{-1}$$

At 272 nm: $0.178 = \epsilon_{o,272}(1.00 \text{ cm})(4.63 \times 10^{-4} M)$

$$\epsilon_{o,272} = 384 \; M^{-1} \cdot cm^{-1}$$

For the para isomer:

At 264 nm: $0.548 = \epsilon_{p,264}(1.00 \text{ cm})(4.63 \times 10^{-4} M)$

$$\epsilon_{p,264} = 1184 \; M^{-1} \cdot cm^{-1}$$

At 272 nm: $0.433 = \epsilon_{p,272}(1.00 \text{ cm})(4.63 \times 10^{-4} M)$

$$\epsilon_{p,272} = 935 \; M^{-1} \cdot cm^{-1}$$

Using these values with the data given, Equations 16-1 and 16-2 can be solved:

$$C_{o\text{-cresol}} = \frac{(0.716)(935) - (0.318)(1184)}{(1577)(935) - (1184)(384)} = 2.87 \times 10^{-4} \; M$$

$$C_{p\text{-cresol}} = \frac{(0.318) - (384)(2.87 \times 10^{-4})}{935} = 2.22 \times 10^{-4} \; M$$

Automatic and Repetitive Determinations

When a large number of similar samples must be analyzed, automatic analyzers may be especially useful because they can provide analytical results rapidly and with minimal operator assistance. Nowhere has the need for automation been greater than in the clinical laboratory, where more than 30 constituents are determined routinely in blood and urine samples. It is not unusual for a large clinical laboratory to average one determination every 30 seconds every day of the year. Three types of automatic analyzers commonly found in clinical laboratories, all using spectrophotometric measurements, are described in the following sections.

Discrete analyzers

These devices treat each sample as a separate entity in its own cell or container. In one system, both the sample and the required reagents are pumped or pipetted automatically into separate chambers of a reaction vessel shown in Figure 16-5. The vessels are placed on a turntable or conveyor belt that carries them sequentially through the necessary steps, such as mixing, heating (optional), and measurement. In the Technicon Single Test Analyzer (STAC), mixing is accomplished using a motor-driven piston to move the liquid back and forth between adjacent chambers (Figure 16-5). Systems of

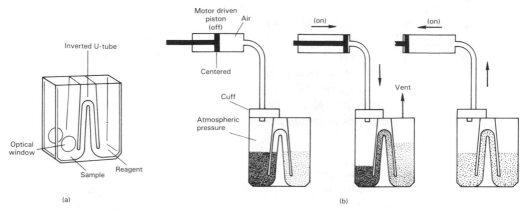

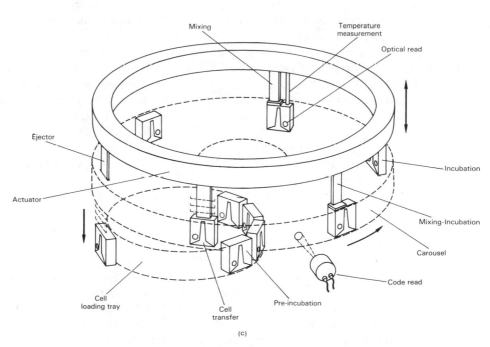

Figure 16-5 STAC discrete sample analyzer: (a) STAC cell; (b) mixing procedure; (c) analyzer system. (Courtesy of Technicon Instruments Corporation)

Techniques of Spectrophotometric Determinations

this type usually include wash and rinse stations which permit the reaction vessels to be recycled into the analysis train.

The STAC and other similar analyzers generally can perform a variety of analyses because they use a computer to monitor and control the various operations. The computer can be programmed for different reagent additions, mixing times, heating temperatures, and so forth, thereby allowing each substance to be determined at a unique set of conditions.

Discrete analyzers perform automatically those operations that are normally done manually. All of the chemistry and knowledge of a particular manual method is retained in the automated procedure. With a computer-controlled system, only the desired determination is performed on a given sample and that determination can be changed for the next sample in sequence. Unlike the segmented flow analyzer described in the next section, the discrete analyzer uses relatively small amounts of reagents, which can be a significant advantage when the reagents are expensive. The primary disadvantage of these analyzers is their mechanical complexity, which makes them more expensive and susceptible to mechanical failure than other types of automatic analyzers.

Segmented-flow analyzers

In this type of analyzer, successive samples are pumped continuously through a small-diameter tube, and at various points, mixed with reagents that react with the analyte to produce the absorbing species. The Technicon AutoAnalyzer, shown in Figure 16-6, uses a series of individual modules, each performing one specific function. Ordinarily, modules can be rearranged or even omitted for different analytical methods. Samples are fed into the analyzer via a small tube that periodically dips into each sample on a large tray or turntable. The samples are mixed with a diluent and then segmented with air bubbles. When blood samples are being analyzed, the segmented sample and a similarly prepared reagent flow into a dialyzer containing membranes that allow small ions and molecules in the sample to pass into the reagent stream. The interfering large protein molecules in the blood remain in the sample stream and are ultimately discarded. Depending on the analyte being determined and the method used, the sample–reagent mixture may pass through a heating bath to speed up the color-developing reaction. Finally, the bubbles are removed and the solution enters a flow-through cell positioned in the optical path of a spectrophotometer.

The sampling sequence of flow analyzers usually can be programmed by the operator. Typically, each sample is run in triplicate and standards are run after every 10 unknowns. In many cases, the slowest step in the overall analysis is the drawing of the blood sample or transporting it to the laboratory.

Segmented-flow analyzers are available in configurations other than that shown in Figure 16-5. For example, a dialyzer may be unnecessary with urine samples because they do not contain the interfering proteins present in blood samples. In such cases, the reagent is mixed directly with the sample. Multichannel flow analyzers often contain dozens of independently operating units, each optimized in terms of sample size, reagent concentrations, flow rates, mixing times, and temperature for a particular analyte determination.

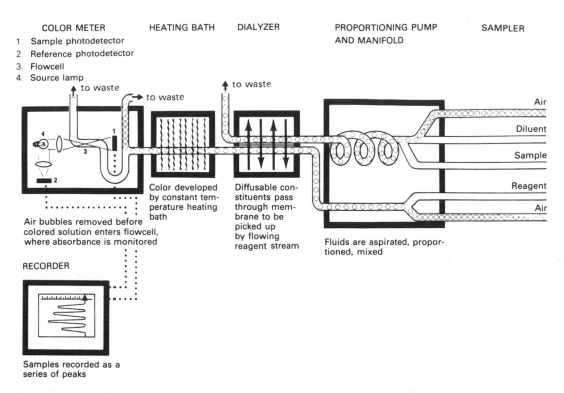

COLOR METER
1. Sample photodetector
2. Reference photodetector
3. Flowcell
4. Source lamp

HEATING BATH

DIALYZER

PROPORTIONING PUMP
AND MANIFOLD

SAMPLER

to waste

to waste

to waste

Air

Diluent

Sample

Reagent

Air

Color developed by constant temperature heating bath

Diffusable constituents pass through membrane to be picked up by flowing reagent stream

Fluids are aspirated, proportioned, mixed

Air bubbles removed before colored solution enters flowcell, where absorbance is monitored

RECORDER

Samples recorded as a series of peaks

Figure 16-6 Single-channel Technicon AutoAnalyzer. (Courtesy of Technicon Instruments Corporation)

Centrifugal analyzers

Whereas discrete and segmented-flow analyzers treat samples sequentially, this analyzer effectively "multiplexes" the detector and examines a number of samples simultaneously. A large number of sample disks, as shown in Figure 16-7, each containing a sample, reagent, and observation chamber, are arranged on a round tray. The tray is rotated so that each disk passes a sample and reagent filling station. When the filling operations are completed, the tray rotation becomes continuous and is accelerated until the resulting centrifugal force pushes the sample first into the reagent chamber and then into the observation chamber. As the disk rotates, each sample passes through the optical beam, where its absorbance is measured. A typical disk rotation is 600 rpm, which means that all of the disks on the tray are measured once each 100 ms. Ordinarily, values from 6 to 10 consecutive rotations are averaged for each sample. As is the case with the other analyzers, most of the operations are monitored and controlled by a computer.

Spectrophotometric Titrations

The spectrophotometer is a useful device for detecting the equivalence point in a titration. As absorbance is directly proportional to concentration, we may use it to follow

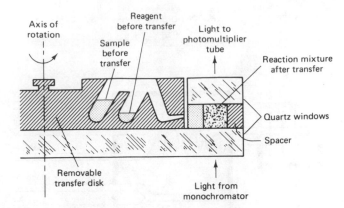

Figure 16-7 Cross-sectional view of a centrifugal analyzer sample disk. (Reprinted from American Laboratory, volume 3, number 7, page 26, 1971. Copyright 1971 by International Scientific Communications, Inc.)

the progress of any titration in which the analyte, titrant, or product absorbs. The usual procedure is to acquire data well before and after the equivalence point and plot the absorbance (corrected for dilution) as a function of the volume of titrant added. The basis of this approach was discussed in Chapter 5. The titration curve will consist of two straight-line portions of differing slopes, one before and one after the equivalence point. The end point is taken as the extrapolated intersection of the two straight lines.

The titration of $Cu(II)$ with EDTA is shown by curve (a) in Figure 16-8. At 745 nm only the product of the reaction, CuY^{2-}, absorbs. Consequently, the absorbance is zero initially but increases as product is formed. No additional product is formed *after* the equivalence point, so the absorbance remains constant. The titration of iron(II) with permanganate is illustrated by curve (b) in Figure 16-8. Here the measurement wavelength is 520 nm and only the permanganate titrant absorbs. Since no titrant remains unreacted in the solution prior to the equivalence point (other than a negligible equilibrium amount), the absorbance remains at zero and rises only after the equivalence point is passed.

It is possible to apply the spectrophotometric technique to a stepwise titration of two components. Bismuth(III) and copper(II) have been determined this way using EDTA as the titrant. Bismuth forms the more stable complex and reacts first. At 745 nm, only CuY^{2-} absorbs. Thus the absorbance is initially zero and remains so until just after the bismuth end point where it begins to rise as CuY^{2-} is formed. When the copper is completely titrated, the absorbance reaches a maximum and remains constant on the addition of excess titrant. The titration curve is shown in Figure 16-9.

Instrumentation

Spectrophotometric titrations can be performed using regular, commercial spectrophotometers that have been modified to accept the titration vessel in place of the normal sample cell. It is usually unnecessary to measure the true absorbance, because end-point detection is based on how the absorbance *changes* (relative values). As a result, reagent blanks and matching reference containers are unnecessary.

Instruments designed specifically for spectrophotometric titrations are available. Usually, they are able to accommodate automatic titrant delivery systems, which are useful when large numbers of samples are being titrated.

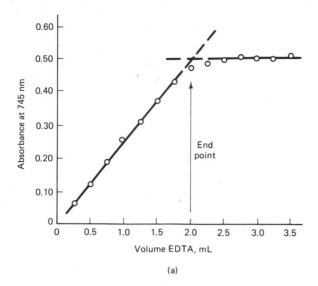

(a)

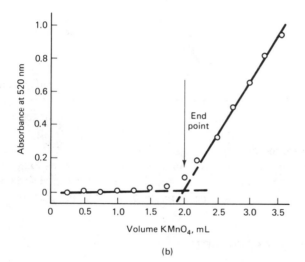

Figure 16-8 Spectrophotometric titration curves:
(a) 20.0 mL of 5.00×10^{-3} M Cu^{2+} titrated with 5.00×10^{-4} M EDTA;
(b) 20.0 mL of 1.00×10^{-3} M Fe^{2+} titrated with 2.00×10^{-3} M $KMnO_4$.

(b)

Applications

Spectrophotometric end-point detection has been applied to all types of reactions, including precipitation reactions in which the suspended solid product diminishes the radiant power by scattering rather than absorption. Although stepwise titration of mixtures is possible, it is limited in scope. In addition to the basic requirements of a stepwise titration, only certain combinations of reactants, titrant, and products can absorb at a particular wavelength if more than one end point is to be detected.

It was pointed out in Chapter 5 that end-point detection using data well away from the equivalence point is especially useful with reactions whose equilibrium constants are not particularly large. Incompleteness of the titration reaction is manifested as curvature

Techniques of Spectrophotometric Determinations

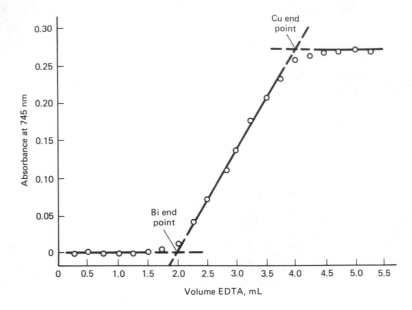

Figure 16-9 Curve for the spectrophotometric titration of 50.0 mL of a solution 4.00 $\times 10^{-3}$ M in both Bi^{3+} and Cu^{2+} with 0.100 M EDTA.

in the vicinity of the equivalence point of the spectrophotometric titration curve. This curvature is not observed away from the equivalence-point region because of the presence of either excess analyte (before) or titrant (after), which forces the reaction to completion.

In contrast to direct spectrophotometry, background absorption due to other constituents in the sample does not interfere in spectrophotometric titrations, as long as it remains constant throughout the titration. This is so because the *change* in absorbance rather than its absolute value is used to locate the equivalence point of the reaction.

METHODS FOR OBTAINING THE STOICHIOMETRY OF COMPLEXES

Over the years, several methods employing spectrophotometric measurements have been developed to determine the composition of complexes. Spectrophotometry is a valuable tool for this purpose because absorbance measurements can be made without perturbing the equilibrium of the system being examined. Three of the most common methods are described briefly in the following sections.

Continuous-Variations Method

This method is based on the measurement of a series of solutions in which molar concentrations of the two reactants vary but their sum remains constant. The absorbance of each solution is measured at a suitable wavelength, corrected for any absorbance the solution would have if no reaction occurred, and plotted versus the mole fraction of one reactant. Ordinarily, solutions are prepared in which the mole fraction varies from 0 to 1. A typical plot is shown in Figure 16-10. A maximum in the absorbance occurs at the mole ratio corresponding to the combining ratio of the reactants. If the complex absorbs less strongly than the reactants, a minimum rather than a maximum will be observed in the plot. The maximum absorbance is taken as the intersection of the extrapolated straight-line portions of the curve to avoid uncertainty caused by rounding of the curve near the maximum absorbance.

The extrapolated maximum for curve A in Figure 16-10 occurs when the mole fraction of M $[C_M/(C_M + C_L)]$ is 0.33 or the mole fraction of L $[C_L/(C_M + C_L)]$ is 0.67. Thus C_M/C_L is $0.33/0.67$ or $1/2$, implying a reacting ratio of 1 metal to 2 ligands or a complex-ion composition of ML_2. Similarly, curve B is maximum when the mole fraction of M and L is 0.50, which suggest a composition of ML.

The curvature near the maximum results from incompleteness of the formation reaction. Reactions that are substantially incomplete produce curves that are so rounded that accurate extrapolations are impossible. The difference between the theoretical and experimental maxima can be used to determine the formation constant of the complex.

A continuous-variations plot generally will not produce a valid result if more than

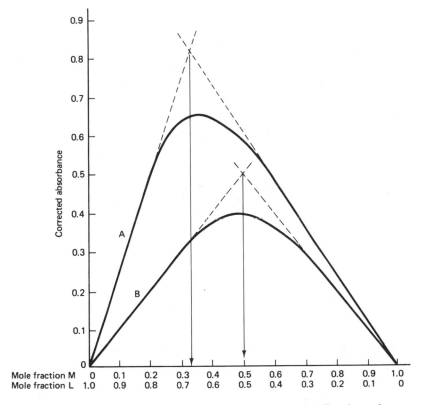

Figure 16-10 Continuous variations plots for 1:2 and 1:1 metal-to-ligand complexes.

Methods for Obtaining the Stoichiometry of Complexes

one complex is formed. If a single complex is formed, the maximum of a continuous-variations plot should be independent of wavelength. Consequently, it is common practice to measure the absorbance of the prepared solutions at several wavelengths. A maximum that varies with wavelength suggests the presence of more than one complex.

Mole-Ratio Method

In the mole-ratio method, solutions containing the same amount of metal are treated with increasing amounts of ligand. The measured absorbance is plotted against the molar ratio of ligand to metal, as shown in Figure 16-11. Again, if the reaction is sufficiently complete, two straight lines of different slope are obtained. The intersection of the extrapolated lines corresponds to the ligand-to-metal ratio in the complex. As was the case with the method of continuous variations, the degree of curvature between the two straight lines is indicative of the incompleteness of the formation reaction. Unlike the method of continuous variations, the measured absorbance does not have to be corrected by subtracting the absorbance the solution would have had if no reaction occurred.

The mole-ratio method generally is superior to the method of continuous variations for complexes having large ligand-to-metal ratios. For example, the relative difference between the position of the maximum for ML_5 and ML_6 is 20% in a mole ratio plot and 3% in a continuous-variations plot, as shown by the data in Table 16-3.

Slope-Ratio Method

The slope-ratio method, used mainly in studying weak complexes, requires that the formation reaction can be forced to completion with a large excess of either metal or ligand. Two sets of solutions are prepared: The first contains various amounts of metal ion each

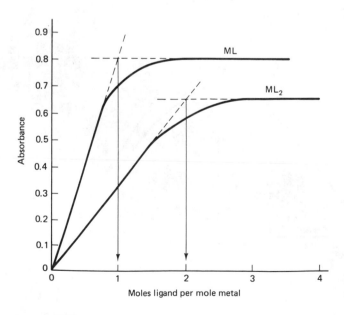

Figure 16-11 Mole-ratio plots for 1 : 1 and 1 : 2 metal-to-ligand complexes.

TABLE 16-3 COMPARISON OF CONTINUOUS VARIATIONS AND
MOLE-RATIO DATA FOR DETERMINING METAL-LIGAND RATIOS

	Method	
	Continuous variations	Mole ratio
Measured ratio for ML_5	0.833	5.00
Measured ratio for ML_6	0.856	6.00
Absolute difference	0.023	1.00
Relative difference (%)	3	20

with the same large excess of ligand, while the second consists of various amounts of ligand each with the same large excess of metal. For the reaction

$$x M + y L \rightleftharpoons M_x L_y$$

when L is present in large excess (driving the reaction to completion), the concentration of product formed is limited by the concentration of the metal, or

$$[M_x L_y] = \frac{C_M}{x}$$

If the system conforms to Beer's law,

$$A = \epsilon b [M_x L_y] = \frac{\epsilon b C_M}{x}$$

and a plot of A versus C_M will yield a straight line with a slope of $\epsilon b / x$. Similarly, for the solutions containing M in large excess,

$$[M_x L_y] = \frac{C_L}{y}$$

and

$$A = \epsilon b [M_x L_y] = \frac{\epsilon b C_L}{y}$$

A similar plot of A versus C_L will produce a straight line with a slope of $\epsilon b / y$. The ratio of the two slopes is the combining ratio for the reaction

$$\frac{\epsilon b / x}{\epsilon b / y} = \frac{y}{x}$$

PROBLEMS

16-1. A 10.00-mL sample containing benzene was diluted to 250.0 mL with hexane and the resulting solution had an absorbance of 0.364 at 204 nm in a 5.00-cm cell. If the molar absorptivity is 7.9×10^3, calculate the concentration of benzene (C_6H_6) in units of mg/mL in the original solution.

16-2. The amino acid tyrosine has an absorption peak at 290 nm. After appropriate treatment of a 0.2162-g sample to isolate the tyrosine, the sample volume was 250.0 mL. This solution had an absorbance of 0.118 in a 2.00-cm cell. A sample of pure tyrosine (MW = 181), weighing 126.4 mg, was dissolved in 100.0 mL of solution. A 10.00-mL aliquot of this solution, diluted to 250.0 mL, had an absorbance of 0.473 at 290 nm in a 1.00-cm cell. Calculate the percentage of tyrosine in the sample.

16-3. A 200-mL sample of well water was treated with excess hydroxylamine hydrochloride to reduce the iron to ferrous ion. After addition of an acetic acid–sodium acetate buffer, excess 4,7-diphenyl-1,10-phenanthroline was added and the volume brought to 250.0 mL with water. The solution had a percent transmittance of 63.1 at the absorption maximum of 533 nm. Another solution, prepared by dissolving 0.0725 g of pure iron wire in acid, was treated similarly, except that the final volume was 1.00 L. A 10.00-mL aliquot diluted to 100.0 mL had an absorbance of 0.288 in the same cell used for the first measurement. Calculate the parts per million of iron in the well water.

16-4. Nickel forms a complex with 1-(2-pyridylazo)-2-naphthol (PAN) that can be represented as NiL_2. A series of solutions of known nickel concentration were treated with an excess of PAN and had the following absorbances at 575 nm:

$C_{Ni^{2+}}$ (ppm)	A_{575}
0.20	0.172
0.35	0.300
0.50	0.438
0.65	0.557
0.80	0.694
0.95	0.830

A 10.00-mL sample treated similarly and diluted to 25.00 mL had an absorbance of 0.417. Calculate the parts per million of nickel in the unknown from a standard curve of the data.

16-5. Fluoride can be determined by its ability to destroy the colored zirconium eriochrome cyanine R complex, ZrL_2:

$$ZrL_2 + H_2O + 2F^- \longrightarrow ZrOF_2 + 2HL^-$$

When a 10.00-mL sample (density 1.00 g/mL) containing fluoride ion was added to 25.00 mL of ZrL_2, the absorbance due to the complex decreased from 0.800 to 0.379 in a 1.00-cm cell. If the molar absorptivity of ZrL_2 is 2.17×10^5, calculate the parts per million of F^- in the sample. *Hint:* Do not forget about dilution.

16-6. A 50.00-mL sample containing inorganic phosphate ions was treated with excess sodium molybdate in perchloric acid, resulting in the formation of a compound called 12-molybdophosphoric acid (MPA):

$$H_3PO_4 + 12MoO_4^{2-} + 24H^+ \longrightarrow H_3PMo_{12}O_{40} + 12H_2O$$

The MPA was separated from excess sodium molybdate by extraction into *n*-butanol. The extract was shaken with aqueous sodium hydroxide, which destroyed the MPA by the reverse of the reaction shown above, and the simple ions transferred to the aqueous phase which was diluted to 25.00 mL with water. The absorbance of this solution was 0.663 in a 1.00-cm cell at 210 nm, a wavelength at which the only absorbing substance was MoO_4^{2-}. If the molar absorptivity of Na_2MoO_4 is 1.6×10^3 and the density of the original sample was 1.00 g/mL, calculate the ppm P in the sample.

16-7. Bilirubin in blood serum absorbs strongly at 461 nm but not at all at 551 nm. On the other hand, the absorbance due to hemoglobin and solution turbidity is about the same at both wavelengths. Thus a simple method for determining bilirubin is based on the difference in absorbance at the two wavelengths. A 20.0-μL sample of blood serum, diluted to 1.00 mL with sodium citrate buffer of pH 8.8, had absorbances of 0.871 and 0.637 at 461 and 551 nm, respectively. When 20.00 μL of a bilirubin standard (50.0 μg/mL) was treated similarly, the absorbances were 0.337 and 0.014 at 461 and 551 nm. Calculate the concentration of bilirubin in units of μg/mL in the serum.

16-8. The copper complex with 2,9-dimethyl-1,10-phenanthroline has a molar absorptivity of 7.95×10^3 at 455 nm. If a sample contains about 0.1% Cu and the final volume containing all of the sample will be 250 mL, what are the minimum and maximum sample weights that will keep the measured absorbance between 0.2 and 0.8 in a 2.00-cm cell?

16-9. Sodium nitrite is often added to certain types of meat to retard oxidation reactions that cause red meat to take on a gray color. Suppose that you are directed to plan a spectrophotometric method for determining nitrite in meat based on its color-forming reaction with N-1-naphthylenediamine and sulfanilic acid ($\epsilon = 2.0 \times 10^4$). If your sample will be diluted to 250 mL and the absorbance measured in 1.00-cm cells, what weight do you recommend so that a sample of average concentration, about 1300 ppm NO_2^-, will have an absorbance of about 0.4?

16-10. Glucose in blood serum can be determined spectrophotometrically by formation of a colored complex with o-tolidine. Six identical 50.0-μL serum samples were treated with various concentrations of glucose and taken through the procedure, producing the following results:

Amount glucose added (μg)	Absorbance
0	0.230
10	0.272
25	0.340
40	0.416
60	0.507
75	0.568

Prepare a standard addition calibration plot and calculate the glucose concentration in the serum in units of mg/100 mL.

16-11. A solution had an absorbance of 0.471 in a 1.00-cm cell when measured against distilled water as the reference. A reagent blank in the same cell had an absorbance of 0.065 against distilled water. What absorbance will the solution have when measured against the reagent blank as the reference?

16-12. The following data were obtained in a determination of mercury using dithizone as the color-forming reagent.

	Cell length (cm)	% Transmittance
Sample vs. H_2O	1.00	13.4
Reagent blank vs. H_2O	5.00	71.6

Calculate the expected absorbance of the sample when measured against the reagent blank in 1.00-cm cells.

16-13. Suppose that the benzene sample in Problem 16-1 was measured against pure water as a reference. If the absorbance of hexane measured against pure water was 0.013 in a 5.00-cm cell, calculate the following.

(a) expected absorbance of the benzene solution when measured against hexane as the reference

(b) absolute and percent relative error in the determination by not using hexane as the reference solution

16-14. The molar absorptivities of M^{2+}, L^{2-}, and ML_2^{2-} at 720 nm are 70.0, 410, and 8.93 $\times$ 10^3, respectively, and the formation constant of ML_2^{2-} is large. If a solution was prepared by mixing 10.00 mL of 1.50×10^{-5} M M^{2+} with 5.00 mL of 5.00×10^{-4} M L^{2-} and diluting to 25.00 mL with water, calculate the expected absorbance at 720 nm in the following.

(a) 5.00-cm cell versus water as a reference

(b) 2.00-cm cell versus 2.5×10^{-4} M L^{2-} as a reference.

16-15. The molar absorptivities of the complexes NiL_2 and ZnL_2 at their respective absorption peaks are

	553 nm	637 nm
NiL_2	1.22×10^3	4.73×10^2
ZnL_2	0	6.29×10^3

A sample weighing 412.6 mg was dissolved, treated with excess L, and diluted to 50.00 mL. This solution had absorbances in 1.00-cm cells of 0.416 and 0.923 at 553 and 637 nm, respectively. Calculate the % Ni and % Zn in the sample.

16-16. *Ortho-* and *meta-*cresol have similar but not identical ultraviolet absorption spectra. A 5.00-mL sample was dissolved in enough isooctane to make the final volume 25.00 mL. Calculate the molar concentration of each isomer in the unknown from the following data.

Substance	Concentration (M)	Absorbance in 1.00-cm cell	
		λ_1	λ_2
o-Cresol	4.63×10^{-4}	0.743	0.466
m-Cresol	4.27×10^{-4}	0.339	0.609
Unknown	—	0.884	0.760

16-17. Suppose that A can be titrated with T to give P and each substance absorbs only in the spectral regions shown below.

Substance	Wavelengths absorbed (nm)
A	275–450
T	400–600, 700–800
P	250–325, 500–700

Sketch the general shape of the titration curve (absorbance versus volume of titrant added) when the absorbance is measured at each of the following wavelengths. Assume that the molar absorptivities of the absorbing substances are approximately the same at these wavelengths.

(a) 300 nm   (c) 425 nm   (e) 650 nm

(b) 375 nm   (d) 550 nm   (f) 750 nm

16-18. Suppose that X and Y in a mixture react stepwise with T, forming XT first and YT last. Sketch the expected spectrophotometric titration curve if the absorbance is measured at a wavelength where only X and YT absorb.

16-19. Strongly acidic and strongly basic solutions of an acid-base indicator HIn^+ ($pK_a = 8.150$) at an analytical concentration of $7.50 \times 10^{-5} M$ had the following absorbances in a 1.00-cm cell:

	Absorbance				Absorbance	
λ (nm)	pH = 1.0	pH = 13.0	λ (nm)	pH = 1.0	pH = 13.0	
400	0.020	0.018	510	0.188	0.475	
420	0.065	0.023	520	0.160	0.728	
440	0.300	0.030	530	0.138	0.876	
450	0.458	0.042	535	0.133	0.898	
455	0.497	0.055	540	0.126	0.840	
460	0.521	0.069	550	0.120	0.587	
465	0.501	0.087	570	0.111	0.195	
475	0.379	0.131	590	0.097	0.101	
490	0.272	0.232	600	0.085	0.063	

(a) Plot A versus λ for the two pH values.

(b) What arc the colors of HIn^+ and In?

(c) Suggest a suitable wavelength for measuring HIn^+ with little interference from In.

(d) Sketch the spectrum expected at a pH of 8.15.

(e) At what wavelength is the absorbance of the indicator independent of pH?

16-20. The complex formed between Ni^{2+} and L was investigated in the following manner. Different volumes of $1.31 \times 10^{-4} M$ L were added to ten 25.00-mL volumetric flasks containing 5.00 mL of $2.46 \times 10^{-4} M Ni^{2+}$, and after dilution to the mark with water, the following absorbances were measured at 470 nm (the absorbance maximum for the complex) in a 2.00-cm cell:

Volume of L added (mL)	Absorbance	Volume of L added (mL)	Absorbance
0	0.006	10.00	0.697
2.00	0.152	12.00	0.722
4.00	0.295	14.00	0.729
6.00	0.461	16.00	0.727
8.00	0.608	18.00	0.730

Calculate the number of ligands for each nickel in the complex.

16-21. A 5.00-mL aliquot of $1.84 \times 10^{-4} M Zn^{2+}$ was added to each of ten 25.00-mL volumetric flasks containing various amounts of L. When the flasks were diluted to volume, the resulting solutions had the following absorbances at 632 nm, the wavelength of maximum absorbance of the complex:

Amount L (mmol × 10⁴)	Absorbance	Amount L (mmol × 10⁴)	Absorbance
0	0.010	20.3	0.806
4.25	0.195	24.0	0.842
7.75	0.353	27.8	0.857
12.0	0.560	32.0	0.860
15.5	0.707	36.3	0.864

Calculate the combining ratio of Zn to L in the complex.

16-22. Calculate the value of x in the complex NiL_x from the following data. The final solution volumes were 25.00 mL and the absorbances were measured in 1.00-cm cells at a wavelength where only the complex absorbed.

Volume of 2.50 × 10⁻⁵ M Ni (mL)	Volume of 2.50 × 10⁻⁵ M L (mL)	Absorbance
0	10.00	0.004
1.00	9.00	0.237
2.00	8.00	0.490
3.00	7.00	0.738
4.00	6.00	0.712
5.00	5.00	0.599
6.00	4.00	0.475
7.00	3.00	0.356
8.00	2.00	0.240
9.00	1.00	0.122
10.00	0	0.002

16-23. The following data were collected in a continuous variations study of a colored complex. Determine the combining ratio of M and L in this complex.

Concentration (M × 10³)		Absorbance of ML_x
M	L	
0	10.00	0.002
0.50	9.50	0.202
1.00	9.00	0.397
1.50	8.50	0.596
2.50	7.50	0.828
3.50	6.50	0.783
4.50	5.50	0.704
5.50	4.50	0.597
6.50	3.50	0.460
7.50	2.50	0.326
8.50	1.50	0.194
9.50	0.50	0.059
10.00	0	0.001

16-24. Use the slope-ratio technique with the following data to determine the stoichiometry of the complex formed between M and L.

$C_L = 2.96 \times 10^{-3} M$		$C_M = 4.83 \times 10^{-3} M$	
C_M (M)	Absorbance	C_L (M)	Absorbance
8.50×10^{-5}	0.482	1.80×10^{-5}	0.034
9.70×10^{-5}	0.550	5.31×10^{-5}	0.100
1.18×10^{-4}	0.669	8.37×10^{-4}	0.158
1.30×10^{-4}	0.737	2.80×10^{-4}	0.529
1.55×10^{-4}	0.878	4.57×10^{-4}	0.862

16-25. A 5.00-mL aliquot of a $3.13 \times 10^{-3} M$ Co^{2+} solution was added to each of five 25.00-mL volumetric flasks containing various amounts of L and a 5.00-mL aliquot of a $3.13 \times 10^{-3} M$ L solution was added to each of five 25.00-mL volumetric flasks containing various amounts of Co^{2+}. When the samples were diluted to volume, the resulting solutions had the following absorbances at the wavelength of maximum absorbance of the complex.

Amount Co^{2+} (mmol × 10^4)	Absorbance	Amount L (mmol × 10^4)	Absorbance
2.50	0.035	2.50	0.013
7.50	0.106	12.5	0.061
12.5	0.173	25.0	0.128
17.5	0.241	75.0	0.389
25.0	0.352	125.0	0.646

Calculate the combining ratio of Co to L in the complex.

17

MOLECULAR FLUORESCENCE SPECTROSCOPY

Although the terms "fluorescence" and "phosphorescence" have long been used to describe certain photon emission processes, many chemists now prefer the more general term of *luminescence*. Addition of a prefix to this term permits us to distinguish between the different ways of producing the electronically excited molecules that eventually undergo photon emission. When the excited molecules are the result of photon absorption, the process is called *photoluminescence*, and when they result from a chemical reaction it is called *chemiluminescence*. The firefly is one of several living organisms that produce light by a chemiluminescence process. This interesting insect makes chemicals that react to yield an electronically excited product which undergoes photon emission in the process of returning to the ground state. Although phosphorescence and chemiluminescence have important applications in analytical chemistry, the remainder of this chapter is devoted to the more often encountered technique of fluorescence spectroscopy. The energetics of the photoluminescence process is described in Chapter 15. The discussion here focuses on the factors affecting fluorescence intensity, the instrumentation used to measure fluorescence, and the methodology of fluorometric determinations.

PRINCIPLES

All molecules have the inherent capability to fluoresce, but relatively few actually do so. The occurrence of fluorescence and its intensity are determined principally by molecular structure and chemical environment. The effects of a few important variables are considered in this section.

Photoluminescence Efficiency

It was pointed out in Chapter 15 that fluorescence is but one of several mechanisms by which molecules may lose absorbed energy. To observe fluorescence, its rate must be competitive with the rates of the other relaxation processes. The fraction of excited molecules that relax (deactivate) via fluorescence is called the *photoluminescence efficiency* or *quantum yield*, ϕ.

$$\phi = \frac{\text{number of photons emitted}}{\text{number of photons absorbed}}$$

The maximum value of ϕ is 1, which would occur when the fluorescence process is so efficient that all of the excited molecules relax this way. When no fluorescence occurs, ϕ has its minimum value of 0. Very few molecules have a photoluminescence efficiency greater than 0.5.

Fluorescence and Structure

Fluorescence in organic molecules is due largely to the presence of the aromatic functional group. Fused ring compounds are especially fluorescent, their photoluminescence efficiency increasing with the number of rings. Simple heterocyclic compounds such as pyridine do not fluoresce to any appreciable extent. Fused to a benzene ring, however, they may become highly fluorescent. Thus, whereas pyridine is nonfluorescent, quinoline (pyridine fused to a benzene ring) is highly fluorescent.

Substituents on the aromatic ring may affect both the fluorescence intensity and the wavelength region where fluorescence occurs as summarized in Table 17-1. The decrease in fluorescence (and shift to longer wavelengths) with increasing atomic weight of substituted halogen is a general phenomenon observed with almost every class of compound.

It has been demonstrated experimentally that structural rigidity in a molecule favors fluorescence. Thus fluorescein is a highly fluorescent molecule, while the very similar phenolphthalein is essentially nonfluorescent.

TABLE 17-1 SUBSTITUENT EFFECTS ON THE FLUORESCENCE OF BENZENE

Substituent	Change in wavelength of fluorescence	Change in intensity of fluorescence
Alkyl	None	None
OH, OCH_3, OC_2H_5	Decrease	Increase
CO_2H	Decrease	Large decrease
NH_2, NHR, NR_2	Decrease	Increase
NO_2, NO	—	Total quenching
CN	None	Increase
SH	Decrease	Decrease
F, Cl, Br, I	Decrease (F → I)	Decrease (F → I)
SO_3H	None	None

Principles

fluorescein

phenolphthalein

The bridge oxygen in fluorescein keeps the molecule planar and inhibits internal molecular motions that lead to internal conversion and vibrational relaxation.

Fluorescence and Environment

Temperature, pH, dissolved oxygen, and solvent commonly affect the fluorescence of molecules. A rise in temperature almost always is accompanied by a decrease in fluorescence because the greater frequency of collisions between molecules increases the probability for deactivation by internal conversion and vibrational relaxation. The pH is important with molecules containing acidic or basic functional groups. Changes in pH influence the degree of ionization, which, in turn, may affect the extent of conjugation or the aromaticity of the compound. Dissolved oxygen often decreases fluorescence dramatically and is an interference in many fluorometric methods. Molecular oxygen is paramagnetic (has a triplet ground state), which promotes intersystem crossing from singlet to triplet states in other molecules. The longer lifetimes of the triplet states increase the opportunity for radiationless deactivation to occur. Other paramagnetic substances, including most transition metals, exhibit this same effect. Solvents affect fluorescence through their ability to stabilize ground and excited states differently, thereby changing the probability and the energy of both absorption and emission.

Fluorescence and Concentration

The fluorescence radiant power F can be obtained by multiplying the absorbed radiant power by the photoluminesence efficiency, which, you will recall, is the fraction of absorbed energy that is emitted as fluorescence. Taking the absorbed radiant power to be the *difference* between the incident and transmitted power, we find that

$$F = \phi(P_0 - P) \tag{17-1}$$

The relationship between the absorbed radiant power and concentration can be obtained from Beer's law. In exponential form, Beer's law (Equation 15-6) can be written

$$\frac{P}{P_0} = 10^{-A} = 10^{-\epsilon bC} \tag{17-2}$$

or

$$P = P_0 10^{-\epsilon bC} \tag{17-3}$$

Substituting Equation 17-3 for P in Equation 17-1 gives

$$F = \phi P_0(1 - 10^{-\epsilon bC})$$

(17-4)

A plot of fluorescence versus concentration for Equation 17-4 is shown in Figure 17-1. The relationship appears linear at low concentrations but reaches a limiting value when the concentration gets very large. As the fluorescence approaches its limiting value, the curve loses its analytical usefulness because solutions of different concentrations will have virtually the same fluorescence.

The portion of Equation 17-4 in parentheses can be represented by an expansion series to give

$$F = \phi P_0\left[\frac{(2.3\epsilon bC)^1}{1!} - \frac{(-2.3\epsilon bC)^2}{2!} - \frac{(-2.3\epsilon bC)^3}{3!} - \cdots - \frac{(-2.3\epsilon bC)^n}{n!}\right]$$

(17-5)

When ϵbC (which equals the absorbance) *is less than about 0.05*, the first term of the expansion series is much larger than the remaining terms and equation 17-5 reduces to

$$F = \phi P_0(2.3\epsilon bC) - K bC$$

(17-6)

where K is a constant equal to $2.3\phi P_0\epsilon$. A proportionality constant arising from the particular instrument design may be included in K as well. This limiting form of Equation 17-4 shows a linear dependence of fluorescence on concentration. It is important to remember that experimental data can be expected to follow Equation 17-6 only when ϵbC is less than about 0.05 and the incident radiant power is constant.

Equation 17-4 has another limiting form when the value of ϵbC is large (greater than about 1.5). Under this condition, $10^{-\epsilon bC}$ is much less than 1 and the equation reduces to a concentration-independent form:

$$F = \phi P_0$$

(17-7)

It is not too surprising that there is an upper limit to F, since the number of photons fluoresced cannot possibly exceed the number absorbed. Unlike absorbance, fluorescence depends directly on the incident radiant power. According to Equation 17-6, F

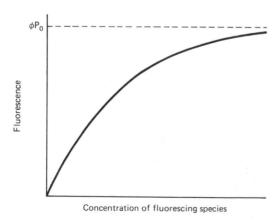

Figure 17-1 Theoretical behavior of fluorescence as a function of concentration.

should double if P_0 is increased by a factor of 2. This provides us with a convenient way to increase the sensitivity of fluorescence measurements.

Self-quenching and self-absorption

Quenching is a word used to describe a decrease in fluorescence caused by a decrease in the photoluminescence efficiency. *Self-quenching* is the result of collisions between excited- and ground-state molecules, which lead to an increase in the amount of radiationless relaxation. In this way it alters the ratio of excited molecules that relax via the fluorescence pathway. Self-quenching depends on the rate at which collisions occur and therefore increases with an increase in concentration.

In many cases, the absorption band of the analyte overlaps the wavelength of the emitted (fluoresced) photon, and as a result, some of the emitted photons are reabsorbed before they can escape the solution. This is called *self-absorption* or the *inner-cell effect* and leads to a decrease in the observed fluorescence. Like self-quenching, it is most serious at high analyte concentrations.

Equation 17-6 does not account for the effects of self-quenching or self-absorption. Self-quenching causes the photoluminescence efficiency to vary with concentration, making the equation virtually useless from an analytical point of view. Self-absorption is a physical artifact that, in theory, can be avoided by the proper cell design and measurement conditions. The decrease in fluorescence due to self-quenching and self-absorption can be so great as to cause the observed fluorescence actually to decrease as the concentration increases.

INSTRUMENTATION FOR MEASURING FLUORESCENCE

The instruments used to measure fluorescence are comprised of essentially the same components as those used in absorption spectrophotometers. The geometric arrangement of the components is somewhat different, to avoid measuring any transmitted radiation along with the fluorescence. A typical arrangement is shown in Figure 17-2. Absorption

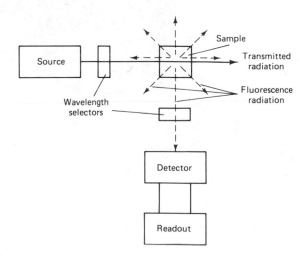

Figure 17-2 Schematic diagram of an instrument to measure fluorescence radiation.

Chapter 17 Molecular Fluorescence Spectroscopy

and transmission of radiant energy occur only along the horizontal optical path, but fluorescence radiation emanates in all directions. The detection of transmitted radiation is avoided by placing the detector at a right angle to the transmitted beam.

Sources

According to Equation 17-6, the fluorescence power, F, is directly proportional to the source power, P_0, and an increase in P_0 will produce a larger signal for a given concentration and thereby improve sensitivity. The tungsten filament and deuterium discharge lamps used in absorption spectrophotometers are not generally suitable sources for fluorescence instruments because they lack the desired intensity. One additional consideration affects the choice of sources. Absorption bands are broad, covering a range of wavelengths, and irradiation at any of these absorbing wavelengths usually is sufficient to initiate fluorescence. For this reason, it is not essential that the source produce a continuum of radiation.

Most simple fluorometers use a mercury discharge lamp as the radiation source. This lamp produces a very intense line spectrum superimposed on a broad, lower-intensity continuum in the near ultraviolet. High-pressure lamps, containing mercury vapor at about 8 atm, have intense lines at 365, 405, 436, 546, 577, 691, and 773 nm. Low-pressure lamps produce intense lines at 254 and 313 nm as well but require a quartz or fused silica window for their transmission. It is relatively uncommon to find an absorption band that does not overlap one of these intense emission lines.

Instruments designed to scan the incident wavelengths use a xenon arc lamp. This lamp produces an intense continuum between about 250 and 600 nm. Unfortunately, it requires water cooling and an expensive power supply to operate, restricting its use to instruments designed for nonroutine work.

Wavelength Selectors

The wavelength selectors shown in Figure 17-2 may be grating monochromators or simple filters. Grating monochromators are found in instruments designed to scan both the incident and fluorescence radiation. Low-cost instruments designed for routine applications use one of several types of filters to select the desired wavelengths. *Interference filters* consist of a thin transparent layer of CaF_2 or MgF_2 sandwiched between two parallel, partially reflecting metal films, as shown in Figure 17-3. A portion of the radiation entering the filter is reflected back and forth between the two metallic films before it escapes and combines with the transmitted portion. The path of the reflected beam is longer, and when the beams combine, constructive interference occurs only for radiation that is a half-wavelength multiple of the spacing. The equation relating the fully reinforced wavelength to the spacing is

$$\lambda = \frac{2\eta b}{m}$$

where η is the dielectric constant of the sandwiched layer, b the thickness of the layer, and m the order of interference. Since destructive interference is incomplete for wavelengths that are close to a half-wavelength multiple of the spacing, the filter actually

Instrumentation for Measuring Fluorescence

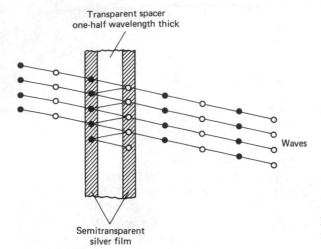

○ Represents wave maximum
● Represents wave minimum

Transparent spacer
one-half wavelength thick

Waves

Semitransparent
silver film

Figure 17-3 Cross-sectional view of an
interference filter.

transmits a band of radiant energies (normally about 10 to 15 nm in spectral width at
50% peak transmittance).

  Absorption filters are comprised of a suitably absorbing substance or substances
dispersed in gelatin, glass, or plastic. A *sharp-cut* or cutoff filter may be designed to
block short or long wavelengths as shown in Figure 17-4. The best sharp-cut filters are
made from substances whose absorption bands rise or fall very steeply. A pair of op-

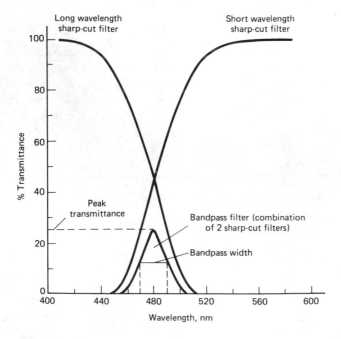

Figure 17-4 Transmittance characteristics
of sharp-cut and bandpass filters.

posing sharp-cut filters can be combined so that only a narrow band of wavelengths is passed. The percent transmittance at the central wavelength of these *bandpass* filters is seldom greater than 30 and spectral bandwidths are typically from 20 to 60 nm.

Sample Cells

Much of the discussion about absorption cells in Chapter 15 applies to fluorescence cells. The one significant difference is that all four faces of a rectangular fluorescence cell must be of optical quality because the measured fluorescence radiation exits the cell at a right angle to the transmitted radiation. The cells may be constructed either of glass or quartz, the latter being required if the incident radiation used is below 350 nm.

Detectors

Fluorescence signals are not very large, partly because of low photoluminescence efficiencies and partly because only a small fraction of the fluorescence radiation reaches the detector. As a result, photomultiplier tubes with their superior sensitivity are preferred over simple phototubes.

Filter Fluorometers

Instruments using filters to isolate the desired incident and fluorescence wavelengths are called filter fluorometers. Because of their simplicity of construction, ease of operation, and low cost, they are the workhorses of fluorescence instrumentation. The optical design of a popular double-beam fluorometer is shown in Figure 17-5. Part of the radiation passes through a primary filter that selects the band of radiation used to excite the analyte molecules in the sample. The resulting fluorescence radiation passes through a secondary

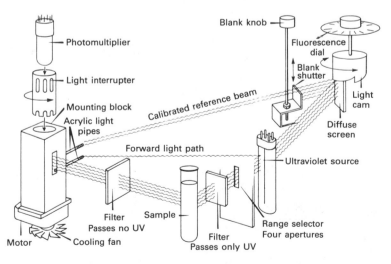

Figure 17-5 Schematic diagram of the Sequoia-Turner model 110 filter fluorometer. (Courtesy of Sequoia-Turner Corporation)

Instrumentation for Measuring Fluorescence

filter selected to prevent any scattered incident radiation (which will be of shorter wavelength than the fluorescence radiation) from reaching the detector. A reference beam, reflected off a mirror which is raised or lowered by a dial on the instrument, is directed to the same detector. The rotating interruptor causes the sample and reference beams to strike the detector alternately and the difference signal is fed to a null meter. The operator adjusts the fluorescence dial until the amount of radiation in the reference beam equals that in the sample beam, as indicated by no deflection on the null meter. The range selector is simply a movable metal plate with four different-size windows. When the fluorescence signal becomes too small to measure accurately, a larger window is used to increase P_0 and therefore F.

Fluorescence measurements with a filter fluorometer often are made by reference to some arbitrarily chosen fluorescence standard. With the standard in the cell compartment, the circuit is balanced with the fluorescence dial set to some arbitrary value. The fluorescence standard is replaced with the unknown and then with each reference solution, the fluorescence being read from the dial after rebalancing the circuit. A plot of fluorescence versus concentration of the reference solutions constitutes the standard curve from which the concentration of the unknown is determined.

Spectrofluorometers

Spectrofluorometers are based on the same general optical design as filter fluorometers, but they use grating monochromators rather than filters to isolate the incident and emitted radiation. The optical layout of a double-beam spectrofluorometer is shown schematically in Figure 17-6. The use of gratings in place of filters greatly improves the resolu-

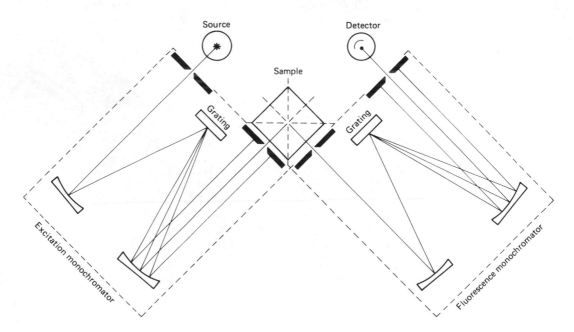

Figure 17-6 Schematic diagram of a typical spectrofluorometer.

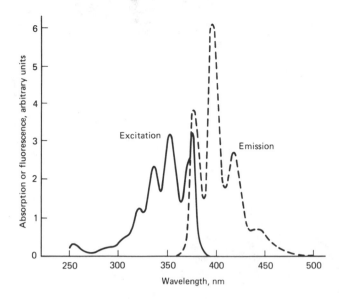

Figure 17-7 Excitation and fluorescence spectra of anthracene.

tion, giving the chemist better control over spectral interferences. Spectrofluorometers employing a continuum source are capable of providing both excitation and fluorescence spectra. Both spectra are actually plots of fluorescence versus wavelength, but they are obtained differently and convey different information. An *excitation spectrum* is a recording of fluorescence versus the wavelength of the exciting or incident radiation and it is obtained by setting the emission monochromator to a wavelength where fluorescence occurs and scanning the excitation monochromator. The resultant spectrum shows which wavelengths the sample must absorb in order to fluoresce. A recording of fluorescence versus the wavelength of the fluorescence radiation is called a *fluorescence spectrum* and it is obtained by setting the excitation monochromator to a wavelength that the sample absorbs and scanning the emission monochromator. These spectra of anthracene are shown in Figure 17-7.

METHODOLOGY

The factors to be considered in developing or evaluating a fluorescence method are very similar to those described for ultraviolet-visible absorption methods in Chapter 16. Perhaps the most significant difference is the need to select both an excitation and fluorescence wavelength (or band of wavelengths), which requires familiarity with the emission spectrum of the source and the excitation and fluorescence spectra of the analyte. The excitation band is isolated by choosing a primary filter whose wavelength transmission range falls in the overlap region of the source emission and analyte excitation spectra, as shown in Figure 17-8. The fluorescence radiation to be detected is isolated by a secondary filter whose function is twofold: (1) to prevent any scattered incident radiation from reaching the detector, and (2) to pass radiation that is not subject to reabsorption by the analyte or other substances in the solution. The first function is accomplished by ensuring that the wavelength transmission ranges of the primary and secondary filters do

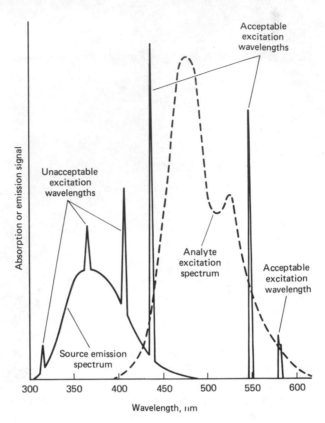

Figure 17-8 Analyte excitation spectrum superimposed on source emission spectrum for the purpose of identifying acceptable excitation wavelengths.

not overlap. Thus any excitation radiation transmitted by the primary filter is blocked by the secondary filter. To perform the second function, the transmission range of the filter should not overlap (at least not severely) the excitation spectrum of any constituents in the solution.

The correct selection of filters may also prevent interference by the fluorescence of other constituents in the sample. If the analyte and another substance absorb in different spectral regions, a primary filter can be selected such that only the analyte is excited. Obviously, if the other constituent is not excited, it cannot fluoresce. If, on the other hand, the analyte and other substances absorb in the same wavelength region but fluoresce in different regions, the interference can be avoided by selecting a secondary filter that transmits only the analyte fluorescence. Both situations are illustrated in Figure 17-9. Note that it is not only acceptable but preferable to use an excitation wavelength where the absorbance is small (less than 0.05).

APPLICATIONS

Fluorescence and absorption methods generally are considered to be more complementary than competitive. Fluorescence methods are best applied to analyte concentrations in the sub-part-per-million range, which is below the detection limit of most absorption

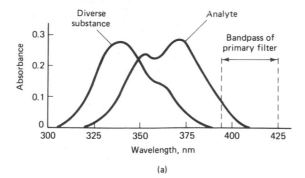

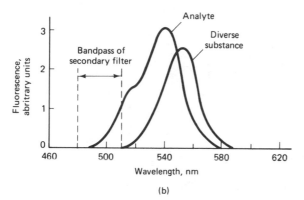

Figure 17-9 Proper choice of primary and secondary filters to avoid interference from another substance: (a) excitation spectra (both substances fluoresce over same wavelength region); (b) fluorescence spectra (both substances absorb in same wavelength region).

methods. The main reason for the lower detection limits lies in the manner by which fluorescence radiation is measured. Because the detector is at a right angle to the incident beam, the background signal (in the absence of scattering) is essentially zero. Even a small signal can be detected against a nearly zero background. In absorption spectrophotometry, the background signal is P_0, a large value, and the detector must recognize a small decrease in this signal due to absorption. A small difference between two large signals is much more difficult to detect accurately. Attempts to apply fluorescence measurements to the determination of larger concentrations are generally unsuccessful due to self-absorption and self-quenching.

Inorganic Substances

Only a few simple inorganic ions fluoresce to any appreciable extent. The uranyl ion, UO_2^{2+}, is probably the best known of this group. Cerium(III), thallium(I), and some of the lanthanides also fluoresce but not as strongly. Because so few ions are naturally fluorescent, most inorganic substances are converted to suitably fluorescent compounds by reaction with a *fluorophore* or are determined indirectly.

Direct determination of metal cations as fluorescent chelates

The use of fluorescent chelates is most successful in the determination of nontransition metals. Although transition metals form many stable chelates with aromatic

TABLE 17-2 SELECTED CHELATING FLUOROPHORES

Name	Structure	Metal(s) determined
1-Amino-4-hydroxyanthraquinone		B, Be, Th
Benzoin (2-hydroxy-1,2-diphenylethanone)		Al, B, Ge, Zn
8-Hydroxyquinoline		Ag, Al, Ca, Cd, Ga, Mg, Mn, Sn, Zn
2,2'-Methylenedibenzothiazole		Zn
Morin (3,5,7,2',4'-pentahydroxyflavone)		Al, B, Be, Cd, Ga, Pb, Sn, Th, U, Zr
Rhodamine B		Al, Au, Ga, Tl, W
Superchrome Blue (2,2'-dihydroxyazonaphthalene-4-sodium sulfonate)	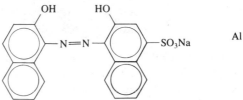	Al

TABLE 17-2 SELECTED CHELATING FLUOROPHORES *(cont.)*

Name	Structure	Metal(s) determined
2-Thenoyltrifluoroacetone		Rare earths
p-Tosylaminoquinoline		Cd, Zn

ligands, very few of these are fluorescent. One reason for this is the paramagnetic nature of the coordinated metal ions. Paramagnetism is known to increase the rate of intersystem crossing between singlet and triplet energy states, and subsequent deactivation from the triplet state is most often via nonradiative processes. A second reason lies in the fact that most transition-metal complexes are characterized by closely spaced energy levels, which enhance the likelihood of deactivation by vibrational relaxation.

In addition to containing the functional groups responsible for the fluorescence, fluorophores also must be excellent chelating agents, as they will be expected to react quantitatively, and perhaps selectively, with a metal ion whose concentration may be only a few parts per billion or less. In many cases, a single fluorophore can be used to determine several different metals. Selectivity can be obtained by carrying out the reaction at a certain pH or in the presence of masking agents such that a single metal ion reacts, or it may be obtained by selecting excitation and fluorescence wavelengths such that only the analyte fluorescence is observed. An abbreviated list of chelating fluorophores and their applications is given in Table 17-2.

Indirect determination by quenching

It is interesting to note that the same properties responsible for the failure of direct fluorometric determinations (unpaired spins and fast vibrational relaxation) are highly desirable for indirect determinations. You may recall that indirect methods rely on measuring the *decrease* in signal produced by adding the sample to a fixed excess of some reagent. Since transition metals and many anions are excellent "quenchers" of fluorescence, they often are determined by indirect fluorometric methods. As little as 1 ppb of iron has been determined by its quenching of the fluorescence of 4,4'-diamino-2,2'-disulfostilbene-N,N,N',N'-tetraacetic acid, and oxygen concentrations in the sub-ppb range have been determined by a similar quenching of the phosphorescence of trypaflavin.

Applications

TABLE 17-3 EXAMPLES OF NATURALLY FLUORESCENT ORGANIC COMPOUNDS

Compound	Wavelength or range of maximum fluorescence (nm)
Aromatic hydrocarbons	
Naphthalene	300–365
Anthracene	370–460
Pyrene	370–400
1,2-Benzopyrene	400–450
Heterocyclic compounds	
Quinoline	385–490
Quinine sulfate	410–500
7-Hydroxycoumarin	450
3-Hydroxyindole	380–460
Dyes	
Fluorescein	510–590
Rhodamine B	550–700
Methylene Blue	650–700
Naphthol AS	516
Coenzymes, nucleic acids, pyrimidines	
Adenine	380
Adenosine triphosphate (ATP)	390
Nicotinamide adenine dinucleotide (NADH)	460
Purine	370
Thymine	380
Drugs	
Aspirin	335
Codeine	350
Diethylstilbestol	435
Estrogens	546
Lysergic acid diethylamide (LSD)	365
Phenobarbital	440
Procaine	345
Steroids	
Aldosterone	400–450
Cholesterol	600
Cortisone	580
Prednisolone	570
Testosterone	580
Vitamins	
Riboflavin (B_2)	565
Cyanocobalamin (B_{12})	305
Tocopherol (E)	340

Organic Substances

Aromatic compounds of almost every type are naturally fluorescent and therefore amenable to direct fluorometric determination. It is well beyond the scope of this book to discuss in any detail the important organic substances that are routinely determined by direct fluorescence measurements, but the great diversity of applications can be seen in any compilation of standard fluorometric methods. A few selected examples are given in Table 17-3.

PROBLEMS

17-1. Calculate the percent relative error in F calculated from Equation 17-6 instead of Equation 17-4 for the following absorbances of the solution.

(a) 0.050   (b) 0.100   (c) 0.250

17-2. Suppose that a solution had a fluorescence of 28 (in arbitrary units) in a 1.00-cm cell when the incident radiant power was P_0. What fluorescence would the solution have in a 5.00-cm cell if the incident radiant power were decreased by 30%?

17-3. Calculate the magnitude and direction of the change in F of a solution if the incident wavelength is changed so that ϕ decreases by 30%, ϵ increases by 50%, and P_0 doubles.

17-4. If the incident wavelength is changed in such a way that ϕ decreases by 12% and ϵ decreases by 8%, how must P_0 change if the fluorescence is to remain constant?

17-5. Calculate the fully reinforced first- and second-order wavelengths (in nanometers) transmitted by an interference filter using a 5.00×10^{-4}-mm-thick layer of MgF_2 ($\eta = 1.378$) as the dielectric material.

17-6. What is the required thickness (in cm) of a MgF_2 ($\eta = 1.378$) interference filter that will transmit fully reinforced radiation of 2000 nm in the second order?

17-7. Anthracene, whose excitation and fluorescence spectra are shown in Figure 17-7, is to be determined by fluorescence measurements with a filter fluorometer equipped with a low-pressure mercury lamp (see the section ''Sources''). Suggest the appropriate wavelengths to be isolated by the primary and secondary filters.

17-8. A 10.00-mL sample containing the reduced form of nicotinamide adenine dinucleotide (NADH) had a fluorescence of 26.0 relative to a reagent blank. When 1.00 μmol of NADH was added to the sample solution (without volume change), the fluorescence increased to 78.3 relative to the blank. Calculate the concentration of NADH in the sample in units of μmol/mL.

17-9. Thorium was determined fluorometrically with 1-amino-4-hydroxyanthraquinone using a standard addition technique. The following volumes of a 2.00×10^{-5} M ThO^{2+} solution were added to six 0.250-g samples, which, after dissolution, treatment with excess reagent, and dilution to 100 mL, had the following fluorescence values.

Volume ThO^{2+} added (mL)	Relative fluorescence
0	23.7
1.00	31.0
3.00	45.6
5.00	60.0
7.00	74.7
10.00	96.3

Calculate the parts per million ThO_2 in the sample.

17-10. A method for determining cyanide ion is based on its reaction with nonfluorescent quinone to form highly fluorescent dicyanohydroquinone. To each of five 50.00-mL volumetric flasks containing 1.00, 2.50, 5.00, 7.50, and 10.00 mL of 2.35×10^{-5} M KCN were added 10 mL of 2×10^{-3} M quinone and sufficient water to bring the volume to the mark. The fluorescence of the resulting solutions were 9.3, 22.9, 45.8, 68.8, and 91.6, respectively. A 1.00-mL aliquot of unknown treated similarly had a fluorescence of 77.2. Calculate the ppm CN^- in the unknown if its density was 1.00 g/mL.

17-11. Glucose can be determined by its rather selective ability to quench the fluorescence of anthranilic acid. The fluorescence of solutions prepared by treating 5.00 mL of 1.50×10^{-3} M anthranilic acid with increasing volumes of 3.30×10^{-4} M glucose and diluting to 50.00 mL with water are given below.

Volume glucose added (mL)	Fluorescence
0	94.4
2.00	79.0
4.00	63.0
6.00	47.1
8.00	30.8
10.00	15.2

A 0.500-mL sample of glucose added to the same volume of anthranilic acid and diluted to 25.00 mL had a fluorescence of 24.1. Calculate the concentration of glucose in units of mg/mL.

18

ATOMIC SPECTROSCOPY

Atomic spectroscopy deals with the absorption and emission of radiation by atoms or atomic ions. In terms of basic principles, atomic and molecular spectroscopy are quite similar. However, there are significant differences in the instrumentation, stemming mainly from the need to convert the analyte to free, unbound atoms or ions—a process called *atomization*. A review of the sections of Chapter 15 dealing with atomic energy levels is good preparation for the material introduced in this chapter. As atoms have no rotational or vibrational energy, transitions occur only between electronic levels and bandwidths in atomic spectra are very narrow. These narrow bandwidths create special instrumental problems, but also provide unique advantages over the broad bands obtained in molecular spectroscopy.

Atomic spectroscopic methods normally are classified according to the type of spectral process involved and the method of atomization used, as shown in Table 18-1. Two spectral processes are discussed in this chapter, absorption and emission, together with several different methods of atomization. Some of these methods are grouped under a single heading in the following discussion.

TABLE 18-1 CLASSIFICATION OF ATOMIC SPECTROSCOPIC METHODS

Atomization method	Atomization temperature (K)	Type of measurement
Flame	2300–3400	Absorption, emission
Electrothermal	2000–3300	Absorption, emission
Inductively coupled plasma	6000–8500	Emission
Electrical arc	4000–5000	Emission

ATOMIC ABSORPTION SPECTROSCOPY

Atomic absorption spectroscopy is similar to molecular absorption spectroscopy, the major difference being that unbound atoms rather than molecules are the absorbing species. The basic instrumental components needed to make absorption measurements are shown in Figure 18-1. In contrast to a molecular absorption spectrophotometer, the monochromator in an atomic absorption or emission instrument is placed *after* the sample (compare with Figure 15-11). This arrangement is necessary to remove unwanted radiation created during the atomization process.

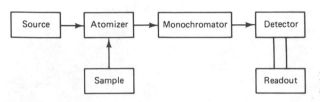

Figure 18-1 Basic components of an atomic absorption spectrophotometer.

Instrumentation

Sources

Atomic absorption bandwidths are so narrow, generally in the range 0.002 to 0.005 nm, they are often referred to as "lines" rather than bands. The narrowest band of wavelengths that can be isolated from a continuum with our *best* monochromators is about 0.5 nm. Thus most of the incident radiation is outside the range of wavelengths the analyte can absorb, as depicted in Figure 18-2. Imagine your eye to be the detector as you look at the figure and consider how little effect the analyte absorption has on the amount of radiant energy you "see." An analyte absorbing *all* the radiation within its 0.002-nm bandwidth removes less than 0.4% of the total radiation in a beam whose bandwidth is 0.5 nm. Obviously, this is an extremely unsatisfactory situation in terms

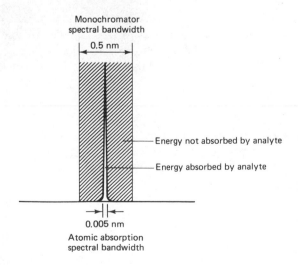

Figure 18-2 Comparison of atomic absorption and monochromator spectral bandwidths.

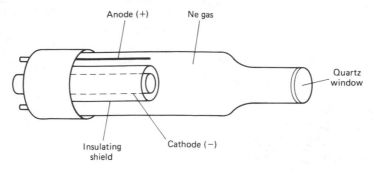

Figure 18-3 Schematic diagram of a hollow cathode lamp.

of how absorption measurements normally are made. The principles of atomic absorption were well understood long before routine measurements could be made. The technique did not become practical until 1955, when Alan Walsh succeeded in making radiation sources with very narrow bandwidths whose wavelengths matched exactly the absorption wavelengths of different analytes.

Hollow cathode lamps. These lamps consist of a cylindrical metallic cathode and tungsten anode sealed in a glass tube containing neon or argon at a pressure of about 1 to 5 torr (Figure 18-3). A voltage is applied to the electrodes sufficient to ionize the neon and the resulting cations are accelerated toward the cathode where on impact they dislodge some of the metal atoms comprising the cathode surface producing an atomic "cloud." This process is called *sputtering*. Some of the metal atoms in the cloud collide with incoming neon ions and become electronically excited. The energy (wavelength) of the photons emitted by these atoms in returning to the ground state is the same energy that can be absorbed by gaseous, unbound atoms of the same element. At the proper operating conditions, the bandwidth of the emitted radiation is even narrower than the atomic absorption bandwidth, as shown in Figure 18-4.

The cathode and anode are shaped and positioned to concentrate the atom cloud in the center of the metal cylinder. A compact atom cloud will concentrate the emitted

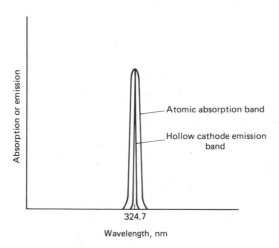

Figure 18-4 Relative line widths for copper emission and absorption.

Atomic Absorption Spectroscopy

radiation in a small area, thereby keeping the intensity high and making it easier to focus the beam on the entrance slit of the monochromator. Some fraction of the emitted radiation is always absorbed by other atoms as it passes through the cloud, thereby decreasing the output power. This self-absorption is reduced by keeping the size of the atom cloud small, thereby shortening the distance the emitted photons must travel to escape the cloud. Also, confining the atomic cloud to the area of the cylinder ensures that most of the atoms ultimately condense onto the cathode rather than the tube walls, thereby extending the useful lifetime of the lamp. Because the cathode must be a solid electrical conductor, hollow cathode lamps for nonmetals are not available.

It should be clear that a different lamp will be needed for each element whose atomic absorption is to be measured, and this is a disadvantage of the atomic absorption technique. A few multielement lamps with cathodes made from alloys or mixtures of metals are available. In these lamps, the atom cloud contains atoms of several elements, and lines characteristic of each element will be present in the output. Only certain combinations of elements work well together, however. If one of the metals is sputtered much more easily than the others, its concentration in the atom cloud is greatest, and after several cycles of sputtering and condensing, the metal cathode becomes coated with this metal. The result is a decrease in intensity of the emission lines for the other elements.

Atomizers

Samples measured in atomic absorption spectroscopy are usually liquids. The purpose of the atomizer is to get a representative portion of the sample into the optical path and convert it to free neutral, ground-state atoms. Chemists have developed numerous ways to atomize samples, two of which are used routinely in atomic absorption spectroscopy: flame atomization and electrothermal atomization.

Flame atomization: The laminar flow burner. In this burner, shown schematically in Figure 18-5, oxidant flows past the tip of a small tube as it enters a large mixing chamber. The pressure drop at the tube opening causes sample solution to be aspirated (sprayed) into the chamber, where it mixes with the incoming fuel. This arrangement is

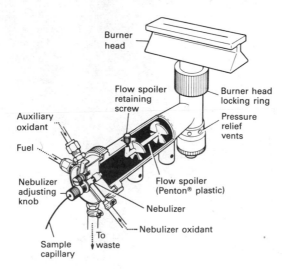

Figure 18-5 Laminar flow burner. (Courtesy of The Perkin-Elmer Corp.)

called a nebulizer. The resulting aerosol then passes over a series of baffles, causing the larger droplets to settle out and drain into a waste container. The finer droplets continue on, exiting the burner via a 5- or 10-cm slot, where the fuel is burned. The burner is designed to be used at conditions where the gas flow velocity at the slot is faster than the burning velocity. Otherwise, the flame will move *into* the mixing chamber and cause an explosion. A pressure relief valve usually is present to prevent damage to the burner or the operator should such an event accidentally occur.

The fuel–oxidant combination used most often with this burner is acetylene–air, which produces a maximum flame temperature of about 2500 K. At this temperature, most compounds are quickly converted to free atoms through a series of steps in which the solvent is evaporated, the remaining solute is vaporized, and the molecules are decomposed. In situations where a higher temperature is desired, nitrous oxide may be used as the oxidant. Acetylene–nitrous oxide flames may reach a temperature of 3000 K.

Chemists are keenly interested in both the chemical and physical processes that occur during atomization. Even in the best atomizers, only a small fraction of the analyte is converted to neutral, ground-state atoms in the portion of the flame through which the source radiation passes. It has been estimated that the present detection limits would be decreased more than 100,000-fold if the atomization process were perfectly efficient, that is, if all the analyte in the flame at a given instant existed as neutral, ground-state atoms in the optical path. Obviously, the potential exists for significant improvement.

Factors such as the oxidant, the fuel/oxidant ratio, and the solvent usually influence both the efficiency of atom production and the spatial distribution of atoms in the flame. Figure 18-6 shows how the distribution changes with the fuel/oxidant ratio. These distribution patterns or atom-population profiles, as they are often called, are very important because they indicate where in the flame the optical path should be located in order to get the largest signal.

Electrothermal atomization: The graphite furnace. As chemists began to understand better the processes controlling atomization, certain deficiencies of flame atomizers became apparent. First, the volume of sample that can be nebulized and carried into the flame is quite small. If large amounts of sample are aspirated, most of the flame energy goes into solvent evaporation and not enough remains to volatilize and decompose the sample. An extreme case is akin to pouring a beaker of water on the flame! Second, the atoms formed in the flame do not remain in the optical path very long, being swept away by the flowing gases. As a result, there is little or no accumulation or concentration of absorbing atoms in the measurement region. Finally, flames require the use of an oxidant, which can react with analyte atoms to form stable oxides, thereby reducing the number of free atoms available for absorption.

Electrothermal atomizers overcome some of these deficiencies and provide greater sensitivity than flame atomizers. The most common type of electrothermal atomizer is the graphite furnace, one variety of which is shown in Figure 18-7. The main part of the atomizer is a small graphite tube about 5 cm in length and 1 cm in diameter. This tube is fitted inside a larger, hollow cylinder and separated by an electrical insulator. A small opening in the top of the graphite tube allows liquid samples to be inserted with a microsyringe. This atomizer is also capable of accepting solid samples which usually are introduced through one end with a special sampling spoon. The graphite tube is connected to a low-voltage high-current power supply and becomes very hot (up to

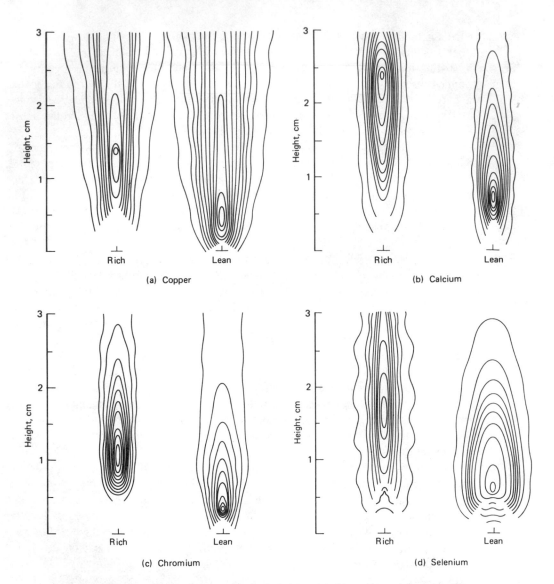

Figure 18-6 Atom-population profiles in fuel-rich and fuel-lean 10-cm air/acetylene flames. Contours are drawn at intervals of 0.1 absorbance unit with the maximum absorbance in the center. (Reprinted with permission from C.S. Rann and A.N. Hambly, *Anal. Chem.* **1965**, *37*, 879–884. Copyright 1965 American Chemical Society.)

3000°C) when current is passed through the circuit. The outer housing is water-cooled to enable a quick return to ambient temperature after the atomization and measurement is completed.

One of the most significant features of electrothermal atomizers is their flexibility in preparing samples for atomization. The treatment normally involves three steps: drying, ashing, and atomizing. To dry liquid samples, the furnace is heated to a tem-

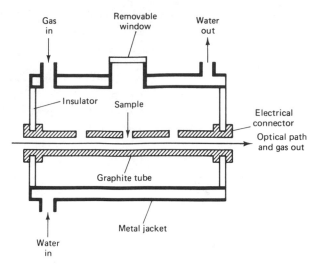

Gas in

Removable window

Water out

Insulator

Sample

Electrical connector

Optical path and gas out

Graphite tube

Metal jacket

Water in

Figure 18-7 Cross-sectional view of a graphite furnace atomizer.

perature slightly above the boiling point of the solvent and held there for perhaps half a minute. The ashing cycle, carried out at a higher temperature (perhaps 600 to 1000°C), is designed to volatilize components of the sample whose boiling points are below that of the analyte and to pyrolyze (thermally decompose) organic compounds that would interfere later by producing carbon particles (smoke) in the optical beam. Finally, atomization is brought about by rapidly increasing the current through the graphite furnace, causing the temperature to rise to between 2000 and 3000°C. Generally, the total atomization time is between 5 and 10 seconds. Most furnaces are designed so that an inert gas can be passed through the graphite tube to aid in removal of the substances volatilized during the drying and ashing cycles and to remove oxygen from the sample area during atomization, thereby preventing the formation of stable metal oxides and minimizing oxidation of the graphite tube.

Electrothermal atomizers provide excellent detection limits, often 1000-fold better than flame atomizers. The ability of the furnace to create the gaseous atoms quickly and keep them in the optical path longer are the major reasons for this improvement. Also, much smaller volumes of sample are consumed by electrothermal analyzers (1 to 10 μL versus 1 to 5 mL for flame atomizers). The relative precisions of methods using electrothermal atomization generally are around 5 to 10%, compared to 1 to 2% with flame atomizers, although direct comparisons probably are not fair because the atomizers are used at different levels of analyte concentration. Graphite furnaces are expensive attachments to atomic absorption instruments and require more care and maintenance than does the laminar flow burner.

Monochromator and detector

Despite the fact that line sources are used, monochromators still are a necessary component of atomic absorption instruments. For one reason, the output of a hollow cathode lamp normally consists of several lines, some of which are due to the filler gas. Furthermore, there will be spectral emission from various sample components during atomization. The function of the monochromator is to isolate the desired or "analytical" wavelength for passage on to the detector. The monochromators and detectors used in molecular absorption instruments are equally suitable to atomic absorption instruments.

Optical systems

The schematic diagram of a single-beam atomic absorption spectrometer is shown in Figure 18-8. The duplication of an atomizer for use in a reference beam is difficult and expensive. As a result, the double-beam design loses much of the advantage it enjoys over the single-beam arrangement in molecular absorption instruments. It is necessary to prevent radiation *originating* in the flame or furnace atomizer from reaching the detector. Most of this radiation has a wavelength different from the measurement wavelength and is removed by the monochromator, which is placed between the atomizer and detector for this very purpose. To distinguish between source and flame radiation of the same wavelength, a technique called *modulation* is used. The intensity of the source radiation is caused to fluctuate at a constant frequency. As a result, the detector receives an alternating signal, corresponding to source radiation transmitted by the sample, plus a continuous signal, corresponding to emission from the atomizer. It is a relatively simple matter to electronically remove the unmodulated or dc signal and pass the alternating or ac signal on to the amplifier.

Modulation is accomplished by using an ac power supply that effectively has the hollow cathode lamp turning on and off at a constant frequency or by periodically interrupting the radiation beam with a rotating chopper, a circular disk with alternating segments removed. The operation of a beam chopper and the resulting modulated signal is depicted in Figure 18-9.

Methodology

In this section we describe some of the procedural details that must be considered in carrying out atomic absorption determinations. Because of their similarity in both principles and instrumentation, atomic and molecular absorption spectroscopy share many problems, and it may be worthwhile to review those sections of Chapter 16 that would logically apply to atomic absorption.

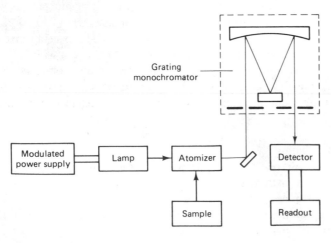

Figure 18-8 Schematic optical diagram of a single-beam atomic absorption spectrophotometer.

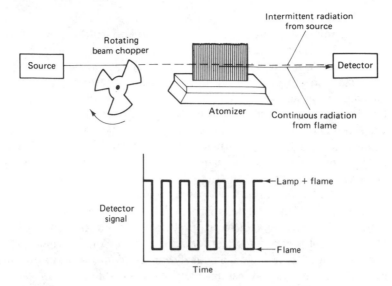

Figure 18-9 Use of a beam chopper to modulate the signal of interest and discriminate against a continuous background.

Sample preparation

The required sample preparation varies somewhat depending on which type of atomizer is to be used. Flame atomizers require liquid samples of relatively low viscosity and with no suspended particulate matter that could clog the nebulizer. Highly viscous liquids are not readily nebulized and must be diluted with a low-viscosity solvent. Relatively few solid samples are directly soluble in common solvents and require some chemical pretreatment to convert the various constituents to soluble forms. Not only are these decomposition and solution steps time consuming, they are potential sources of error. When the analyte concentration is very low, it is especially important to ensure that the reagents used in the pretreatment steps do not contaminate the sample with traces of the analyte. While the graphite furnace can handle both liquid and solid samples, matrix effects generally are more severe with this atomizer and more sample pretreatment may be necessary. Chemists use the term ''matrix'' as a general reference to the non-analyte part of the sample or the environment in which the analyte exists.

A few elements, including tin, lead, arsenic, antimony, bismuth, and selenium, form volatile hydrides that can be fed directly into a flame atomizer. Rapid generation of the hydride is brought about by adding the acidified sample to a small volume of dilute (about 1%) sodium borohydride. An inert gas such as argon sweeps the resulting hydride from the reaction vessel into a quartz tube heated to several hundred degrees centigrade in a flame or tube furnace. This temperature is sufficient to atomize the hydride molecules. Hydride generation techniques usually improve the detection limits by a factor of 10 to 100 over flame or furnace atomization techniques and, in addition, offer the significant advantage of separating the analyte from many potentially interfering substances in the sample matrix.

Atomization

Atomization is the most complicated and difficult-to-control step in an atomic absorption determination, and because the potential exists for a great deal of improvement, it continues to be an area of intense study by many research groups. Instruments are reasonably effective at discriminating against unwanted radiation originating in the sample area during atomization and, as a result, the conditions employed for the atomization process are designed mostly to overcome possible chemical interferences. For example, when refractory metals such as chromium, molybdenum, tungsten, and vanadium are being determined, fuel-rich (reducing) flames are used to minimize the formation of stable oxides and the optical path is kept low in the flame, where oxides have not yet had a chance to form. When electrothermal atomizers are used, the times and temperatures of the drying, ashing, and atomization cycles are individually tailored to each sample. Thus, if the analyte is fairly volatile, relatively low drying and ashing temperatures will be used to avoid loss of the analyte prior to atomization. As pointed out in Chapter 16, it is very important for the chemist to know as much as possible about the composition of the sample being analyzed in order to make correct decisions about the sample pretreatment and instrument operating conditions.

Standards

Ideally, standards for atomic absorption determinations should closely resemble the sample in terms of overall composition in addition to containing a known concentration of the analyte element. The complexity and uncertain composition of many samples makes this ideal difficult to attain in many cases. Thus it becomes especially important to adopt sample pretreatment steps and operating conditions that will minimize chemical interferences and other undesirable matrix effects. Since atomic absorption is used to measure very low concentrations, the reagents used to prepare standards must be free of trace amounts of the analyte. In some cases, the singly distilled water used for dilutions can contain more analyte than the weighed standard. It does not take much to contaminate a 1-ppb solution.

Establishing a relationship between absorbance and concentration

Several methods of relating the measured value for an unknown to the concentration of analyte are used in atomic absorption spectroscopy. Two of these, the standard curve and standard addition methods, have been described previously in connection with other techniques.

Standard curve method. Ideally, atomic absorption can be expected to follow Beer's law, with absorbance being directly proportional to concentration. Unfortunately, the efficiency with which atomization occurs often decreases with an increase in analyte concentration, making calibration plots of absorbance versus concentration nonlinear, particularly at higher concentrations, as shown in Figure 18-10. It is especially important that the standards used to prepare the calibration curve cover the entire concentration range within which the unknown is expected to fall because it is very difficult to extrapolate nonlinear data.

Standard addition method. This method is used frequently in atomic absorption spectroscopy because it is effective in compensating for spectral and chemical interferences caused by the sample matrix. The basis for the method is described in Chapter 13. It is important to remember that the standard addition method requires an extrapo-

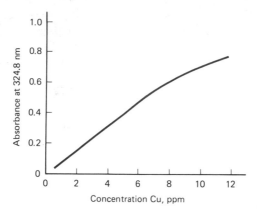

Figure 18-10 Standard calibration curve for copper.

lation of the analytical data, and as just pointed out, many calibration plots are nonlinear and, as such, cannot be extrapolated safely. On the other hand, nonlinear curves can appear to be almost linear over a small range of concentration. Consequently, if the standard additions are kept small, an acceptable extrapolation such as that shown in Figure 16-3 may be possible.

Internal standard method. This method relies on a comparison of two signals: one due to the analyte and the other to an internal standard. An internal standard is an element, not present originally, that is added in known amount to the sample. A series of solutions is prepared with different, known concentrations of analyte and a single concentration of internal standard. The absorbances due to analyte and standard are measured at the appropriate wavelengths and the calibration curve is constructed plotting their ratio versus the concentration of analyte, as shown in Figure 18-11. The unknown is treated with the same amount of internal standard, the absorbances measured, and the analyte concentration determined from the calibration curve.

The internal standard method is most useful when some sample loss or nonspecific matrix interference is expected. In the case of some sample being lost during pretreatment, it is assumed that the fractions of analyte and internal standard lost are the same.

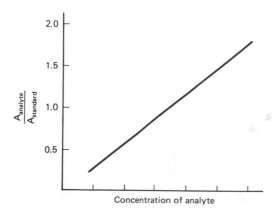

Figure 18-11 Internal standard calibration curve.

Although the absorbance due to each component is less, the absorbance ratio will be the same as if no loss occurred. Occasionally, a type of matrix interference is encountered that tends to affect the signal of all elements in the same way. For example, increased viscosity may decrease all measured absorbances by roughly the same amount because less sample gets aspirated into the flame. In situations where it is difficult or impossible to match the viscosity of the standards to that of the unknowns, the internal standard method may be the method of choice. Unfortunately, use of this method requires measurements at two different wavelengths, which may be a considerable inconvenience with instruments not designed for two optical paths through separate monochromators.

Applications

The scope of atomic absorption methods is quite impressive. Virtually every metallic element can be determined by atomic absorption, but not all with the same sensitivity or detection limit. Although direct determinations are limited to those elements for which hollow cathode lamps are available, other substances can be determined by indirect procedures. For example, PO_4^{3-} reacts with MoO_4^{2-} in acidic solution, forming 12-molybdophosphoric acid, $H_3P(Mo_3O_{10})_4$. This compound can be extracted into any of several organic solvents, freeing it from excess molybdate ion, and the organic extract analyzed for molybdenum. On a molar basis, the phosphorus or phosphate concentration is one-twelfth the measured molybdenum concentration.

The concentration range over which a suitable calibration plot can be prepared is seldom more than an order of magnitude and often is less. In many cases, samples are subjected to a crude but quick analysis to determine the approximate concentration of

TABLE 18-2 COMPARISON OF DETECTION LIMITS FOR SELECTED ELEMENTS[a]

| Element | AAS | | AES | |
	Flame	Electrothermal	Flame	ICP
Al	30	0.005	5	2
As	100	0.02	0.0005	40
Ca	1	0.02	0.1	0.02
Cd	1	0.0001	800	2
Cr	3	0.01	4	0.3
Cu	2	0.002	10	0.1
Fe	5	0.005	30	0.3
Hg	500	0.1	0.0004	1
Mg	0.1	0.00002	5	0.05
Mn	2	0.0002	5	0.06
Mo	30	0.005	100	0.2
Na	2	0.0002	0.1	0.2
Ni	5	0.02	20	0.4
Pb	10	0.002	100	2
Sn	20	0.1	300	30
V	20	0.1	10	0.2
Zn	2	0.00005	0.0005	2

[a]Parts per billion ($\mu g/L$). AAS, atomic absorption spectroscopy; AES, atomic emission spectroscopy.

the analyte. This information is used to decide what sample size and dilutions are necessary to yield a final concentration falling within the desired analytical range.

The detection limits vary greatly for different elements and for different methods of atomization (flame versus electrothermal). As illustrated by the data in Table 18-2, the detection limits obtained with electrothermal atomizers are typically two to three orders of magnitude smaller than those obtained with flame atomizers. As mentioned earlier, the lower detection limits are achieved generally at the expense of precision and accuracy. Overall relative uncertainties of 5 to 10% are common with an electrothermal atomizer, versus 1 and 2% when using flames.

Flame atomizers have the advantage over their electrothermal counterparts in terms of cost, speed, and ease of automation. The latter characteristic is of considerable importance in clinical applications where a large number of samples must be analyzed each day.

FLAME EMISSION SPECTROSCOPY

Flame emission spectroscopy (FES) is the oldest of the atomic spectroscopic methods, dating back to the late nineteenth century, when certain metals were detected by burning the sample and observing the color of the flame. The relative importance of flame emission as an analytical technique became greatly diminished as a result of the development of atomic absorption. Although one is an emission and the other an absorption method, there are certain features common to both: in particular, the use of a flame to atomize the sample. In atomic absorption spectroscopy, the primary purpose of the atomizer is to produce the highest possible concentration of neutral, ground-state atoms, capable of absorbing characteristic radiation from an external source. The atomizer has an additional responsibility in flame emission: to excite the analyte atoms so they may emit their characteristic radiation upon relaxing to a lower electronic energy state. This "thermal excitation" occurs as a result of high-energy collisions between the analyte atoms and particles formed during combustion of the flame gas.

Flame emission spectra consist of fairly simple series of narrow lines whose wavelengths are characteristic of the emitting element. The wavelength (or energy) of the most intense line usually corresponds to the energy of the transition between the first excited electronic state and the ground state. According to Figure 15-5, this transition for the sodium atom is from the $3p$ to the $3s$ state, with an energy corresponding to a wavelength of 589.3 nm. Sodium is the element responsible for the yellow color observed in many flames.

Instrumentation

The schematic diagram of a simple flame emission spectrophotometer looks very much like Figure 18-8 without the lamp and its power supply. Since the flame serves as both atomizer and "exciter," no external radiation source is required. The laminar-flow burner (Figure 18-5) used in atomic absorption has proven to be an effective burner for flame emission as well. The monochromator removes background radiation from the flame, whose wavelength is different from that being measured. Unfortunately, the flame is likely to produce some background radiation whose wavelength is the same as that being

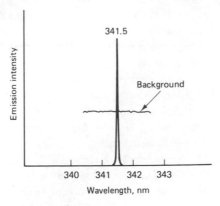

Figure 18-12 Emission line of Ni in the presence of broadband background emission.

measured (passed by the monochromator). The practice of modulation, used in atomic absorption spectroscopy to discriminate against this radiation, cannot be used in flame emission because both analyte and background emission originate in the flame. This is a major limitation of the flame emission technique. In cases where the background radiation is known to be of a broad-band type, and the desired emission line is fairly well isolated from other emission lines, a correction may be possible by making two measurements at slightly different wavelengths. According to Figure 18-12, the signal measured at 341.5 nm includes analyte and background. By changing the monochromator setting slightly to 341.0 nm, the background signal alone can be measured. A simple subtraction yields the pure analyte signal.

Methodology

Many of the problems associated with sample preparation and atomization for atomic absorption are encountered also in flame emission methods. That is not to say, however, that the optimum conditions and procedures used are always the same for both methods. For example, the spectral emission of the flame itself is of much greater consequence in flame emission than in atomic absorption because background correction is more difficult (no modulation). As a result, the various flame conditions selected depend partly on the spectral characteristics of the flame gases and certain organic solvents may have to be avoided due to their high spectral emission during the burning process.

It is important to remember that a successful flame emission method depends not only on the ability of the flame to atomize the analyte but also to excite the atoms. Temperature is a critical factor in determining the extent of both processes. At thermal equilibrium, the number of atoms in an excited level (N^*) relative to those in the ground level (N_0) is given by the Maxwell–Boltzmann equation:

$$\frac{N^*}{N_0} = \frac{g^*}{g_0} e^{-\Delta E/kT} \qquad (18\text{-}1)$$

where g_0 and $g*$ are the number of energy states in each level (called the degeneracy), ΔE is the energy difference between the two levels, T is the temperature in kelvin, and k is a constant. According to this equation, raising the temperature should increase the number of excited atoms and the radiant power of the emission signal. Ordinarily, less than 0.1% of the available atoms are in an excited level, which means that even a small temperature increase (or decrease) will have a large relative effect on the number of excited atoms and almost no effect on the number of unexcited atoms. Suppose, for example, that there are 100 excited atoms and 100,000 ground-state atoms in a sample, and a rise in temperature of 10 K increases the number of excited atoms to 104. This represents a 4% increase in the emission signal. At the same time, it represents only a 0.004% *decrease* in the number of ground-state atoms and should have no measurable effect on the atomic absorption signal. The lesson here is simple: Flame parameters that might affect the temperature must be controlled much more carefully in flame emission than in atomic absorption spectroscopy.

The high percentage of ground-state atoms in flames is the source of a problem called *self-absorption*. The outer portion of a flame is cooler than the center and therefore contains a higher concentration of unexcited atoms. Some of the photons emitted from the center of the flame are absorbed as they attempt to pass through this region, causing a decrease in the signal. Since the line width for absorption is narrower than that for emission in a flame, self-absorption tends to alter the center of a line more than its edges. Self-absorption becomes more severe as the analyte concentration increases and ultimately causes the signal–concentration relationship to become nonlinear.

Applications

In theory, the scope of FES is somewhat broader than that of AAS because there is no need for a line source characteristic of the element being determined. However, the atomic emission lines of most nonmetals fall in the ultraviolet spectral range below 200 nm, where special, expensive equipment is required for their measurement. Presently, the most important applications of flame emission spectroscopy are for the determination of the alkali and alkaline earth metals, especially in biological fluids and tissues. It has found use also in the determination of the lanthanum group elements (rare earths).

The analytical concentration range for most determinations is about an order of magnitude, which is quite similar to the range in atomic absorption spectroscopy. The detection limits for a number of metals are given in Table 18-2. Although individual values vary quite a bit, they are, with the exception of the alkali and alkaline earth metals, not as low as those obtained for atomic absorption.

Unlike atomic absorption, flame emission spectroscopy has the inherent capability for simultaneous, multielement determinations, and many research groups are presently exploring ways to turn this capability into a practical reality. One approach involves removing the exit slit of the monochromator and using a detector capable of monitoring many different wavelengths simultaneously. Photodiode arrays show considerable promise in this regard. One of the limitations to simultaneous determinations lies in the fact that the optimum atomization and excitation conditions vary considerably from element to element.

PLASMA AND ELECTRICAL DISCHARGE EMISSION SPECTROSCOPY

These techniques are similar in principle to flame emission spectroscopy but use much more energetic atomization-excitation processes. The energy available in these atomizers is sufficient to excite atoms to many upper levels. In addition, a substantial number of ions are formed, which also become excited. As a result, the emission spectra are complex, containing dozens of lines characteristic of each element. This complexity can be an advantage in identifying which elements are present in a sample. These high-energy atomizers also are more efficient than flames at breaking down the stable oxides formed by refractory elements such as boron, phosphorus, niobium, zirconium, and tungsten.

The optimum conditions for excitation with plasmas and electrical discharges do not vary much from element to element, and good spectra can be obtained for most elements at a single set of atomization-excitation conditions. Consequently, the techniques are conducive to simultaneous, multielement determinations.

Emission Spectroscopy with Plasma Sources

The plasma source is a relatively recent innovation that is rapidly replacing the older electrical discharge source. A plasma consists of a gas or mixture of gases in which a significant fraction of the atoms are ionized. As an electrical conductor, it can be heated inductively by coupling it with an oscillating magnetic field. A schematic diagram of an inductively coupled plasma (ICP) torch is shown in Figure 18-13. The sample is nebulized in much the same way as in a laminar flow burner and the sample aerosol and argon nebulizing gas are fed into the torch through a small quartz tube above which is placed a high-power radio-frequency induction coil. A spark from a Tesla coil initiates ionization in the flowing gas, and the resulting ions and electrons rapidly acquire enough energy from the oscillating field of the induction coil to sustain a high degree of ionization. A second stream of argon gas flows up and around the sample tube, helping to aspirate the sample into the plasma and to thermally isolate the outer wall from the hot plasma. The geometry of the torch causes the ions and electrons to move in closed annular paths depicted in Figure 18-14. The plasma consists of a brilliant white core (shaded area) and a flame-like tail. The volume just above the core can reach a temperature of 7000 K and is remarkably free from background radiation, making it an ideal location from which to measure the analyte emission.

Figure 18-14 also shows how the temperature varies throughout the plasma. Sample fragments reaching the observation point have spent about 2 ms in the plasma at temperatures from 6000 to 8000 K. This is two to three times longer and hotter than in a typical combustion flame. The consequence is obvious: Atomization is much more complete and the concentration of emitting atoms is larger. In addition, the atomization process takes place in a chemically inert atmosphere (argon) which largely prevents the formation of stable oxides.

Plasma emission instruments

The real power of plasma and electrical discharge spectroscopy lies in their ability to provide multielement determinations on a single sample. There are two types of plasma emission instruments, and both are designed with this capability in mind. Figure 18-15

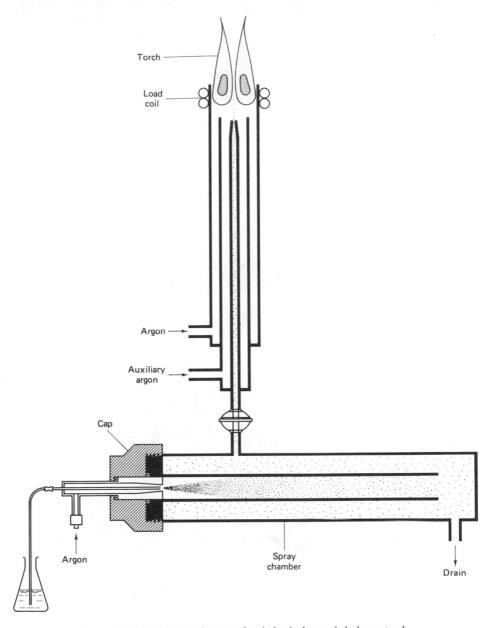

Torch

Load
coil

Argon →

Auxiliary
argon →

Cap

Argon

Spray
chamber

Drain

Figure 18-13 Schematic diagram of an inductively coupled plasma torch.

shows the optical layout of an instrument using a single detector with a scanning monochromator to select the desired emission wavelength. The monochromator is programmed to move sequentially to the wavelengths characteristic of each element being
determined, pausing briefly (perhaps a few seconds) at each wavelength to allow a satisfactory emission signal to be recorded. A successful instrument of this type requires
steady signals from the source during the entire scanning procedure.

Plasma and Electrical Discharge Emission Spectroscopy

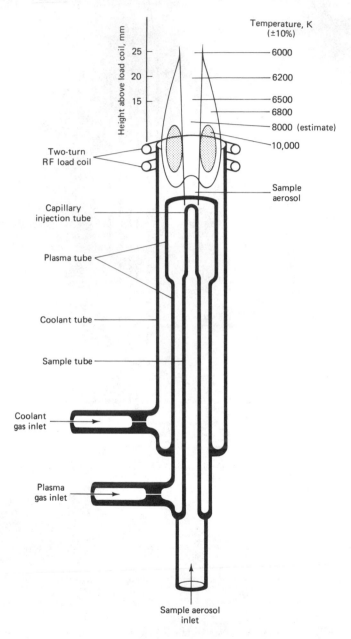

Figure 18-14 Temperature profile of an inductively coupled plasma. (Adapted with permission from V.A. Fassel, *Anal. Chem.* **1979,** *51*, 1290A–1308A and R.N. Savage and G.M. Hieftje, *Anal. Chem.* **1979,** *51*, 408–413. Copyright 1979 American Chemical Society.)

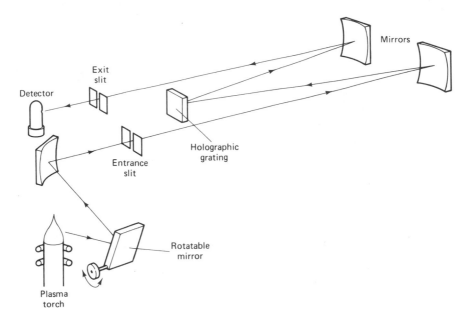

Figure 18-15 Schematic optical diagram of a sequential ICP emission spectrometer.

The other type of instrument, shown in Figure 18-16, uses multiple detectors along the focal plane of a fixed, concave diffraction grating. The various signals are measured at the same time, making this a true *simultaneous, multichannel* instrument. The movable entrance slit enables the optical path to be moved slightly with respect to the detector, creating a small scanning effect. This permits each detector to record a background signal at a wavelength very close to the analytical wavelength, which can be used to correct for broad-band radiation. Since the detectors are operating simultaneously, their signals must be stored for a later, sequential readout. Newer instruments use computers with multichannel analog-to-digital converters to acquire and save the detector signals. Instruments with as many as 50 photomultiplier-tube detectors are available. The great advantage of the true multichannel instrument is that all of the data are collected simultaneously and the source output does not have to be constant for as long a period as with the scanning instruments.

Applications

The performance of inductively coupled plasma sources is quite remarkable. Excellent stability coupled with low background and insensitivity to sample matrix makes it the source of choice in a great many applications. A significant feature of ICP spectroscopy is the large concentration range over which the calibration curves are linear: in some cases, as many as five orders of magnitude (Figure 18-17). Self-absorption, one of the major causes of nonlinear standard curves in emission spectroscopy, is largely absent in plasmas because the temperature, and therefore the concentration of absorbing atoms, is quite uniform throughout each horizontal plane.

Plasma and Electrical Discharge Emission Spectroscopy

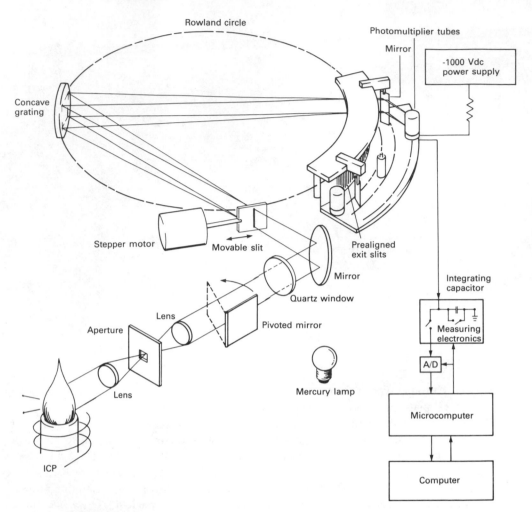

Figure 18-16 Schematic diagram of a multichannel, direct-reading ICP spectrometer. (Courtesy of Baird Corp.)

Emission Spectroscopy with Electrical Discharges

Several types of electrical discharges are used as atomization and excitation sources in emission spectroscopy, two of which are described in this section. The *direct-current arc* is a high-current low-voltage discharge taking place between two electrodes. The passage of a large current through the air space separating two electrodes produces temperatures of 4000 to 6000 K, which are sufficiently high to volatilize, atomize, and excite most substances. One of the main problems with the arc is its tendency to selectively volatilize certain constituents in the sample. This results in the various emission signals being somewhat sequential rather than simultaneous in appearance and causes some problems in detection. Direct-current arc discharges tend to wander over the electrode surface, causing different parts of the discharge to move in and out of the optical path, which is a major factor contributing to the relatively poor precision obtained with this source.

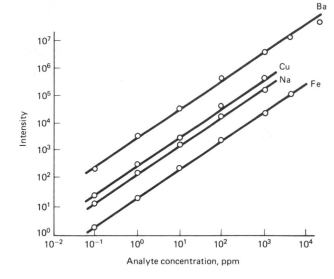

Figure 18-17 ICP emission calibration curves for several metals. Note that a log-log plot is used due to the large range of intensity and concentration values. (Reprinted with permission from R.N. Savage and G.M. Hieftje, *Anal. Chem.* **1979**, *51*, 408–413. Copyright 1979 American Chemical Society.)

Some of the problems caused by selective volatility and wandering are overcome by using an intermittent discharge that cycles on and off at a high frequency, called an *alternating-current spark* discharge. The duration of a single spark is short compared to the time between sparks, allowing the electrodes and sample time to cool between discharges. As a result, the heating effects responsible for selective volatilization in the dc arc discharge are greatly reduced. In addition, wandering is much less of a problem because each discharge is most likely to be initiated between the same two points of least resistance on the electrodes. The *average* current in a spark, and therefore the average heating, is much less than in a dc arc, but the *initial* current is very large (perhaps 1000 A). This current is carried by a very narrow column of ions between the electrodes and, within this region, the temperature may momentarily reach values as high as 40,000 K, which produces a high concentration of ions. Because the *average* temperature in the spark is much lower than in a dc arc and the overall atomization and excitation efficiency is not as great, the limits of detection are not as low with this source.

Arc and spark source instruments

The major difference between instruments using plasma and electrical discharge sources is the manner in which samples are introduced. Plasmas are best suited for liquid samples, whereas electrical discharge sources can handle a variety of sample types. Conducting metals or alloys can be fashioned into an appropriate shape and made one of the electrodes (Figure 18-18a). Nonmetallic samples that are not easily dissolved may be ground to powders and placed in a graphite cup electrode (Figure 18-18b). Powdered graphite is usually added to nonconducting solid samples to increase the conductivity and aid the atomization process. Liquids are placed in a small porcelain boat under a rotating graphite wheel (Figure 18-18c). As the wheel turns, it becomes wet and brings fresh sample into the discharge area. Graphite is the preferred electrode material because it is a good electrical conductor, can withstand the high temperatures produced in the discharge, and is available in a high state of purity.

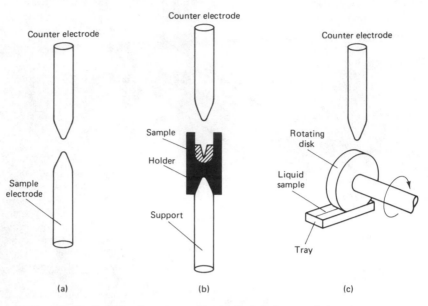

Figure 18-18 Graphite electrodes used in electrical discharge emission spectroscopy.

Arc and spark sources are used only with simultaneous, multichannel instruments, similar in design to that shown in Figure 18-16. To obtain sufficient reproducibility and sensitivity, the emission signal for an element must be integrated over a 15- to 60-second interval. The time required and amount of sample consumed in determining several elements *sequentially* are unacceptably large.

Although the importance of arc and spark sources in quantitative analysis has been diminished greatly by the development of ICP sources, their use in qualitative analysis remains high. Instruments designed for qualitative analysis are similar in optical design to that shown in Figure 18-16 but use a strip of photographic emulsion (unexposed film) as the detector in place of the individual photomultiplier tubes. During the arc or spark discharge, emissions from the various elements strike the film, thereby exposing it at discrete locations commensurate with their individual wavelengths. To complete the analysis, the film is developed and placed in an optical comparator with a master film

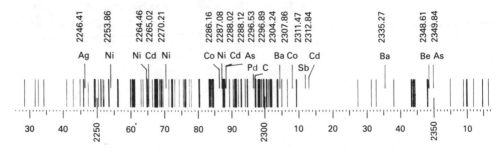

Figure 18-19 Portion of an emission spectroscopy master film used for qualitative analysis. (Courtesy of Spex Industries, Inc.)

Element	Wavelength (nm)	Detection limit (ppm)	Element	Wavelength (nm)	Detection limit (ppm)
Ag	328.1	1	K	404.7	3000
Al	396.2	2	Li	323.3	40
As	228.8	20	Mg	285.2	0.4
Au	267.6	5	Mn	257.6	1.5
B	249.8	4	Mo	313.3	30
Ba	455.4	10	Na	589.6	1
Be	234.9	1	Ni	341.5	2
Bi	306.8	3	P	253.6	20
Ca	422.7	1	Pb	405.8	3
Cd	228.8	10	Sb	260.0	20
Co	345.4	4	Si	251.6	2
Cr	425.4	2	Sn	317.5	10
Cu	324.8	0.8	Ti	337.3	10
Fe	372.0	1	V	318.5	20
Hg	253.7	30	Zn	334.5	30

containing a wavelength scale and markers indicating the exact position of the most characteristic lines for each element. A portion of a master film is shown in Figure 18-19. After aligning the two films, the operator looks for matches of important lines. Normally, three or more matches are required for a positive identification. Typical detection limits for qualitative analysis with a dc arc are given in Table 18-3.

PROBLEMS

18-1. The iron content of a municipal water supply was determined by an atomic absorption procedure. The absorbance of the water, after a five-fold dilution, was 0.646 at 248.3 nm. A standard solution, prepared by dissolving 0.1483 g of pure iron wire in acid, diluting to 250.0 mL, and then making a 100-fold dilution, had an absorbance of 0.813. Calculate the ppm Fe in the water sample.

18-2. In a more careful analysis of the water sample described in Problem 18-1, a series of iron standards was prepared by taking various volumes of an iron solution whose concentration was 0.05932 mg Fe/mL and diluting to 100 mL. The absorbances of these solutions was as follows:

Volume of Fe taken (mL)	$A_{248.3}$
1.00	0.113
3.00	0.334
5.00	0.530
7.00	0.672
10.00	0.813

(a) Calculate the ppm Fe using the calibration plot.
(b) Calculate the percent error in the answer for Problem 18-1.
(c) What assumption was made in Problem 18-1 that was not valid?

18-3. Sodium and potassium in blood serum commonly are determined by a flame emission technique using lithium as an internal standard. Standard solutions were prepared using various amounts of Na^+ and K^+ with a fixed amount amount of Li^+, all diluted to 25.0 mL. A 10.0-μL serum sample was treated in a similar fashion. The following data were obtained:

Sample	Volume of 8.00 μg/μL Na^+ std. (μL)	Volume of 0.450 μg/μL K^+ std. (μL)	Volume of 5.00 μg/μL Li^+ std. (μL)	Emission signal, arbitrary units		
				Na^+	K^+	Li^+
1	1.00	1.00	5.00	51	3.2	64.0
2	3.00	3.00	5.00	165	11.0	69.0
3	5.00	5.00	5.00	247	16.1	62.2
4	7.00	7.00	5.00	340	20.9	61.1
5	9.00	9.00	5.00	481	30.0	66.8
Unknown			5.00	275	18.8	67.9

(a) Prepare calibration plots for sodium and potassium using their absolute signals and using their signals relative to the corresponding lithium signal.

(b) Determine the concentration of Na^+ and K^+ in the serum, in units of mg/100 mL, from each pair of graphs.

(c) Which of the two plots in part a is the most linear? Why?

(d) Calculate the absolute error in the determination of both metals using the two graphs.

18-4. A dietary supplement capsule weighing 4.84 g was ground into a fine powder. Two portions of the solid, both weighing 0.137 g, were dissolved in dilute acid and transferred to 50-mL volumetric flasks. To one of these, 5.00 mL of 40.0 ppm Mn^{2+} was added, then both flasks were diluted to the mark with distilled water. When aspirated into the flame of an atomic absorption spectrometer set at a manganese absorption wavelength, the absorbances of the unknown and unknown plus standard were 0.374 and 0.641, respectively. Calculate the percentage of Mn in the capsule.

18-5. Six 5.00-mL samples of raw milk were treated with 0, 5.00, 10.00, 15.00, 20.00, and 25.00 μg of Zn^{2+} and diluted to 50.00 mL. The absorbances of these solutions were 0.235, 0.292, 0.346, 0.397, 0.458, and 0.510. Prepare a standard calibration plot and calculate the ppm Zn in the sample of milk.

18-6. A 25.0-mL sample of water from Lake Erie was acidified and treated with excess sodium molybdate, leading to the formation of 12-molybdophosphoric acid (12-MPA):

$$H_3PO_4 + 12Na_2MoO_4 + 24H^+ \longrightarrow H_3PMo_{12}O_{40} + 24Na^+ + 12H_2O$$

The 12-MPA was separated from the excess sodium molybdate by extraction with two successive 10.0-mL portions of iso-amyl alcohol. When the combined extract was aspirated into the flame of an atomic absorption spectrometer set up for measuring molybdenum, an absorbance of 0.662 was obtained. The absorbance of a 5.00-ppm Mo standard in the same volume of iso-amyl alcohol was 0.203. Calculate the ppm PO_4^{3-} in the sample.

19

INTRODUCTION TO SEPARATIONS

Despite chemists' best efforts to the contrary, analytical methods are never completely specific for a single substance. The ways in which substances can interfere in a determination are numerous and, consequently, so are the specific procedures used to overcome or avoid interferences. There are three general approaches to dealing with interferences: adjust conditions to *prevent* their occurrence; *compensate* for their effect; and *separate* the analyte from the interfering constituent. The chemist is most likely to attempt the alternatives in the order listed above because, generally, that is the order of increasing difficulty.

Adjusting the solution and measurement conditions to avoid an interference can be quite simple and highly effective. Consider, for example, the determination of Fe^{3+} in water containing Ca^{2+} and Mg^{2+} by an EDTA titration. Both Ca^{2+} and Mg^{2+} can be made sufficiently unreactive by adjusting the pH to 3.0 before titrating. Compensation techniques also are quite simple and used widely. They involve correcting the measured value for the effect of another constituent. Thus, in a spectrophotometric determination, the absorbance of the sample can be corrected for the background absorption of the solvent and other reagents by subtracting the absorbance of a reagent blank. Separation as a means of avoiding an interference generally is considered only as a last resort, because it is more time consuming and requires more sample manipulation than do the other methods. Despite this fact, separations are an extremely important part of modern analytical chemistry.

A chemical separation is characterized by a change in the *relative* concentration of two or more substances within a defined region of space. This chapter is devoted only to separations that are the result of equilibrium processes. All equilibrium-based separations involve partitioning of components between two distinct phases that can be separated mechanically. The extent to which the components partition differently between

TABLE 19-1 REPRESENTATIVE SEPARATION METHODS BASED ON PHASE EQUILIBRIA

Gas-liquid	Gas-solid	Liquid-liquid	Liquid-solid
Distillation	Sublimation	Extraction	Precipitation
Gas chromatography	Adsorption	Liquid chromatography	Adsorption
	Molecular sieves		Ion exchange

the phases determines the ease with which a given degree of separation can be achieved. For example, chloride ion can be separated quite completely and easily from nitrate ion in water by addition of excess silver ion. The insoluble nature of silver chloride ensures that almost all of the silver ion is transferred to the solid phase, whereas the nitrate ions remain dissolved in the liquid phase. In situations such as this, where two solutes have opposite strong preferences for the two phases, a high degree of separation is possible with a single partitioning. More often, however, the substances to be separated have similar chemical and physical properties, including the manner in which they partition between two phases. As a result, the partitioning process may have to be repeated many times to effect a given degree of separation.

Separation methods are often classified according to the phases used. Since matter ordinarily exists in one of three physical phases, six two-phase systems are possible. The difficulty of mechanically separating two solids and the miscibility of all gases makes these two-phase systems unsuitable for separations, leaving the four classifications shown in Table 19-1. A discussion of all of the separation methods is beyond the scope of this book. The remainder of this and the three succeeding chapters is devoted to the principles and practices associated with the separation methods currently of most widespread interest in analytical chemistry.

PRECIPITATION

Precipitation is used occasionally as a means of separating constituents prior to a determination. In a gravimetric procedure the analyte is determined by weighing the precipitate, but there is no fundamental reason why it cannot be determined by other types of measurements as well. Our interest in this section is the use of precipitation as a general separation tool.

A separation based on precipitation results from a difference in the solubility of the various substances. Although solubility product constants in conjunction with other equilibrium constants can be used to determine both the theoretical feasibility and the conditions necessary to achieve a given degree of separation, other variables not reflected in the values of the equilibrium constants often play a vital role in the success or failure of a given separation. For example, in Chapter 4 we discussed several nonequilibrium processes by which precipitates become contaminated with unwanted substances whose solubility has not been exceeded. Additionally, it was noted that some precipitates tend to be colloidal and are not easily separated from the liquid solution.

Precipitating Reagents

Quantitative separations based on precipitation are restricted largely to inorganic substances, mainly because organic precipitates tend to be impure and difficult to filter. Although many precipitating agents have found use in organic separations, the discussion here is limited to a few representative reagents of general applicability. You should refer to Chapter 4 for a discussion of specific precipitants.

Hydrogen or hydroxide ion

Large differences exist in the solubilities of acid, base, and oxide forms of the elements. Couple this with the fact that the concentration of hydrogen or hydroxide ion can easily be varied by as much as 15 orders of magnitude and it becomes clear that pH control may be quite useful as a means of effecting certain separations. In practice, pH control is used mainly to isolate groups of substances rather than single components. As illustrated in Table 19-2, the separations can be divided into four categories according to the acidity of the solution: (1) those occurring in solutions of strong, oxidizing acids; (2) those occurring in weakly acidic solution; (3) those occurring in weakly basic solution; and (4) those occurring in strongly basic solution. Table 19-2 is only representative of group separations that can be achieved through pH control; it is not an exhaustive compilation.

Sulfide ion

Cations of most elements, except the alkali and alkaline earth metals, form insoluble sulfides. Separations are possible because the solubilities differ greatly and it is an easy matter to control the concentration of the precipitating sulfide ion by pH control (S^{2-} is a weak base). A theoretical treatment of the ionic equilibria influencing the solubility of sulfide precipitates does not always provide correct conclusions regarding the feasibility of a particular separation because of coprecipitation and slow rates of precipitation. At one time, however, sulfide precipitation schemes for separating and identifying cations were extremely important in analytical chemistry, and a great deal

TABLE 19-2 PRECIPITATION SEPARATIONS BASED ON CONTROL OF ACIDITY

Reagent	Substances precipitated	Substances not precipitated
Concentrated HNO_3 or $HClO_4$	SiO_2, Nb_2O_5, H_2SnO_3, Sb_2O_5, Ta_2O_5, WO_3, PbO_2	Most other metal ions and anions
CH_3CO_2H/CH_3CO_2Na buffer	$Al(OH)_3$, $Cr(OH)_3$, $Fe(OH)_3$	Most divalent metal cations
NH_3/NH_4Cl buffer	BeO, $Al(OH)_3$, $Sc(OH)_3$, $Cr(OH)_3$, $Fe(OH)_3$, $Ga(OH)_3$, $Zr(OH)_4$	Alkali and alkaline earth metals, Mn^{2+}, Co^{2+}, Ni^{2+}, Cu^{2+}, Zn^{2+}, Ag^+, Cd^{2+}
Concentrated $NaOH$	Most divalent metal ions, rare earths, Fe^{3+}	Alkali metals, Al^{3+}, Zn^{2+}, oxyanions

Precipitation

TABLE 19-3 ELEMENTS PRECIPITATED
BY SULFIDE ION

Reagent	Substances precipitated
H_2S in HCl	HgS, PbS, Bi_2S_3, CuS, CdS, As_2S_3, Sb_2S_3, SnS
H_2S in NH_3	FeS, MnS, ZnS, NiS, CoS, $Al(OH)_3$, $Cr(OH)_3$, $Ti(OH)_4$

of practical information is known about the properties of sulfide precipitates. Table 19-3 shows how groups of ions can be separated based on sulfide and hydroxide precipitation at controlled pH.

8-Hydroxyquinoline

With the exception of the alkali metals, cations of almost every metal in the periodic table can be precipitated with 8-hydroxyquinoline. Trivalent cations form $1:3$ complexes as shown below with Al^{3+}, whereas $1:2$ complexes are the rule with divalent cations.

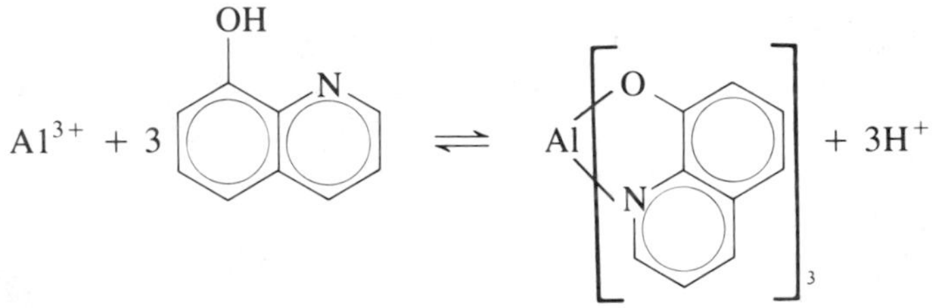

The reaction indicates that the equilibrium constant will be pH dependent and that some selectivity may be gained by careful control of the pH during precipitation. It is common

TABLE 19-4 METALS PRECIPITATED
BY 8-HYDROXYQUINOLINE

Solution condition	Substances precipitated
pH 4–5	Al^{3+}, Ti^{4+}, Fe^{3+}, Co^{2+}, Cu^{2+}, Ni^{2+}, Zn^{2+}, Ga^{3+}, Zr^{4+}, Pd^{2+}, Ag^+, Cd^{2+}, Hf^{4+}
pH 11 (NH_3)	Be^{2+}, Mg^{2+}, Al^{3+}, Sc^{3+}, Ti^{4+}, Mn^{2+}, Fe^{3+}, Cu^{2+}, Zn^{2+}, Ga^{3+}, Zr^{4+}, Pd^{2+}, Cd^{2+}, In^{3+}, Hf^{4+}, rare earths

practice to use various masking agents in conjunction with 8-hydroxyquinoline to effect group precipitations. One such example with ammonia is shown in Table 19-4.

SOLVENT EXTRACTION

Solvent extraction is a method of separation based on the distribution of solutes between two immiscible liquid solvents. In its simplest application, the only apparatus needed is a separatory funnel. The solution containing the solutes is placed in the funnel along with a second, immiscible solvent. After shaking for a short time to hasten the distribution process, the contents are allowed to settle and the lower layer is drained from the funnel. If the extraction does not produce an adequate separation, it may be repeated using a fresh portion of the extracting solvent.

Organic chemists use solvent extraction in preparative work, that is, isolating and purifying compounds for further use or study. In analytical chemistry, solvent extraction is generally an integral part of an analysis scheme, being used to isolate the analyte from possible interferences and, at the same time, converting it to a measurable form. In this context, the majority of its applications are in atomic and molecular spectrophotometric determinations. In this section we examine the mathematical description of the separation process and then review a few important applications.

Distribution Equilibria

The distribution of a solute between two immiscible liquids is an equilibrium process. Accordingly, for a solute S, partitioning between aqueous and organic solvents, we may write

$$S_{aq} \rightleftharpoons S_{org}$$

for which

$$K_D = \frac{[S]_{org}}{[S]_{aq}} \tag{19-1}$$

where K_D is an equilibrium constant called the distribution coefficient. Technically, Equation 19-1 should be written in terms of activities, but as we have done in the past, concentrations will be considered acceptable estimates of the activities.

Knowing the distribution coefficient makes it possible to determine the extent to which a solute is transferred from one solvent to another. To see how this is done, let m be the amount in millimoles of solute S and q the fraction of S remaining in the *aqueous* phase when equilibrium is achieved. Then

$$[S]_{aq} = \frac{(\text{fraction remaining})(\text{total amount})}{\text{volume aqueous phase}} = \frac{qm}{V_{aq}} \tag{19-2}$$

$$[S]_{org} = \frac{(\text{fraction transferred})(\text{total amount})}{\text{volume organic phase}} = \frac{(1-q)m}{V_{org}} \tag{19-3}$$

Substituting these concentrations in Equation 19-1 gives

$$K_D = \frac{(1-q)m/V_{org}}{qm/V_{aq}} = \frac{(1-q)V_{aq}}{qV_{org}}$$

which can be solved for q:

$$q = \frac{V_{aq}}{K_D V_{org} + V_{aq}} \tag{19-4}$$

The fraction of solute transferred to the organic phase is $1 - q$, or

$$1 - q = 1 - \frac{V_{aq}}{K_D V_{org} + V_{aq}} = \frac{K_D V_{org}}{K_D V_{org} + V_{aq}} \tag{19-5}$$

If the phases are separated and the extraction repeated with the same volume of fresh organic solvent, the fraction of solute remaining in the aqueous phase *for this extraction* is also given by Equation 19-4. In terms of the *original* solution, the fraction remaining after two extractions is $q \times q$, or

$$\text{fraction remaining after two extractions} = q \times q = \left(\frac{V_{aq}}{K_D V_{org} + V_{aq}}\right)^2$$

The fraction remaining in the aqueous phase after n extractions with V_{org} portions of organic solvent is

$$\text{fraction remaining after } n \text{ extractions} = q^n = \left(\frac{V_{aq}}{K_D V_{org} + V_{aq}}\right)^n \tag{19-6}$$

The percentage of solute transferred to the combined organic phases after n extractions is

$$\% \ E = (1 - q^n) \times 100$$

or

$$\% \ E = \left[1 - \left(\frac{V_{aq}}{K_D V_{org} + V_{aq}}\right)^n\right] \times 100 \tag{19-7}$$

For most analytical purposes, an extraction must be at least 99.9% complete to be considered quantitative. According to Equation 19-7, the number of extractions necessary to effect a quantitative transfer of solute from one phase to another depends mainly on the value of the distribution coefficient.

Example 19-1

If the distribution coefficient for a metal chelate partitioning between water and chloroform is 6.4, calculate the fraction of chelate extracted when 25.0 mL of 4.3×10^{-2} M ML is shaken with (a) one 10.0-mL portion of chloroform and (b) two successive 10.0-mL portions of chloroform.

Solution

(a) According to Equation 19-6, the fraction remaining after one extraction is

$$q = \frac{25.0 \text{ mL}}{(6.4)(10.0 \text{ mL}) + 25.0 \text{ mL}} = 0.281$$

The fraction extracted is

$$1 - q = 1 - 0.281 = 0.719$$

(b) Again, according to Equation 19-6,

$$q^2 = \left[\frac{25.0 \text{ mL}}{(6.4)(10.0 \text{ mL}) + 25.0 \text{ mL}} \right]^2 = 0.0790$$

and the fraction extracted is

$$1 - q^2 = 1 - 0.0790 = 0.921$$

Note that the concentration was not used to solve this problem. Sometimes more information is available than is needed and the chemist must know when to ignore superfluous data.

Effect of other solute equilibria

It is important to recognize that the distribution coefficient as defined in Equation 19-1 describes the ratio of *equilibrium concentrations* of solute in two phases. Ordinarily, the chemist is more interested in the total or *analytical concentration* of the solute in each phase. The ratio of these concentrations is called the *distribution ratio*, D:

$$D = \frac{C_{\text{ML,org}}}{C_{\text{ML,aq}}} \tag{19-8}$$

If there are no competing equilibria in either phase, the analytical and equilibrium concentrations are the same and D equals K_D. It is not unusual, however, for a solute to be involved in other equilibria in one or both phases. As a result, the solute will exist in more than one form and D will not be the same as K_D. Consider a common example—the extraction of a metal complex that can dissociate in the aqueous phase. The equilibrium between a complex and its ions in aqueous solution is

$$M^{n+} + L^{n-} \rightleftharpoons ML$$

for which

$$K_{\text{ML}} = \frac{[\text{ML}]_{\text{aq}}}{[\text{M}]_{\text{aq}}[\text{L}]_{\text{aq}}} \tag{19-9}$$

where the charges are omitted for the sake of clarity. The analytical concentration of metal in the aqueous phase is the sum of the equilibrium concentrations of its two forms:

$$C_{\text{ML,aq}} = [\text{ML}]_{\text{aq}} + [\text{M}]_{\text{aq}} \tag{19-10}$$

If we assume that no competing equilibrium occurs in the organic phase, Equation 19-8 can be rewritten as

$$D = \frac{[\text{ML}]_{\text{org}}}{[\text{ML}]_{\text{aq}} + [\text{M}]_{\text{aq}}}$$

Solving Equation 19-9 for $[M]_{aq}$ and substituting in the equation above gives

$$D = \frac{[ML]_{org}}{[ML]_{aq} + [ML]_{aq}/K_{ML}[L]_{aq}}$$

which can be rearranged to yield

$$D = \frac{K_D K_{ML}[L]_{aq}}{K_{ML}[L]_{aq} + 1} \qquad (19\text{-}11)$$

Thus we have an expression which says that the distribution of the complex depends not only on K_D but also on the formation constant of the complex and the concentration of the ligand. Equations may be derived in a similar fashion for systems in which the solute participates in other competing equilibria. The extraction of metal complexes almost always varies with pH because of its effect on the formation of the complex.

Extractants

The large majority of substances are converted to neutral chelates or ion-association complexes in extraction procedures. Usually, the complex is formed in the aqueous phase prior to extraction, but sometimes the complexing agent is dissolved in the organic phase, in which case complexation and extraction take place simultaneously when the two phases are shaken together.

In *chelate extraction systems*, anionic ligands displace coordinated water from metal ions forming neutral covalent chelates that are more soluble in organic solvents than in water. In attempting to separate metals, the chemist uses a number of solution variables such as pH and auxiliary complexing agents to control which metal ions form extractable complexes. In *ion-association extraction systems*, large organic ions form stable ion pairs with oppositely charged analyte ions. Overall, the complex is neutral, and the large organic group makes it soluble in many organic solvents. Our ability to manipulate solution conditions to gain selectivity is more limited with ion-association complexes than with chelates. The formation constant of an ion pair depends partly on the closeness of approach of the two ions, which is somewhat dependent on the size and structure of the organic ion.

Solvents

There are two primary considerations in the selection of an extraction solvent: the relative extractability of the substance of interest, and the degree of immiscibility with the initial solvent. In many analytical applications, extraction is used both to isolate the analyte from potential interferences and simultaneously, to convert it to a form suitable for spectrophotometric measurement. To accomplish this, various reagents may be added to the aqueous phase and the chemist must ensure that these reagents are not extracted along with the analyte, or if they are, that they will not interfere in later measurements. Solvent immiscibility is important in this regard. Solvents are seldom totally immiscible; that is, they exhibit some solubility in one another. If, for example, water dissolves slightly in an extracting organic solvent, it can "carry along" various solutes that could interfere in the later measurement of the analyte.

When there is a choice of solvents suitable for the extraction, the analyst will consider other factors, such as specific gravity, toxicity, flammability and the tendency to form emulsions. Emulsions are the result of one immiscible liquid becoming dispersed in a continuum of another. The stability of the dispersion is of interest because it is necessary to separate the phases to complete the analytical procedure. In general, emulsions separate faster as the viscosity of the continuous phase decreases and the difference in density of the two phases increases. Occasionally, solvents are mixed to achieve the desired characteristics. For example, mixtures of alcohols and diethyl ether are commonly used in the extraction of certain metal thiocyanates from aqueous solutions. The properties of selected solvents employed in extractions are summarized in Table 19-5.

Applications

The ease with which solvent extractions are performed and the possibility of simultaneously converting the analyte to a measurable form make this a popular technique. Specific examples are far too numerous to discuss. Instead, a brief summary of the different types of extraction systems is given in the remainder of this section.

Extraction of metal chelates

Table 19-6 lists a few of the many chelating reagents reported to be useful extractants for metal ions. The solubility differences among metal chelates is seldom large enough to enable their separation by extraction. Instead, we rely on an adjustment of solution conditions to prevent certain metals from forming extractable complexes. For example, solution acidity has a dramatic effect on the formation and extraction of metal diphenylthiocarbazones, as shown in Figure 19-1. Such figures are extremely useful in defining the conditions necessary to effect a desired separation. Thus copper can be separated quantitatively from zinc by carrying out the extraction at a pH between 3.5

TABLE 19-5 SELECTED SOLVENTS USED IN EXTRACTIONS

Name	Formula	Specific gravity	Boiling point (°C)	Solubility in water (g/L)
Cyclohexane	C_6H_{12}	0.779	81.4	0.1
Hexane	C_6H_{14}	0.660	69.0	0.14(16°C)
Toluene	$C_6H_5CH_3$	0.867	110.6	0.47(16°C)
1,3,5-Trimethylbenzene	$(CH_3)_3C_6H_3$	0.865	164.7	Very small
Carbon tetrachloride	CCl_4	1.595	76.8	0.8
Chloroform	$CHCl_3$	1.498	61.3	10(15°C)
Trichloroethylene	C_2HCl_3	1.456	87.0	1
1-Butanol	$CH_3(CH_2)_3OH$	0.810	117.7	79
2-Methyl-1-propanol	$(CH_3)_2CHCH_2OH$	0.801	108.3	95(18°C)
3-Methyl-1-butanol	$(CH_3)_2CHCH_2CH_2OH$	0.812	130.5	26.7(22°C)
Diethyl ether	$C_2H_5OC_2H_5$	0.714	34.6	75
4-Methyl-2-pentanone	$(CH_3)_2CHCH_2COCH_3$	0.802	119.0	19
Ethyl acetate	$CH_3CO_2C_2H_5$	0.901	77.2	86
Tri(n-butyl)phosphate	$(C_4H_9)_3PO_4$	0.973	177.5[a]	6

[a]At 25 mmHg.

Solvent Extraction

TABLE 19-6 SOME CHELATING REAGENTS USED IN SOLVENT EXTRACTION

Name (abbreviation)	Formula	Metals extracted[a]
Acetylacetone	$CH_3C-CH_2-CCH_3$ (with two $=O$ groups)	Al^{3+}, Be^{2+}, Cr^{3+}, Co^{3+}, Cu^{2+}, Fe^{3+}, Mo(VI), U(VI), Zn^{2+}
Thenoyltrifluoroacetone (TTA)		>40 metals
Morin		Al^{3+}, Be^{2+}, Ga^{3+}, Sb^{3+}, Sc^{3+}, Sn^{2+}, Ti^{4+}, Zr^{4+}
8-Hydroxyquinoline (oxine)		>50 metals
Cupferron		>30 metals
1-(2-Pyridylazo)-2-naphthol (PAN)		Bi^{3+}, Cd^{2+}, Cu^{2+}, Pd^{2+}, Sn^{2+}, Hg^{2+}, Co^{2+}, Pb^{2+}, Fe^{3+}, Ni^{2+}, Zn^{2+}, Sc^{3+}
Diphenylthiocarbazone (dithizone)		>25 metals
Sodium diethyldithiocarbamate	$(C_2H_5)_2N-C-S^-Na^+$ (with $=S$ group)	>30 metals
N-Benzoyl-N-phenylhydroxylamine (BPHA)		>25 metals

[a] Not an exhaustive list.

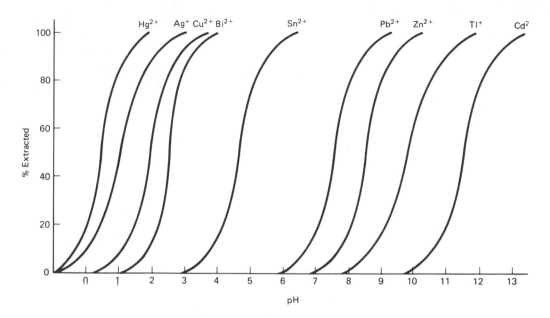

Figure 19-1 Effect of pH on efficiency of extraction of metal ions into chloroform by diphenyl-thiocarbazone. [Reprinted by permission of the publisher from G.H. Morrison and H. Freiser in C.L. Wilson and D.W. Wilson (eds.): *Comprehensive Analytical Chemistry*, vol. IA (Amsterdam: Elsevier Science Publishers B.V., 1959), p. 151.]

and 6.0. When a desired separation cannot be achieved by pH regulation alone, masking agents such as cyanide, thiocyanate, and thiosulfate may be employed.

Unlike transition metals, the alkali metals do not form stable, neutral chelates, and their separation and extraction into organic solvents is more difficult. Recently, analytical chemists have developed an interest in macrocyclic molecules called *crown ethers* which can *envelop* certain metal ions in a protective pocket of oxygen atoms, thereby enabling their transfer from an aqueous to a nonaqueous solvent. The structure of a typical crown ether is shown in Figure 19-2. The nonpolar nature of the exposed (outward) part of the molecule makes it quite soluble in organic solvents such as chloroform. Because there is no formal covalent bonding between the ether oxygens and metal cation, these molecules are not true chelates and the crown ethers are sometimes referred to as *phase transfer catalysts*. To satisfy the laws of electroneutrality, an anion must be transferred together with the cation, or the crown ether must have an acidic functional group that can transfer a proton to the aqueous phase to replace the metal cation.

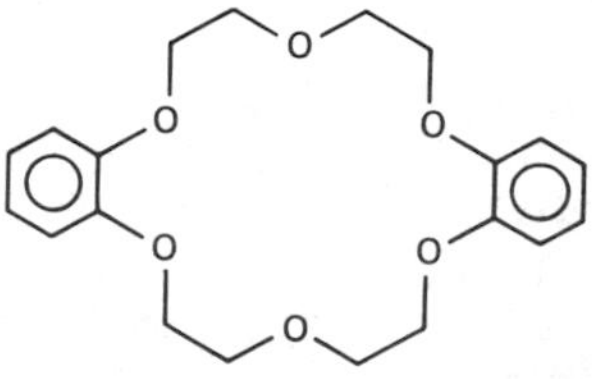

Figure 19-2 Formula of dibenzo-18-crown-6.

Solvent Extraction

There are many natural compounds that behave similarly to synthetic crown ethers. Living cells use such molecules to transport ions and hydrophilic molecules through hydrophobic membranes. A class of antibiotics called ionophores kill certain bacteria by carrying metal ions through the cell wall and disrupting their metabolism. One such antibiotic, valinomycin, is so specific for K^+ that it is used as the active membrane material in a potassium ion-selective electrode (Chapter 13).

Extraction of ion-association complexes

Ion-association systems are, in the majority of cases, better suited than chelate systems for the extraction of large *quantities* of metals. This should not be interpreted to mean that ion-association reagents cannot be used for the extraction of low *concentrations* of metal. Many metals form anionic halide complexes that, when paired with a suitable cation, can be extracted into organic solvents. Ion pairs with hydronium ion often are soluble in oxygenated solvents such as alcohols, ethers, esters, and ketones.

TABLE 19-7 DIETHYL ETHER EXTRACTION OF METALS WITH VARIOUS ANIONS

Metal	Percent extracted				
	6.0 M HCl[a]	6.0 M HBr[b]	6.9 M HI[c]	7.0 M NH$_4$SCN[b]	8.0 M HNO$_3$[b]
Ag	0	0	—	0.1	2.4
Al	0	0	0	49	0
Au(III)	95	73	100	—	97
As(III)	68	73	62	0.4	—
Be	0	—	0	92	1.4
Bi(III)	0	0	34	0.1	6.8
Ca	0	0	0	0	0
Cd	0	0.9	100	0.2	0.3
Co(II)	0	0.08	0	75	0.2
Cr(III)	0	0	0	3.4	—
Cu(II)	0.05	6.2	—	—	0
Fe(III)	99	95	—	53	0.13
Ga	97	95	0	99	<0.2
Hg(II)	0.2	1.5	100	0.65	4.7
Mg	0	0	—	0	—
Mn(II)	0	0	0	0.17	0.2
Mo(VI)	85	54	6.5	97	0.6
Ni	0	0	0	0	0
Pb(II)	0	0	0	<0.1	0.5
Sb(III)	6	6.1	100	2.2	—
Sn(II)	22	36	100	—	—
U(VI)	0	0	0	7	65
V(V)	0	0	0	—	2
W(VI)	0	0	0	—	—
Zn	0.2	3.6	10.6	93	0

[a] 3:2 volume ratio of organic to aqueous phase.

[b] 1:1 volume ratio of organic to aqueous phase.

[c] 4:1 volume ratio of organic to aqueous phase.

Thus, in the presence of 6 M HCl, iron(III) is extracted quantitatively into diethyl ether as $H_3O^+FeCl_4^-$. The selectivity of such an extraction is determined primarily by the magnitude of the difference in the haloanion and ion-pair formation constants for the various metals in a sample. Large differences in these equilibrium constants make it much easier to find conditions that will permit the quantitative formation and extraction of a single ion pair. The extracting solvent also plays a role in determining the selectivity, mainly through its effect on the formation constant of the ion pair. It was pointed out in Chapter 8 that ion-pair formation is favored by a decrease in the dielectric constant of the solvent. Unfortunately, solvents of low dielectric constant often lack sufficient solvating ability to dissolve small inorganic ion pairs.

A fairly large number of metal thiocyanates and nitrates are extractable into oxygenated organic solvents. At low concentrations of hydrogen and thiocyanate ion, simple thiocyanates such as $Sc(SCN)_3$, $Fe(SCN)_3$, $Zn(SCN)_2$, $Ga(SCN)_3$, and $UO_2(SCN)_2$ are extracted into diethyl ether. At higher concentrations of thiocyanate, complex anion salts such as $(NH_4)_2Co(SCN)_4$ and $(NH_4)_2Zn(SCN)_4$ are extracted. Table 19-7 summarizes the extractability of various metals with different anions.

Large, bulky organic ions comprise a second type of ion-association extractant used widely in analytical separations. As a group, these extractants are more versatile

TABLE 19-8 SOME ORGANIC ION-ASSOCIATION REAGENTS USED IN SOLVENT EXTRACTION

Name	Formula	Substances extracted[a]
Di-*n*-butylphosphoric acid ester (DBPA)	$(C_4H_9O)_2\overset{\displaystyle O}{\overset{\displaystyle \|}{P}}-OH$	In, Nb, Sn(IV), Ta, Zr
Rhodamine B		Ag, Au(III), Bi(III), Ga, Hg(II), Mo(VI), Sb(V), Tl(III)
Sodium tetraphenylborate	$(C_6H_5)_4B^-Na^+$	K, Rb, Cs
Tetraphenylarsonium chloride	$(C_6H_5)_4As^+Cl^-$	Au(III), Cr(VI), Fe(III), Ga, Hg(II), Mn(VII), Sb(III), ClO_3^-, ClO_4^-, NO_3^-, I^-, SCN^-
Triisooctylamine	$[(CH_3)_2CH(CH_2)_4CH_2]_3N$	Ag, Au(III), Be, Cd, Co(II), Fe(III), Ga, Pt, U(VI), Zn
Tri-*n*-octylphosphine oxide (TOPO)	$(C_8H_{17})_3P=O$	Au(III), Cr(VI), Fe(III), Mo(VI), Sn(IV), Ta, Th, U(VI)

[a]Substances listed are only representative. Others may also be extracted.

Solvent Extraction

than their inorganic counterparts. Reagents are available that can form ion pairs with either anions or cations, and often they are excellent chromophores, which allows the concentration of the extracted substance to be determined by a simple spectrophotometric measurement. As a rule, these complexes are most soluble in solvents of low polarity, such as chloroform, toluene, or hexane. A selected list of organic ion-association reagents is given in Table 19-8.

ION-EXCHANGE SEPARATIONS

A variety of natural and synthetic substances exhibit the ability to exchange one type of ion for another of the same sign. The ion-exchange properties of clays and zeolites are widely recognized as being very important to the retention of various metals in soil. Soils with a very low clay content often cannot retain certain metal ions needed to support the growth of various plants. Synthetic ion-exchange compounds were first made about 50 years ago and have since found many applications in ion separations, reagent preparation, and water purification.

Ion-Exchange Resins

The most widely used synthetic ion-exchange resins are made by the copolymerization of styrene and divinylbenzene while suspended as tiny droplets in water. The resulting beads are porous and swell as they absorb water, but they do not dissolve in water or common hydrocarbons. The divinylbenzene acts to "cross-link" the linear polystyrene chains, creating a three-dimensional polymeric network. The degree of swelling depends on the amount of cross-linking agent present, which is generally about 8 to 12%.

The resin is made into a cation exchanger by sulfonating the benzene rings (Figure 19-3a). The $-SO_3^-$ group is covalently bound to the polymer, but the associated cation is free to move about and can be replaced with another cation. Such resins are called *strong* cation-exchange resins because the sulfonate functional group is always completely ionized. Anion-exchange resins are made by treating the beads with chloromethyl ether and then trimethylamine, which places a quaternary ammonium group on the benzene rings (Figure 19-3b). In this case it is the associated anion that is free to move about and can be replaced. Quaternary ammonium groups usually are completely ionized; thus we have a *strong* anion-exchange resin. Weak cation- and anion-exchange resins containing carboxylic and triamine functional groups, respectively, are also available but are not discussed in this chapter.

The *capacity* of an ion-exchange resin is the amount, in milliequivalents, of exchangeable ion per gram of resin. Sulfonated cation-exchange resins have a capacity of about 5 meq/g of dry resin in the hydrogen-ion form. The capacity of quaternary ammonium anion-exchange resins usually is slightly less, about 3 to 3.5 meq/g of dry resin in the chloride-ion form.

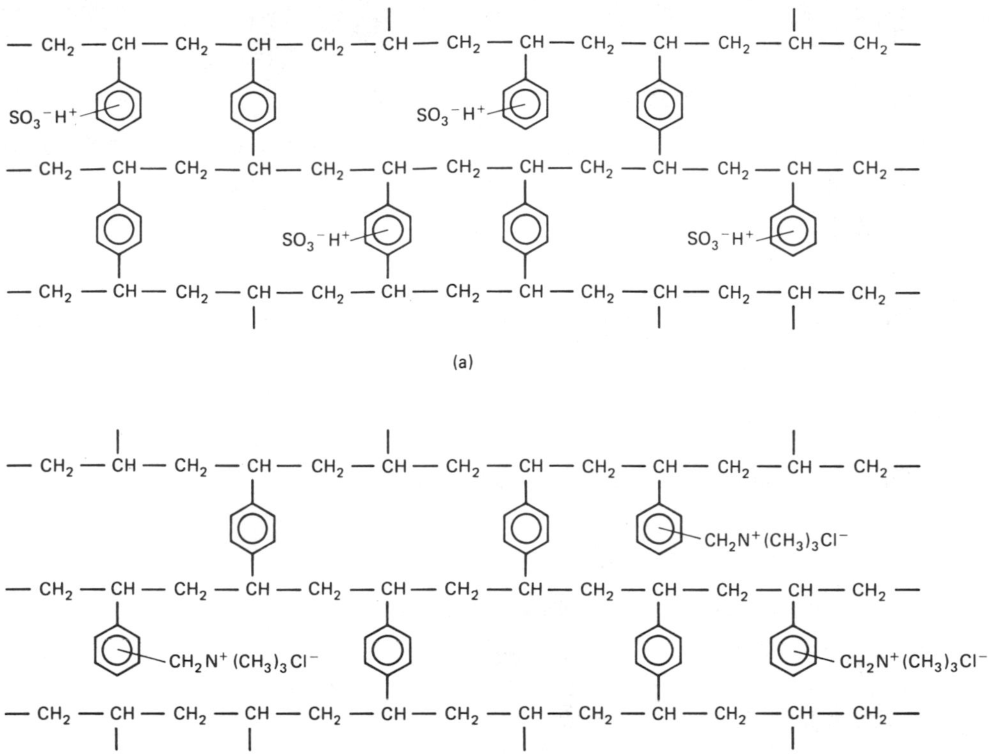

(a)

(b)

Figure 19-3 Chemical structure of styrene-divinylbenzene ion-exchange resins: (a) strong cation exchanger; (b) strong anion exchanger.

Ion-Exchange Equilibria

Ion-exchange reactions are reversible and may be treated in terms of the law of mass action. If a solution of sodium chloride is brought into contact with a strong cation-exchange resin in the hydrogen-ion form, the following reaction occurs:

$$RSO_3{}^-H^+ + Na^+ \rightleftharpoons RSO_3{}^-Na^+ + H^+$$

for which the equilibrium constant is

$$K_{eq} = \frac{[RSO_3{}^-Na^+][H^+]}{[RSO_3{}^-H^+][Na^+]}$$

The value of the equilibrium constant depends on the specific nature of the ion-exchange resin, particularly the degree of cross-linking. In spite of the variability of the equilibrium constants, it is possible to arrange ions in the order of their increasing preference for a given resin, by making comparisons to a common standard ion. In terms of charge, the order of preference is $1+ < 2+ < 3+ < 4+$. The orders within the monovalent and divalent groups are

$$Li^+ < H^+ < Na^+ < NH_4^+ < K^+ < Rb^+ < Cs^+ < Ag^+ < Tl^+$$

$$Mg^{2+} < Zn^{2+} < Co^{2+} < Cu^{2+} < Cd^{2+} < Ni^{2+} < Ca^{2+} < Sr^{2+} < Pb^{2+} < Ba^{2+}$$

A strong anion-exchange resin in the hydroxide-ion form in contact with a solution containing chloride ions undergoes an exchange represented by the reaction

$$R_4N^+OH^- + Cl^- \rightleftharpoons R_4N^+Cl^- + OH^-$$

Between charge groups, the order of preference is $1- < 2- < 3- < 4-$ and among monovalent anions the order is

$$F^- \simeq OH^- < CH_3CO_2^- < Cl^- < NO_2^- < CN^- < Br^- < NO_3^- < I^-$$

It is important to note that other competing reactions can drastically alter the position of an ion-exchange equilibrium. Consider, for example, the following two exchange reactions:

$$RSO_3^-H^+ + (Na^+ + Cl^-) \rightleftharpoons RSO_3^-Na^+ + (H^+ + Cl^-) \quad (19\text{-}12)$$

$$RSO_3^-H^+ + (Na^+ + OAc^-) \rightleftharpoons RSO_3^-Na^+ + HOAc \quad (19\text{-}13)$$

Despite the fact that the actual exchange is between H^+ and Na^+ in both cases, Reaction 19-13 proceeds to a much greater extent because the hydrogen ions released in the exchange reaction combine covalently with the acetate ion.

Applications

Since ion-exchange equilibrium constants are not very large, a single equilibration is seldom sufficient to effect a "quantitative" exchange. Consequently, the usual procedure is to pour the sample solution through the ion-exchange resin in a long column (about 5 to 50 cm). To help understand how this works, consider a column containing a cation-exchange resin in the hydrogen-ion form and divide it into imaginary layers. When a sodium chloride solution is poured onto the column an equilibrium is established in the first "layer" and some of the sodium ions in the solution are replaced by hydrogen ions from the resin. As the solution moves down the column into the next "layer," another equilibrium is established and the concentration of sodium ions in the solution is reduced further. If this process is repeated a sufficient number of times, the solution exiting the column will contain a negligible amount of sodium ions, as they will have been replaced almost entirely with hydrogen ions. It is important to realize that the column does not *actually* consist of discrete layers in which equilibrium is established. This is merely a convenient concept that helps us to understand the results we obtain experimentally.

Ion-exchange resins serve many useful purposes in analytical chemistry. A few of them are discussed below. One very important application, chromatography, is discussed in Chapter 22.

Replacing interfering ions

There are many instances where a specific ion or group of ions interferes in a determination or other chemical process and must be removed. Most of the applications of this type involve replacement of ions that are oppositely charged from the analyte.

For example, high concentrations of ferric, aluminum, and potassium ions interfere in the gravimetric determination of sulfate by coprecipitating with barium sulfate. Passing the sample through a cation-exchange resin in the hydrogen-ion form will result in these interfering cations being exchanged for noninterfering hydrogen ions. As an anion, the sulfate will pass through the resin unretained.

Tetramethylammonium hydroxide is used in many electrochemical studies and nonaqueous titrations. Unfortunately, it is expensive and often not available in a highly pure state. On the other hand, the much less expensive tetramethylammonium chloride is commonly available. The desired reagent can be prepared first by converting a cation-exchange resin to the tetramethylammonium-ion form by passing a solution of the chloride salt through the column, then rinsing with water and passing a known quantity of sodium hydroxide through the resin. Sodium ions exchange with the tetramethylammonium ions on the resin to produce the desired reagent. The chemist must ensure that the column contains enough quaternary ammonium ion to exchange with all the sodium ions, or else the final solution will be contaminated with sodium hydroxide. Anion-exchange resins in the hydroxide-ion form are somewhat unstable, which makes a direct hydroxide-for-chloride exchange less feasible.

Commercial water softeners use a cation resin to exchange sodium ions for calcium and magnesium ions, the two substances primarily responsible for "hard" water. The reaction can be written as

$$2RSO_3^-Na^+ + M^{2+} \rightleftharpoons (RSO_3^-)_2M^{2+} + 2Na^+ \qquad (19\text{-}14)$$

where M^{2+} represents Ca^{2+} or Mg^{2+}. When the resin is depleted of sodium ions, it can be regenerated by pouring concentrated sodium chloride solution through the canister or container, thereby causing Reaction 19-14 to proceed in the reverse direction.

Determining total salt

Determining the total number of equivalents of salt in a solution is difficult by any procedure other than one using ion exchange. The sample is passed through a cation-exchange resin in the hydrogen-ion form with sufficient capacity to exchange hydrogen ions for all of the cations in the sample. The cation content of the sample is determined by titrating the acid solution collected from the column. Note that one hydrogen ion is produced for each monovalent cation present, two for each divalent cation, and so forth.

Deionizing water

The preparation of deionized water is among the most common uses of ion-exchange resins. The water is passed through a mixture of a cation-exchange resin in the hydrogen-ion form and an anion-exchange resin in the hydroxide-ion form. The hydrogen and hydroxide ions released from the resins combine to form water. When the resins need to be regenerated; that is, put back into their original hydrogen-ion and hydroxide-ion forms, they are first separated by placing them in a salt solution where the less dense anion resin floats to the top.

Concentrating trace electrolytes

Ion-exchange resins have significant utility in their ability to isolate and retain ions. Occasionally, the analytical chemist may find it inconvenient to transport samples in their original form to the laboratory for analysis. For example, suppose that you were fortunate enough to be researching a pollution problem that required information about the metal-ion concentrations of certain alpine lakes in the Colorado Rocky Mountains. If the determinations require 500-mL samples, which you would like to do in triplicate, and there are 20 lakes involved, you should be able to figure out that you are dealing with about 60 lb of water, which is certainly more weight than anyone would like to have in their backpack while climbing around in the mountains. One solution is to pass each 500-mL sample through a 10-cm column of cation-exchange resin in the hydrogen-ion form, transport the columns back to the lab, and displace the metal ions from the resin with concentrated hydrochloric acid.

In addition to solving certain transportation problems, ion-exchange resins can be used to concentrate ions from very dilute solutions. The cations collected by passing several liters of sample solution through a resin may be liberated by a much smaller volume of concentrated acid.

PROBLEMS

19-1. After shaking 25.0 mL of aqueous solute with 10.0 mL of hexane, the amounts of solute in the two solvents were determined to be 3.58 μmol in hexane and 1.09 μmol in water. Calculate the distribution coefficient for the solute.

19-2. When 50.0 mL of aqueous 8.50×10^{-4} M acetylsalicylic acid was shaken with 20.0 mL of diethyl ether, its concentration in the aqueous phase was reduced to 5.45×10^{-5} M. Calculate the distribution coefficient for acetylsalicylic acid.

19-3. If a solute has a distribution coefficient of 8.5, calculate the fraction remaining in 50.0 mL of aqueous phase after one, two, and three successive extractions with 10.0 mL of toluene.

19-4. If the distribution coefficient of 12-molybdophosphoric acid, $H_3PMo_{12}O_{40}$, between n-butanol and water is 11.4, calculate the fraction of this substance remaining in 25.0 mL of water after extraction with the following portions of n-butanol.
(a) one 50.0-mL portion
(b) two 25.0-mL portions
(c) five 10.0-mL portions
(d) ten 5.00-mL portions

19-5. Suppose that a solute has a distribution coefficient of 8.8 between water and chloroform. What volume of chloroform will be needed for a single extraction if the fraction of solute remaining in 25.0 mL of water is to be the same as the fraction remaining after three successive extractions with 10.0-mL portions of chloroform?

19-6. When 25.0 mL of an aqueous solution of metal chelate was extracted with 10.0 mL of cyclohexane, the fraction extracted was 0.71. How many successive extractions with 10.0-mL portions of cyclohexane will be required to extract at least 99.5% of the metal chelate?

19-7. Calculate the volume of toluene needed to remove 99% of a solute from 50.0 mL of water in a single extraction for the following values of K_D.
(a) 50
(c) 10
(d) 5.0
(b) 25

19-8. What is the minimum value of K_D that will permit 99% of a solute to be removed from 20.0 mL of water in a single extraction with the following volumes of the organic solvent?
(a) 10.0 mL
(b) 25.0 mL
(c) 100.0 mL

19-9. What is the minimum value of K_D that will permit 99.8% of a solute to be removed from

25.0 mL of water with the following number of successive extractions with 10.0 mL portions of organic solvent?

(a) 1   (c) 4   (d) 10

(b) 2

19-10. Using the necessary data in Table 19-7, calculate the percentage of As(III) and Cd(II) extracted from 25.0 mL of aqueous 6.0 M HBr with two successive 10.0-mL portions of diethyl ether.

19-11. If the original aqueous solution in Probem 19-10 contained As(III) and Cd(II) in equimolar concentrations, calculate the percent relative purity of As(III) in the ether and Cd(II) in the water after the first and second extractions. Relative purity can be taken as the amount of one solute divided by the total amount of both solutes *in the solvent*.

19-12. A 10.00-mL aliquot of seawater was added to the top of a column of cation-exchange resin in the hydrogen-ion form and washed through with distilled water. The solution exiting the column was collected and required 38.74 mL of 0.1437 M NaOH for titration. Calculate the total salt concentration of the seawater as % NaCl. Assume that the sample had a density of 1.020 g/mL.

19-13. A 0.4097-g sample containing NaOH, Na_2CO_3, and nonionic material was dissolved and diluted to 100.0 mL with water. A 25.00-mL aliquot of the dissolved sample was added to a column of anion-exchange resin in the chloride-ion form and washed through with distilled water. The collected solution required 12.86 mL of 0.01522 M HCl for titration to the phenolphthalein end point. Another 25.00-mL aliquot of the dissolved sample required 16.47 mL of 0.01522 M HCl for titration to the phenolphthalein end point. Calculate the % NaOH and % Na_2CO_3 in the sample.

INTRODUCTION TO CHROMATOGRAPHY

The first applications of chromatography are attributed to a Russian botanist named Mikhail Tswett, who in 1903 began publishing papers describing the separation of colored plant pigments by passing a solution of the compounds through a glass column packed with finely divided calcium carbonate. The separated pigments appeared as colored bands on the column, which inspired him to coin the term "chromatography" from the Greek words *chroma* meaning "color," and *graphein* meaning "to write." The technique was largely forgotten for many years, not resurfacing until the 1930s.

The period of modern chromatography dates to the 1940s, when the first comprehensive theories were developed. Present-day chromatography is comprised of a diverse group of techniques and methods with impressive applications not only in chemistry but in biology, medicine, environmental science, and agriculture as well. This is an area in which theory and practice have benefited one another in a most effective manner. Theoretical developments during the 1940s spurred many practical applications. As chemists gained familiarity with the technique, they began to experiment with new materials and methods, which led to new theoretical developments and to new applications.

To the analytical chemist, chromatography is much more than a means of separating substances; it is a means for analyzing very complex samples. This chapter is devoted to the principles by which solutes become separated on a chromatographic column. In Chapters 21 and 22 we describe specific types of chromatography and how the separated solutes can be determined as they exit the column.

GENERAL DESCRIPTION OF CHROMATOGRAPHY

Modern chromatography encompasses a diverse group of separation methods, all of which are based on components partitioning between a *stationary phase* and a *mobile phase*. It is the responsibility of the mobile phase to "carry" the sample components through the stationary phase; that is, solutes move only while they are in the mobile phase. The fraction of time a solute spends in each phase is determined by its distribution coefficient, which must be different from those of the other solutes if a separation is to occur.

Classification of Methods

The diversity of chromatographic methods makes classification difficult using a single set of criteria. Table 20-1 summarizes several different classifications that are used. The terms *gas* and *liquid* in gas chromatography and liquid chromatography refer to the nature of the mobile phase or carrier, as it is sometimes called. Each of these can be subdivided according to the type of stationary phase used and the manner in which it is immobilized. The specific characteristics of some of these techniques are discussed in Chapters 21 and 22.

Chromatographic methods can also be subdivided according to the mechanism by which solutes are retained by the stationary phase. *Partition* is a bulk-phase distribution process in which the solute forms homogeneous solutions in each phase. *Adsorption* involves interactions at a surface or fixed sites on a normally solid stationary phase. *Exclusion* relies on the ability of a porous solid stationary phase to discriminate on the basis of size by admitting small molecules to its pores but excluding larger ones.

A third, very broad method of classification is based on the physical configuration of the stationary phase. In most chromatographic methods, the stationary phase is held in a column through which the mobile phase is pushed under pressure or drawn by gravity. *Column chromatography* is divided into two types according to how the stationary phase is immobilized. *Packed columns* use a stationary phase of small solid particles (often coated with a thin layer of liquid) contained in an open tube. *Open tubular* or *capillary columns* are made by forcing a liquid through a small-diameter tube. A thin layer of the liquid coats the inside wall of the tube and is held there by capillary forces or by chemical bonding to the tube surface.

In *planar chromatography*, the stationary phase is a flat strip of paper or a solid coated onto a glass plate. The liquid mobile phase moves through the stationary phase by capillary wetting or a combination of wetting and gravity. This type of chromatography has limited applications in quantitative analysis and is not discussed in this book.

Yet another means of classification is made according to how the sample is introduced to the stationary phase and moves through it. In *frontal chromatography* the sam-

TABLE 20-1 CLASSIFICATION OF CHROMATOGRAPHIC METHODS

According to phases		According to mechanism of retention	According to physical configuration	According to sample development
Mobile phase	Stationary phase			
Gas	Liquid	Partition	Column	Frontal
Gas	Solid	Adsorption	Planar	Displacement
Liquid	Liquid	Exclusion		Elution
Liquid	Solid			

General Description of Chromatography

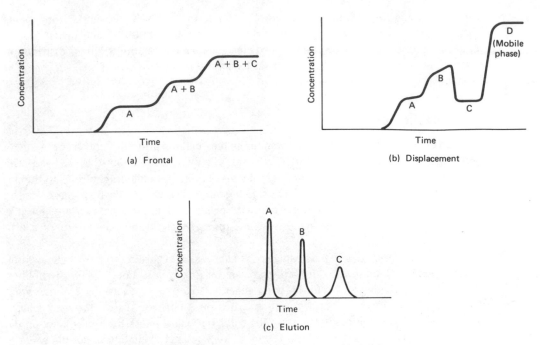

Figure 20-1 Types of chromatographic development.

ple is fed onto the column continuously and acts as the mobile phase. Sample components emerge from the column as *fronts* (Figure 20-1a). The least retained component A begins to emerge first, followed by a mixture of A and the next least retained component B, and so on. A complete recovery of pure components cannot be achieved with this technique; its main utility is for concentrating trace compounds and purifying large volumes of liquid or gas.

Displacement chromatography is characterized by a mobile phase that is strongly attracted to the stationary phase, causing the sample components to be "pushed" through the column by the advancing solvent. The displacement technique generally produces poorer separations than elution (Figure 20-1b) but can tolerate larger samples, thereby making it more useful in preparative and industrial-scale separations.

In *elution* chromatography the sample components are carried along the column by a mobile phase, and separation is the result of their spending different fractions of time in that phase. In addition to producing good separations (Figure 20-1c), elution chromatography has the considerable advantage of leaving the column in its original condition, ready for another sample. The remainder of this chapter is devoted to the elution technique.

LINEAR ELUTION CHROMATOGRAPHY

In elution chromatography a discrete amount of sample is added to the mobile phase at the top or beginning of the column. As the sample flows onto the column the solutes partition between the two phases. That portion of a solute in the mobile phase moves down the column, bringing it in contact with a different section of stationary phase, where further partitioning occurs. At the same time, the portion of that solute in the stationary phase comes in contact with a different section of mobile phase and therefore undergoes additional partitioning. This process repeats itself many times as fresh mobile phase continuously flows onto the column. Solutes move through the column only while they are in the mobile phase. The rate at which they move depends on the fraction of time spent in the mobile phase, which is a function of the distribution coefficient. Thus solutes with different distribution coefficients should move through a column at different rates, causing them to be separated into bands, as illustrated in Figure 20-2. Eventually, the solutes are eluted from the column, where they may be detected and/or collected.

The mobile phase flowing into the column is called the *eluent*; the solution emerging from the column is the *eluate*; and the process by which the solutes move and separate is called *elution*. A detector at the end of the column monitors the eluate and produces a signal as each solute emerges. A plot of this signal versus the volume of eluate is called a *chromatogram*. In most cases, the flow rate of the mobile phase is held constant, enabling time to be substituted for volume in plotting the chromatogram. Elution chromatograms contain information useful in both qualitative and quantitative anal-

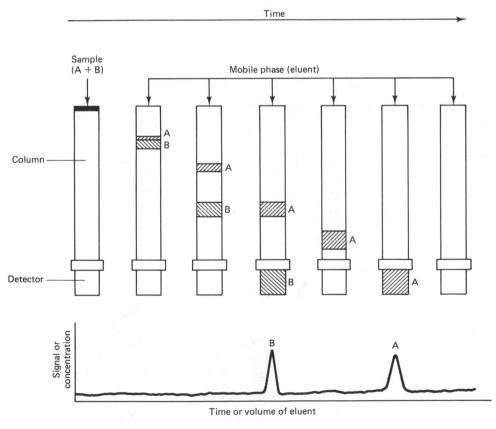

Figure 20-2 Schematic diagram of a chromatographic separation by the elution technique.

Linear Elution Chromatography

ysis. The time required for a solute to be eluted is characteristic of the substance and the height or integrated area of the signal can be related to the concentration of the solute.

The basis of elution chromatography is a partitioning of solutes between the stationary and mobile phases, which is an equilibrium process described by a distribution coefficient, K_D. It is common to find that K_D is not strictly constant over a large range of concentrations, which affects the shape of the eluted peaks and complicates the theoretical description of the chromatographic process. Fortunately, at the low concentrations generally encountered in elution chromatography, K_D is often nearly concentration independent, so that a *linear* relationship exists between the solute concentrations in each phase.

Solute Retention (or Migration)

The volume of mobile phase that must flow through a column to elute a solute to its maximum concentration is called the retention volume, V_R (see Figure 20-4). This value is a measure of the extent to which a solute is retained by the column. Two compounds must have different retention volumes in order to be separated on a chromatographic column. The retention time, t_R, is the *time* required to elute a solute to its maximum concentration and can be obtained from V_R if the volumetric flow rate, F, of the mobile phase is known:

$$t_R = \frac{V_R}{F}$$

where F has units of volume per time. As a matter of convenience, most chromatograms are recorded as a function of time and yield t_R directly.

The degree to which a solute is retained on a column is represented by the *retention ratio*, R, which is defined as

$$R = \frac{\text{average velocity of solute through column}}{\text{average velocity of eluent through column}} \tag{20-1}$$

The volumes required to pass eluent and solute through the column are inversely proportional to their average velocities and Equation 20-1 can be written as

$$R = \frac{V_M}{V_R} \tag{20-2}$$

where V_M is the volume of mobile phase in the column (sometimes called the dead volume). This is the same volume that must be added to push an eluent molecule completely through the column. If the solute is not retained at all, it moves through the column at the same rate as the eluent and R has its maximum value of 1.0. Since a solute moves through the column only when it is in the mobile phase, the ratio of velocities equals the fraction of time the solute spends in the mobile phase, which, in turn, is the mole fraction of solute in that phase. Thus Equation 20-1 can be written as

$$R = \frac{\text{amount solute in mobile phase}}{\text{total amount solute}} = \frac{C_M V_M}{C_M V_M + C_S V_S} \tag{20-3}$$

where C_M and C_S are the concentrations of the solute in the mobile and stationary phases, respectively, and V_S is the volume of the stationary phase. By dividing both numerator and denominator by C_M, Equation 20-3 can be rearranged to

$$R = \frac{V_M}{V_M + K_D V_S} \tag{20-4}$$

where K_D is the distribution coefficient (C_S/C_M).

Equations 20-2 and 20-4 can be combined to produce one of the fundamental equations of chromatography:

$$\frac{V_M}{V_R} = \frac{V_M}{V_M + K_D V_S}$$

which rearranges to

$$V_R = V_M + K_D V_S \tag{20-5}$$

Thus, for a given column where V_M and V_S are fixed, the degree to which solutes move apart (as measured by V_R) is determined by the values of K_D.

Example 20-1

Calculate the difference in retention volumes for solutes A and B that are passed through a column containing 1.8 mL of stationary phase and 2.7 mL of mobile phase. The distribution coefficients for A and B are 8.5 and 16.8, respectively.

Solution According to Equation 20-5,

$$V_{R,A} = 2.7 + (8.5)(1.8) = 18 \text{ mL}$$

and

$$V_{R,B} = 2.7 + (16.8)(1.8) = 33 \text{ mL}$$

Then

$$\Delta V_R = 33 \text{ mL} - 18 \text{ mL} = 15 \text{ mL}$$

Column Efficiency: The Plate Theory

Figure 20-2 shows us that two important processes occur during elution: (1) the solutes move through the column at different rates, and (2) the solutes spread from a very short to a much longer length of the column, a process called band broadening. The first process acts to move the solutes apart, whereas the second process pushes them together. Separation is possible only if the solute bands move apart faster than they broaden. The best chromatography columns are those producing the least amount of band broadening for a given retention volume. This fact is illustrated in Figure 20-3. Even though the retention times are the same on both columns, a better separation is achieved on column 1 because the solute bands are not broadened into one another as they are on column 2.

The first comprehensive theory of chromatography that related band broadening to solute migration through the column was developed in the 1940s by Martin and Synge. It was not long before the importance of their work was recognized by the scientific community, as both men were awarded the 1952 Nobel Prize in Chemistry. Martin and Synge suggested that a chromatographic column can be viewed as consisting of a series

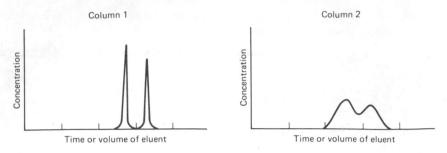

Figure 20-3 Elution chromatograms of a sample on two columns with different efficiencies.

plate = thin each to allow equil betw 2 phases

of thin, contiguous sections or plates, each thick enough to permit a partitioning solute to equilibrate between the two phases. Movement of the solute is viewed simply as the stepwise transfer from one plate to the next. Since these plates are not a *physical* reality, chemists prefer to call them *theoretical plates*. In essence, this theory treats the chromatographic process as a series of many discrete equilibrations, each of which can be described mathematically like a single extraction.

Column efficiency is characterized by the "thinness" or height of a theoretical plate, H, which is calculated by dividing the length of the column by the total number of theoretical plates, n:

h = L/N

$$H = \frac{L}{n} \tag{20-6}$$

The term *height equivalent to a theoretical plate* (HETP), also used by chemists, is synonymous with H. Since n is a unitless number, H has a unit of length. For the peak shapes obtained in linear elution chromatography, the number of theoretical plates can be calculated from two easily measured quantities, the retention time (or retention volume) and the extrapolated base width of the peak:

$$n = 16\left(\frac{t_R}{W}\right)^2 \quad \text{or} \quad n = 16\left(\frac{V_R}{W}\right)^2 \tag{20-7}$$

The extrapolated base width is determined by drawing tangents to the rising and falling portions of the peak and extrapolating them to the baseline as shown in Figure 20-4. Alternatively, n can be calculated using the following equation if the width is measured at one-half the peak height (see Figure 20-4):

$$n = 5.55\left(\frac{t_R}{W_{1/2}}\right)^2 \quad \text{or} \quad n = 5.55\left(\frac{V_R}{W_{1/2}}\right)^2 \tag{20-8}$$

In both cases the units of W and t_R or V_R must be the same.

Band Broadening: The Rate Theory

The mathematics of the plate theory is relatively straightforward and adequately describes the migration of solute through a column, but it does not explain any of the factors *responsible* for band broadening. Chromatographic bands are broadened by three

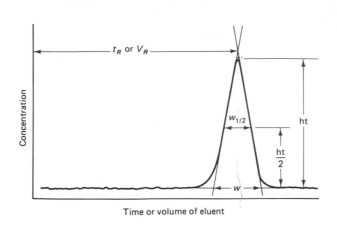

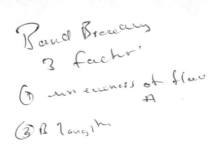

Figure 20-4 Elution chromatogram showing how the retention time, extrapolated base width, and width at half-height are measured.

kinetically controlled processes: *unevenness of flow* (eddy diffusion), *longitudinal diffusion*, and *resistance to mass transfer*. The rate theory of chromatography is a sophisticated attempt to describe the elution process in terms of the forces responsible for moving solute molecules within and between phases. The development of this theory represents a milestone in the evolution of chromatography because it tells *how* various column parameters and operating conditions can be altered to increase column efficiency.

Unevenness of flow

Packed columns have many irregular channels which cause solute molecules to travel different overall paths through the column, as illustrated in Figure 20-5a. Since the paths are not equal in length, solute molecules reach the end of the column at different times even though they travel at the same velocity, which is another way of saying that the solute band has been broadened or spread out. The term *eddy diffusion* arises from the observation that some molecules may become trapped temporarily in swirling pools of mobile phase whose forward progress has been deflected by groups of solid particles. The variation in path length is proportional to the average diameter of the particles and the packing irregularity. In turn, packing irregularity is influenced by such factors as particle shape, particle size, uniformity of size, and packing tightness or density. In theory, none of these factors are appreciably affected by the velocity of the mobile phase. Thus band broadening due to the unevenness of flow is essentially independent of the flow velocity.

Longitudinal diffusion

You should recall that a natural tendency exists for molecules to diffuse from regions of high to low concentration. As a result, solute bands in chromatography columns are constantly being broadened as diffusion pushes the leading and trailing edges farther apart (Figure 20-5b). Diffusion can occur in either phase, and it is much more significant in gases than in liquids. The rate of diffusion is governed by the diffusion coefficient, which depends on a variety of factors, including solvent viscosity and temperature. Since the amount of diffusion increases with time, broadening of this type increases with decreasing flow velocity.

Linear Elution Chromatography

(a) Unevenness of flow

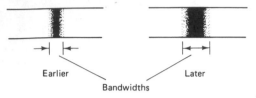

Earlier   Later

Bandwidths

(b) Longitudinal diffusion

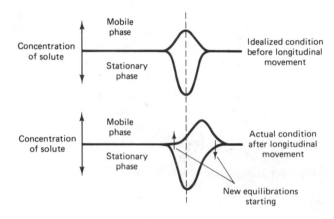

(c) Resistance to mass transfer

Figure 20-5 Kinetic factors contributing to band broadening in chromatography.

Resistance to mass transfer

As a solute moves through the column it undergoes a series of transfers between the two phases. Although commonly called mass transfer, this may be viewed as diffusion from one phase to the other. Broadening occurs because the phases are not in static contact with one another. Consider a small section of column within which the solute is attempting to equilibrate between the mobile and stationary phases. Equilibrium cannot be established "instantaneously" because molecules must diffuse to the phase boundary before they can transfer to the opposite phase. In the meantime, the mobile phase carries the solute down the column, where the leading edge of the band comes into contact with new stationary phase, starting a new equilibration (Figure 20-5c). Similarly, new mobile phase begins to flow over the trailing edge of the solute band, initiating a new equilibrium. Thus the leading and trailing edges of the band are spread apart more than they would be if the equilibrium was established before any movement of the mobile phase.

Band broadening of this type is diminished by anything that increases the speed with which the partitioning process approaches equilibrium. High temperatures and low viscosities (of both solvents) favor rapid attainment of equilibrium. Similarly, a high

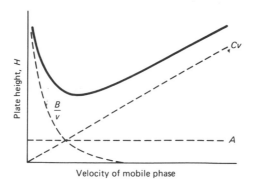

Figure 20-6 Effect of flow velocity on plate height according to the van Deemter equation.

contact area per unit volume of phase (thin phases) is desirable because it means that solute molecules will not have to diffuse very far to reach the phase boundary. Obviously, mass transfer broadening becomes more significant as the flow velocity increases.

Van Deemter equation   *Column efficiency*

The first equation successfully relating column efficiency to the processes responsible for band broadening, known as the van Deemter equation, was derived for gas chromatography using packed columns. The equation in its general form is

(Gas chromatography using packed)

$$H = A + \frac{B}{v} + Cv \qquad H = A + B/v + Cv \qquad (20\text{-}9)$$

where v is the velocity of the mobile phase and A, B, and C are quantities representing the amount of broadening due to the unevenness of flow, longitudinal diffusion, and resistance to mass transfer, respectively. The flow velocity affects longitudinal diffusion and resistance to mass transfer oppositely and the optimum velocity (minimum broadening) can be found experimentally by plotting plate height versus flow velocity, as shown in Figure 20-6. Note that at slow flow velocities band broadening or plate height is due almost entirely to longitudinal diffusion, whereas at high flow velocities, resistance to mass transfer is the major cause of broadening.

Band broadening
① ↓v B↑
↑m̄ C

Resolution of Eluted Bands

In chromatography, resolution is a term used to describe how well two elution bands are separated from one another. Chromatograms of a sample containing two solutes obtained with four different columns are shown in Figure 20-7. The separations range from poor to excellent. It would be inconvenient if we always had to draw a figure to illustrate the quality of a given separation, so chemists have devised a simple mathematical description that conveys the desired information. The resolution R_s is taken as the distance between two peaks divided by the average extrapolated base width of the bands:

$$R_s = \frac{2(t_{R,\mathrm{B}} - t_{R,\mathrm{A}})}{W_\mathrm{A} + W_\mathrm{B}} \qquad (20\text{-}10)$$

The symbol R_s should not be confused with the retention ratio, R. A large value of R_s

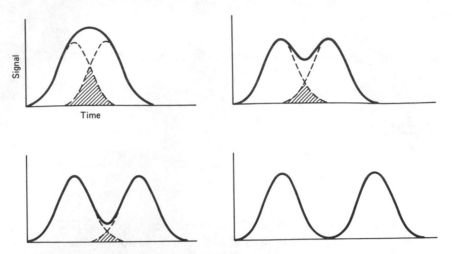

Figure 20-7 Significance of resolution values for two components of approximately equal concentration.

means that the bands are well separated. For two solutes of equal concentrations, the overlap area of the two bands is 16% at $R = 0.5$, 4.4% at $R = 0.75$, 2.3% at $R = 1.0$, and 0.1% at $R = 1.5$. These values will change somewhat if the concentrations of the two solutes are not the same.

Equation 20-10 defines resolution in terms of experimental results rather than fundamental quantities and as such, provides no information on how to improve the separation. It can be shown that resolution depends on the number of theoretical plates and the distribution coefficients for the solutes as given by the equation

$$R_s = \frac{\sqrt{n}}{4}\left(\frac{\alpha - 1}{\alpha}\right)\frac{k_2'}{1 + k_2'} \tag{20-11}$$

where

$$\alpha = K_{D,2}/K_{D,1}$$

$$k_2' = K_{D,2}V_S/V_M \text{ (often called the capacity factor)}$$

and 1 and 2 refer to first eluted and second eluted solute. The parameters n, α, and k_2' can be adjusted somewhat independently of one another. The number of theoretical plates is determined primarily by kinetic processes that occur during separation and can be increased by lengthening the column, adjusting the flow velocity of the mobile phase closer to its optimum value, or changing various column parameters to minimize the A, B, and C terms of the van Deemter equation. On the other hand, α and k_2' are thermodynamic properties of the system and are changed by using different stationary and mobile phases or by varying the temperature.

PROBLEMS

20-1. It takes 2 min 24 s for a solute to be eluted from a chromatographic column to its maximum concentration. Calculate its retention volume if the flow rate is 24.0 mL·min^{-1}.

20-2. Calculate the expected retention volume for a solute whose distribution coefficient is 0.68 on a column containing 58.0 mL of stationary phase and 6.3 mL of mobile phase.

20-3. It took 16.4 s for a solute with absolutely no retention on the stationary phase of a column to be eluted to its maximum concentration. What will be the retention time of a solute whose distribution coefficient is 1.24 on this column? Assume that the flow rate is a constant 25.0 mL·min^{-1} and the volume of the stationary phase is 7.30 mL.

20-4. What is the distribution coefficient of a solute whose retention time is 128 s on the column described in Problem 20-3?

20-5. Suppose that it is necessary for two retention times to differ by at least 5.0 s in order to prevent any overlap of the elution peaks. Calculate the minimum allowable difference in the distribution coefficients if the volume of the stationary phase is 6.02 mL and the flow rate is 30.0 mL·min^{-1}.

20-6. What is the approximate minimum number of theoretical plates a column must have so that the extrapolated base widths of two symmetrical elution peaks ($t_{R,1}$ = 83.4 s and $t_{R,2}$ = 86.6 s) do not overlap one another?

20-7. A column produces an elution peak after 148 s that is 3.30 s wide at one-half its height.
(a) Calculate the number of theoretical plates in the column.
(b) Calculate the column efficiency (H) if the column is 6 ft 4 in. long.

20-8. What will be the width at one-half the height of a peak for a solute whose retention time is 213 s on a column with 4.5×10^3 theoretical plates? What will be its extrapolated base width?

20-9. If the width of a solute peak is 6.5 mm at one-half its height, what is its extrapolated base width?

20-10. A 4.20-m column has a height equivalent to a theoretical plate of 0.70 mm. If the flow rate is 32.5 mL·min^{-1}, calculate the extrapolated base width (in seconds) of a peak for a solute having the given retention time.

(a) 38 s   (c) 3 min 28 s   (d) 0.0833 h
(b) 1 min 4 s

20-11. Calculate the height equivalent to a theoretical plate for a 2.50-m column operated at a flow velocity of 9.7 cm·s^{-1} if the A, B, and C terms of Equation 20-9 are 0.037 cm, 0.29 cm^2·s^{-1}, and 0.0093 s, respectively.

20-12. Calculate the predicted height equivalent to a theoretical plate for the column described in Problem 20-11 if the flow velocity is increased to 12.0 cm·s^{-1}.

20-13. Calculate the resolution factor R_s for the separation of the following pairs of peaks.
(a) $t_{R,A}$ = 283 s, $t_{R,B}$ = 291 s, W_A = 4.6 s, W_B = 6.5 s
(b) $t_{R,A}$ = 161 s, $t_{R,B}$ = 163 s, W_A = 3.1 s, W_B = 2.9 s
(c) $t_{R,A}$ = 145 s, $t_{R,B}$ = 151 s, $W_{1/2,A}$ = 3.5 s, $W_{1/2,B}$ = 3.6 s

20-14. What length of column is necessary to produce a resolution of 1.0 between two solute peaks? The distribution coefficients for the two solutes are 1.83 and 1.97; the volumes of stationary and mobile phases are 74.1 mL and 7.6 mL, respectively; and the height equivalent to a theoretical plate for the column is 0.174 cm.

20-15. Two solutes with distribution ratios of 1.47 and 1.86 are to be separated on a column whose volume ratio of stationary phase to mobile phase (V_S/V_M) is 13.6.

(a) How many theoretical plates are needed to ensure a resolution of 1.35?

(b) What length of column is required for part (a) if H is 0.250 cm?

20-16. A solute whose distribution coefficient is 2.63 has a retention time of 116 s on a column with a value of V_S/V_M of 6.37. What will be the retention time of a solute whose distribution coefficient is 2.31?

20-17. (a) For a column with 2.00×10^4 theoretical plates, calculate the extrapolated base width of the elution peak for a solute whose retention time is 3.50 min.

(b) Repeat the calculation in part (a) for a solute whose retention time is 3.00 min.

(c) Using the average of these two widths as the nominal base width, calculate the maximum number of peaks with no overlapping extrapolated bases that can occur between 3.00 and 3.50 min, inclusive.

21

GAS CHROMATOGRAPHY

[handwritten notes:]
Gas mp
s or 2 sr

Ga SC

GSC — retention based on physical adsorption
of solute by solid phase

Gas chromatography is characterized by the use of a *gaseous* mobile phase with either a solid or a liquid stationary phase. In *gas-solid chromatography* (GSC), retention is based on the physical adsorption of solutes by the solid phase. The technique is used mainly for the separation of low-molecular-weight gases and does not have many applications in analytical chemistry. This chapter is devoted to *gas-liquid chromatography* (GLC), in which retention is based on the solubility of a solute in a liquid stationary phase. Gas-liquid chromatography has experienced phenomenal growth since its inception in the 1950s and certainly is among the most powerful and often used tools available for separation and analysis. Only in the last five years or so has this growth began to slow, mainly as a result of the rapid development of a companion technique, high-performance liquid chromatography, which is discussed in Chapter 22.

There are several important advantages associated with the use of a gas as the mobile phase in a chromatographic separation: *[handwritten: low viscosity]*

1. The low viscosity of gases permits the use of very long columns, thereby improving the efficiency of separations, and the use of high flow rates, which means that separations can be performed quickly.
2. Gases are considered "inert" in terms of their interactions with the solute, and the equilibrium of solute distribution between the phases is largely independent of the gas.
3. Many simple, sensitive, and fast-responding detectors are available for measuring concentrations of substances in a mobile gas phase.

Because there is little interaction between the solute and the gas phase, gas chromatography is limited to the separation of relatively volatile substances (at the operating temperature of the column).

INSTRUMENTATION

The essential components of a typical gas chromatograph consist of a source of carrier gas with associated flow controllers and meters, a sample inlet system, a column, a detector, and a readout device. A block diagram of these components is shown in Figure 21-1. The sample inlet system and detector often have to be heated, but they may not be encased in an oven like the column.

Carrier Gas

The carrier gas must be chemically unreactive at the temperatures encountered in the various components of the chromatograph and should be of low viscosity. The gases used most frequently are helium, nitrogen, and argon. Although hydrogen also is an adequate carrier for many applications, it is not used much because of the extra precautions that must be observed due to its explosive nature. The choice of carrier gas generally is dictated by the type of detector used, as will become evident in the discussion of detectors.

All of the common carrier gases are available in large, pressurized tanks that can be fitted with an appropriate pressure regulator to control the gas flow rate. In addition, many commercial instruments contain a needle valve to further regulate the flow through the column. Volumetric flow rates generally are in the range of 20 to 100 mL/min at the column exit. The inlet gas pressure needed to sustain this flow depends on the dimensions of the column and the nature of the packing material, but usually it is between 10 and 50 psi. Flow rates are usually measured with a soap bubble flow meter attached either to the end of the column or to the detector. The flow meter is a simple device that measures the time required for flowing gas to push a soap film between two graduations on a buret or calibrated tube.

Sample Inlet System

Sample introduction is probably the most exacting operation encountered in gas chromatography. To avoid any appreciable spreading of the solute band, the sample must

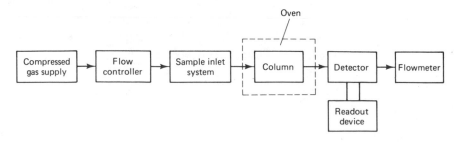

Figure 21-1 Block diagram of the basic components of a gas chromatograph.

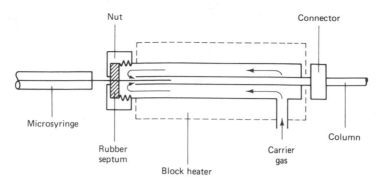

Figure 21-2 Simplified diagram of a flash vaporizer injection port.

enter the column very quickly as a thin "plug" of vapor. Furthermore, no decomposition or fractionation should occur during introduction. Liquid samples are commonly injected with a microsyringe through a silicone rubber septum into a small heated chamber attached to the beginning of the column (Figure 21-2). In order to vaporize the entire sample "instantaneously," the temperature of the chamber must exceed the boiling points of all the sample components. However, it must not be so high as to cause any decomposition. Figure 21-3 shows the effect of inlet chamber temperature on the separation of a mixture containing three components, whose boiling points range from 65 to 82°C. As general rule, the temperature of the injection chamber should exceed the boiling point of the highest-boiling component by at least 10°C. Special gastight syringes are available for injecting gas samples.

Although syringe injection is simple, it cannot match sampling valves such as those shown in Figure 21-4 for accuracy, precision, or ease of automation. Rotary valves allow a sample to be stored in a volume-calibrated loop and switched into the carrier gas stream at the appropriate time. The valves generally are available with interchangeable sample loops, each calibrated to hold a different volume.

Very small samples are used in gas chromotography. Ideally, the injected sample occupies a single theoretical plate at the beginning of the column. If the sample is too

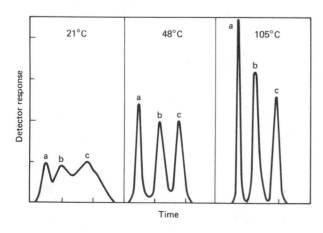

Figure 21-3 Effect of injection port temperature on resolution. The same sample size and column conditions were used in all three cases. a, Methanol; b, ethanol; c, isopropanol. [From C.E. Bennett, S. Dal Nogare, and L.W. Safranski, in Part I, Vol. 3, Chap. 37, of I.M. Kolthoff and P.J. Elving, eds. *Treatise on Analytical Chemistry* (New York: John Wiley & Sons, Inc.), p. 1681. Copyright © 1961 by Interscience Publishers, Inc. Reprinted by permission of John Wiley & Sons, Inc.]

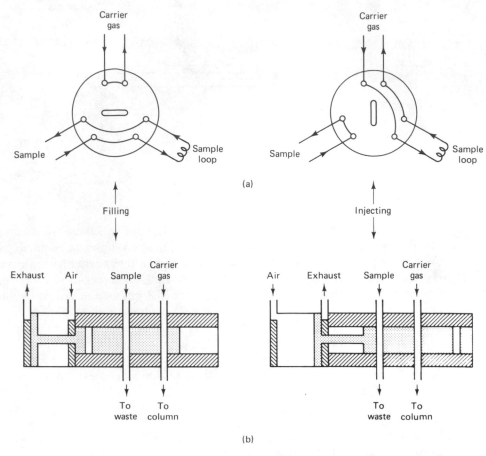

Figure 21-4 Valve injection systems for gas samples: (a) rotary; (b) sliding.

large, it simply spreads over several plates, thus broadening the solute bands before the separation even begins. Also, a large liquid sample may not vaporize "instantaneously" and therefore will flow onto the column over a period of time rather than all at once. Liquid samples usually are from 0.1 to 10 μL in volume, whereas gas samples may vary from 0.5 to 10 mL. Open tubular columns require only a few nanoliters of sample, a volume too small for direct injection. In such cases, a splitter is used to select as little as 0.1% of the sample for injection onto the column.

Packed Columns

Packed columns are made from 3- to 20-ft lengths of metal (copper, aluminum, or stainless steel) or glass tubing ranging from $\frac{1}{16}$ to $\frac{1}{4}$ in. in inside diameter (i.d.). Ordinarily, they are coiled for compactness. The smaller-diameter columns are used most often for analytical determinations and the larger ones for preparative work, where the separated components are collected and used in additional experiments.

Solid supports

The solid support serves to hold the liquid phase in place in the column and ideally should:

1. Consist of small, uniformly sized spheres
2. Be inert at high temperature
3. Have good mechanical strength
4. Be wetted uniformly by the liquid phase
5. Hold the liquid phase strongly

A high uniformity of shape and size minimizes the packing irregularities responsible for band broadening (unevenness of flow effect). Small particles provide a large surface area for contact with the mobile phase, which increases the number of equilibrations or theoretical plates per given length of column. The solid particles must be chemically inert toward the sample components and carrier gas at the operating temperature of the column (which may exceed 300°C), and they must be able to withstand the high pressure without fracturing. A strong physical attraction between the solid support and liquid phase is necessary to prevent the liquid from being stripped away by the flowing mobile phase.

The supports used most widely for gas chromatography are derived from diatomaceous earth, a naturally occurring substance consisting of the skeletal remains of ancient single-cell plants. Untreated diatomaceous earth has considerable surface activity, which must be reduced before it is suitable as a support material. Several techniques are used to deactivate the surface, including washing with acid and chemically bonding a compound such as dimethyldichlorosilane to the surface (silanizing).

Liquid phases

The purpose of the immobilized liquid phase is to dissolve the solutes (temporarily) as they migrate through the column. The properties of a good liquid phase include:

1. High thermal stability
2. Low chemical reactivity
3. Low vapor pressure at the temperature used
4. Good (but not too good) dissolving power
5. Strong attachment to the solid support

In evaluating liquids as potential stationary phases it is important to remember that gas chromatographic columns are operated at high temperatures. Compounds unreactive at room temperature may not behave similarly at 300°C. If the liquid has an appreciable vapor pressure at the operating temperature, it will "bleed" off the column, giving a false detector signal and ultimately ruining the column. As a general rule, the liquid should have a boiling point at least 200°C above the temperature at which it will be used. The liquid phase cannot be too good a solvent for the types of compounds being separated because that would result in inappropriately long retention times.

Although a very large number of substances have the necessary properties of a stationary liquid phase, as few as 25 have become accepted as unofficial industry standards, suitable for most applications. The choice among these "standards" depends largely on the nature of the solutes to be separated and the maximum temperature to which the column will be subjected (see Table 21-3). These and other factors are discussed in a later section.

Open Tubular (Capillary) Columns

Open tubular columns are made from 25- to 500-ft lengths of narrow-bore (0.1 to 0.4 mm i.d.) glass or fused silica tubing. In *wall-coated open tubular* (WCOT) columns, a thin (ca. 0.5 μm) layer of liquid stationary phase is coated on the inside surface of the glass tube, where it is held by capillary forces. The inside surface of the tube may be roughened by etching to help it bond the liquid phase more strongly. Also, the inner surface of the glass may be deactivated by treating it with a silanizing agent similar to that used to deactivate the solid supports for packed columns. Open tubular columns also are made with a thin (ca. 30 μm) lining of liquid-coated solid support. These *support-coated open tubular* (SCOT) columns can accommodate larger samples than WCOT columns. *Fused silica open tubular* (FSOT) columns, which first appeared in 1979, have very thin walls and are quite flexible, allowing them to be coiled very tightly. The compactness, flexibility, and physical strength of these columns are superior to those of glass columns and, as a result, they are beginning to dominate the market.

The liquid phases used in open tubular columns are the same as those used in packed columns. For a time, chemists were unable to coat open tubular columns with very polar stationary phases, but this problem has largely been overcome in recent years and columns covering a wide range of polarities are available from commercial suppliers. Some of the important properties and characteristics of open tubular and packed columns are compared in Table 21-1.

Ovens

Column temperature is an important variable requiring careful control in gas chromatography. Columns are contained in a thermostatted oven capable of reaching tempera-

TABLE 21-1 PROPERTIES AND CHARACTERISTICS OF DIFFERENT TYPES OF GAS CHROMATOGRAPHIC COLUMNS

	FSOT	WCOT	SCOT	Packed
Length (m)	10–150	10–150	10–25	1–6
Inside diameter (mm)	0.1–0.3	0.2–0.8	~0.5	1.5–6
Efficiency (plates/m)	2000–4000	1000–4000	500–1500	1000–3000
Capacity (μg sample)	0.01–0.1	0.01–1	0.01–1	1–100
Pressure drop	Low	Low	Low	High
Rate of thermal equilibrium	Very fast	Fast	Fast	Slow
V_M/V_S	—	100–300	20–70	15–20
Stability	Excellent (bonded) Good (coated)	Fair	Good	Good
Cost (dollars)	300–400 (bonded) 200–300 (coated)	200–300	200–300	50–100

tures up to 400°C and maintaining them to within a few tenths of a degree. Small ovens are more efficient in this respect, which is the major reason why coiled, compact columns are preferred.

Detectors

The purpose of the detector is to sense and measure the small amount of each solute present in the carrier gas as it leaves the column. Since the column separates the sample components into discrete bands, interferences during detection are not a problem and the detectors of greatest utility are those that respond to a wide variety of substances. Partially selective detectors that respond well only to certain groups of compounds are useful in situations where the column is unable to separate completely all the components of a sample. Both types of detectors are described in the following sections.

Thermal conductivity detector (TCD)

This detector consists of a heated metal block with two cylindrical channels, each containing a pair of thin, wire filaments made of tungsten or a tungsten–rhenium alloy (Figure 21-5). Pure carrier gas flows through one channel (reference side), and column effluent flows through the other channel (sample side). Passage of an electrical current heats the filaments to a temperature above that of the block. At the same time, the filaments are being cooled by the gases flowing through each channel. The efficiency of the cooling process and, therefore, the filament temperature depend on the thermal conductivity of the gas. Normally, a carrier gas of very high thermal conductivity is used so that the reference filaments are kept at a relatively low temperature. When carrier gas containing an eluted solute flows over the sample filaments, their temperature rises due to the lower thermal conductivity of the gas, causing their electrical resistance to increase. This change in resistance is converted to an electrical signal by wiring each pair of filaments in opposing arms of a Wheatstone bridge circuit like that shown in Figure 21-6. With pure carrier gas flowing over both sets of filaments (before sample injection), the bridge is balanced by adjusting a variable resistance (the "zero" control). When a sample component flows into the detector the bridge becomes unbalanced and the resulting voltage is divided by an attenuator and sent to a recorder or other readout device.

The sensitivity of this detector is proportional to the difference in thermal conduc-

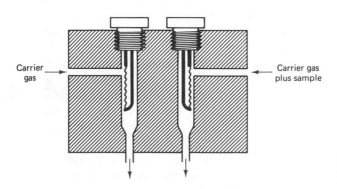

Carrier gas ——→   ←—— Carrier gas plus sample

Figure 21-5 Mounted filaments for a thermal conductivity detector.

Instrumentation

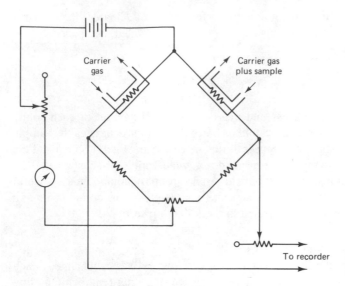

Figure 21-6 Wheatstone bridge circuit for a thermal conductivity detector.

tivity between the carrier gas and the sample components. Only two gases, hydrogen and helium, have thermal conductivities vastly greater (about one to two orders of magnitude) than the bulk of organic compounds. Hydrogen is seldom used because of its reactivity toward certain organic molecules and the need for extra safety precautions when working with a highly combustible gas.

The desirable properties of the thermal conductivity detector lie in its simplicity, large linear dynamic range (about five orders of magnitude), ability to respond to virtually every substance, and nondestructive character. The fact that this detector senses without destroying the solute makes it very useful in preparative applications, where the separated substances are to be collected. The most severe limitation of the detector is its relatively poor sensitivity, which makes it generally unsuitable for use with the low-capacity open tubular columns.

Flame ionization detector (FID)

This detector burns the column effluent in a hydrogen–air flame and measures the current produced from the ions formed during combustion. As shown in Figure 21-7, the column effluent is mixed with hydrogen gas and burned in the presence of air. A positively charged (ca. 400 V) metallic cylinder is mounted just above the negatively charged burner tip. Ions and electrons formed during combustion migrate to these electrodes, causing a current to flow in the external circuit. The currents are very small and require an electrometer for their measurement.

Nearly all carbon compounds produce ions when burned in hydrogen, although the number of ions produced varies with the nature of the compound. Certain functional groups, including carbonyl, hydroxyl, halogen, and amine, produce few or no ions. Also, the detector is not responsive toward most inorganic compounds, including stable gases such as H_2O, CO, CO_2, CS_2, NO_x, NH_3, and H_2S.

The flame ionization detector, like the TCD, is considered a nonselective detector, capable of sensing all but a very few substances. Perhaps its most important characteristic is excellent sensitivity, about five orders of magnitude better than that of the thermal

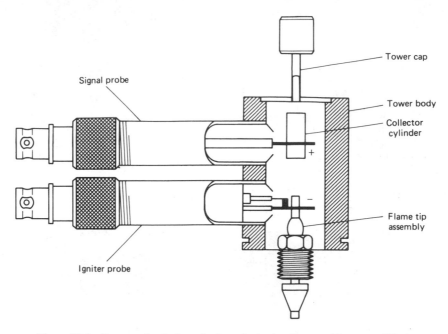

Figure 21-7 Cross-sectional view of a flame ionization detector. (Courtesy of Varian Associates, Inc.)

conductivity detector, making it very compatible with the low-capacity open tubular columns. Also important are its large linear dynamic range (about 10^7) and relative insensitivity to temperature variations of the carrier gas or detector chamber. Its insensitivity to water often is used to advantage in analyzing samples which contain water that is not completely separated from other solutes by the column. Since this detector places no restriction on the carrier gas, nitrogen is sometimes used in place of the much more expensive helium. The FID is destructive and cannot be used in situations where the separated components are to be saved.

Thermionic emission detector (TED)

A simple modification of the flame ionization detector makes it almost solely responsive toward organic phosphorus and nitrogen compounds. The modification involves placing a nonvolatile rubidium silicate bead just above the burner tip and burning the sample in a hydrogen-poor flame. For reasons that currently are not well understood, phosphorus- and nitrogen-containing compounds produce unusually large numbers of ions in such a flame, whereas other organic compounds produce fewer than normal numbers of ions. Consequently, the thermionic emission detector is particularly well suited to the determination of phosphorus-containing pesticides.

Flame photometric detector (FPD)

This detector, shown in Figure 21-8, is essentially a flame emission photometer (Chapter 18). Column effluent flows into a hydrogen-rich, low-temperature flame where solute molecules are partially combusted. The relatively simple fragments from this

Instrumentation

515

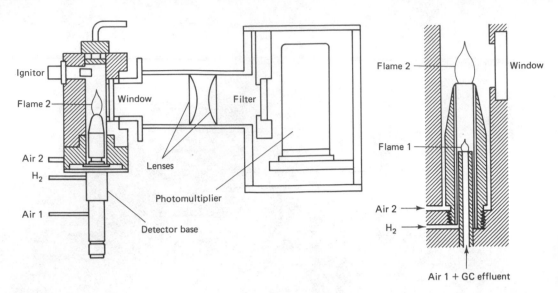

Figure 21-8 Flame photometric detector. (Courtesy of Varian Associates, Inc.)

combustion are passed into a second flame, where the combustion process is completed. Flame conditions are such that phosphorus and sulfur compounds are combusted to molecular species (HPO and S_2), which become thermally excited and emit radiant energy (λ_{max} at 526 and 394 nm, respectively) as they deactivate. An optical fiber carries the emitted radiant energy to a photomultiplier tube for detection. The double-flame arrangement minimizes background emission and quenching of the sulfur emission by other organic compounds in the flame. Like the thermionic emission detector, the FPD is a highly selective detector used in the determination of pesticides and pesticide residues.

Electron capture detector (ECD)

The basis of this detector is the ability of certain types of molecules to "capture" free electrons. As shown in Figure 21-9, the column effluent is passed over two electrodes, one of which is coated with a radioactive material (^{3}H or ^{63}Ni) that emits beta particles of sufficient energy to ionize the carrier gas, which is usually nitrogen or a mixture of argon and methane. In a properly designed cell, a large number of free electrons are produced from each emitted beta particle. These electrons migrate to the positive electrode, giving rise to a steady current. As solute molecules enter the cell, they "capture" some of these electrons, thereby decreasing the current. The magnitude of the decrease can be related to the concentration of solute. Ordinarily, the voltage is applied in short, rapid pulses rather than continuously. In the period between pulses, thermal equilibrium is established between the free electrons and the gas molecules and electron capture is maximized, both of which are necessary for a stable, sensitive response.

The electron capture detector is very sensitive toward electrophilic substances such as halogenated compounds, anhydrides, peroxides, conjugated carbonyls, nitriles, nitrates, ozone, sulfur-containing compounds, and a variety of organometallic compounds.

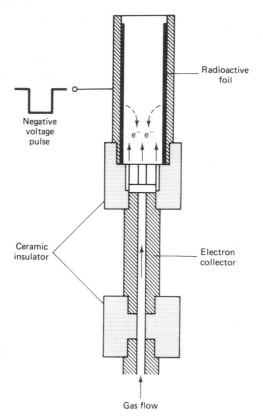

Negative
voltage
pulse

Radioactive
foil

e^- e^-

Ceramic
insulator

Electron
collector

Gas flow

Figure 21-9 Electron capture detector.
(Courtesy of Varian Associates, Inc.)

It also responds well to traces of oxygen, making leak-free systems and oxygen-free carrier gases essential.

METHODOLOGY

Before the applications of gas-liquid chromatography are discussed, it is worthwhile to consider some of the steps to be performed and decisions that must be made in carrying out a determination. The ultimate success of a GLC determination depends in large part on how well the analyst evaluates the sample and selects the column and operating conditions conducive to a good separation.

Forming Volatile Substances

Generally, a GLC column must be operated at a temperature above the boiling point of the highest-boiling substance in the sample; otherwise, this substance will spend most of the time condensed in the stationary phase and will not be eluted in a reasonable period of time. The maximum operating temperature of a column depends largely on the nature of the stationary liquid, but it is seldom above 300°C. Samples containing solutes that are not sufficiently volatile to be chromatographed directly may be treated with a reagent known to form volatile derivatives of the solutes. A favorite method, called *silanization*, involves the replacement of acidic hydrogens with di- or trimethylsilyl groups. A number of metals can be separated as volatile trifluoroacetylacetone chelates. Many reagents are known that form volatile derivatives with various groups of compounds, but to be useful in this type of application, the reactions must be simple to perform, fast, and quantitative, even at the very low concentrations normally encountered with GLC samples. A few of the more successful reagents are listed in Table 21-2.

Liquids and solids that are not very volatile can sometimes be decomposed thermally into characteristic gaseous products that can be separated and determined chromatographically. This procedure is known as *pyrolysis gas chromatography*. Special techniques are used to reproducibly heat the sample to its decomposition temperature in a few milliseconds. When both the heating rate and final temperature are well controlled, the pyrolysis process is quite reproducible. This is a particularly good method of studying and characterizing polymeric materials.

Selecting a Column

It is common to see half a dozen or more columns hanging on the wall next to a gas chromatograph. In selecting the right column for a particular application, the characteristics most likely to be considered are polarity of the stationary-phase liquid, maximum operating temperature, capacity, and efficiency. Let us examine each of these in terms of how they can affect a given separation.

Polarity of the stationary-phase liquid

For a chromatographic separation to occur, the solutes must have different distribution coefficients, which will in turn lead to different retention times. As a practical matter, the distribution coefficients must be within the range of about 0.1 to 10. Solutes with very large coefficients spend a disproportionate amount of time in the stationary phase, leading to undesirably long retention times, while solutes with very small coefficients spend most of their time in the mobile phase and pass through the column too rapidly for any appreciable separation to occur. Since the carrier gas is considered "inert," the distribution coefficient of a solute depends almost entirely on the column temperature and the dissolving power of the stationary liquid. Solubility and temperature

TABLE 21-2 SELECTED REAGENTS USED IN FORMING VOLATILE DERIVATIVES FOR GAS-LIQUID CHROMATOGRAPHY

Derivative	Reagent	Application
Methyl esters	Diazomethane, BF_3-methanol	Acids
Silyl ethers	N,O-Bis(trimethylsilyl)acetamide, trimethylsilylimidazole	Alcohols, acids, amides, amines
Fluoroacetylacetonates	Trifluoroacetylacetone, hexafluoroacetylacetone	Metals
Fluoroacetates	Trifluoroacetic anhydride	Alcohols, amines, amino acids

must be played off one another to obtain a satisfactory distribution coefficient and retention time. Columns with a liquid stationary phase that attracts solutes very strongly will have to be operated at relatively high temperatures to obtain a satisfactory distribution of solute between the phases. On the other hand, a successful separation of relatively volatile substances that are not strongly attracted to the stationary phase may be possible only at temperatures low enough to ensure that the solutes spend a reasonable fraction of time in the condensed state.

Of all the properties of a solvent that affect solubility, *polarity* is probably the most important. Columns are often classified according to the polarity of their stationary-phase liquid, which enables chemists to apply the "like dissolves like" rule in selecting a column for a particular job. Although hundreds of substances have been reported as suitable stationary liquids for specific separations, only about 25 have gained widespread acceptance. A few of these are listed in Table 21-3.

Solutes of similar polarity are separated mainly as a result of differences in their boiling points, the assumption being that their solubilities in the stationary phase are about the same. Conversely, solutes with nearly identical boiling points are separated as a result of differences in their solubility in the stationary phase. A good understanding of the attractive and repulsive forces that operate between molecules is essential in making informed decisions about which type of column should be used to effect a particular separation.

Maximum operating temperature

Gas-liquid chromatographic columns are so efficient that unless the separation is a particularly difficult one, the nature of the stationary phase may be of secondary importance to the maximum temperature at which the column can be used. The point has been made that the temperature must be high enough to keep the sample components properly volatilized in order to ensure reasonable elution times. Consequently, columns

TABLE 21-3 NATURE AND TEMPERATURE RANGE OF SOME STATIONARY-PHASE LIQUIDS

Type	Material	Temperature Range (°C)
Nonpolar	Apiezon L	50–300
	C-87	30–280
	Dexsil-300	50–400
	Silicone DC-200	50–350
	Squalene	20–150
Somewhat polar	Dinonylphthalate	20–150
	Poly-A 103	70–275
	Silicone OV-17	0–300
	Tricresylphosphate	20–125
Very polar	Carbowax 20M	60–225
	Ethylene glycol adipate	20–200
	Neopentyl glycol succinate	50–230
	Silar 7CD	25–275

whose maximum operating temperatures are below the boiling point of the least volatile sample component will be unsuitable and can be discarded from further consideration.

Column capacity

The capacity of a column determines how large a sample can be accommodated without degrading the quality of the separation. Sample size and column capacity should be fairly closely matched: Small samples tend to get "lost" on large-capacity columns and large samples overload small capacity columns, causing broad elution bands and incomplete separations. Column capacities vary greatly, depending primarily on the type of column but also on the amount of liquid phase used (called the *loading*). A $\frac{1}{4}$-in. packed column may tolerate as much as 0.5 to 1 mg of each solute producing a separate peak, whereas a WCOT column can seldom handle more than 5 μg of solute per peak.

Column efficiency

Although the efficiencies, in terms of plate thickness, of packed and open tubular columns are about the same (see Table 21-1), the much greater length of the open tubular columns gives them superior separating power. For example, a 150-m WCOT column may have as many as 500,000 theoretical plates, compared to 18,000 plates for a 6-m packed column. Open tubular columns have so many theoretical plates that satisfactory separations can be achieved even when the stationary phase is not optimum in terms of its interaction with the solutes. Hence some laboratories find it necessary to keep on hand only two columns, one polar and one nonpolar, to handle their applications.

Temperature Programming

Now that you are aware of the importance of column temperature on retention time, consider the problem presented by a sample that contains solutes of similar polarity but very different volatilities. If the column is kept at a low temperature best suited for separating the lower-boiling components, an inordinate amount of time may be required to elute the higher-boiling components. Conversely, if the column is operated at a high temperature suitable for elution of the higher-boiling components, the lower-boiling ones pass through the column too quickly for adequate separation to occur. The solution to this problem is to increase the column temperature *during* the separation, a process called *temperature programming*. The result of such an experiment is illustrated in Figure 21-10. As the sample is introduced to the column, the higher-boiling components condense in the stationary liquid and do not migrate at all (nor do they spread very much). As the temperature rises, these components are vaporized according to their boiling points and begin to migrate down the column. Temperature programming is a very powerful technique for the analysis of complex mixtures, and most instruments can be purchased with this capability.

APPLICATIONS

The applications of GLC are extensive. Indeed, entire books are devoted to its applications in one specific area, such as pesticide, petroleum, food, or flavor analysis. This section is devoted to a general discussion of how qualitative and quantitative data are acquired and used.

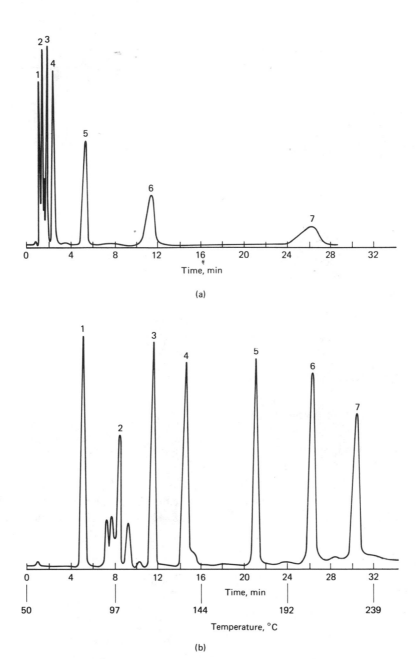

Figure 21-10 Chromatogram of a mixture of seven *n*-alkanes: (a) constant temperature (isothermal) of 168°C; (b) programmed temperature from 50 to 250°C at 6°C per minute. (Reprinted with permission from S. Dal Nogare and C.E. Bennett, *Anal. Chem.* **1956**, *30*, 1157–1158. Copyright 1956 American Chemical Society.)

Applications

Qualitative Analysis

With the use of a nonselective detector, the only qualitative information obtained from a gas chromatogram are the retention times. For a given column and set of operating conditions, the retention time is characteristic of the eluting solute, and it may be possible to identify a substance by comparing its retention time with tabulated retention times for a group of known compounds. Ordinarily, the same column and operating conditions must be used in collecting all the retention data. Even so, it generally is not possible to reproduce retention times to better than about 1 to 2%. Thus all of the compounds in the "known" group must have unique retention times; that is, no two times can be within 2% of one another if unequivocal identifications are to be made. To minimize the effect of variations in the operating conditions, the retention times may be measured relative to that of an added internal standard.

Compound identification is greatly enhanced by using retention data from two or more different columns. The likelihood of two compounds having the same retention times on different types of columns is small. This situation is illustrated by the data in Table 21-4. Using the nonpolar Apiezon L column, four different compounds have retention times within 2% of the retention time of the unknown, but only one (substance A) has a similar retention time on the polar Carbowax 20M column.

Gas chromatography has other qualitative applications, such as determining the effectiveness of a purification process and monitoring the progress of a chemical reaction. In purifying an organic compound, the removal of a volatile contaminant can be verified by the absence of its peak in the gas chromatogram. Similarly, a chemical reaction might be declared complete when the GC peak for the product has stopped getting larger.

Quantitative Analysis

Modern gas chromatographs are capable of producing signals that rival spectrophotometry in terms of accuracy and precision. This, coupled with their inherent speed, versatility, and ease of automation, has made them an extremely powerful tool in quantitative analysis. When certain operating conditions are properly controlled, the height and area of an analyte peak vary linearly with concentration. Quantitative determinations are based on a comparison of one of these values with those from a series of standards.

TABLE 21-4 TYPICAL TWO-COLUMN RETENTION DATA FOR IDENTIFYING AN UNKNOWN

Substance	Retention time (s)	
	Apiezon L (nonpolar)	Carbowax 20M (polar)
Unknown	163	386
A	160	390
B	161	129
C	164	178
D	167	312

Although peak heights are extremely easy to measure, they are quite sensitive to changes in flow rate, column temperature, sample inlet temperature, and sample injection rate. Consequently, these parameters must be controlled carefully to obtain reproducible peak heights. Because of the broadening that occurs with increased retention times, the sensitivity of peak height as an indicator of concentration is greatest for early occurring peaks. Peak area is more difficult to measure than peak height without the assistance of electronic signal integrators, but it is somewhat less sensitive to small changes in the operating conditions, especially the rate of sample introduction. Unlike peak height, peak area remains constant with increasing retention time as long as the detector operates with a constant sensitivity. Most modern gas chromatographs are equipped with electronic signal integrators that compute the area under each peak and relay this information to a printer or other readout device. Two other simple methods of measuring peak areas are worth mentioning. If the peaks are symmetrical, the area can be approximated by multiplying the peak height by the peak width at one-half height. Unfortunately, the elution bands for solutes with short retention times are often so narrow that their widths cannot be measured accurately. Some laboratory chart recorders are equipped with a second pen driven by an electromechanical integrator that represents the recorded area in terms of the number of oscillations between two chart lines.

The sensitivity of peak height and, to a lesser extent, peak area to a host of operating conditions creates some difficulties in using external standards to determine the concentration of a solute. These difficulties can largely be overcome by adding an *internal standard* to the sample and measuring the peak height of each analyte relative to this substance. It is expected that minor variations in one or more of the operating conditions will affect the height of both peaks by the same relative amount. Data for a calibration plot are acquired from mixtures of various known amounts of analyte with a constant amount of internal standard. The ratio of analyte peak height to internal standard peak height is plotted against the amount or concentration of analyte. The internal standard must be selected with considerable care: It should be a substance not present in the original sample; it should be eluted close to the analyte yet be completely resolved; and the ratio of its peak height to that of the analyte should be as close to 1 as possible.

PROBLEMS

21-1. The gas chromatogram of a mixture containing only the ortho, meta, and para isomers of cresol had three peaks whose integrated areas were 24.6, 30.8, and 9.3, respectively. Assuming that the detector responds equally to the isomers, calculate the percentage of each isomer in the mixture.

21-2. A 5.00-μL sample containing aniline ($C_6H_5NH_2$) and anisole ($C_6H_5OCH_3$) together with other substances was injected into a gas chromatograph. The heights for the peaks of these two solutes in the resulting chromatogram were 4.22 (aniline) and 7.60 (anisole) chart divisions. Another 5.00-μL sample was injected together with 0.25 μL of pure aniline (all in the same syringe), producing aniline and anisole peak heights of 8.73 and 7.60 chart divisions. Calculate the concentration, in volume percent, of the two components under the following assumptions.
 (a) The detector responds equally to both compounds.
 (b) The detector response (on a volume basis) is 1.35 times greater for anisole than for aniline.

21-3. The relative detector responses (on a molar basis) of a gas chromatograph for 1,3-dichlorobenzene (I), 3-methylcyclohexanol (II), diphenyl ether (III), and 3,5-dibromopyridine (IV) were determined to be 1.00, 0.83, 0.91, and 0.66, respectively. From the following data, calculate the concentration of the four components in the unknown, in units of $\mu mol/\mu L$. The density of I is 1.28 g/mL.

| | | Peak area (cm^2) | | | |
		I	II	III	IV
Sample	Volume (μL)				
Unknown	5.00	3.61	2.99	1.70	2.46
I	0.25	1.13			

21-4. The bromobenzene in a sample was determined by an internal standard technique. A series of standards, prepared by mixing various volumes of pure bromobenzene with 1.00 μL of pure n-propylbenzene, yielded the following results:

| Volume of | Peak area (cm^2) | |
bromobenzene (μL)	Bromobenzene	n-Propylbenzene
0.50	0.74	4.07
1.00	1.50	4.11
2.00	2.87	3.94
3.50	5.17	4.06
5.00	7.25	3.99

When 5.00 μL of sample (containing no n-propylbenzene) was mixed with 1.00 μL of n-propylbenzene and chromatographed, the ratio of bromobenzene to n-propylbenzene peak area was 0.873. Prepare an internal standard calibration plot and calculate the volume percent of bromobenzene in the sample.

21-5. Seven identical portions of a substance were chromatographed on the same column at different flow velocities, with the following results:

Sample	Flow velocity ($cm \cdot s^{-1}$)	Retention time (s)	Peak width at half-weight (s)
1	7.0	625	7.9
2	10.0	438	5.2
3	15.0	292	3.2
4	25.0	175	1.9
5	40.0	110	1.2
6	60.0	73	0.8
7	80.0	61	0.7

(a) Prepare a van Deemter plot.
(b) Determine the optimum flow velocity.
(c) Calculate the number of theoretical plates at the optimum flow velocity.

21-6. The chromatogram of a sample containing 1- and 2-naphthol had two peaks with retention times of 228 and 235 s. Their widths at half-height were 2.4 and 2.5 s, respectively. Calculate the resolution of these two peaks.

21-7. The following table is a summary of the retention times of a series of alcohols and ketones on two different columns:

	Retention time (s)	
	Carbowax 20M	Silicone OV-17
n-Butanol	46.3	16.1
n-Hexanol	96.0	35.0
n-Octanol	201	73.4
n-Decanol	450	164

(a) Plot the retention time versus the number of carbons in the molecule for both columns.
(b) Plot the logarithm of the retention time versus the number of carbon atoms in the molecule for both columns.
(c) Which normal alcohol will have a retention time of about 49 s on the Silicone OV-17 column?
(d) What will be the retention time of *n*-propanol on the Carbowax 20M column?
(e) Did you use the plot from part (a) or part (b) to answer parts (c) and (d)? Why?
(f) Plot log t_R on the Silicone OV-17 column versus log t_R on the Carbowax 20M column.
(g) Suppose that a sample produced two peaks with the following retention times on the two columns:

	$t_R(1)$ (s)	$t_R(2)$ (s)
Silicone OV-17	110	22.9
Carbowax 20M	302	158

Are either of the solutes a straight-chain alcohol, and if so, which one?

HIGH-PERFORMANCE LIQUID CHROMATOGRAPHY

It is natural that we should want to compare liquid chromatography (LC) in columns with gas chromatography. Despite the fact that liquid chromatography is the older technique, it is just beginning to realize its considerable potential in chemical analysis and already may have surpassed GC in the diversity of applications. In some respects the constraints on liquid chromatography are less severe than on gas chromatography. For example, sample constituents do not have to be volatile or thermally stable at high temperatures since most liquid chromatographic separations are performed at or near room temperature. In many forms of liquid chromatography the mobile phase liquid interacts with the sample constituents and therefore has a direct effect on the distribution coefficients. This is in contrast to gas chromatography, where solute distribution is largely independent of the carrier gas.

Two difficulties had to be overcome to bring liquid chromatography to its present competitive state. The chief difficulty was slow speed. In early liquid chromatography the mobile phase moved through the column under the force the gravity, and flow rates typically were a few tenths of a milliliter per minute. To achieve even these flow rates, relatively large packing particles were required, which, in turn, limited column efficiency. The second difficulty was a lack of detectors comparable in speed and sensitivity to those used in gas chromatography. It has been known for some time that column efficiency (as indicated by the height equivalent to a theoretical plate) improves dramatically with a decrease in the diameter of the particles comprising or supporting the stationary phase, as illustrated in Figure 22-1. Unless the applied pressure is increased, the flow rate decreases with a decrease in the size of the particles. Thus, while the efficiency problem can be solved by using smaller packing particles, it is done so at the expense of flow rate.

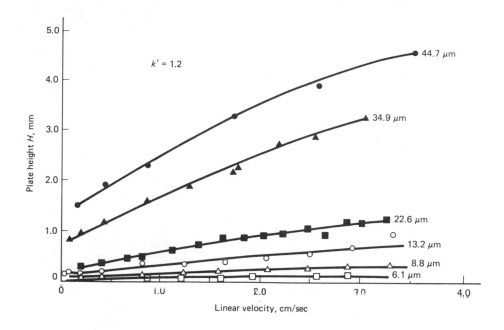

Figure 22-1 Effect of particle size and flow rate on column efficiency of a packed column. (From R.E. Majors, *J. Chrom. Sci.* **1973**, *11*, 88–95. Reprinted with permission.)

The problem of slow flow rates was overcome by designing systems to operate at high pressures, up to 600 atm in some cases. At these pressures, adequate flow rates are obtained even with packing particles as small as 2 to 3 μm in diameter. Once the high-pressure technology had been developed, several detection systems were adapted for use with the chromatographic columns. Liquid chromatography using these high-efficiency columns at high pressures has come to be known as *high-performance liquid chromatography* (HPLC).

INSTRUMENTATION

The flow diagram for a typical liquid chromatograph is shown in Figure 22-2. The mobile phase, which may be a single liquid or a variable mixture of two or more liquids, is pumped at high pressures into a controlled-temperature oven, where it passes first through an open coil to bring it to the operating temperature and then through a guard column designed to protect the analytical column from impurities and extend its lifetime. If a differential type of detector is used, the flow may be split at this point, with part going directly to the reference side of the detector and part to the analytical column. As in GC, the sample is added to the mobile phase just prior to the column. Ultimately, the column effluent passes through the sample side of the detector and on to a collection device or to waste.

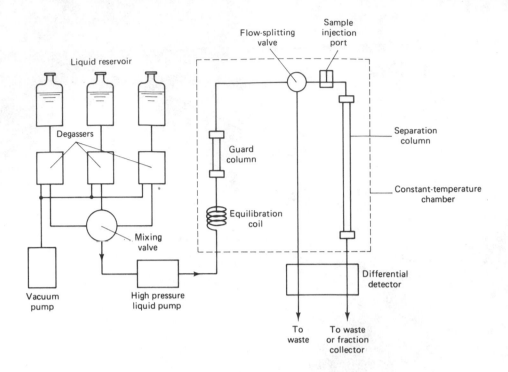

Figure 22-2 Flow diagram for a liquid chromatograph.

Eluent Delivery System

Delivery systems consist of a reservoir of the eluent, a degassing chamber, and a pump. Many liquids dissolve appreciable amounts of atmospheric gases that must be removed to avoid bubbles forming in the column and detector when the pressure is released. Most instruments remove gases by vacuum pumping, distillation, heating, or sparging with a fine spray of an inert gas of low solubility such as argon or helium.

The demands on the pump are quite severe. It should be capable of pulseless delivery of up to 10 mL of liquid per minute at a pressure of at least 1000 psi and preferably much higher. Three types of pumps currently dominate the instrument market. *Reciprocating pumps* use an oscillating piston in a small cylinder to compress the liquid and push it through the system. To minimize the pulsating nature of such a device, dual pistons and cylinders are used in alternating cycles, one filling as the other empties. By overlapping the initial and final stages of the displacement strokes of the pistons, the pressure drop caused by the slowing of one piston at the end of its compression stroke is partially offset by the pressure rise caused by the other piston. This technique reduces, but does not eliminate, pulses in the liquid flow, and most good chromatographs have some sort of damping system to improve the flow regularity further. The advantages of reciprocating pumps are their high output pressures (up to 10,000 psi), adaptability to eluent programming (discussed in a later section), and independence of the reservoir volume. Disadvantages include a tendency to form gas bubbles and cavitate with some liquids, the need for damping devices to minimize pulses in the liquid flow, and the need for monitoring and feedback circuits to obtain reproducible flow rates over a range of liquid types.

Syringe pumps consist of a large, motor-driven piston which pushes on the liquid in a reservoir cylinder and forces it through the column. They are capable of generating a pulseless flow at pressures equal to those of reciprocating pumps, but their maximum flow volumes are not as great. Syringe pumps have a fixed, maximum reservoir volume of 250 to 500 mL, which may not be sufficient to carry out some separations. Sometimes, two pumps are used in tandem to increase the maximum reservoir volume. These are not desirable types of pumps for eluent programming work.

Pneumatic pumps use gas pressure to drive a piston or diaphragm acting on the eluent. One type of pneumatic pump uses a piston with a large area on the gas side and a much smaller area on the eluent side. Since the pressure on the eluent is proportional to the *ratio* of the two areas (which can be 50:1), a gas pressure of 5 atm can produce a liquid pressure of 250 atm. These pumps are inexpensive and free of pulses, but they have a limited reservoir capacity.

Sample Inlet System

In general, the problems of sample introduction are the same in LC as in GC and they are handled in much the same way. The simplest and least expensive method is *syringe injection* through a septum. This approach is limited to systems operating at relatively low pressures (<40 atm) unless a double-septum design is used. Special, high-pressure syringes are used to prevent plunger "blowback." The major problem with septums is their tendency to leak at the high operating pressures. Some instruments avoid using a septum by sealing the syringe needle to the sample inlet with a tight-fitting Teflon collar. A valve opens the injector chamber to the high-pressure column, allowing the needle to slide into the moving liquid, where the sample is injected. Since vaporization of the sample is unnecessary, no special heating device is required.

The problems of blowback due to the high pressure can be overcome with a technique called *stop-flow injection*. A high-pressure valve just ahead of the inlet chamber momentarily closes, stopping the flow and relieving the pressure on the injector. The sample is injected at this low-pressure condition, after which the valve opens, restoring the pressure and flow.

Sampling valves such as those shown in Figure 21-4 are common in higher-quality instruments. Samples are drawn into the valve core or loop at low pressure, then switched into the high-pressure, flowing liquid. Interchangeable calibrated loops allow a choice of sample volumes from perhaps 0.5 to 500 μL. Good sampling valves can operate at pressures up to 7000 psi, with a volume reproducibility of a few tenths of a percent.

Columns

Most HPLC columns are fabricated from stainless steel tubing because it is chemically inert and can withstand the high pressures involved. Typically, the columns are 10 to 50 cm long with an inside diameter of 2 to 10 mm. The inside diameter must be very constant throughout the length of the tube and the inside wall very smooth to prevent

Instrumentation

packing irregularities that would allow the mobile phase to channel along the wall/packing material interface. Channeling is a major source of band broadening and loss of column efficiency in liquid chromatography. Connections to the column, especially those involving a change in diameter, must be made with very low dead-volume fittings that leave little or no room for the formation of stagnant pools or eddys, which are also a source of band broadening.

The packing particles currently available range from 3 to 10 μm in diameter, which is an order of magnitude smaller than those used in the first HPLC columns some 20 years ago. The pressure drop required to maintain a given volumetric flow rate increases with the *square* of the decrease in the diameter of the packing particles or of the column: thus each twofold decrease in *either* diameter requires about a fourfold increase in the applied pressure to maintain the flow rate. Of course, the increased pressure required can be offset partially by reducing the length of the column.

Columns with internal diameters of around 5 mm packed with 5-μm particles offer a good compromise between sample capacity, column efficiency, applied pressure, and volume of mobile phase used. Such columns may be 25 cm in length and contain 12,000 to 15,000 theoretical plates. Smaller-diameter columns packed with smaller particles are available in somewhat shorter lengths. They are about as efficient as the larger columns and have the advantage of using a smaller volume of eluent to complete a separation. This can be an important advantage because the high-purity mobile phase liquids required for HPLC are quite expensive. The major disadvantage of these small columns is their limited sample capacity.

Careful temperature control is important in some types of HPLC. Columns generally are housed in circulating air or water baths capable of maintaining the temperature within 0.2°C over the range 30 to 150°C.

A short *guard column* usually precedes the analytical column in the flow path. The purpose of this column is to extend the lifetime of the analytical column by removing particulate matter and other contaminants from the eluent and saturating it with stationary-phase liquid. If the mobile phase is not presaturated with the stationary-phase liquid, the analytical column may slowly lose its liquid phase and ultimately become useless. Guard columns are considered expendable, and are reconditioned or replaced periodically. Generally, they are similar in composition to the analytical column except that larger packing particles are used to keep the pressure drop small.

Detectors

Unfortunately, there are no HPLC detectors that compare favorably in terms of reliability, sensitivity, versatility, and applicability to the thermal conductivity and flame ionization detectors used in gas chromatography. Most of the detectors used in HPLC respond to a physical property of the solute that is different from or not exhibited by the pure eluent. These detectors are selective, responding only to solutes that exhibit the particular physical property. Only a very few detectors function on the basis of a bulk property that is largely independent of the nature of the solute. Although more universal in their applicability, detectors of this type are limited in other ways. The characteristics of a number of different types of HPLC detectors are summarized in Table 22-1. The three most popular detectors are discussed in the following sections.

TABLE 22-1 CHARACTERISTICS OF SOME HPLC DETECTORS

Detector	Type[a]	Maximum sensitivity ($\mu g/mL$)	Sensitivity to flow rate	Sensitivity to temperature	Suitable with solvent programming?
UV absorption	S	10^{-4}	Low	Low	Yes
Fluorescence	S	10^{-5}	Low	Low	Yes
Refractive index	N	10^{-1}	Low	High	No
Conductance	S	10^{-2}	High	$2\%/°C$	No

[a]S, selective; N, nonselective.

Ultraviolet absorption detectors

The most common HPLC detectors are based on the absorption of radiant energy by the eluted solutes as they pass through a flow cell attached to the end of the column. Several types of absorption detectors can be found in commercial instruments: Some use one or two fixed wavelengths from an intense UV line source such as a mercury vapor lamp, while others employ a continuous source and scanning monochromator capable of acquiring absorbances at a single wavelength or entire absorption spectra. The cell for a fixed-wavelength, differential-absorption detector is shown in Figure 22-3. Radiant energy from a low-pressure mercury vapor lamp is divided into two parallel beams, one passing through the column eluate and the other through pure eluent (or other reference solution) in a dual, flow-through cell. The volume of the cell is very small (1 to 10 μL), to minimize band broadening. To obtain path lengths of 2 to 10 mm, which are necessary to provide adequate sensitivity for many compounds, very small diameter passages are used. A filter in the optical path selects the very intense 254-nm mercury emission line for transmission to the matched photodetectors. Some detectors of this type employ a

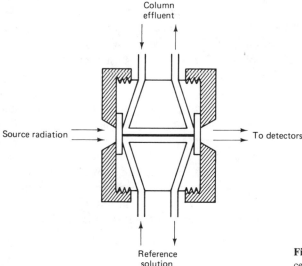

Figure 22-3 Double-beam, flow-through cell for an HPLC absorption detector.

special "phosphor converter" which absorbs *some* of the primary 254-nm radiation and emits it as fluorescence with a maximum intensity at 280 nm. This is done in such a way that both wavelengths pass through the cell simultaneously, thereby creating a dual-wavelength detector. Medium-pressure mercury lamps have numerous intense emission lines throughout the ultraviolet and lower visible portions of the spectrum that can be individually selected with an appropriate filter.

The advantages of this type of detector relative to other HPLC detectors are considerable: It is inexpensive, has a high sensitivity for many organic compounds, and is largely unaffected by small changes in flow rate, temperature, and the composition of the eluent. Absorption detectors place some limitations on the eluents that can be used, as they must not absorb at the wavelength to which the detector is set.

Fluorescence detector

Fluorescence detectors in HPLC are essentially filter fluorometers or spectrofluorometers (Chapter 17) with a flow-through observation cell. Spectrofluorometers have an advantage in versatility but are much more expensive. The already excellent sensitivity is increased further by using a half-sphere mirror to collect the fluorescence energy over a large solid angle and reflect it onto the photodetector.

Highly fluorescent compounds are encountered frequently in pharmaceutical preparations, clinical samples, petroleum products, and natural products, such as plant pigments, vitamins, alkaloids, and flavorings. Fluorescent derivatives of nonfluorescent compounds often are made either before the sample is chromatographed or on a small reaction column placed immediately before the detector. The eluent must be transparent to both the exciting and fluorescence radiation. Quenching is not as severe a limitation in the HPLC detector as in regular fluorometry because the column separates the potentially interfering components. In most cases the solution passing through the detector consists of a single solute dissolved in the mobile-phase liquid.

Refractive index detector

Changes in the refractive index of the mobile phase caused by the presence of a solute can alter the amount of radiant energy reflected or transmitted along a particular optical path. A *reflection-type* refractive index detector, shown in Figure 22-4a, measures the change in the radiant power reflected at and transmitted through the interface of the eluted liquid and a glass prism. As the refractive index of the liquid flowing through the cell changes, the fraction of radiant power reflected off the solution–prism interface (dashed lines in Figure 22-4a) changes. In turn, this alters the amount of radiation passing through the liquid, which is reflected off the steel cell wall onto the photodetector. If the angle of incident radiation is chosen correctly, the power of the observed radiation is a linear function of the liquid's index of refraction. A somewhat different approach is taken in the *deflection-type* detector shown in Figure 22-4b. Column eluent and eluate flow through opposite compartments of a two-chamber cell. The glass dividing plate is set at an angle causing an incident beam to be bent or deflected when the two solutions have different indexes of refraction. A deflected beam does not fully illuminate the photosensitive surface of the detector and a smaller output signal is recorded.

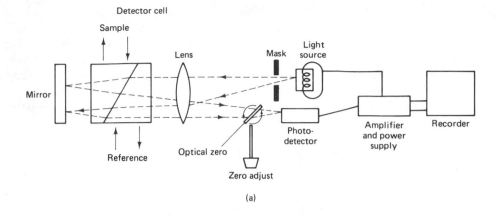

(a)

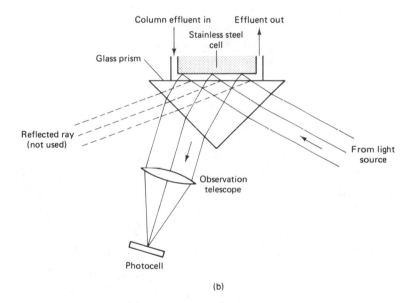

(b)

Figure 22-4 Refractive index detectors for HPLC: (a) deflection type; (b) reflection-type. (Courtesy of Waters Chromatography division of Millipore Corporation)

The most important feature of the refractive index detectors is their ability to respond to nearly all solutes. Only in the rare event that the solute and eluent have almost exactly the same refractive index (to within 1 part in 10^7) is detection theoretically impossible. The response of these detectors is unaffected by minor variations in flow rate but is extremely sensitive to changes in temperature (see Table 22-1). Overall, the deflection-type detector is more convenient to use, but it has a smaller linear range than that of its reflection-type counterpart. A major limitation of this detector is its incompatibility with the technique of eluent programming.

Instrumentation

LIQUID-LIQUID PARTITION CHROMATOGRAPHY

In the earliest example of liquid-liquid chromatography, the stationary liquid was retained on the solid support by physical adsorption. There are numerous problems with these coated supports when used in modern, high-pressure columns. The loading (amount of stationary-phase liquid for a given amount of support) is critical and different for each liquid and support used. If too little liquid is used, coverage will be incomplete and the exposed support may adsorb solutes from the mobile phase in the manner of liquid-solid chromatography. The adsorption forces binding the stationary liquid to the support are weak and if too much liquid is used, some will be sheared away by the mobile phase at the high operating pressures.

Bonded-Phase Supports

Bonded-phase supports overcome many of the problems encountered with adsorbed liquid phases. Here the molecules comprising the stationary phase are covalently bonded to a silica or silica-based support particle. The most popular bonded phases, siloxanes, are made by heating the silica particles in dilute acid for a day or two to generate the reactive silonal group:

$$-\underset{|}{\overset{\overset{\displaystyle OH}{|}}{Si}}-O-\underset{|}{\overset{\overset{\displaystyle OH}{|}}{Si}}-O-\underset{|}{\overset{\overset{\displaystyle OH}{|}}{Si}}-$$

silica particle

which then is treated with an organochlorosilane:

$$\left\{-\underset{|}{\overset{|}{Si}}-OH + Cl-\underset{\underset{\displaystyle CH_3}{|}}{\overset{\overset{\displaystyle CH_3}{|}}{Si}}-R \longrightarrow \left\{-\underset{|}{\overset{|}{Si}}-O-\underset{\underset{\displaystyle CH_3}{|}}{\overset{\overset{\displaystyle CH_3}{|}}{Si}}-R + HCl\right.\right.$$

These bonded phases are quite stable between pH 2 and 9 and up to temperatures of about 80°C.

The surface polarity of the bonded phase is determined largely by the nature of the R group of the silane. A fairly common bonded phase is made with a linear C_{18} hydrocarbon in which the groups protrude out from the particle surface much like the bristles on a brush. Exactly how these surfaces retain solute molecules is not clear and is being studied by various research groups.

Mobile Phases

Unlike gas chromatography, there are many mobile phases available for liquid-liquid partition chromatography, and because they interact with the sample components, they have a direct influence on partition coefficients and solute retention. To be suitable for use as a mobile phase, liquids must have the following characteristics:

1. Low viscosity to facilitate moving the liquid through the column
2. High purity to avoid contaminating the sample
3. High chemical stability so that it will not react with solutes or the stationary phase
4. Low volatility at the operating temperature to avoid the formation of bubbles in the column or detector
5. Immiscibility with adsorbed-type stationary phases so that it does not dissolve the stationary liquid and carry it off the column; usually not a problem with bonded-phase supports
6. Compatibility with the detector

The characteristics of a few liquids widely used as mobile phases in liquid chromatography are summarized in Table 22-2. The polarity index P' is of particular interest in liquid-liquid partition chromatography. It is an empirical measure of the polarity of a liquid based on its ability to dissolve in certain "standard" solvents. The values are used by chemists as an aid in selecting the eluent (or eluent mixture) that in conjunction with the stationary liquid will produce the solute distribution ratios needed for a given separation.

Normal- and Reversed-Phase Chromatography

The early applications of liquid chromatography were based on the use of highly polar stationary phases and relatively nonpolar mobile phases. This phase combination became known as *normal-phase chromatography*. With such an arrangement, the least polar solute is the first to be eluted and increasing the polarity of the mobile-phase liquid

TABLE 22-2 PROPERTIES OF LIQUIDS USED AS MOBILE PHASES IN LIQUID CHROMATOGRAPHY

Liquid	Polarity index, P'	Eluent strength,[a] ϵ^0	Viscosity (cP)[b]	Boiling point (°C)	Refractive index
Cyclohexane	0.04	0.03	1.02	81	1.426
n-Hexane	0.1	0.01	0.326	69	1.375
Carbon tetrachloride	1.6	0.14	0.969	77	1.460
Diisopropyl ether	2.4	0.22	0.38	68	1.368
Toluene	2.4	0.22	0.590	110	1.494
Diethyl ether	2.8	0.29	0.233	35	1.353
Tetrahydrofuran	4.0	0.35	0.46	66	1.407
Chloroform	4.1	0.31	0.58	61	1.443
Ethanol	4.3	0.68	1.200	78	1.359
Ethyl acetate	4.4	0.45	0.455	77	1.370
1,4-Dioxane	4.8	0.43	1.2	101	1.422
Methanol	5.1	0.73	0.597	65	1.329
Acetonitrile	5.8	0.50	0.345	82	1.344
Nitromethane	6.0	0.49	0.620	101	1.380
Water	10.2	Large	1.00	100	1.333

[a]For silica.

[b]Centipoise.

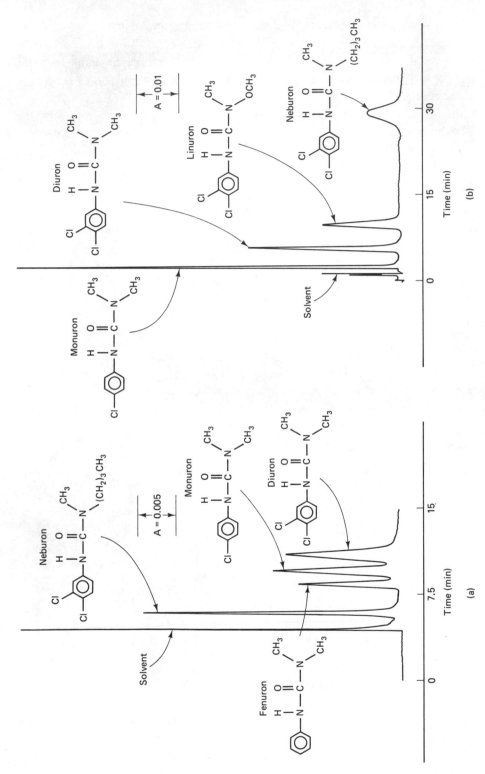

Figure 22-5 Separation of ureas on ETH-Permaphase recorded with a UV detector: (a) nonpolar mobile phase of 1% dioxane in hexane; (b) polar mobile phase of 35% methanol in water. (Reprinted with permission from J.J. Kirkland, *Anal. Chem.* **1971**, *43* (Oct.), 36A-48A. Copyright 1971 American Chemical Society.)

TABLE 22-3 CHARACTERISTICS OF NORMAL- AND REVERSED-PHASE LIQUID CHROMATOGRAPHY

	Normal phase	Reversed phase
Stationary phase	Polar	Nonpolar
Mobile phase	Nonpolar	Polar
Elution order	Least polar first	Most polar first
Effect on retention of *increasing* mobile-phase polarity	t_R decreases	t_R increases

usually causes the retention times to decrease. *Reversed-phase chromatography* is performed using a polar mobile phase and a nonpolar stationary phase. Polar substances prefer the mobile phase and are eluted first followed by the nonpolar solutes. This reversed order can be seen in the separation of a mixture of ureas shown in Figure 22-5. Somewhat surprisingly, an increase in the polarity of the mobile phase usually results in longer retention times. Apparently, the driving force for retention is derived less from the interaction between solute and stationary phase than from the ability of the mobile-phase liquid to force the solute into the nonpolar stationary phase. The characteristics of normal- and reversed-phase chromatography are summarized in Table 22-3. Actually, the term "reversed phase" is a misnomer because it implies that something is abnormal, which is not the case. Furthermore, the development of new types of bonded-phase supports has made available stationary phases covering a wide range of polarities, which have lessened the distinction between the two techniques.

LIQUID-SOLID ADSORPTION CHROMATOGRAPHY

Retention in liquid-solid chromatography (LSC) is due to adsorption of solute molecules on a solid stationary phase. Polar molecules are held more strongly and therefore elute more slowly than do nonpolar molecules. Since the polarity of organic molecules is determined largely by the nature of their functional groups, we can predict with reasonable accuracy the order of retention times for classes of compounds:

aliphatic hydrocarbons < olefins < aromatic hydrocarbons, organohalides < organosulfides < ethers < nitro compounds < aldehydes, ketones, esters < alcohols, amines < sulfones < sulfoxides < amides < carboxylic acids

This elution order must not be taken too literally because functional group polarity is not the *only* factor affecting the strength of adsorption. Molecular structure and steric factors also play a role, as evidenced by the fact that cis-trans isomers can sometimes be separated.

Stationary-Phase Materials

Silica and alumina are the principal substances used as stationary phases in high-performance LSC. Silica is preferred because of its higher sample capacity and our ability to prepare it in several different forms. *Porous microparticles* are the most popular form of stationary-phase material for analytical work. These are small particles, with diameters in the range 3 to 10 μm, and containing a high concentration of surface hydroxyl groups, which are responsible for the retention of solutes. The porous nature of these particles gives them an extremely large surface area per unit weight (100 to 900 m^2/g). The activity or adsorption power of silica is a direct function of its surface area. Columns made with small particles of high surface area will be effective at separating solutes having polarities that lie in a narrow range close to the polarity of the mobile phase. Columns packed with particles of low surface area will separate solutes having a wider range of polarities but will lack the high resolving power needed to separate solutes of very similar polarity.

A major difficulty encountered in LSC is the variation in distribution coefficient with the concentration of solute, which is another way of saying that the amount of solute adsorbed is not directly proportional to its concentration in the mobile phase. This causes retention to be dependent on concentration, especially with strongly adsorbed solutes. Naturally, this has made chemists interested in how the surface activity of silica and alumina might be moderated to gain some control over the adsorption characteristics. Through research it has been learned that a polar substance added in small amounts to the mobile phase will become selectively adsorbed on the most active sites, thereby decreasing the overall surface activity. Such substances are called *modifiers*. Distribution coefficients on silica and alumina that have been conditioned in this manner tend to be relatively independent of concentration. It is convenient that the adsorption of modifiers is reversible which enables the surface activity to be altered by changing the concentration or type of modifier in the mobile phase.

Porous-layer or *pellicular* particles represent a second type of stationary phase, consisting of spherical glass beads, 30 to 40 μm in diameter, covered with a porous layer of adsorbent about 1 to 3 μm thick. The adsorbent may be silica, a collection of small spherical particles bonded to the larger glass bead, or a bonded organic polymer. Porous-layer particles have a much lower surface area per weight (about 5 to 20 m^2/g) than porous microparticles and cannot accommodate large samples. Consequently, they are not especially useful for preparative work or with the less sensitive detectors.

Mobile Phases

The general requirements for liquids to be used as eluents in LSC are essentially the same as those described earlier for liquid-liquid partition chromatography. Because adsorption rather than solubility in the stationary phase is responsible for retention, the polarity index developed for liquid-liquid partition chromatography is not equally applicable to LSC. The *eluent strength* $\epsilon°$, which is the adsorption energy per unit area of liquid, is considered to be a better index of retention. The values of $\epsilon°$ depend on the type of adsorbent and are about 20% larger with alumina than with silica. Values are given with the eluents in Table 22-2. Even with this useful index, the eluent selection process begins with an experienced guess. If retention times are too long, a more polar eluent is tried, and vice versa. The addition of a polar modifier to the eluent has the *effect* of increasing the polarity of the eluent. What it actually does is decrease the adsorption power of the stationary phase toward the solute. It is common practice to mix two solvents to gain just the right effective polarity or eluent strength.

ION-EXCHANGE CHROMATOGRAPHY

The basic principles of ion exchange are described in Chapter 19. This section is devoted exclusively to the chromatographic applications of ion-exchange resins. The large porous resin beads used in batch separations are not satisfactory for HPLC because of their compressibility and the long times required for ions to diffuse through the micropores of the bead. Both pellicular and porous microparticles have been adapted for use with ion-exchange chromatography by coating the rigid particles with a thin layer of relatively nonporous ion-exchange resin. In some cases, the resin is bonded to the silica microparticles by means of silation reactions. By using thin layers of ion-exchange resin on a rigid support, the problems of compressibility and long diffusion times are largely overcome, although it is at the expense of sample capacity, which is about a hundred-fold less than for the resins used in batch separations.

Mobile Phase

In the types of chromatography discussed thus far, the eluent always consisted of a pure solvent or mixture of solvents. The eluents in ion-exchange chromatography are aqueous solutions, sometimes mixed with other miscible solvents, but always containing ionic solutes. You may recall from the discussion in Chapter 19 that an ion-exchange resin will bind all ions of the opposite charge type. It is the role of the ionic solute in the mobile phase to compete for the binding sites, thereby causing the sample ions to migrate through the column. An ionic solute that is bound too strongly will allow the sample ions very little time on the column, resulting in very short retention times and a poor separation. Conversely, if the ionic solute is not competitive for the active sites, there will be little or no migration of the sample ions.

Hydrogen ion is a favorite eluting agent for the separation of cations because its binding capability can be adjusted easily through the use of pH buffers. For example, when a resin with sodium ions at the active sites is exposed to an eluent containing hydrogen ions, it will strive to establish the following equilibrium:

$$\text{Resin}^-\text{Na}^+ + \text{H}^+(\text{aq}) \rightleftharpoons \text{Resin}^-\text{H}^+ + \text{Na}^+(\text{aq})$$

If the pH of the eluent (mobile phase) is increased, the hydrogen ion becomes less competitive for the binding sites (equilibrium shifts to the left). The opposite effect is observed when the pH is decreased. The utility of hydrogen ion as the active eluting agent is illustrated by the separation of a complex mixture of amino acids shown in Figure 22-6.

The applications of high-performance ion-exchange chromatography are expanded by the fact that most metal cations can be converted to anions through complexation with a negatively charged ligand. The anions potentially can be separated on an anion-exchange resin. Variation of the ligand concentration in the eluent changes the fraction

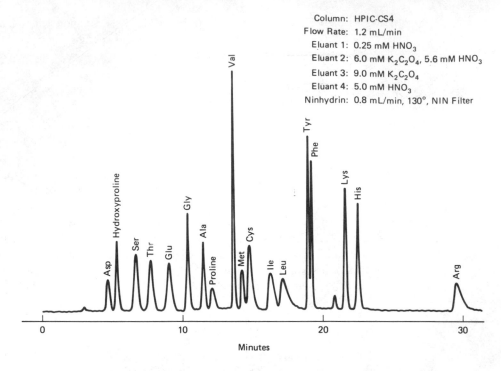

Column: HPIC-CS4
Flow Rate: 1.2 mL/min
Eluant 1: 0.25 mM HNO_3
Eluant 2: 6.0 mM $K_2C_2O_4$, 5.6 mM HNO_3
Eluant 3: 9.0 mM $K_2C_2O_4$
Eluant 4: 5.0 mM HNO_3
Ninhydrin: 0.8 mL/min, 130°, NIN Filter

Figure 22-6 Separation of amino acids on an ion-exchange column. (Courtesy of Dionex Corp.)

of the metal existing in the complexed, anionic form and becomes an excellent means of controlling the degree to which a metal is retained on the column. Several such complexing agents were discussed in Chapters 9 and 19.

Suppressor Columns

Since, by definition, ion exchange deals with ionic constituents, detection can be based on conductivity measurements. A conductivity detector has the potential for good sensitivity, is responsive to any charged species, and it is simple to construct, inexpensive, and easy to maintain. The major difficulty that must be overcome is the presence of the high concentration of electrolyte required to elute most sample ions with reasonable speed. The conductance of the eluted ions tends to get "lost" in the high conductance of the eluent itself.

This problem is solved by placing a second ion-exchange column, called a *suppressor column*, immediately after the analytical column. The suppressor column is packed with an ion-exchange resin capable of converting the eluent ions (but *not* the sample ions) to neutral molecules. For example, if cations are being separated using aqueous hydrochloric acid as the eluent, the suppressor column would consist of an anion-exchange column in the hydroxide form. As the eluent passes through this column, chloride ions are replaced by hydroxide ions, which immediately combine with the hydrogen ions to form water:

$$Resin^+OH^- + (H^+ + Cl^-) \longrightarrow Resin^+Cl^- + H_2O$$

The separated cations are unaffected by this column unless they happen to form covalent hydroxides, which is not very probable at a neutral pH and the low concentrations normally encountered.

EXCLUSION CHROMATOGRAPHY

In *liquid exclusion* or *gel permeation chromatography* substances are separated according to their size and to a lesser extent, shape. The stationary-phase material consists of small polymeric or silica-based particles that are characterized by a network of uniformly sized pores into which certain molecules can diffuse. Molecules that enter the pores effectively become a part of the stationary phase, much as if they had become adsorbed or dissolved. The average time a substance spends in the pores is determined by its size. Molecules too large to enter any of the pores are not retained at all and move through the column at the same rate as the eluent. Very small molecules enter virtually every pore they encounter and are retained the most (exit the column last). The extent to which intermediate-size molecules can penetrate the pores is determined by their size and, to some extent, shape. Since the porous packing particles generally do not interact chemically or physically with the solute molecules, a column of such particles can be regarded as having a variable path length. It is a short column for molecules too large to "go exploring in the various pores or caves," a longer column for smaller molecules that "explore some of the caves," and the longest for molecules small enough to "explore every cave."

For a given shape, the size of a molecule usually can be related directly to its molecular weight. It is not unusual to hear biochemists discussing the separation of complex protein samples into different molecular weight fractions using gel permeation chromatography. Shape can be important, as it is entirely possible for a linear compound to enter pores that would not admit a spherical compound of lower molecular weight.

Column Packings

Packing materials are either semirigid organic polymers (called gels) or rigid silica-based particles. The rigidity of the porous silica-based particles makes them especially suitable for use with different solvents and at high pressures. Their principal drawbacks are a tendency to retain solutes by adsorption and to catalyze the decomposition of many biochemical molecules.

The porous polymers have very little chemical or physical interaction with solute molecules but suffer the disadvantages of slight swelling in some solvents and a certain degree of compressibility. Several different polymers are used, including polystyrene cross-linked with divinylbenzene (the base polymer for ion-exchange resins, which is shown in Figure 19-3), polyacrylamide cross-linked with *N,N'*-methylene-*bis*-acrylamide, and dextran cross-linked with glycerine (Figure 22-7). The average pore size,

Figure 22-7 Structure of polyacrylamide and polydextran gels used in exclusion chromatography.

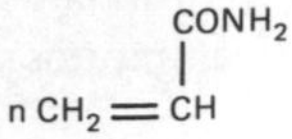

Acrylamide

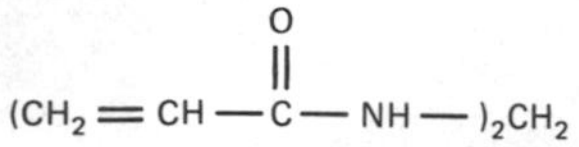

$$n\ CH_2 = CH - CONH_2$$

N, N'-methylene-*bis*-acrylamide

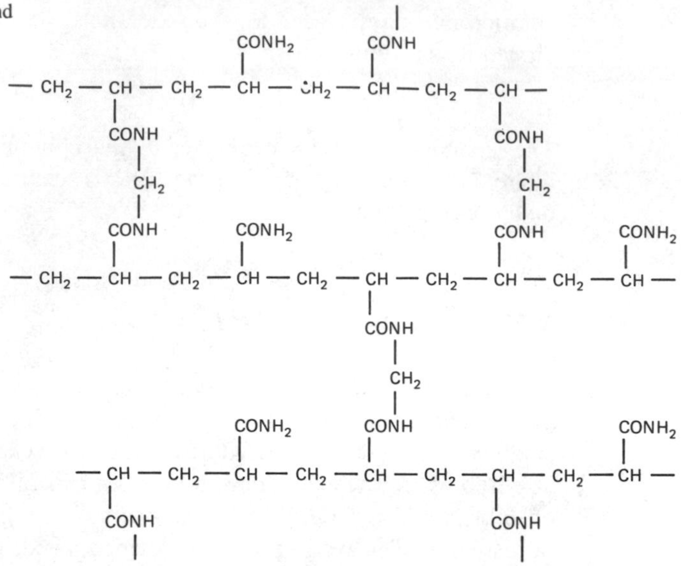

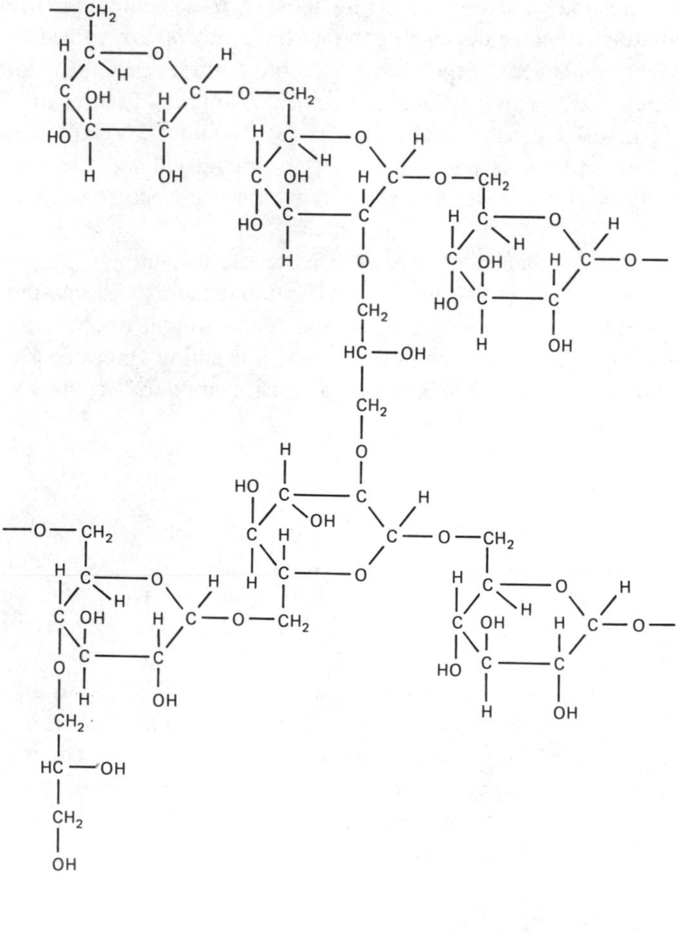

$$(CH_2 = CH - \overset{\displaystyle O}{\overset{\|}{C}} - NH -)_2 CH_2$$

TABLE 22-4 SPECIFICATIONS OF A CONTROLLED-PORE GLASS SIZE-EXCLUSION PACKING MATERIAL

Average pore diameter (nm)	Solute diameter (nm)	Approximate molecular weight	Surface area (m²/g)
750	800 – 350	$1 \times 10^4 - 2 \times 10^3$	340
1,200	1,200 – 450	$3 \times 10^4 - 3 \times 10^3$	210
1,700	1,700 – 550	$6 \times 10^4 - 6 \times 10^3$	150
2,400	2,400 – 650	$2 \times 10^5 - 9 \times 10^3$	110
3,500	3,500 – 800	$3 \times 10^5 - 1 \times 10^4$	75
5,000	5,000 – 900	$6 \times 10^5 - 2 \times 10^4$	50
7,000	7,000 – 1,200	$1 \times 10^6 - 3 \times 10^4$	36
10,000	10,000 – 1,500	$3 \times 10^6 - 4 \times 10^4$	25
14,000	14,000 – 1,800	$5 \times 10^6 - 6 \times 10^4$	18
20,000	20,000 – 2,000	$2 \times 10^7 - 9 \times 10^4$	13

which can vary from 10^1 to 10^5 nm, is determined by the degree of cross-linking in the polymer.

Each type of particle is available in a wide range of pore sizes typical of those shown in Table 22-4 for a controlled-pore glass (silica). The chemist must select a packing material whose average pore diameter is appropriate for the size or molecular weight of the substances to be separated. In the event that the sample contains solutes with molecular weights so different that a separation cannot be effected on a single column, two or more different columns can be joined in series.

The useful molecular weight range for a given packing material often is shown with a calibration curve like the one in Figure 22-8. The molecular weight on a *logarithmic scale* is plotted against the retention volume V_R. The *exclusion limit* represents the molecular weight beyond which no retention occurs; that is, any molecule with a molecular weight equal to or exceeding this value is too large to get into the pores of the packing material. Such molecules travel the shortest paths through the column and are eluted the earliest. Obviously, if no retention occurs, the distribution coefficient must be zero. The *permeation limit* is the molecular weight below which maximum retention occurs; that is, any molecule with a molecular weight equal to or less than this value is small enough to get into every pore of the packing material. On the average, these small molecules move through the same number of pores and therefore elute together. Total permeation is equivalent to a distribution coefficient of unity. Molecules falling in the selective permeation range have distribution coefficients between 0 and 1. If the values are different enough, a separation is possible.

Since K_D is restricted to values between 0 and 1, the number of peaks that can be resolved is limited. As a result, the total separation of complex samples is seldom possible using exclusion chromatography. The technique has many applications in biochemistry and polymer chemistry, where mixtures of large molecules are common. A typical separation of four compounds on a small-particle gel column is illustrated in Figure 22-9. At a flow rate of 1.0 mL/min, the separation required about 20 min to complete.

Mobile Phase

There are very few restrictions on the mobile phase used in exclusion chromatography. Like any mobile phase used in HPLC, it must have a low viscosity and be capable of

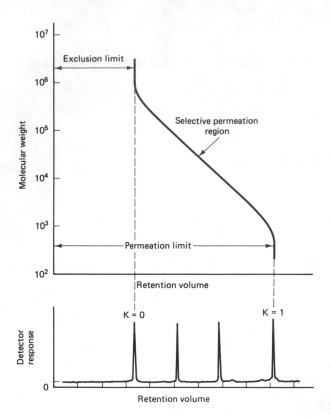

Figure 22-8 Calibration curve for a size exclusion chromatographic column.

dissolving the sample. Some liquids may cause excessive swelling of certain semirigid polymeric particles, which will limit the maximum operating pressure. Polar eluents are preferred with silica-based stationary phases where adsorption can occur, because they act as modifiers and deactivate the surface toward the adsorption of the sample constituents.

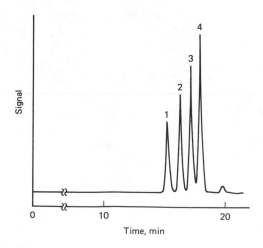

Figure 22-9 Exclusion chromatographic separation of phthalates on a 2 × 30-cm column of LiChrogel on PS4 (10 μm). Mobile phase was tetrahydrofuran at a flow rate of 1.0 mL/min: (1) dioctylphthalate; (2) dibutylphthalate; (3) diethylphthalate; (4) dimethylphthalate. (Courtesy of EM Science)

SPECIAL TECHNIQUES

A number of special techniques are used by chemists to broaden the applications of the different types of liquid chromatography. Some of these, such as the use of modifiers and suppressor columns, were discussed with the individual chromatographic method. A few additional techniques of somewhat more general application are discussed in the following sections.

Derivatization

It was pointed out earlier that one of the major shortcomings of LC compared to GC is the lack of a good, sensitive, universal detector. Chemists are discovering that the range of substances to which a given detector will respond can be broadened substantially through the formation of derivatives. Off-line derivatization is carried out before the separation and the treated sample then is introduced into the chromatograph. On-line derivatization is carried out in a unit that is an integral part of the chromatograph; it may be done before or after the sample passes through the column. Because the physical processes involved are specific and generally complicated, they are not discussed here.

The most common derivatization techniques are designed to enhance ultraviolet absorption or fluorescence by combining the analyte with a suitable chromophore or fluorophore. A few reagents used for this purpose are listed in Table 22-5.

Gradient Elution (Eluent Programming)

When an eluent of constant composition is used throughout a chromatographic separation, the process is called *isocratic elution*. It is an acceptable process when the substances to be separated are of a similar type because usually it is possible to find a single eluent that will provide the desired separation in a reasonable length of time. Isocratic elution will not give satisfactory results with samples containing substances whose distribution coefficients vary widely. Consider a sample with two distinct groups of sub-

TABLE 22-5 DERIVATIVE REAGENTS USED TO ENHANCE DETECTION BY ABSORPTION OR FLUORESCENCE MEASUREMENTS

Reagent	Analytes
For ultraviolet absorption	
3,5-Dinitrobenzoyl chloride	Alcohols, phenols, amines
p-Nitrobenzyloxyamine hydrochloride	Aldehydes, ketones
O-p-Nitrobenzyl-*N*,*N'*-diisopropylisourea	Carboxylic acids
N-Succinimidyl-*p*-nitrophenylacetate	Amines, amino acids
For fluorescence	
4-Bromomethyl-7-methoxycoumarin	Carboxylic acids
7-Chloro-4-nitrobenzyl-2-oxa-1,3-diazole	Amines, thiols
1-Dimethylaminonaphthalene-5-sulfonyl chloride	Amines, phenols
1-Dimethylaminonaphthalene-5-sulfonyl hydrazine	Carbonyls

stances: Group 1 compounds have relatively small distribution coefficients and are the least strongly retained by the stationary phase, while group 2 compounds have much larger distribution coefficients and are retained more strongly by the stationary phase. An eluent effective at separating and eluting group 1 substances will not be able to move group 2 molecules through the column at any reasonable speed. On the other hand, an eluent with sufficient strength to elute group 2 compounds will surely move group 1

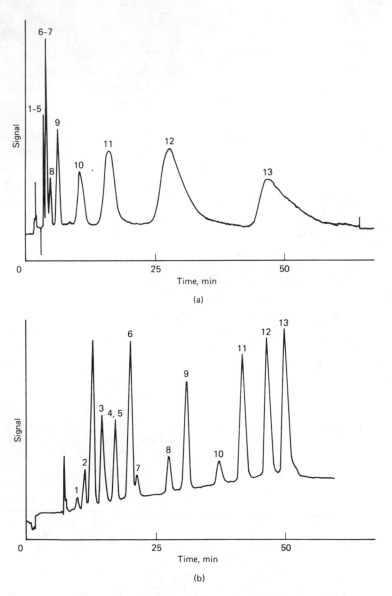

Figure 22-10 Effect of eluent programming on the ion-exchange separation of a 13-component mixture of carboxylic acids: (a) isocratic elution; (b) gradient elution. (From J. Aurenge, *J. Chromatogr.* **1973**, *84*, 285–298. Reprinted with permission of Elsevier Science Publishers B.V., Amsterdam)

compounds through the column so rapidly that they have little chance to become separated.

The solution to this problem lies with a technique called *gradient elution* or *eluent programming*. The technique involves changing the composition of the mobile phase, either stepwise or continuously, to increase its eluting strength during the separation. One might say that eluent programming is to liquid chromatography what temperature programming is to gas chromatography. To see how this might work, consider the problem described in the preceding paragraph. If the separation is begun with a "weak" mobile phase, the group 2 compounds will remain fixed in the stationary phase while the group 1 compounds begin to separate. As the separation continues, the composition of the mobile phase is gradually changed to make it a "stronger" eluent. This causes the group 2 compounds to begin partitioning and move through the column. The new composition may be less than ideal for the group 1 compounds, but they have already been separated and will remain so as they continue to move through the column. Chromatograms of a thirteen-component mixture with and without gradient elution are shown in Figure 22-10. After an eluent programming experiment, the column contains the final rather than the original eluent and must be reconditioned before its next use.

Closed-Loop Recycling

In gas chromatography when solute peaks are not separated as well as one might like, switching to a longer column will often solve the problem. This solution seldom is possible in HPLC because a longer column requires a much higher inlet pressure to maintain a suitable flow rate. Recycling the partially separated components through the same column has the effect of increasing the column length without changing the pressure. Figure 22-11 shows the progress of a separation of naphthalene and biphenyl as it passes through the same column three times. The bands are broadened after each pass but the resolution is improved. The extent of broadening is somewhat greater than expected from the normal on-column processes, due to the extra connectors and valves required to recycle the liquid. The optimum number of cycles depends on the degree of broadening and resolution being sought.

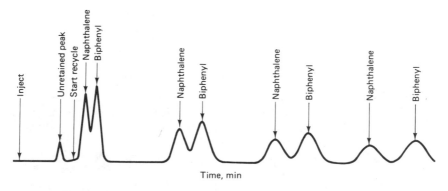

Figure 22-11 Three-cycle chromatographic separation of naphthalene and biphenyl. (Courtesy of Waters Chromatography division of Millipore Corporation)

Special Techniques

PROBLEMS

22-1. When 1.00-μL volumes of pure benzene ($d = 0.879$ g/mL), toluene ($d = 0.867$ g/mL), and ethylbenzene ($d = 0.867$ g/mL) were chromatographed, the areas of their elution peaks were 27.3, 25.5, and 23.1 cm², respectively.

   (a) Calculate the detector molar response factors of toluene and ethylbenzene relative to benzene.

   (b) Calculate the volume percent of toluene in a 2.00-μL sample if the area of its elution peak was 30.7 cm².

22-2. In determining trace amounts of cresols in water using a chromatographic technique, a 10.00-μL sample produced peaks with areas of 36.7 cm² for *o*-cresol, 12.0 cm² for *m*-cresol, and 26.9 cm² for *p*-cresol. A 0.105-μg sample of pure *o*-cresol yielded a peak with an area of 15.3 cm². Assuming that the detector responds equally to all three isomers, calculate the parts per million of each isomer in the sample.

22-3. It was pointed out in the chapter that significant band broadening can occur in off-column chambers such as a UV detector cell. Suppose that we desire to limit the cell volume to 5.00 μL.

   (a) What must be the diameter of a cylindrical UV cell whose path length is 2.00 cm?

   (b) What is the detection limit (in nanograms) of benzene in such a cell if the minimum detectable absorbance is 0.0002? Benzene (C_6H_6) has a molar absorptivity of 204 at the measurement wavelength of 254 nm.

22-4. A 30-cm column is packed with 45.0 g of a strong cation-exchange resin whose exchange capacity is 0.402 meq/g resin. What is the largest volume of a sample containing 200 ppm Ca^{2+} and 125 ppm Mg^{2+} that should be injected into the column if the sample is to use no more than 0.5% of the column's capacity?

22-5. The following elution data were obtained from a 2.5 × 50-cm Sephadex G-200 gel permeation column:

Compound	Molecular weight	Retention volume (mL)
Sucrose	342	242
Glucagon	5,500	233
Cytochrome *c*	11,000	214
Chymotripsinogen	24,000	188
Bovine serum albumin	80,000	149
Aldolase	153,000	127
α-Conarachin	486,000	92
α-Crystallin	825,000	78
Blue dextran	2,000,000	75
Unknown	?	119

   (a) Plot the data in an appropriate manner.

   (b) Estimate the molecular weight of the unknown from the calibration plot.

   (c) Explain why the retention volume is nearly independent of MW at both very low and very high molecular weights.

   (d) What will be the retention volume of an enzyme whose molecular weight is 39,000?

   (e) How many theoretical plates must the column have to provide a resolution of 1.0 for two identically shaped peaks whose retention volumes are 200 and 205 mL?

LABORATORY OPERATIONS AND PRACTICES

The material presented in this chapter relates primarily to the laboratory experience in quantitative analysis. Chemistry is an experimental science and the development of good laboratory skills is fundamental to its practice and perhaps even to its understanding and appreciation. The particular items chosen for discussion represent the fundamental tools and techniques common to almost all introductory analysis problems.

ANALYTICAL BALANCE

A typical chemistry laboratory may contain a variety of balances, each designed for a particular type of application. The most important specifications of a balance are its accuracy, precision, capacity, sensitivity, and readability. The term *analytical balance* is used to describe a balance capable of weighing objects with a very high degree of accuracy and precision. *Capacity* refers to the maximum weight that the balance can measure. Macro analytical balances can weigh objects of up to several hundred grams, while semimicro analytical balances have a capacity of perhaps 100 to 160 g. The terms "sensitivity" and "readability" are sometimes confused when examining balance specifications. *Sensitivity* is generally taken as the smallest weight that will produce a certain measurable response, typically 0.1 or 0.01 mg. *Readability* refers to the smallest discernible scale division, which may or may not be the same as the sensitivity of a balance.

Construction and Operation

Modern analytical balances employ a single pan suspended from an unequal-arm beam and are capable of displaying the final weight on the front panel. A cutaway view of a typical balance is shown in Figure 23-1. The balance beam contains sets of weights in

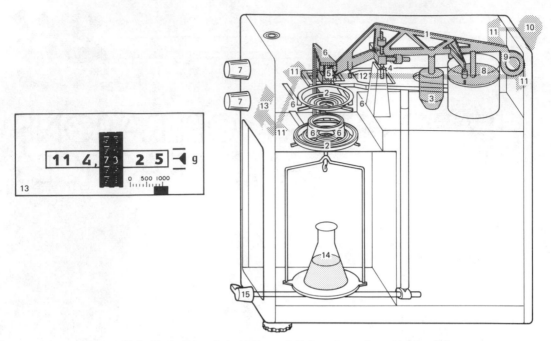

Figure 23-1 Single-pan analytical balance. 1, Balance beam; 2, removable weights; 3, counterweight; 4, knife-edge beam fulcrum; 5, pan and weight suspension knife edge; 6, weight-lifting mechanism; 7, weight-control dials; 8, air damper; 9, scale graduation plate; 10, lamp; 11, mirrors; 12, light path; 13, readout panel; 14, sample; 15, beam release dial. (Courtesy of Mettler Instrument Corp.)

the shape of concentric rings together with a suspended weighing pan on one end and a large counterweight on the other end. The beam is balanced when the weighing pan is empty. When an object is placed on the pan, it causes the beam to pivot and the approximate balance point is reestablished by lifting some of the weights from the beam using mechanical levers controlled by calibrated dials on the front panel of the balance. Generally, the total weight removed is within 1 g of the weight of the object. The remaining weight needed to restore the exact balance point is estimated by a simple optical measurement of the distance by which the beam is displaced from that point.

Analytical balances have a lever that can release the beam either partially or completely. The partial release position is used while "dialing in" the correct weights in order to prevent sudden, extreme fluctuations of the beam that might wear or otherwise damage the knife-edge finish of the fulcrum. The full release position is used only in making the final reading. The beam should be completely supported whenever an object is being placed on or removed from the weighing pan.

TECHNIQUES OF WEIGHING

Objects are weighed either directly or by difference. In direct weighing, the receiving vessel or a sheet of weighing paper is placed on the pan and weighed, the sample is transferred, and the container plus sample is reweighed. Some balances can be set to read zero with the empty receiving vessel on the pan, a process called *taring*. Thus, when the sample is transferred, only its weight is displayed.

Weighing by difference involves weighing the container and its contents, transferring a portion of the contents to another vessel, and reweighing the container. The difference equals the weight of the transferred material. This technique is especially useful when weighing hygroscopic or volatile substances because the weighing container can be kept closed during the weighing process.

Errors

Most analytical balances are of sufficiently high quality that any inherent weighing error is usually much less than the errors from other sources in a determination. When the object being weighed has a density that is considerably different from the density of the weights used by the balance, a *buoyancy* error may be significant. An object on a balance pan displaces a certain amount of air. Since the balance is set to zero with air on the pan, the object will appear lighter than its actual weight. The correct weight can be calculated using the equation

$$w_{\text{corr}} = w + wd_{\text{air}}\left(\frac{1}{d_o} - \frac{1}{d_w}\right)$$

where w_{corr} is the corrected weight of the object, w the mass of the weights, d_{air} the density of the air (0.0012 g/mL at $25\,^\circ$C), d_o the density of the object, and d_w the density of the weights (7.76 g/mL for the stainless steel weights used in most modern single-pan balances). In general, buoyancy corrections are necessary only when weighing gases, liquids, or low-density solids.

Two other sources of error not inherent in the balance itself are worth mentioning: The object being weighed should not be touched with bare hands because the resulting fingerprints can change the weight, and objects should be at room temperature before being weighed because heat transfer between an object and its surroundings will cause air currents within the balance chamber that can lead to highly erroneous measurements.

VOLUMETRIC GLASSWARE

Volumetric glassware refers to glass containers that are designed to measure the volume of liquids with a high degree of accuracy. Three such containers—the buret, the pipet, and the volumetric flask—are found in every introductory quantitative analysis laboratory. Volumetric containers are marked by the manufacturer as to the manner of calibration and the temperature at which the calibration was made. Items marked TD are designed *to deliver* the specified volume at the calibration temperature. Items marked TC and/or which have a frosted ring near the top are designed *to contain* the specified volume at the calibration temperature.

Glassware expands and contracts with rising and falling temperatures. However, an object that has expanded on heating does not always return to the same volume on cooling—a phenomenon called hysteresis. For this reason, volumetric glassware should not be heated much above its calibration temperature. *It is especially inappropriate to dry such glassware in a laboratory drying oven.*

Cleaning Glassware

Glassware should be soaked and/or scrubbed with warm detergent, then rinsed, first with large quantities of tap water to remove detergent and finally with several small portions of distilled water. Remember that impurities may not be visible to the eye. Water will wet clean glassware, forming an unbroken, uniform film on the surface. Grease and other organic material that survives a vigorous cleaning with detergent usually can be removed with a cleaning solution consisting of sodium dichromate dissolved in concentrated sulfuric acid.* The cleaning action may be slow at room temperature but becomes quite rapid at 60 to 70°C. Warm the cleaning solution in a beaker or flask and transfer a *small* portion to the glassware being cleaned. If too large a volume is used, it may cause undesirable heating of the glassware. Dichromate cleaning solution should be handled with *extreme caution*, especially when it is warm. It will attack skin and clothing with the same vigor and speed with which it attacks grease on the walls of a glass container. Spills should be diluted immediately with a large volume of water.

Glassware in which solutions have been allowed to stand for long periods or which have evaporated to dryness may be especially hard to clean. Nobody likes to do dishes, but remember that the longer the job is put off, the more difficult it is likely to be.

Burets

A buret consists of a long tube of highly uniform diameter marked with volume graduations and fitted with a ground-glass or Teflon stopcock and a short delivery tip, as shown in Figure 23-2. A ground-glass stopcock must be lightly greased so that it forms a liquid-tight seal. In applying the grease, make sure that the stopcock is dry, and avoid the area immediately adjacent to the hole. Teflon stopcocks do not require lubrication. Class A burets are certified to meet the specifications listed in Table 23-1.

Rather than drying a buret to remove traces of water, it is simply rinsed several times with a few milliliters of the solution it will contain. Each portion of rinse solution should be expelled through the stopcock to ensure that it and the tip are also rinsed. It is common practice to overfill a buret and open the stopcock fully to dislodge any small air bubbles that may collect at the junction of the stopcock and the tip. Once the entire buret is free of bubbles, the liquid is drained slowly until the level reaches the graduations. Any solution clinging to the outside of the buret tip is removed by touching the tip to the inside wall of a beaker. Wiping the tip is not recommended because of the risk of some of the liquid "wicking out."

Aqueous solutions wet the wall of the buret, giving a shallow curve to the surface of the liquid, which is called the *meniscus*. The bottom of the meniscus is taken as the level of the solution. In estimating the volume, the eye should be level with this point to avoid a reading error called *parallax*. If the meniscus is below eye level, it will appear to coincide with a smaller volume marking on the buret. If the meniscus is above eye level, the opposite effect occurs. When the stopcock is opened fully, the layer of liquid

*Cleaning solution can be prepared by dissolving about 5 g of sodium dichromate in 5 mL of water in a large Pyrex flask and *slowly* adding about 100 mL of concentrated sulfuric acid with constant stirring. After cooling, the solution is transferred to a labeled storage bottle. The solution may be reused until it acquires the green color indicative of chromic ion.

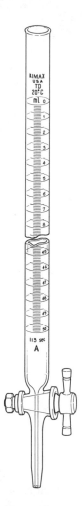

Figure 23-2
A 50-mL buret
with stopcock.

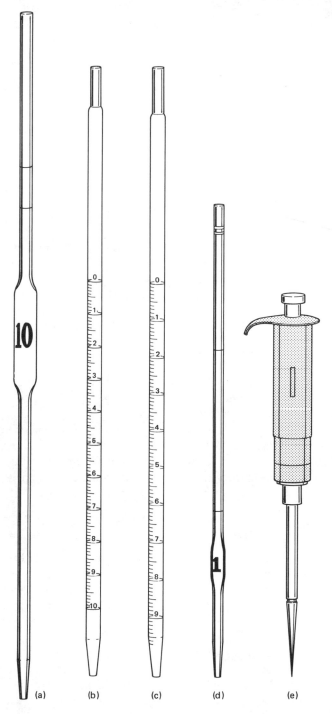

Figure 23-3 Common pipets: (a) volumetric or transfer;
(b) measuring (Mohr); (c) measuring (serological);
(d) Ostwald-Folin; (e) syringe.

TABLE 23-1 TOLERANCES FOR CLASS A VOLUMETRIC GLASSWARE

Volume (mL)	Tolerance (mL)		
	Buret	Pipet	Volumetric flask
0.5		±0.006	
1		±0.006	±0.02
2		±0.006	±0.02
3		±0.01	
4		±0.01	
5	±0.01	±0.01	±0.02
10	±0.02	±0.02	±0.02
15		±0.03	
20		±0.03	
25	±0.03	±0.03	±0.03
50	±0.05	±0.05	±0.05
100	±0.10	±0.08	±0.08
200			±0.10
250			±0.12
500			±0.20
1000			±0.30
2000			±0.50

wetting the inside wall cannot drain as rapidly as can the main body of liquid. Therefore, you should allow 20 to 30 s to elapse between the time the stopcock is closed and the time the volume is measured.

As the end point of a titration is approached, it is often desirable to deliver less than one drop of liquid at a time from the buret. A fraction of a drop is delivered by opening the stopcock only slightly until a portion of drop is clinging to the tip, closing the stopcock, and rinsing the partial drop off the tip and into the receiving container with a stream of water from a plastic wash bottle.

Pipets

Pipets are devices used to accurately transfer known volumes of liquid. They are available in a variety of types, as illustrated in Figure 23-3, each designed for certain types of applications. Pipets marked with a frosted band at the top are calibrated to deliver a specified volume when the last drop is blown out. The absence of a frosted band means that an allowance has been made for the small amount of liquid remaining inside after the pipet has drained freely. *Volumetric* or *transfer* pipets are designed to deliver a definite specified volume of between 0.5 and 100 mL. The tolerances for class A volumetric pipets are given in Table 23-1. *Ostwald-Folin* pipets (available in volumes from 0.5 to 10 mL) are similar to volumetric pipets except that they are calibrated to deliver a specified volume when the last drop is blown out. Both *measuring* and *serological* pipets are used to deliver a variable volume, generally between 1 and 25 mL. The only difference between them is that serological pipets are calibrated to deliver their maximum volume when the last drop is blown out.

Syringe pipets are available that can deliver either fixed or variable volumes in the range of 1 to 1000 μL. They employ a disposable polypropylene tip into which the liquid is drawn by the action of a spring-operated piston. The liquid is dispensed simply by reversing the action of the piston.

Before filling a pipet, it should be rinsed two to three times with small portions of the liquid to be transferred. This is done by drawing in the liquid through the tip, rotating the pipet while holding it in a horizontal position to ensure that the rinse solution contacts the entire inside surface, and draining through either the tip or the top opening. Pipets are filled by drawing liquid in through the tip by the sucking action of an evacuated rubber bulb held over the top opening (Figure 23-4). Liquid is drawn in above the top graduation and the bulb is replaced quickly with the index finger. At this point, the tip is wiped to remove any droplets of liquid clinging to the outside and the pipet is allowed to drain slowly until the level reaches the graduation mark. This is accomplished best by reducing the pressure of the index finger and rotating the pipet back and forth between the thumb and middle finger. The tip of the pipet should be touched to the side of a waste container to remove any partial drops that may have formed and are clinging to the outside. Finally, the index finger is removed and the pipet is allowed to drain freely, usually onto the side of the receiving vessel to minimize splashing and possible sample loss. *Pipets should never be filled by using your mouth instead of a rubber bulb.* Because they are difficult to clean, it is good practice to rinse pipets with distilled water immediately after use.

Volumetric Flasks

Volumetric flasks (Figure 23-5) are calibrated to contain a particular volume of liquid at 20°C. A careful examination of the writing on the flask will reveal the letters TC, which

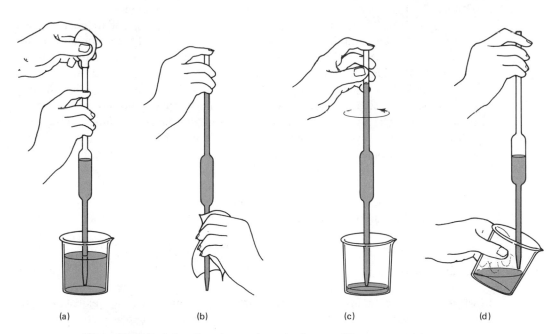

(a)   (b)   (c)   (d)

Figure 23-4 Technique for using a volumetric pipet: (a) filling; (b) wiping; (c) adjusting to the mark; (d) emptying.

Volumetric Glassware

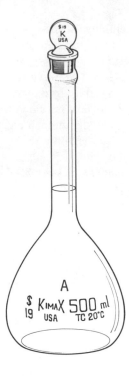

Figure 23-5 A 500-mL volumetric flask.

stand for "to contain," as opposed to the letters TD for "to deliver" that are found on burets and many pipets. These flasks are available with capacities ranging from 1 to 2000 mL. The tolerances for class A volumetric flasks are given in Table 23-1.

Volumetric flasks are used to prepare solutions of known concentration. A weighed sample is dissolved in a minimum volume of solvent, transferred to the appropriate size flask, and diluted to the mark with solvent. If the sample is not dissolved prior to being transferred to the flask, it should be allowed to dissolve completely before filling the volumetric flask with solvent. This permits easy mixing by swirling. Once full, mixing is more difficult and is accomplished by holding the stopper firmly in place while repeatedly inverting the flask. The large air bubble rising through the solution causes mixing. It is very easy to overfill a volumetric flask while trying to make the final volume adjustment. This problem can be avoided by using a disposable dropping pipet or medicine dropper to add the last milliliter or so of solvent.

Volumetric flasks should not be used as storage bottles. Once the flask has been filled and its contents mixed, the solution should be transferred to an ordinary bottle for long-term storage, and the flask should be rinsed and cleaned.

SAMPLE PREPARATION

Most solids absorb atmospheric moisture (are hygroscopic) to some extent, thereby undergoing a change in composition. The extent of absorption depends not only on the chemical nature of the substance but also on the amount of exposed surface area and the atmospheric humidity. In such instances, it is usually necessary to dry the solid in an

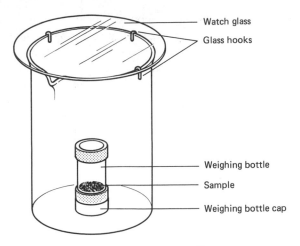

Watch glass
Glass hooks

Weighing bottle

Sample

Weighing bottle cap

Figure 23-6 Arrangement for drying samples in an oven.

oven before weighing, in order to achieve a uniform and reproducible composition. Generally, a temperature of about 110°C for 1 to 2 hours is sufficient for small quantities of samples. The material to be dried is placed in an open weighing bottle, which, in turn, is placed inside a small beaker covered with a watch glass supported by glass hooks on the edge of the beaker, as illustrated in Figure 23-6.

When the sample is removed from the oven, the weighing bottle and its top must be cooled to room temperature (about 30 min) in a low-humidity environment to prevent reabsorption of moisture. The device used to accomplish this, shown in Figure 23-7, is a *desiccator*. A drying agent or *desiccant*, such as calcium sulfate (Drierite), calcium chloride, magnesium perchlorate, silica gel, or phosphorus pentoxide, is put in the bottom of the chamber and the sample is placed on a perforated plate above the desiccant. The lid of the desiccator seals to the base by means of a flat, ground-glass rim that is kept lightly greased. The most effective and safest way to open and close a desiccator is to *slide* the lid gently off and on. When a very hot object is placed in a desiccator, it may warm the air and increase the pressure enough to break the seal, which, on occasion,

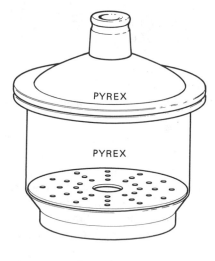

PYREX

PYREX

Figure 23-7 Desiccator for keeping samples dry.

Sample Preparation

causes the lid to slide off and break. On cooling, the opposite effect takes place and the reduced pressure inside the desiccator may make it difficult to open, and upon opening, the inrushing air may blow sample material and/or desiccant around the chamber. The simplest way to minimize this problem is to leave the lid open just a bit for the first few minutes of cooling. Once the sample has cooled to room temperature, the desiccator can be opened and the lid of the weighing bottle can be replaced.

On those few occasions when samples must be weighed in their hydrated form, the desiccator can be converted into a high-humidity chamber by using a shallow tray of water in place of the desiccant. Samples stored in such an environment will tend to hydrate fully and reach a uniform, reproducible composition.

DISSOLUTION AND DECOMPOSITION

Although many samples may arrive at the laboratory in solid form, analysis procedures usually involve measurement or manipulation of the analyte in aqueous solution. The process of dissolving a sample often is far from simple or routine and may even represent the majority of the total analysis time. Ordinarily, the entire sample must be dissolved. Attempts to dissolve just the analyte often result in incomplete separation from the undissolved residue. Since the decomposition–dissolution process involves addition of other chemicals, the analyst must keep in mind the limitations of the method chosen and be certain that no reagents are added that would interfere in later steps. In addition, care must be exercised to avoid loss of the analyte through the formation of volatile compounds.

Concentrated Acids

Strong inorganic acids are simple to use, relatively safe, and can dissolve a wide variety of materials. Because the vapors are very corrosive, acid solutions should be heated in a well-ventilated fume hood.

Hydrochloric acid
Although normally referred to as a nonoxidizing acid, this reagent does oxidize and dissolve metals that lie below hydrogen on the activity scale, such as magnesium, iron, and zinc. Aqueous hydrochloric acid is an excellent solvent for metal oxides. The concentrated reagent is about 37% HCl by weight or about 12 M. At this concentration, it is fairly volatile and the vapors can give a severe stinging sensation in the eyes and nose. Hydrogen chloride gas is easily volatilized from aqueous solution by heating until a constant-boiling 6 M solution is formed.

Nitric acid
With the exception of aluminum and chromium, all common metals will dissolve in hot, concentrated nitric acid. Much of the acid's dissolving power is derived from the ability of the nitrate ion to act as an oxidizing agent. Concentrated nitric acid is about 70% HNO_3 by weight or 16 M. Concentrated acid that has stood for a long time may decompose slightly, forming yellow-brown nitrogen dioxide. There is no reason to dis-

card such a solution, however, since nitrogen dioxide is a normal reduction product formed whenever metals are dissolved in nitric acid.

Sulfuric acid

Unlike concentrated nitric and hydrochloric acids, concentrated sulfuric acid contains very little water (98% H_2SO_4 by weight or 18 M) and has a high boiling point (340°C). As a result, hot sulfuric acid is a very effective oxidizing agent for many metals and for organic compounds as well. Concentrated sulfuric acid has a very strong affinity for water and generates a great deal of heat when it hydrates. Extreme care should be used in diluting sulfuric acid solutions.

Aqua regia

A mixture of three parts of concentrated hydrochloric acid to one part of concentrated nitric acid (by volume) is an extremely powerful oxidizing agent known as *aqua regia* or "royal water." The name was chosen for its ability to dissolve noble metals such as platinum and gold, which are insoluble in either acid alone. When mixed, the two acids react to yield elemental chlorine and nitrosyl chloride:

$$3HCl + HNO_3 \longrightarrow Cl_2(g) + NOCl(g) + 2H_2O$$

The dissolving power of the mixture is derived mainly from the ability of chlorine and nitrosyl chloride to convert the metals to metal chlorides, which are then transformed to stable complex anions by reaction with chloride ion. The overall reaction for gold is

$$Au(s) + 4HCl + HNO_3 \longrightarrow HAuCl_4 + NO(g) + 2H_2O$$

Hydrofluoric acid

Many ores and minerals contain appreciable quantities of metal silicates which will not dissolve completely in oxidizing acids. Aqueous hydrofluoric acid has been found to be a particularly effective solvent for such substances, due to its ability to form volatile silicon tetrafluoride. Since fluoride ion is such an effective complexing agent, excess hydrofluoric acid often must be removed completely after the dissolution process. Usually, this is done by repeated addition of concentrated sulfuric acid followed by evaporation to near dryness. The silicon tetrafluoride and hydrofluoric acid are volatilized during the evaporation steps. Hydrofluoric acid reacts slowly with silica in glass and given sufficient time, it will produce a frosted surface. Whenever possible, hydrofluoric acid solutions should be kept in plastic containers. When glass must be used, the solutions should not be allowed to remain in the container any longer than is absolutely necessary.

Concentrated hydrofluoric acid is about 49% HF by weight or 29 M. It can cause painful burns and serious injuries if allowed to come in contact with the skin. Unfortunately, there may be no discomfort or indication of a burn for several hours after the exposure. For these reasons, this acid demands great respect and care during handling. Many chemists choose never to handle this reagent without the use of protective gloves.

Fusions with Molten Fluxes

The term *fusion* refers to the dissolving of a sample in a molten solvent, called a *flux*. Fusions are performed by mixing the finely ground sample with an excess of the solid flux in an inert metal or porcelain crucible and heating to a temperature above the melting point of the flux. During this stage, the sample components are decomposed and transformed into water-soluble salts. After an appropriate period of time, the melt is cooled and the salts are dissolved in water or dilute aqueous acid.

There are three reasons why molten solvents exhibit much greater dissolving power than aqueous solvents. First, the temperature of the melt is usually much higher, typically 400 to 1200°C, than can be achieved with aqueous solutions, making the rate at which the decomposition reactions occur much faster. Second, the molten solvent itself is the reactive agent in a fusion and its concentration is much greater than it would be as a solute dissolved in water. Third, the absence of water can greatly increase the acidity or basicity of the dissolving reagent.

Fluxes often are categorized according to whether they are acids, bases, or simply strong oxidants. Acidic fluxes, such as potassium pyrosulfate, are used primarily to dissolve metal oxides. Fusions with this reagent are performed between 300 and 400°C, at which highly acidic sulfur trioxide is slowly evolved:

$$K_2S_2O_7(s) \xrightarrow{\Delta} K_2SO_4 + SO_3$$

Sodium carbonate is a basic flux commonly used to dissolve silicates and phosphates. Generally, it is heated to a few hundred degrees above its melting point of 851°C. Because of the very high temperatures used and the strong basicity of the flux, the fusions normally are done in highly inert platinum crucibles. Sodium peroxide is an example of a powerful oxidizing flux used to dissolve sulfides and a number of alloys.

GRAVIMETRIC TECHNIQUES

Gravimetry probably requires more sample manipulation and manual dexterity than any of the classical analytical methods. This section is a very brief description of some of the fundamental operations that must be performed in most gravimetric methods.

Weighing

All gravimetric procedures rely on the difference in weight of a container with and without the precipitate. The empty container is heated to the same temperature at which the precipitate will be heated, cooled to room temperature in a desiccator, and weighed. This process is repeated until a constant weight (within a few tenths of a milligram) is achieved. Ultimately, the container with the precipitate is treated similarly.

Filters

Precipitates are collected on paper, sintered glass, or unglazed porcelain filters. Collecting on paper is slower and more tedious but is necessary with gelatinous and very finely divided crystalline precipitates because they tend to clog the tiny pores of other filters.

Sintered glass filters consist of a porous disk of small glass particles sintered (partially melted) together and set into the bottom of a glass crucible. The crucibles are labeled C, M, or F, for coarse, medium, and fine, depending on the porosity of the sintered glass disk. Because the glass particles can soften and fuse together, thereby changing the porosity, these filters should not be heated above 200°C. Unglazed porcelain is similar in principle to sintered glass but has the advantage of tolerating much higher temperatures.

Paper filters are circular in shape and available in several degrees of porosity. As might be expected, fine-porosity paper passes liquid more slowly and clogs more easily than coarse-porosity paper. Because it is quite hygroscopic, paper cannot be dried to a constant weight and must be burned away after filtering. As a result, it can be used only to filter precipitates that can tolerate the high temperature encountered during burning. High-quality filter paper is manufactured in such a way that it produces a negligible amount of ash on ignition.

Heating Equipment

Some precipitates can be weighed after drying at 110°C, while others must be ignited to temperatures as high as 1000°C. Temperatures to about 300°C can be achieved in a simple electric oven. Sometimes the efficiency of the drying process in such ovens is improved by forcing through predried air or by maintaining a partial vacuum in the heating chamber.

Ordinary heat lamps are finding numerous applications in drying precipitates. They are capable of producing temperatures of several hundred degrees, which is sufficient to char filter paper.

Laboratory burners, such as the Meker burner, and heavily insulated electric ovens, called muffle furnaces, are used when precipitates must be heated above 300°C. Burners are inexpensive and simple but provide little control over the exact temperature to which the material is heated. Muffle furnaces are expensive by comparison but can reach temperatures above 1000°C that can be controlled to within a few percent of the set value.

Filtering Techniques

The actual filtering process is essentially the same with both filter crucibles and paper, although the physical setup is a little different. Filter crucibles are set into a rubber adapter mounted on a filter flask which is connected to a water aspirator through a trap, as shown in Figure 23-8. The purpose of the trap is to isolate the filter flask from an accidental backup of water through the aspirator. Filter paper is supported in a narrow-stem funnel. The circular paper is folded in half twice and one corner is torn off. Then it is opened to form a cone, pressed into the bottom of the filter funnel, and wetted with water. All air bubbles between the paper and the funnel are worked out by gently pressing with the fingers. The removal of one corner of the folded paper allows a better seal of the opened paper to the glass funnel, thereby preventing air from channeling between the paper and the glass, which can slow the filtering process considerably.

The filtration process is pictured in Figure 23-9. The first step of the process is to decant (gently pour) the bulk of the liquid onto the filter. Allowing the liquid to flow down a glass rod onto the filter prevents splashing and allows the analyst to direct the flow. Depending on the nature of the precipitate, it may be desirable to wash it with successive portions of a wash liquid, decanting each portion onto the filter. Finally, a stream of water from a plastic wash bottle is used to transfer the bulk of the precipitate

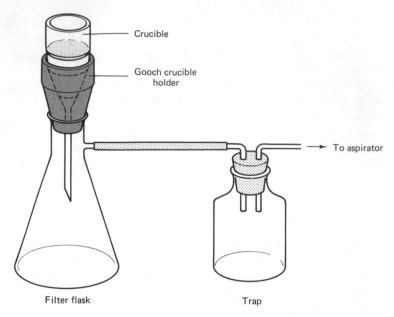

Crucible

Gooch crucible
holder

To aspirator

Filter flask

Trap

Figure 23-8 Arrangement for filtering with crucibles.

to the filter. The precipitate is transferred last, to minimize clogging of the pores of the filter, which can slow the filtering process considerably. In some cases it may be necessary to use a rubber policeman (a small rubber squeegee fitted on the end of a glass rod) to loosen particles of precipitate that stick to the walls of the container.

When all of the precipitate has been transferred to the filter, it is washed perhaps five or six times with small portions of an appropriate wash liquid. Each portion of liquid should be directed onto the upper portion of the filter medium so that it tends to wash the precipitate toward the bottom, where it is less likely to be lost during handling. Each portion of the wash liquid should be drained or drawn through completely before the next portion is added.

Ashing Filter Paper

Since filter paper cannot be dried to a constant weight, it must be burned off. If the filter paper is allowed to dry partially, it can be removed from the funnel easily without tearing. Once removed, it is flattened, the sides are folded in, the top edge is folded down, and the paper is placed in a weighed porcelain crucible with the bulk of the precipitate on the bottom.

To ash the paper with a heat lamp, the crucible is placed on a noncombustible surface with the lamp about $\frac{1}{2}$ in. above. No further attention is necessary until charring is complete. To ash the paper with a bunsen burner, the crucible is mounted on a wire triangle at a 45° angle, as shown in Figure 23-10. The process is started with a very low flame directed toward the bottom edge of the crucible. The object is to char the paper without allowing it to ignite. A flame will expel moisture too fast and set up strong air currents, both of which can carry minute amounts of precipitate out of the crucible.

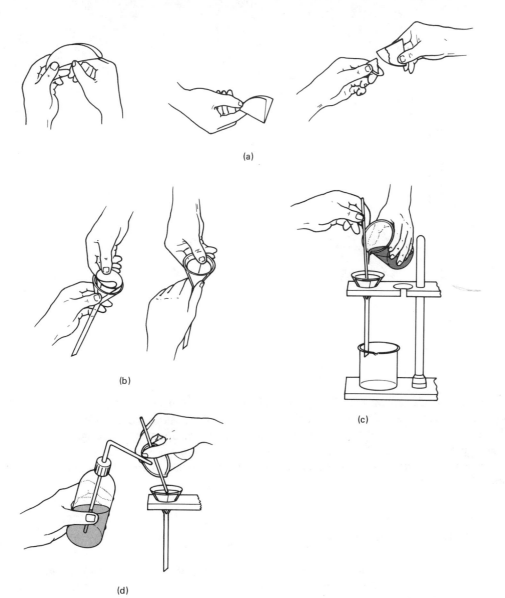

Figure 23-9 Arrangement for filtering with paper: (a) preparing the paper; (b) mounting the paper; (c) decanting the liquid; (d) transferring the precipitate.

Ideally, only thin wisps of smoke should be present. If the smoke gets heavy or the paper bursts into flame, remove the burner immediately and hold the crucible cover over the opening for 10 to 15 s. Continue the charring process until no smoke is observed, then increase the heat slowly until the residual carbon is vaporized.

Preparing the Precipitate for Weighing

If the precipitate was collected on a fritted-glass filter crucible, it is dried in an oven (usually at 100 to 200°C). Some precipitates collected on porous ceramic filters and

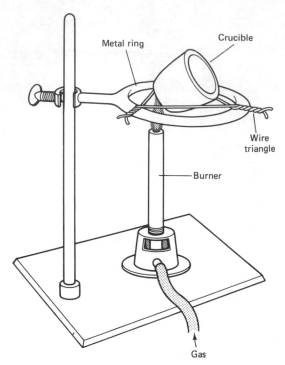

Metal ring

Crucible

Wire triangle

Burner

Gas

Figure 23-10 Arrangement for ashing filter paper in porcelain crucibles.

precipitates collected on paper must be ignited at very high temperatures in order to achieve a constant weight. Ignition using a flame is carried out much like ashing except that a Meker burner with the maximum flame is used. Precipitates can be ignited in a muffle furnace as well. Once ignition is complete, the crucible is allowed to cool for a few minutes before being tranferred to a desiccator for further cooling to room temperature.

SELECTION AND HANDLING OF REAGENTS

In this section we provide a very brief description of the grade classifications of common laboratory chemicals and precautions that should be observed in their handling.

Grade Classifications of Chemicals

Most chemicals are available in several grades of purity. To obtain the greatest accuracy in a quantitative determination, the reagents used should be of the highest purity available. However, in most cases, some balance must be struck between the accuracy of the determination and the cost of the reagents. The cost of a reagent rises sharply with its purity, and examples can be found where improving the purity from 99.90% to 99.98% more than doubles the cost of the reagent.

There are no national or international labeling laws that govern the purity of chemicals. However, several scientific organizations have established minimum standards for a large number of common chemicals, and most manufacturers use the designations suggested by these organizations.

Practical or technical grade

Chemicals carrying these designations are of relatively low purity, typically 70 to 95%. As a result, they should be used only when high purity is not important, such as in the preparation of a cleaning solution. Such chemicals are seldom used in the analysis laboratory.

Reagent grade

Chemicals carrying this designation meet the minimum specifications of the Reagent Chemical Committee of the American Chemical Society. This committee is charged with establishing practical concentration limits for all of the impurities normally associated with a given reagent. Chemical suppliers label their products with either the actual results of determinations for the various impurities or the maximum amount of each impurity allowed by the specifications. Reagent-grade chemicals are of high purity and are the most common grade of chemical used in quantitative analysis.

Primary-standard grade

This designation is reserved for chemicals of exceptionally high purity. The concentrations of primary standards are determined experimentally by the most accurate methods and the results are printed on the label. Not all reagents are available in this grade of purity. Primary standards are available from chemical supply companies and from the National Bureau of Standards.

USP grade

Chemicals carrying a USP designation must conform to the tolerances set forth in the *United States Pharmacopoeia*. Because this document is concerned primarily with contaminants that pose a health hazard to humans, chemicals may meet the USP specifications but still contain considerable amounts of nonhazardous impurities. This grade of chemical is encountered most often in the medical, pharmaceutical, and food-additive fields.

Specific-use chemicals

Numerous applications require chemicals with very stringent specifications with regard to a specific impurity or physical property. For example, reagents used for the determination of water must be scrupulously anhydrous but may contain larger amounts of other impurities, and solvents used in spectrophotometry may have to be absolutely free of any impurities that absorb above a certain wavelength. Manufacturers of such chemicals usually provide the desired information on the label or in their catalog.

Handling Reagents

Perhaps the first rule to observe in handling reagents is to read the label, even before opening the bottle. Make sure that you have selected the appropriate grade of chemical necessary for the job at hand. Be sure to note any special warnings on the label, such as *poison* or *flammable* and handle the chemical accordingly. The reagent in a newly opened bottle can be expected to meet the purity specifications on the label. Whether this remains true after repeated use depends on how careful we are not to contaminate the unused reagent. There are three important "*nevers*" to remember:

1. Never leave the bottle open longer than is necessary to remove the desired portion of reagent.

2. Never return unused reagent to the bottle. The potential savings is usually not worth the risk of contamination that might ruin the entire bottle.

3. Never insert spoons, knives, or spatulas into a reagent bottle unless they are known to be clean and totally unreactive toward the reagent.

In removing a reagent, the best approach is to shake the bottle or tap it gently on a soft surface (wood) to loosen the contents, then pour the desired quantity into a suitable container. If shaking or tapping does not loosen the contents, a clean porcelain or stainless steel spatula may be used to loosen and remove the material.

Despite one's best efforts to the contrary, an occasional spill is going to occur. There is little harm associated with most spills *unless* they are not cleaned up immediately. Innumerable balance pans and benchtops have been pitted and marked by stray sodium hydroxide pellets left by unthinking analysts. Consult with your laboratory instructor on how to clean up and dispose of spilled materials.

SAFETY

Safety in the laboratory should always be of the utmost concern to everyone. It is not possible to give a definitive discourse on safe laboratory practices in a few paragraphs or even a few pages. Indeed, entire books have been written on this subject. The large majority of safe laboratory practices do not have to be *learned*; they are a matter of common sense. It is our responsibility to remember at all times that the potential for dangerous mistakes exists. Mistakes are avoided when we *think* about what we are doing and *anticipate* what will happen rather than just blindly follow a procedure. For example, anytime you intend to mix two substances, you should ask yourself the question "Will these substances react, and if so, how?" Reading labels is an excellent practice to cultivate. If you handle a toxic chemical, use rubber gloves or wash your hands as soon as you are finished. We do not have to be exceptionally bright to figure out that it is not wise to handle a toxic substance if we have an open sore or cut on our hand. What we have to do is *read* the label and *think* about what we are doing. Following are a few of the more important do's and don'ts of lab safety:

1. Know the location of all fire extinguishers, showers, and eye wash fountains and how to use them. Ask your laboratory instructor to demonstrate their use.

2. Label all chemicals and solutions that are stored outside their original containers. Include your name and the date of preparation.

3. Tie up long hair, especially in the presence of open flames.

4. Keep work areas clean and uncluttered.

5. Never begin working until your instructor arrives.

6. Never bring food, drinks, or smoking materials into the laboratory.

7. Never work without an approved laboratory coat or apron and never wear open-toed or open-top shoes.

8. Never work without approved eye protection.

9. Never heat flammable liquids over an open flame or outside a hood.

Your laboratory instructor almost certainly will have a great deal more to say on the matter of safety. Pay especially close attention when safety is discussed.

LABORATORY NOTEBOOK

The validity of an experimental result depends on the ability of another scientist to duplicate that result. Obviously, some written record of what was done, how it was done, and what was observed is needed. The laboratory notebook is this written record. The manner in which information is recorded in a notebook may vary considerably between scientists and from experiment to experiment. Check with your instructor to find out whether you must conform to a particular format or whether you will be allowed some latitude in devising your own style of recording data and observations. The details of simple procedures and the data for semi- or nonquantitative measurements can be omitted as long as doing so does not introduce any uncertainty as to what and how something was done. For example, it may be quite sufficient to record that a 0.10 M solution of sodium hydroxide was prepared by dissolving approximately 4.0 g of reagent-grade NaOH in about 1.0 L of distilled water, without specifying the exact weights used and the nature of the weighing and mixing process. It is especially important to label all data entered in the notebook. After a few weeks, even the writer may have difficulty interpreting a notebook's contents when items are not labeled or identified in some way.

Another important reason for keeping an accurate, up-to-date record of your work is to protect yourself and your employer against challenges and infringements by others. For example, if the validity of your results is challenged in a court of law, your notebook can be valuable evidence that you performed the work correctly and drew legitimate scientific conclusions from the data. The decision in a patent suit may hinge on whether a chemist or company can demonstrate precisely *when* certain discoveries were made. For these reasons, scientists often must date and sign their laboratory notebooks each day. In some cases, another person may also sign the book to verify its validity.

Most instructors have somewhat different rules regarding laboratory notebooks, so no effort is made here to present a detailed set of instructions that must be followed. However, a few widely used rules are listed below.

1. Use a bound notebook.

2. Number all pages consecutively.

3. Enter all data *directly* into the book *in ink*.

4. Delete unused or incorrect data by drawing a single line through it. Do not erase,

Laboratory Notebook

obliterate, or remove data from the notebook. Occasionally, such data become important at a later date.

5. Write balanced equations for all important reactions that took place.

6. Include one complete example of every calculation made with the data.

7. Date and sign your notebook at the end of each laboratory period.

24

REPRESENTATIVE
EXPERIMENTS

Experiment 18	Comparison of spectrophotometric methods for determining the stoichiometry of a complex
Experiment 19	Determination of salicylic acid in aspirin by fluorescence
Experiment 20	Determination of copper by atomic absorption spectroscopy
Experiment 21	Determination of total salt content by ion exchange and acid-base titration
Experiment 22	Determination of the optimum flow rate in gas chromatography
Experiment 23	Determination of chlorinated hydrocarbons in a mixture by gas chromatography

The experiments described in this chapter were selected to illustrate the important methods and techniques discussed in the book. In general, they are well tested and can be performed inexpensively with glassware and equipment available in any modern quantitative analysis laboratory. Although the instructions provided are sufficiently specific to avoid ambiguity, students should not expect to be able to perform these experiments without first studying the pertinent background information. The reagents and apparatus required for each experiment are listed for the convenience of the instructor. The minimum amount of each reagent needed per student is given in parentheses.

EXPERIMENT 1: GRAVIMETRIC DETERMINATION OF SULFUR AS BARIUM SULFATE

This method for determining sulfur is based on the precipitation, collection, and weighing of insoluble barium sulfate. Freshly precipitated barium sulfate consists of very finely divided crystalline particles that must be digested in order to facilitate the washing and filtering processes. The digested precipitate, collected on fine-porosity filter paper, is ignited in a porcelain crucible prior to weighing. The most serious source of error is the coprecipitation of cations, especially potassium and iron(III). Although it results in the formation of smaller particles initially, *rapid* addition of the precipitating reagent to a hot solution of the sample has been shown to produce a more accurate result.

Reaction

$$Na_2SO_4 + BaCl_2 \longrightarrow BaSO_4(s) + 2NaCl$$

Reagents

HCl, concentrated (6 mL)
$BaCl_2$, 0.05 *M* (300 mL)
$AgNO_3$, 0.1 *M* (3 drops)
HNO_3, concentrated (6 drops)

Procedure

1. Transfer the unknown sample to a marked weighing bottle and dry in an oven at 110°C for at least 1 h. Cool to room temperature in a desiccator.

2. Clean and mark three porcelain crucibles. Mount them at a 45° angle on wire triangles and ignite for 20 min at the highest temperature attainable with Meker burners. Cool to room temperature in a desiccator (at least 30 min) and weigh. Repeat the ignition, cooling, and weighing until a constant weight (± 0.3 mg) is achieved.

3. Weigh accurately three 0.4- to 0.6-g samples of unknown and transfer to clean, numbered 600-mL beakers.

4. Dissolve each sample in 200 mL of water, add 2 mL of concentrated HCl, and heat to near boiling.

   Steps 5 through 10 should be performed during a single laboratory period.

5. Heat 100 mL of 0.05 M $BaCl_2$ to near boiling and add it quickly and with vigorous stirring to the hot sample solution. Leave the stirring rod in the beaker. Repeat this step with the remaining two samples.

6. Cover each beaker with a clean watch glass and heat to near boiling for 1 h.

7. While the solutions are heating, fit three marked filter funnels with fine-porosity ashless filter paper (Whatman No. 42, Schleicher and Schuell No. 598, or equivalent).

8. Decant the hot supernate through the filter, wash the precipitate in the beaker three times with small portions of hot water, and decant the washings through the filter.

9. Transfer the precipitate to the filter paper. Use a rubber policeman to loosen any particles that are sticking to the wall of the beaker.

10. Wash the precipitate in the filter with 5-mL portions of hot water until the filtrate, after acidification with 2 drops of concentrated HNO_3, gives only a faint turbidity on the addition of 1 drop of 0.1 M $AgNO_3$.

11. Remove each filter paper, fold, and place in the appropriate marked crucible. Char the paper. When no carbon remains, ignite the precipitate for 20 min using the hottest flame possible with the Meker burner.

12. Cool to room temperature in a desiccator (at least 30 min) and weigh. Repeat the ignition, cooling, and weighing process until a constant weight (± 0.3 mg) is attained.

13. Report the amount of sulfur as % SO_3.

EXPERIMENT 2: GRAVIMETRIC DETERMINATION OF NICKEL AS NICKEL DIMETHYLGLYOXIMATE

In ammoniacal solution, nickel forms an insoluble red coordination compound with dimethylglyoxime ($C_4H_8O_2N_2$). Only palladium(II) reacts similarly, and then only in weakly acidic solution. The coprecipitation of iron(III) and aluminum is prevented by addition of sodium tartrate, which forms soluble complex ions with these metals. It is necessary to know the approximate amount of nickel in the sample, to avoid adding a large excess of dimethylglyoxime, which is not very soluble in water and may precipitate along with the nickel complex. Nickel dimethylglyoximate is quite bulky when first precipitated and tends to creep up the walls of its container. As a result, the sample taken for analysis should not contain more than about 50 mg of nickel. The procedure is appropriate for the determination of nickel in steel.

Reaction

$$Ni^{2+} + 2C_4H_8O_2N_2 \xrightarrow{NH_3} Ni(C_4H_7O_2N_2)_2 + 2H^+$$

Reagents

HCl, 6 M (90 mL)

HNO$_3$, 6 M (35 mL)

Tartaric acid (3 g)

Methyl red indicator, 0.02% (w/v) in 60% (v/v) ethanol (1 mL)

NH$_3$, 6 M (90 mL)

Dimethylglyoxime, 1% (w/v) in ethanol (80 mL)

CH$_3$CO$_2$NH$_4$, 3 M (120 mL)

AgNO$_3$, 0.1 M (3 drops)

Procedure

1. Clean and mark three medium-porosity sintered-glass filter crucibles. Heat each crucible in an oven at 110°C for 1 h, cool in a desiccator, and weigh. Repeat the heating, cooling, and weighing process until a constant weight ($\pm$0.3 mg) is achieved.

2. Weigh accurately three samples of steel, each containing between 30 and 40 mg of Ni (see your instructor), and transfer to separate 400-mL beakers. Add 30 mL of 6 M HCl and warm on a hot plate in the hood until all or most of the sample dissolves.

3. Add 10 mL of 6 M HNO$_3$ and boil *gently* until the yellow-brown oxides of nitrogen are no longer evolved.

4. Dilute each sample with 150 mL of water and heat to near boiling. Add 1 g of tartaric acid and 5 drops of methyl red indicator, followed by 6 M NH$_3$ until a permanent color change is observed. The solution may be removed from the hood.

5. Heat the solution to about 70°C and add slowly 20 mL of the dimethylglyoxime reagent. With constant stirring, add 3 M ammonium acetate until a permanent red precipitate of nickel dimethylglyoximate appears. Slowly add an additional 20 mL of the ammonium acetate and mix. Leave the stirring rod in the beaker.

6. Add 1 mL of the dimethylglyoxime reagent to the supernate and note if it causes additional precipitation to occur. If so, add 5 mL more.

7. Digest the precipitates on a hot plate at 60°C for 1 h. Cool to room temperature and filter through the previously weighed sintered-glass crucibles.

8. Decant the supernate into the filter crucible and wash the precipitate in the beaker three times with 5 mL of a solution prepared by adding 1 mL of 6 M NH$_3$ to 500 mL of distilled water.

9. Transfer the precipitate to the filter crucible, using a rubber policeman to loosen particles sticking to the wall of the beaker.

10. Wash the precipitate in the crucible with successive 5-mL portions of the wash solution described in step 8 until the filtrate, after acidification with a few drops of 6 M HNO$_3$, gives only a faint turbidity on the addition of 1 drop of 0.1 M AgNO$_3$.

11. Dry the crucible plus precipitate to constant weight in exactly the same manner as used in step 1.

12. Report the amount of nickel in the steel as % Ni.

EXPERIMENT 3: DETERMINATION OF THE TOTAL ACIDITY OF VINEGAR BY ACID-BASE TITRATION

Vinegar contains several acids whose total concentration can be determined by titration with a strong base such as sodium hydroxide. The total acidity generally is reported as percent acetic acid, which is the principal acid present in most commercial vinegars. Since the pH at the equivalence point of the titration is basic, phenolphthalein is the indicator of choice.

Reactions

Standardization:

$$KHC_8H_4O_4 + NaOH \longrightarrow NaKC_8H_4O_4 + H_2O$$

Titration of unknown:

$$CH_3CO_2H + NaOH \longrightarrow CH_3CO_2Na + H_2O$$

Reagents

NaOH, 50% (w/v), carbonate-free (7 mL)

$KHC_8H_4O_4$, primary-standard grade (5 g)

Phenolphthalein indicator, 0.05% (w/v) in 50% (v/v) ethanol (12 drops)

Procedures

Preparation and standardization of 0.1 *M* carbonate-free sodium hydroxide

1. A carbonate-free, 50% (w/v) solution of NaOH should be available in the laboratory or prepared ahead of time. A convenient storage-delivery system is shown in Figure 24-1. Transfer approximately 7 mL of this reagent to a 1-L storage bottle containing about 500 mL of freshly boiled and cooled distilled water. Mix, add another 500 mL of the boiled water, and mix again. Keep the bottle tightly closed with a *rubber* stopper.

2. Dry about 5 g of primary-standard-grade potassium hydrogen phthalate at 110°C for 2 h and cool in a desiccator.

3. Weigh accurately three 0.8- to 0.9-g samples of the dried potassium hydrogen phthalate and transfer to numbered 250-mL conical flasks.

4. Dissolve each sample just prior to titration in 25 mL of CO_2-free water and add 2 drops of phenolphthalein indicator solution.

5. Titrate the sample with the NaOH solution until the first appearance of a pink color that persists for at least 20 s. Swirl the flask continuously while adding titrant and

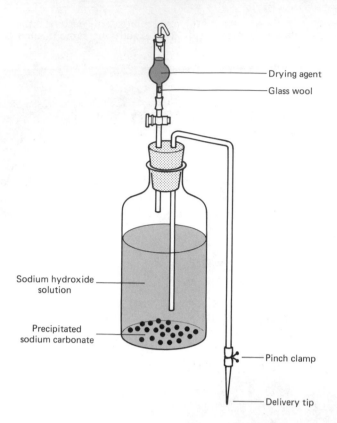

Drying agent

Glass wool

Sodium hydroxide solution

Precipitated sodium carbonate

Pinch clamp

Delivery tip

Figure 24-1 Storage-delivery system for carbonate-free base.

observe the color development against a white background. The approach of the end point is indicated by the slowness with which the pink color fades as the titrant mixes with the sample.

6. Calculate the molarity for each standardization and the average molarity.

Analysis of the unknown (vinegar)

1. If the vinegar sample has not been diluted previously by the instructor, pipet a 50-mL portion into a 500-mL volumetric flask and dilute to the mark with CO_2-free water. Mix thoroughly.

2. Pipet three 50-mL portions of the diluted sample into separate, numbered, 250-mL conical flasks. If the sample has a noticeable color, it may be diluted with 50 mL of CO_2-free water prior to titration.

3. Add 2 drops of phenolphthalein indicator and titrate each sample with standard 0.1 M NaOH to the first appearance of a pink color that persists for at least 20 s.

4. Calculate the total acidity as grams of acetic acid per 100 mL of sample.

EXPERIMENT 4: DETERMINATION OF CARBONATE AND BICARBONATE IN A MIXTURE BY ACID-BASE TITRATION

There are a number of ways in which these two bases can be determined in a mixture. Two procedures are presented here. Instructors may choose to have the students perform either or both. In the *two-indicator method*, the amount of carbonate present is taken as the amount of hydrochloric acid required to titrate to the phenolphthalein (Ph) end point. The total amount of the two analytes equals the amount of hydrochloric acid required to continue the titration to the methyl red (MR) end point. The amount of bicarbonate is found by difference. In the *two-titration method*, the total is determined by titration of a sample to the methyl red end point. A second sample is treated with a known volume of standard sodium hydroxide which converts the bicarbonate to carbonate. After removal of the carbonate by precipitation with barium chloride, the excess sodium hydroxide is titrated with hydrochloric acid. The amount of bicarbonate is taken as the difference between the amount of sodium hydroxide added and the amount left. The amount of carbonate is found by difference.

Reactions

Standardization:

$$Na_2CO_3 + 2HCl \xrightarrow{MR} H_2O + CO_2(g) + 2NaCl$$

Two-indicator method:

$$Na_2CO_3 + HCl \xrightarrow{Ph} NaHCO_3 + NaCl$$

$$NaHCO_3 + HCl \xrightarrow{MR} H_2O + CO_2(g) + NaCl$$

Two-titration method:

$$Na_2CO_3 + 2HCl \xrightarrow{MR} H_2O + CO_2(g) + 2NaCl$$

$$NaHCO_3 + NaOH \longrightarrow Na_2CO_3 + H_2O$$

$$Na_2CO_3 + BaCl_2 \longrightarrow BaCO_3(s) + 2NaCl$$

$$NaOH + HCl \xrightarrow{Ph} NaCl + H_2O$$

Reagents

HCl, concentrated (9 mL)
Na_2CO_3, primary-standard grade (2 g)
Methyl red indicator, 0.02 % (w/v) in 60 % (v/v) ethanol (1 mL)
Phenolphthalein indicator, 0.05 % (w/v) in 50 % (v/v) ethanol (1 mL)
NaOH (4 g)
$BaCl_2$, 10 % (w/v) (60 mL)

Procedures

Preparation and standardization of 0.1 *M* hydrochloric acid

1. Transfer 8 to 9 mL of concentrated HCl to a labeled storage bottle and dilute to about 1.0 L with distilled water. Mix thoroughly.
2. Dry about 2 g of primary-standard-grade sodium carbonate at 110°C for 2 h and cool in a desiccator.

3. Weigh *by difference* three 0.20- to 0.25-g samples of the dried reagent to the nearest 0.1 mg and transfer to numbered 250-mL conical flasks.

4. Dissolve each sample in 25 mL of distilled water and add 2 drops of methyl red indicator.

5. Titrate each sample with the 0.1 M HCl solution until the indicator has changed gradually from its initial yellow color to a definite red color. Stop the titration and boil the solution *gently* for 2 min. Cover the flask with a watch glass or inverted beaker (to retard absorption of CO_2 from the air) and allow the solution to cool. The yellow color should return. Continue the titration until a sharp yellow-to-red color change is observed.

6. Calculate the molarity of the titrant.

Two-indicator method

1. Dry the unknown at 110°C for 2 h and cool in a desiccator.

2. Ask your instructor for the approximate sample weight to be used. Weigh accurately three samples, dissolve in about 25 mL of distilled water, and transfer to 250-mL volumetric flasks. Dilute to volume, and mix. Transfer the solutions to clean, numbered storage bottles.

3. Pipet 25 mL of each solution into numbered 250-mL conical flasks and add 2 drops of phenolphthalein indicator.

4. Cool the solution for 5 min in an ice-water bath and titrate *slowly*, with constant, vigorous swirling, until the pink color just disappears. Record the volume of titrant used.

5. Add 2 drops of methyl red indicator and continue to titrate as in step 5 of the standardization procedure. Repeat the titration with the two remaining samples.

6. Calculate the % Na_2CO_3 and % $NaHCO_3$ for each sample.

Two-titration method

1. Same as step 1 in the two-indicator method.

2. Same as step 2 in the two-indicator method.

3. Prepare a 0.1 M NaOH solution by dissolving 4.0 g of the solid in 1.0 L of distilled water. Store in a bottle with a *rubber* stopper.

4. To determine the total amount of the two substances, pipet 25 mL of each sample solution into numbered 250-mL conical flasks and add 2 drops of methyl red indicator. Titrate as in step 5 of the standardization procedure.

5. To determine the amount of bicarbonate, pipet 25 mL of each sample solution into numbered 250-mL conical flasks. Treat each sample individually from this point.

6. Pipet 50 mL of the 0.1 M NaOH solution into one of the flasks, mix, and immediately add 10 mL of the $BaCl_2$ solution and 2 drops of phenolphthalein indicator.

7. Without delay, titrate the excess NaOH with standard 0.1 M HCl to the disappearance of the pink color. Add the titrant slowly with constant, vigorous swirling of the flask. Repeat steps 6 and 7 with the two remaining samples.

8. Prepare three blanks, each consisting of 25 mL of water and 10 mL of the $BaCl_2$ solution.

9. Add 50 mL of the standard NaOH solution and 2 drops of phenolphthalein indicator, then titrate with the standard HCl to the disappearance of the pink color.

10. The difference between the amount of HCl needed for the titration of the sample and the blank is the amount of $NaHCO_3$ present in the titrated solution. Calculate the % Na_2CO_3 and % $NaHCO_3$ in the sample. Do not forget that only part of the weighed sample was titrated.

EXPERIMENT 5: DETERMINATION OF NITROGEN BY THE KJELDAHL METHOD

This procedure is suitable for the determination of nitrogen (or protein) in a variety of samples, including dried blood, flour, and grain cereals. The nitrogen is converted to ammonia, which is distilled into a known volume of standard hydrochloric acid. The excess acid is determined by titration and the amount of nitrogen equals the difference between the amount of hydrochloric acid taken and the amount left in excess.

Reactions

Sample preparation:

$$\text{Organic N} + H_2SO_4 \longrightarrow NH_4HSO_4 + CO_2(g) + H_2O$$

$$NH_4HSO_4 + 2NaOH \longrightarrow NH_3 + Na_2SO_4 + 2H_2O$$

$$NH_3 + HCl \longrightarrow NH_4Cl$$

Titration:

$$HCl + NaOH \longrightarrow NaCl + H_2O$$

Reagents

HCl, concentrated (9 mL)
Na_2CO_3, primary-standard grade (2 g)
Methyl red indicator, 0.02% (w/v) in 60% (v/v) ethanol (1 mL)
NaOH, 50% (w/v), carbonate-free (7 mL)
$KHC_8H_4O_4$, primary-standard grade (5 g)
Phenolphthalein indicator, 0.05% (w/v) in 50% (v/v) ethanol (6 drops)
Na_2SO_4, anhydrous (30 g)
$CuSO_4$ (0.1 g)
H_2SO_4, concentrated (75 mL)
NaOH, 10 M (300 mL)
Zn, mossy (6 g)

Procedures

Preparation and standardization of 0.1 *M* hydrochloric acid solution
Use the procedure in Experiment 4.

Preparation and standardization of 0.1 *M* sodium hydroxide solution
Use the procedure in Experiment 3.

Analysis of the unknown

1. Weigh accurately three samples (of a size determined by the instructor) and transfer each to the center of a 9-cm non-acid-washed filter paper.
2. Wrap the samples in the filter papers and place in dry 500-mL Kjeldahl flasks. Wrapping the samples prevents any solid from sticking to the neck of the flask.
3. Add about 10 g of anhydrous Na_2SO_4, a crystal of $CuSO_4$ (catalyst), and about 25 mL of concentrated H_2SO_4.
4. Clamp the flask at a 45° angle in the hood and heat gently until the initial frothing diminishes. Then gradually increase the heat until the solution reaches a vigorous boil. Maintain boiling for 15 min after the solution turns colorless or a pale yellow (this can take several hours).
5. Cool the contents of the flask, *cautiously* add 100 mL of distilled water, and cool in an ice bath. Add another 100 mL of water and cool again.
6. Assemble the Kjeldahl distillation apparatus (Figure 7-13). Pipet 50 mL of the standard 0.1 *M* HCl into the 250-mL receiver flask. Position the flask so that the tip of the condenser is completely immersed in the acid solution.
7. Begin water circulation through the condenser. Add 2 drops of methyl red indicator followed by about 100 mL of 10 *M* NaOH. Pour the NaOH down the side of the Kjeldahl flask *slowly* so that it does not mix with the solution and forms a second layer.
8. Quickly add 1 to 2 g of mossy zinc, to prevent bumping during the distillation, and *immediately* connect the flask to the spray trap.
9. Mix the two layers very slowly by gently swirling the flask. Be especially careful not to point the opening of the flask toward yourself or another person.
10. Quickly bring the solution to a boil and continue distilling until at least 100 mL has been transferred to the receiving flask. You may wish to mark the receiving flask for this volume in advance.
11. Continue to heat while lowering the receiving flask until the tip of the condenser is well above the solution. Then disconnect the Kjeldahl flask and turn off the burner. Remove the spray trap and rinse the inside of the condenser with several small portions of distilled water, collecting the rinsings in the receiving flask.
12. Add 2 drops of methyl red indicator and titrate with the standard 0.1 *M* NaOH solution.
13. Calculate the % N in the sample.

EXPERIMENT 6: DETERMINATION OF ANILINE BY NONAQUEOUS ACID-BASE TITRATION

Many amines, including aniline, cannot be titrated in water because they are too weakly basic. However, in a solvent of much lower basicity such as acetic acid, their base strength is enhanced sufficiently to enable their titration with perchloric acid. Although samples may be dissolved directly in pure acetic acid, the perchloric acid titrant must be treated in some way to remove its water. Concentrated perchloric acid contains about 28% water. Most of this water is converted to acetic acid by reaction with acetic anhydride. The perchloric acid is standardized with potassium hydrogen phthalate, which acts as a base in acetic acid solvent.

Although it is a weak acid, pure acetic acid can cause severe skin irritation. Be careful handling this reagent. Another problem with acetic acid is its large temperature coefficient of expansion ($0.11\%/°C$). Care must be taken to ensure that the temperature of the titrant is essentially the same during both the standardization and determination steps. The following procedure is written for aniline, but it can be used for other amines as well.

Reactions

Standardization:

$$KHC_8H_4O_4 + HClO_4 \longrightarrow H_2C_8H_4O_4 + KClO_4$$

Titration:

$$C_6H_5NH_2 + HClO_4 \longrightarrow C_6H_5NH_3{}^+ClO_4{}^-$$

Reagents

$HClO_4$, concentrated, 72% (4.3 mL)
CH_3CO_2H, concentrated (655 mL)
$(CH_3CO_2)_2O$ (10 mL)
$KHC_8H_4O_4$, primary-standard grade (5 g)
Methyl violet, 0.2% (w/v) in CH_3CO_2H (12 drops)

Procedures

Preparation and standardization of 0.1 *M* perchloric acid

1. With a measuring pipet, transfer 4.3 mL of concentrated $HClO_4$ to a 250-mL conical flask containing 100 mL of concentrated acetic acid. Add 10 mL of acetic anhydride, mix, and allow to stand for at least an hour and preferably overnight.
2. Transfer the solution to a storage bottle and dilute to about 500 mL with pure acetic acid. Allow the mixture to cool to room temperature before using.

3. Dry about 5 g of potassium hydrogen phthalate at 110°C for 1 h and cool in a des-iccator.

4. Weigh accurately three samples of dried $KHC_8H_4O_4$, 0.5 to 0.7 g each, and transfer to numbered 250-mL conical flasks.

5. Add 60 mL of concentrated acetic acid to each flask and heat to near boiling in a hood until dissolution is complete. Cool to room temperature.

6. Add 2 drops of methyl violet indicator to each flask and titrate with the perchloric acid solution until a violet-to-blue (not green or yellow) color change occurs.

7. Consult with your instructor about how to dispose of the solutions. Calculate the molarity of the perchloric acid solution.

Analysis of the unknown

1. Transfer three portions of sample, each containing between 3 and 4 mmol of aniline (ask your instructor for appropriate weights or volumes), to numbered 250-mL con-ical flasks and dilute each to a total volume of about 75 mL with pure acetic acid.

2. Add 2 drops of methyl violet and titrate with 0.1 M perchloric acid to the same color change observed in the standardization titration.

3. Report the concentration of aniline as mmol aniline/mL sample.

EXPERIMENT 7: DETERMINATION OF THE TOTAL HARDNESS OF WATER BY COMPLEXATION TITRATION

The two ions in water that are primarily responsible for its "hardness," Ca^{2+} and Mg^{2+}, can be titrated together with the complexing agent, EDTA. The procedure presented here works well with either Eriochrome Black T or Calmagite indicator. Neither indi-cator functions well in the titration of Ca^{2+} alone. Consequently, a small amount of Mg^{2+} is added to the titrant before its standardization with pure, primary standard cal-cium carbonate. Although the sum of the calcium and magnesium ions is determined, the hardness is usually calculated as parts per million $CaCO_3$ (mg $CaCO_3$/L solution).

Reactions

$$Ca^{2+} + HY^{3-} \xrightarrow{pH\ 10} CaY^{2-} + H^+$$

$$Mg^{2+} + HY^{3-} \xrightarrow{pH\ 10} MgY^{2-} + H^+$$

Reagents

$Na_2H_2Y \cdot 2H_2O$ (1 g)

NaOH (0.5 g)

$MgCl_2 \cdot 6H_2O$ (0.05 g)

$CaCO_3$, primary-standard grade (2 g)

HCl, concentrated (15 mL)

NH$_4$Cl/NH$_3$ buffer, 3.2 g NH$_4$Cl + 28.5 mL concentrated NH$_3$ diluted to 500 mL (60 mL)

Eriochrome Black T, 0.5% (w/v) in C$_2$H$_5$OH (fresh each month) (2 mL)

Procedures

Preparation and standardization of 0.005 *M* EDTA

1. Dissolve about 1.0 g of Na$_2$H$_2$Y·2H$_2$O in 450 mL of distilled water containing 0.5 g of NaOH and 0.05 g of MgCl$_2$·6H$_2$O. Dilute to 500 mL and mix well. Transfer to a labeled storage bottle.

2. Dry 2 g of primary-standard-grade CaCO$_3$ at 110°C for 1 h and cool in a desiccator.

3. Weigh accurately 0.20 to 0.25 g of the dry CaCO$_3$ and transfer to a 250-mL conical flask.

4. Carefully add 5 mL of concentrated HCl solution. After dissolution is complete, add about 50 mL of distilled water, rinsing down the sides of the flask in the process. Boil for 5 min to remove carbon dioxide. Transfer the cooled solution and rinsings to a 500-mL volumetric flask and dilute to the mark with distilled water. Mix well.

5. Pipet three 25-mL portions of the calcium solution into 250-mL conical flasks. Add 10 mL of the ammonia buffer to each flask.

6. To the first flask, add 5 drops of Eriochrome Black T indicator and titrate with the 0.005 *M* EDTA to the first appearance of a clear blue color. Ordinarily, the initial wine-red color turns slowly to purple and then sharply to blue. Repeat the titration with the remaining two solutions.

7. Calculate the average molarity of the EDTA.

Analysis of the unknown

1. Pipet three 50-mL portions of the water sample into 250-mL conical flasks and add 10 mL of the ammonia buffer to each sample.

2. Titrate as in step 6 of the standardization procedure.

3. Calculate the total hardness as ppm CaCO$_3$.

EXPERIMENT 8: DETERMINATION OF MAGNESIUM, ZINC, AND MERCURY IN A MIXTURE BY COMPLEXATION TITRATION

This procedure uses the data from three titrations to determine the concentration of three metal ions in a mixture. Although magnesium forms the weakest complex with EDTA, it can be titrated in the presence of the other two metals if they are first complexed with cyanide ion. Once this titration has been completed, zinc can be released from its cyanide complex by addition of excess formaldehyde. The mercury cyanide complex is too stable to react with formaldehyde. The liberated zinc then can be titrated with EDTA. Mercury is determined by taking the difference between the total amount and the amounts of magnesium and zinc. The total cannot be determined by direct titration with EDTA because mercury forms too strong a complex with the EBT indicator. This problem is overcome by using a back titration in which excess EDTA is added to a mixture of the three ions before the indicator is added, and the remaining EDTA is titrated with a standard magnesium solution. Once the mercury–EDTA complex is formed, it reacts only slowly with free EBT indicator.

This experiment calls for the use of cyanide. Students must remember that cyanide is toxic and HCN gas is easily evolved from acidic solutions of cyanide salts. Be especially careful when handling this reagent. Clean up any spills immediately, following the directions of your instructor, and then wash your hands. Work in the hood and make sure that you never add a cyanide salt to an acidic solution or pour a cyanide solution down the drain, where it may come in contact with an acidic solution.

Reactions

Masking:

$$M^{2+} + 4CN^- \longrightarrow M(CN)_4^{2-} \quad \text{where M = Zn or Hg}$$

Titration of magnesium:

$$Mg^{2+} + HY^{3-} \longrightarrow MgY^{2-} + H^+$$

Demasking:

$$Zn(CN)_4^{2-} + 4HCHO + 4H_2O \longrightarrow Zn^{2+} + 4HOCH_2CN + 4OH^-$$

Titration of zinc:

$$Zn^{2+} + HY^{3-} \longrightarrow ZnY^{2-} + H^+$$

Back titration for total:

$$M^{2+} + \text{excess } HY^{3-} \longrightarrow MY^{2-} + H^+ \quad \text{where M = Mg, Zn, or Hg}$$

$$HY^{3-} + Mg^{2+} \longrightarrow MgY^{2-} + H^+$$

Reagents

$Na_2H_2Y \cdot 2H_2O$ (2 g)

NaOH (0.5 g)

$MgCl_2 \cdot 6H_2O$ (0.55 g)

$CaCO_3$, primary-standard grade (2 g)

HCl, 6 M (15 mL)

NH_4Cl/NH_3 buffer, 3.2 g NH_4Cl + 28.5 mL concentrated NH_3 diluted to 500 mL (105 mL)

Eriochrome Black T indicator, 0.5% (w/v) in C_2H_5OH (fresh each month) (3 mL)

KCN (1.5 g)

CH_3CO_2H, concentrated (5 mL)

HCHO, 40% (15 mL)

Procedures

Preparation and standardization of 0.005 *M* EDTA

Use the procedure given in Experiment 7 except adjust the amounts to make 1 L.

Preparation and standardization of 0.005 *M* MgCl$_2$

1. Dissolve about 0.5 g of MgCl$_2$·6H$_2$O in water and dilute to 500 mL. Transfer to a storage bottle.
2. Pipet three 25-mL aliquots of the standard 0.005 *M* EDTA solution into separate 250-mL conical flasks. Add 25 mL of distilled water to each flask followed by 10 mL of the ammonia buffer and 5 drops of Eriochrome Black T indicator. Titrate each solution with the 0.005 *M* MgCl$_2$ until the first appearance of a pink color.
3. Calculate the molarity of the MgCl$_2$ solution.

Determination of magnesium and zinc

1. Pipet 10 mL of your unknown, containing Mg^{2+}, Zn^{2+}, and Hg^{2+}, into a 250-mL conical flask. Add about 50 mL of distilled water and 10 mL of the ammonia buffer. Check the solution with pH paper to ensure that it is about pH 10.
2. Weigh about 0.5 g of KCN and add it to the basic sample **in the hood**. Again, check the solution with pH paper to ensure that it is still about pH 10.
3. Add 5 drops of Eriochrome Black T indicator and titrate with the 0.005 *M* EDTA to the first appearance of a clear blue color.
4. Calculate the amount of magnesium in the sample.
5. To the solution from step 3, add 3 mL of a freshly prepared mixture of 5 mL of concentrated acetic acid, 15 mL of 40% formaldehyde, and 20 mL of distilled water. Check to make sure that the pH is still 10. If not, add 2 mL of ammonia buffer and check again. The indicator will turn red.
6. Titrate immediately with the 0.005 *M* EDTA to the first appearance of a clear blue color.
7. Calculate the amount of zinc in the sample.
8. Obtain directions from your instructor regarding the proper way to dispose of your solution.
9. Repeat steps 1 to 8 with two more aliquots of sample.

Determination of magnesium, zinc, and mercury total

1. Pipet 10 mL of sample into a 250-mL conical flask and add exactly 75 mL of the 0.005 *M* EDTA, followed by 5 mL of ammonia buffer.
2. Add 5 drops of Eriochrome Black T indicator. The solution should be clear blue due to an excess of EDTA. If it is not, pipet another 25 mL of the EDTA into the flask. Back titrate the excess EDTA with standard 0.005 *M* MgCl$_2$ to the appearance of a purple color. Do not try to titrate to a pink color because before enough pink magnesium-indicator complex forms, it will react irreversibly with HgY^{2-}.

3. Calculate the total amount of the three metals present in the sample and then determine the amount of mercury by difference. Report the results of all three metals in units of parts per million.

EXPERIMENT 9: DETERMINATION OF IRON IN ORE BY REDOX TITRATION

Common iron ores, such as hematite (Fe_2O_3), limonite ($Fe_2O_3 \cdot 3H_2O$), and magnetite (Fe_3O_4), generally dissolve completely in concentrated hydrochloric acid, yielding ferric ions. Iron ores containing silicates will leave a residue of white, hydrated silica, but this does not interfere with the determination. This method uses stannous chloride to prereduce iron(III) to iron(II), which subsequently is titrated with standard potassium dichromate. A Jones or Walden reductor may also be used for prereduction of iron(III). The excess stannous chloride is removed by the addition of mercuric chloride:

$$SnCl_2 \underset{\text{(small amount)}}{} + 2HgCl_2 \longrightarrow SnCl_4 + Hg_2Cl_2(s)$$

The silky, white, insoluble mercurous chloride does not react with the potassium dichromate titrant, and the excess mercuric chloride does not reoxidize iron(II).

The addition of too great an excess of stannous chloride may lead to a different reaction when mercuric chloride is added:

$$SnCl_2 \underset{\text{(large amount)}}{} + HgCl_2 \longrightarrow SnCl_4 + Hg(l)$$

The metallic mercury appears as finely divided black particles that can be oxidized by the titrant. This reaction is avoided by careful control of the amount of excess stannous chloride.

Reactions

Pretreatment:

$$FeCl_3 + SnCl_2 \longrightarrow FeCl_2 + SnCl_4$$

$$SnCl_2 + 2HgCl_2 \longrightarrow SnCl_4 + Hg_2Cl_2(s)$$

Titration:

$$6FeCl_2 + K_2Cr_2O_7 + 14HCl \longrightarrow 6FeCl_3 + 2CrCl_3 + 2KCl + 7H_2O$$

Reagents

$K_2Cr_2O_7$, primary-standard grade (5 g)

HCl, concentrated (30 mL)

$SnCl_2$, 0.25 M in 4 M HCl (6 mL)

$HgCl_2$, 0.25 M (30 mL)

H_2SO_4, concentrated (6 mL)

H$_3$PO$_4$, concentrated (15 mL)

Sodium diphenylaminesulfonate indicator, 0.2% (w/v) (1 mL)

Procedures

Preparation of standard 0.017 *M* potassium dichromate

1. Dry about 5 g of primary-standard-grade K$_2$Cr$_2$O$_7$ at 150°C for 2 h and cool in a desiccator.

2. Weigh accurately about 2.5 g of the dry K$_2$Cr$_2$O$_7$, transfer to a small beaker and dissolve in about 50 mL of distilled water. Transfer to a 500-mL volumetric flask and dilute to the mark with distilled water. Mix well and transfer to a labeled storage bottle.

3. Calculate the molarity of this solution.

Analysis of the unknown

1. Dry the ore at 110°C for at least 2 h and cool in a desiccator.

2. Weigh three ore samples of appropriate size (see your instructor) and transfer to numbered 500-mL conical flasks.

3. Add 10 mL of concentrated HCl, cover each flask with a watch glass, and heat gently (in the hood). To the hot solution, add stannous chloride dropwise until the yellow color of FeCl$_4$$^-$ just disappears.

4. Continue heating, and if the yellow color reappears, add more stannous chloride as in step 3. Continue this treatment until dissolution of the ore seems complete. Examine the residue. A white, flocculent residue is probably silica and complete dissolution of the iron can be assumed. Treat each sample individually from this point.

5. If a yellow color has reappeared on standing, add stannous chloride dropwise to the hot sample until the yellow color just disappears. Avoid adding more than 1 to 2 drops in excess. Wash the bottom of the watch glass, allowing the water to fall into the flask. Also wash down the sides of the flask.

6. Cool the solution and add about 50 mL of water, followed rapidly by 10 mL of the HgCl$_2$ solution. A small quantity of a silky, white precipitate should appear. If no precipitate appears, insufficient SnCl$_2$ was added. If a gray precipitate forms, too much excess SnCl$_2$ was present. In either case, the mistake cannot be corrected and the solution should be discarded.

7. Immediately add 200 mL of distilled water, 2 mL of concentrated H$_2$SO$_4$, 5 mL of concentrated H$_3$PO$_4$, and 8 drops of the diphenylaminesulfonate indicator.

8. Titrate with standard K$_2$Cr$_2$O$_7$ until the indicator changes from green to violet.

EXPERIMENT 10: DETERMINATION OF CALCIUM IN LIMESTONE BY REDOX TITRATION

Calcium is determined indirectly by precipitation with oxalate ion, dissolution in acid, and titration of the redissolved oxalate with potassium permanganate. The method works well with samples that contain alkali metals or magnesium but not other metals, most of which interfere either by forming insoluble oxalates or coprecipitating with the calcium oxalate.

The most serious source of error in this method stems from the formation of an impure calcium oxalate precipitate. One of the best ways to improve the purity and filterability of this precipitate is to carry out the precipitation under conditions that favor crystal growth over nuclei formation. In this procedure, the calcium oxalate is precipitated homogeneously from a solution made basic gradually by the hydrolysis of urea. The precipitate is filtered to separate it from excess oxalate and redissolved in acid for the titration with permanganate.

Reactions

Precipitation:

$$CO(NH_2)_2 + H_2O \xrightarrow{\Delta} CO_2 + 2NH_3$$
$$H_2C_2O_4 + 2NH_3 \longrightarrow 2NH_4^+ + C_2O_4^{2-}$$
$$Ca^{2+} + C_2O_4^{2-} + H_2O \longrightarrow CaC_2O_4 \cdot H_2O(s)$$

Dissolution:

$$CaC_2O_4 \cdot H_2O + H_2SO_4 \longrightarrow CaSO_4 + H_2C_2O_4 + H_2O$$

Titration:

$$5H_2C_2O_4 + 2KMnO_4 + 3H_2SO_4 \longrightarrow 2MnSO_4 + 10CO_2(g) + K_2SO_4 + 8H_2O$$

Reagents

$KMnO_4$ (3.2 g)
$CaCO_3$, primary-standard grade (2 g)
HCl, concentrated (140 mL)
$(NH_4)_2C_2O_4$, 0.5 M (180 mL)
NH_3, 6 M (120 mL)
Methyl red indicator, 0.02% (w/v) in 60% (v/v) ethanol (12 drops)
$CO(NH_2)_2$ (90 g)
$(NH_4)_2C_2O_4$, 0.01 M (200 mL)
H_2SO_4, 4 M (120 mL)
HCl, 0.1 M (135 mL)
Br_2 water, saturated solution (15 mL)
NH_4NO_3, 0.1 M (60 mL)

Procedures

Preparation and standardization of 0.02 *M* potassium permanganate

1. Dissolve about 3.2 g of $KMnO_4$ in 1 L of distilled water.
2. Heat to just below boiling for 1 h, cover, and let stand overnight (or longer).

3. Remove $MnO_2(s)$ by filtering through a fine-porosity sintered-glass filter crucible.

4. Store the filtered solution in a glass-stoppered, amber bottle that has been cleaned with dichromate cleaning solution and rinsed thoroughly with distilled water and several small portions of the permanganate solution. Place a small beaker over the stopper to prevent dust from collecting on the lip of the stopper and getting into the solution.

5. Dry about 2 g of primary-standard-grade $CaCO_3$ at 110°C for 1 h and cool in a desiccator.

6. Weigh accurately three 0.20- to 0.25-g samples of dry $CaCO_3$ and transfer to separate 400-mL beakers.

7. To each beaker, carefully add 15 mL of concentrated HCl, taking care to avoid losing any sample due to spattering. Quickly cover with a watch glass. When the effervescence ceases, wash down the sides of the beakers and the bottom of the watch glasses. Place the beakers on a hot plate and heat until the samples dissolve completely.

8. Dilute to 200 mL with distilled water, add 30 mL of 0.5 M ammonium oxalate, and reheat to boiling.

9. Add 6 M NH_3 slowly until a precipitate just begins to form. Then add 2 drops of methyl red indicator, 5 drops of concentrated HCl, and 15 g of urea. Continue heating until the indicator changes from red to yellow.

10. Cool to room temperature and filter through a medium-porosity sintered-glass filter crucible. Wash the calcium oxalate precipitate 5 to 10 times with small portions of 0.01 M $(NH_4)_2C_2O_4$ and 3 times with hot water.

11. Transfer the crucibles and contents to separate 600-mL beakers and add 200 mL of hot 4 M H_2SO_4. Leave the crucibles in the beakers.

12. Heat the solutions to about 85°C and titrate with 0.02 M $KMnO_4$. Use a stirring rod to mix the solutions. A faint pink color of excess MnO_4^- that persists for at least 30 s is taken as the end point. The temperature of the solution should be kept above 60°C during the entire titration, and it may take a minute or two for the first portion of titrant to react.

13. Calculate the molarity of the $KMnO_4$.

Analysis of the unknown

1. Dry the unknown at 110°C for 1 h and cool in a desiccator.

2. Weigh accurately three samples of approximately 0.5 g each and transfer to numbered 250-mL beakers.

3. Moisten each sample with a few drops of water and carefully add 15 mL of concentrated HCl, taking care to avoid losing sample due to spattering. Quickly cover the beakers with watch glasses.

4. When the effervescence ceases, wash down the sides of the beakers and the bottom of the watch glasses. Place the beakers on a hot plate (in the hood) and evaporate to dryness. Elevate the watch glasses on glass hooks to speed this process. Transfer to an oven at 110°C and bake for 1 h.

5. To each baked residue, add 10 ml of concentrated HCl, 90 mL of hot water, and

mix well. Filter through medium-porosity filter paper. Wash 15 times with 3-mL portions of hot 0.1 M HCl and twice with hot water. Collect the filtrate and washings in a clean 250-mL beaker. Discard the residues.

6. Add 6 M NH$_3$ dropwise until a precipitate just begins to form. Add 5 mL of bromine water to oxidize ferrous and manganous ions and boil for 5 min.

7. Add 2 drops of methyl red indicator and just enough 6 M NH$_3$ to make the solutions *slightly* basic (yellow). After 5 min, filter the hot solutions through coarse, ashless filter paper. Wash each precipitate six times with 3-mL portions of 0.1 M NH$_4$NO$_3$, and collect the filtrate and washings in clean 400-mL beakers. Discard the precipitates.

8. Add 5 mL of concentrated HCl and 30 mL of 0.5 M (NH$_4$)$_2$C$_2$O$_4$ to each filtrate. Heat to just below boiling and add 6 M NH$_3$ slowly until a precipitate just begins to form. Then add 5 drops of concentrated HCl and 15 g of urea. Continue heating until the indicator changes from red to yellow.

9. Proceed as in steps 10 to 12 of the standardization procedure.

10. Calculate the % CaO in the limestone sample.

EXPERIMENT 11: DETERMINATION OF COPPER IN ORE BY IODOMETRIC REDOX TITRATION

This procedure, which is suitable for the determination of copper in ores, involves the reaction of cupric ions with iodide to produce iodine, which is titrated with standard sodium thiosulfate. Oxides of nitrogen produced during dissolution of the ore sample are removed by reaction with urea. The reaction between cupric and iodide ions is carried out in a solution buffered with NH$_4$HF$_2$, which produces a pH of about 3.5. The choice of this particular buffer is quite propitious, serving to avoid three problems. Above pH 4, cupric ion begins to hydrolyze and the hydrolysis products react slowly and incompletely with iodide ion, leading to low results. Below about pH 3, arsenic(V), if present, is reduced by iodide ion and the iodine produced leads to high results. Finally, the fluoride ion complexes any iron(III) present and prevents it from oxidizing iodide ion.

The solid cuprous iodide that is formed during the oxidation of iodide absorbs some of the iodine formed, releasing it only slowly at the end point. This can cause the color change to be gradual at the equivalence point. Potassium thiocyanate is added to displace the iodine, resulting in a sharper color change and more accurate estimate of the equivalence point.

Reactions

Pretreatment:

$$3Cu(s) + 8HNO_3 \longrightarrow 3Cu(NO_3)_2 + 2NO + 4H_2O$$
$$\downarrow$$
$$HNO_2$$

$$2HNO_2 + CO(NH_2)_2 \longrightarrow 3H_2O + CO_2(g) + 2N_2(g)$$

$$2Cu^{2+} + 4I^- \longrightarrow 2CuI(s) + I_2$$

Titration:

$$I_2 + 2S_2O_3^{2-} \longrightarrow S_4O_6^{2-} + 2I^-$$

Reagents

$Na_2S_2O_3 \cdot 5H_2O$ (25 g)

NaOH (1 pellet)

Cu wire, pure, No. 14 or 16 (2 in.)

HNO_3, 6 M (25 mL)

Urea (1.5 g)

NH_3, 6 M (60 mL)

CH_3CO_2H, concentrated (30 mL)

KI (18 g)

Starch indicator, 1% (w/v), boiled and filtered (30 mL)

KSCN (9 g)

HCl, concentrated (30 mL)

HNO_3, concentrated (15 mL)

H_2SO_4, concentrated (15 mL)

Br_2 water, saturated (15 mL)

NH_4HF_2 (6 g)

Procedures

Preparation and standardization of 0.1 *M* sodium thiosulfate

1. Dissolve about 25 g of $Na_2S_2O_3 \cdot 5H_2O$ in 1 L of distilled water that has just been boiled for 5 min and cooled. Transfer to a clean storage bottle that was rinsed with the boiling water and add 1 pellet of NaOH as a preservative.

2. Dip a 2-in. length of pure copper wire (14 or 16 gauge) in dilute nitric acid for 5 s, rinse with water, and wipe dry with a piece of filter paper.

3. Weigh accurately three 0.20- to 0.25-g samples of wire and transfer to numbered, 250-mL conical flasks. In the hood, add 5 mL of 6 M HNO_3 and warm until the copper is dissolved.

4. Dilute with 25 mL of water, add 0.5 g of urea, and boil again for 5 min.

5. Add 6 M NH_3 until deep blue $Cu(NH_3)_4^{2+}$ starts to form. A precipitate of $Cu(OH)_2$ will form and redissolve before the deep blue color appears. Add 5 mL of concentrated acetic acid. Treat each sample individually from this point.

6. Add 3 g of KI and titrate the liberated iodine immediately with the 0.1 M thiosulfate solution, continuing until the solution is a very pale yellow.

7. Interrupt the titration to add 5 mL of starch indicator, then continue the addition of titrant dropwise until the blue color just disappears with the addition of 1 drop.

8. Interrupt the titration again and add 1.5 g of KSCN. The blue color will reappear. Continue to add titrant dropwise until the blue color disappears for at least 20 s.

9. Calculate the molarity of the titrant.

Analysis of the unknown

1. Dry the sample at 110°C for 2 h and cool in a desiccator.

2. Weigh accurately three samples of an appropriate weight so that about 35–45 mL of titrant is used (see your instructor) and transfer to numbered 250-mL conical flasks.

3. *In the hood*, add 10 mL of concentrated HCl and 5 mL of concentrated HNO_3. Mix and warm until the ore dissolves. Do not mistake yellow sulfur, which may be present in some samples, for undissolved ore.

4. When dissolution is complete, add 5 mL of concentrated sulfuric acid and boil until thick white fumes of SO_3 are evolved. Cool and add 50 mL of water and 5 mL of bromine water. Mix well and boil to remove the excess, unreacted bromine.

5. Cool and add 6 M NH_3 until either deep blue $Cu(NH_3)_4^{2+}$ or insoluble $Fe(OH)_3$ begins to form. Add 5 mL of concentrated CH_3CO_2H and 2 g of NH_4HF_2. Mix well and treat each sample individually from this point.

6. Continue as indicated in steps 6 to 8 of the standardization procedure.

7. Calculate the % Cu in the ore.

EXPERIMENT 12: DETERMINATION OF CHLORIDE BY PRECIPITATION TITRATION

The amount of chloride is determined by measuring the amount of silver required to precipitate it as silver chloride. Dextrin is added to the solution to prevent the precipitate from coagulating. Prior to the equivalence point, the colloidal silver chloride particles are negatively charged, due to their adsorption of excess chloride ion. When the equivalence point is past, the particles become positively charged due to adsorption of excess silver ion. As a result, the indicator anion is attracted to the counter-ion layer, where its electronic configuration is altered slightly, causing it to change from colorless to red.

Reaction

$$NaCl + AgNO_3 \longrightarrow AgCl(s) + NaNO_3$$

Reagents

$AgNO_3$, primary-standard grade (1 g)

Dextrin, 0.1% (w/v) (15 mL)

Dichlorofluorescein indicator, 0.1% (w/v) in 75% (v/v) ethanol (1.5 mL)

Procedures

Preparation of standard 0.02 *M* silver nitrate

1. Dry 1 g of primary-standard-grade silver nitrate at 110°C for 1 h (not longer) and cool in a desiccator. **Caution:** Silver nitrate will stain skin and fingernails a dark brown. Avoid direct contact with the solid or its solutions.
2. Weigh accurately about 0.85 g of the dried reagent, transfer to a small beaker, and dissolve in about 25 mL of distilled water. Transfer the solution to a 250-mL volumetric flask and dilute to the mark with distilled water. Mix thoroughly and transfer the solution to a clean amber bottle for storage.

Analysis of the unknown

1. Dry the sample at 110°C for 1 h and cool in a desiccator.
2. Weigh accurately three samples of appropriate weight (see your instructor) and transfer to numbered, 250-mL conical flasks.
3. Add 50 mL of distilled water, 5 mL of the dextrin solution, and 0.5 mL of dichlorofluorescein indicator. Mix well.
4. Titrate slowly with the silver nitrate solution, swirling vigorously, until a permanent pink color is observed. The end-point color change is much sharper when the titration vessel is protected from direct illumination of sunlight.
5. Report the % Cl in the sample.

EXPERIMENT 13: DETERMINATION OF FLUORIDE BY POTENTIOMETRIC TITRATION

Fluoride ion is titrated with lanthanum nitrate to form insoluble lanthanum fluoride. The progress of the titration is followed by measuring the potential of a fluoride ion-selective electrode and the end point is determined graphically. There are few interferences in the method if the pH is kept between 4 and 10. Below pH 4, significant amounts of molecular hydrofluoric acid form, to which the electrode does not respond. Above pH 10, the electrode response toward hydroxide becomes significant.

Reaction

$$La(NO_3)_3 + 3NaF \longrightarrow LaF_3(s) + 3NaNO_3$$

Apparatus

Fluoride ion-selective electrode
Calomel reference electrode
pH meter
Magnetic stirrer

Reagents

La$(NO_3)_3 \cdot 6H_2O$ (1.1 g)

NaF (2 g)

CH_3CO_2H/CH_3CO_2Na buffer, pH 5.0, total acetate = 0.10 M (40 mL)

Procedures

Preparation and standardization of 0.01 M lanthanum nitrate

1. Dissolve about 1.1 g of La$(NO_3)_3 \cdot 6H_2O$ in 250 mL of distilled water and transfer to a storage bottle.
2. Dry 2 g of reagent-grade NaF at 110°C for 2 h and cool in a desiccator.
3. Weigh accurately 0.8 to 0.9 g of dry NaF, dissolve in a small amount of distilled water, and transfer to a 1-L volumetric flask. Dilute to volume with distilled water. Mix and transfer immediately to a labeled plastic storage bottle.
4. Pipet four 25-mL aliquots of the standard NaF solution into separate 125-mL beakers and add 5 mL of the acetic acid–sodium acetate buffer to each solution.
5. Position the first beaker on the magnetic stirrer and place the fluoride-ion and calomel electrodes in the solution. Make sure there is sufficient clearance for the magnetic stirring bar to turn without striking the electrodes.
6. Set the meter to the millivolt range, turn on the stirrer, and titrate slowly with the 0.01 M La$(NO_3)_3$. Watch the meter and note the approximate volume of titrant required to reach the point where a large rapid change in potential occurs. This is the approximate end-point volume.
7. Repeat the titration with each of the three remaining solutions. Record the electrode potential at exactly 0.5-mL intervals until 1 mL prior to the estimated end point. Then take readings at exactly 0.05-mL intervals until 1 mL past the large rapid change in potential. Continue to titrate until another 5 mL of titrant has been added, again taking readings at exactly 0.5-mL intervals. Rinse the electrodes and stirring bar with distilled water between each titration.
8. Discard the solutions and rinse the electrodes and beakers immediately after you are finished. Do not let fluoride solutions stand in contact with glass for long periods.
9. Plot regular, first-, and second-derivative titration curves and determine the end-point volume from each plot. Show the plots to your instructor and be prepared to discuss which end point should be used to calculate the molarity of the lanthanum nitrate.

Analysis of the unknown

1. Transfer your sample to a 250-mL volumetric flask and dilute to the mark with distilled water.
2. Titrate the unknown as in steps 4 through 8 of the standardization procedure.
3. Plot regular, first-, and second-derivative titration curves and determine the end-point volume from each. Calculate the concentration of fluoride as milligrams of F per 25 mL of sample.

EXPERIMENT 14: DETERMINATION OF THE IONIZATION CONSTANT OF A WEAK ACID BY POTENTIOMETRIC TITRATION

The potentiometric titration curve of a weak acid is recorded using a glass pH electrode. The pK_a for each titratable hydrogen is taken as the pH when 50% of that hydrogen has been titrated. The number of observed end points and the calculated pK_a values may be used to identify the acids in an unknown.

Reaction

$$HA + NaOH \longrightarrow NaA + H_2O$$

Apparatus

pH meter
Glass pH electrode
Calomel reference electrode
Magnetic stirrer

Reagents

Standard buffer, pH 7.00 (20 mL)
Potassium hydrogen tartrate standard buffer, saturated solution (20 mL)
NaOH, 50% (w/v), carbonate-free (7 mL)

Procedures

Preparation of 0.1 M sodium hydroxide

Prepare this reagent according to the procedure in Experiment 3. It does not have to be standardized.

Determination of ionization constant

1. Transfer about 20 mL of the pH 7 standard buffer to a clean and dry 125-mL beaker. Set the pH meter to standby and place the pH and calomel electrodes in the solution, making sure that the tips of both electrodes are fully immersed. Swirl the solution gently.
2. Set the meter to the pH scale and adjust the standardization dial until a value of 7.00 is displayed.
3. Return the meter to the standby position. Remove the electrodes, rinse with distilled water, and pat dry with a tissue. Place the electrodes in about 20 mL of standard potassium hydrogen tartrate buffer and adjust either the slope or temperature dial (see your instructor or the instrument manual) until the meter displays a value of 3.56.
4. Return the meter to the standby position. Remove and rinse the electrodes with distilled water.

5. Your sample may be a liquid or a solid. Follow the directions from your instructor for the amounts to be used. Weigh or pipet three portions of sample and transfer to separate 125-mL beakers. Add 25 mL of distilled water to each beaker.

6. Mount the first beaker on the magnetic stirrer and place the pH and calomel electrodes in the solution, making sure that they are immersed but high enough for the stirring bar to rotate freely.

7. Set the meter to the pH position and titrate with the 0.1 M NaOH solution. Add the titrant in about 0.5-mL increments, taking the pH reading after each increment, until the pH begins to rise sharply. Throughout the region of sharply rising pH, add the titrant in increments of about 0.05 mL. If the pH does not rise to nearly 13 after the end point, continue titrating until the second break has been passed and the pH is nearly constant.

8. Return the meter to the standby position. Remove the electrodes, rinse, and pat dry. Repeat steps 6 and 7 for the remaining samples.

9. Plot the titration curve of pH versus volume of NaOH added and determine the K_a for each titratable hydrogen.

10. From the experimental K_a value(s), identify your acid. Look up the accepted value or values in the literature and calculate the relative error of your results.

EXPERIMENT 15: DETERMINATION OF PHENOL BY COULOMETRIC TITRATION

Phenol is determined by measuring the amount of bromine needed for its reaction. The bromine is generated electrolytically in the titration vessel by the oxidation of bromide ion at a platinum anode. By carrying out the electrolysis at constant current, the amount of bromine required to reach the end point can be obtained from the measured time of electrogeneration.

The end point may be determined amperometrically or potentiometrically. In the amperometric method, a potential difference of 200 to 300 mV is applied across two small, identical platinum electrodes. This voltage is not large enough to cause electrolysis in the analyte solution and no current flows (electrodes are polarized). Bromine, however, is reduced at the positive electrode, and when it appears in the solution just past the equivalence point, the electrodes become depolarized and a small current flows. In the potentiometric method, a calomel reference electrode is substituted for one of the small platinum electrodes and the electrode potential of the Br_2/Br^- half-cell is measured.

Reactions

$$2Br^- \xrightarrow{\text{electrolysis}} Br_2 + 2e^-$$

Apparatus

Platinum generating electrode, at least 10 cm² surface area

Platinum counter electrode, No. 14 wire

Platinum indicator electrode (2), amperometry

Calomel reference electrode

Coulometer or constant current source with timer

Amperometer

Magnetic stirrer

pH meter

Reagent

NaBr, 1.0 M (80 mL)

Procedure

1. Assemble the cell as shown in Figure 24-2 and place on a magnetic stirrer. Add 1.0 M NaBr to the tube containing the counter electrode until the level is 1 to 2 cm below the cap. Connect the platinum generating electrode to the positive output terminal and the counter electrode to the negative output terminal of the coulometer. Connect the platinum indicator electrodes to the amperometer. Leave the calomel reference electrode unconnected for the time being.

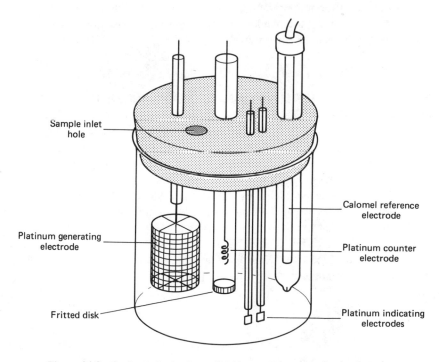

Figure 24-2 Coulometric titration cell with amperometric end-point detection.

Experiment 15

2. Transfer about 50 mL of water and 5 mL of the NaBr solution to the cell. Set the amperometer output to 200 mV and start the stirrer. Turn on the coulometer and generate Br_2 for 1 to 2 min. This conditions the electrodes. Turn off the stirrer and the applied voltage to the indicator electrodes.

3. Discard the solution and rinse the cell with distilled water. Transfer a 25-mL aliquot of phenol sample to the cell. Add 5 mL of 1.0 M NaBr and dilute with distilled water until the electrodes are fully immersed. Turn on the stirrer and amperometer. Record the amperometric current. Switch the coulometer to the generate position. Periodically, interrupt the generation of titrant, wait 5 to 10 s for complete reaction, and record the amperometric current and the time or amount of titrant generated (ask your instructor for the appropriate interval).

4. When the current begins to increase steadily, interrupt the generation of the titrant more frequently and record the values. Acquire at least 10 measurements after the equivalence point. Data in the vicinity of the equivalence point are not important. Switch the coulometer to standby and turn off the stirrer and amperometer.

5. Repeat steps 3 and 4 using 25 mL of distilled water as a blank in place of the phenol sample.

6. Disconnect the indicator electrodes from the amperometer. Connect *one* of these electrodes and the calomel electrode to the pH meter. The pH meter is used in the *millivolt* setting.

7. Repeat steps 3 through 5, but take measurements frequently in the vicinity of the equivalence point and less frequently before and after. Continue the titration until the potential becomes nearly constant.

8. Plot both the amperometric and potentiometric titration curves. Determine the end points, subtract the blanks, and calculate the concentration of phenol in the unknown as mg phenol/25 mL sample.

EXPERIMENT 16: DETERMINATION OF MANGANESE IN STEEL BY SPECTROPHOTOMETRY

Small quantities of manganese can be determined as highly colored permanganate ion. Potassium periodate is effective at oxidizing lower oxidation states of manganese to MnO_4^-, which exhibits a maximum absorbance at 525 nm. A reagent blank, consisting of the unoxidized sample, is used to compensate for other constituents in the sample that absorb at this wavelength. Large quantities of cerium(III) and chromium(III) interfere because they are oxidized by the periodate to cerium(IV) and $Cr_2O_7^{2-}$, both of which absorb at 525 nm. Ferric ion also absorbs at the measurement wavelength, but its interference is avoided by adding phosphoric acid, which reacts with the iron, converting it to a nonabsorbing complex ion.

Reaction

$$2Mn^{2+} + 5IO_4^- + 3H_2O \longrightarrow 2MnO_4^- + 5IO_3^- + 6H^+$$

Apparatus

Colorimeter or spectrophotometer

Reagents

$MnSO_4 \cdot H_2O$ (2 g)

HNO_3, 6 M (150 mL)

$(NH_4)_2S_2O_8$ (3 g)

Na_2SO_3 (0.3 g)

H_3PO_4, concentrated (45 mL)

KIO_4 (3.0 g)

Procedures

Preparation of a standard manganese(II) solution

1. Dry 2 g of $MnSO_4 \cdot H_2O$ at 110°C for 1 h and cool in a desiccator.
2. Weigh accurately 0.30 to 0.31 g of the dried reagent, dissolve in a small volume of distilled water, and transfer to a 1-L volumetric flask.
3. Dilute to the mark with distilled water, mix, and transfer to a storage bottle.

Analysis of the unknown

1. Weigh accurately three identical steel samples of between 0.5 and 1.0 g each and transfer to separate 250-mL conical flasks.
2. In the hood, add 50 mL of 6 M HNO_3 to dissolve the sample and boil to remove the yellow-brown nitrogen oxides.
3. In small portions, add about 1 g of $(NH_4)_2S_2O_8$ and boil gently for 10 to 15 min. If a permanganate color develops or brown, insoluble MnO_2 begins to form, add 0.1 g of Na_2SO_3 and boil for another 5 min to decolorize the solution and remove sulfur dioxide.
4. Cool and transfer the solutions and rinsings to numbered 100-mL volumetric flasks and dilute to the mark with distilled water. Mix well.
5. Pipet 25-mL aliquots of the first sample into three separate 125-mL beakers. To the first beaker (unknown) add 5 mL of concentrated H_3PO_4 and 0.5 g of KIO_4. To the second beaker (unknown + standard) add the same amount of H_3PO_4 and KIO_4 plus 5.00 mL (pipet) of the standard manganese solution. To the third beaker (blank) add only 5 mL of H_3PO_4.
6. Gently boil each solution for 5 min. When cool, transfer to numbered 50-mL volumetric flasks and dilute to the mark with distilled water.
7. Measure the absorbance of solutions 1 and 2 at 525 nm versus solution 3 as the reference. Repeat steps 5 through 7 with the two remaining samples.
8. Use the standard addition equation to calculate the amount of Mn in each measured solution and then calculate the % Mn in the sample.

EXPERIMENT 17: DETERMINATION OF PHOSPHATE IN SERUM BY SPECTROPHOTOMETRY

Phosphate and molybdate ions combine in acidic solution to form 12-molybdophosphoric acid, which, upon treatment with a suitable reducing agent such as hydrazine sulfate, yields a highly colored blue product called heteropoly blue. Acidic molybdate can also be reduced to a blue substance, but only in less acidic solutions. Although the blue heteropoly compound has not been characterized completely, it appears to have a molecular composition similar to that of the unreduced species, differing only in that some of the covalently bound molybdenum atoms are in a +5 rather than +6 oxidation state. The absorbance can be measured at 650 or 830 nm, but the method is more sensitive and subject to fewer interferences when the longer wavelength is used.

Reactions

$$H_3PO_4 + 12H_2MoO_4 \xrightarrow{H^+} H_3P(Mo_3O_{10})_4 + 12H_2O$$

$$H_3P(Mo_3O_{10})_4 + N_2H_2SO_4 \longrightarrow \text{Blue product}$$

Apparatus

Colorimeter or spectrophotometer
Absorption cells (2), 1.00 cm

Reagents

KH_2PO_4 (2 g)
Cl_3CCO_2H, 0.60 M (660 mL)
Na_2MoO_4, 5.0 × 10^{-3} M in 5 M H_2SO_4 (10 mL)
$N_2H_4SO_4$, 6.0 × 10^{-3} M (10 mL)

Procedures

Preparation of standard phosphate solution

1. Dry 2 g of reagent-grade KH_2PO_4 at 110°C for 1 h and cool in a desiccator.
2. Weigh and dissolve 0.4395 g of the dry reagent in distilled water and dilute to volume in a 1-L volumetric flask. Store in a plastic bottle. Each milliliter of this solution contains 0.1000 mg of phosphorus.

Preparation of the calibration graph

1. Transfer 0, 1, 2, 4, 6, 8, and 10 mL of the standard phosphate solution to each of seven 100-mL volumetric flasks. Add 90 mL of the trichloroacetic acid reagent to each flask and dilute to volume with distilled water. Mix each solution thoroughly.
2. Pipet 5 mL of each solution into separate 10-mL volumetric flasks. Add 1 mL of the

molybdate and 1 mL of the hydrazine sulfate reagents and dilute to the mark with distilled water. Mix each solution thoroughly.

3. Suspend the volumetric flasks in a boiling-water bath for 10 min. Remove and cool rapidly in an ice-water bath. Check the volumes in the flasks and, if necessary, add water to the mark and mix.

4. Measure the absorbance of each solution at 830 nm in a 1-cm cell against a reagent blank (the solution with no phosphate).

Analysis of the unknown

1. Pipet 1.0 mL of blood serum and 9.0 mL of the trichloroacetic acid reagent into a small centrifuge tube. Mix and centrifuge for 5 min.

2. Pipet 5 mL of the supernate into a 10-mL volumetric flask and continue as in steps 2 through 4 of the procedure for the preparation of the calibration curve.

3. Prepare a graph of absorbance versus concentration of phosphate and determine the concentration of phosphate in the serum unknown. Report the results in mg-% of P.

EXPERIMENT 18: COMPARISON OF SPECTROPHOTOMETRIC METHODS FOR DETERMINING THE STOICHIOMETRY OF A COMPLEX

The stoichiometry of the complex formed between iron (II) and 1,10-phenanthroline is determined by three methods: continuous variations, mole ratio, and slope ratio. Although each method has unique strengths and weaknesses, all yield acceptable results with this complex. The iron-phenanthroline complex forms easily and is quite stable. The complex absorbs strongly in the visible region where neither iron (II) nor 1,10-phenanthroline absorb. The experiment is designed to illustrate the acquisition and treatment of data for each method.

Reaction

$$x\text{Fe}^{2+} + y\text{C}_{12}\text{H}_8\text{N}_2 \longrightarrow \text{Fe}_x(\text{C}_{12}\text{H}_8\text{N}_2)_y^{2+}$$

Apparatus

Colorimeter or spectrophotometer (recorder optional)
Absorption cells (2), 1.00-cm

Reagents

Fe(NH$_4$)$_2$(SO$_4$)$_2 \cdot$6H$_2$O, 7.00×10^{-4} M (110 mL)
1,10-Phenanthroline, 7.00×10^{-4} M (140 mL)
1,10-Phenanthroline, 2.10×10^{-3} M (50 mL)
CH$_3$CO$_2$H/CH$_3$CO$_2$Na buffer, pH 4.0, total acetate 0.10 M (155 mL)
Hydroxylamine hydrochloride, 0.7 M (31 mL)

Procedures

Continuous variations method

1. Pipet 0, 1, 2, 3, 4, 5, 6, 7, 8, 9, and 10 mL of the iron solution into separate 25-mL volumetric flasks. Add 5 mL of the acetate buffer and 1 mL of the hydroxylamine hydrochloride solution to each flask.

2. Into each flask, respectively, pipet 10, 9, 8, 7, 6, 5, 4, 3, 2, 1, and 0 mL of the 7.00 $\times$ 10^{-4} M phenanthroline solution. Dilute to the mark with distilled water and mix thoroughly.

3. After 10 min, measure the absorbance of each solution at 508 nm using distilled water as a reference. As an option, your instructor may have you record the spectrum of one or all of the solutions.

4. Plot absorbance versus mole fraction of iron(II). Extrapolate the linear portions of the plot until they intersect and compute the indicated stoichiometry.

Mole-ratio method

1. Pipet 2 mL of the standard iron solution into ten 25-mL volumetric flasks. Add 5 mL of the acetate buffer, 1 mL of the hydroxylamine hydrochloride solution, and mix.

2. Pipet 1, 2, 3, 4, 5, 6, 8, 10, 12, and 15 mL of the 7.00 $\times$ 10^{-4} M phenanthroline solution into the flasks. Dilute to the mark with distilled water and mix thoroughly.

3. After 10 min, measure the absorbance of each solution at 508 nm versus distilled water as a reference.

4. Plot absorbance versus the ratio of the moles of phenanthroline to the moles of iron. Extrapolate the linear portions of the plot until they intersect and report the indicated stoichiometry.

Slope-ratio method

1. Pipet 5 mL of the iron solution into each of five 25-mL volumetric flasks. Add 5 mL of the acetate buffer, 1 mL of the hydroxylamine hydrochloride solution, and mix.

2. Pipet 1, 2, 3, 4, and 5 mL of the 7.00 $\times$ 10^{-4} M phenanthroline solution into the flasks. Dilute to the mark with distilled water and mix.

3. After 10 min, measure the absorbance of each solution at 508 nm versus distilled water as a reference.

4. Repeat steps 1 through 3 except use 10 mL of the 2.10 $\times$ 10^{-3} M phenanthroline solution and 0.5, 1.0, 1.5, 2.0, and 2.5 mL of the iron solution.

5. Plot absorbance versus concentration of iron and absorbance versus concentration of phenanthroline. From the slopes of these two graphs, calculate the stoichiometry.

EXPERIMENT 19: DETERMINATION OF SALICYLIC ACID IN ASPIRIN BY FLUORESCENCE

Acetylsalicylic acid, the active ingredient in aspirin, hydrolyzes to salicylic and acetic acids. Both hydrolysis products are usually present to some extent in aspirin, and a measure of their concentration can be indicative of the extent of the hydrolysis. Both salicylic and acetylsalicylic acid fluoresce, but their excitation and emission spectra are sufficiently different that conditions can be selected such that neither substance interferes in the determination of the other.

Salicylic acid has excitation and fluorescence maxima near 310 and 450 nm, respectively. Acetylsalicylic acid is not excited at wavelengths longer than about 300 nm, and when excited at shorter wavelengths, its fluorescence occurs in the 300- to 400-nm wavelength region. Thus salicylic acid can be selectively excited using the 313-nm emission line from a low-pressure mercury discharge lamp. Even if some acetylsalicylic acid becomes excited, its fluorescence is easily prevented from reaching the detector by using a sharp-cut secondary filter that blocks radiation below about 460 nm.

Reaction

acetylsalicylic acid + H_2O → salicylic acid + CH_3CO_2H

Apparatus

Filter fluorometer with low-pressure mercury discharge lamp
Primary filter: No. 7-54 plus Wratten No. 34A narrow-pass (isolates 313-nm line)
Secondary filter: No. 4 sharp-cut (465 nm)

Reagents

Salicylic acid (0.25 g)
Acetic acid, concentrated (5 mL)
Chloroform, Spectro-quality (500 mL)

Procedures

Preparation of standard 25-μg/mL salicylic acid

1. Weigh 250 mg of pure salicylic acid and transfer to a 100-mL volumetric flask. Dissolve and dilute to volume with a solution prepared by mixing 5 mL of concentrated acetic acid with 500 mL of Spectro-quality chloroform. Mix thoroughly.
2. Pipet 1 mL of this solution into a 100-mL volumetric flask and dilute to volume with the acetic acid–chloroform mixture. Each milliliter of this solution contains 25.0 μg of salicylic acid.

Preparation of the standard curve

1. Pipet 1, 2, 3, 4, 6, 8, and 10 mL of the 25-μg/mL salicylic acid standard into 25-mL volumetric flasks and dilute each to the mark with the acetic acid–chloroform mixture.

2. Measure the fluorescence of each solution using the acetic acid–chloroform solvent as a reference.
3. Plot fluorescence versus concentration (in $\mu g/mL$).

Analysis of the unknown

1. Either acquire a sample from your instructor or grind an aspirin tablet in a mortar until it is powdered. Weigh the powered sample (about 400 mg) and transfer to a 250-mL beaker.
2. Add about 75 mL of the acetic acid–chloroform solvent and stir. Filter rapidly through filter paper (Whatman No. 1 or equivalent), catching the filtrate in a 100-mL volumetric flask. Rinse twice with the solvent and add to the flask. Dilute to the mark with the solvent and mix.
3. Measure the fluorescence versus the solvent as a blank. Determine the concentration of salicylic acid from the standard curve and report the results as milligrams of salicylic acid per aspirin tablet.

EXPERIMENT 20: DETERMINATION OF COPPER BY ATOMIC ABSORPTION SPECTROSCOPY

Atomic absorption spectroscopy is an excellent technique for determining a large number of metals, including copper. This experiment is written for water samples, but it can be adapted easily for use with other types of samples, such as algaecides, used oil or grease, and foods. If time permits, the instructor may wish to include a brief study of the effects of varying certain instrument parameters, such as burner height and flame composition.

Apparatus

Atomic absorption spectrophotometer
Copper hollow cathode lamp

Reagents

Cu wire, No. 14 or 16 (1 g)
HNO_3, 6 M (50 mL)

Procedures

Preparation of standard 100-ppm copper solution

1. Weigh exactly 1 g of pure copper wire and dissolve in 50 mL of 6 M HNO_3 in the hood. Transfer to a 250-mL volumetric flask and dilute to volume with distilled water.
2. Pipet 25 mL of this solution into a 1-L volumetric flask and dilute to the mark with distilled water. This solution contains 100 ppm Cu.

Preparation of the standard curve

1. Consult the instrument manual or directions from your instructor regarding the instrument settings to be used.
2. Pipet 1, 3, 5, 7, and 10 mL of the standard 100-ppm copper solution into separate 100-mL volumetric flasks and dilute each to the mark with distilled water.
3. Zero the instrument while aspirating distilled water into the flame.
4. Aspirate each standard into the flame and record the absorbance.
5. Plot absorbance versus concentration and fit the best straight line to the data.

Analysis of the unknown

1. Prepare your sample using directions provided by the instructor.
2. With the same instrument settings used for measuring the standards, aspirate the unknown and record the absorbance.
3. Determine the concentration of copper from the standard curve.

EXPERIMENT 21: DETERMINATION OF TOTAL SALT CONTENT BY ION EXCHANGE AND ACID-BASE TITRATION

The total concentration of dissolved salts in water and certain commercial products is a quantity often sought. Instead of determining the amount of each ionic constituent, it is usually easier to replace all the cations with hydrogen ions, which can then be titrated with a standard base. The replacement is accomplished by passing the sample solution through a strong acid ion-exchange resin in the hydrogen form. Since the exchange is stoichiometric and the resin is insoluble in water, the effluent from the column contains an amount of H^+ equivalent to the total amount of cations in the sample.

Reactions

Exchange:

$$R-SO_3^-H^+ + (M^{n+} + nX^-) \longrightarrow (R-SO_3^-)_n M^{n+} + n(H^+ + X^-)$$

Titration:

$$HX + NaOH \longrightarrow NaX + H_2O$$

Reagents

Strong acid cation-exchange resin, BioRad Dowex 50W-X8 (50 to 100 mesh) or equivalent
NaOH, 50%(w/v), carbonate-free (7 mL)
KCl, primary-standard grade (2 g)
$KHC_8H_4O_4$, primary-standard grade (5 g)
Phenolphthalein indicator, 0.05%(w/v) in 50%(v/v) ethanol (1 mL)

HCl, 4 *M* (50 mL)

Methyl red indicator, 0.02%(w/v) in 60%(v/v) ethanol (2 drops)

Procedures

Preparation of the ion-exchange column

1. Push a small plug of glass wool into the bottom of an empty 20 to 30-cm column (the lower half of an old or broken buret makes a fine column). Do not make the plug so large that it fits too tightly. Remember: What goes in must eventually come out.

2. Pass distilled water through the tube until the glass-wool plug is free of air bubbles. Close the stopcock and leave the tube about two-thirds filled with water.

3. Make a slurry of the resin using about 2 mL of water per 1 g of resin (see your instructor for the quantity of resin to be used). Pour this slurry into the column, draining some of the water if necessary. A bed of about 15 to 20 mL is adequate. At no time during this or later steps should the liquid level in the tube be allowed to fall below the top of the resin bed.

4. Clamp the column upright to a ring stand and place a beaker under the tip. Adjust the stopcock for a flow rate of about 5 mL/min. When the water level is just above the top of the resin bed, add a few milliliters of 4 *M* HCl, allowing it to drain down the side of the tube so as to not disturb the top of the resin bed. Repeat this procedure until 50 mL of HCl has been passed through the column.

5. Rinse the column with distilled water, using the procedure in step 4, until the effluent is not acidic toward methyl red indicator.

Preparation and standardization of 0.1 *M* sodium hydroxide

1. Transfer approximately 7 mL of the carbonate-free NaOH solution to a 1-L storage bottle and dilute with 1 L of cool, freshly boiled distilled water. Keep the bottle tightly closed with a *rubber* stopper.

2. Dry about 2 g of pure potassium chloride at 110°C for 1 h and cool in a desiccator.

3. Weigh accurately three 0.20- to 0.25-g samples of dry KCl, transfer to small beakers, and dissolve in 10 to 15 mL of distilled water.

4. Adjust the flow rate of the column to about 1 mL/min and pour the first sample onto the column. Rinse the beaker three times with small amounts of water and transfer the rinsings to the column. Collect the effluent in a clean, 250-mL conical flask.

5. When the liquid level approaches the top of the resin bed, begin a rinsing operation with 75 mL of distilled water, adding about 5 mL at a time as the liquid level reaches the top of the resin bed. Increase the flow rate to about 5 mL/min for this step. Collect the rinsings in the same flask with the previous effluent. The resin will generally have the capacity to handle at least six samples before it must be reconditioned by performing steps 4 and 5 of the column preparation procedure.

6. Add 2 drops of phenolphthalein indicator and titrate with the 0.1 *M* NaOH to the appearance of a pink color that persists for at least 20 s.

7. Repeat steps 4 to 6 for the remaining two samples.
8. Calculate the molarity of the sodium hydroxide solution.

Analysis of the unknown

1. Pipet an appropriate aliquot of sample (see your instructor) onto the column. Adjust the flow rate to about 1 mL/min and catch the effluent in a clean 250-mL conical flask.
2. Repeat steps 4 through 6 of the standardization procedure.
3. Repeat steps 1 and 2 of this procedure with two additional aliquots of sample.
4. Calculate the amount of total salt in units of meq/mL sample.

EXPERIMENT 22: DETERMINATION OF THE OPTIMUM FLOW RATE IN GAS CHROMATOGRAPHY

According to the van Deemter equation, the efficiency of a chromatography column, as measured by its height equivalent to a theoretical plate (H), gets worse at both very high and low flow rates. At high flow rates, band broadening is due mainly to the resistance to mass transfer of the solute between the phases, while at low flow rates, it is due mainly to molecular diffusion of the solute. The optimum flow rate can be determined experimentally from a plot of H versus flow rate, where H is determined from the band width and retention time of a given solute.

The following experiment can be performed with a gas chromatograph equipped with virtually any column and standard detector. The particular solute used may depend on the column and detector employed.

Apparatus

    Gas chromatograph with thermal conductivity or flame ionization detector
    Column appropriate for the instrument
    Soap-bubble flowmeter

Reagents

    Carrier gas appropriate for the instrument
    Solute to be chromatographed (see your instructor)

Procedure

A wide variety of instruments, equipped with different detectors and columns, can be used to perform this experiment. As a result, only a general procedure is given here. Operating conditions consistent with the equipment being used will be specified by the instructor.

1. Using the instrument manual or directions from your instructor, turn on the instru-

ment and adjust the various operating controls, such as injection port temperature, column temperature, and detector sensitivity.

2. Obtain from your instructor the range of flow rates to be used and adjust the flow rate to the smallest value. To determine the flow rate, squeeze the bulb on the bottom of the soap-bubble flowmeter and get a few bubbles moving up the tube. Measure the time it takes for one of these bubbles to move 10 mL.

3. Set the recorder baseline and turn the recorder to standby. Mark the point on the chart paper where the recording will start. Simultaneously inject a sample and start the recorder. Continue to record until the solute has been eluted.

4. Increase the flow rate by about 10% of the range to be studied and repeat step 3. Continue until the maximum flow rate has been reached.

5. Calculate H from each chromatogram and plot it versus flow rate. Determine the optimum flow rate from the graph.

EXPERIMENT 23: DETERMINATION OF CHLORINATED HYDROCARBONS IN A MIXTURE BY GAS CHROMATOGRAPHY

This experiment compares the more common method of external calibration with that of internal calibration. The use of an ordinary calibration plot based on external standards does not always give satisfactory results in gas chromatography, owing partly to short-term variations in operating conditions such as temperature and flow rate, and partly to errors in measuring the small volumes of sample used. Sample volumes in gas chromatography seldom exceed a few microliters. The measuring of such small volumes can be done fairly reproducibly, but the accuracy is often quite poor. An internal standard technique, in which the peak height of an analyte is compared to that of an introduced standard, eliminates much of the error from these sources. The expectation is that any variation will affect the height of both analyte and internal standard peak by the same relative amount and the ratio of their heights will remain a constant function of their relative concentrations.

There are three requirements for a good internal standard: (1) It should yield a peak completely resolved from any other peak; (2) it should have a retention time close to the retention time of the analyte peak(s); and (3) the ratio of the peak heights should be fairly close to 1. It is desirable, but not always feasible, to use an internal standard that is similar structurally to the analyte.

This experiment is written for the separation and determination of three simple chlorinated hydrocarbons and can be performed with almost any gas chromatograph. It can be adapted for use with a great many other chemical systems as well.

Apparatus

    Gas chromatograph with thermal conductivity or flame ionization detector

    Packed column with nonpolar stationary phase such as dinonylphthalate on Celite 545

    Chart recorder

    Soap-bubble flowmeter

Reagents

CCl$_4$ (7.5 mL)

CHCl$_3$ (7.5 mL)

CH$_2$Cl$_2$ (7.5 mL)

CH$_3$COCH$_3$ (12 mL)

Procedures

Preparation of the calibration plot

1. If necessary, turn on the instrument and set the various controls, such as column temperature, detector sensitivity, and flow rate, according to directions supplied by the instructor. The particular settings depend on the type of instrument and the column used.

2. Prepare standard solutions by mixing the following volumes (mL) of pure reagents in a test tube.

CCl$_4$	CHCl$_3$	CH$_2$Cl$_2$	CH$_3$COCH$_3$
0.5	1.5	2.5	2.0
1.0	2.0	1.5	2.0
1.5	2.5	0.5	2.0
2.0	0.5	2.0	2.0
2.5	1.0	1.0	2.0

Mix well and stopper tightly with a rubber septum.

3. Set the recorder baseline and turn the recorder to standby. Mark the point on the chart paper where the recording will start. Simultaneously inject 1.0 μL of the first mixture and start the recorder. Continue to record until all four solutes have been eluted. Repeat this step with each of the remaining solutions.

4. Measure the peak height of each component in each chromatogram and plot the peak height of each analyte versus the volume percent in the mixture. Plot the ratio of the peak height of each analyte to the peak height for acetone versus the volume percent in the mixture.

Analysis of the unknown

1. Obtain an unknown from your instructor and pipet 4.5 mL into a test tube. Add exactly 2 mL of the acetone as an internal standard. Stopper the tube with a rubber septum and mix.

2. Set the recorder baseline and turn the recorder to standby. Mark the point on the chart paper where the recording will start. Simultaneously start the chart recorder and inject 1.0 μL of the sample into the chromatograph. Record until all components have been eluted.

3. Repeat step 2 with two additional portions of the sample.

4. Use the retention times to identify the component(s) present in your unknown and determine the volume percent of each from both sets of calibration plots.

APPENDIX

A
EXPONENTS AND LOGARITHMS

Exponents

Most people prefer to avoid both very large and very small numbers. Often, this can be done by selection of an appropriate unit. For example, the length of a football field is reported in yards rather than miles or inches. Sometimes, the necessary units are dictated by other considerations and the very large or small numbers must be used without change. Scientists have devised a method of representing numbers, called scientific notation, which eliminates the use of leading and trailing zeros to place the decimal point. In this notation, numbers generally are written with one digit to the left of the decimal, and the proper position of the decimal is indicated by multiplying the number with 10 raised to the appropriate power (called the exponent). Thus

$$1,200,000 = 1.2 \times 10^6$$
$$416.3 = 4.163 \times 10^2$$
$$0.0000188 = 1.88 \times 10^{-5}$$

In 1.2×10^6, the exponent 6 means that the decimal point in the nonexponential number belongs six places to the *right* of its present location. In 1.88×10^{-5}, the minus sign means it belongs five places to the left of its present location. It is customary, but not absolutely necessary, to maintain only one digit to the left of the decimal (thus 1.2×10^6 is preferred to 12×10^5).

The addition and subtraction of numbers written in scientific notation must be carried out cautiously. Numbers to be added or subtracted must have the *same* exponent, and that exponent becomes part of the result.

$$\begin{array}{rcr} 8.1 \times 10^4 & \longrightarrow & 0.81 \times 10^5 \\ + \ 1.21 \times 10^5 & \longrightarrow & + \ \underline{1.21 \times 10^5} \\ & & 2.02 \times 10^5 \end{array}$$

Multiplying and dividing numbers written in scientific notation is quite easy. The nonexponential parts of the numbers are multiplied or divided in the usual way, but the exponents are *added* when multiplying and *subtracted* when dividing.

$$(2.34 \times 10^3) \times (4.08 \times 10^5) = (2.34 \times 4.08) \times 10^{(3+5)} = 9.55 \times 10^8$$

$$(8.11 \times 10^{-4}) \times (2.06 \times 10^2) = (8.11 \times 2.06) \times 10^{(-4+2)}$$

$$= 16.7 \times 10^{-2} = 1.67 \times 10^{-1}$$

$$\frac{9.06 \times 10^{12}}{3.02 \times 10^3} = \frac{9.06}{3.02} \times 10^{(12-3)} = 3.00 \times 10^9$$

$$\frac{6.84 \times 10^{-6}}{3.17 \times 10^{-2}} = \frac{6.84}{3.17} \times 10^{[-6-(-2)]} = 2.16 \times 10^{-4}$$

Exponents of exponents are also easily evaluated. Again, the nonexponential and exponential parts of the numbers are handled separately. The exponents are *multiplied* in all cases. When the second exponent is *positive*, the nonexponent is multiplied with itself a number of times equal to the second exponent.

$$(2.0 \times 10^4)^3 = (2.0 \times 2.0 \times 2.0) \times 10^{(4 \times 3)} = 8.0 \times 10^{12}$$

$$(3.0 \times 10^{-6})^2 = (3.0 \times 3.0) \times 10^{(-6 \times 2)} = 9.0 \times 10^{-12}$$

When the second exponent is *negative*, the reciprocal of the nonexponent is multiplied with itself a number of times equal to the second exponent.

$$(2.0 \times 10^4)^{-3} = \left(\frac{1}{2.0} \times \frac{1}{2.0} \times \frac{1}{2.0}\right) \times 10^{[(4)(-3)]}$$

$$= \left(\frac{1}{2.0 \times 2.0 \times 2.0}\right) \times 10^{-12} = \left(\frac{1}{8.0}\right) \times 10^{-12}$$

$$= 0.125 \times 10^{-12} = 1.25 \times 10^{-13}$$

$$(3.0 \times 10^{-2})^{-2} = \left(\frac{1}{3.0 \times 3.0}\right) \times 10^{[(-2)(-2)]} = \left(\frac{1}{9.0}\right) \times 10^4$$

$$= 1.1 \times 10^3$$

Square roots, cube roots, and so on, are merely fractional exponents and are treated the same way as whole-number exponents.

$$\sqrt{9.0 \times 10^4} = (9.0 \times 10^4)^{1/2} = \sqrt{9.0} \times 10^{(4 \times 1/2)} = 3.0 \times 10^2$$

$$\sqrt[3]{8.0 \times 10^{-12}} = (8.0 \times 10^{-12})^{1/3} = \sqrt[3]{8.0} \times 10^{[(-12)(1/3)]} = 2.0 \times 10^{-4}$$

In taking roots of exponential numbers, it is convenient to adjust the exponent to a value exactly divisible by the root. Thus

$$\sqrt{1.6 \times 10^3} = (1.6 \times 10^3)^{1/2} = (16 \times 10^2)^{1/2} = \sqrt{16} \times 10^{(2 \times 1/2)} = 4.0 \times 10^1$$

Logarithms

Having just reviewed exponential notation, it should not be surprising to learn that all numbers can be expressed in exponential form:

$$N = b^x \qquad (1)$$

where N is the number, b the base, and x the exponent. The logarithm of the number N is the power x to which the base b must be raised to satisfy the equality. That is,

$$x = \log_b N \qquad (2)$$

Only two bases are used in chemistry: 10 and e. Base 10 logarithms are called common logarithms and are written as log or occasionally $\log_{10}$. Base e logarithms are called natural or Naperian logarithms and are written as ln or occasionally $\log_e$.

$$\log N = 2.3 \ln N$$

Only base 10 logs are discussed in this section.

Pocket calculators, with their log functions, have eliminated the need for lengthy tables of log values. There are, however, some manipulations of logs that are still worthwhile learning. One of these is to find the *antilog* of a value; that is, to find N for a known x in Equation 2. Some calculators do not have an antilog function key, but N can be found by solving Equation 1. The calculator's exponential function key is used to raise b to the x power to give N.

It is also worthwhile to note that the log of a number raised to some power is equal to the power times the log of the number.

$$\log N^y = y \log N$$

The exponent y can be positive, negative, whole, or fractional.

$$\log 4^2 = \log 16 \text{ or } (2)(\log 4) = 1.20$$

$$\log 9.4^{-3} = (-3)(\log 9.4) = -2.92$$

$$\log \sqrt{16} = \log (16)^{1/2} = (\tfrac{1}{2})(\log 16) = 0.60$$

B

BALANCING REDOX REACTIONS USING THE METHOD OF HALF-REACTIONS

Every redox reaction can be divided into two half-reactions: one representing the oxidation and the other representing the reduction. The advantage of this method of balancing is based on the premise that each half-reaction can be balanced more easily than the entire redox reaction. In aqueous solution, H^+, OH^-, and H_2O often are reactants or products of redox reactions and must be added in appropriate quantities to construct a balanced equation. Although free electrons will not appear in the balanced net reaction, they are always necessary in the half-reactions to obtain a charge balance. A complete, net redox reaction can be obtained by combining the balanced half-reactions. The individual steps are illustrated below for the reaction

$$KMnO_4 + H_3AsO_3 \xrightleftharpoons{HCl} MnCl_2 + H_3AsO_4$$

1. Using oxidation numbers, identify the substances being reduced and oxidized.
   a. $\overset{+7}{K}MnO_4$ is reduced to $\overset{+2}{Mn}Cl_2$.
   b. $H_3\overset{+3}{As}O_3$ is oxidized to $H_3\overset{+5}{As}O_4$.
2. Write a half-reaction for the reduction process *using ionic formulas*.

$$MnO_4^- \rightleftharpoons Mn^{2+}$$

3. Balance the half-reaction chemically by addition of H^+, OH^-, and H_2O as necessary.
   a. In acidic solution:
      (1) balance hydrogen by adding H^+ to the hydrogen-deficient side.
      (2) balance oxygen by adding H_2O to the oxygen-deficient side and twice as much H^+ to the other side.

$$MnO_4^- \rightleftharpoons Mn^{2+} + 4H_2O$$
$$MnO_4^- + 8H^+ \rightleftharpoons Mn^{2+} + 4H_2O$$

   b. In basic solution:
      (1) Balance hydrogen by adding H_2O to the hydrogen-deficient side and twice as much OH^- to the other side.
      (2) Balance oxygen by adding OH^- to the oxygen-deficient side and half as much H_2O to the other side.

4. Balance the half-reaction electrically by adding electrons.

$$MnO_4^- + 8H^+ + 5e^- \rightleftharpoons Mn^{2+} + 4H_2O$$

5. Repeat steps 2 through 4 for the substance undergoing oxidation. The half-reaction may be written as an *oxidation*, in which case it will be *added* to the other half-reaction, or as a *reduction*, in which case it will be *subtracted* from the other half-reaction.*

   Step 2: $H_3AsO_4 \rightleftharpoons H_3AsO_3$

   Step 3: $H_3AsO_4 \rightleftharpoons H_3AsO_3 + H_2O$

   $\qquad H_3AsO_4 + 2H^+ \rightleftharpoons H_3AsO_3 + H_2O$

   Step 4: $H_3AsO_4 + 2H^+ + 2e^- \rightleftharpoons H_3AsO_3 + H_2O$

6. Normalize the number of electrons by multiplying each half-reaction with the appropriate whole numbers:

$$2 \times (MnO_4^- + 8H^+ + 5e^- \rightleftharpoons Mn^{2+} + 4H_2O)$$

$$5 \times (H_3AsO_4 + 2H^+ + 2e^- \rightleftharpoons H_3AsO_3 + H_2O)$$

7. Subtract (or add if the half-reactions are written in opposite directions) the two half-reactions to get the net ionic reaction. Cancel where possible.

$$2MnO_4^- + \overset{6}{\cancel{16}}H^+ + \cancel{10e^-} + 5H_3AsO_3 + \cancel{5H_2O}$$

$$\rightleftharpoons 2Mn^{2+} + \overset{3}{\cancel{8}}H_2O + 5H_3AsO_4 + \cancel{10H^+} + \cancel{10e^-}$$

8. Check the net equation *both* chemically and electrically. Add the spectator ions to convert to a molecular equation, if necessary.

$$2KMnO_4 + 6HCl + 5H_3AsO_3 \rightleftharpoons 2MnCl_2 + 3H_2O + 5H_3AsO_4 + 2KCl$$

*This author prefers writing the half-reactions as reductions and subtracting to obtain the net reaction because it is consistent with the practices commonly used in calculating electrode potentials and cell voltages.

C

SOLVING QUADRATIC EQUATIONS

An exact solution to many equilibrium problems involves solving a quadratic equation. In such cases, the expression should be arranged to the form

$$ax^2 + bx + c = 0$$

for which

$$x = \frac{-b \pm \sqrt{b^2 - 4ac}}{2a}$$

This equation has two roots, but in all equilibrium problems, one root will be chemically impossible.

Consider, for example, the equilibrium expression for the dissociation of chlorous acid, where the equilibrium constant and analytical concentration of acid are known and the equilibrium concentration of hydrogen ion is sought.

$$HClO_2 \rightleftharpoons H^+ + ClO_2^-$$

$$K_{eq} = \frac{[H^+][ClO_2^-]}{[HClO_2]} = \frac{[H^+]^2}{C_{HClO_2} - [H^+]}$$

Putting this in the appropriate form gives

$$[H^+]^2 + K_{eq}[H^+] - K_{eq}C_{HClO_2} = 0$$

If K_{eq} is 1.12×10^{-2} and C_{HClO_2} is $0.100\ M$,

$$[H^+]^2 + 1.12 \times 10^{-2}[H^+] - 1.12 \times 10^{-3} = 0$$

Substituting in the quadratic formula yields

$$[H^+] = \frac{-(1.12 \times 10^{-2}) \pm \sqrt{(1.12 \times 10^{-2})^2 - (4)(1)(-1.12 \times 10^{-3})}}{2(1)}$$

$$= \frac{-(1.12 \times 10^{-2}) \pm (6.79 \times 10^{-2})}{2}$$

and the two roots are

$$[H^+] = \frac{-(1.12 \times 10^{-2}) + (6.79 \times 10^{-2})}{2} = \frac{5.67 \times 10^{-2}}{2} = 0.0284 \, M$$

$$[H^+] = \frac{-(1.12 \times 10^{-2}) - (6.79 \times 10^{-2})}{2} = \frac{-7.91 \times 10^{-2}}{2} = -0.0396 \, M$$

Since a negative concentration has no physical meaning, the second root is discarded.

D

MULTIPLICATION AND DIVISION WITH SIGNIFICANT FIGURES

In carrying out a multiplication or division with experimental values, the result should be written in such a way that its *relative* uncertainty is "comparable" to the relative uncertainty of the least accurate value. That is, the answer should not be written to appear significantly more or less accurate than the least accurate value used in its calculation. One widely used rule states that *the relative uncertainty of the answer must fall between 0.2 to 2 times the largest relative uncertainty in the data used for its calculation.*

According to the general rule adopted in Chapter 3, the absolute uncertainty of a properly expressed experimental value is taken to be ± 1 in the last digit, unless otherwise specified. Relative uncertainty is obtained by dividing the absolute uncertainty by the value being considered. With numbers containing three or four significant figures, it is convenient to represent the relative uncertainty in parts per thousand, as shown in the following examples.

Experimental value	Absolute uncertainty	Relative uncertainty (ppt)
26.07	± 0.01	$\dfrac{0.01}{26.07} \times 1000 = 0.4$
0.0988	± 0.0001	$\dfrac{0.0001}{0.0988} \times 1000 = 1$
164.27	± 0.01	$\dfrac{0.01}{164.27} \times 1000 = 0.06$
7.2×10^{-3}	$\pm 0.1 \times 10^{-3}$	$\dfrac{0.1 \times 10^{-3}}{7.2 \times 10^{-3}} \times 1000 = 14$

The following arithmetic operation, carried out with a pocket calculator, produced a result with obviously too many digits:

$$\frac{(26.07)(7.2 \times 10^{-3})}{(0.0988)(164.27)} = 0.0115653379$$

The value, 7.2×10^{-3}, has the greatest relative uncertainty (14 ppt) and, according to the rule, the relative uncertainty of the answer must lie between 0.2 to 2 times 14 ppt, or between 2.8 and 28 ppt. If the answer is rounded off to 0.01157, a value with four significant figures, it will have a relative uncertainty of

$$\frac{0.00001}{0.01157} \times 1000 = 0.86 \text{ ppt}$$

This uncertainty is *less* than the smallest permissible value (2.8 ppt), which means that the answer is too accurate and more rounding off is necessary. The answer rounded off to 0.0116, a value with three significant figures, has a relative uncertainty of

$$\frac{0.0001}{0.0116} \times 1000 = 8.6 \text{ ppt}$$

which falls within the acceptable range, making it the correct answer. Rounding off the answer to two significant figures, or 0.012, produces a value with too much uncertainty:

$$\frac{0.001}{0.012} \times 1000 = 83 \text{ ppt}$$

The value, 83, is greater than the largest permissible relative uncertainty (28 ppt).

Using the simple rule stated in Chapter 3 of keeping the same number of digits in the final answer as is found in the value with the least number of significant digits, the answer would be written as 0.012 (same number of significant figures as 7.2×10^{-3}). Obviously, the simple rule does not always give the same result as the more complicated (but better) rule.

FOUR-PLACE LOGARITHMS OF NUMBERS

No.	0	1	2	3	4	5	6	7	8	9
10	0000	0043	0086	0128	0170	0212	0253	0294	0334	0374
11	0414	0453	0492	0531	0569	0607	0645	0682	0719	0755
12	0792	0828	0864	0899	0934	0969	1004	1038	1072	1106
13	1139	1173	1206	1239	1271	1303	1335	1367	1399	1430
14	1461	1492	1523	1553	1584	1614	1644	1673	1703	1732
15	1761	1790	1818	1847	1875	1903	1931	1959	1987	2014
16	2041	2068	2095	2122	2148	2175	2201	2227	2253	2279
17	2304	2330	2355	2380	2405	2430	2455	2480	2504	2529
18	2553	2577	2601	2625	2648	2672	2695	2718	2742	2765
19	2788	2810	2833	2856	2878	2900	2923	2945	2967	2989
20	3010	3032	3054	3075	3096	3118	3139	3160	3181	3201
21	3222	3243	3263	3284	3304	3324	3345	3365	3385	3404
22	3424	3444	3464	3483	3502	3522	3541	3560	3579	3598
23	3617	3636	3655	3674	3692	3711	3729	3747	3766	3784
24	3802	3820	3838	3856	3874	3892	3909	3927	3945	3962
25	3979	3997	4014	4031	4048	4065	4082	4099	4116	4133
26	4150	4166	4183	4200	4216	4232	4249	4265	4281	4298
27	4314	4330	4346	4362	4378	4393	4409	4425	4440	4456
28	4472	4487	4502	4518	4533	4548	4564	4579	4594	4609
29	4624	4639	4654	4669	4683	4698	4713	4728	4742	4757
30	4771	4786	4800	4814	4829	4843	4857	4871	4886	4900
31	4914	4928	4942	4955	4969	4983	4997	5011	5024	5038
32	5051	5065	5079	5092	5105	5119	5132	5145	5159	5172
33	5185	5198	5211	5224	5237	5250	5263	5276	5289	5302
34	5315	5328	5340	5353	5366	5378	5391	5403	5416	5428
35	5441	5453	5465	5478	5490	5502	5514	5527	5539	5551
36	5563	5575	5587	5599	5611	5623	5635	5647	5658	5670
37	5682	5694	5705	5717	5729	5740	5752	5763	5775	5786
38	5798	5809	5821	5832	5843	5855	5866	5877	5888	5899
39	5911	5922	5933	5944	5955	5966	5977	5988	5999	6010
40	6021	6031	6042	6053	6064	6075	6085	6096	6107	6117
41	6128	6138	6149	6160	6170	6180	6191	6201	6212	6222
42	6232	6243	6253	6263	6274	6284	6294	6304	6314	6325
43	6335	6345	6355	6365	6375	6386	6395	6405	6415	6425
44	6435	6444	6454	6464	6474	6484	6493	6503	6513	6522
45	6532	6542	6551	6561	6571	6580	6590	6599	6609	6618
46	6628	6637	6646	6656	6665	6675	6684	6693	6702	6712
47	6721	6730	6739	6749	6758	6767	6776	6785	6794	6803
48	6812	6821	6830	6839	6848	6857	6866	6875	6884	6893
49	6902	6911	6920	6928	6937	6946	6955	6964	6972	6981
50	6990	6998	7007	7016	7024	7033	7042	7050	7059	7067
51	7076	7084	7093	7101	7110	7118	7126	7135	7143	7152
52	7160	7168	7177	7185	7193	7202	7210	7218	7226	7235
53	7243	7251	7259	7267	7275	7284	7292	7300	7308	7316
54	7324	7332	7340	7348	7356	7364	7372	7380	7388	7396
	0	1	2	3	4	5	6	7	8	9

No.	0	1	2	3	4	5	6	7	8	9
55	7404	7412	7419	7427	7435	7443	7451	7459	7466	7474
56	7482	7490	7497	7505	7513	7520	7528	7536	7543	7551
57	7559	7566	7574	7582	7589	7597	7604	7612	7619	7627
58	7634	7642	7649	7657	7664	7672	7679	7686	7694	7701
59	7709	7716	7723	7731	7738	7745	7752	7760	7767	7774
60	7782	7789	7796	7803	7810	7818	7825	7832	7839	7846
61	7853	7860	7868	7875	7882	7889	7896	7903	7910	7917
62	7924	7931	7938	7945	7952	7959	7966	7973	7980	7987
63	7992	8000	8007	8014	8021	8028	8035	8041	8048	8055
64	8062	8069	8075	8082	8089	8096	8102	8109	8116	8122
65	8129	8136	8142	8149	8156	8162	8169	8176	8182	8189
66	8195	8202	8209	8215	8222	8228	8235	8241	8248	8254
67	8261	8267	8274	8280	8287	8293	8299	8306	8312	8319
68	8325	8331	8338	8344	8351	8357	8363	8370	8376	8382
69	8388	8395	8401	8407	8414	8420	8426	8432	8439	8445
70	8451	8457	8463	8470	8476	8482	8488	8494	8500	8506
71	8513	8519	8525	8531	8537	8543	8549	8555	8561	8567
72	8573	8579	8585	8591	8597	8603	8609	8615	8621	8627
73	8633	8639	8645	8651	8657	8663	8669	8675	8681	8686
74	8692	8698	8704	8710	8716	8722	8727	8733	8739	8745
75	8751	8756	8762	8768	8774	8779	8785	8791	8797	8802
76	8808	8814	8820	8825	8831	8837	8842	8848	8854	8859
77	8865	8871	8876	8882	8887	8893	8899	8904	8910	8915
78	8921	8927	8932	8938	8943	8949	8954	8960	8965	8971
79	8976	8982	8987	8993	8998	9004	9009	9015	9020	9025
80	9031	9036	9042	9047	9053	9058	9063	9069	9074	9079
81	9085	9090	9096	9101	9106	9112	9117	9122	9128	9133
82	9138	9143	9149	9154	9159	9165	9170	9175	9180	9186
83	9191	9196	9201	9206	9212	9217	9222	9227	9232	9238
84	9243	9248	9253	9258	9263	9269	9274	9279	9284	9289
85	9294	9299	9304	9309	9315	9320	9325	9330	9335	9340
86	9345	9350	9355	9360	9365	9370	9375	9380	9385	9390
87	9395	9400	9405	9410	9415	9420	9425	9430	9435	9440
88	9445	9450	9455	9460	9465	9469	9474	9479	9484	9489
89	9494	9499	9504	9509	9513	9518	9523	9528	9533	9538
90	9542	9547	9552	9557	9562	9566	9571	9576	9581	9586
91	9590	9595	9600	9605	9609	9614	9619	9624	9628	9633
92	9638	9643	9647	9652	9657	9661	9666	9671	9675	9680
93	9685	9689	9694	9699	9703	9708	9713	9717	9722	9727
94	9731	9736	9741	9745	9750	9754	9759	9763	9768	9773
95	9777	9782	9786	9791	9795	9800	9805	9809	9814	9818
96	9823	9827	9832	9836	9841	9845	9850	9854	9859	9863
97	9868	9872	9877	9881	9886	9890	9894	9899	9903	9908
98	9912	9917	9921	9926	9930	9934	9939	9943	9948	9952
99	9956	9961	9965	9969	9974	9978	9983	9987	9991	9996
	0	1	2	3	4	5	6	7	8	9

F

SOLUBILITY PRODUCT CONSTANTS

According to Anion

Substance	K_{sp}	pK_{sp}
Bromates		
$AgBrO_3$	5.5×10^{-5}	4.26
$Pb(BrO_3)_2$	7.9×10^{-6}	5.10
$TlBrO_3$	1.7×10^{-4}	3.78
Bromides		
$AgBr$	5.0×10^{-13}	12.30
$CuBr$	5×10^{-9}	8.3
$HgBr_2$	1×10^{-19}	19.0
Hg_2Br_2	5.6×10^{-23}	22.25
$PbBr_2$	2.1×10^{-6}	5.68
$TlBr$	3.6×10^{-6}	5.44
Carbonates		
Ag_2CO_3	8.1×10^{-12}	11.09
$BaCO_3$	5.0×10^{-9}	8.30
$CaCO_3$	4.5×10^{-9}	8.35
$CdCO_3$	1.8×10^{-14}	13.74
Hg_2CO_3	8.9×10^{-17}	16.05
$La_2(CO_3)_3$	4×10^{-34}	33.4
$MgCO_3$	3.5×10^{-8}	7.46
$PbCO_3$	7.4×10^{-14}	13.13
$SrCO_3$	9.3×10^{-10}	9.03
Chlorides		
$AgCl$	1.8×10^{-10}	9.74
$CuCl$	1.9×10^{-7}	6.73
Hg_2Cl_2	1.2×10^{-18}	17.91
$PbCl_2$	1.7×10^{-5}	4.78
$TlCl$	1.8×10^{-4}	3.74
Chromates		
Ag_2CrO_4	1.2×10^{-12}	11.92
$BaCrO_4$	2.1×10^{-10}	9.67
$CuCrO_4$	3.6×10^{-6}	5.44
Hg_2CrO_4	2.0×10^{-9}	8.70
Tl_2CrO_4	9.8×10^{-13}	12.01
Cyanides		
$AgCN$	2.2×10^{-16}	15.66
$Hg_2(CN)_2$	5×10^{-40}	39.3
$Zn(CN)_2$	3×10^{-16}	15.5

According to Anion (*Cont.*)

Substance	K_{sp}	pK_{sp}
Fluorides		
CaF_2	3.9×10^{-11}	10.41
LaF_3	1×10^{-29}	29.0
MgF_2	6.6×10^{-9}	8.18
ThF_4	5×10^{-29}	28.3
Hydroxides		
Ag_2O ($\rightleftharpoons 2Ag^+ + 2OH^-$)	3.8×10^{-16}	15.42
$Ca(OH)_2$	6.5×10^{-6}	5.19
$Cd(OH)_2$	4.5×10^{-15}	14.35
$Cr(OH)_3$	2×10^{-30}	29.8
$Cu(OH)_2$	4.8×10^{-20}	19.32
Cu_2O ($\rightleftharpoons 2Cu^+ + 2OH^-$)	4×10^{-30}	29.4
$Fe(OH)_2$	8×10^{-16}	15.1
$Fe(OH)_3$	2×10^{-39}	38.8
$La(OH)_3$	2×10^{-21}	20.7
$Mg(OH)_2$	7.1×10^{-12}	11.15
PbO ($\rightleftharpoons 2Pb^{2+} + 2OH^-$)	8×10^{-16}	15.1
Iodates		
$AgIO_3$	3.1×10^{-8}	7.51
$Ba(IO_3)_2$	1.5×10^{-9}	8.81
$Ca(IO_3)_2$	7.1×10^{-7}	6.15
$Hg_2(IO_3)_2$	1.3×10^{-18}	17.89
$Sr(IO_3)_2$	3.3×10^{-7}	6.48
$Th(IO_3)_4$	2.4×10^{-15}	14.62
$TlIO_3$	3.1×10^{-6}	5.51
Iodides		
AgI	8.3×10^{-17}	16.08
CuI	1×10^{-12}	12.0
Hg_2I_2	1.1×10^{-28}	27.95
PbI_2	7.9×10^{-9}	8.10
SnI_2	8.3×10^{-6}	5.08
TlI	5.9×10^{-8}	7.23
Oxalates		
BaC_2O_4	1×10^{-6}	6.0
CaC_2O_4	1×10^{-8}	8.0
$La_2(C_2O_4)_3$	1×10^{-25}	25.0
SrC_2O_4	4×10^{-7}	6.4
Phosphates		
Ag_3PO_4	2.8×10^{-18}	17.55
$Ca_3(PO_4)_2$	2.0×10^{-29}	28.70
$FePO_4$	4×10^{-27}	26.4
$Fe_3(PO_4)_2$	1×10^{-36}	36.0
$MgNH_4PO_4$	3×10^{-13}	12.5
Sulfates		
Ag_2SO_4	1.5×10^{-5}	4.83
$BaSO_4$	1.1×10^{-10}	9.96
$CaSO_4$	2.4×10^{-5}	4.62
Hg_2SO_4	7.4×10^{-7}	6.13
$PbSO_4$	1.6×10^{-8}	7.79
Sulfides		
Ag_2S	8×10^{-51}	50.1
CdS	1×10^{-27}	27.0
CuS	8×10^{-37}	36.1

Appendix F

According to Anion (*Cont.*)

Substance	K_{sp}	pK_{sp}
Sulfides, (*Cont.*)		
Cu_2S	3×10^{-49}	48.5
FeS	8×10^{-19}	18.1
HgS	2×10^{-53}	52.7
NiS	4×10^{-20}	19.4
PbS	3×10^{-28}	27.5
SnS	1×10^{-26}	26.0
ZnS	2×10^{-25}	24.7
Thiocyanates		
AgSCN	1.1×10^{-12}	11.97
CuSCN	4.0×10^{-14}	13.40
$Hg(SCN)_2$	2.8×10^{-20}	19.56
$Hg_2(SCN)_2$	3.0×10^{-20}	19.52
TlSCN	1.6×10^{-4}	3.79

According to Cation

Substance	Formula	K_{sp}	pK_{sp}
Barium carbonate	$BaCO_3$	5.0×10^{-9}	8.30
Barium chromate	$BaCrO_4$	2.1×10^{-10}	9.67
Barium iodate	$Ba(IO_3)_2$	1.5×10^{-9}	8.81
Barium oxalate	BaC_2O_4	1×10^{-6}	6.0
Barium sulfate	$BaSO_4$	1.1×10^{-10}	9.96
Cadmium carbonate	$CdCO_3$	1.8×10^{-14}	13.74
Cadmium hydroxide	$Cd(OH)_2$	4.5×10^{-15}	14.35
Cadmium sulfide	CdS	1×10^{-27}	27.0
Calcium carbonate	$CaCO_3$	4.5×10^{-9}	8.35
Calcium fluoride	CaF_2	3.9×10^{-11}	10.41
Calcium hydroxide	$Ca(OH)_2$	6.5×10^{-6}	5.19
Calcium iodate	$Ca(IO_3)_2$	7.1×10^{-7}	6.15
Calcium oxalate	CaC_2O_4	1×10^{-8}	8.0
Calcium phosphate	$Ca_3(PO_4)_2$	2.0×10^{-29}	28.70
Calcium sulfate	$CaSO_4$	2.4×10^{-5}	4.62
Chromium(III) hydroxide	$Cr(OH)_3$	2×10^{-30}	29.8
Copper(I) bromide	CuBr	5×10^{-9}	8.3
Copper(I) chloride	CuCl	1.9×10^{-7}	6.73
Copper(I) hydroxide	$CuOH (Cu_2O)$	4×10^{-30}	29.4
Copper(I) iodide	CuI	1×10^{-12}	12.0
Copper(I) sulfide	Cu_2S	3×10^{-49}	48.5
Copper(I) thiocyanate	CuSCN	4.0×10^{-14}	13.40
Copper(II) chromate	$CuCrO_4$	3.6×10^{-6}	5.44
Copper(II) hydroxide	$Cu(OH)_2$	4.8×10^{-20}	19.32
Copper(II) sulfide	CuS	8×10^{-37}	36.1
Iron(II) hydroxide	$Fe(OH)_2$	8×10^{-16}	15.1
Iron(II) phosphate	$Fe_3(PO_4)_2$	1×10^{-36}	36.0
Iron(II) sulfide	FeS	8×10^{-19}	18.1

According to Cation (Cont.)

Substance	Formula	K_{sp}	pK_{sp}
Iron(III) hydroxide	Fe(OH)$_3$	2×10^{-39}	38.8
Iron(III) phosphate	FePO$_4$	4×10^{-27}	26.4
Lanthanum carbonate	La$_2$(CO$_3$)$_3$	4×10^{-34}	33.4
Lanthanum fluoride	LaF$_3$	1×10^{-29}	29.0
Lanthanum hydroxide	La(OH)$_3$	2×10^{-21}	20.7
Lanthanum oxalate	La$_2$(C$_2$O$_4$)$_3$	1×10^{-25}	25.0
Lead(II) bromate	Pb(BrO$_3$)$_2$	7.9×10^{-6}	5.10
Lead(II) bromide	PbBr$_2$	2.1×10^{-6}	5.68
Lead(II) carbonate	PbCO$_3$	7.4×10^{-14}	13.13
Lead(II) chloride	PbCl$_2$	1.7×10^{-5}	4.78
Lead(II) hydroxide	Pb(OH)$_2$ (PbO)	8×10^{-16}	15.1
Lead(II) iodide	PbI$_2$	7.9×10^{-9}	8.10
Lead(II) sulfate	PbSO$_4$	1.6×10^{-7}	7.79
Lead(II) sulfide	PbS	3×10^{-28}	27.5
Magnesium ammonium phosphate	MgNH$_4$PO$_4$	3×10^{-13}	12.5
Magnesium carbonate	MgCO$_3$	3.5×10^{-8}	7.46
Magnesium fluoride	MgF$_2$	6.6×10^{-9}	8.18
Magnesium hydroxide	Mg(OH)$_2$	7.1×10^{-12}	11.15
Mercury(I) bromide	Hg$_2$Br$_2$	5.6×10^{-23}	22.25
Mercury(I) carbonate	Hg$_2$CO$_3$	8.9×10^{-17}	16.05
Mercury(I) chloride	Hg$_2$Cl$_2$	1.2×10^{-18}	17.91
Mercury(I) chromate	Hg$_2$CrO$_4$	2.0×10^{-9}	8.70
Mercury(I) cyanide	Hg$_2$(CN)$_2$	5×10^{-40}	39.3
Mercury(I) iodate	Hg$_2$(IO$_3$)$_2$	1.3×10^{-18}	17.89
Mercury(I) iodide	Hg$_2$I$_2$	1.1×10^{-28}	27.95
Mercury(I) sulfate	Hg$_2$SO$_4$	7.4×10^{-7}	6.13
Mercury(I) thiocyanate	Hg(SCN)$_2$	3.0×10^{-20}	19.52
Mercury(II) bromide	HgBr$_2$	1.3×10^{-19}	18.89
Mercury(II) sulfide	HgS	2×10^{-53}	52.7
Mercury(II) thiocyanate	Hg(SCN)$_2$	2.8×10^{-20}	19.56
Nickel sulfide	NiS	4×10^{-20}	19.4
Silver bromate	AgBrO$_3$	5.5×10^{-5}	4.26
Silver bromide	AgBr	5.0×10^{-13}	12.30
Silver carbonate	Ag$_2$CO$_3$	8.1×10^{-12}	11.09
Silver chloride	AgCl	1.8×10^{-10}	9.74
Silver chromate	Ag$_2$CrO$_4$	1.2×10^{-12}	11.92
Silver cyanide	AgCN	2.2×10^{-16}	15.66
Silver hydroxide	AgOH (Ag$_2$O)	3.8×10^{-16}	15.42
Silver iodate	AgIO$_3$	3.1×10^{-8}	7.51
Silver iodide	AgI	8.3×10^{-17}	16.08
Silver phosphate	Ag$_3$PO$_4$	2.8×10^{-18}	17.55
Silver sulfate	Ag$_2$SO$_4$	1.5×10^{-5}	4.83
Silver sulfide	Ag$_2$S	8×10^{-51}	50.1
Silver thiocyanate	AgSCN	1.1×10^{-12}	11.97
Strontium carbonate	SrCO$_3$	9.3×10^{-10}	9.03
Strontium iodate	Sr(IO$_3$)$_2$	3.3×10^{-7}	6.48
Strontium oxalate	SrC$_2$O$_4$	4×10^{-7}	6.4

Substance	Formula	K_{sp}	pK_{sp}
Thallium(I) bromate	$TlBrO_3$	1.7×10^{-4}	3.78
Thallium(I) bromide	$TlBr$	3.6×10^{-6}	5.44
Thallium(I) chloride	$TlCl$	1.8×10^{-4}	3.74
Thallium(I) chromate	Tl_2CrO_4	9.8×10^{-13}	12.01
Thallium(I) iodate	$TlIO_3$	3.1×10^{-6}	5.51
Thallium(I) iodide	TlI	5.9×10^{-8}	7.23
Thallium(I) thiocyanate	$TlSCN$	1.6×10^{-4}	3.79
Thorium fluoride	ThF_4	5×10^{-29}	28.3
Thorium iodate	$Th(IO_3)_4$	2.4×10^{-15}	14.62
Tin(II) iodide	SnI_2	8.3×10^{-6}	5.08
Tin(II) sulfide	SnS	1×10^{-26}	26.0
Zinc cyanide	$Zn(CN)_2$	3×10^{-16}	15.5
Zinc sulfide	ZnS	2×10^{-25}	24.7

G

ACID DISSOCIATION CONSTANTS

Name	Formula	K_a	pK_a
Acetic acid	CH_3CO_2H	1.76×10^{-5}	4.754
Adipic acid	$HO_2C(CH_2)_4CO_2H$	3.8×10^{-5}	4.42
		3.8×10^{-6}	5.42
Arsenic acid	H_3AsO_4	5.8×10^{-3}	2.24
		1.1×10^{-7}	6.96
		3.2×10^{-12}	11.49
Arsenious acid	H_3AsO_3	5.1×10^{-10}	9.29
Benzoic acid	⬡—CO_2H	6.28×10^{-5}	4.202
Boric acid	H_3BO_3	5.81×10^{-10}	9.236
		1.8×10^{-13}	12.74
		1.6×10^{-14}	13.80
Carbonic acid	H_2CO_3	4.45×10^{-7}	6.352
		4.69×10^{-11}	10.329
Chloroacetic acid	$ClCH_2CO_2H$	1.36×10^{-3}	2.866
Chlorous acid	$HClO_2$	1.1×10^{-2}	1.96
Citric acid	$\begin{array}{c} CO_2H \\ | \\ HO_2CCH_2CCH_2CO_2H \\ | \\ OH \end{array}$	7.45×10^{-4}	3.128
		1.73×10^{-5}	4.762
		4.02×10^{-7}	6.396
Dichloroacetic acid	Cl_2CHCO_2H	5.0×10^{-2}	1.30
Ethylenediaminetetraacetic acid	$(HO_2CCH_2)_2NCH_2CH_2N(CH_2CO_2H)_2$	1×10^{-2}	2.0
		2.1×10^{-3}	2.68
		7.8×10^{-7}	6.11
		6.8×10^{-11}	10.17
Formic acid	HCO_2H	1.80×10^{-4}	3.745
Fumaric acid	$\begin{array}{c} HO_2C \quad\quad H \\ \quad C=C \\ H \quad\quad CO_2H \end{array}$	8.85×10^{-4}	3.053
		3.21×10^{-5}	4.493
Hydrazoic acid	HN_3	2.2×10^{-5}	4.66

Name	Formula	K_a	pK_a
Hydrocyanic acid	HCN	6.2×10^{-10}	9.21
Hydrofluoric acid	HF	6.8×10^{-4}	3.17
Hydrogen peroxide	H_2O_2	2.2×10^{-12}	11.66
Hydrosulfuric acid	H_2S	9.5×10^{-8}	7.02
		1×10^{-14}	14.0
Hydrothiocyanic acid	HSCN	1×10^{-1}	1.0
Hydroxyacetic acid	$HOCH_2CO_2H$	1.48×10^{-4}	3.830
Hypochlorous acid	HClO	3.0×10^{-8}	7.52
Hypophosphorous acid	$\overset{\displaystyle O}{\overset{\displaystyle \|}{HOPH_2}}$	5.9×10^{-2}	1.23
Iodic acid	$\overset{\displaystyle O}{\overset{\displaystyle \|}{HOI=O}}$	1.7×10^{-1}	0.77
Lactic acid	$\overset{\displaystyle OH}{\overset{\displaystyle \|}{CH_3CHCO_2H}}$	1.35×10^{-4}	3.870
Maleic acid	$\begin{array}{c} HO_2C \quad\quad CO_2H \\ \diagdown \quad\quad \diagup \\ C=C \\ \diagup \quad\quad \diagdown \\ H \quad\quad\quad H \end{array}$	1.23×10^{-2} 4.66×10^{-7}	1.910 6.332
Malic acid	$\overset{\displaystyle OH}{\overset{\displaystyle \|}{HO_2CCH_2CHCO_2H}}$	3.48×10^{-4} 8.00×10^{-6}	3.458 5.097
Malonic acid	$HO_2CCH_2CO_2H$	1.42×10^{-3} 2.01×10^{-6}	2.848 5.697
Nitrilotriacetic acid	$N(CH_2CO_2H)_3$	2.24×10^{-2} 1.15×10^{-3} 4.63×10^{-11}	1.650 2.939 10.334
Nitrous acid	HNO_2	7.1×10^{-4}	3.15
Oxalic acid	HO_2CCO_2H	5.60×10^{-2} 5.42×10^{-5}	1.252 4.266
Periodic acid	H_5IO_6	2.6×10^{-2} 5.1×10^{-9}	1.59 8.29
Phenol	⬡—OH	1.05×10^{-10}	9.979
Phenylacetic acid	⬡—CH_2CO_2H	4.90×10^{-5}	4.310

Name	Formula	K_a	pK_a
Phosphoric acid	H_3PO_4	7.11×10^{-3}	2.148
		6.32×10^{-8}	7.199
		4.5×10^{-13}	12.35
Phosphorus acid	$\overset{\displaystyle O}{\overset{\|}{HP(OH)_2}}$	3×10^{-2}	1.5
		1.6×10^{-7}	6.80
Phthalic acid		1.12×10^{-3}	2.951
		3.91×10^{-6}	5.408
Salicylic acid		1.1×10^{-3}	2.96
		1.8×10^{-14}	13.74
Succinic acid	$HO_2CCH_2CH_2CO_2H$	6.21×10^{-5}	4.207
		2.31×10^{-6}	5.636
Sulfamic acid	H_2NSO_3H	1.03×10^{-1}	0.987
Sulfuric acid	H_2SO_4	—	—
		1.0×10^{-2}	2.00
Sulfurous acid	H_2SO_3	1.2×10^{-2}	1.92
		6.6×10^{-8}	7.18
Tartaric acid	$\overset{\displaystyle OH}{\overset{\|}{\underset{\underset{\displaystyle OH}{\|}}{HO_2CCHCHCO_2H}}}$	9.20×10^{-4}	3.036
		4.31×10^{-5}	4.366
Trichloroacetic acid	Cl_3CCO_2H	2.2×10^{-1}	0.66

Name	Formula	K_b	pK_b
Ammonia	NH_3	1.75×10^{-5}	4.757
Aniline	⬡—NH_2	3.99×10^{-10}	9.399
Benzylamine	⬡—CH_2NH_2	2.2×10^{-5}	4.66
2,2'-Bipyridine	(N, N ring structure)	2.2×10^{-10}	9.66
Butylamine	$CH_3(CH_2)_3NH_2$	4.37×10^{-4}	3.360
Cyclohexylamine	⬡—NH_2	4.4×10^{-4}	3.36
Diethylamine	$(CH_3CH_2)_2NH$	8.57×10^{-4}	3.067
Dimethylamine	$(CH_3)_2NH$	5.94×10^{-4}	3.226
Ethanolamine	$HOCH_2CH_2NH_2$	3.15×10^{-5}	4.502
Ethylamine	$CH_3CH_2NH_2$	4.33×10^{-4}	3.364
Ethylenediamine	$H_2NCH_2CH_2NH_2$	8.47×10^{-5} 7.05×10^{-8}	4.072 7.152
Glycinate	$H_2NCH_2CO_2^-$	6.00×10^{-5} 2.24×10^{-12}	4.222 11.650
Hydrazine	H_2NNH_2	9.5×10^{-7}	6.02
Hydroxylamine	$HONH_2$	9.1×10^{-9}	8.04
8-Hydroxyquinolinate	(ring structure with O^- and N)	6.5×10^{-5} 8.1×10^{-10}	4.19 9.09

Name	Formula	K_b	pK_b
4-Methoxyaniline (p-anisidine)	CH_3O—⬡—NH_2	2.28×10^{-9}	8.642
Methylamine	CH_3NH_2	4.4×10^{-4}	3.36
2-Methylaniline (o-toluidine)	⬡ with CH_3 and NH_2	2.80×10^{-10}	9.553
4-Methylaniline (m-toluidine)	CH_3—⬡—NH_2	1.21×10^{-9}	8.917
1,10-Phenanthroline	(phenanthroline structure)	7.2×10^{-10}	9.14
Propylamine	$CH_3CH_2CH_2NH_2$	3.68×10^{-4}	3.434
Pyridine	(pyridine structure)	1.69×10^{-9}	8.772
Triethanolamine	$(HOCH_2CH_2)_3N$	5.78×10^{-7}	6.238
Triethylamine	$(CH_3CH_2)_3N$	5.19×10^{-4}	3.285
Trimethylamine	$(CH_3)_3N$	6.31×10^{-5}	4.200
Tris(hydroxymethyl) aminomethane (Tris or THAM)	$(HOCH_2)_3CNH_2$	1.19×10^{-6}	5.924

I

STEPWISE FORMATION CONSTANTS

Ligand/cation	log K_1	log K_2	log K_3	log K_4	log K_5	log K_6
NH$_3$						
Ag$^+$	3.31	3.91				
Cd^{2+}	2.55	2.24	1.34	0.84	−0.36	−1.61
Co^{2+}	1.99	1.51	0.93	0.64	0.06	−0.74
Cu^{2+}	4.04	3.43	2.80	1.48		
Ni^{2+}	2.72	2.17	1.66	1.12	0.67	−0.03
Zn^{2+}	2.21	2.29	2.36	2.03		
Br$^-$						
Ag$^+$	4.30	2.34	1.5	0.8		
Bi^{3+}	3.06	2.5	1.8	1.2		
Cd^{2+}	2.14	0.9	0.0	−0.1		
Hg^{2+}	9.40	8.58	2.7	1.5		
Pb^{2+}	1.77	0.8	0.2	−0.7		
Cl$^-$						
Ag$^+$	3.70	1.92	0.8	−0.3		
Bi^{3+}	2.2	1.3	2.3	1.0	0.5	0.1
Cd^{2+}	1.98	0.6	−0.2	−0.7		
Fe^{2+}	0.4	0.0				
Fe^{3+}	1.48	0.65				
Hg^{2+}	6.74	6.48	0.9	1.0		
Pb^{2+}	1.59	0.2	−0.1	−0.3		
Sn^{2+}	1.51	0.74	−0.25	−0.50		
CN$^-$						
Ag$^+$	—	20.48[a]	0.9			
Cd^{2+}	5.6	5.1	4.8	3.5		
Hg^{2+}	18.00	16.71	3.83	3.0		
Ni^{2+}	—	—	—	30.2[b]		
Zn^{2+}	5.3	6.4	5.0	4.9		
EDTA						
See Appendix J						
F$^-$						
Al^{3+}	6.11	5.01	3.9	3.0	1.4	0.4
Fe^{3+}	5.18	3.95	2.8			
Sn^{2+}	4.08	2.60	2.8			
Th^{4+}	8.4	6.7	4.7	3.4		
OH$^-$						
Al^{3+}	9.01	—	—	24[d]		
Cd^{2+}	4.1	3.6	2.6	1.7		
Cu^{2+}	6.3	—	—	10.1[d]		
Fe^{2+}	4.5	2.9	2.6	−0.4		

Stepwise Formation Constants (*Cont.*)

Ligand/cation	$\log K_1$	$\log K_2$	$\log K_3$	$\log K_4$	$\log K_5$	$\log K_6$
OH^-, (*Cont.*)						
Fe^{3+}	11.1	11.0				
Hg^{2+}	10.6	11.2	−0.9			
Ni^{2+}	4.1	4	3			
Pb^{2+}	6.3	4.6	3.0			
Sn^{2+}	10.4	13.5				
Zn^{2+}	5.0	—	8.6[c]			
I^-						
Ag^+	6.6	5.1	1.4	1.3		
Bi^{3+}	3.63	—	—	11.4[d]	1.8	2.0
Cd^{2+}	2.28	1.64	1.1	1.0		
Hg^{2+}	12.87	10.95	3.8	2.2		
Pb^{2+}	1.26	1.5	0.6	0.5		
$C_2O_4{}^{2-}$						
Al^{3+}	—	13[a]	3.8			
Cd^{2+}	3.5	1.8				
Co^{2+}	4.7	2.4	0.8			
Cu^{2+}	6.2	1.8				
Fe^{2+}	4.7	3.0				
Fe^{3+}	9.4	6.8	4.0			
Ni^{2+}	5.3	2.3				
Pb^{2+}	—	6.5[a]				
Zn^{2+}	4.9	2.7				
SCN^-						
Ag^+	4.8	3.4	1.3	0.2		
Bi^{3+}	2.21	0.5	1.4	0.8	0.6	−0.4
Cd^{2+}	0.89	0.89	0.02	−0.50		
Co^{2+}	0.98	0.34				
Cu^{2+}	1.74	1.00				
Fe^{3+}	2.21	1.43	1.4	1.3		
Hg^{2+}	9.08	7.78	2.84	2.0		
Ni^{2+}	1.13	0.45	0.0			
Pb^{2+}	1.08	0.40				

[a] $\log K_1 K_2$

[b] $\log K_1 K_2 K_3 K_4$

[c] $\log K_2 K_3$

[d] $\log K_2 K_3 K_4$

J

METAL–EDTA FORMATION CONSTANTS

Metal	K_{MY}	$\log K_{MY}$
Ag^+	2.1×10^7	7.32
Al^{3+}	1.3×10^{16}	16.11
Ba^{2+}	5.8×10^7	7.76
Bi^{3+}	6.3×10^{27}	27.80
Ca^{2+}	5.0×10^{10}	10.70
Cd^{2+}	2.9×10^{16}	16.46
Co^{2+}	2.0×10^{16}	16.30
Cu^{2+}	6.3×10^{18}	18.80
Fe^{2+}	2.1×10^{14}	14.32
Fe^{3+}	1.3×10^{25}	25.11
Hg^{2+}	6.3×10^{21}	21.80
Mg^{2+}	4.9×10^8	8.69
Mn^{2+}	6.2×10^{13}	13.79
Ni^{2+}	4.2×10^{18}	18.62
Pb^{2+}	1.1×10^{18}	18.04
Sn^{2+}	2.0×10^{18}	18.30
Sr^{2+}	4.3×10^8	8.63
Zn^{2+}	3.2×10^{16}	16.51

STANDARD ELECTRODE POTENTIALS

According to Element

Half-reaction	$E°\,(V)^a$
$Ag^+ + e^- \rightleftharpoons Ag(s)$	0.800
$AgBr(s) + e^- \rightleftharpoons Ag(s) + Br^-$	0.0713
$AgCl(s) + e^- \rightleftharpoons Ag(s) + Cl^-$	0.222
$Ag(CN)_2^- + e^- \rightleftharpoons Ag(s) + 2CN^-$	−0.31
$AgI(s) + e^- \rightleftharpoons Ag(s) + I^-$	−0.152
$Ag_2S(s) + 2e^- \rightleftharpoons 2Ag(s) + S^{2-}$	−0.691
$Al^{3+} + 3e^- \rightleftharpoons Al(s)$	−1.662
$AlF_6^{3-} + 3e^- \rightleftharpoons Al(s) + 6F^-$	−2.069
$Al(OH)_4^- + 3e^- \rightleftharpoons Al(s) + 4OH^-$	−2.33
$H_3AsO_4 + 2H^+ + 2e^- \rightleftharpoons H_3AsO_3 + H_2O$	0.560
$H_3AsO_3 + 3H^+ + 3e^- \rightleftharpoons As(s) + 3H_2O$	0.248
$Ba^{2+} + 2e^- \rightleftharpoons Ba(s)$	−2.912
$Bi^{3+} + 3e^- \rightleftharpoons Bi(s)$	0.200
$Br_2(aq) + 2e^- \rightleftharpoons 2Br^-$	1.087
$Br_2(l) + 2e^- \rightleftharpoons 2Br^-$	1.066
$Br_3^- + 2e^- \rightleftharpoons 3Br^-$	1.051
$2BrO_3^- + 12H^+ + 10e^- \rightleftharpoons Br_2(l) + 6H_2O$	1.495
$Ca^{2+} + 2e^- \rightleftharpoons Ca(s)$	−2.868
$C_6H_4O_2(quinone) + 2H^+ + 2e^- \rightleftharpoons C_6H_4(OH)_2$	0.700
$C_6H_6O_6 + 2H^+ + 2e^- \rightleftharpoons C_6H_8O_6(ascorbic\ acid)$	0.390
$2CO_2(g) + 2H^+ + 2e^- \rightleftharpoons H_2C_2O_4$	−0.49
$Cd^{2+} + 2e^- \rightleftharpoons Cd(s)$	−0.403
$Ce^{4+} + e^- \rightleftharpoons Ce^{3+}$	1.70 (1 M HClO$_4$)
	1.61 (1 M HNO$_3$)
	1.44 (1 M H$_2$SO$_4$)
	1.28 (1 M HCl)
$ClO_4^- + 2H^+ + 2e^- \rightleftharpoons ClO_3^- + H_2O$	1.19
$ClO_3^- + 3H^+ + 2e^- \rightleftharpoons HClO_2 + H_2O$	1.21
$2ClO_3^- + 12H^+ + 10e^- \rightleftharpoons Cl_2(g) + 6H_2O$	1.46
$HClO_2 + 2H^+ + 2e^- \rightleftharpoons HOCl + H_2O$	1.65
$2HOCl + 2H^+ + 2e^- \rightleftharpoons Cl_2(g) + 2H_2O$	1.61
$Cl_2(g) + 2e^- \rightleftharpoons 2Cl^-$	1.358
$Cl_2(aq) + 2e^- \rightleftharpoons 2Cl^-$	1.395

According to Element (*Cont.*)

Half-reaction	$E°\,(V)^a$
$Co^{3+} + e^- \rightleftharpoons Co^{2+}$	1.83
$Co^{2+} + 2e^- \rightleftharpoons Co(s)$	-0.28
$Cr_2O_7^{2-} + 14H^+ + 6e^- \rightleftharpoons 2Cr^{3+} + 7H_2O$	1.33
$Cr^{3+} + e^- \rightleftharpoons Cr^{2+}$	-0.407
$Cr^{3+} + 3e^- \rightleftharpoons Cr(s)$	-0.744
$Cr^{2+} + 2e^- \rightleftharpoons Cr(s)$	-0.913
$Cu^{2+} + e^- \rightleftharpoons Cu^+$	0.153
$Cu^{2+} + 2e^- \rightleftharpoons Cu(s)$	0.342
$Cu^{2+} + I^- + e^- \rightleftharpoons CuI(s)$	0.86
$Cu^+ + e^- \rightleftharpoons Cu(s)$	0.521
$CuCl(s) + e^- \rightleftharpoons Cu(s) + Cl^-$	0.137
$F_2(g) + 2e^- \rightleftharpoons 2F^-$	2.87
$Fe^{3+} + e^- \rightleftharpoons Fe^{2+}$	0.771
	0.762 (1 M HClO$_4$)
	0.746 (1 M HNO$_3$)
	0.732 (1 M HCl)
$Fe(phen)_3^{3+} + e^- \rightleftharpoons Fe(phen)_3^{2+b}$	1.06
$Fe(CN)_6^{3-} + e^- \rightleftharpoons Fe(CN)_6^{4-}$	0.358
$Fe^{2+} + 2e^- \rightleftharpoons Fe(s)$	-0.447
$Ga^{3+} + 3e^- \rightleftharpoons Ga(s)$	-0.560
$2H^+ + 2e^- \rightleftharpoons H_2(g)$	0.00
$2H_2O + 2e^- \rightleftharpoons H_2(g) + 2OH^-$	-0.828
$2Hg^{2+} + 2e^- \rightleftharpoons Hg_2^{2+}$	0.920
$Hg^{2+} + 2e^- \rightleftharpoons Hg(l)$	0.854
$Hg_2^{2+} + 2e^- \rightleftharpoons 2Hg(l)$	0.797
$Hg_2Cl_2(s) + 2e^- \rightleftharpoons 2Hg(l) + 2Cl^-$	0.268
$H_5IO_6 + H^+ + 2e^- \rightleftharpoons IO_3^- + 3H_2O$	1.601
$2IO_3^- + 12H^+ + 10e^- \rightleftharpoons I_2(s) + 6H_2O$	1.195
$IO_3^- + 3H_2O + 6e^- \rightleftharpoons I^- + 6OH^-$	0.26
$I_2(aq) + 2e^- \rightleftharpoons 2I^-$	0.615
$I_2(s) + 2e^- \rightleftharpoons 2I^-$	0.536
$I_3^- + 2e^- \rightleftharpoons 3I^-$	0.536
$K^+ + e \rightleftharpoons K(s)$	-2.931
$Li^+ + e^- \rightleftharpoons Li(s)$	-3.040
$Mg^{2+} + 2e^- \rightleftharpoons Mg(s)$	-2.372
$MnO_4^- + 4H^+ + 3e^- \rightleftharpoons MnO_2(s) + 2H_2O$	1.679
$MnO_4^- + 8H^+ + 5e^- \rightleftharpoons Mn^{2+} + 4H_2O$	1.507
$MnO_2(s) + 4H^+ + 2e^- \rightleftharpoons Mn^{2+} + 2H_2O$	1.224
$Mn^{2+} + 2e^- \rightleftharpoons Mn(s)$	-1.185
$MoO_4^{2-} + 4H^+ + 2e^- \rightleftharpoons MoO_2(s) + 2H_2O$	0.606
$2NO_3^- + 4H^+ + 2e^- \rightleftharpoons N_2O_4(g) + 2H_2O$	0.803
$NO_3^- + 3H^+ + 2e^- \rightleftharpoons HNO_2 + H_2O$	0.934
$NO_3^- + 4H^+ + 3e^- \rightleftharpoons NO(g) + 2H_2O$	0.957
$HNO_2 + H^+ + e^- \rightleftharpoons NO(g) + H_2O$	0.957

Half-reaction	$E°$ (V)[a]
$2NO(g) + 2H^+ + 2e^- \rightleftharpoons N_2O(g) + H_2O$	1.59
$N_2O(g) + 2H^+ + 2e^- \rightleftharpoons N_2(g) + H_2O$	1.77
$Na^+ + e^- \rightleftharpoons Na(s)$	−2.71
$Ni^{2+} + 2e^- \rightleftharpoons Ni(s)$	−0.257
$O_3(g) + 2H^+ + 2e^- \rightleftharpoons O_2(g) + H_2O$	2.08
$O_2(g) + 2H^+ + 2e^- \rightleftharpoons H_2O_2$	0.695
$O_2(g) + 4H^+ + 4e^- \rightleftharpoons 2H_2O$	1.229
$H_2O_2 + 2H^+ + 2e^- \rightleftharpoons 2H_2O$	1.776
$H_3PO_4 + 2H^+ + 2e^- \rightleftharpoons H_3PO_3 + H_2O$	−0.276
$H_3PO_3 + 2H^+ + 2e^- \rightleftharpoons H_3PO_2 + H_2O$	−0.499
$PbO_2(s) + 4H^+ + SO_4^{2-} + 2e^- \rightleftharpoons PbSO_4(s) + 2H_2O$	1.691
$PbO_2(s) + 4H^+ + 2e^- \rightleftharpoons Pb^{2+} + 2H_2O$	1.455
$PbSO_4(s) + 2e^- \rightleftharpoons Pb(s) + SO_4^{2-}$	−0.359
$Pb^{2+} + 2e^- \rightleftharpoons Pb(s)$	−0.126
$PtCl_6^{2-} + 2e^- \rightleftharpoons PtCl_4^{2-} + 2Cl^-$	0.68
$PtCl_4^{2-} + 2e^- \rightleftharpoons Pt(s) + 4Cl^-$	0.755
$Pt^{2+} + 2e^- \rightleftharpoons Pt(s)$	1.188
$S_2O_8^{2-} + 2e^- \rightleftharpoons 2SO_4^{2-}$	2.01
$SO_4^{2-} + 4H^+ + 2e^- \rightleftharpoons H_2SO_3 + H_2O$	0.172
$2H_2SO_3 + 2H^+ + 4e^- \rightleftharpoons S_2O_3^{2-} + 3H_2O$	0.400
$H_2SO_3 + 4H^+ + 4e^- \rightleftharpoons S(s) + 3H_2O$	0.449
$S_4O_6^{2-} + 2e^- \rightleftharpoons 2S_2O_3^{2-}$	0.08
$S(s) + 2H^+ + 2e^- \rightleftharpoons H_2S(aq)$	0.142
$S(s) + 2H^+ + 2e^- \rightleftharpoons H_2S(g)$	0.171
$SbO^+ + 2H^+ + 3e^- \rightleftharpoons Sb(s) + H_2O$	0.230
$Sb_2O_3(g) + 6H^+ + 6e^- \rightleftharpoons 2Sb(s) + 3H_2O$	0.167
$H_2SeO_3 + 4H^+ + 4e^- \rightleftharpoons Se(s) + 3H_2O$	0.74
$Se(s) + 2H^+ + 2e^- \rightleftharpoons H_2Se(g)$	−0.369
$SiO_2(s) + 4H^+ + 4e^- \rightleftharpoons Si(s) + 2H_2O$	−0.807
$SiF_6^{2-} + 4e^- \rightleftharpoons Si(s) + 6F^-$	−1.24
$Sn^{4+} + 2e^- \rightleftharpoons Sn^{2+}$	0.151 0.139 (1 M HCl)
$Sn^{2+} + 2e^- \rightleftharpoons Sn(s)$	−0.136
$TiO^{2+} + 2H^+ + e^- \rightleftharpoons Ti^{3+} + H_2O$	0.100
$Ti^{3+} + e^- \rightleftharpoons Ti^{2+}$	−0.368
$Tl^{3+} + 2e^- \rightleftharpoons Tl^+$	1.252 0.77 (1 M HCl)
$Tl^+ + e^- \rightleftharpoons Tl(s)$	−0.336
$UO_2^{2+} + e^- \rightleftharpoons UO_2^+$	0.052
$UO_2^{2+} + 4H^+ + 2e^- \rightleftharpoons U^{4+} + 2H_2O$	0.327
$UO_2^+ + 4H^+ + e^- \rightleftharpoons U^{4+} + 2H_2O$	0.612
$U^{4+} + e^- \rightleftharpoons U^{3+}$	−0.607
$VO_2^+ + 2H^+ + e^- \rightleftharpoons VO^{2+} + H_2O$	0.991

According to Element (Cont.)

Half-reaction	$E°$ (V)[a]
$VO^{2+} + 2H^+ + e^- \rightleftharpoons V^{3+} + H_2O$	0.337
$V^{3+} + e^- \rightleftharpoons V^{2+}$	-0.255
$Zn^{2+} + 2e^- \rightleftharpoons Zn(s)$	-0.762

[a]A number of these potentials cannot be determined experimentally and are calculated from theory.
[b]Phen stands for 1,10-phenanthroline.

According to Value

Half-reaction	$E°$ (V)[a]
$F_2(g) + 2e^- \rightleftharpoons 2F^-$	2.87
$O_3(g) + 2H^+ + 2e^- \rightleftharpoons O_2(g) + H_2O$	2.08
$S_2O_8^{2-} + 2e^- \rightleftharpoons 2SO_4^{2-}$	2.01
$Co^{3+} + e^- \rightleftharpoons Co^{2+}$	1.83
$H_2O_2 + 2H^+ + 2e^- \rightleftharpoons 2H_2O$	1.776
$N_2O(g) + 2H^+ + 2e^- \rightleftharpoons N_2(g) + H_2O$	1.77
$Ce^{4+} + e^- \rightleftharpoons Ce^{3+}$	1.70 (1 M HClO$_4$)
$PbO_2(s) + 4H^+ + SO_4^{2-} + 2e^- \rightleftharpoons PbSO_4(s) + 2H_2O$	1.691
$MnO_4^- + 4H^+ + 3e^- \rightleftharpoons MnO_2(s) + 2H_2O$	1.679
$HClO_2 + 2H^+ + 2e^- \rightleftharpoons HOCl + H_2O$	1.65
$2HOCl + 2H^+ + 2e^- \rightleftharpoons Cl_2(g) + 2H_2O$	1.61
$Ce^{4+} + e^- \rightleftharpoons Ce^{3+}$	1.61 (1 M HNO$_3$)
$H_5IO_6 + H^+ + 2c^- \rightleftharpoons IO_3^- + 3H_2O$	1.601
$2NO(g) + 2H^+ + 2e^- \rightleftharpoons N_2O(g) + H_2O$	1.59
$MnO_4^- + 8H^+ + 5e^- \rightleftharpoons Mn^{2+} + 4H_2O$	1.507
$2BrO_3^- + 12H^+ + 10e^- \rightleftharpoons Br_2(l) + 6H_2O$	1.495
$2ClO_3^- + 12H^+ + 10e^- \rightleftharpoons Cl_2(g) + 6H_2O$	1.46
$PbO_2(s) + 4H^+ + 2e^- \rightleftharpoons Pb^{2+} + 2H_2O$	1.455
$Ce^{4+} + e^- \rightleftharpoons Ce^{3+}$	1.44 (1 M H$_2$SO$_4$)
$Cl_2(aq) + 2e^- \rightleftharpoons 2Cl^-$	1.395
$Cl_2(g) + 2e^- \rightleftharpoons 2Cl^-$	1.358
$Cr_2O_7^{2-} + 14H^+ + 6e^- \rightleftharpoons 2Cr^{3+} + 7H_2O$	1.33
$Ce^{4+} + e^- \rightleftharpoons Ce^{3+}$	1.28 (1 M HCl)
$Tl^{3+} + 2e^- \rightleftharpoons Tl^+$	1.252
$2O_2(g) + 4H^+ + 4e^- \rightleftharpoons 2H_2O$	1.229
$MnO_2(s) + 4H^+ + 2e^- \rightleftharpoons Mn^{2+} + 2H_2O$	1.224
$ClO_3^- + 3H^+ + 2e^- \rightleftharpoons HClO_2 + H_2O$	1.21
$2IO_3^- + 12H^+ + 10e^- \rightleftharpoons I_2(s) + 6H_2O$	1.195
$ClO_4^- + 2H^+ + 2e^- \rightleftharpoons ClO_3^- + H_2O$	1.19
$Pt^{2+} + 2e^- \rightleftharpoons Pt(s)$	1.188
$Br_2(aq) + 2e^- \rightleftharpoons 2Br^-$	1.087
$Br_2(l) + 2e^- \rightleftharpoons 2Br^-$	1.066
$Fe(phen)_3^{3+} + e^- \rightleftharpoons Fe(phen)_3^{2+}$[b]	1.06
$Br_3^- + 2e^- \rightleftharpoons 3Br^-$	1.051
$VO_2^+ + 2H^+ + e^- \rightleftharpoons VO^{2+} + H_2O$	0.991

Half-reaction	$E°$ (V)[a]
$HNO_2 + H^+ + e^- \rightleftharpoons NO(g) + H_2O$	0.957
$NO_3^- + 4H^+ + 3e^- \rightleftharpoons NO(g) + 2H_2O$	0.957
$NO_3^- + 3H^+ + 2e^- \rightleftharpoons HNO_2 + H_2O$	0.934
$2Hg^{2+} + 2e^- \rightleftharpoons Hg_2^{2+}$	0.920
$Cu^{2+} + I^- + e^- \rightleftharpoons CuI(s)$	0.86
$Hg^{2+} + 2e^- \rightleftharpoons Hg(l)$	0.854
$2NO_3^- + 4H^+ + 2e^- \rightleftharpoons N_2O_4(g) + 2H_2O$	0.803
$Ag^+ + e^- \rightleftharpoons Ag(s)$	0.800
$Hg_2^{2+} + 2e^- \rightleftharpoons 2Hg(l)$	0.797
$Fe^{3+} + e^- \rightleftharpoons Fe^{2+}$	0.771
$Tl^{3+} + 2e^- \rightleftharpoons Tl^+$	0.77 (1 M HCl)
$Fe^{3+} + e^- \rightleftharpoons Fe^{2+}$	0.762 (1 M HClO$_4$)
$PtCl_4^{2-} + 2e^- \rightleftharpoons Pt(s) + 4Cl^-$	0.755
$Fe^{3+} + e^- \rightleftharpoons Fe^{2+}$	0.746 (1 M HNO$_3$)
$H_2SeO_3 + 4H^+ + 4e^- \rightleftharpoons Se(s) + 3H_2O$	0.74
$Fe^{3+} + e^- \rightleftharpoons Fe^{2+}$	0.732 (1 M HCl)
$C_6H_4O_2 \text{(quinone)} + 2H^+ + 2e^- \rightleftharpoons C_6H_4(OH)_2$	0.700
$O_2(g) + 2H^+ + 2e^- \rightleftharpoons H_2O_2$	0.695
$PtCl_6^{2-} + 2e^- \rightleftharpoons PtCl_4^{2-} + 2Cl^-$	0.68
$I_2(aq) + 2e^- \rightleftharpoons 2I^-$	0.615
$UO_2^+ + 4H^+ + e^- \rightleftharpoons U^{4+} + 2H_2O$	0.612
$MoO_4^{2-} + 4H^+ + 2e^- \rightleftharpoons MoO_2(s) + 2H_2O$	0.606
$H_3AsO_4 + 2H^+ + 2e^- \rightleftharpoons H_3AsO_3 + H_2O$	0.560
$I_2(s) + 2e^- \rightleftharpoons 2I^-$	0.536
$I_3^- + 2e^- \rightleftharpoons 3I^-$	0.536
$Cu^+ + e^- \rightleftharpoons Cu(s)$	0.521
$H_2SO_3 + 4H^+ + 4e^- \rightleftharpoons S(s) + 3H_2O$	0.449
$2H_2SO_3 + 2H^+ + 4e^- \rightleftharpoons S_2O_3^{2-} + 3H_2O$	0.400
$C_6H_6O_6 + 2H^+ + 2e^- \rightleftharpoons C_6H_8O_6 \text{(ascorbic acid)}$	0.390
$Fe(CN)_6^{3-} + e^- \rightleftharpoons Fe(CN)_6^{4-}$	0.358
$Cu^{2+} + 2e^- \rightleftharpoons Cu(s)$	0.342
$VO^{2+} + 2H^+ + e^- \rightleftharpoons V^{3+} + H_2O$	0.337
$UO_2^{2+} + 4H^+ + 2e^- \rightleftharpoons U^{4+} + 2H_2O$	0.327
$Hg_2Cl_2(s) + 2e^- \rightleftharpoons 2Hg(l) + 2Cl^-$	0.268
$IO_3^- + 3H_2O + 6e^- \rightleftharpoons I^- + 6OH^-$	0.26
$H_3AsO_3 + 3H^+ + 3e^- \rightleftharpoons As(s) + 3H_2O$	0.248
$SbO^+ + 2H^+ + 3e^- \rightleftharpoons Sb(s) + H_2O$	0.230
$AgCl(s) + e^- \rightleftharpoons Ag(s) + Cl^-$	0.222
$Bi^{3+} + 3e^- \rightleftharpoons Bi(s)$	0.200
$SO_4^{2-} + 4H^+ + 2e^- \rightleftharpoons H_2SO_3 + H_2O$	0.172
$S(s) + 2H^+ + 2e^- \rightleftharpoons H_2S(g)$	0.171
$Sb_2O_3(s) + 6H^+ + 6e^- \rightleftharpoons 2Sb(s) + 3H_2O$	0.167
$Cu^{2+} + e^- \rightleftharpoons Cu^+$	0.153
$Sn^{4+} + 2e^- \rightleftharpoons Sn^{2+}$	0.151
$S(s) + 2H^+ + 2e^- \rightleftharpoons H_2S(aq)$	0.142
$Sn^{4+} + 2e^- \rightleftharpoons Sn^{2+}$	0.139 (1 M HCl)
$CuCl(s) + e^- \rightleftharpoons Cu(s) + Cl^-$	0.137
$TiO^{2+} + 2H^+ + e^- \rightleftharpoons Ti^{3+} + H_2O$	0.100
$S_4O_6^{2-} + 2e^- \rightleftharpoons 2S_2O_3^{2-}$	0.08
$AgBr(s) + e^- \rightleftharpoons Ag(s) + Br^-$	0.0713
$UO_2^{2+} + e^- \rightleftharpoons UO_2^+$	0.052
$2H^+ + 2e^- \rightleftharpoons H_2(g)$	0.00

According to Value (Cont.)

Half-reaction	$E°$ $(V)^a$
$Pb^{2+} + 2e^- \rightleftharpoons Pb(s)$	-0.126
$Sn^{2+} + 2e^- \rightleftharpoons Sn(s)$	-0.136
$AgI(s) + e^- \rightleftharpoons Ag(s) + I^-$	-0.152
$V^{3+} + e^- \rightleftharpoons V^{2+}$	-0.255
$Ni^{2+} + 2e^- \rightleftharpoons Ni(s)$	-0.257
$H_3PO_4 + 2H^+ + 2e^- \rightleftharpoons H_3PO_3 + H_2O$	-0.276
$Co^{2+} + 2e^- \rightleftharpoons Co(s)$	-0.28
$Ag(CN)_2^- + e^- \rightleftharpoons Ag(s) + 2CN^-$	-0.31
$Tl^+ + e^- \rightleftharpoons Tl(s)$	-0.336
$PbSO_4(s) + 2e^- \rightleftharpoons Pb(s) + SO_4^{2-}$	-0.359
$Ti^{3+} + e^- \rightleftharpoons Ti^{2+}$	-0.368
$Se(s) + 2H^+ + 2e^- \rightleftharpoons H_2Se(g)$	-0.369
$Cd^{2+} + 2e^- \rightleftharpoons Cd(s)$	-0.403
$Cr^{3+} + e^- \rightleftharpoons Cr^{2+}$	-0.407
$Fe^{2+} + 2e^- \rightleftharpoons Fe(s)$	-0.447
$2CO_2(g) + 2H^+ + 2e^- \rightleftharpoons H_2C_2O_4$	-0.49
$H_3PO_3 + 2H^+ + 2e^- \rightleftharpoons H_3PO_2 + H_2O$	-0.499
$Ga^{3+} + 3e^- \rightleftharpoons Ga(s)$	-0.560
$U^{4+} + e^- \rightleftharpoons U^{3+}$	-0.607
$Ag_2S(s) + 2e^- \rightleftharpoons 2Ag(s) + S^{2-}$	-0.691
$Cr^{3+} + 3e^- \rightleftharpoons Cr(s)$	-0.744
$Zn^{2+} + 2e^- \rightleftharpoons Zn(s)$	-0.762
$SiO_2(s) + 4H^+ + 4e^- \rightleftharpoons Si(s) + 2H_2O$	-0.807
$2H_2O + 2e^- \rightleftharpoons H_2(g) + 2OH^-$	-0.828
$Cr^{2+} + 2e^- \rightleftharpoons Cr(s)$	-0.913
$Mn^{2+} + 2e^- \rightleftharpoons Mn(s)$	-1.185
$SiF_6^{2-} + 4e^- \rightleftharpoons Si(s) + 6F^-$	-1.24
$Al^{3+} + 3e^- \rightleftharpoons Al(s)$	-1.662
$AlF_6^{3-} + 3e^- \rightleftharpoons Al(s) + 6F^-$	-2.069
$Al(OH)_4^- + 3e^- \rightleftharpoons Al(s) + 4OH^-$	-2.33
$Mg^{2+} + 2e^- \rightleftharpoons Mg(s)$	-2.372
$Na^+ + e^- \rightleftharpoons Na(s)$	-2.71
$Ca^{2+} + 2e^- \rightleftharpoons Ca(s)$	-2.868
$Ba^{2+} + 2e^- \rightleftharpoons Ba(s)$	-2.912
$K^+ + e^- \rightleftharpoons K(s)$	-2.931
$Li^+ + e^- \rightleftharpoons Li(s)$	-3.040

[a]A number of these potentials cannot be determined experimentally and are calculated from theory.

[b]Phen stands for 1,10-phenanthroline.

L

THE LITERATURE OF ANALYTICAL CHEMISTRY

Journals

American Laboratory

Analusis
  (formerly Chimie Analytique)

Analyst

Analytica Chimica Acta

Analytical Biochemistry

Analytical Chemistry

Analytical Instrumentation

Analytical Letters

Applied Spectroscopy

Clinica Chimica Acta

Clinical Chemistry

Fresenius' Zeitschrift für analytische Chemie

Journal of Analytical Chemistry of the USSR

Journal of Chromatographic Science

Journal of Chromatography

*Journal of Electroanalytical Chemistry and
  Interfacial Electrochemistry*

Microchemical Journal

Microchimica Acta (in 2 parts)

Separation Science and Technology

Spectrochimica Acta

Talanta

Reviews

Analytical Chemistry: Fundamental Reviews, American Chemical Society, Washington. Appears
  biennially in even-numbered years.

Analytical Chemistry: Applications, American Chemical Society, Washington. Appears biennially
  in odd-numbered years.

Critical Reviews in Analytical Chemistry, CRC Press, Boca Raton. Appears quarterly.

Standard Methods

ASTM Book of Standards, American Society for Testing and Materials, Philadelphia. Revised
  annually.

Official Methods of Analysis, 14th ed., Association of Official Analytical Chemists, Washington,
  1984.

Standard Methods for the Examination of Water and Wastewater, 16th ed., American Public
  Health Association, New York, 1985.

Compilations of Methods

D.F. Boltz and J.A. Howell, Eds., *Colorimetric Determination of Nonmetals*, 2nd. ed., John
  Wiley & Sons, Inc., New York, 1978.

G. Charlot, *Colorimetric Determination of Elements: Principles and Methods*, Elsevier Science Publishers B.V., Amsterdam, 1965.

Z. Marczenko, *Spectrophotometric Determination of Elements*, John Wiley & Sons, Inc., New York, 1976.

E.B. Sandell, *Colorimetric Determination of Traces of Metals*, 3rd ed., John Wiley & Sons, Inc., New York, 1959.

E.B. Sandell and H. Onishi, "Photometric Determination of Traces of Metals," 4th ed. of Part I of *Colorimetric Determination of Traces of Metals*, John Wiley & Sons, Inc., New York, 1978.

F.D. Snell, *Photometric and Fluorometric Methods of Analysis*, John Wiley & Sons, Inc., New York, 1978 and 1981. In 2 volumes.

L.C. Thomas and G. J. Chamberlin, *Colorimetric Chemical Analytical Methods*, 9th ed., Tintometer, Salisbury, 1980.

W.J. Williams, *Handbook of Anion Determination*, Butterworths, London, 1979.

Treatises and General Reference Books

A.J. Bard, R. Parsons, and J. Jordan, *Standard Potentials in Aqueous Solution*, Marcel Dekker Inc., New York, 1985.

A.J. deBethune and N.A.S. Loud, *Standard Aqueous Electrode Potentials and Temperature Coefficients at $25°C$*, Clifford A. Hampel, Skokie, 1964.

D. Glick, Ed., *Methods of Biochemical Analysis*, John Wiley & Sons, Inc., New York, 1954–. A continuing series.

E. Hoegfeldt and D.D. Perrin, *Stability Constants of Metal–Ion Complexes*, The Chemical Society, London, 1979 and 1981. In 2 volumes.

I.M. Kolthoff and P.J. Elving, Eds., *Treatise on Analytical Chemistry*, 2nd ed., John Wiley & Sons, Inc., New York, 1978–. A continuing, multivolume series in three parts.

J.J. Lingane, *Analytical Chemistry of Selected Metallic Elements*, Van Nostrand Reinhold Co., New York, 1966.

A.E. Martell and R.M. Smith, *Critical Stability Constants*, Vol. 4, Plenum Publishing Corp., New York, 1976.

L. Meites, Ed., *Handbook of Analytical Chemistry*, McGraw-Hill, New York, 1963.

G. Milazzo, S. Caroli, and V.K. Sharma, *Tables of Standard Electrode Potentials*, John Wiley & Sons, Inc., New York, 1978.

J. Mitchell, Jr., I.M. Kolthoff, E.S. Proskauer, and A. Weissberger, Eds., *Organic Analysis*, John Wiley & Sons, Inc., New York, 1953.

D.D. Perrin, *Ionisation Constants of Inorganic Acids and Bases in Aqueous Solution*, 2nd ed., Pergamon Press, Oxford, 1982.

M. Pinta, *Modern Methods for Trace Element Analysis*, Ann Arbor Science, Ann Arbor, 1978.

E.P. Serjeant and B. Dempsey, *Ionization Constants of Organic Acids in Aqueous Solution*, Pergamon Press, Oxford, 1979.

S. Svehla, Ed., *Wilson and Wilson's Comprehensive Analytical Chemistry*, Elsevier Science Publishers B.V., Amsterdam, 1959–. A continuing, multivolume series.

A. Weissberger and B.W. Rossiter, Eds., *Techniques of Chemistry*, Vol. I: *Physical Methods of Chemistry, Part IIIB*, John Wiley & Sons, Inc., New York, 1972.

Advanced Analytical and Instrumental Analysis Texts

G.D. Christian, J.E. O'Reilly, Eds., *Instrumental Analysis*, 2nd ed., Allyn & Bacon, Inc., New York, 1986.

G.W. Ewing, *Instrumental Methods of Chemical Analysis*, 5th ed., McGraw-Hill, New York, 1985.

G.G. Guilbault and L.G. Hargis, *Instrumental Analysis Manual*, Marcel Dekker Inc., New York, 1970.

H.A. Laitinen and W.E. Harris, *Chemical Analysis*, 2nd ed., McGraw-Hill, New York, 1975.

K.A. Rubinson, *Chemical Analysis*, Little, Brown & Company, Boston, 1987.

D.T. Sawyer, W.R. Heineman, and J.M. Beebe, *Chemistry Experiments for Instrumental Methods*, John Wiley & Sons, Inc., New York, 1984.

D.A. Skoog, *Principles of Instrumental Analysis*, 3rd ed., W.B. Saunders Company, New York, 1985.

H.H. Willard, L.L. Merritt, Jr., J.A. Dean, and F.A. Settle, Jr., *Instrumental Methods of Analysis*, 6th ed., Van Nostrand Reinhold Co., New York, 1981.

Monographs

Gravimetry and titrimetry

M.R.F. Ashworth, *Titrimetric Organic Analysis*. Part I, "Direct Methods"; Part II, "Indirect Methods," John Wiley & Sons, Inc., New York, 1964.

L. Erdey, *Gravimetric Analysis*, Vol. 1, Pergamon Press, Oxford, 1963; Vol. 2, Macmillan Publishing Co., Inc., New York, 1963; Vol. 3, Macmillan Publishing Co., Inc., New York, 1965.

J.S. Fritz, *Acid-Base Titrations in Nonaqueous Solvents*, Allyn & Bacon, Inc., Boston, 1973.

I.M. Kolthoff, V.A. Stenger, and R. Belcher, *Volumetric Analysis*, John Wiley & Sons, Inc., New York, 1942–57. In 3 volumes.

T.S. Ma and R.C. Rittner, *Modern Organic Elemental Analysis*, Marcel Dekker, Inc., New York, 1979.

R. Pribil, *Applied Complexometry*, Pergamon Press, Oxford, 1982.

G. Schwartzenbach and H. Flaschka, *Complexometric Titrations*, 2nd ed., English translation by H.M.N.H. Irving, Methuen and Co., Ltd., London, 1969.

W. Wagner and C.J. Hull, *Inorganic Titrimetric Analysis*, Marcel Dekker, Inc., New York, 1971.

Electroanalytical chemistry

A.J. Bard and R.L. Faulkner, *Electrochemical Methods*, John Wiley & Sons, Inc., New York, 1980.

H. Freiser, Ed., *Ion-Selective Electrodes in Analytical Chemistry*, Plenum Publishing Corp., New York, Vol. 1, 1978, Vol. 2, 1980.

J. Koryta and K. Stulik, *Ion-Selective Electrodes*, 2nd ed., Cambridge University Press, Cambridge, 1983.

H.W. Nurnberg, Ed., *Electroanalytical Chemistry*, John Wiley & Sons, Inc., New York, 1974.

P.H. Rieger, *Electrochemistry*, Prentice Hall, Englewood Cliffs, NJ, 1987.

D.T. Sawyer and J.L. Roberts, Jr., *Experimental Electrochemistry for Chemists*, John Wiley & Sons, Inc., New York, 1974.

E.P. Serjeant, *Potentiometry and Potentiometric Titrations*, John Wiley & Sons, Inc., New York, 1984.

Spectrometry

C. Burgess and A. Knowles, Eds., *Techniques in Visible and Ultraviolet Spectrometry*, Chapman and Hall, New York, 1981–. A continuing, multivolume series.

J.A. Dean and T.C. Rains, Eds., *Flame Emission and Atomic Absorption Spectroscopy*, Marcel Dekker, Inc., New York, 1974. In 3 volumes.

J. Ingle and S. Crouch, *Spectrochemical Analysis*, Prentice Hall, Englewood Cliffs, NJ, 1988.

E.D. Olsen, *Modern Optical Methods of Analysis*, McGraw-Hill, New York, 1975.

S.G. Schulman, Ed., *Molecular Luminescence Spectroscopy, Methods and Applications: Part I*, John Wiley & Sons, Inc., New York, 1985.

J.C. Van Loon, *Analytical Atomic Absorption Spectroscopy: Selected Methods*, Academic Press, New York, 1980.

Separations

E. Heftmann, *Chromatography*, 3rd ed., Van Nostrand Reinhold Co., New York, 1975.

E.L. Johnson and R. Stevenson, *Basic Liquid Chromatography*, Varian Associates, Inc., Palo Alto, 1978.

B.L. Karger, L.R. Snyder, and C. Horvath, Eds., *An Introduction to Separation Science*, John Wiley & Sons, Inc., New York, 1973.

J.M. Miller, *Separation Methods in Chemical Analysis*, John Wiley & Sons, Inc., New York, 1975.

F.C. Smith, Jr. and R.C. Chang, *The Practice of Ion Chromatography*, John Wiley & Sons, Inc., New York, 1983.

L.R. Snyder and J.J. Kirkland, *Introduction to Modern Liquid Chromatography*, 2nd ed., John Wiley & Sons, Inc., New York, 1979.

W.W. Yau, J.J. Kirkland, and D.D. Bly, *Modern Size-Exclusion Liquid Chromatography*, John Wiley & Sons, Inc., New York, 1979.

Miscellaneous

A. Albert and E.P. Serjeant, *The Determination of Ionisation Constants: A Laboratory Manual*, 3rd ed., Chapman & Hall, London, 1984.

R. Caulcutt and R. Boddy, *Statistics for Analytical Chemists*, Chapman & Hall, London, 1983.

H.A. Flaschka and A.J. Barnard, Jr., *Chelates in Analytical Chemistry*, Marcel Dekker, Inc., New York, 1967–72. In 4 volumes.

J.G. Grasselli, *The Analytical Approach*, American Chemical Society, Washington, 1983.

H. Irving, T.S. West, and H. Freiser, *Compendium of Analytical Nomenclature: Definitive Rules 1977*, Pergamon Press, Inc., New York, 1978.

D.D. Perrin, *Masking and Demasking of Chemical Reactions*, John Wiley & Sons, Inc., New York, 1970.

R.W. Ramette, *Chemical Equilibrium and Analysis*, Addison-Wesley Publishing Co., Inc., Reading, 1981.

ANSWERS TO SELECTED PROBLEMS

CHAPTER 2, page 31

2-1. (a) Soluble: Na^+, Br^- **(c)** Soluble: H^+, NO_3^- **(e)** Insoluble **(g)** Insoluble
(i) Soluble: Ni^{2+}, SO_4^{2-}

2-3. (a) 10.4 mmol **(b)** 4.88 mmol **(c)** 5.00×10^{-2} mmol **(d)** 9.00 mmol
(e) 3.02×10^{-3} mmol **(f)** 1.38 mmol

2-5. (a) 182 mg **(b)** 51.0 mg **(c)** 842 mg **(d)** 6.30×10^3 mg **(e)** 0.0457 mg **(f)** 34.1 mg

2-7. (a) 0.217 M **(b)** 0.0144 M **(c)** 0.0672 M **(d)** 1.50 M **(e)** 8.86×10^{-3} M
(f) 9.83×10^{-6} M

2-8. (a) 37.05 g/eq **(c)** 49.04 g/eq **(e)** 126.9 g/eq **(g)** 62.95 g/eq

2-9. (a) $2KMnO_4 + 3H_2SO_4 + 5H_2C_2O_4 \rightleftharpoons 2MnSO_4 + 8H_2O + 10CO_2(g) + K_2SO_4$
(c) $K_2Cr_2O_7 + 14HCl + 6KI \rightleftharpoons 2CrCl_3 + 7H_2O + 3I_2 + 8KCl$
(e) $H_2O_2 + 2K_4Fe(CN)_6 \rightleftharpoons 2KOH + 2K_3Fe(CN)_6$ **2-11. (a)** 0.100 N **2-13.** 6.71 mg

2-15. (a) 7.62×10^{-3} mol/L **(b)** 4.88×10^{-2} mol/L **(c)** 0.117 mol/L **2-17.** 3.26×10^3 mg

2-19. (a) 0.0840 M **(b)** 0.0840 M **(c)** 0.168 M **2-21. (a)** 0.10 M **(c)** 1.5 M **(e)** 0.60 M

2-23. 8.95 g/L **2-25.** 0.117 M **2-27.** 10.9 **2-29.** 45.1 g

2-31. (a) Dissolve 47.1 g in water and dilute to 1.00 L.
(c) Dissolve 1.50 g in water and dilute to 2.00 L. **(e)** Dilute 5.66 mL to 300 mL with water.

2-33. $K_{eq} = \dfrac{[Ce^{3+}]^2[Sn^{4+}]}{[Ce^{4+}]^2[Sn^{2+}]}$; $K_{eq} = \dfrac{[Ce^{3+}][Sn^{4+}]^{1/2}}{[Ce^{4+}][Sn^{2+}]^{1/2}}$; the values are different.

2-35. With assumption: $[H^+] = 0.0283$ M Without assumption: $[H^+] = 0.0259$ M **2-37.** 5.6 mg

2-39. 7.09×10^{-4} **2-41.** 5.0 g **2-43.** 0.083 M **2-45.** Concentration of N_2O_4 quadruples

2-47. 1.65×10^{-9} M

CHAPTER 3, page 62

3-1. (a) 4 **(b)** 4 **(c)** 3 **(d)** 4 **(e)** 2
3-3. (a) 2.00×10^3 **(b)** 0.0156 **(c)** 3.14 **(d)** 1.00×10^{-6}

3-5. (a) 1; 0.06   (b) 0.001; 5×10^{-4} (c) 0.01×10^{-3}; 0.01   (d) 0.000001; 2×10^{-5}
3-7. (a) 1.62   (b) 0.0316   (c) 1.44×10^3 (d) 8.62×10^{-3}
3-9. (a) 5.6   (b) 3.78×10^3 (c) 11   (d) 3.18×10^{-4} (e) 4.652×10^2 (f) 0.38   (g) 7.69
(h) -1.903 **3-11.** (a) 4.19   (b) 4.20   (c) 0.7   (d) 0.7
3-13. Absolute error $= 0.04\%$; percent error $= 1.5$ **3-15.** 72
3-17. C, because it has the smallest relative average deviation.   **3-19.** (a) 0.11   (b) 0.090
3-21. 0.0064   **3-23.** No significant difference
3-25. A significant difference exists at both 90% and 99%.
3-27. $Q_C = 0.56$ and $Q_T = 0.73$. Retain the value.
3-29. No. Q_C is independent of the magnitude of the divergent value.   **3-31.** (a) 3.0   (b) 0.60
3-33. (b) Slope $= 0.0293$; intercept $= 1.03 \times 10^{-3}$
(c) $\beta = 0.0293 \pm 5.6 \times 10^{-5}$; $\alpha = 1.03 \times 10^{-3} \pm 0.017$

CHAPTER 4, page 90

4-1. (a) $\dfrac{2Al}{Al_2O_3}$ (c) $\dfrac{2Zn}{Zn_2P_2O_7}$ (e) $\dfrac{Na_2B_4O_7 \cdot 10H_2O}{2B_2O_3}$ (g) $\dfrac{Pb(C_2H_5)_4}{PbCrO_4}$ **4-3.** 300 mg   **4-5.** 588 mg
4-7. 3.5% P; 6.6% K   **4-9.** 99.98%
4-11. 9.20% Sn; 78.6% Cu; 3.05% Zn; 3.48% Pb. No, the total is less than 100%.
4-13. (a) 16.26%   (b) 138.2 mg   (c) 0.8 ppt   (d) 1.4 ppt   **4-15.** 0.87%   **4-17.** 0.388 g
4-19. (a) 4.57%   (b) 4.37%   (c) 45.8 ppt   **4-21.** -0.56% **4-23.** 1.812 g   **4-25.** 171 mg
4-27. 6.36 mL   **4-29.** 39.1%   **4-31.** 27.4% $BaCl_2 \cdot 2H_2O$; 19.4% KCl
4-33. 8.009% Al; 4.994% Mg

CHAPTER 5, page 111

5-1. (a) 0.01700 M (c) 0.01282 M (e) 0.02449 M (g) 0.009121 M **5-2.** (a) 5.00 g   (c) 6.09 g
(d) 10.6 g   **5-3.** 4.945 g   **5-5.** (c) End point occurs at 9.60 mL; 34.5% Fe   **5-7.** 0.1035 M
5-9. 0.07297 M **5-11.** 0.1411 M **5-13.** $CaCl_2 \cdot 2H_2O$ **5-15.** 18.0%
5-17. 1.29%; the molecular formula or the % Br in the pure antihistamine   **5-19.** 4.94%   **5-21.** 0.147

CHAPTER 6, page 146

6-1. (a) H_2S (c) H_2NSO_3H **6-2.** (a) CH_3CO_2Na (c) Na_3PO_4 **6-3.** (a) HCO_3 (c) NH_3
6-4. (a) H_2CO_3 (c) $C_6H_5NH_3{}^+$ **6-5.** (a) 1.6×10^{-5} (c) 1.58×10^{-7}
6-6. (a) 5.71×10^{-10} (c) 5.92×10^{-6} **6-7.** (a) 2.664   (c) 1.986   (e) 12.875   (g) 11.11
6-9. 1.1 M **6.11.** (a) 7.000   (b) 8.54   (c) 1.954   (d) 0.778   (e) 1.255   (f) 1.255
6-13. (a) 8.95 g   (b) 0.639 g   (c) 0.84 g   **6-15.** (a) 0.58   (b) 0.16   (c) 0.20   (d) 0.19
6-17. 8.4×10^{-6} **6-19.** 1.18×10^{-6} **6-21.** (a) 0.48   (b) 0.53   (c) 0.96   (d) 2.03   (e) 2.67
6-23. (a) Trichloroacetic acid/sodium trichloroacetate   (b) Acetic acid/sodium acetate
(c) Ammonium chloride/ammonia   **6-25.** (a) 8.70   (b) 13.60   (c) 3.166   (d) 3.17   **6-27.** 2.20
6-29. (a) 1.54 (via quadratic formula)   (b) 10.08   (c) $H_2SO_3 + Na_2SO_3 \rightarrow 2NaHSO_3$ (d) 4.55

CHAPTER 7, page 187

7-1. 0 mL: 13.314   15.6 mL: 12.738   27.4 mL: 7.000   38.1 mL: 1.498
7-3. 95%: 2.59   100%: 7.00   105%: 11.39
7-5. 0 mL: 11.127   10 mL before e.p.: 9.177   At e.p.: 5.276   10 mL past e.p.: 1.738
7-7. 50%: 3.870   95%: 5.149   100%: 8.281   105%: 11.387
7-9. (a) 11.04   (b) 9.08   (c) 8.48   (d) 7.56   (e) 6.64   (f) 6.04   (g) 3.67   (h) 1.74

7-11. (a)

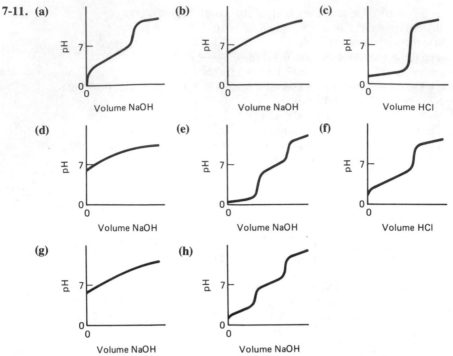

7-15. 1.72 pH units **7-17.** 0.04811 M **7-19.** 0.2469 M **7-21.** 4.13% **7-23.** 10.10%
7-25. 658.4 mg **7-27. (b)** 21.42% Na_2CO_3; 3.66% NaOH **7-29.** 64.19% Na_2CO_3; 25.27% $NaHCO_3$
7-31. 9.41% **7-33.** 23.4% **7-35.** 97.46%

CHAPTER 8, page 204

8-1. (a) $2H_2O \rightleftharpoons H_3O^+ + OH^-$ $K_w = [H_3O^+][OH^-]$
(c) $2CH_3CO_2H \rightleftharpoons CH_3CO_2H_2^+ + CH_3CO_2^-$ $K_{HS} = [CH_3CO_2H_2^+][CH_3CO_2^-]$
(e) $2H_2NCH_2CH_2NH_2 \rightleftharpoons H_2NCH_2CH_2NH_3^+ + H_2NCH_2CH_2NH^-$ $K_{HS} =$
$[H_2NCH_2CH_2NH_3^+][H_2NCH_2CH_2NH^-]$
8-3. (a) $1.00 \times 10^{-7} M$ **(c)** $5.96 \times 10^{-8} M$ **(e)** $2.24 \times 10^{-8} M$ **8-5. (a)** 2.9×10^{-9}
(b) Ionization may be complete but dissociation of the ion pair may be quite incomplete.
8-7. (a) 2.20 **(b)** 2.48 **(c)** 7.89 **8-9. (a)** 15.74 **(b)** 15.39 **(c)** 8.35 **(d)** 1.821 **8-11.** 16.83

CHAPTER 9, page 233

9-1. (a) 6.3×10^6 **(b)** 9.0×10^{10} **(c)** 1.1×10^{13} **9-3.** 5.6×10^{-3}
9-5. (a) Undefined **(b)** 14.37 **(c)** 17.02 **(d)** 2.79
9-7. (a) No **(b)** No **(c)** Yes **(d)** Yes **(e)** Yes **9-9.** 0.1086 M
9-11. (a) 178.9 mg/L **(b)** 100.2 mg/L **9-13.** 8.751% **9-15.** 50.92%
9-17. 18.2 ppm Mg; 41.60 ppm Ca **9-19.** 0.8669% Mg; 3.94% Zn; 19.4% Hg

CHAPTER 10, page 256

10-1. (a) $Fe^{3+} + e^- \rightleftharpoons Fe^{2+}$ **(c)** $I_2 + 2e^- \rightleftharpoons 2I^-$ **(e)** $MnO_4^- + 8H^+ + 5e^- \rightleftharpoons Mn^{2+} + 4H_2O$
(g) $S_4O_6^{2-} + 2e^- \rightleftharpoons 2S_2O_3^{2-}$ **(i)** $VO_2^+ + 2H^+ + e^- \rightleftharpoons VO^{2+} + H_2O$
10-3. $E = E° - \dfrac{0.0661}{n} \log \dfrac{[C]^c[D]^d}{[A]^a[B]^b}$ **10-5. (a)** 0.323 V **(b)** -0.53 V **(c)** 0.658 V **(d)** 0.435 V

10-7. (a) 1.412 V **(b)** 1.318 V **(c)** 1.223 V **(d)** 1.128 V **10-9.** 3.943
10-11. O_3, H_3AsO_4, $S_4O_6^{2-}$, V^{3+}, Cr^{2+} **10-13. (a)** Decreases **(b)** Decreases **(c)** Unchanged
(d) Unchanged **(e)** Decreases **10-15.** 1.883
10-17. (a) Left to right **(b)** Right to left **(c)** Left to right **(d)** Right to left **(e)** Right to left
(f) Left to right
10-19. 1.1×10^{83}; the reaction is too slow to measure at room temperature in the absence of a catalyst.
10-21. (a) 0.974 V **(b)** $2Sb_2O_3(s) + 3Si(s) \rightleftharpoons 4Sb(s) + 3SiO_2(s)$ **(c)** 0.974 V
(d) The potential of each electrode depends only on $[H^+]$, which remains constant.
10-23. 4.66×10^{-15} **10-25.** 5.96×10^9 **10-27.** 1.32×10^{21}

CHAPTER 11, page 295

11-1. (a) +3 **(c)** +4 **(e)** +3 **11-3. (a)** 0.807 V **(b)** 0.828 V **(c)** 1.305 V **(d)** 1.400 V
(e) 1.404 V
11-5. (a) $pFe^{2+} = 1.44$; $E = 0.735$ V **(b)** $pFe^{2+} = 1.81$; $E = 0.781$ V
(c) $pFe^{2+} = 8.40$; $E = 1.19$ V **(d)** $pFe^{2+} = 15.32$; $E = 1.59$ V **(e)** $pFe^{2+} = 15.67$; $E = 1.60$ V
11-7. (a) 4.17 mL **(b)** 20.8 mL **(c)** 29.0 mL **(d)** 203 mL
11-9. (a) -0.150 V **(b)** -0.109 V **(c)** 0.763 V **(d)** 1.06 V **11-11. (a)** 0.390 V **(b)** 0.331 V
11-13. (a) 0.68 V **(b)** 0.68 V **(c)** 0.68 V
11-15. (a) $FeCl_3 + Ag(s) \rightleftharpoons FeCl_2 + AgCl(s)$
(c) $K_2Cr_2O_7 + 6Ag(s) + 14HCl \rightleftharpoons 2CrCl_3 + 6AgCl(s) + 7H_2O + 2KCl$
(e) $MnCl_3 + Ag(s) \rightleftharpoons MnCl_2 + AgCl(s)$
11-16. (a) $K_2Cr_2O_7 + 4Zn(s) + 7H_2SO_4 \rightleftharpoons 2CrSO_4 + 4ZnSO_4 + 7H_2O + K_2SO_4$
(c) $2KMnO_4 + 5Zn(s) + 8H_2SO_4 \rightleftharpoons 2MnSO_4 + 5ZnSO_4 + 8H_2O + K_2SO_4$
(e) $Fe_2(SO_4)_3 + Zn(s) \rightleftharpoons 2FeSO_4 + ZnSO_4$ **11-17.** 0.08807 M **11-19.** 7.732% **11-21.** 5.018%
11-23. 1057 ppm **11-25.** 3.26% **11-27.** 12.17% **11-29.** 445.1 mg **11-31.** 0.02839 M
11-33. 0.974 M **11-35.** 35.52% **11-37.** 0.1613 M **11-39.** 62.18%

CHAPTER 12, page 320

12-1. (a) 1.3×10^{-2} M **(c)** 4×10^{-6} M **(e)** 7×10^{-7} M
12-2. (a) 1.10×10^{-12} **(c)** 1.50×10^{-9} **(e)** 2.40×10^{-15}

12-3.

	(a)	(b)	(c)
For $AgIO_3$:	1.8×10^{-4} M	1.8×10^{-4} M	2.1×10^{-7} M
For Ag_2CrO_4:	8.4×10^{-5} M	4.2×10^{-5} M	5.3×10^{-11} M
For Ag_3PO_4:	3.1×10^{-5} M	1.0×10^{-5} M	8.3×10^{-16} M

12-5. 0.54% **12-7. (a)** No **(b)** Yes **(c)** Yes **(d)** No **12-9.** 2.3×10^{-6} M
12-11. (a) 0.884 **(b)** 1.228 **(c)** 1.434 **(d)** 6.00 **(e)** 10.33 **12-13.** 73.85 mL
12-15. 6.4×10^{-6} M; after the equivalence point **12-17.** 28.06% **12-19.** 0.07532 M
12-21. 3.562% **12-23.** 378.8 mg **12-25.** -0.07 mL; -3×10^{-3}

CHAPTER 13, page 348

13-1. (a) 0.269 V **(b)** 0.331 V **(c)** 0.392 V **13-3. (a)** 0.174 V **(b)** 0.174 V **(c)** 0.174 V
(d) 0.174 V Not at all selective **13-5. (a)** 11 mV; 0.19 pH unit **(b)** 46 mV; 0.78 pH unit
13-7. 2.38; to determine the value of K **13-9.** 0.014 **13-11.** 0.176 V **13-13.** 1.16×10^{-4} M
13-15. 323 ppm **13-17.** 3.98 ppm **13-19.** 300.9 ppm

CHAPTER 14, page 374

14-1.

	Anode	*Cathode*
(a)	$2Cl^- \rightleftharpoons Cl_2(g) + 2e^-$	$2H^+ + 2e^- \rightleftharpoons H_2(g)$
(b)	$2H_2O \rightleftharpoons 4H^+ + O_2(g) + 4e^-$	$Ni^{2+} + 2e^- \rightleftharpoons Ni(s)$
(c)	$2Br^- \rightleftharpoons Br_2(l) + 2e^-$	$2H^+ + 2e^- \rightleftharpoons H_2(g)$
(d)	$2H_2O \rightleftharpoons 4H^+ + O_2(g) + 4e^-$	$2H^+ + 2e^- \rightleftharpoons H_2(g)$

14-3. 684 s   **14-5.** 12.5%   **14-7.** 0.9916%   **14-9.** 5.997% Cu; 1.351% Ni; 2.572% Zn
14-11. 0.942 mg/mL   **14-13.** 7.884×10^{-4} M   **14-15.** 1.63×10^{-4} M

CHAPTER 15, page 404

15-1. (a) 700.0 nm; visible   **(c)** 224.9 nm; ultraviolet   **(e)** 11.99 nm; x-ray
15-2. (a) 5.59×10^{14} s^{-1} **(c)** 1.22×10^9 s^{-1} **(e)** 4.67×10^{19} s^{-1} **15-3. (a)** 396.1 nm
(b) 403.4 nm   **(c)** 442.5 nm   **(d)** 433.5 nm
15-5.

	Wavelength	*Frequency*
(a)	496.6 nm	6.037×10^{14} s^{-1}
(b)	198.7 nm	1.509×10^{15} s^{-1}
(c)	2484 nm	1.207×10^{14} s^{-1}
(d)	824.3 nm	3.637×10^{14} s^{-1}

15-6. (a) 3.000   **(c)** 1.000   **(e)** 0.383   **15-7. (a)** 99%   **(c)** 48.5%   **(e)** 10.00%
15-9. 852 L/mol·cm in both cells   **15-11.** 6.55×10^{-5} M
15-13. No. The ratios of the absorbances at the two wavelengths are different.   **15-15.** 6.62×10^{-9} M
15-17. (a) 0.503   **(b)** 7.25×10^4 L/mol·cm   **(c)** 1.26   **(d)** 1.99 cm   **15-19.** 0.501
15-21. (a) 0.574   **(b)** -0.86% **(c)** 2.61×10^{-5} M   **(d)** 0.76%

CHAPTER 16, page 427

16-1. 0.0180 mg/mL   **16-3.** 6.30 mg/L   **16-5.** 0.165 ppm   **16-7.** 36.2 μg/mL   **16-9.** 0.18 g
16-11. 0.406   **16-13. (a)** 0.351   **(b)** 0.037   **16-15.** 0.243% Ni; 0.0959% Zn
16-17. (a)

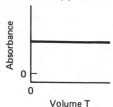

(c)

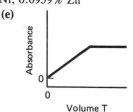

(e)

16-19. (b) HIn^+ is orange-red; In is violet   **(c)** About 440 nm
(d) The absorbance at each wavelength is the average of the absorbances at pH 1.0 and 13.0.
(e) 493 nm   **16-21.** 1:1.95 or ZnL_2 **16-23.** 1:3 or ML_3 **16-25.** 1:2.93 or CoL_3^{2+}

CHAPTER 17, page 449

17-1. (a) 5.7   **(b)** 11.8   **(c)** 31.4   **17-3.** $F_2 = 2.1F_1$ **17-5.** First order: 1.38×10^3 nm.   Second
order: 689 nm   **17-7.** Primary (excitation): 313 or 365 nm.   Secondary (fluorescence): 400 or 420 nm
17-9. 64.5 ppm   **17-11.** 0.480 mg/mL

CHAPTER 18, page 473

18-1. 23.6 ppm
18-3. (b) From graph of absolute signals: 424 mg Na^+/100 mL; 25.7 mg K^+/100 mL. From graph of relative signals: 408 mg Na^+/100 mL; 24.5 mg K^+/100 mL.
(c) The plot of the relative signal versus concentration; instrumental variations affect *both* signals proportionately. **(d)** 16 mg Na^+/100 mL; 1.2 mg K^+/100 mL **18-5.** 4.36 ppm

CHAPTER 19, page 492

19-1. 8.21 **19-3.** 1 extraction: 0.37. 2 extractions: 0.14. 3 extractions: 0.051. **19-5.** 260 mL
19-7. (a) 99 mL **(b)** 198 mL **(c)** 495 mL **(d)** 990 mL
19-9. (a) 1.25×10^3 **(b)** 53.4 **(c)** 9.32 **(d)** 2.15
19-11. After first extraction: 0.99 for As and 0.67 for Cd. After second extraction: 0.99 for As and 0.81 for Cd. **19-13.** 20.25% Na_2CO_3; 2.15% NaOH

CHAPTER 20, page 505

20-1. 57.6 mL **20-3.** 38.1 s **20-5.** 0.415 **20-7. (a)** 1.12×10^4 plates **(b)** 6.79×10^{-3} in./plate
20-9. 11.0 mm **20-11.** 0.16 cm **20-13. (a)** 1.44 **(b)** 0.67 **(c)** 1.0
20-15. (a) 697 plates **(b)** 174 cm **20-17. (a)** 5.93 s **(b)** 5.08 s **(c)** 5

CHAPTER 21, page 523

21-1. 38.0% ortho; 47.6% meta; 14.4% para
21-3. 1.4 μmol I/μL; 1.4 μmol II/μL; 0.72 μmol III/μL; 1.5 μmol IV/μL
21-5. (a) Plot H versus v **(b)** v_{opt} = 35 cm/s **(c)** 4.81×10^4 plates
21-7. (a) Plots are nonlinear. **(b)** Plots are linear. **(c)** C_7 alcohol (*n*-heptanol) **(d)** 31.6 s
(e) Part (b); because they are linear and the line can be fit more accurately and easily.
(f) Plot is linear. **(g)** Compound 1 is a C_9, alcohol (*n*-nonanol).

CHAPTER 22, page 548

22-1. (a) For toluene: 1.12; for ethylbenzene: 1.17 **(b)** 60.0%
22-3. (a) 5.64×10^{-2} cm **(b)** 0.191 ng
22-5. (a) Plot V_R versus log MW. **(b)** 2.09×10^5
(c) At low MW, the small molecules get into every pore and end up retained about the same amount; at high MW, the large molecules are excluded from every pore and are not retained at all. **(d)** 172 mL
(e) 2.6×10^4 plates

INDEX

retention, 498
  solid support, 511, 529, 534, 539, 541
  stationary phase, 511, 534, 537, 539, 541
  temperature programming, 520
  theoretical plate, 500
  unevenness of flow broadening, 501
  van Deemter equation, 503
Chromophore, 408
Chromotropic acid as spectrophotometric
    reagent, 416
Cleaning glassware, 552
Cleaning solution, 552
Closed-loop recycling in LC, 547
Cobalt, spectrophotometric determination,
    416
Coefficient of variation, 43
Collisional deactivation, 386
Colloid:
  coagulation, 72
  formation, 68
Colloidal suspension, 68
Color, 388
Color transition range:
  acid-base indicator, 170
  redox indicator, 273
Column:
  capillary, 495
  efficiency, 499
  liquid chromatography, 529
  open tubular, 495
  packed, 495
  selection, 518
  solid support in exclusion
      chromatography, 541
  solid support in GLC, 511
  solid support in LC, 530
  stationary phase in GLC, 511
  unevenness of flow, 501
Column chromatography, 495
Common ion:
  effect on equilibrium, 28
  effect on solubility, 302
Comparator for emission spectroscopy film,
    472
Complexation titration:
  calculation of shape, 219–23
  effect of conditional constant, 216–18
  indicators, 227–29
Complexing agent:
  effect on metal-ion concentration, 214–16
  effect on solubility, 306–8

prevent precipitation of metal hydroxides,
    214–16
Concentration, 10–18
  analytical, 18
  effect on absorbance, 389
  effect on acid-base titration, 152, 158–60
  effect on electrode potential, 245
  effect on fluorescence, 436–38
  equilibrium, 18
  overpotential, 352
  polarization, 352
  standard state, 20, 242
Conditional equilibrium constant, 216–18,
    304
  minimum for successful titration, 225
Conductivity detector, 540
Confidence interval, 50
Confidence level, 50
Constant-boiling hydrochloric acid, 174
Constant-current coulometry, 363–67
Constant error, 47
Constant-voltage coulometry, 361–63
Constructive interference, 379
Continuous variations method, 424, 600
Controlled-potential electrolysis, 361–63
Convection, 369
Coordination compound:
  chemical masking, 210
  as colored substances, 209
  titrating metal ions, 211
Coordination number, 206
Copper:
  atomic absorption determination, 602
  electrode, 336
  iodometric titration, 291, 588
  photometric titration, 421
  primary standard, 295
  spectrophotometric determination, 417
  standard for sodium thiosulfate, 589
Coprecipitation, 69
Coulomb, 360
Coulometric titration, 363–67
  current efficiency, 365
  table of applications, 365
Coulometry, 359–67
  constant current, 363–67
  constant potential, 361–63
  external generation of titrant, 366
Counter-ion layer, 69
  reduction of size, 72
Cresol purple indicator, 170

Drierite, 557
Dropping mercury electrode, 367, 371
Drying:
  precipitates, 563
  samples, 5, 557
Dynode, 399

Eddy diffusion in chromatography, 501
Efficiency of chromatography column, 499,
  520
Electrical discharge emission spectroscopy,
  470–73
Electrical double layer:
  coagulating colloids, 72
  colloidal particle, 69
Electric force field, 377
Electroanalytical chemistry texts, 642
Electrochemistry:
  coulometry, 359–67
  electrode potentials, 241
  electrogravimetry, 357
  electrolysis, 350–56
  galvanic cell, 238
  ion-selective electrodes, 328–38
  Nernst equation, 245
  polarography, 367–74
  potentiometry, 323
Electrode:
  calcium, 332
  calibration, 339–42
  carbon dioxide, 337
  constant-current coulometry, 363
  constant-potential coulometry, 363
  crystalline membrane, 335
  definition, 238
  direct potentiometry, 339–45
  dropping mercury, 367
  electrical discharge atomizer, 471
  first-order, 326
  fluoride, 335
  gas sensing, 336–38
  glass membrane, 328–32
  indicating-reference combination, 332
  inert, 328
  ion-selective, 328–38
  junction potential, 339
  liquid membrane, 332–35
  membrane, 328–38
  metal indicator, 326–28
  non-Nernstian response, 340

  pH, 328–32
  polarization, 352
  polymer membrane, 332–35
  reference, 243, 323–26
  saturated calomel, 324
  second-order, 327
  silver, 326
  silver-silver chloride, 325
  standard hydrogen, 323
Electrode polarization:
  concentration, 352
  kinetic, 353
Electrodeposition, 357–59
Electrode potential:
  definition, 241
  effect of concentration, 245
  equivalence point, 263–65
  irreversible electrode reactions, 255
  IUPAC sign convention, 244
  limitations, 254–56
  measurement, 243, 338
  predicting reaction direction, 244
  standard, 243
  table of values, 634–39
Electrogravimetry, 357
Electrolysis, 350–56
  constant applied voltage, 355
  constant current, 356
  current-voltage behavior, 355
  voltage necessary, 351–53
Electrolysis cell:
  mercury cathode, 359
  three electrode, 361
Electrolyte, 9
Electrolytic cell, 238, 350
Electromagnetic radiation, 377–82 (*see also*
  Radiant energy)
  interference, 379
  particle properties, 380
  polarization, 379
  power and intensity, 379
  wave properties, 377–80
Electromagnetic spectrum, 381
Electron capture detector, 516
Electronic energy, 383
Electrothermal atomization, 455–57
Eluate, 497
Eluent, 497
  delivery in LC, 528
  programming, 545–47
  strength in LC, 538, 535